Michael Kaschke,
Karl-Heinz Donnerhacke,
and Michael Stefan Rill

Optical Devices in Ophthalmology and Optometry

Related Titles

Smith, F., King, T.A., Wilkins, D.

Optics and Photonics – An Introduction

2007
2nd edn, ISBN: 978-0-470-01784-5,
Also available in digital formats.

Murphy, D.B.

Fundamentals of Light Microscopy and Electronic Imaging

2002
ISBN: 978-0-470-24202-5

Abdel-aleem, S.M.

Design, Execution, and Management of Medical Device Clinical Trials

2009
ISBN: 978-0-470-47426-6,
Also available in digital formats.

Valberg, A.

Light Vision Color

2005
ISBN: 978-0-470-84902-6,
Also available in digital formats.

Porter, J., Queener, H.M., Lin, J.E., Thorn, K., Awwal, A.A. (eds)

Adaptive Optics for Vision Science
Principles, Practices, Design, and Applications

2006
ISBN: 978-0-471-91464-8

Snell, R., Lemp, M.

Clinical Anatomy of the Eye

1998
2nd edn, ISBN: 978-0-632-04344-6,
Also available in digital formats.

Still, D., Fletcher, R. (eds)

Eye Examination and Refraction

1998
2nd edn, ISBN: 978-0-632-05141-0

Gross, H., Gross, H., Blechinger, F., Achtner, B.

Handbook of Optical Systems
Volume 4: Survey of Optical Instruments

2008
ISBN: 978-3-527-40380-6

Michael Kaschke, Karl-Heinz Donnerhacke,
and Michael Stefan Rill

Optical Devices in Ophthalmology and Optometry

Technology, Design Principles, and Clinical Applications

WILEY-VCH
Verlag GmbH & Co. KGaA

Authors

Dr. Michael Kaschke
Carl Zeiss AG
Oberkochen
Germany

Dr. Karl-Heinz Donnerhacke
Jena
Germany

Dr. Michael Stefan Rill
Carl Zeiss AG
Oberkochen
Germany

Cover
Background illustration:
© Hammerstudio/Fotolia

Other illustrations:
authors.

Library of Congress Card No.:
applied for

British Library Cataloguing-in-Publication Data:
A catalogue record for this book is available from the British Library.

Bibliographic information published by the Deutsche Nationalbibliothek
The Deutsche Nationalbibliothek lists this publication in the Deutsche Nationalbibliografie; detailed bibliographic data are available on the Internet at http://dnb.d-nb.de.

Cover Design Simone Benjamin, Bluesea Design, McLeese Lake, Canada
Typesetting le-tex publishing services GmbH, Leipzig
Printing and Binding Markono Print Media Pte Ltd, Singapore

ISBN 978-3-527-41068-2
ePDF ISBN 978-3-527-64899-3
ePub ISBN 978-3-527-64898-6
Mobi ISBN 978-3-527-64897-9
obook ISBN 978-3-527-64896-2

Printed on acid-free paper

To Sylvia without her support nothing would have been possible.

Michael Kaschke

To Christel for her abundance of understanding all the time.

Karl-Heinz Donnerhacke

To my family and friends for their steady support.

Michael Stefan Rill

Contents

Preface

This book is based on lectures "Optical Systems in Medical/Ophthalmic Technology" held by two of the authors since 2007 at the Karlsruhe Institute of Technology (Germany) and since 2003 at the Ernst Abbe University of Applied Sciences in Jena (Germany). The idea behind these lectures was to create a link between fundamental physical methods in optics, photonics, and measurement technology on the one hand, and to communicate their applications in medical sciences, in particular ophthalmology and optometry, to graduate students in physics and in electrical and mechanical engineering on the other. As this book is essentially based on these lectures, the structure, motivation, and target group of readers for this is quite similar. However, this book is intended as a textbook and compendium for the engineer and physicist in academic and industrial research and development. It will also be useful for the teaching and practicing ophthalmologist and optometrist with an interest in optical technology. To cover the broad spectrum of readers, we have not only described the application of optical methods to the design of ophthalmic diagnostic and laser treatment systems, but have also discussed clinical applications in such a way that the advantage of using a particular optical system and design becomes evident.

A large number of excellent reference books and textbooks exist on ophthalmology and optometry. The same applies to applied and technical optics. However, most of these books focus on a certain aspect (e.g., disease diagnosis, treatment procedures, etc.) or distinct optical and photonics techniques (e.g., OCT, aberrometry, etc.). Consequently, they often address experts with a certain scientific background. Students looking for a systematic and relatively complete introduction to the topic of optical methods and systems in ophthalmology and optometry are thus confronted with a number of books from which the necessary information has to be extracted with considerable effort. The situation is actually very similar for engineers and scientists of academic and research institutions, who often look for a compendium or reference book which gives them a general overview of available solutions in medical technology associated with their area of expertise. To close this gap, we provide an interdisciplinary overview of the currently most relevant (and most often used) techniques and technologies in ophthalmology and optometry, their underlying optical principles, as well as their corresponding diagnostic and therapeutic application.

The eye is our most important sensory organ and any reduction or loss of vision is a major impairment of our quality of life. Although the eye is quite accessible for optical examination methods, the diagnostic options available to the ophthalmologist were very limited until the middle and end of the nineteenth century. It was not until 1850 that Herrmann von Helmholtz invented the ophthalmoscope affording a view of the inside of the living eye for the very first time. This can be considered as the advent of modern ophthalmology and the birth of ophthalmic equipment making. Over the years, it has been demonstrated that no other organ necessitates the use of as many different optical-medical devices as the eye. It is also no surprise that ophthalmology has become by far the most successful application area of lasers in medicine since the invention of the laser in the 1960s. Ophthalmic and optometric methods and technologies have rapidly grown and matured during the last couple of decades and have actually seen an acceleration, but this is still a very active field of research today. Presenting an in-depth coverage of all the ongoing activities is certainly beyond the scope of this book. Consequently, we will try to walk the line between covering the more general and principal approaches in design, development and application of ophthalmic systems, and providing detailed background information on exciting current research topics.

As this book is intended to bridge technology and clinical domains, we will discuss modern optical technologies alongside their clinical deployment. In this way, it addresses graduate and postgraduate students in physics, electrical, mechanical, and biomedical engineering who want to gain a general insight into the principles and concepts of ophthalmic systems. We have also added some topic-related application-oriented "Problems" at the end of each chapter. The problems are presented with fully elaborated solutions, which can be downloaded from the reserved website of Wiley (http://www.wiley.com). These problems demonstrate how basic design parameters of an ophthalmic device are calculated.

Ophthalmologists and optometrists who want to gain a profound understanding of how the diagnostic and therapy systems work will also greatly benefit from the application oriented approach used in our book. We also give references to the most current and relevant literature throughout the chapters and in corresponding "Recommended Reading" sections. These references might be particularly useful for specialists or students who want to acquire further expertise in a special subject.

The book has a modular structure so that it can be used by readers with different backgrounds and interests. In the first part, a basic introduction to key aspects of ophthalmology and optometry is given, including a brief introduction to anatomy, optical properties, as well as refractive errors and diseases of the human eye. The second part is dedicated to ophthalmic diagnosis and imaging devices and techniques. Within this part, Chapter 4 gives as an overview of the link-up between common eye diseases and clinical conditions as well as relevant ophthalmic devices and methods. In the third part of this book, we focus on the therapeutic aspects of ophthalmology in which the use of laser systems is of particular importance. The appendix of this book provides the basics of optics and lasers, which are relevant to understanding the physical concepts of the presented ophthalmic and optometric systems. Here, the intention was not to present the entire content of textbooks on

Table 1 Examples of structured courses.

Background	Electrical Engineer	Biomedical Engineer	Physicist	Engineer/Scientist in practice	Ophthalmologist/ Optometrist
Chapter 1	2	R	1	R	R
Chapter 2	3	R	2	R	R
Chapter 3	4	R	3	R	R
Chapter 4	5	2	4	R	1
Chapter 5	6	3	5	1	2
Chapter 6	7	4	6	2	3
Chapter 7	8	5	7	3	4
Chapter 8	9	6	8	4	5
Chapter 9	11	8	9	5	6
Chapter 10	12	9	10	6	7
Appendix A	1	1	R	R	X
Appendix B	10	7	R	R	X

optics and lasers, but rather to focus on the topics relevant to ophthalmic devices and to provide a consistent reference base and notation.

The chapters of this book may be combined in various ways for use in semester courses. Representative examples of such structured courses are shown in Table 1 in which we also suggest a potentially beneficial sequence for reading the chapters of this book. Topics which should have already been treated during a previous course or are considered to be already known are marked with an "R", standing for "revision".

Commonly accepted notation and symbols have been used whenever possible. However, as this book covers a number of different topics, a number of symbols exist that have multiple meanings. To avoid confusion, we have added an overview of abbreviations and symbols to the Appendix C.

Acknowledgments
We would like to acknowledge all the inspiration and motivation we received from many people while writing this book. We would like to thank Herbert Gross and Rudolf von Bünau in particular for their useful hints, support and extensive discussions. We are also indebted to Michael Vincent Patella for valuable help and clarifications on the topic of functional imaging in Chapter 8 and Michael Kempe for detailed discussions on OCT in Chapter 7 and the laser Sections 9 and 10. We also greatly acknowledge further help by (in alphabetical order) Peter Andrews, Detlef Biernat, Mark Bischoff, Tobias Bühren, Tobias Damm, Manfred Dick, Matt Everett, Gerald Kunath-Fandrei, Werner Nahm, and Clive Poole. Our thanks go to all the people who supported us by providing relevant data and pictures, in particular Thomas Behling, Jens Dawczynski, Peter Esser, Herbert Gross, and all the organizations and companies who consented to the inclusion of their graphic ma-

terial throughout this book. We also would like to thank the students and academic staff at the Karlsruhe Institute of Technology (Germany) and the Ernst Abbe University of Applied Sciences in Jena (Germany).

We are grateful to our editors Valérie Molière and Anja Tschörtner at WILEY-VCH Verlag GmbH (Germany).

The book would not have been possible without the encouragement, support and interest of our families and friends. We therefore also see it as their book.

Part One

1
Structure and Function

In the following chapters of this book, we will deal with the examination and treatment of the human eye. As the majority of readers are expected to have an engineering or scientific background, we would like to provide a background in human ocular anatomy. This chapter shall also serve as an introduction and general reference for the more technical chapters. Of course, these chapters cannot claim to cover the entire anatomy and physiology of the eye, but they should be sufficient to gain an understanding of the interaction between the eye and the ophthalmic devices under consideration.

The human eye is a sophisticated sensory organ through which 80% of the sensory information we receive is processed. It is indeed our most important connection to the outside world. Any reduction or loss of vision means a major impairment of our quality of life.

In principle, the eye works like an artificial optical imaging system. To create an image, optical components focus light rays onto a photosensitive detector. But vision is more than just a projection of the surroundings onto a passive screen. The optical data is "preprocessed" by the photosensitive and neuronal tissue before it is sent to the brain for final "image analysis". Even with efficient preprocessing, the eye sends 10 × more data to the brain than all other sensory organs altogether. To analyze this huge amount of information flow with almost no latency[1)], 30 different parts of the brain are involved simultaneously. During image processing, relevant information is filtered by recognition of known patterns.

The brain's important role for vision is illustrated in Figure 1.1. In Figure 1.1a, an animated image is shown as it would be directly projected onto photosensitive tissue. Here, a sharp image exists only in the center of the visual field. The eye now automatically changes the viewing direction in a fast manner and, for a moment, the margins of the image are sharply imaged as well. All the image segments are then merged by the brain so that the perceived field of sharp vision extends to the margins (Figure 1.1b).

1) The average period of latency to transmit a light stimulus from the retina to the visual cortex takes about 95–115 ms, and it takes about 300 ms to perceive the signal. However, the brain "simulates" a real-time perception.

Optical Devices in Ophthalmology and Optometry, First Edition. Michael Kaschke, Karl-Heinz Donnerhacke, and Michael Stefan Rill.

(a) (b)

Figure 1.1 Image processing by the brain. (a) Simulated "raw" image as it would be detected by photosensitive tissue. (b) Due to fast changes of the viewing direction and preprocessing of the "detected data", the perceived field of clear vision is considerably extended. Taken with permission from [1].

We can trick the brain's hard-wired processing algorithms by looking at visual illusions (Figure 1.2). It is an interesting and currently not fully answered question how much visual illusions are based purely on innate factors, or to what extent they are also based on experience and adaption.

1.1
Anatomy of the Human Eye

The human eye can be divided into the *anterior* and *posterior segments* (Figure 1.3b and c, respectively). The anterior segment (Figure 1.4) is the optical window to the environment. It mainly consists of optical components, such as the cornea, eye lens, and iris. The posterior segment of the eye (Figure 1.5) is referred to as the *fundus*. It is connected to the visual cortex of the brain via the optic nerve (Figure 1.6).

Sclera and cornea The spherical outer shell of the human eye consists of the white, opaque *sclera*. It serves as a mechanical support and protects the eye from injuries caused by mechanical force. The collagen fibers in the sclera are randomly distributed. Consequently, the incident light is strongly scattered (Section 9.2) so that the tissue appears white and opaque, which is why this part is also called "the whites of the eyes".

In the anterior segment of the eye, the sclera passes into the transparent *cornea*. The cornea is composed of multiple functional layers (Figure 1.7) and is covered by the 4–7 μm thick *tear film*. The tear film consists of a viscous, aqueous fluid that smooths the surface roughness of the corneal surface. A smooth surface reduces light scattering (Section 9.2), and thus improves the clarity of vision. The corneal *epithelium* is a 50 μm thick chemical barrier of the outer cornea which protects the eye against water, large molecules, and toxic substances. This is followed by *Bowman's membrane* which is a thin layer (8–14 μm) above the 500 μm thick *stroma*. The stroma is composed of approximately 250 stacked collagen layers termed *lamellae*. Each lamella has a thickness of about 2 μm and contains ordered, cylindrical-

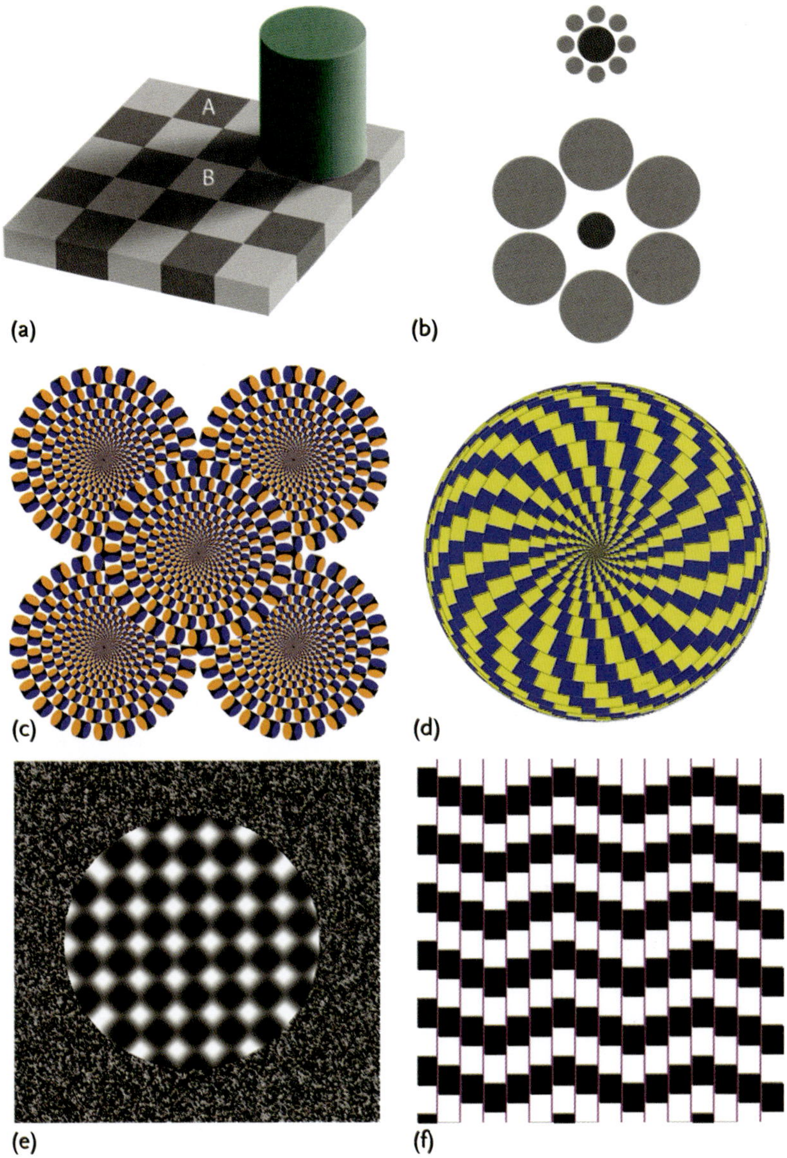

Figure 1.2 Visual illusions which trick the image processing capability of the brain. (a) The checkerboard field A seems to be darker than B although the gray scales are equal in both cases. (b) The big dark dot at the top appears larger than the lower one, even though both dots are exactly the same size. (c) The circles seem to rotate. (d) Does the image show spirals? A closer look reveals that the structures are closed rings. (e) Fixate the pattern in the center. When moving your head, the circle seems to move independently from the background. When bringing your head closer, the pattern inside the circle seems to approach. (f) The lines appear to be bent although they are perfectly straight. See also [2].

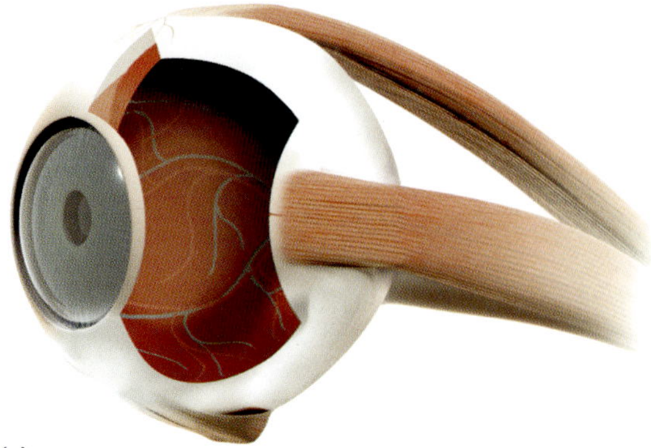

(a)

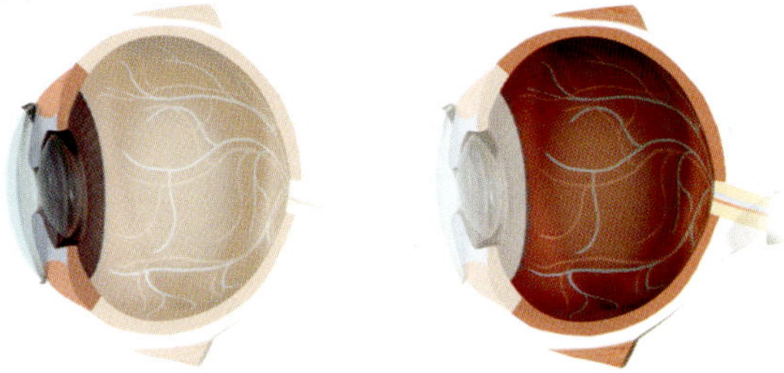

(b) (c)

Figure 1.3 Anatomy of the human eye. (a) Oblique view of the human eye ball. (b) Side view of the eye with highlighted anterior segment. (c) Side view of the eye with highlighted posterior segment.

shaped collagen *fibrils* with diameters of 25–35 nm and a spacing of 20–50 nm [3]. Underneath the stroma we have the *Descemet's membrane* (approximately 10 μm thick) which forms the basement layer of *endothelial cells*. The innermost corneal layer is the 5 μm thick *endothelium*. It is composed of hexagonal cells arranged in a honeycomb lattice and allows leakage of nutrients to the upper layers of the cornea. At the same time, the endothelium actively pumps water out of the cornea to keep it clear and transparent.

Uvea, choroid, iris, and ciliary body The *uvea* forms the middle shell of the eye. In the posterior part of the eye (Figure 1.5), the uvea forms the so-called choroid, that is, a blood-rich tissue supplying nutrients to the retina (Section 1.2). The *choroid* has a total thickness of 350–450 μm. In the anterior segment (Figure 1.4), the uvea has evolved into the *iris*. In optical terms, the iris is an adjustable aperture stop whose diameter can be modified by two antagonistic muscles (*sphincter* and *dilator pupillae*). The hole of the iris is called the *pupil*. Note that the "anatomic pupil" does *not* correspond to the optical entrance or exit pupils (Sections 2.1.1 and A.1.4). The color of the iris depends on the amount of pigmentation in the anterior limiting layer of iris and stroma.

Between iris and choroid, the uvea has formed into the *ciliary body* which has two important functions. On the one hand, it produces the aqueous humor. On the

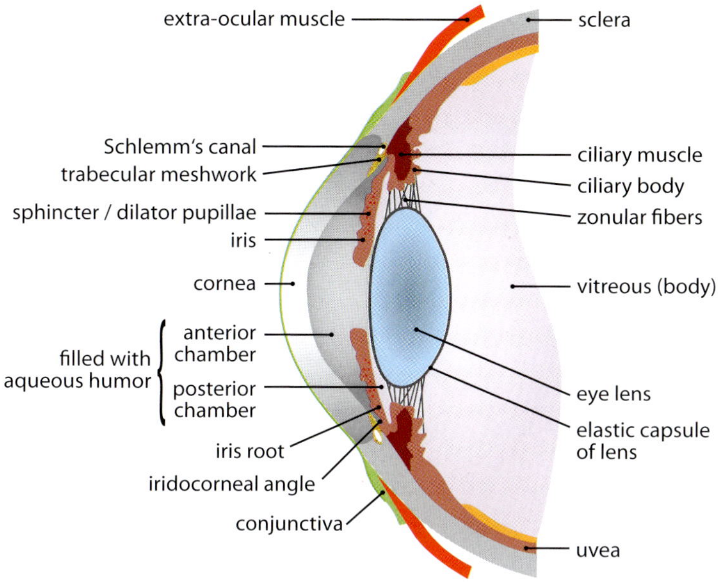

Figure 1.4 Scheme of the eye's anterior segment (see also Figure 1.3b). The depicted eye components are not to scale.

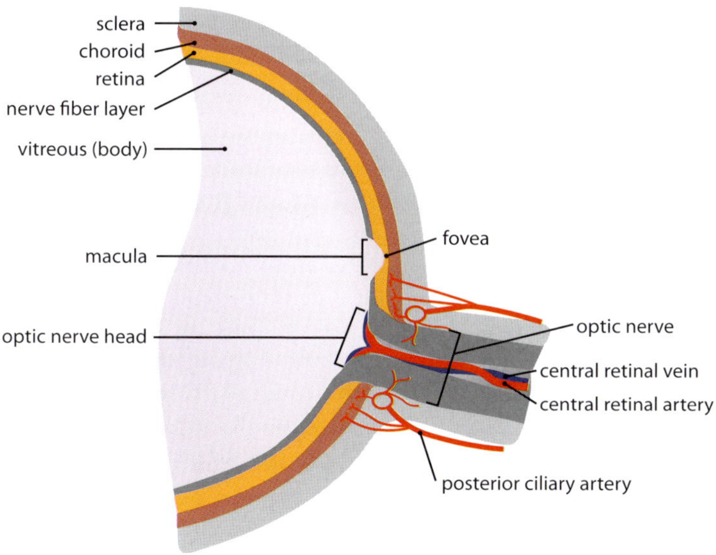

Figure 1.5 Scheme of the eye's posterior segment (see also Figure 1.3c). The depicted eye components are not to scale.

other hand, it comprises the ciliary muscle which may relax the tension on the eye lens so that near vision is possible (Section 2.1.4).

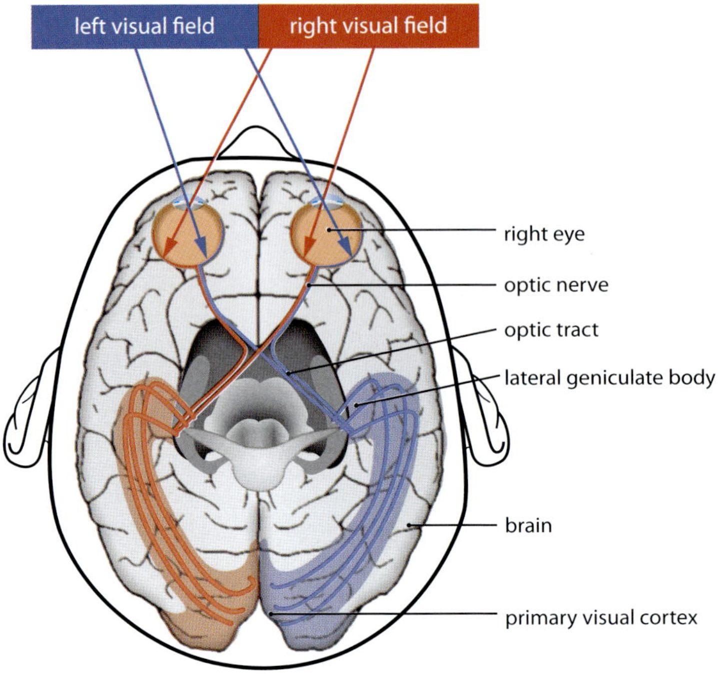

Figure 1.6 Transverse cross-section of the visual pathway including the primary visual cortex. The left and right areas of the retina are connected to different parts of the visual cortex. Hence, if one side of the visual cortex is impaired, the visual information of both eyes can still be used.

Eye lens Similar to the cornea, the eye lens is a transparent tissue which contains no nerve fibers or blood vessels. The required nutrients are supplied by the aqueous humor, which is a clear fluid. The eye lens is embedded into an elastic capsule which is again attached to the ciliary body via *zonular fibers*. The capsule is composed of collagen and varies from 2–28 μm in thickness. The lens itself consists of an *epithelial layer*, which is only located in the anterior part of lens, and the *lens fibers*. The cells of the epithelium are located between the lens capsule and the outermost layer of lens fibers. Lens fibers form the bulk of the interior of the lens. They are long (up to 12 mm), thin, and transparent cells whose diameter ranges between 4 and 7 μm. The eye lens consists of two kinds of fiber. The inner core, the so-called *nucleus*, is formed by primary lens fibers. The nucleus is surrounded by the *cortex* which is formed by secondary lens fibers. The major purpose of the lens is the refractive change (i.e. accommodation; Section 2.1.4) to focus nearby objects.

Eye chambers The interior of the eye is divided into three chambers. The space between cornea and iris is called the *anterior chamber*, and between iris and eye lens we have the *posterior chamber*. The remaining space between lens and retina is referred to as the *vitreous*. The anterior and posterior chambers are filled with aque-

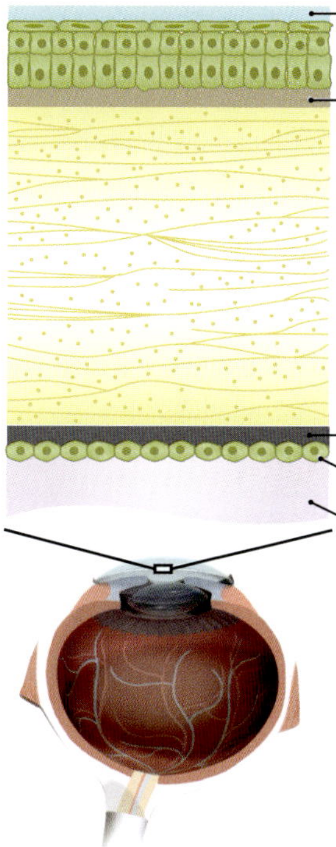

Figure 1.7 Detailed view of the corneal layer structure. The layer thicknesses are not to scale. Corresponding mean values are shown in parentheses. For reference, the cross-section of the whole eye is also shown. Adapted from [4].

ous humor which contains nutrients for the eye lens and inner cornea. The aqueous humor is produced by the epithelial cells of the ciliary body in an amount of approximately 2.4 mm^3/min and dispensed to the posterior chamber. It then flows continuously through the pupil to the *trabecular meshwork*, where it is eventually drained through *Schlemm's canal* (Figure 1.4). The relationship between production and discharge of aqueous humor determines the *intraocular pressure*, which is slightly higher than atmospheric pressure and approximately matches the intercranial pressure. The vitreous is filled with a transparent, colorless, gelatinous mass. In contrast to the aqueous humor, the vitreous does not flow through the eye.

Retina The inner layer of the human eye is referred to as the *retina*. The retina is an upstream part of the brain which "detects" light and converts the stimulus to neuronal signal, which are then transmitted by the optic nerve to the visual centers of the brain, where the image is eventually "formed". The retina will be highlighted in detail in the following section because of its high structural and functional complexity as well as its importance for vision.

Most parts of the eye are optically not accessible. When looking at the eye, we can only see the colored iris with the pupil and the white sclera. For optical imaging and structural analysis of the other parts, special optical techniques are required which we will discuss in Part Two of this book.

1.2 Retina: The Optical Sensor

The retina has evolved from the central nervous system and is actually part of the brain. It consists of approximately 127 million photoreceptors which convert incident light to electrical signals by a light-induced chemical reaction. These signals are than to some extent preprocessed by the retinal neural network and the ganglion cells. The nerve fibers of the approximately 1.2 million ganglion cells (called axons) merge on the *optic nerve head* (also called *optic disk*) into the optic nerve (Figure 1.5). From there, the neuronal signals are transmitted to the *primary visual cortex* of the brain (Figure 1.6).

1.2.1 Retinal Structure

The human retina is a multilayered tissue with a total thickness of about 180 μm in the fovea and between 200 and 400 μm elsewhere. The retinal structure is schematically shown in Figure 1.8. The retina contains two types of photoreceptors which are named after their shape: *rods* and *cones* (Figure 1.9). The photoreceptors are separated into three subregions, that is, outer segment, inner segment, and synaptic terminal. The outer segment is built up of a densely-packed stack of membranes which include the chromophores rhodopsin (rods) or iodopsin (cones)[2]. This part of the receptor is amazingly sensitive to incident light. If the eye is totally adapted to a dark environment, five to eight photons can be perceived when they hit the membrane within 20 ms[3]. The inner segment of a photoreceptor consists of the cell nucleus, mitochondria, and the endoplasmic reticulum. The synaptic terminal is a fiber-like extension of the nerve cell which conducts electric nerve impulses to the horizontal and bipolar cells of the retinal neural network. After some preprocessing there, these electrical signals are then conducted via the axons of the ganglion cells to the brain.

The spatial distribution of the photoreceptors across the retina is highly nonuniform (Figure 1.10b). The density of the about 6–7 million cones is highest in the fovea. The peak density of the about 120 million rods is located at about 15° from the eye's optical axis. In the fovea, no rods are present. Cones are only used under

2) Each pigment molecule consists of two components, a large protein molecule (*opsin*) and a small molecule derived from vitamin A (*retinal*). The latter is responsible for light absorption.

3) Light perception (also referred to as *transduction*) takes place in less than 1 ms. During this short period, the retinal molecule changes its shape and dissociates from its binding site on the opsin. This process, known as *bleaching*, is the only step in vision which depends on light. Then, an electrically conductive sodium ion (Na^+) channel is closed. As long as the channel is open in the dark, the excitatory neurotransmitter *glutamate* is steadily released.

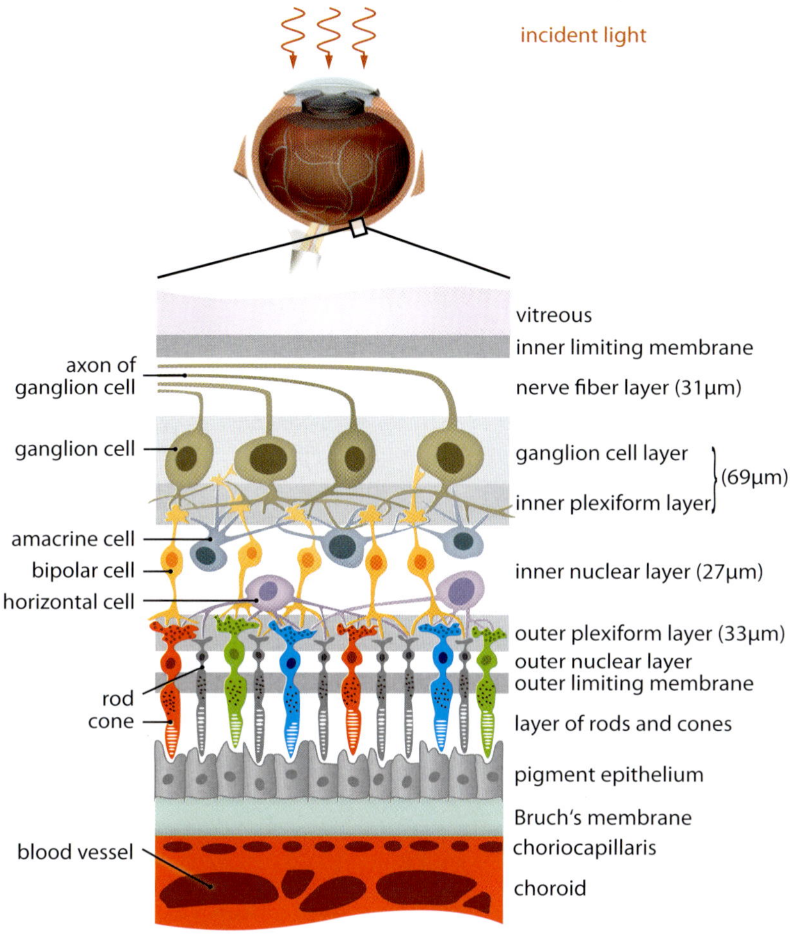

Figure 1.8 Cross-section of the functional retinal layers. At the top, the cross-section of the whole eye is added as a reference. The layer thicknesses and the relative distance (and density) of the photoreceptors are *not* to scale. Measured (mean) thickness values [5] of some retinal layers are shown in parentheses. The different colors of the cones refer to their distinct spectral absorption of light (Section 2.3). *S*, *M*, and *L* cones are shown in blue, green, and red, respectively. Adapted from [4].

sufficient ambient light conditions, that is, *daylight* or *photopic vision* (Section 2.1.6). In this case, three types of cones are used with different spectral absorptions so that we are able to perceive colors (Section 2.3). *S*, *M*, and *L* cones (color coding in Figure 1.8) contain pigments which absorb in the blue, green, and red spectral range, respectively. The human eye is thus *trichromatic* in daylight. The nervous system combines the signals of all types of cones and assigns a respective color.

In contrast to cones, rods are needed when the ambient light conditions are poor, that is, *night* or *scotopic vision* (Section 2.1.6). All rods contain the same chromophore so that the color impression decreases progressively with the light level.

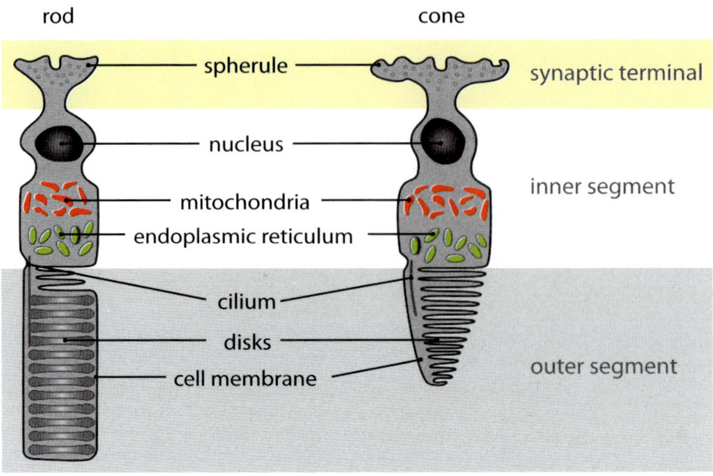

Figure 1.9 Segmentation and structure of retinal photoreceptors.

To achieve sufficient light sensitivity, the output signal of about 100 rods is combined on the way to the brain. As a consequence of this sampling the spatial resolution (Section 2.1.5) for scotopic vision is much lower compared to photopic vision, as in the fovea the density of cones is highest (Figure 1.10b), and each cone is connected to one nerve fiber.

1.2.2 Functional Areas

The *macula* (anatomic fovea centralis) is located in the center of the retina and is about 1.5 mm in diameter (Figures 1.5 and 1.10a). The most central part of the macula, called fovea (anatomic foveola), has a depression with a diameter of about 0.35 mm located at a field angle of approximately 5° from the eye's optical axis (Section 2.1.3). In this region we find the highest density of cone photoreceptors (Figure 1.10b). As a consequence, visual acuity and resolution for photopic vision (Section 2.1.5, Figure 2.8) have their maximum values in this region.

In the fovea, the cone nerve fibers are spread radially from the center and nearly parallel to the surface of the retina and form the so-called *Henle fiber layer* (Info Box 6.5). The first connection between the cone fibers and the bipolar cells of the neuronal network occurs outside the fovea. Consequently, there are no tissue layers such as inner plexiform, ganglion cell, and nerve fiber layers (Figure 1.8) above the foveal cones, which could result in an impaired image quality due to scattering of light (see Section 9.2) by these layers. For the same reason, the central part of the macula (500 μm in diameter) contains no retinal capillaries (so-called *foveal avascular zone*). Since rods are not present in the fovea, this area is basically night-blind. The sharpest vision in the case of scotopic vision is achieved at about 15° from the eye's optical axis because there we find the highest of the rod density (Figure 1.10b).

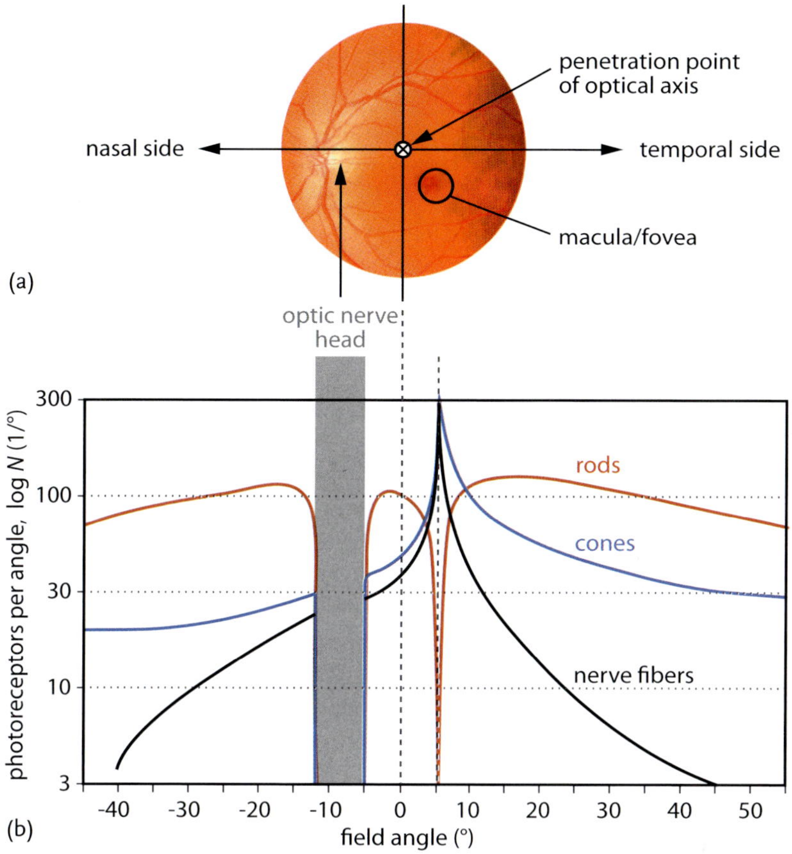

Figure 1.10 (a) Image of the eye fundus as a reference for the diagram in (b). The penetration point of the optical axis as well as the locations of fovea and optic nerve head are highlighted. (b) Distribution of photoreceptors and nerve fibers in the retina relative to the optical axis (0°). The logarithm of the number of photoreceptors *N* per degree of field angle is taken as the vertical axis. Data taken from [1].

It is worth noting that the human retina is an inverted detector (Figure 1.8). Incident light is converted to an electric signal by photoreceptors which form the lowest layer. After being preprocessed, the electric signal is transmitted by nerve fibers of the ganglion cell layer that form the top layer (towards the vitreous). This kind of structure has the advantage that photoreceptors can be directly supplied with nutrients by the choroid. At one common exit point, the nerve fibers are connected to the brain and, thus, have to penetrate all other layers (*not* shown in Figure 1.8). Here, at the optic nerve head, the retina is 600 μm thick, but does not comprise any photoreceptors (Figure 1.10b). The optic nerve head is located about 10° nasally relative to the optical axis and 1.5° upwards relative to the fovea (Figure 1.10a).

1.3 Recommended Reading

Further details about the anatomy, structure, and function of the human eye are presented in [1, 4, 6–10]. Further examples of visual illusions can be found in [11].

References

1 Gross, H., Blechinger, F., and Achtner, B. (2008) *Handbook of Optical Systems – Survey of Optical Instruments*, vol. 4, Wiley-VCH Verlag GmbH, Weinheim.

2 Ditzinger, T. (2006) *Illusion des Sehens*, Spektrum, Heidelberg.

3 Komai, Y. and Ushiki, T. (1991) The three-dimensional organization of collagen fibrils in the human cornea and sclera. *Invest. Ophthalmol. Vis. Sci.*, **32**, 2244–2258.

4 Grehn, F. (2008) *Augenheilkunde*, Springer.

5 Bagci, A.M., Shahidi, M., Ansari, R., Blair, M., Blair, N.P., and Zelkha, R. (2008) Thickness profiles of retinal layers by optical coherence tomography image segmentation. *Am. J. Ophthalmol.*, **146**, 679–687.

6 Remington, L. (2011) *Clinical Anatomy and Physiology of the Visual System*, Butterworth–Heinemann.

7 Snell, R.S. and Lemp, M.A. (1997) *Clinical Anatomy of the Eye*, Wiley–Blackwell.

8 Lemp, M.A. (1998) *Clinical Anatomy of the Eye*, Blackwell Publishers.

9 Tasman, W. (ed.) (1992) *Duane's Foundations of Clinical Ophthalmology*, vol. 2, Lippincott Williams & Wilkins.

10 Atchison, D.A. and Smith, G. (2006) *Optics of the Human Eye*, Butterworth–Heinemann.

11 Seckel, A. (2009) *Optical Illusions*, Firefly Books.

2
Optics of the Human Eye

In all ophthalmic and optometric devices to be presented, the eye is an essential part of the entire optical system. Because of this, the functional principles of these devices cannot be fully understood without an understanding of basic optics of the human eye. In this chapter, we will see that the eye can be described in a similar way as other optical systems. This finding is very important for further discussions in this book, so that this chapter serves as a basic reference.

The anatomy of the human eye is readily comparable to the design of a photo camera (Figure 2.1). We can thus identify the iris as an aperture stop (Section 2.1.1), that the cornea and eye lens form the objective lens (Section 2.1.4), that the retina is a photo sensor, and that the brain acts as a very sophisticated image processing computer with intelligent algorithms. In contrast to a photo camera, the eye is *not* a centered optical system as its refractive components and aperture stop are not centered at a common optical axis (Section 2.1.3). The performance of the "light sensors" are also different for the photo camera and the human eye. The resolution of the photo camera's detector is equal for the whole area, whereas the retinal resolution is inhomogeneous (Section 2.1.5). In the central part of the retina, the resolution is high and decreases at the margins.

2.1
Optical Imaging

When light is incident to the eye, first of all it enters the cornea. As our eyes are usually surrounded by air, the refractive power (Section A.1.2.1) at the air–cornea interface[1] is as high as 42 diopters (D). According to the Fresnel[2] equation (A5), the transmittance of the air–cornea interface is 98%. However, this value does not take scattering (Section 9.2) and absorption (Section 9.1) of the ocular media into

1) Strictly speaking, we have to consider the interface between air and the tear film (Section 1.1). However, the refractive indices of the tear film and the corneal layers are very similar. Thus, for the following discussions, we will regard the tear film as being a part of the cornea.
2) Augustin Fresnel (1788–1827).

Optical Devices in Ophthalmology and Optometry, First Edition. Michael Kaschke, Karl-Heinz Donnerhacke, and Michael Stefan Rill.

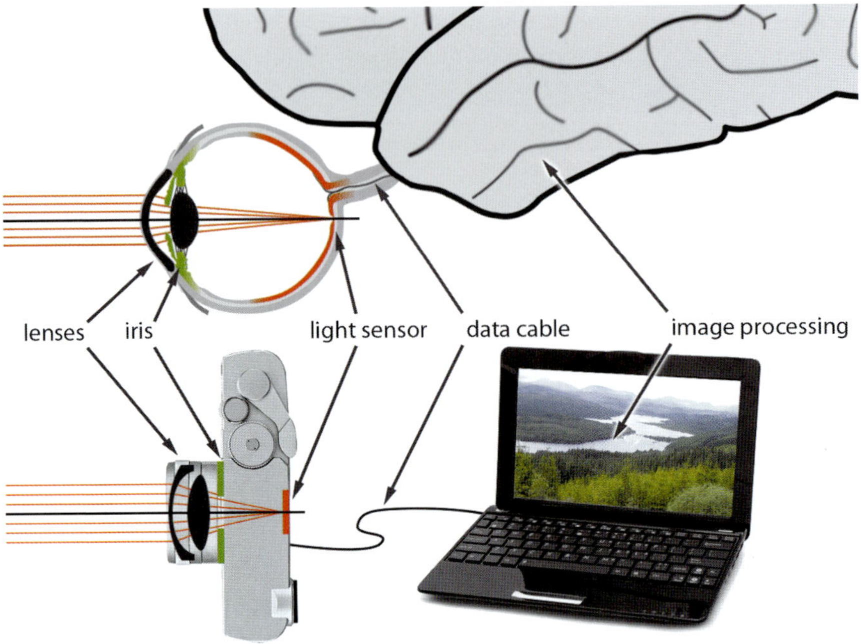

Figure 2.1 Comparison of human eye and photo camera. The optical system of the photo camera is reduced to the most necessary components. Usually, the arrangement is much more complicated since optical aberrations (Section A.1.7) have to be corrected.

account. Nevertheless, healthy corneal tissue is remarkably transparent. This is due to the ordered collagen fibrils[3] which are weak scatterers, as their radius is much smaller than the wavelength of visible light (fibril diameter: 25–35 nm; wavelength of visible light: 380–780 nm). In addition, the spatial distribution of fibrils reduces scattering because of destructive interference (Section A.2.3).

After the incident light rays have been refracted (Section A.1.1) by the cornea, they travel through the anterior chamber and cross the iris. For the eye as an optical system, the iris forms an aperture stop with a variable inner diameter (Figure 2.1). It limits the maximum acceptance angle, that is, the so-called *visual field*, for incident light rays to about 105°. On the nasal side, this angle is further reduced to 60° by the nose. For an iris diameter of 8 mm (scotopic vision), the eye has a maximum numerical aperture (Section A.1.4) of about 0.23 [1]. For an iris diameter of 3 mm, the numerical aperture reduces to 0.1.

When the light rays have passed the iris, they travel through the posterior chamber and enter the eye lens. The shape of the eye lens and thus its refractive power can be adjusted depending on the distance of the object being fixated. The lens consists of multiple shells which are stacked layer-by-layer. Each shell has a different refractive index, where a maximum refractive index of 1.42 is found in the core.

3) The human corneal stroma is composed of stacked lamellae. Within each lamella, collagen fibrils run parallel to each other and show a regular spacing (Section 1.1).

Behind the lens, light passes through the vitreous and is eventually "detected" by the retina. The image formed on the retina is inverted, that is, upside down, which is analogous to the imaging of a single lens (Figure A.6). Another inversion process happens in the brain, which results in the correct visual perception of our environment.

2.1.1
Entrance and Exit Pupils

In ophthalmology and optometry, the term "pupil" is often referred to as the hole of the iris (iris aperture). But technically, the iris is actually an aperture stop (Section A.1.4).[4] The cornea forms an image of this aperture stop which is in optical terms the entrance pupil of the eye. The exit pupil of the eye is the image of the same aperture stop formed by the eye lens.

As shown in Figure 2.2, we follow the paths of a marginal ray and the chief ray to determine the location and diameter of the entrance and exit pupils. The optical design of the eye has been simplified in this scheme, because we assume that the iris is centered on an optical axis. The chief ray emanates from the outermost off-axis point O_1 of the focused object. It is then refracted by the cornea, crosses the center of the iris, and is again refracted by the eye lens. Eventually, the chief ray hits the retina at point I_1'. As explained in Section A.1.4, the extensions of the chief ray define the positions of the pupil centers on the optical axis. The detailed view in Figure 2.2b shows the resulting pupil centers, E (entrance pupil) and E′ (exit pupil), relative to the ocular parts. The optical design in this figure is not to scale, but reveals correctly that the entrance pupil is larger and in front of the iris.

The marginal ray emanates from an on-axis object point O_0 and grazes the inner edge of the iris. When we extend the object- and image-side parts of the marginal ray to the pupil planes, we obtain the inner diameters of the entrance and exit pupils, respectively.

We can now use the chief ray to determine the image size on the retina. In paraxial approximation (Section A.1.2), the absolute value of the image size follows as[5] (Figure 2.2):

$$\left|h_{\mathrm{I}}'\right| = \varphi' \overline{\mathrm{E}'\mathrm{I}_0'}\,. \tag{2.1}$$

The overbar in (2.1) symbolizes the length between the points.[6] The iris of each human eye is decentered nasally by about 0.5 mm relative to the optical axis (formed by cornea and eye lens). In contrast to centered optical systems, the angles φ and φ' are *not* equal, but it can be shown that the relation $\varphi'/\varphi = m$ is constant[7]. With

4) In a very crude approximation, iris and entrance pupil coincide.

5) Throughout this book, primed variables describe optical design parameters and points within the image space, that is, between the first refracting surface and the image (see also Section A.1). Unprimed variables are used for object space quantities.

6) For example, $\overline{\mathrm{E}'\mathrm{I}_0'}$ stands for the distance between the center of the exit pupil E′ and the on-axis image point I_0'.

7) However, m depends on the state of accommodation (Section 2.1.4).

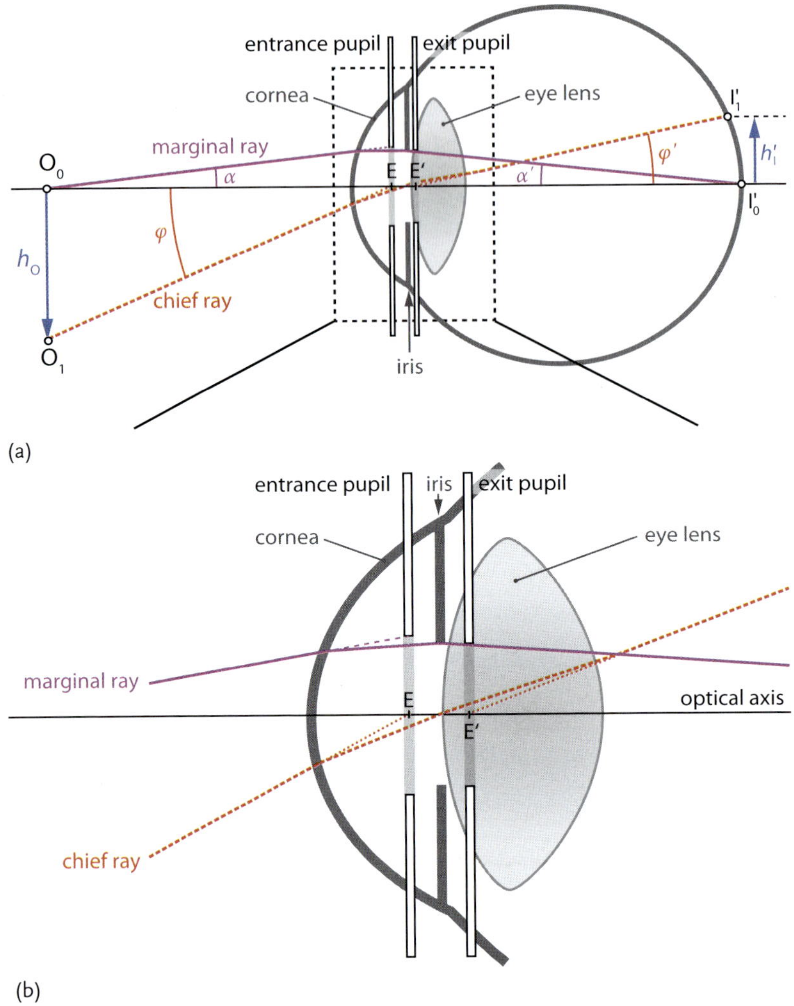

Figure 2.2 Location of entrance and exit pupils of the human eye. The indication of parameters is comparable to Figure A.13. (a) Path of the chief ray which starts at the outermost point O_1 of the object and passes through the center of the iris. The extensions of the chief ray on the object and image side define the centers of the entrance and exit pupils, respectively. E is the center of the entrance pupil and E′ the center of the exit pupil. φ and φ' denote the included angles between optical axis and chief ray on the object and image side, respectively. The corresponding angles between marginal ray and optical axis are α and α'. (b) Detailed view (dashed box of (a)) of the optical design in the anterior segment. Adapted from [2].

the object height h_O and $\varphi = |h_O|/\overline{O_0E}$, we may thus rewrite (2.1) as

$$|h'_I| = m\,|h_O|\,\frac{\overline{E'I'_0}}{\overline{O_0E}}\;. \tag{2.2}$$

2.1.2 Cardinal Points

For human eyes, the description of optical imaging can be simplified by introducing three types of *cardinal points.*[8] In centered optical systems, these points represent special locations on the optical axis which determine the basic imaging properties like image size, image location, and orientation. As the eye is not a centered optical system, we still use the concept of cardinal points but should understand that these points can merely be used as an approximate reference.

Focal points When incident light rays cross the object-side focal point F (Section A.1.2.1) and pass into the eye, they propagate parallel after refraction at cornea and lens. For an emmetropic eye (i.e., without refractive errors; Section 3.1), the image is formed at an infinite distance on the image side. If the incident light rays are parallel, they will be focused on the retina (at point F′) after refraction at cornea and (unaccommodated) eye lens.

Principal points The two principal points P and P′ are defined by the intersections of the principal planes (see, e.g., planes K and K′ in Figure A.8 of Section A.1.2.2) with the optical axis. These points are of interest if the combination of cornea and lens is considered to be one "thick" lens. In this case, the optical design may be simplified by assuming that incident rays are effectively refracted at the two principal planes.

Nodal points We consider a light ray emanated from the off-axis object point O_1 in Figure 2.3 which travels towards nodal point N. After refraction by cornea and lens, the same ray seems to originate from image-side nodal point N′. The special feature of this ray is that its angle to the optical axis is equal for the ray's incident (between O and the corneal front surface) and refracted part (between rear surface of lens and retina). As this ray passes through both nodal points, it is referred to as the *nodal ray*. It also defines the visual axis (Section 2.1.3) of the eye if I_1' marks the center of the fovea. In centered optical systems, the nodal and chief rays coincide so that the nodal points are actually located in the pupil centers.

Relation between cardinal points From geometrical considerations (Figure 2.3), we may derive some useful relations between the cardinal points [2]. We have

$$\mathcal{D}_{\text{eye}} = -\frac{n}{\overline{\text{PF}}} = \frac{n'}{\overline{\text{P}'\text{F}'}} , \tag{2.3}$$

$$\overline{\text{PN}} = \overline{\text{P}'\text{N}'} = \frac{n' - n}{\mathcal{D}_{\text{eye}}} , \tag{2.4}$$

8) In geometric optics (Appendix A), we also use cardinal points to simplify the description of optical imaging. However, the artificial systems discussed in that chapter are centered so that only focal and principal points must be considered. In decentered optical systems like the human eye, we also have to introduce nodal points.

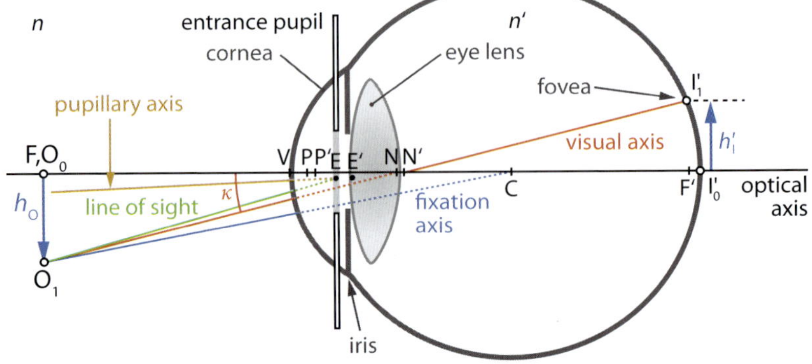

Figure 2.3 Axes and cardinal points of the eye with corresponding inclination angles. V is the point of intersection of the optical axis with the cornea (*corneal vertex*). E and E′ represent the centers of the entrance and exit pupils. N and N′ denote the nodal points of the eye, and C is the rotation center of the eye. P and P′ are the principal points of the eye. *n* and *n*′ represent the refractive indices outside and inside the eye, and κ is the angle between visual axis and optical axis. Adapted from [2].

$$\overline{\mathrm{FN}} = \overline{\mathrm{P'F'}}\,, \tag{2.5}$$

$$\overline{\mathrm{FP}} = \overline{\mathrm{N'F'}}\,. \tag{2.6}$$

$\mathcal{D}_{\text{eye}}$ denotes the total refractive power of the human eye, n is the refractive index of the object space (usually air), and n' the refractive index of the image space (i.e., the refractive index of the vitreous). Values for $\mathcal{D}_{\text{eye}}$ and n' will be specified in Section 2.2.

2.1.3
Eye Axes

In centered optical systems, the optical axis is usually determined by the line which intersects with the centers of curvature of all refracting and reflecting surfaces. Since the ocular parts are decentered, it is useful to "redefine" the optical axis of an eye as the best-fit line between the centers of curvature of all refracting surfaces (black line in Figure 2.3). In addition, we introduce some other axes that help us to describe the eye's optical geometry (see also [2]).

Visual axis The line between the fixated point O_1 and fovea by way of nodal points N and N′ is referred to as the *visual axis*. The visual axis thus consists of the two line segments $\overline{\mathrm{O}_1\mathrm{N}}$ and $\overline{\mathrm{N}'\mathrm{I}'_1}$ (red line in Figure 2.3) and forms the actual imaging axis of the eye. On average, the optical axis and the visual axis enclose an angle of $\kappa \approx 5°$ on the object side.

With the visual axis, we can once again determine the retinal image size (compare with Section 2.1.1). In the case of paraxial optics, the absolute value of the retinal image size is given by

$$|h'_{\mathrm{I}}| = \kappa \overline{\mathrm{N'I'_0}} \tag{2.7}$$

$$= |h_{\mathrm{O}}| \frac{\overline{\mathrm{N'I'_0}}}{\overline{\mathrm{O_0N}}}, \tag{2.8}$$

where $\kappa = -h_{\mathrm{O}}/\overline{\mathrm{O_0N}}$.

Line of sight The *line of sight* is given by the line between a fixated object point O_1 and the center of entrance pupil E (green line in Figure 2.3). On average, the angle between the line of sight and the pupillary axis is approximately 2.5°. The position at which the line of sight crosses the cornea is referred to as the *corneal sight center*.

Pupillary axis The *pupillary axis* passes through the center of entrance pupil E and is perpendicular to the corneal surface (orange line in Figure 2.3). It is used as an objective measure to judge the amount of eccentric fixation. As the eye is not a centered optical system, the entrance pupil is often not concentric to the cornea. The cornea may also have an irregular shape. Both factors cause the pupillary axis to be different from the optical axis. However, for the following discussions, we assume the center of the entrance pupil to lie on the optical axis.

Fixation axis The *fixation axis* is the reference axis for eye movements. It is determined by the line between object point O_1 and center of eye rotation C (blue line in Figure 2.3).

2.1.4 Accommodation

In healthy eyes, the refractive power of the eye lens is at maximum $\mathcal{D}_{\mathrm{l}} = 20\,\mathrm{D}$ and thus contributes only $\leq$ 30% to the total eye refraction.[9] However, within a certain limit, the lens is able to change the refractive power so that nearby as well as distant objects can be sharply imaged on the retina. This process is referred to as *accommodation*. The range over which the refractive power can be changed depends on age.

Mechanism of accommodation If the eye focuses on nearby objects, the ciliary muscle is contracted and the zonular fibers are relaxed (*accommodated eye*). When the tension on the lens is decreased, the elasticity of the lens capsule keeps it in a more spherical shape (upper part of Figure 2.4). As the lens becomes more strongly curved in this case,[10] the eye's total refractive power increases (*near vision*). To focus objects which are located far away from the eye (*far vision*), the deformable, elastic

9) The lower refractive power of the eye lens results from the smaller difference of refractive indices at the aqueous humor–lens and lens–vitreous interfaces.
10) This corresponds to a reduced radius of curvature.

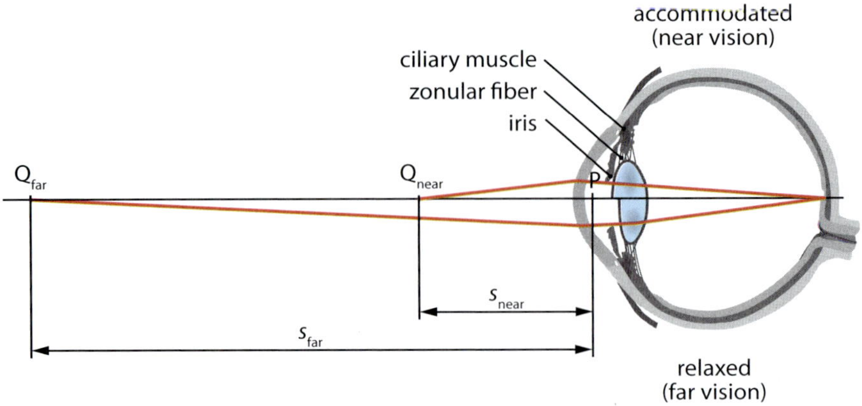

Figure 2.4 Physiology of accommodation. Upper half of figure: If the ciliary muscle is contracted, the zonular fibers are relaxed and the elasticity of the lens capsule keeps the lens in a more spherical shape. In this case, the refractive power of the lens is higher so that nearby objects can be imaged (near vision). Lower half of figure: To fixate objects which are located far away from the eye (far vision), the deformable eye lens is brought to an elliptical shape by pulling on the lens capsule. The pulling force which acts on the zonular fibers is generated by a relaxed ciliary muscle. For reference, the near point Q_{near}, far point Q_{far}, principal point P, and the corresponding distances (s_{near} and s_{far}) are shown.

lens is brought to a more elliptical shape by pulling on the lens capsule. The pulling force acting on the zonular fibers is generated by a *relaxed* ciliary muscle (*relaxed eye*). This situation is illustrated in the lower part of Figure 2.4.

Accommodation is an unconscious process that is not yet fully understood. But it is a common belief that chromatic aberrations (Section A.1.9) may deliver the required optical stimulus [3, 4].

Range of accommodation The refractive power of the eye lens can be changed only within certain limits. The upper and lower limits of attainable refractive power determine the *range of accommodation* within which sharp vision is possible. The endpoints of the range of accommodation are called *far* and *near point*, respectively. The far point Q_{far} is the object point imaged by the eye when the total refractive power is minimal. The near point Q_{near} is the object point imaged by the eye when the total refractive power is maximal. The corresponding distances of Q_{far} and Q_{near} from the object-side principal point P (Section 2.1.2) of the eye are referred to as the *far point distance* s_{far} and *near point distance* s_{near}, respectively. If s_{far} or s_{near} are situated in front of the eye, the distances are negative. If they lie (virtually) behind the eye, the distances are set positive. The inverse distances are called the *far point refraction* $A_{far} = 1/s_{far}$ and *near point refraction* $A_{near} = 1/s_{near}$ ($[A_{far}] = [A_{near}] = \mathrm{D}$).

The difference between far and near point refraction is referred to as the *amplitude of accommodation*

$$\Delta A_{max} = A_{far} - A_{near} \,. \tag{2.9}$$

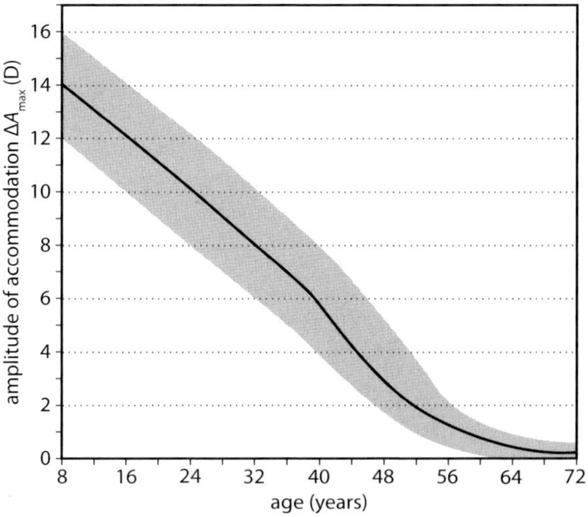

Figure 2.5 Age-dependence of the amplitude of accommodation ΔA_{max}. The typical range of deviation from the mean values (black curve) is shown in gray. Data taken from [6].

For example, if Q_{far} lies at optical "infinity"[11] ($A_{far} = 0$ D) and Q_{near} at a distance of $s_{near} = 0.2$ m ($A_{near} = -5$ D), the amplitude of accommodation is $\Delta A_{max} = 5$ D. The amplitude of accommodation is not constant during life. With advancing age, the elasticity of the lens and thus the range of accommodation decreases. As a consequence, humans usually need eye glasses for near vision at the age of 50. This reduction of ΔA_{max} (Figure 2.5), called *presbyopia,* is *not* a refractive error or an eye disease (Chapter 3), but a usual consequence of aging. According to [5], about 1.04 billion people suffered from presbyopia in 2005, 67% of which (696 million people) live in less- and least-developed regions of the world.

2.1.5
Resolution

The resolution of an optical system can be defined as the smallest distance between two Airy disks which can be perceived as being separated (Section A.2.1.6). Let us consider the resolution of the human eye by using the same approach. At a wavelength of 550 nm and a pupil diameter of 3 mm, we obtain from (A77) a minimum angle of resolution (MAR) for the eye of $\theta_{eye,min} = \text{MAR} = 48''$ (arcsec) $\approx 1'$ (arcmin). This value merely reveals the normal maximum possible optical resolution of a diffraction-limited eye (Section A.2.1.6). To evaluate the eye's maximum possible anatomic resolution, we have to consider the photoreceptor spacing as well. In the fovea, the density of cones is maximal (Figure 1.10b), and a nearly one-to-one connection exists between cones and retinal ganglion cells. Cones in the fovea have

11) This means that the far point Q_{far} is infinitely far away from the eye. An incident bundle of rays emanated from this point is thus parallel (i.e. *collimated light*).

a center-to-center distance of about 2 μm. When the images of two adjacent point sources stimulate two adjacent cones, they are perceived as being only one point source. If, however, there is one cone unstimulated in between those stimulated by the point sources, then their images are perceived as being separated. As a consequence, a minimum separation of $\approx 4\ \mu m$ between image points is required. This value corresponds well to the angle of $48''$ at the nodal point, taking into account that the image-side nodal point lies about 17 mm in front of the retina (Figure 2.13). The angular limit of photoreceptor resolution is thus in agreement with the minimum angle of resolution of a diffraction-limited eye. In the case of scotopic vision (Section 1.2.1) however, the eye's resolution is much lower ($\theta_{eye} > 10'$). As the output signal of about 100 rods is combined into one ganglion cell, the effective size of one "detector pixel" is much larger and sensitivity is much higher, too.

2.1.5.1 **Visual Performance**

The values we have calculated for MAR can only be achieved with healthy eyes under ideal ambient light conditions. Normal vision may be impaired by refractive errors (Section 3.1), higher-order aberrations (Section 5.4), eye diseases (Sections 3.2–3.7), and/or problems with the processing of visual signals.

To quantify the visual performance of a patient, we could directly determine the minimum angle of resolution (MAR). In practice, however, this quantity could be a bit confusing, since a large angle means low vision and vice versa. Thus, the visual performance is usually expressed by the inverse of the minimum angle of resolution

$$V = \frac{1}{\text{MAR}}\,, \tag{2.10}$$

where V is the so-called *visual acuity*, which we will preferably use in this book. A patient with $V = 1$ ($[V] = 1/' = (\text{arcmin})^{-1}$) is considered to have normal vision. By definition, the visual acuity scale is divided into intervals such that the quotient of two adjacent values of V is constant[12] (e.g., $1/1'$, $1/1.3'$, $1/1.6'$, $1/2.0'$, $1/2.5'$). As a consequence, the visual acuity scale has 22 divisions ranging from $V = 0.020$ to $V = 2.5$. After three scale divisions the visual acuity has doubled.

Another common measure for the visual performance is specified by the common logarithm of the minimum angle of resolution ($\log_{10}$ MAR). This so-called *logMAR* scale is particularly used in scientific publications. The visual acuity scale is related to the logMAR scale via

$$V = 10^{-\text{logMAR}} \tag{2.11}$$

and can also be defined via distances. As a reference parameter, a distance is chosen at which two object points or lines can be clearly distinguished under an angle of $1'$. For example, two points separated by 1.75 mm are placed at a distance of 6 m so that they appear at an angle of $1'$. If a test person can perceive these as two points only from a distance of 3 m, then his or her visual acuity is consequently $V = 0.5$ (which corresponds to 0.3 logMAR).

12) This actually corresponds to a decibel (dB) scale.

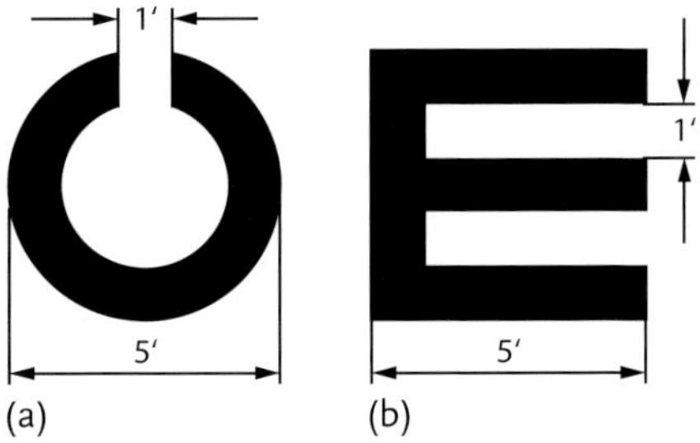

Figure 2.6 Two typical symbols which are used to determine the visual acuity. For subjective measurements, the symbols are displayed in different sizes and in various states of rotation. The patient then has to state in which direction the respective feature of interest shows. This type of chart can thus also be used for patients who are illiterate or too young to read. (a) Landolt ring. The feature of interest is the gap of the "C"-shaped symbol. (b) Snellen E. The feature of interest is the limb.

2.1.5.2 **Determination of the Visual Performance**

Visual acuity is measured by subjective methods. The smallest feature size which can be clearly resolved by the patient determines the visual acuity. For this purpose, a wall chart with *Landolt rings*[13] (Figure 2.6a) is used as a test target. The symbols are displayed in different sizes and orientations at a defined distance from the patient. The patient is now asked to state in which direction the corresponding feature of interest shows. The smallest feature which can be clearly recognized by the patient determines the visual acuity. For example, if a patient is able to recognize the orientation of Landolt ring gaps (top, right, bottom, left) with a gap size of 1.75 mm at a distance of 6 m, he or she has a visual acuity of 1. Patients with a lower visual acuity see a blurred image (e.g., the Landolt ring is perceived as a closed ring or dot) and thus cannot find the right feature orientation.

In clinical practice, the so-called *Snellen chart* is used as an alternative measure for the visual acuity. It consist of letters of the alphabet (see e.g. the "Snellen E" in Figure 2.6b) which are arranged in rows. In each row, the size of the letters is different so that each row can be used to test a different level of acuity. The rating of the acuity relates to the distance at which an emmetropic test person ("normal" visual acuity) is able to recognize the letters in that line. The (Snellen) visual acuity is defined by

$$V_S = \frac{\text{testing distance (in m)}}{\text{distance (in m) at which test line letters subtend an angle of } 5'} \,. \tag{2.12}$$

The 6/6 acuity line represents the "normal" line and contains letters that subtend an angle of 5′ (with a minimum feature size of 1′) at a distance of 6 m.

Subjective methods do not only check the performance of the pure optical system, but also determine the image processing capability of the sensory organ on the whole (i.e., the combined eye–brain imaging system).

13) Edmund Landolt (1846–1926).

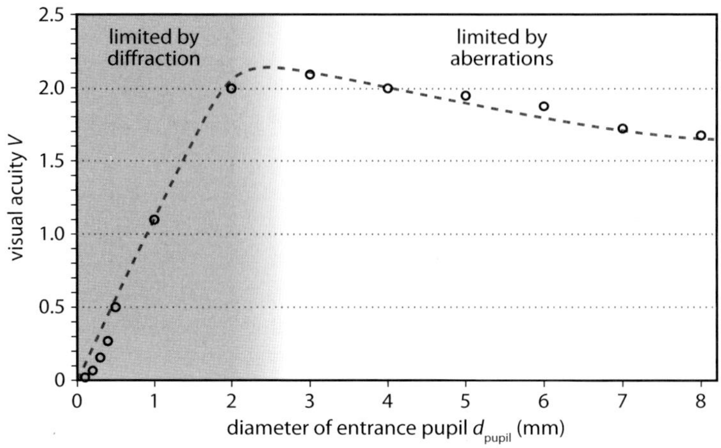

Figure 2.7 Diameter of the eye's entrance pupil versus visual acuity. A maximum visual acuity is attained for a pupil diameter of about 2.5 mm. Below 2.5 mm (gray area), the optical performance is limited by diffraction. Above 2.5 mm, aberrations deteriorate the optical resolution of the eye. Data points are taken from [7]. The dashed line is a best-fit curve through the data points and meant as a guidance.

2.1.5.3 **Influence Factors on the Visual Performance**

As already mentioned, refractive errors and insufficient ambient light conditions decrease the visual acuity of human eyes. In addition, a number of other factors may influence vision, such as the diameter of the eye's entrance pupil (Figure 2.7). For photopic vision (Section 2.1.6), the best resolution is given for a pupil diameter of about $d_{pupil} \approx 2.5$ mm. A smaller pupil diameter deteriorates the resolution as diffraction (Section A.2.1.6) comes into play. If the pupil diameter is larger, optical aberrations (Section A.1.6) reduce the resolution so that the visual acuity eventually "saturates".

As the density of cones and ganglion cells rapidly decreases outside the fovea, the visual performance for photopic vision also depends on the field angle at which the image is projected on the retina (Figure 2.8). At the optic nerve head, no photoreceptors exist at all so that the visual acuity equals to zero in this area (the so-called blind spot). Other influence factors for the visual performance are the shape, brightness, and color of considered objects as well as the degree of attention (psychological influence factors).

2.1.6 **Adaption**

The eye is able to maintain a high sensitivity to small changes in light intensity across a broad range of ambient light levels. Full operation of human vision is possible for a luminance between 10^{-6} and 10^8 cd/m^2 [8]. For this purpose, the eye uses the following mechanisms to adapt to the given ambient light conditions:

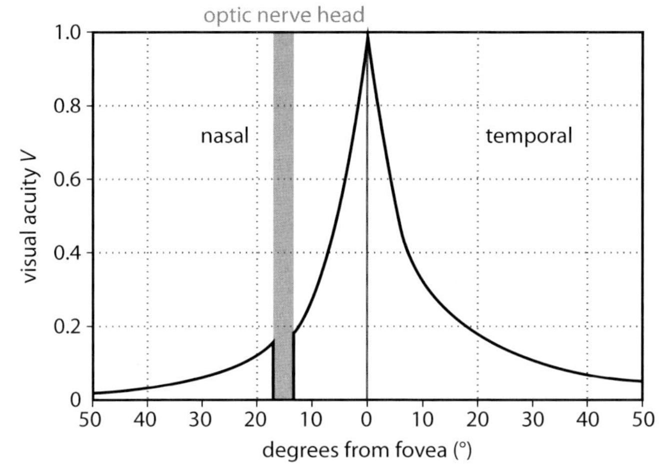

Figure 2.8 Dependence of the visual acuity on the field angle for photopic vision. The density of cones is maximal at the fovea (0°), but rapidly decreases outside this retinal region. As a consequence, the resolution is gradually reduced for increasing field angles. Adapted from [6].

1. As discussed in Section 1.2, the retina has two types of photoreceptors (rods and cones), which are used at different illumination levels. The operational range of rods – which ends at rod saturation – spans a remarkable 8 orders of magnitude in luminance (Figure 2.9). In the range of *scotopic vision*, only rods are used. Cones operate even over a range of 11 orders of magnitude in luminance, at which the range of operation partly overlaps with the rods (*mesopic vision*). If the luminance is higher than 10 cd/m^2, only cones are used for vision (*photopic vision*). The "operational" range of each type of photoreceptor can be regulated by biochemical (Section 1.2.1) and neuronal processes in the pigments within the outer segment (Figure 1.9). Since these processes are relatively slow, the adaption of the photoreceptors takes several minutes. Therefore, a faster adaption process is required which protects the retina from overload and damage.
2. The inner diameter of the iris diaphragm can be changed by the so-called *pupil reflex*[14]. This mechanism is the first stage of sensitivity regulation when the ambient light level is changed. When the iris diameter reduces from 8 mm (maximum diameter at total darkness) to 2 mm (minimum diameter at very high light levels), the area of the aperture stop decreases by a factor of 16. As a consequence, the iris can regulate the amount of light which enters the eye only by one order of magnitude. This is obviously not sufficient to account for the manageable luminance range of 14 orders of magnitude. The variation of the iris diameter is thus only responsible for the first stages of adaption, before the photoreceptors adjust their sensitivity according to the present luminance.

14) The term "pupil" can be misleading. In optics, the hole in the iris acts as an aperture stop. Only the image of this hole formed by the cornea should be referred to as a "pupil".

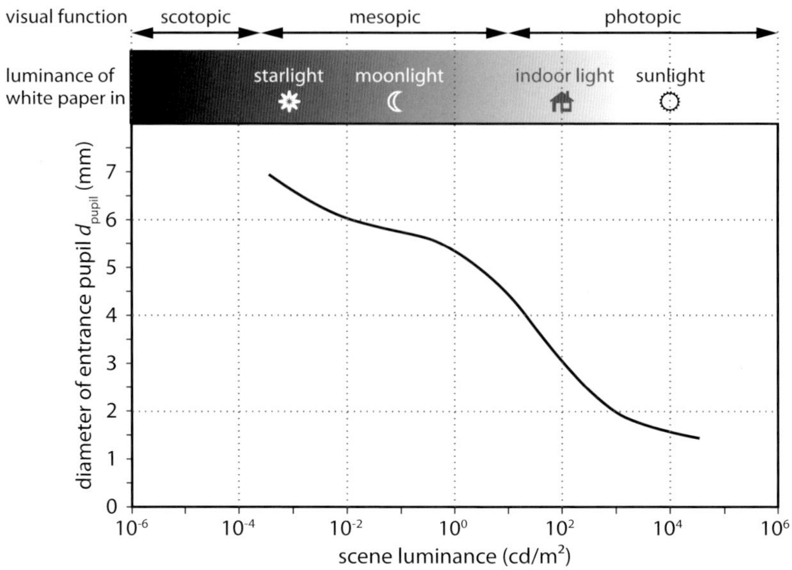

Figure 2.9 Range of illumination levels which can be handled by the eye. The luminance magnitudes of a white paper in starlight, in moonlight, indoors, and in sunlight are also shown for reference. Data taken from [9].

Let us consider the process when the ambient light level becomes low (transition from photopic to scotopic vision). As discussed, the inner diameter of the iris increases, the rods are activated, and the brain starts to increase the neural detection time of the photoreceptors. The adaption process to darkness takes about 40 min to complete. During the first 7 min, new pigments are generated for the cones so that the threshold sensitivity for the smallest detectable visual stimulus is increased by a factor of 50. During the following 30 min, the rod pigment rhodopsin is produced which increases the sensitivity of the retina by a factor of 1000 [6]. At low light levels, the sensitivity is high, but the spatial resolution of the eye is reduced (Section 2.1.5) and colors cannot be distinguished (Section 2.3).

In the reverse case of high ambient light conditions (photopic vision), the iris diameter decreases, the cones are activated, and the neural detection time is reduced. As a consequence, the eye is no longer sensitive to small changes in light intensity. The biochemical adaption of cones takes about 3–4 min. The spatial resolution and the color perception is enhanced compared to scotopic vision.

2.1.7
Stiles–Crawford Effect

Due to the tapered shape of cones (Figure 1.9), incident light is guided towards the end of the photoreceptors by total internal reflection (Section A.1.1), similar to an optical waveguide (Section 10.2.4.2). Light rays which pass through the center of the eye's entrance pupil are incident at a small angle to the cones of the curved

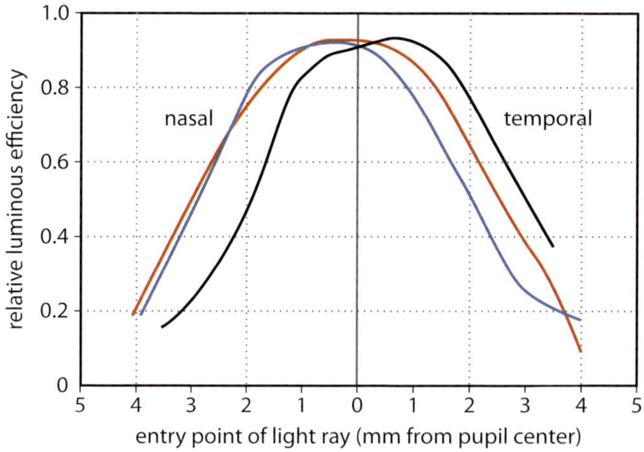

Figure 2.10 The relative luminous efficiency of a focused target versus the horizontal pupillary position of the beam. The values were measured under photopic conditions for three test persons (different curve colors). Apparently, the maximum efficiency is not measured in the center of the entrance pupil. Data taken from [8].

retina and stimulate the cones more "effectively" than rays which pass through the outer zone of the pupil. If the angle of incidence is large, the total reflection is suppressed and not all rays can be absorbed. This phenomenon is referred to as the *Stiles–Crawford effect*. The relative luminous efficiency, that is, the amount of light which is detected by the photoreceptors versus the incident luminance, is shown for three different test persons in Figure 2.10. We see that the effect varies slightly for every person and is not always symmetric about the center of the entrance pupil. However, a simple function to fit the data in Figure 2.10 can be given by [8]

$$\log_{10}\left(\frac{\eta_{\text{SC}}}{\eta_{\text{SC,max}}}\right) \approx -0.07 r^2 \,, \tag{2.13}$$

where $\eta_{\text{SC}}/\eta_{\text{SC,max}}$ is the relative luminous efficiency and r the distance of the ray from the pupil center. The Stiles–Crawford effect has a positive effect on the image quality of the eye, because the influence of spherical aberration (Section A.1.6.1) decreases for a large pupil diameter.

2.1.8
Depth of Field

If an object is moved slightly away from the focal point, this has only a little effect on the quality of the retinal image. But the image quality deteriorates progressively as we move further away from the focal point. The depth of field Δz_{dof} is defined as the maximum distance the object can be shifted and the image still reveals an acceptable sharpness. Δz_{dof} depends on the state of adaption (Section 2.1.6) and is affected by retinal, neural, and psychophysical factors. A commonly used empirical formula to determine the depth of field for the human eye is given by [10]

$$\Delta z_{\text{dof}} = \frac{\lambda}{2\text{NA}^2} + \frac{1}{7\text{NA}\beta} , \tag{2.14}$$

where λ denotes the wavelength of the incident light and NA the numerical aperture (Section A.1.4) of the eye. β is the magnification (Section A.1.2.1) of the eye that depends on the state of accommodation. Similar to camera systems, a large diameter of the aperture stop – which corresponds to a high NA – thus means a small depth of field. In (2.14), accommodation was taken into account by ways of β.

2.1.9
Binocular Vision

Both eyes are arranged in a common plane and separated by an *interpupillary distance* PD (i.e., the distance between both pupil centers) of 50–75 mm. This special arrangement allows humans to get a three-dimensional impression of the environment. For example, we can estimate depth and distance of objects which are placed in a row. This property of binocular vision is referred to as *stereopsis*. It is based on the comparison of two slightly different retinal images by the brain. In more concrete terms, both images received by the eyes are two-dimensional but horizontally shifted. The brain is now able to combine this information to "generate" a three-dimensional image and can distinguish between any objects which are located in different planes along the viewing direction (i.e., the z direction in Figure 2.11).

Figure 2.11 shows a top-view scene of two eyes of a person who fixates object point O_{fix}. O_{fix} is located at a distance L from the nodal plane[15] of the eyes (Section 2.1.3) and is sharply imaged onto each fovea. Object point O_{f}, which lies in front of O_{fix}, is imaged onto the temporal sides of both eyes (image point I'_{f}). Object point O_{b}, which lies behind O_{fix}, is imaged onto the nasal sides of the eyes (image point I'_{b}). The retinal images of O_{b} are closer together than the images of O_{f} which is "translated" to a different position in depth.

In Figure 2.11, the temporal ray of the right eye crosses the nasal ray of the left eye at point A. Similarly, the temporal ray of the left eye crosses the nasal ray of the right eye at point B. Points A and B are both located on the object plane and separated by the *parallax* distance s_{p} with which we can write the *stereo angle* as

$$\varepsilon = 2 \arctan\left(\frac{s_{\text{p}}}{2L}\right) . \tag{2.15}$$

The minimum stereo angle determines the smallest angle that can be resolved by the eye and still allows stereoscopic perception. Under appropriate conditions, the human eye has a minimum stereo angle of $\varepsilon_{\text{min}} = 10''$ [6].

A related quantity is the stereoscopic depth perception ΔL [12]. This quantity specifies the distance between two objects in a row which can be perceived as being

15) At the nodal point (Section 2.1.3), the incident light rays are crossing the optical axis of an eye. The nodal points of both eyes lie on one common *nodal plane*.

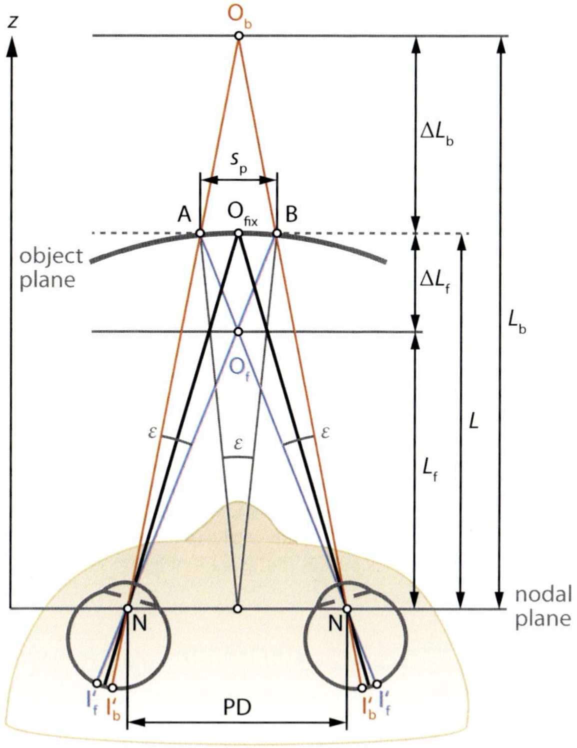

Figure 2.11 Geometric consideration of stereopsis. O_{fix} is an object point focused by both eyes. Hence, an image is created at both foveae. The object points O_b and O_f are projected onto the nasal and temporal sides of the retina, respectively. ε is the stereo angle. Adapted from [11].

separated in depth. For objects which lie in front of O_{fix}, it is given by [11]

$$\Delta L_f = \frac{s_p L}{PD + s_p} . \tag{2.16}$$

For objects behind O_{fix}, we have

$$\Delta L_b = \frac{s_p L}{PD - s_p} . \tag{2.17}$$

If the parallax distance s_p is small compared to the interpupillary distance PD, the stereo angle becomes also very small and can be approximated by $\varepsilon \approx s_p/L$. In this case, we may use the simplified relation

$$\Delta L_f \approx \Delta L_b = \frac{s_p L}{PD} = \frac{\varepsilon L^2}{PD} . \tag{2.18}$$

For a typical reading distance of $s_{nv} = 25$ cm (i.e., the *typical near viewing distance* of healthy eyes of young adults), the minimum stereoscopic depth perception of a human eye can be calculated as

$$\Delta L_{\min} = \frac{\varepsilon_{\min} s_{\mathrm{nv}}^2}{\mathrm{PD}} = 45\ \mu\mathrm{m}\,. \tag{2.19}$$

With adequate visual aids (e.g., surgical microscopes; Section 6.2), $\Delta L_{\min}$ can be further reduced. For this purpose, we can either enlarge the *stereo base* (i.e., the effective interpupillary distance), e.g., by mirrors or prisms, or increase the magnification of the image so that the three-dimensional image impression is considerably enhanced. Without visual aids, stereopsis breaks down at object distances > 500 m.

Besides stereopsis, limited depth perception of distant objects can be attained with just one eye as well. In this case, the distinction in depth is based on recognition patterns and experience. Some typical determining factors are:

- recognition of perspective, that is, smaller objects appear to be further away than larger objects,
- partial overlap of objects,
- distribution of light and shadows, and
- parallax of motion, that is, a moving object appears to be slower if it is further away.

2.1.10
Spectral Properties

Since water is the major substance of the eye's optical components, the spectral properties (Figure 9.3) such as absorption and scattering are determined by those of water.[16] In particular for wavelengths $\lambda > 600$ nm, the absorption of water dominates the spectral properties of ocular media. For wavelengths $\lambda < 600$ nm, however, proteins and chromophores become important.

Figure 2.12 shows the transmittance spectrum at different locations in the eye. We can see that wavelengths below 400 nm and above 1400 nm are totally absorbed by the eye's ocular media, whereas the transmittance remains higher than 0.6 in the spectral range between 420 and 920 nm (see transmittance at retina (black) in Figure 2.12).

With regard to dispersion, the properties of ocular media are determined by the Abbe number (see Eq. (A6)) $\nu_{\mathrm{eye}} \approx 50.2$. Hence, eye tissue shows a higher degree of dispersion than most silica glasses.

16) This is certainly true for the aqueous humor. In cornea and lens, proteins are included so that the absorption properties are slightly different for them. Nevertheless, it is reasonable to model the spectral properties of an eye by a water depth of 16 mm.

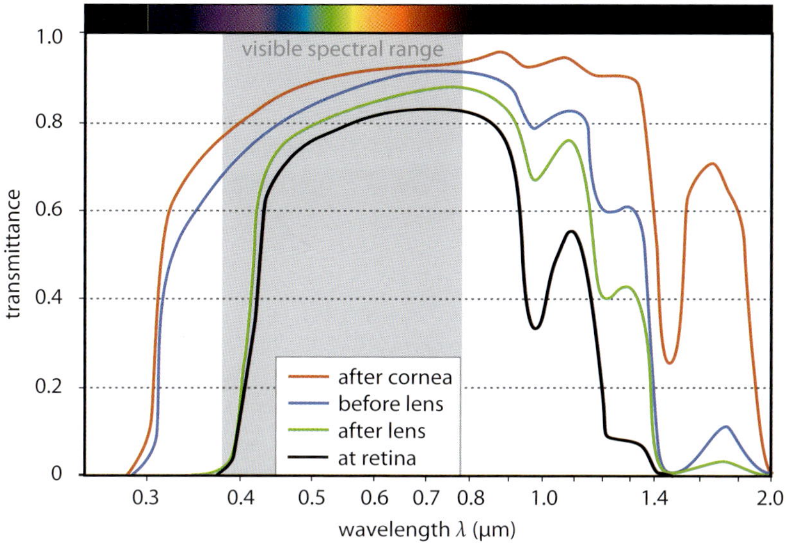

Figure 2.12 Spectral transmittance inside the eye after the incident light has passed the cornea (red), the aqueous humor (blue), the lens (green), and the vitreous (black). The wavelength scale is not linear. For reference, a color bar is added at the top of the diagram. Data taken from [6].

2.2 Schematic Eye Models

Both the optical properties and the geometry of eyes that we have discussed so far vary between human beings. This obviously makes it difficult to design and simulate optical devices where the eye is part of the whole system – regardless of whether the eye acts as a detector or as the examined object. To deal with this issue in a somewhat more rigorous manner, schematic *eye models* [2] have been developed which emulate real eyes under certain boundary conditions as closely as possible. The eye parameters included in the eye models are basically average values that have been determined from measurements on many test eyes. Depending on the intended applicative situation (relaxed eye, accommodated eye; Section 2.1.4), we can utilize more or less accurate and complicated eye models[17]. In this context, we distinguish between *paraxial models* (Section 2.2.1), which are only valid in the case of paraxial approximation, and *finite wide-angle models* (Section 2.2.2), which are also able to describe aberrations (Sections A.1.6–A.1.9) quantitatively.

17) An eye model is called *exact* if it includes at least four refracting surfaces, i.e., two corneal and two lenticular surfaces.

2.2.1
Paraxial Model: The Gullstrand Eye

In optometry, the most common paraxial eye model is the *Exact Gullstrand*[18] *Eye #1*, which is based on measured data from accommodated and relaxed eyes[19]. Consequently, two versions of this model exist for different states of accommodation of the eye lens:[20] 0 D and 10.878 D.

2.2.1.1 Optical Properties

In the Exact Gullstrand Eye, we assume the refracting surfaces to be spherical and centered at a common optical axis. Since all ocular media of the model have a constant refractive index, the refractive power of the lens is varied by implementing a lens core (central nucleus) of high refractive index which is surrounded by an outer shell (cortex) with lower refractive index. All in all, the Exact Gullstrand Eye thus consists of six (i.e., two corneal and four lenticular) refracting surfaces (Figure 2.13). The geometric and optical parameters for the relaxed and accommodated state are listed in Table 2.1.

In the Exact Gullstrand Eye, the total refractive power of the cornea $\mathcal{D}'_\mathrm{c}$ can be calculated with the *Gullstrand formula* (or *thick lens formula*)

$$\mathcal{D}'_\mathrm{c} = \mathcal{D}'_\mathrm{a} + \mathcal{D}'_\mathrm{p} - \frac{L_\mathrm{c}}{n_\mathrm{c}} \mathcal{D}'_\mathrm{a} \mathcal{D}'_\mathrm{p} \,. \tag{2.20}$$

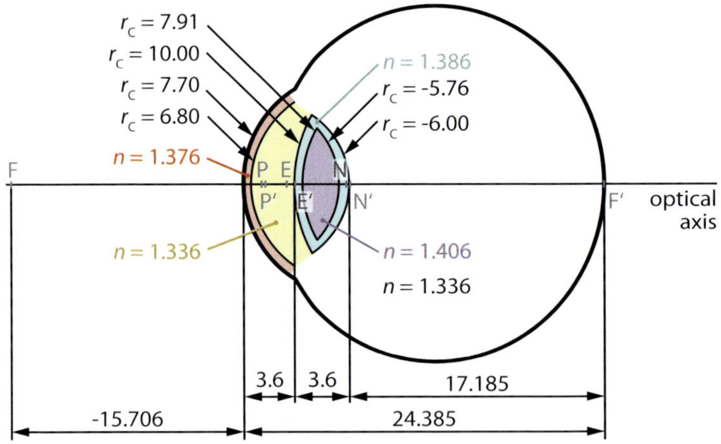

Figure 2.13 Scheme (2 : 1 scale) of the Exact Gullstrand Eye #1 for relaxed vision based on data from Table 2.1. The radii of curvature and lengths are given in millimeters.

18) Alvar Gullstrand (1862–1930).

19) Other paraxial eye models are discussed in-depth in [2, 6].

20) The original version of the Exact Gullstrand Eye cannot be used to model variable accommodation. However, with some modifications, continuous accommodation may be included [2].

Table 2.1 Parameters of the Exact Gullstrand Eye #1 for relaxed (0 D accommodation) and accommodated vision (10.878 D accommodation). The locations refer to the vertex of the cornea (point V in Figure 2.3). r_C denotes the radius of curvature, L is the thickness, and n the refractive index of the respective eye component. Data taken from [2, 13].

Parameter	Relaxed vision	Accommodated vision
Location of object-side focal point F (mm)	−15.706	−12.397
Location of image-side focal point F′ (mm)	24.385	21.016
Location of object-side nodal point N (mm)	7.078	6.533
Location of image-side nodal point N′ (mm)	7.331	6.847
Location of object-side principal point P (mm)	1.348	1.772
Location of image-side principal point P′ (mm)	1.601	2.086
Location of entrance pupil E (mm)	3.047	2.668
Diameter of entrance pupil (mm)	8.000	8.000
Location of exit pupil E′ (mm)	3.665	3.212
Diameter of exit pupil (mm)	7.276	7.524
Refractive power of cornea (D)	43.053	43.053
Refractive power of lens (D)	19.111	33.057
Refractive power of eye (D)	58.636	70.576
Total eye length (mm)	24.385	24.385

	Relaxed vision			Accommodated vision		
	r_C (mm)	L (mm)	n	r_C (mm)	L (mm)	n
Corneal front surface	7.700	–	–	7.700	–	–
Cornea	–	0.500	1.376	–	0.500	1.376
Corneal back surface	6.800	–	–	6.800	–	–
Anterior chamber	–	3.100	1.336	–	2.700	1.336
Front surface of lens cortex	10.000	–	–	5.333	–	–
Anterior lens cortex	–	0.546	1.386	–	0.673	1.386
Front surface of lens core	7.911	–	–	2.655	–	–
Lens core	–	2.419	1.406	–	2.655	1.406
Back surface of lens core	−5.760	–	–	−2.655	–	–
Posterior lens cortex	–	0.635	1.386	–	0.673	1.386
Back surface of lens cortex	−6.000	–	–	−5.333	–	–
Vitreous	–	17.185	1.336	–	17.185	1.336

$\mathcal{D}'_a$ denotes the refractive power of the corneal front surface (air–cornea interface) and $\mathcal{D}'_p$ the refractive power of the corneal back surface (interface between cornea and anterior chamber). n_c and L_c are the refractive index and the thickness of the cornea, respectively. In optometry, one often uses the *corneal back vertex power* $\mathcal{D}'_{cv}$ instead of $\mathcal{D}'_c$. $\mathcal{D}'_{cv}$ represents the inverse distance from the corneal back vertex to its image-side focus and is given by [13]

$$\mathcal{D}'_{\text{cv}} = \frac{\mathcal{D}'_{\text{c}}}{1 - L_{\text{c}}\mathcal{D}'_{\text{a}}/n_{\text{c}}} . \tag{2.21}$$

In paraxial approximation, the refractive power of a spherical surface (also referred to as the *surface power*)[21] is determined by

$$\mathcal{D}' = \frac{n' - n}{r_{\text{C}}} , \tag{2.22}$$

where n and n' are the refractive indices of the media on the incident and the refracted side, respectively. Equation (2.22) corresponds to the right side of (A11) in Section A.1.2 for which we now use the dioptric equivalents.

Let us now apply (2.20)–(2.22) to the Exact Gullstrand Eye model. With the values from Table 2.1, we obtain

$$\mathcal{D}'_{\text{a}} = \left(\frac{1.376 - 1}{0.0077}\right) \text{D} = 48.83 \text{ D} , \tag{2.23}$$

$$\mathcal{D}'_{\text{p}} = \left(\frac{1.336 - 1.376}{0.0068}\right) \text{D} = -5.88 \text{ D} . \tag{2.24}$$

With (2.20), the total refractive power of the cornea follows as

$$\mathcal{D}'_{\text{c}} = \left(48.83 - 5.88 - \frac{0.0005}{1.336} \cdot 48.83 \cdot (-5.88)\right) \text{D} = 43.06 \text{ D} , \tag{2.25}$$

and for the corneal back vertex power, we obtain $\mathcal{D}'_{\text{cv}} = 43.86$ D.

For many applications, the Exact Gullstrand Eye is still too "complicated". In these cases, it is possible to simplify the model without introducing noticeable errors. In the so-called *Simplified Gullstrand Eye #1* (Table 2.2), the cornea is represented by a single refracting surface. In this simplified model, the lens has just two refracting surfaces instead of four.

2.2.1.2 **Treatment of Aberrations**

In the first instance, the Gullstrand Eye model is meant to adequately describe the optics of the human eye only for paraxial rays. In real eyes, we also have to deal with nonparaxial or obliquely incident rays which are affected by optical aberrations (Section A.1.5). For example, the pupil diameter allows oblique rays to be projected onto the retina, and the ocular media show considerable dispersion. Refractive errors and diseases (Chapter 3) intensify the naturally given aberrations even further. In the following, we examine if, and to what extent, the Gullstrand Eye can "handle" optical aberrations.

21) The surface power quantifies the ability of the lens surface to change the direction of an incident light ray, i.e., the degree of divergence or convergence.

Table 2.2 Parameters of the Simplified Gullstrand Eye #1 for relaxed (0 D accommodation) and accommodated vision (8.599 D accommodation). The locations refer to the vertex of the cornea. r_C denotes the radius of curvature, L is the thickness, and n the refractive index of the respective eye component. Data taken from [13].

Parameter	Relaxed vision	Accommodated vision
Location of object-side focal point F (mm)	−14.983	−12.561
Location of image-side focal point F′ (mm)	23.896	21.252
Location of object-side nodal point N (mm)	7.062	6.562
Location of image-side nodal point N′ (mm)	7.363	6.909
Location of object-side principal point P (mm)	1.550	1.782
Location of image-side principal point P′ (mm)	1.851	2.128
Location of entrance pupil E (mm)	3.052	2.674
Diameter of entrance pupil (mm)	8.000	8.000
Location of exit pupil E′ (mm)	3.687	3.249
Diameter of exit pupil (mm)	7.334	7.532
Refractive power of cornea (D)	42.735	42.735
Refractive power of lens (D)	21.755	32.295
Refractive power of eye (D)	60.483	69.721
Total eye length (mm)	23.896	23.896

	Relaxed vision			Accommodated vision		
	r_C (mm)	L (mm)	n	r_C (mm)	L (mm)	n
Corneal front surface	7.800	–	–	7.800	–	–
Anterior chamber	–	3.600	1.333	–	3.200	1.333
Front surface of eye lens	10.000	–	–	5.000	–	–
Eye lens	–	3.600	1.416	–	4.000	1.416
Back surface of eye lens	−6.000	–	–	−5.000	–	–
Vitreous	–	16.696	1.333	–	16.696	1.333

Spherical aberration In the Gullstrand Eye model, 60% of the total spherical aberration (Section A.1.6.1) is caused by the anterior corneal surface and 30% by the posterior surface of the lens. The accommodated Gullstrand Eye shows 3 × more spherical aberration than the relaxed version. Interestingly, real eyes have much less spherical aberration than calculated from the paraxial model. For example, the spherical aberration of the Gullstrand Eye is 6 × higher than corresponding experiments have shown for light rays entering the pupil at a ray height of 4 mm (Figure 2.16).

Astigmatism For real eyes, astigmatism (Section A.1.6.3) occurs when the eye's refractive parts are irregular (Section 3.1.2) or in the case of obliquely incident rays. Compared to real eyes, the Gullstrand Eye shows more astigmatism for oblique rays. For example, we have a deviation by a factor of 2 when the light rays are incident at angles $< 50°$.

Field curvature Because of the curved shape of the retina, the field curvature (Section A.1.6.4) has nearly no influence on imaging. We can understand this by looking at Figure A.19 in Section A.1.6.4. In the eye, the image surface is curved so that the incident off-axis rays are focused on the retina. Up to an angle of incidence of 30°, this behavior is well reproduced by the Gullstrand Eye. For angles $> 30°$, field curvature is still irrelevant for real eyes, whereas it becomes important for the Gullstrand Eye.

Distortion Distortion (Section A.1.6.5) mainly depends on both position and diameter of the entrance pupil. With regard to distortion, the Gullstrand model predicts the imaging behavior of real eyes quite well.

Chromatic aberration As dispersion (Section A.1.1) is not included to the Gullstrand model, chromatic aberration (Section A.1.9) cannot be modeled. Nevertheless, chromatic effects play an important role for the imaging with real eyes.

In summary, the Gullstrand Eye model is *not* able to describe aberrations caused by nonparaxial rays with sufficient accuracy. In addition, chromatic aberrations are not considered at all. It is thus useful to develop more sophisticated schematic eye models which are able to describe nonparaxial optics and dispersion as well.

2.2.2 Finite Wide-Angle Models

Paraxial eye models can be further improved when we try to reproduce the optical and geometric parameters of real eyes as exactly as possible. For example, we may include the facts that the surfaces of cornea and lens are not spherical (aspheric lenses) and that the refracting surfaces are not centered at a common axis. We can also add dispersion to the models and/or describe the lens with a gradient refractive index, instead of the shell approach used for the Exact Gullstrand Eye. But we have to bear in mind that every generalization increases the complexity of the model, which makes calculations even more challenging. In the following, we present two prominent examples for so-called *finite wide-angle models* which allow the description of the imaging of human eyes for nonparaxial rays and off-axis object points quite well.

2.2.2.1 Navarro Eye Model

In the Navarro Eye model (Figure 2.14), the refractive parts of the eye are represented by four aspheric surfaces (so-called *conicoids*) which are centered to a common optical axis. Each surface is determined by the condition

$$x^2 + y^2 + (1 + Q)z^2 - r_C z = 0 , \qquad (2.26)$$

Table 2.3 Parameters of the Navarro Eye. The locations refer to the vertex of the cornea. *A* is the state of accommodation in diopters (D). r_C denotes the radius of curvature, *L* is the thickness, and *Q* the asphericity parameter of the respective eye component. *n* is the refractive index for a wavelength of 589.3 nm. The additional parameters r_{C3}, r_{C4}, L_2, L_3, n_3, Q_3, and Q_4 depend on the state of accommodation and are defined in the text. Data taken from [2, 6, 14].

Parameter	Relaxed vision ($A = 0$ D)	Accommodated vision ($A = 10$ D)
Location of object-side focal point F (mm)	−14.969	−12.051
Location of image-side focal point F′ (mm)	24.004	21.172
Location of object-side nodal point N (mm)	7.145	6.727
Location of image-side nodal point N′ (mm)	7.452	7.116
Location of object-side principal point P (mm)	1.583	2.005
Location of image-side principal point P′ (mm)	1.890	2.393
Location of entrance pupil E (mm)	3.042	2.928
Location of exit pupil E′ (mm)	3.682	3.551
Refractive power of cornea (D)	42.882	42.882
Refractive power of lens (D)	21.779	34.548
Refractive power of eye (D)	60.416	71.145
Total eye length (mm)	24.004	24.000

	r_C (mm)	L (mm)	n	Q
Corneal front surface	7.72	–	–	–
Cornea	–	0.55	1.3670	−0.2600
Corneal back surface	6.50	–	–	–
Anterior chamber	–	$3.05 - L_2$	1.3374	0
Front surface of eye lens	$10.20 - r_{C3}$	–	–	–
Eye lens	–	$4.00 + L_3$	$1.4200 + n_3$	$-3.1316 - Q_3$
Back surface of eye lens	$-6.00 + r_{C4}$	–	–	–
Vitreous	–	16.403 98	1.3360	$-1.000 - Q_4$

where r_C denotes the radius of curvature, Q the asphericity parameter, and z the direction of the optical axis. The refractive index of the eye lens is assumed to be constant for a given wavelength. Analogous to other models, all optical and geometric parameters are based on average data taken from measurements. The corresponding data set is listed in Table 2.3.

Accommodation is fully included in the Navarro Eye model in that the distance between cornea and lens as well as the lens parameters are functions of the state of accommodation. As a consequence, some parameters in Table 2.3 have additional terms which depend on the state of accommodation A (in D), that is,

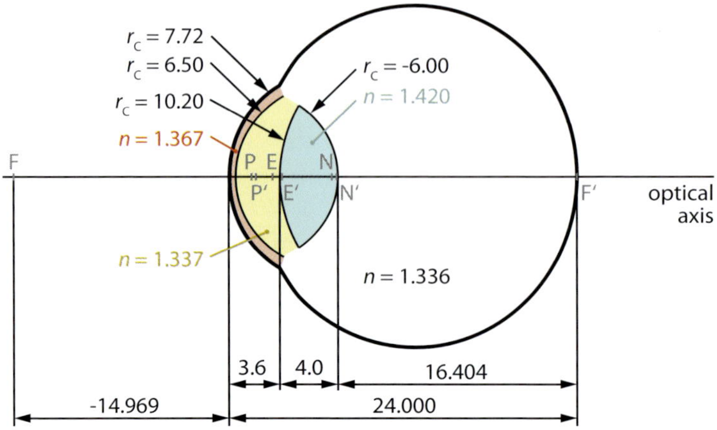

Figure 2.14 Scheme (2 : 1 scale) of the Navarro Eye for relaxed vision based on data from Table 2.3. The radii of curvature and lengths are given in millimeters.

$$r_{C3} = 1.75 \ln(A+1) ,$$

$$r_{C4} = 0.2294 \ln(A+1) ,$$

$$L_2 = 0.05 \ln(A+1) ,$$

$$L_3 = 0.1 \ln(A+1) ,$$

$$n_3 = 9 \times 10^{-5}(10A + A^2) ,$$

$$Q_3 = 0.34 \ln(A+1) ,$$

$$Q_4 = 0.125 \ln(A+1) .$$

In contrast to the Gullstrand Eye model, chromatic dispersion is fully included in the Navarro Eye (see [14]). For this purpose, the ocular media are described by the so-called *Herzberger formula* which is given by

$$n(\lambda) = a_1(\lambda) n^{**}(\lambda = 365\text{nm}) + a_2(\lambda) n_{\mathrm{F}}(\lambda = 486.1\text{ nm}) + a_3(\lambda) n_{\mathrm{c}}(\lambda = 656.3\text{nm}) + a_4(\lambda) n^{*}(\lambda = 1014\text{ nm}) . \quad (2.27)$$

The coefficients are determined by

$$a_1(\lambda) = 0.661\,471\,96 - 0.040\,352\,796\ \mu\text{m}^{-2}\lambda^2 - \frac{0.280\,467\,9\ \mu\text{m}^2}{\lambda^2 - \lambda_0^2} + \frac{0.033\,859\,79\ \mu\text{m}^4}{(\lambda^2 - \lambda_0^2)^2} , \quad (2.28)$$

$$a_2(\lambda) = -4.201\,463\,83 + 2.735\,089\,56\ \mu\text{m}^{-2}\lambda^2 + \frac{1.505\,437\,84\ \mu\text{m}^2}{\lambda^2 - \lambda_0^2} - \frac{0.115\,932\,35\ \mu\text{m}^4}{(\lambda^2 - \lambda_0^2)^2} , \quad (2.29)$$

Table 2.4 Refractive indices of ocular media used in the Herzberger formula (2.27). Data taken from [14].

Ocular Medium	n^{**} ($\lambda = 365$ nm)	n_F ($\lambda = 486.1$ nm)	n_C ($\lambda = 656.3$ nm)	n^* ($\lambda = 1014$ nm)
Cornea	1.3975	1.3807	1.37405	1.3668
Aqueous humor	1.3593	1.3422	1.3354	1.3278
Lens	1.4492	1.42625	1.4175	1.4097
Vitreous	1.3565	1.3407	1.3341	1.3273

$$a_3(\lambda) = 6.298\,342\,37 - 4.694\,099\,35\ \mu\text{m}^{-2}\lambda^2 - \frac{1.575\,086\,5\ \mu\text{m}^2}{\lambda^2 - \lambda_0^2} + \frac{0.102\,930\,38\ \mu\text{m}^4}{(\lambda^2 - \lambda_0^2)^2}\,, \tag{2.30}$$

$$a_4(\lambda) = 1.758\,350\,59 + 2.362\,537\,94\ \mu\text{m}^{-2}\lambda^2 + \frac{0.350\,116\,57\ \mu\text{m}^2}{\lambda^2 - \lambda_0^2} - \frac{0.020\,857\,82\ \mu\text{m}^4}{(\lambda^2 - \lambda_0^2)^2}\,, \tag{2.31}$$

with $\lambda_0^2 = 0.028\ \mu\text{m}^2$ and $[\lambda] = \mu\text{m}$. The refractive indices of the ocular media used in Eq. (2.27) for the various optical components are listed in Table 2.4.

2.2.2.2 Liou–Brennan Eye Model

The Liou–Brennan Eye model (geometry shown in Figure 2.15) reproduces the eye geometry fairly exactly. The optical and geometric parameters are based on measurements of people at the ages of about 45. For example, this model can be used advantageously to simulate the visual performance before and after a refractive surgery (Chapter 10).

In the Liou–Brennan Eye, the decentration of the real human eye is also regarded. The corresponding visual axis (red line in Figure 2.15) includes an angle of $\kappa = 5°$ with the optical axis. The iris is modeled as a circular aperture stop at the front surface of the lens whose center is shifted by 0.5 mm from the optical axis in the nasal direction. Accommodation is, however, not included and parameters are only given for the relaxed eye. Dispersion of the ocular media is set to the dispersion of water, which is approximately

$$n_{\text{water}} \approx 1.3847 - 0.1455\ \mu\text{m}^{-1} \cdot \lambda + 0.0961\ \mu\text{m}^{-2} \cdot \lambda^2 \tag{2.32}$$

with the wavelength of light λ ($[\lambda] = \mu\text{m}$). Similar to the Navarro Eye, the anterior corneal surface and the surfaces of the eye lens are aspheric. In addition, the eye lens has a gradient refractive index. The distribution of the refractive index at the front surface of the lens (Figure 2.15) is given by

$$n_A(\rho, z) = 1.368 + 0.049\,057\ \mu\text{m}^{-1}z - 0.015\,427\mu\text{m}^{-2}z^2 - 0.001\,978\ \mu\text{m}^{-2}\rho^2\,. \tag{2.33}$$

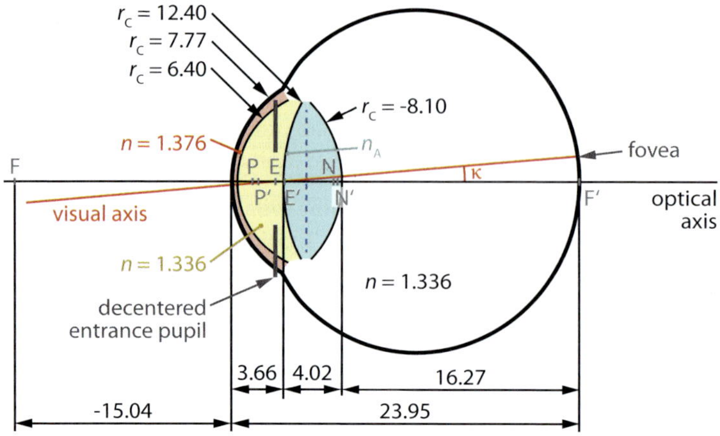

Figure 2.15 Scheme (2:1 scale) of the relaxed Liou–Brennan Eye model. The radii of curvature and lengths are given in millimeters. The geometry, the visual axis, and the decentration of the entrance pupil (by 0.5 mm nasally) are shown. The blue dashed line represents the lens center, which is an imaginary plane separating the eye lens into anterior and posterior segments.

$\rho = \sqrt{x^2 + y^2}$ ($[\rho] = \mu\text{m}$) is the coordinate normal to the traveling direction z of the incident light rays. At the central lens surface (dashed blue line in Figure 2.15), the distribution of the refractive index is determined by

$$n_P(\rho, z) = 1.407 - 0.006\,605\,\mu\text{m}^{-2} z^2 - 0.001\,978\,\mu\text{m}^{-2} \rho^2 \qquad (2.34)$$

with $[z] = [\rho] = \mu\text{m}$.

2.2.2.3 Aberrations in Finite Wide-Angle Models

Finite wide-angle models are more realistic than paraxial models. Thus, we expect that the aberrations calculated by finite wide-angle models come closer to experimental data.

Spherical aberration In Figure 2.16, the distance of incident light rays at the pupil from the optical axis (ray height) is plotted versus the spherical aberration in diopters (D) for the Gullstrand Eye #1, the Navarro Eye, and the Liou–Brennan Eye. When the calculations are compared with experimental data (circles), we can clearly observe that the finite wide-angle models deliver much better results than the paraxial model. In particular, the Liou–Brennan Eye perfectly matches the experiment.

Astigmatism At angles of incidence $< 50°$, the Navarro Eye deviates by a factor of < 1.5 from real eyes. Thus, it fits slightly better to experimental data than the paraxial Gullstrand Eye. The Liou–Brennan Eye shows roughly the same behaviour at larger angles as the Navarro Eye, but the deviations from real eyes at small angles of incidence are not as large as for the Navarro Eye.

Table 2.5 Parameters of the Liou–Brennan Eye. The locations refer to the vertex of the cornea. r_C denotes the radius of curvature, L is the thickness, and Q the asphericity parameter of the respective eye component. n is the refractive index at a wavelength of 555 nm. n_A and n_P are gradient-index distributions of the unaccommodated lens (see Eqs. (2.33) and (2.34)). Data taken from [2, 15].

Parameter	Relaxed Vision
Location of object-side focal point F (mm)	−15.040
Location of image-side focal point F′ (mm)	23.950
Location of object-side nodal point N (mm)	7.100
Location of image-side nodal point N′ (mm)	7.378
Location of object-side principal point P (mm)	1.532
Location of image-side principal point P′ (mm)	1.890
Location of entrance pupil E (mm)	3.098
Location of exit pupil E′ (mm)	3.720
Refractive power of cornea (D)	42.262
Refractive power of lens (D)	22.134
Refractive power of eye (D)	60.314
Total eye length (mm)	23.950

	r_C (mm)	L (mm)	n	Q
Corneal front surface	7.77	–	–	–
Cornea	–	0.50	1.376	−0.18
Corneal back surface	6.40	–	–	–
Anterior chamber	–	3.16	1.336	−0.60
Front surface of eye lens	12.40	–	–	–
Eye lens (front segment)	–	1.59	n_A	−0.94
Lens center (dashed line Figure 2.15)	∞	2.43	n_P	–
Back surface of eye lens	−8.10	–	–	–
Vitreous	–	16.27	1.336	0.96

Field curvature Like for the paraxial Gullstrand Eye, the discussed finite wide-angle models fit exactly to experimental data for small angles ($< 30°$). The finite wide-angle models tend to emulate the field curvature of real eyes better for angles $> 30°$.

Chromatic aberration Since dispersion is included in both finite wide-angle models, chromatic aberrations can be described by these models quantitatively. The dispersion model of the Navarro Eye was chosen especially with the objective to reproduce experimental data as accurately as possible.

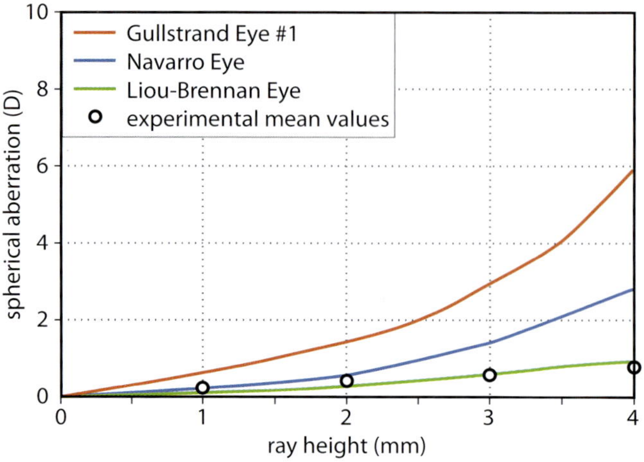

Figure 2.16 Ray height versus spherical aberration for different eye models. Experimental data points are shown for reference. Regarding spherical aberration, the Liou–Brennan Eye can emulate real eyes very well. Data taken from [2].

2.2.3 Applications of Eye Models

The optics of the human eye must often be considered for the design of ophthalmic, optometric, and other optical systems. For this purpose, the presented schematic eye models are used to derive the standard optical properties like the refractive powers, the imaging behavior, the positions of cardinal points, the influence of aberrations, and so on. From this, some typical examples for the application of eye models follow:

- Design of corrective glasses and contact lenses
- Design and layout of optical imaging systems and visual aids such as loupes (Section 6.1) and surgical microscopes (Section 6.2)
- Design and layout of diagnostic devices (Chapter 5)
- Simulation and planning of surgeries for treatment of eye diseases and refractive errors (Chapter 10)
- Calculation of derived eye parameters (see Example 2.1).

Example 2.1

Calculation of the Retinal Image Size As a simple example for the application of eye models, we calculate the retinal image size of a Landolt ring with a height of $h_O = 6\,\text{mm}$. We assume that the test person looks at this symbol from a distance of $L = 4\,\text{m}$. For the calculation, we use (2.7), as the angle κ of the nodal ray (Figure 2.3) does not change after passage through the eye's refractive parts (Section 2.1.3).[22]

The angle at which the Landolt ring appears to the test person is given by

$$\kappa = \arctan\left(\frac{h_O}{L}\right) = 5.157' = 0.0015\,\text{rad}\,. \tag{2.35}$$

The distance between the image-side nodal point N' and the retinal image surface (represented by image point I'_0) can, for example, be taken from the Exact Gullstrand Eye (Table 2.1) for which $\overline{N'I'_0} = 17.054\,\text{mm}$. As a consequence, we obtain for the retinal image size

$$h'_I = \kappa \overline{N'I'_0} = 25.6\,\mu\text{m}\,. \tag{2.36}$$

Note that we inserted the value of angle κ in radiant units. The gap feature of the Landolt ring is $5\times$ smaller than the whole Landolt ring symbol. On the retina, the gap thus forms an image height of $5.1\,\mu\text{m}$.

When we use the parameters of the Navarro Eye (Table 2.3) instead of the Exact Gullstrand Eye model, the distance between image-side nodal point and retinal image changes to $\overline{N'I'_0} = 16.548\,\text{mm}$. Hence, in this case, we obtain a retinal image size of $h'_I = 24.8\,\mu\text{m}$.

2.3 Color Vision

As mentioned in Section 1.2.1, the color perception of eyes mainly depends on the ambient light conditions. For scotopic vision, only rods are used. Rods have a maximum sensitivity at a wavelength of 507 nm (black curve in Figure 2.17). This is the reason why we cannot distinguish colors at night.

For photopic vision, the human eye uses three types of cones (S, M, L cones) with different spectral sensitivity (Figure 2.17). The three detector types allow us to have *trichromatic vision*. Light with wavelengths between 380 and 780 nm can be processed by the retina, which actually determines the *visible spectrum*. From the mean value of the sensitivity curves of all cones in Figure 2.17, we obtain a maximum sensitivity at a wavelength of 555 nm under photopic conditions.

For vision, the signals of S, M, and L cones are combined in the brain, resulting in a visual stimulus interpreted as a *color*.[23] Hence, the color at which an object appears to us is *not* an inherent property of the object itself, but rather depends on

22) In principle, we could also use relation (2.1) to calculate the retinal image size. However, the angle φ of the chief ray is modified by cornea and eye lens (Figure 2.2), and φ' cannot be measured in practice.

23) An unknown number of women may perceive millions of colors invisible to the rest of human beings. It is supposed that they possess 4 (instead of 3) different types of cones which expand the perceivable spectral range [16].

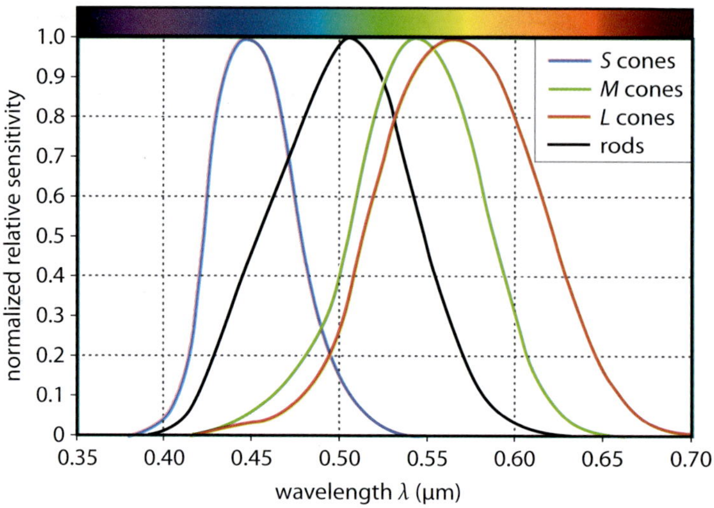

Figure 2.17 Dependence of normalized relative sensitivity of *S*, *M*, *L* cones, and rods on the wavelength. Data taken from [6].

our visual impression. The wavelength-dependent visual stimulus is characterized by three parameters:

1. *Hue:* The hue determines the degree to which a visual stimulus can be described as similar or different from stimuli described as "red", "yellow", and "blue" (i.e., primary hues). It depends largely on what eye and brain perceive to be the predominant wavelength of light reflected or sent from an object.
2. *Saturation:* The saturation characterizes the purity or "richness" of a color. If all the light which is seen by the eye has the same wavelength, the color will appear fully saturated. The more wavelengths are added, the paler (desaturated) the color appears.
3. *Brightness:* The brightness determines the intensity level at which the visual impression is perceived. It is our subjective interpretation of luminance (Section A.2.1.5).

White, gray, and black stimuli cannot be described as a color, as they have no hue and saturation and are completely determined by the brightness.

In contrast to the ear – which is able to distinguish several acoustic frequencies playing at once – the eyes and the brain cannot determine which wavelengths of light are simultaneously present in the observed color. For example, if we look at a monochromatic laser beam which only consist of light with a wavelength of 590 nm, the eye perceives a yellow color. The same impression can be obtained when we spatially overlap two laser beams with wavelengths of 540 and 680 nm and proper intensity. In this case, we do *not* realize that the beam consists of green and red light. This is, in fact, the basis for the concept of primary hues (red, yellow, blue). With these hues, we can "mix" any other color of the visible spectrum[24], for example, cyan is a combination of blue and green.

The full theory of color vision and color formation is beyond the scope of this book. Please refer to references [17–19] for further details about this topic.

2.4 Recommended Reading

Further information about the optics of the human eye can be found in standard ophthalmology textbooks, for example, [2, 6, 11, 13].

References

1 Porter, J., Queener, H., Lin, J., Thorn, K., and Awwal, A.A.S. (eds) (2006) *Adaptive Optics for Vision Science: Principles, Practices, Design and Applications*, John Wiley & Sons, Inc, Hoboken.

2 Atchison, D.A. and Smith, G. (2006) *Optics of the Human Eye*, Butterworth–Heinemann.

3 Aggarwala, K.R., Nowbotsing, S., and Kruger, P.B. (1995) Accommodation to monochromatic and white-light targets. *Invest. Ophthalmol. Vis. Sci.*, **36**, 2695–2705.

4 Graef, K. and Schaeffel, F. (2012) Control of accommodation by longitudinal chromatic aberration and blue cones. *J. Vis.*, **12**, 1–12.

5 Holden, B.A., Fricke, T.R., Ho, S.M., Wong, R., Schlenther, G., Cronjé, S., Burnett, A., Papas, E., Naidoo, K.S., and Frick, K.D. (2008) Global vision impairment due to uncorrected presbyopia. *Arch. Ophthalmol.*, **126**, 1731–1739.

6 Gross, H., Blechinger, F., and Achtner, B. (2008) *Handbook of Optical Systems – Survey of Optical Instruments*, vol. 4, Wiley-VCH Verlag GmbH, Weinheim.

7 Holladay, J.T., Lynn, M.J., Waring, G.O., Gemmill, M., Keehn, G.C., and Fielding, B. (1991) The relationship of visual acuity, refractive error, and pupil size after radial keratotomy. *Arch. Ophthalmol.*, **109**, 70–76.

8 Optical Society of America (ed.) (1994) *Handbook of Optics: Fundamentals, Techniques, and Design*, vol. 1, McGraw–Hill Professional.

9 Tasman, W. (ed.) (1992) *Duane's Foundations of Clinical Ophthalmology*, vol. 2, Lippincott Williams & Wilkins.

10 Gross, H., Zügge, H., Peschka, M., and Blechinger, F. (2007) *Handbook of Optical Systems – Correction of Optical Systems*, vol. 3, Wiley-VCH Verlag GmbH, Weinheim.

11 Diepes, H. (2004) *Refraktionsbestimmung*, DOZ Verlag Optische Fachveröffentlichungen GmbH, 3rd edn.

12 Schober, H. (1964) *Das Sehen*, vol. 1, VEB Fachbuchverlag Leipzig.

13 Smith, G. and Atchison, D.A. (1997) *The Eye and Visual Optical Instruments*, Cambridge University Press.

14 Navarro, R., Santamaría, J., and Bescós, J. (1985) Accommodation-dependent model of the human eye with aspherics. *J. Opt. Soc. Am. A*, **2**, 1273–1281.

15 Liou, H.L. and Brennan, N.A. (1997) Anatomically accurate, finite model eye

24) As the optical resolution of human eyes is limited, colors can also be generated by additive color composition. This means that the color impression is generated by superposition of small areas (e.g., tiny dots) with primary hues. The hue we perceive is then determined by the relative intensity of each color component. This principle of "composite" colors is, e.g., used for digital image projection and color raster prints.

for optical modeling. *J. Opt. Soc. Am. A*, **14**, 1684–1695.

16 Greenwood, V. (2012) The humans with super human vision. http://discovermagazine.com/2012/jul-aug/06-humans-with-super-human-vision (accessed 24 August 2013).

17 Gross, H. (2005) *Handbook of Optical Systems – Fundamentals of Technical Optics*, vol. 1, Wiley-VCH Verlag GmbH, Weinheim.

18 Still, D., Fletcher, R. (eds) (1998) *Eye Examination and Refraction*, Wiley-VCH Verlag GmbH, Weinheim.

19 Valberg, A. (2005) *Light Vision Color*, Wiley-VCH Verlag GmbH, Weinheim.

3
Visual Disorders and Major Eye Diseases

The optical devices presented in this book are used to diagnose or treat disorders and diseases of the human eye. Before we focus on the technical principles, let us introduce the most common visual disorders and eye diseases in brief. As this book is not intended to be a textbook for clinical ophthalmology, the presented explanations do not claim to cover all details. This chapter rather serves as a quick reference for readers without a medical background. References for recommended readings are provided throughout the following sections and at the end of this chapter.

Compared to other sensory organs, the human eye delivers by far the highest portion of information about the environment. If vision is reduced due to refractive errors and/or eye diseases, this means a major impairment for the person concerned. In the following, we will first discuss the widely spread refractive errors (Section 3.1) which can also be considered as optical aberrations of the eye's optical parts. In the second part of this chapter (Sections 3.2–3.7), we will introduce the most common eye diseases and, eventually, provide information about the worldwide statistical distribution of visual impairment and blindness, as well as their impact on the society and economy (Sections 3.8 and 3.10)

3.1
Refractive Errors

Emmetropia The human eye is called *emmetropic* if the far point (Section 2.1.4) is located at infinity for relaxed accommodation (Figure 3.1a). In this case, the image-side focal point of the eye coincides with the retina for far vision.[1] Any deviation from this ideal state of refraction is called *ametropia* (refractive error).

The goal of refractive correction of an ametropic eye is to restore emmetropia. For this purpose, ametropia is at first quantified by using special diagnostic techniques (Chapter 5). Once the degree of ametropia is known, the eye can be corrected with eye glasses (Section 5.5), contact lenses, intraocular lenses, or by corneal refractive surgery (Section 10.3).

1) This means that we have a match between eye length and refractive power.

Optical Devices in Ophthalmology and Optometry, First Edition. Michael Kaschke, Karl-Heinz Donnerhacke, and Michael Stefan Rill.

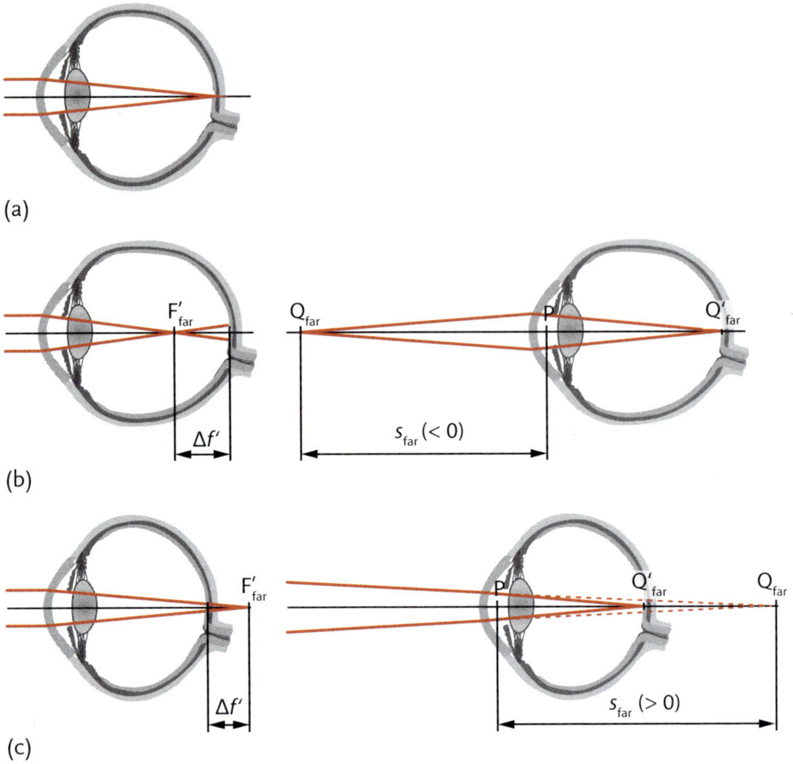

Figure 3.1 Refractive errors of the human eye. (a) In an emmetropic eye, a bundle of parallel rays is focused on the fovea. (b) In myopia, an excess of refractive power exists compared to an emmetropic eye with the same axial eye length. The focal point F'_{far} of a distant object is thus shifted by $\Delta f'$ so that the image plane is located in front of the retina. The far point Q_{far} of the myopic eye is located at a finite distance in front of the eye such that $A_{far} < 0$. (c) In a hyperopic eye, a lack of refractive power exists compared to an emmetropic eye with the same axial length. The focal point F'_{far} of a distant object is shifted by $\Delta f'$ so that the image plane is located behind the retina. The far point Q_{far} is located at a finite distance behind the eye such that $A_{far} > 0$.

Ametropia In an ametropic eye, the far point is *not* located at infinity for relaxed accommodation. It is always based on a mismatch of refractive power and the axial eye length. For ametropia, it is not relevant whether the axial eye length is normal and the refractive power deviates or the refractive power is normal and the axial eye length deviates. The only quantity measured and corrected in an ametropic eye is the lack or excess of the refractive power compared to an emmetropic eye with the same axial length.

The degree of ametropia is quantified by the *far point refraction* A_{far} ($[A] = \mathrm{D} = 1/\mathrm{m}$), that is, the inverse of the far point distance (Section 2.1.4). The far point distance from the eye's principal point is negative if we measure in the opposite direction of light propagation. Accordingly, A_{far} is negative if the far point is located in front of the eye.

3.1.1
Axial-Symmetric Ametropia: Myopia and Hyperopia

For *axial-symmetric ametropia*, the optical system of the eye, also called the *dioptric apparatus*, is rotationally symmetric about the optical axis. However, a discrepancy between the refractive power and the axial eye length exists. Here, we distinguish between the following cases.

Myopia For *myopia*, the far point Q_{far} is situated at a finite distance s_{far} in front of the eye. The image-side focal point for far vision (relaxed eye) thus lies in front of the retina (Figure 3.1b). The far point distance s_{far} and the far point refraction A_{far} are negative. For example, when the far point distance of a myopic eye is $s_{far} = -0.5$ m, the eye has a refractive error of -2 D. Object points which are further away than the far point appear blurred. Thus, the term "near-sightedness" is obvious for myopic eyes. At accommodation for near vision (accomodated eye; Section 2.1.4) to distances closer than the far point, the refractive power, and thus the degree of myopia, is further increased.

Hyperopia For *hyperopia*, the far point is located at a finite distance behind the eye (Figure 3.1c) and is "virtual" (see also Section A.1.2.1). Consequently, the image-side focal point for far vision lies behind the retina. The far point distance as well as the far point refraction are positive in this case. Without accommodation, all objects which are placed at an arbitrary distance in front of the eye appear blurred. As the refractive power of the eye is increased during accommodation, the degree of hyperopia decreases. Young people are thus often able to compensate hyperopia with steady accommodation. But when the range of accommodation decreases with age due to presbyopia (Section 2.1.4), the latent hyperopia becomes more and more manifest.

3.1.2
Astigmatism

An eye is *astigmatic* when the eye's refractive parts are *not* rotationally symmetric about the optical axis. In the case of *regular astigmatism*[2)], two perpendicular principal meridians exist for which we can determine different far point positions. If this is *not* possible, the astigmatism is called *irregular*. Irregular astigmatism is often caused by an abnormal change of the corneal surface, for example, due to the formation of scars or keratoconus (Section 3.1.6).

Regular astigmatism can be corrected with standard toric eye glasses or contact lenses. In irregular astigmatism, a contact lens can be used for correction under certain conditions. For severe irregular astigmatism, corneal refractive surgery (Section 10.3.3) is required in which the cornea is flattened by an excimer laser

2) Regular on-axis astigmatism should not be confused with off-axis astigmatic aberrations (Section A.1.6.3) which also appear for spherical lenses.

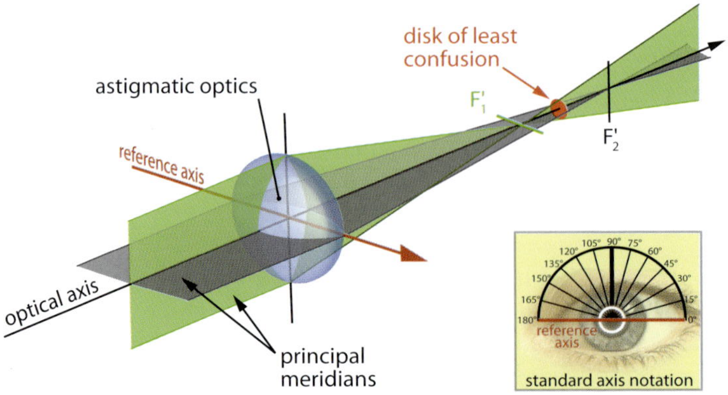

Figure 3.2 With-the-rule astigmatism described by the conoid of Sturm. The reference axis and the principal meridians are shown. F_1' is the focal line formed by rays which pass the vertical meridian. F_2' is the focal line formed by rays which pass the horizontal meridian. Inset: Standard axis notation. A meridian is specified by the counterclockwise angle included with the horizontal reference axis (red). The reference axis is defined as the connecting line between the pupil centers of both eyes. By convention, the horizontal setting is denoted by 180° and not by 0°.

(Section B.5.1.2). In the following, we will restrict ourselves to regular astigmatism, since it is the most often occurring type of astigmatic refractive errors.

The path of rays in an astigmatic eye can be described by the *conoid of Sturm* (Figure 3.2a). A distant object point which is located at infinity forms two focal lines F_1' and F_2' near the retina. Both focal lines are centered on the optical axis and are perpendicular to the corresponding principal meridians. Between the focal lines, we find the *disk of least confusion* which represents the location where the image of the object point becomes rotationally symmetric and at which the image of an extended object appears blurred and distorted, but is very similar to the original.

Depending on the axis orientation of the principal meridians (inset of Figure 3.2), we can distinguish between different types of regular astigmatism:

- *With-the-rule astigmatism*: The meridian with a higher refractive power is vertical ($90° \pm 15°$). The horizontal focal line F_1' thus lies in front of the vertical focal line F_2' (as shown in Figure 3.2).
- *Against-the-rule astigmatism*: The meridian with a higher refractive power is horizontal ($0° \pm 15°$ or $180° \pm 15°$). As a consequence, the vertical focal line F_2' lies in front of the horizontal focal line F_1'.
- *Oblique astigmatism*: The principal meridians of the astigmatic lens are tilted by an angle between 15° and 75° (or 105° and 165°), but are still perpendicular to each other.

Alternatively, astigmatism can also be classified by the location of the focal lines relative to the retina:

- *Simple myopic astigmatism*: The first focal line is located in front of the retina, while the second one is located on the retina.
- *Simple hyperopic astigmatism*: The first focal line is located on the retina, while the rear focal line lies behind the retina.
- *Compound myopic (hyperopic) astigmatism*: Both focal lines are located in front of (or behind) the retina.
- *Mixed astigmatism*: One focal line is situated in front of the retina and the other one behind the retina.

3.1.3 Notations of Spherocylindric Refraction in Astigmatic Eyes

Polar notation In clinical practice, the spherocylindric refractive error A_{far} (see also Section A.1.8.2) of an astigmatic eye is indicated in polar notation by means of the parameters sph (sphere), cyl (cylinder), and axis. Here, sph is the far point refraction along one principal meridian with a certain orientation described by the axis parameter, cyl is determined by the difference between the far point refractions of both orthogonal principal meridians. Depending on the sign of cyl, we distinguish between two comparable spherocylindric notations, referred to as the *plus cylinder* and *minus cylinder* notation. For example, let us assume a far point refraction of −5 D for one principal meridian with an axis orientation of 95°. The far point refraction of the perpendicular axis shall be −3 D. We can then write the refractive error

- in plus cylinder notation as $A_{\text{far}} = \text{sph} - 5\,\text{D}/\text{cyl} + 2\,\text{D}/\text{axis}\ 95°$
- in minus cylinder notation as $A_{\text{far}} = \text{sph} - 3\,\text{D}/\text{cyl} - 2\,\text{D}/\text{axis}\ 5°$

For the conversion from plus to minus cylinder notation, the following rule applies:

1. Add cyl to sph.
2. Change the sign of cyl from + to −.
3. Add 90° to axis. If cyl > 180°, subtract 180°.

For the conversion from minus to plus cylinder notation, the following rule applies:

1. Add cyl to sph.
2. Change the sign of cyl from − to +.
3. Subtract 90° from axis. If axis < 0°, add 180°.

Power vector notation Another possible description of spherocylindric refractive errors is the *power vector notation* [1]. This type of notation is often used in scientific publications, as it is advantageous for the analysis of refraction data. In certain studies, it is necessary to add, subtract, or compute the mean value of several refractive measurements. The traditional polar notation is unsuitable for these kinds of analysis, since cylinders with different axes cannot be directly added or subtracted.

The polar notation can be converted to three orthogonal components of the power vector notation via

$$M = \text{sph} + \frac{\text{cyl}}{2}\,, \tag{3.1}$$

$$J_0 = -\frac{\text{cyl}}{2}\cos(2\,\text{axis})\,, \tag{3.2}$$

$$J_{45} = -\frac{\text{cyl}}{2}\sin(2\,\text{axis})\,, \tag{3.3}$$

with $[M] = [J_0] = [J_{45}] = D$, where M denotes the spherical equivalent power (Section 5.4.1), J_0 (sometimes also referred as J_{180}) the component of astigmatism with a vertical or horizontal axis, and J_{45} the component of astigmatism with oblique axes. In the power vector notation, the orientation of the axes does not have to be specified, and it is possible to directly add cylindric refraction contributions, as they all have the same orientation (axis value).

To convert from power vector notation to polar notation, the following relations can be used:

$$\text{cyl} = -2\sqrt{J_{45}^2 + J_0^2}\,, \tag{3.4}$$

$$\text{sph} = M - \frac{\text{cyl}}{2}\,, \tag{3.5}$$

$$\text{axis} = \frac{1}{2}\arctan\left(\frac{J_{45}}{J_0}\right)\,. \tag{3.6}$$

3.1.4
Anisometropia

Anisometropia is a condition in which both eyes have unequal refractive powers (e.g., different degrees of myopia). The correction of large degrees of anisometropia with eye glasses creates a difference in image magnification between the eyes. This may cause *aniseikonia*. This is a condition in which the perceived image size and/or shape is different for each of both eyes. As long as the deviation of the refractive powers is small, each refractive error can be corrected individually (e.g., by glasses or preferentially by contact lenses) so that binocular vision is restored. For a large difference of refractive powers, however, the brain cannot match the images of both eyes anymore. As a consequence, binocular vision (Section 2.1.9) is lost and the patient sees double images.

3.1.5
Distribution of Refractive Errors

When considering the refractive status of people (Figure 3.3), we realize that most eyes are not emmetropic, but have a refractive error between 0 and 1 D. However,

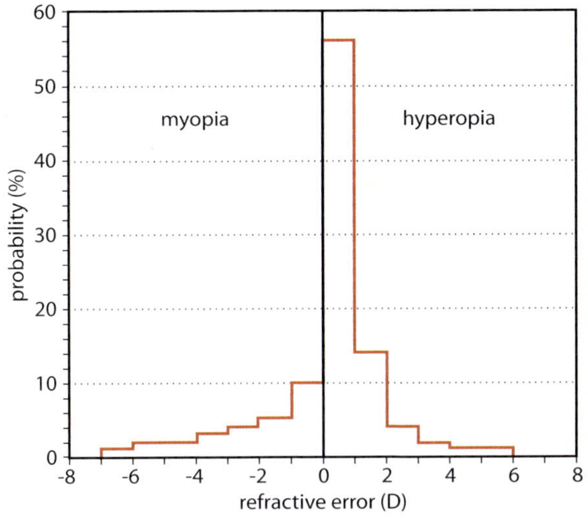

Figure 3.3 Probability distribution for the occurrence of spherical refractive errors. Adapted from [2].

this low degree of hyperopia does not noticeably reduce the visual performance (Section 2.1.5.1), since it can be compensated by accommodation.

We can also deduce from Figure 3.3 that the prevalence of large refractive errors is relatively low. Large degrees of ametropia rarely appear and are mostly related to eye diseases, which are considered next.

3.1.6 Refractive Errors Caused by Diseases

Refractive errors presented so far are caused by biological variations in the refractive power and length of the human eye. In addition, certain eye diseases exist which may lead to strong (and often irregular) refractive errors. Let us now consider the two most relevant of those.

Progressive myopia In patients with *progressive myopia*, the eye grows in length such that the degree of myopia considerably increases. The abnormal growth leads to stresses within the retinal layers which, in turn, increases the risk for retinal detachment and choroidal neovascularization (see also Section 3.4). To date, only the symptoms of progressive myopia can be treated. An adequate therapy to prevent the abnormal eye growth is still an active field of research.

Keratoconus *Keratoconus* is a degenerative disorder of the cornea which may appear in the case of weak corneal collagen. In the progression of this disease, structural changes make the cornea thinner so that it ends up in a conical shape. The resulting irregular myopic astigmatism impairs vision in that patients see multiple images and are often sensitive to light. Keratoconus is inheritable and often diagnosed for young people who have Down's syndrome. It becomes most severe at the ages between 15 and 30 years. A rare subgroup of keratoconus is *keratoglobus* for which the structural changes of the corneal tissue lead to a spherical bulge reaching up to the outer rim of the cornea.

Keratoconus in the early stages can be treated with contact lenses. Advanced stages are treated with corneal surgery (e.g., corneal transplantation, intrastromal corneal ring segments) or by so-called *corneal collagen cross-linking* with riboflavin to stabilize the weak corneal tissue.

3.2 Cataract

So far, we have considered refractive errors and visual disorders which lead to the formation of blurred images. In most cases, these disorders can be corrected with classic optical correction methods (i.e., eye glasses and contact lenses). We will now discuss eye diseases which require a surgical and/or pharmaceutical treatment. First, *cataract* shall be mentioned as the major cause of blindness worldwide (Section 3.9). Cataract leads to an opacity of the eye lens and a change of the refractive power.

The lens is transparent because of its ordered structure (fibers arranged in an array) and composition (mainly crystalline proteins). If, however, proteins in the lens or the lens capsule aggregate in a chaotic manner, the lens becomes gray-yellowish and turbid (Figure 3.4). Incident light is thus strongly scattered, and light intensity reaching the retina is reduced. People who suffer from cataract have an increased sensitivity to light as well as a reduced color and contrast perception. Eventually, cataract may lead to lens opacity and thus to blindness if not treated early enough.

The majority of diagnosed cataracts are age-related. For example, 75% of people older than 75 years suffer from such lens opacities [3]. Cataract can also be congenital, associated with other ocular diseases (e.g., glaucoma), or may have a traumatic

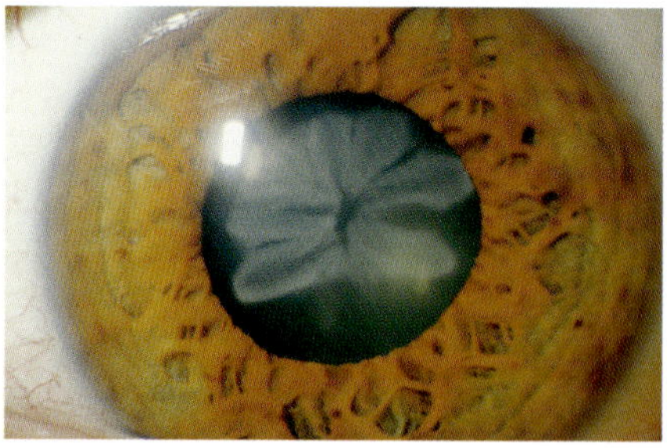

Figure 3.4 Photograph of an eye affected by cataract. The gray fog of the turbid lens is clearly visible. Courtesy of Carl Zeiss.

origin (e.g., mechanical stress). Studies have shown that intensive exposure to radiation (e.g., intensive sunlight or X-rays) considerably increases the likelihood of cataract formation.

Cataracts can be efficiently treated with a surgical intervention at which the opaque lens is replaced by an artificial intraocular lens. Before surgery, the optical properties and dimensions of the affected eye are determined (Section 7.7). From the measured data, the required refractive power of the intraocular lens is deduced.

3.3 Glaucoma

Glaucoma is a progressive neurodegenerative disease which is characterized by an irreversible loss of retinal ganglion cells and their axons (Section 1.1). Under these conditions, the structure of the retinal nerve fiber layer (RNFL) and the topography of the optic nerve head (ONH) change (Figure 1.8 in Section 1.2 and Section 1.2.2). When glaucoma is left untreated, it may lead to a progressive vision impairment – in particular a reduction of the visual field (Section 2.1) – and eventual blindness.

In industrialized countries, glaucoma is the second most frequent cause of blindness (Section 3.9). As the disease is often relatively asymptomatic, especially in the early stages, and as the public awareness of glaucoma and its risk factors is quite low, approximately 50% of individuals with glaucoma remain undiagnosed. For diagnosis and control of progression, a number of techniques are used which either examine structural changes (e.g., shape of the ONH) or functional changes (e.g., shrinkage of the visual field). We will discuss the corresponding techniques in Chapters 6–8.

The most relevant risk factors which favor the development and progression of glaucoma are age, race, family history, and an increased intraocular pressure (IOP). The mean value of the IOP lies in the range of 15–16 mmHg (1 mmHg corresponds to 133.3 Pa). It is considered to be critical for values above 21 mmHg. Based on the origin for high IOP, we distinguish between *primary* and *secondary glaucoma*. In primary glaucoma, no directly associated eye disease can be found which has caused the IOP elevation. In secondary glaucoma, the IOP is increased either due to another ocular disease, systemic disease, or an injury. The most widespread form of glaucoma is primary.

Except for few cases, the increased IOP is related to a high outflow resistance of the aqueous humor (Section 1.1). Depending on the causes of reduced outflow, we categorize glaucoma into *open-angle* and *angle-closure glaucoma*.

Primary open-angle glaucoma In primary open-angle glaucoma, the increased IOP is caused by a reduced drainage of the trabecular meshwork. The origin of the blockage which leads to the increased outflow resistance is however unknown. The primary open-angle glaucoma is the most common type of glaucoma and very often leads to blindness.

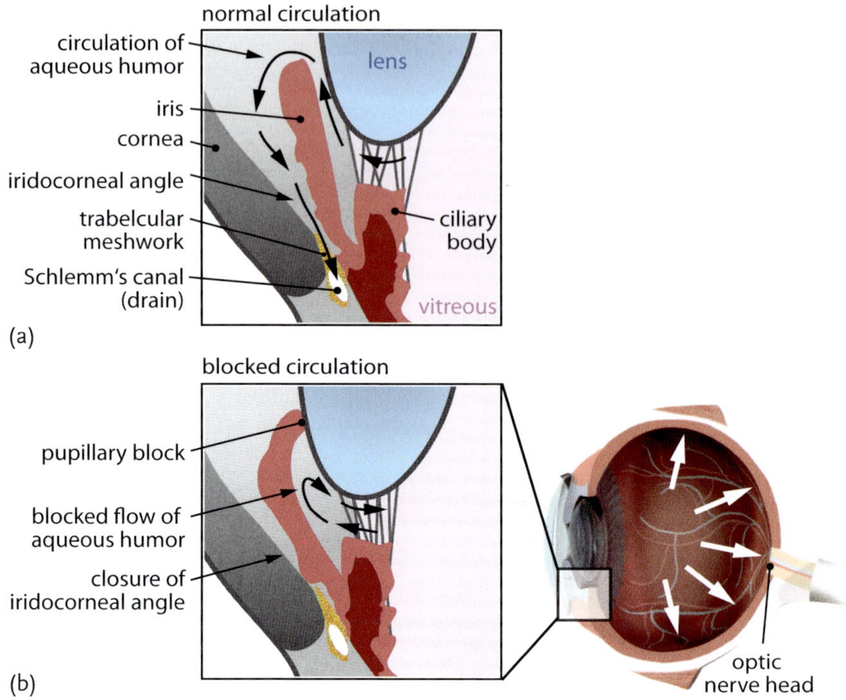

Figure 3.5 Glaucoma occurs when the drainage of aqueous humor is blocked. (a) Normal situation of a healthy eye in which the aqueous humor flows through the trabecular meshwork and is drained at Schlemm's canal. (b) Primary angle-closure glaucoma. The outflow of the aqueous humor is reduced or blocked due to a closure of the iridocorneal angle. Angle closure may be caused by a pupillary block (a contact of iris and lens), a plateau iris (an anterior displacement of the peripheral iris), or a combination thereof (depicted case). The rising intraocular pressure (represented by white arrows in overview) damages ganglion cells and their axons until they die off.

Primary angle-closure glaucoma In primary angle-closure glaucoma, the increased IOP originates from a closure of the iridocorneal angle. This means that the iris touches the trabecular meshwork (*iridotrabecular contact*). A closure of the iridocorneal angle can be caused by a pupillary block (in 75% of all cases), a *plateau iris*, or a combination thereof.

In the case of a pupillary block, the inner part of the iris (pupillary margin) touches the eye lens, which then blocks the flow of the aqueous humor from the posterior to the anterior chamber. Due to the resulting pressure difference, the outer part of the iris is bent such that it touches the trabecular meshwork (Figure 3.5b). As a consequence, the flow of the aqueous humor is blocked both at the pupillary margin and the trabecular meshwork.

In the case of a plateau iris, the iris root[3] is displaced in the direction of the cornea. As a result of this morphological anomaly, the iridocorneal angle is permanently partially or totally closed. For small iridocorneal angles, a total closure is sometimes obtained when the pupil is dilated. In the latter case, the drainage of the aqueous humor is blocked due to a thickening of the iris root caused by dilation.

Primary angle-closure glaucoma can be acute or chronic. In acute angle-closure glaucoma, the iridocorneal angle is suddenly closed (within a few hours) which leads to an extremely high IOP of up to 50–80 mmHg. In contrast to primary open-angle glaucoma, this condition manifests itself in severe pain, redness of the affected eye, corneal edema, blurred vision, and nausea. It is often caused by a pupillary dilation which can be naturally or pharmacologically induced. Even if the periods of increased IOP are relatively short, rapid and significant damage can occur. In the case of an acute primary angle-closure glaucoma, the patient thus needs immediate medical help. First-line treatment is laser peripheral iridotomy (Section 10.4.4.2) at which a hole is created in the peripheral part of the iris to bypass the pupillary block. In chronic angle-closure glaucoma, the elevation of the IOP progresses slowly and gradually, since the closure happens gradually or intermittently. Thus, chronic angle-closure glaucoma is often confused with open-angle glaucoma. Chronic primary angle-closure glaucoma occurs more often and usually causes more serious damage than the acute form.

The major risk factors for primary angle-closure glaucoma are age, hyperopia, and East Asian descent. Although primary angle-closure glaucoma appears only in about a third of all primary glaucoma cases, the number of people becoming blind due to this form of glaucoma is nearly equal to the number of people becoming blind from primary open-angle glaucoma [4].

The role of the IOP in glaucoma Currently, an increased IOP is the only treatable risk factor associated with glaucoma. Hence, the goal of all therapeutic measures concerning glaucoma is to lower the IOP. Besides drug-based therapy methods, surgical interventions are also used. In the latter case, laser-based microsurgical techniques play a major role (Section 10.4.4.2).

The absolute value of the IOP is, however, only of limited importance for each individual patient, since every eye shows a different sensitivity. On the one hand, some patients suffer from *low-tension glaucoma,* in which the typical symptoms of glaucoma also appear for eyes with an IOP below the critical value. On the other hand, some patients whose eyes have an IOP above the critical value do *not* show any glaucoma-related symptoms (*ocular hypertension*). Consequently, IOP values between 21 and 26 mmHg do not necessarily require a glaucoma therapy, but the risk of developing glaucoma considerably rises for an increased IOP.

3) With "iris root" we mean the peripheral part of the iris at the transition to the ciliary body (see also Figure 1.4 in Section 1.1).

3.4
Age-Related Macular Degeneration

The macula is the most sensitive part of the retina, since it represents the area of sharpest vision. A widespread disease which considerably impairs the quality of vision in this area is *age-related macular degeneration* (AMD). As the name suggests, mainly elderly people suffer from this retinal disease.

AMD is the major cause of blindness in industrialized nations (Section 3.9). In most cases, both eyes are affected by this disease. As the peripheral field of view is usually maintained, the patients are not totally blind but have a severe visual impairment. Particularly, the ability to read is limited. The appearance and progression of AMD is strongly related to aging processes around the retinal pigment epithelium (RPE), Bruch's membrane, and the choriocapillaris (inner layer of blood vessels of the choroid; Figure 1.8) which form debris and lead to abnormal structural changes.

The outer segments of the photoreceptors, which are responsible for the conversion of light to neuronal signals, are continuously regenerated. Simultaneously, the "exhausted" parts of the outer segments (near the RPE) are rejected. These "waste products" are absorbed and catabolized by the RPE. The RPE cells then discard a portion of the products of decomposition through Bruch's membrane to the choriocapillaris. Age-related disorders affecting this process lead to an accumulation of debris between the RPE cells and Bruch's membrane and thus to the formation of *drusen*. A share of the waste products remains in the RPE cells and accumulates in the form of *lipofuscin granules* which, in turn, disturbs the cellular metabolism.

Age-related macular diseases are classified into three main stages. The early stage is the *age-related maculopathy* (ARM). During the progression, ARM "transforms" to the advanced stages, called *dry* and *wet AMD*.

3.4.1
ARM

Characteristic manifestations of ARM are changes of the pigment distribution (*hyperpigmentation*) and the formation of drusen (Figures 3.7 and 3.6b). ARM is usually diagnosed by chance, since during this stage the symptoms are marginal and often not recognized by the patient. To date, suitable therapy methods are not available. However, food supplements such as vitamins and essential minerals are often administered in an attempt to slow down progression of ARM.

3.4.2
Dry AMD

A characteristic manifestation of dry AMD is *geographic atrophy* (Figure 3.6c), a condition where the RPE and the adjacent retinal structures (choriocapillaris and photoreceptors) are continuously destroyed. As a consequence, depigmented areas appear which grow in diameter (Figure 3.8). These irreversible structural changes

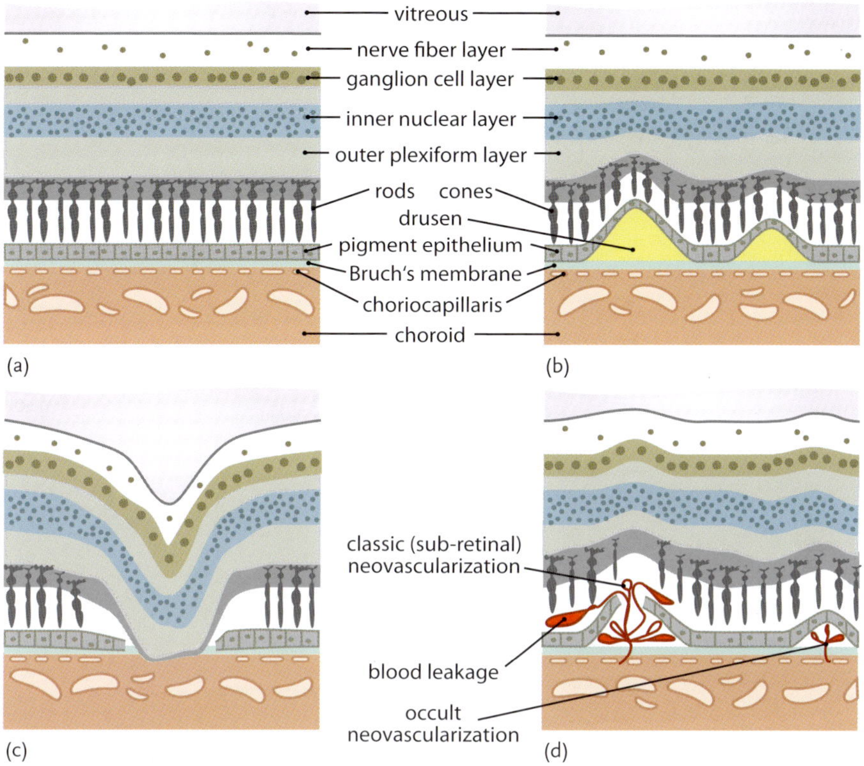

Figure 3.6 Stages of age-related macular diseases. (a) Cross-section of the retina in the case of a healthy eye. (b) Age-related maculopathy: formation of drusen which contain waste products (retinal debris) produced by the retinal pigment epithelium (RPE). (c) Dry age-related macular degeneration: RPE and adjacent retinal structures (choriocapillaris and photoreceptors) are continuously destroyed. As a consequence, depigmented areas occur, which are called *geographic atrophies*. (d) Wet age-related macular degeneration: abnormal vessels develop from the choriocapillaris, penetrate Bruch's membrane, and grow below the retina. Depending on the location of neovascularization (above or below the RPE), we distinguish between the classic and occult form. Adapted from [5].

then impair the quality of vision. As soon as the fovea is affected, the visual acuity is considerably reduced and the ability to read is lost.

About 85% of AMD patients suffer from the dry form. To date, a suitable therapy method does not exist. As with ARM, food supplements are often used in an attempt to slow progression of the disease.

3.4.3 Wet AMD

In wet AMD, new abnormal blood vessels are formed in the choroid (*sub-foveal choroidal neovascularization* (CNV)). The abnormal vessels develop from the choriocapillaris, penetrate Bruch's membrane, and grow below the retina (Figure 3.6d).

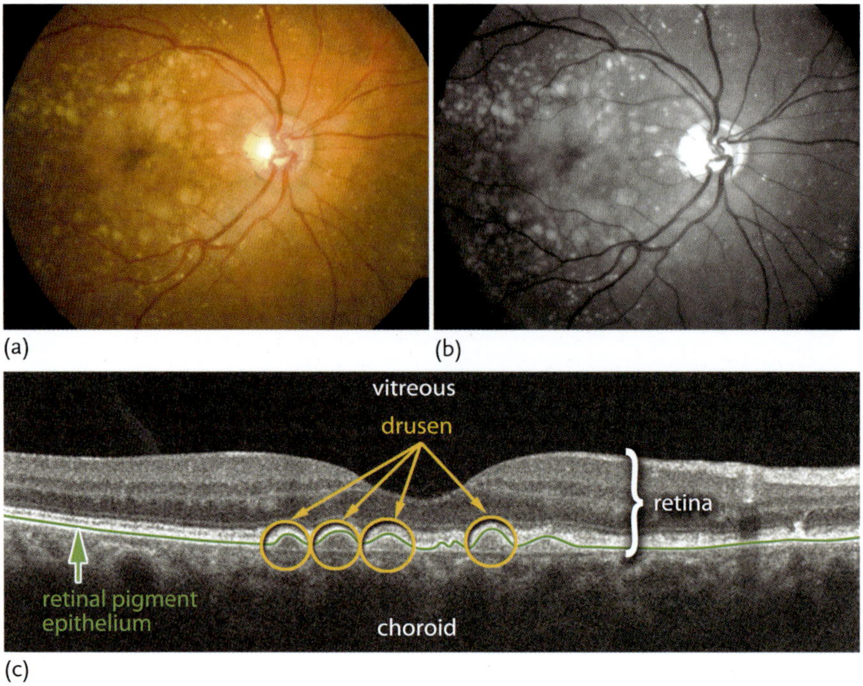

Figure 3.7 Age-related maculopathy (ARM). (a) Color fundus image. Multiple, partially conflating soft drusen (light yellow spots) are distributed over the whole fundus. Courtesy of Dr. Thomas Behling. (b) Monochromatic fundus image (Section 6.7.5). With green light illumination (red-free imaging), the drusen appear as bright spots. Courtesy of Dr. Thomas Behling. (c) Cross-sectional OCT image of the retina (Chapter 7) which shows how the retinal pigment epithelial layer (green line) is locally detached (orange circles) from Bruch's membrane due to the formation of drusen. Courtesy of Carl Zeiss.

Depending on the location of neovascularization, we distinguish between the *classic* and *occult*[4] form. Occult neovascularization is located between the RPE layer and the choriocapillaris. Classic neovascularization is located between the RPE layer and the photoreceptors and can be identified by fluorescence angiography imaging (Figure 3.9b).

The newly grown vessels are very fragile and leaky. Because of the resulting discharge of fluids and blood, edema appear, and in some cases an RPE detachment is induced. Bleedings around the fovea may lead to a rapid and dramatic reduction of the visual acuity. Compared to occult neovascularization, the classic form occurs less frequently but grows faster and rapidly reduces the quality of vision. Generally, long-standing neovascularization leads to degenerative and structural changes in

4) The term "occult" is derived from the fact that this form of neovascularization cannot be made visible with fluorescence angiography (Sections 6.7.6 and 6.8.1.2) due to the absorption properties of the RPE layer.

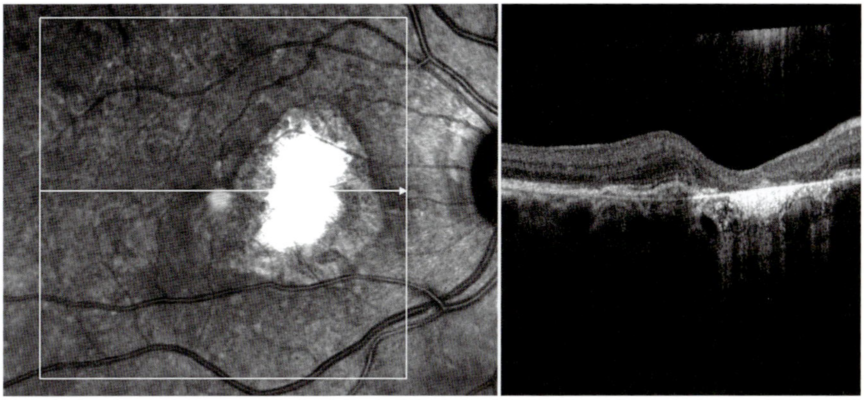

(a) (b)

Figure 3.8 Geographic atrophy in dry AMD. (a) Monochromatic near-infrared fundus image. The geographic atrophy is represented by the bright area, since RPE and photoreceptors are destroyed in this region. (b) Cross-sectional OCT image (Chapter 7) in a plane perpendicular to the arrow in (a). The absence of RPE and photoreceptors as well as the thinning of the retina are clearly visible. Courtesy of Prof. Dr. Jens Dawczynski.

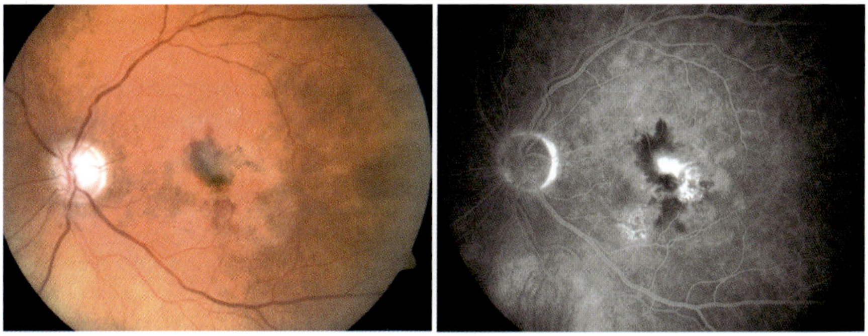

(a) (b)

Figure 3.9 Wet AMD with typical sub-foveal choroidal neovascularization (CNV). (a) The fundus image clearly shows CNV as a dark brown spot in the center of the macula. (b) In the fluorescein angiographic image (Section 6.7.6), the leakage of fluids and blood from CNV is visible as a large bright spot. Courtesy of Dr. Thomas Behling.

both RPE and choriocapillaris so that the photoreceptors eventually die off. In the final stage of wet AMD, the retina shows an extended scar.

Wet AMD appears in only 15% of all cases of advanced macular degeneration. However, it is the major cause of vision loss (in 80% of all cases). For several years, effective medical therapies have been available for treatment of wet AMD. The so-called *anti-VEGF*[5] treatment halts vision loss in more than 90% of patients and leads to vision improvement in 30% of all cases. This therapy, based on intravit-

5) Vascular endothelial growth factor (VEGF) is one of the most important mediators to stop the growth of new vessels.

real injection of anti-VEGF drugs, has widely replaced laser-assisted therapy methods (photodynamic therapy and photocoaguation; Sections 9.4.1 and 9.4.2, respectively).

3.5 Diabetic Retinopathy

Diabetic retinopathy (DR) is one of the most sight-threatening complications of diabetes mellitus (insulin-dependent diabetes) and one of the most important emerging causes of blindness among working-age individuals in developed countries. More than 90% of people with long-standing (more than 20 years) diabetes mellitus suffer from DR. Two main stages exist: *nonproliferative* and *proliferative* DR.

Nonproliferative DR In the early stages of nonproliferative DR, perycites are lost which support the retinal capillary endothelial cells of the blood vessels. Their absence leads to an increased permeability of the capillaries and a decrease of their wall thickness. As a result, punctuate hemorrhages occur, and small patchy vessel dilatations, so-called *microaneurysms*, develop. They appear as small red dots in the color fundus image or as bright dots in the fluorescein angiography image (Figure 3.10a).

During nonproliferative DR, macular edema are the most important clinical manifestation. Macular edema occur when intercellular fluids (which stem from leaking microaneurysms) and capillary leakage accumulate on or under the macula of the eye and causes it to thicken and swell. As an edema separates retinal cells,

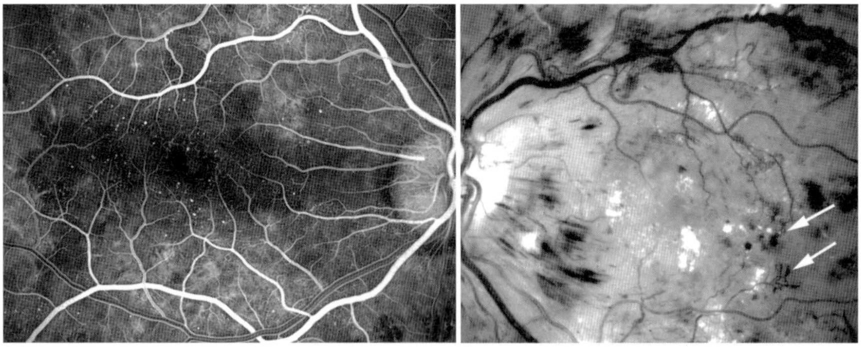

(a) (b)

Figure 3.10 Diabetic retinopathy (DR). (a) Nonproliferative DR: in the fluorescein angiography image, the formed microaneurysms and punctuate hemorrhages appear as small bright dots. (b) Proliferative DR: with green light monochromatic illumination (red-free imaging), hemorrhages appear as extended dark spots. Newly formed abnormal blood vessels (neovascularization) are also clearly visible (see white arrows). The bright spots are cotton-wool spots, that is, infarctions of the retinal nerve fiber layer. Courtesy of Dr. Thomas Behling.

multiple interfaces are created in the retina, which leads to an increase of light scattering. The visual quality is thus considerably reduced down to the threshold of ambulatory vision[6]. Diabetic macular edema is the main cause of moderate vision loss in nonproliferative DR. However, it can develop at any stage of DR and is the leading cause of legal blindness in patients with diabetes.

In advanced stages of nonproliferative DR, multiple bleeding, capillary nonperfusions, venous loops, small abnormal blood vessels, and cotton-wool spots (i.e., infarctions caused by blockage of blood supply) of the retinal nerve fiber layer appear (Figure 3.10b).

Proliferative diabetic retinopathy As DR progresses, blood vessels become occluded which leads to retinal ischemia (undersupply). In response, growth factors are released which induce the formation of new blood vessels (neovascularization; Figure 3.10b). These new blood vessels grow towards the surface of the retina or optic nerve head. As they are fragile and can spontaneously bleed, severe vitreous hemorrhage, vitreous shrinking, and/or retinal detachment occur, any of which may lead to severe loss of vision. If proliferative DR is left untreated, 50% of patients will become blind within several years.

For proliferative DR, diabetic macular edema, and other vitreo-retinal diseases, severe visual loss can be reduced or prevented with laser-induced photocoagulation (Section 10.2).

3.6 Retinal Vein Occlusions

After diabetic retinopathy, retinal vein occlusion (RVO) is the second most common cause of visual loss of all retinal vascular disorders. RVO is characterized by a partial or total obstruction of the central retinal venous outflow caused by a thrombus located in the central retinal vein. Depending on the location of the thrombus, we may distinguish between *central retinal vein occlusion* (CRVO) and *branch retinal vein occlusion* (BRVO).

A CRVO results from a thrombus in the central retinal vein within the optic nerve. The BRVO occurs in the case of obstruction at a branch of the retinal vein. BRVO is approximately 3–5 × more common than CRVO. The severity of RVO depends on the completeness of the occlusion and the size and location of the retinal area drained by the affected vein. Due to the vein occlusion, a pressure forms in the capillaries which, in turn, leads to superficial hemorrhages and leakage of fluid and blood (Figure 3.11). Patients affected by RVO typically complain of an onset of blurred vision or a central visual field defect. Depending on the severity of the disease, a visual loss can be noticed in an acute or more subtle and intermittent manner. The most significant complications of RVO are macular edema and retinal

6) "Ambulatory vision" is defined as being able to see objects and move around a room without tripping over obstacles.

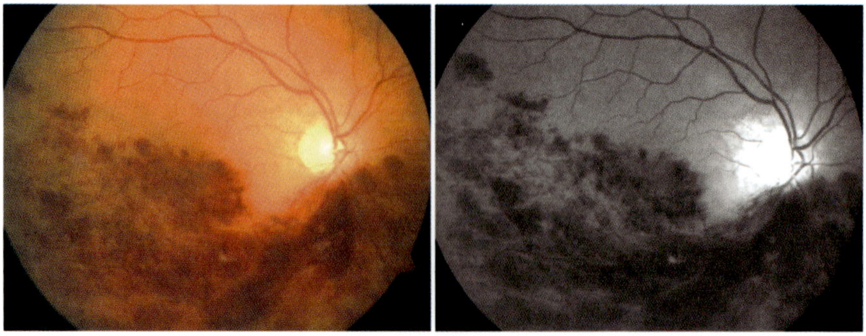

(a) (b)

Figure 3.11 Branch retinal vein occlusion (BRVO). (a) Color fundus image: the large superficial hemorrhages appear as dark areas. (b) Monochromatic fundus image (Section 6.7.5). When illuminated with green light (red-free imaging), the hemorrhages can be more clearly seen. Courtesy of Dr. Thomas Behling.

neovascularization leading to severe vision loss or blindness. Both manifestations of RVO can be treated with laser photocoagulation (Section 10.2).

3.7 Infective Eye Diseases

Let us briefly discuss two other common eye diseases, which are caused by infection. These diseases mainly emerge in developing regions with insufficient hygienic conditions (see also Section 3.8).

3.7.1 Trachoma

Trachoma is the most common infectious cause of blindness. This disease is caused by the *Chlamydia trachomatis* bacterium which produces a characteristic roughening of the inner surface of the eyelids. Children who suffer from trachoma are the reservoir of infection, while blindness rather affects adults. Trachoma mainly appears in communities with poor water supplies and sanitation, where the bacteria are transmitted from person to person through direct or indirect contact and by flies. Blindness due to trachoma is caused by repeated episodes of infection.

Apart from prevention, antibiotic therapy and surgery are possible treatment options to cure advanced stages of the disease.

3.7.2
Onchocerciasis

Onchocerciasis (also known as *river blindness* and *Robles' disease*) is caused by infection with the roundworm *Onchocera volvulus* which is transmitted through the bite of a black fly. A few years ago, it was the world's second-leading infectious cause of blindness, but successful international health programs have tremendously decreased the number of infections. This disease can be treated by drug administration.

3.8
Major Causes for Visual Impairment

In 2010, about 285 million people were visually impaired (i.e., low vision and blindness) [9]. According to a study of the World Health Organization (WHO) [6], the major causes for visual impairment worldwide are uncorrected refractive errors (42%), cataract (33%), and glaucoma (2%; Figure 3.12). The distribution varies for different global regions as it mainly depends on the average age and income situation of the population [7].

Dependence on age As shown in Figure 3.13a, about 65% of visually impaired persons are older than 50 years. Currently, this group of people comprises about 20% of the world population. As the average age (especially in industrial nations) steadily rises, the number of people affected by age-related eye diseases like age-related macular degeneration, cataract, and glaucoma will possibly increase in the next decades.

The same trend is also found for presbyopia. In 2005, about 1.04 billion people were affected by this age-related visual impairment [8]. It is estimated that this number will increase to about 1.4 billion by the year 2020 and to 1.8 billion by 2050.

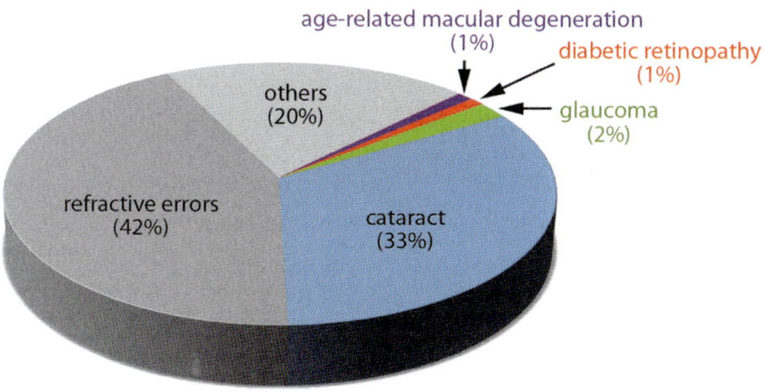

Figure 3.12 Major causes for visual impairment worldwide in 2010. Data taken from [6].

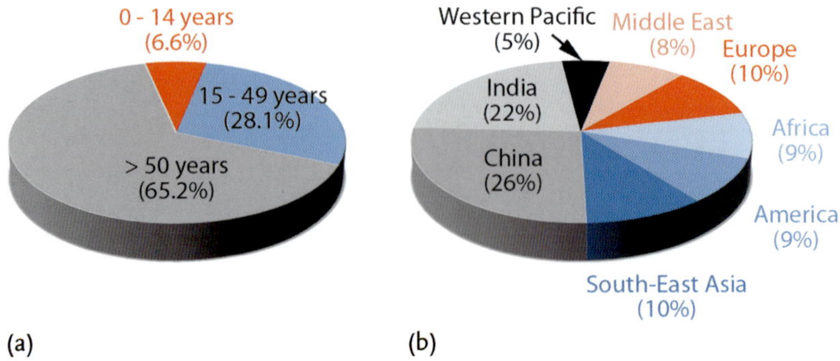

Figure 3.13 Worldwide distribution of visual impairment in 2010. (a) Distribution sorted by age. (b) Distribution sorted by global region. Data taken from [6].

Regional dependence Figure 3.13b shows the regional distribution of visually impaired people. Obviously, China, India, as well as African and Middle East countries are more greatly affected, partly because of insufficient medical care. As the population in developing countries rapidly rises, the total number of people with uncorrected refractive errors, cataract, and hygiene-related diseases (e.g., trachoma and onchocerciasis) will possibly increase. Note that children are especially affected by this trend. The WHO estimated that the vision of 19 million children between 0 and 14 years of age was impaired [9]. 17.5 million of them had low vision and 1.4 million were irreversibly blind for the rest of their lives. In fact, uncorrected refractive errors are the main causes of visual impairment in children between 5 and 15 years.

Counteractive measures In principle, 80% of global visual impairment could be prevented or cured. To address this shortcoming, national and international programs (e.g., Vision 2020) have been established to support low income countries. In some countries, eye care services which are affordable and of sufficient quality have been integrated into standard health care systems. Campaigns (including school-based education) also raise the awareness of eye diseases. In spite of initial achievements, there is still great potential for improvement.

3.9
Major Causes of Blindness

A person is considered to be blind if the (Snellen) visual acuity (Section 2.1.5.2) is below 0.05 and/or the field of view is reduced to less than 10°. In 2010, the WHO estimated that more than 39 million people were blind [9] and that this number grows by about 17 000 each year. Figure 3.14 shows that cataracts are the major cause of blindness, accounting for more than 50% of all cases. This is followed by glaucoma (8%) and age-related macular degeneration (5%). The large percentage

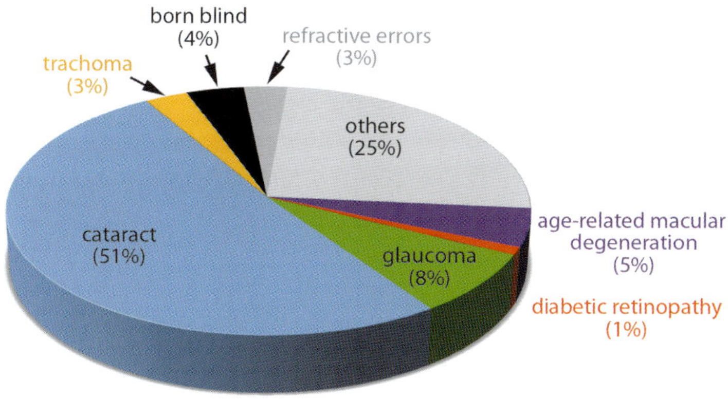

Figure 3.14 Major causes of blindness worldwide in 2010. Data taken from [6].

of "other diseases" reveals that a high percentage of blind people live in developing countries where no reliable statistical surveys exist.

Dependence on age Figure 3.15a shows the worldwide age distribution of blind people. With 81.7%, the share of people older than 50 years is even higher than for visual impairment in general. To a certain degree, this tendency can be understood, since most eye disease are not congenital, but appear for elderly adults. At first, the most common eye diseases lead to a decrease of the visual acuity. If left untreated, they cause blindness after several years of progression.

Regional dependence In Figure 3.15b, the regional distribution of blind people is shown. From this, we conclude that developing countries are more greatly affected by blindness than industrialized countries. The tendency is also apparent in Figure 3.16, which shows the number of blind people per million population. In particular, low and middle income countries in Africa and the Middle East are most seriously affected. In fact, about 75% of blind people live in developing countries of Asia (21.4 million) and Africa (7.1 million) [10].

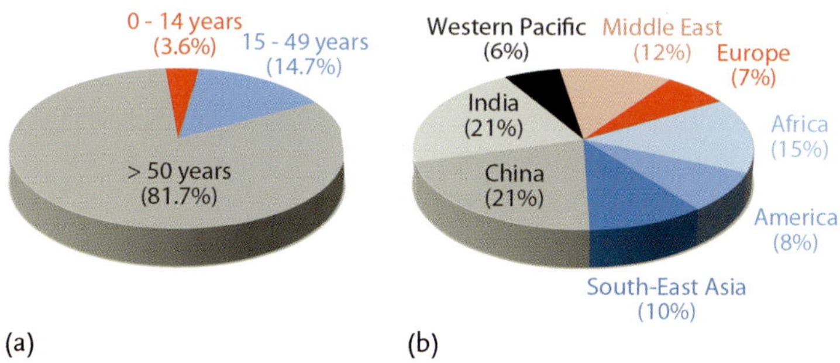

Figure 3.15 Worldwide distribution of blindness in 2010. (a) Distribution sorted by age. (b) Distribution sorted by global region. Data taken from [6].

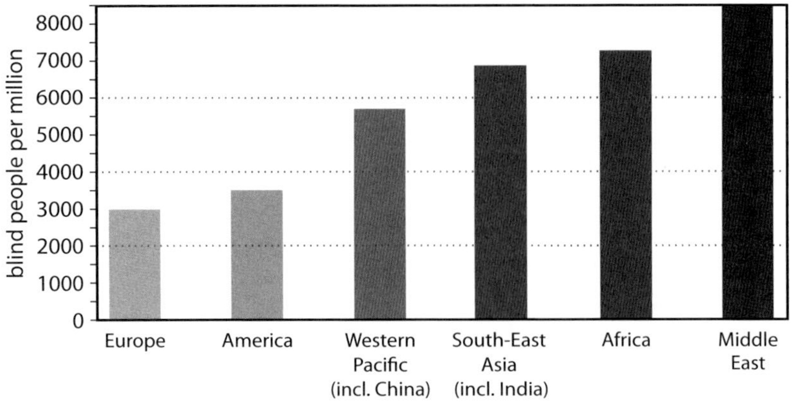

Figure 3.16 Number of blind persons per million in 2010 sorted by global region. Data taken from [6].

Let us highlight cataract in this context. Cataract is strongly related to aging and cannot be prevented, but cataract surgery has been identified by the World Bank as one of the "most highly cost-effective" interventions[7)] that can be offered to the developing countries [7]. However, the poor population often cannot afford the required operations. As a consequence, they go blind despite the fact that their blindness could, in principle, be corrected with relatively low effort by on-time treatments. In industrial countries, however, cataract is successfully treated. There, age-related diseases (e.g., age-related macular degeneration and glaucoma) and diabetes-induced diseases (e.g., diabetic retinopathy and glaucoma) are the major causes of blindness. For example, in the United States, vision loss due to diabetes has been the leading cause of blindness in adults for more than 30 years.

3.10 Socio-Economic Impact of Eye Diseases

Simulations have shown that the age and size of the world population will increase over the next 20 years. If we assume that the current age distribution in Figure 3.13a remains constant, the combination of population growth and increased length of life will rapidly increase the number of diseased (particularly blind) people. This will, in turn, result in serious consequences for the global economy in all societies. The emerging costs of lost productivity, rehabilitation, and special education pose a significant burden to affected persons, their family, and their society [10]. The worldwide economic productivity loss (due to visual impairment) alone was esti-

7) Cataract intervention has been highly optimized during the last years. Under ideal conditions, one ophthalmologist can carry out more than 1000 surgeries each year [15]. In total, there are about 20 million cataract surgeries each year.

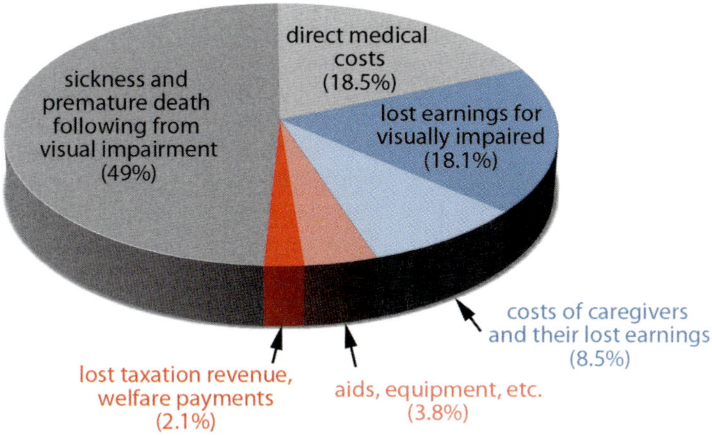

Figure 3.17 Cost structure emerging from visual impairment. With a share of approximately 18.5%, the costs for medical care are relatively low compared to the socio-economic costs resulting as a consequence of visual impairment. Other important (indirect) cost factors are the lost earnings of the concerned person and caregivers as well as the related costs for aids and required equipment. Data taken from [9].

mated to be 42 billion USD per year and will possibly rise to 110 billion USD by the year 2020 [7].

Cost structure The economic impact of visual impairment can be categorized into direct costs, which are related to the treatment of the respective eye disease (e.g., medical services, pharmaceuticals, research, administration), and indirect costs, which refer to the financial burden of the concerned people (e.g., lost earnings, costs for visual aids, welfare payment, lost taxation revenues). The structure of arising costs was surveyed in an Australian study in 2004 (Figure 3.17). Interestingly, the medical care for the actual disease (or refractive error) is *not* the most relevant cost factor. Instead, costs related to following diseases and premature death of visually impaired people have by far the highest share (49%). The loss of earnings of diseased persons is also tremendously high (18.1%), and it should be noted that one blind person usually absorbs the whole manpower of another person [11]. As a consequence, children blinded during childhood or at birth generate a higher economic cost to their family members and society over their lifetime than adults blinded in later life.

Regional impact Let us now have a look at the regional impact of visual impairment. Figure 3.18 shows the estimated loss of the gross domestic product (GDP) of selected global regions (in billion USD). The simulation [7] also considered the steady increase of age and population and provides forecasts up to the year 2020. Apparently, the emerging costs related to visually impaired people will rise in all regions. In particular, industrialized nations will face a serious financial burden due to the drop out and required support of impaired persons.

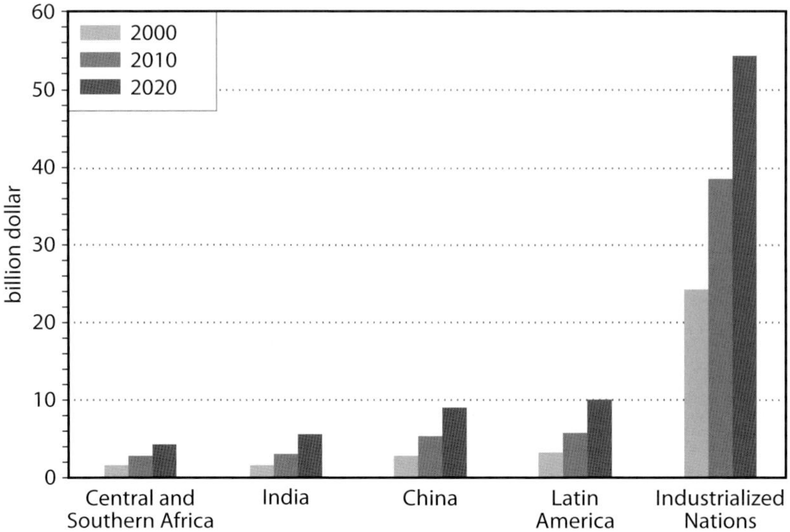

Figure 3.18 Estimated loss of the gross domestic product (GDP) of selected global regions for 2000, 2010, and 2020. The influence of blindness and trends of the demographic development were considered in this simulation. Data taken from [7].

3.11 Recommended Reading

In this chapter, we restricted ourselves to common disorders and eye diseases which are relevant for the ophthalmic systems to be discussed throughout this book. Further information about presented and other kinds of eye disease can be found in standard ophthalmology textbooks, for example, [3, 5, 12–14, 16].

Problems to Chapters 1–3

PI.1. Size of the retinal image The retinal image size can be calculated via

$$|h_{\mathrm{I}}| = \kappa \overline{\mathrm{N}'\mathrm{I}'_0} \tag{3.7}$$

$$= |h_{\mathrm{O}}| \frac{\overline{\mathrm{N}'\mathrm{I}'_0}}{\overline{\mathrm{O}_0\mathrm{N}}}, \tag{3.8}$$

where $\kappa = -h_{\mathrm{O}}/\overline{\mathrm{O}_0\mathrm{N}}$ (Section 2.1.3). For an object which is located at infinity (relaxed eye) and with a mean refractive power of the eye of 60 D, we may derive the relation

$$|h_{\mathrm{I}}| \approx 16.667\,\mathrm{mm} \cdot \kappa\,. \tag{3.9}$$

If κ is given in degrees, the relation can also be expressed by

$$|h_{\mathrm{I}}| \approx 0.291\,\mathrm{mm} \cdot \kappa^{\circ} \tag{3.10}$$

with $\kappa = h_O/\overline{O_0V}$ (Figure 2.3).

1. Please derive Eq. (3.9) or (3.10).
2. Calculate the size of the retinal image for a tower (50 m high at a distance of 1 km), a person (1.8 m high at a distance of 10 m), a thumbnail (diameter of 2 cm at a distance of 60 cm), and for the full moon ($\kappa = 0.5°$).

PI.2. Gullstrand Eye model Calculate the position of the eye's entrance pupil relative to the corneal vertex. Please also calculate the diameter of the entrance pupil relative to the iris aperture by using the Gullstrand Eye model #1.

PI.3. Reflectance of the cornea: Calculate the reflectance of the cornea at the vertex with the Fresnel equation (A4).

PI.4. Radius of curvature Variations of the corneal radius of curvature Δr_C mean a change of the corneal refractive power $\Delta\mathcal{D}$ which can be calculated via

$$\Delta\mathcal{D} \approx -\frac{\mathcal{D}_c'^2 \cdot \Delta r_C}{n_c - 1}, \tag{3.11}$$

where $\mathcal{D}_c'$ is the refractive power of the corneal front surface and n_c the refractive index of the cornea.

1. Please derive Eq. (3.11).
2. Verify the following statement for an emmetropic Gullstrand Eye: A variation of the corneal radius of curvature by ± 0.1 mm changes the eye's refractive power by approximately ± 0.6 D.

PI.5. Eye length A small variation of an emmetropic eye's axial length ΔL_{eye} with a refractive power of $\mathcal{D}_{eye}'$ means a change of refraction by $\Delta\mathcal{D}$. We may approximate this change with

$$\Delta\mathcal{D} \approx -\frac{\mathcal{D}_{eye}' \Delta L_{eye}}{n}. \tag{3.12}$$

1. Please derive Eq. (3.12).
2. Verify the following statement for an emmetropic Gullstrand Eye with $\mathcal{D}_{eye}' = 60$ D and $n = 1.336$: The variation of the eye length by ± 0.37 mm changes the eye's refractive power by approximately ± 1 D.

PI.6. Stereoscopic vision In order to check the stereoscopic vision of an eye, real and virtual test objects are used.

1. A real test object shall be used up to a stereo angle of $\varepsilon = 5''$. What minimum stereoscopic depth perception ΔL_{min} must this object have if it is viewed from a distance of 5 m and the interpupillary distance is PD = 62 mm.
2. A virtual stereoscopic test object consists of two identical test objects T_l and T_r (e.g., stripes or triangles). These objects are horizontally arranged (distance $\Delta y = 20$ mm) above *or* below a central focus object F (e.g., circle). The three objects T_l, T_r, and F lie all in one test plane which is perpendicular to the viewing

direction and located at a distance of 5 m. An optical system ensures that each eye of the patient can only see one test object. In this regard, we can distinguish between the following cases:

- symmetric allocation, that is, T_l (T_r) is seen by the left (right) eye, and
- asymmetric allocation, that is, T_l (T_r) is seen by the right (left) eye.

Due to the small relative shift of both identical images on the retina, the patient perceives a virtual object T which seems to float behind and in front of the test plane at a distance ΔL (virtual stereoscopic effect).

- Is the normal stereoscopic resolution sufficient to have a three-dimensional impression?
- At what distance behind and in front of the test plane does a patient (with normal stereoscopic vision) see the test objects (interpupillary distance PD = 65 mm)?
- Which allocation do we have to choose, if the patient shall perceive a floating test object located in front of the test plane?
- Calculate the relative local shift Δs of the retinal images for an eye with a refractive power of 60 D.

PI.7. Resolving power of the eye

1. In order to determine the refraction of an eye, the standard letter "E" is placed at a distance of 2 ft (= 6096 mm) from the eye. Calculate the size of the letter for a visual acuity of $V = 1$.
2. The retinal image resembles the image shown in the Figure 3.19. Calculate the image size on the retina and compare the result to the distance of the retinal cones (in the fovea).
3. Sometime in the future, we will be visited by aliens from the planet XIR2050 whose star emits light only in the red and infrared spectral range and whose atmosphere allows only near-infrared light (wavelengths between 1 and 1.5 μm) to pass through. The eyes of the aliens are adapted to these conditions and the aliens' visual acuity is similar to that of our eyes. How will they fare on Earth? If you were to be selected to travel to XIR2050, how would you prepare for your visit? What should you expect to be faced with on this planet?

PI.8. Refractive errors

1. In the case of cataracts, it used to be common in the past to simply perforate the turbid lens and sort of remove it in a surgical process. The eye was rendered aphakic (i.e., left without eye lens). Where is the image of the far point in an aphakic eye? Would it have been possible to help this person with spectacles? How strong would these spectacles have to have been?
2. Is it true that myopic people see small things better? How much of a difference is there as compared to a person with normal vision?

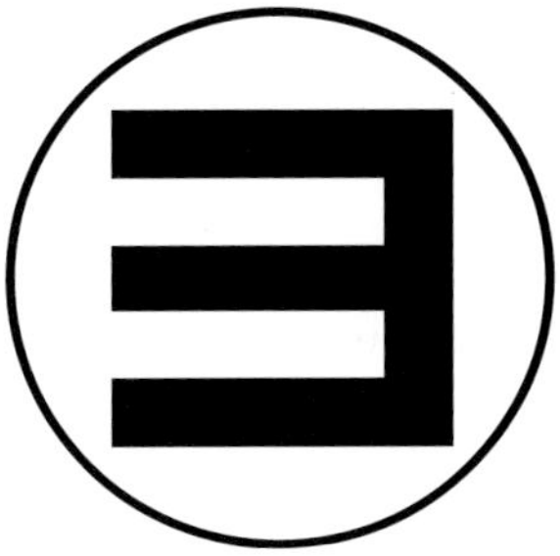

Figure 3.19 Retinal image of Snellen "E".

3. What is your comment regarding the assertion that myopic people can see sharp images under water; meaning that they do not need a pair of goggles?
4. In diving schools, it is taught that you see objects under water 33% bigger and 25% closer. Prove if this is really true.

PI.9. Refractive errors

1. Use the Gullstrand Eye model to calculate the power of spectacles needed to correct an eye if the far point is located 45 cm in front of the eye. Because of the frame of spectacles, the glasses are placed at a distance of 15 mm in front of the corneal vertex.
2. The same spectacle glasses as in a) have inadvertently been mounted in a frame so that we now have a distance from the corneal vertex of only 10 mm. Does this improve or worsen the correction of the refractive error?
3. Draw a conclusion from b) regarding how a contact lens would have to be designed.

PI.10. Chromatic aberration The eye shows some notable chromatic aberration of almost 2 D in the visible spectral range. Why do we generally not notice this, whereas an optical instrument (e.g. a photo camera) with similar chromatic aberration would be unusable? Could the different width of the blue-white-red stripes of the French flag (Tricolore) have anything to do with this?

PI.11. Stereo-camera system Let us consider the special case of a stereo-camera system. Both cameras have equal parameters (in particular equal focal lengths f), the image planes of the two cameras are co-planar, and the x axes of the image planes are parallel to the baseline (Figure 3.20). The distance between the two cameras is $b = 200\,\text{cm}$. The cameras use $1/2''$ CCD chips with 1024 horizontal pixels. A camera system of this type shall be used to track objects (e.g. surgical instruments) in a volume ($a^3 = 50 \times 50 \times 50\,\text{cm}^3$) around a patient's head from a distance $z \approx 3\,\text{m}$.

1. Calculate the stereo-disparity (difference between x_L and x_R) for various objective focal lengths f.
2. Which objective focal length would you recommend to achieve a tracking volume of maximum size or maximum point accuracy?

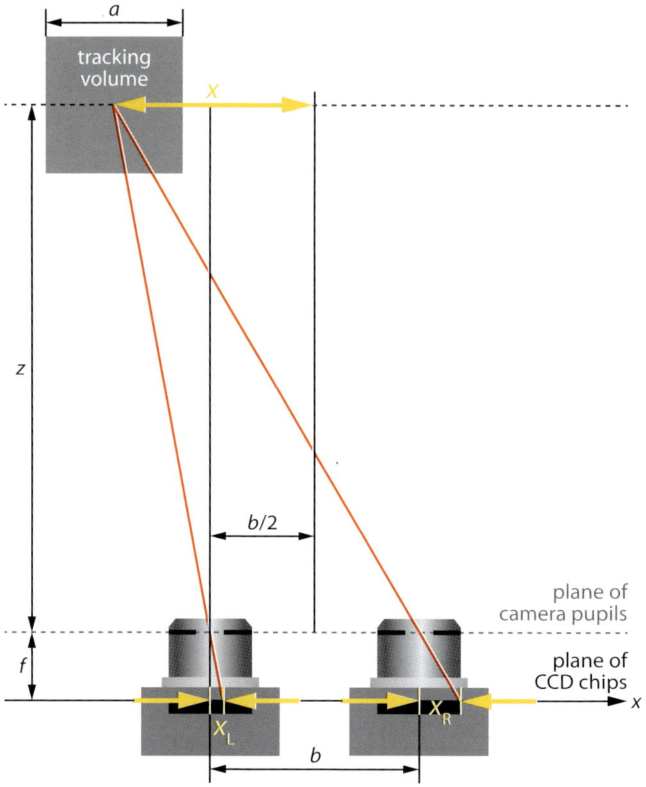

Figure 3.20 Geometry of a stereo-camera system discussed in Problem PI.11.

3. Calculate the maximum attainable (lateral) resolution of the system at the optimal objective focal length. What causes the resolution to be less in reality? How can the resolution be increased?
4. Compare this resolution to that of an acoustical tracking system with a frequency of 50 kHz. Which phase measuring accuracy must be at least attained with an ultrasound system?

References

1 Thibos, L.N., Wheeler, W., and Horner, D. (1997) Power vectors: An application of fourier analysis to the description and statistical analysis of refractive error. *Optom. Vis. Sci.*, **74**, 367–375.

2 Gross, H., Blechinger, F., and Achtner, B. (2008) *Handbook of Optical Systems – Survey of Optical Instruments*, vol. 4, Wiley-VCH Verlag GmbH, Weinheim.

3 Grehn, F. (2008) *Augenheilkunde*, Springer.

4 Quigley, H.A. and Broman, A.T. (2006) The number of people with glaucoma worldwide in 2010 and 2020. *Br. J. Ophthalmol.*, **90**, 262–267.

5 Sachsenweger, M. (2002) *Augenheilkunde*, Thieme.

6 Pascolini, D. and Mariotti, S.P. (2012) Global estimates of visual impairment – 2010. *Br. J. Ophthalmol.*, **96**, 614–618.

7 Frick, K.D. and Foster, A. (2003) The magnitde and cost of global blindness: An increasing problem that can be alleviated. *Am. J. Ophthalmol.*, **135**, 471–476.

8 Holden, B.A., Fricke, T.R., Ho, S.M., Wong, R., Schlenther, G., Cronjé, S., Burnett, A., Papas, E., Naidoo, K.S., and Frick, K.D. (2008) Global vision impairment due to uncorrected presbyopia. *Arch. Ophthalmol.*, **126**, 1731–1739.

9 WHO (2012) Global data on visual impairments 2010. http://www.who.int/blindness/GLOBALDATAFINALforweb.pdf (accessed 24 August 2013).

10 Smith, A. and Smith, J. (1996) The economic burden of global blindness: A price too high! *Br. J. Ophthalmol.*, **80**, 276–277.

11 Javitt, J., Venkataswamy, G., and Sommer, A. (1983) The economic and social aspect of restoring sight, in *ACTA: 24th Int. Congr. Ophthalmol. New York, NY: JP Lippincott*, pp. 1308–1312.

12 Kanski, J.J. (2011) *Clinical Ophthalmology: A Systematic Approach*, Saunders.

13 Cassel, G.H., Billig, M.D., and Randall, H.G. (1998) *The Eye Book: A Complete Guide to Eye Disorders and Health*, The Johns Hopkins University Press.

14 Flammer, J. (2006) *Glaucoma: A Guide for Patients, An Introduction for Care Providers, A Quick Reference*, Hogrefe Verlag GmbH.

15 Bell, C.M., Hatch, W.V., Cernat, G., and Urbach, D.R. (2007) Surgeon volumes and selected patient outcomes in cataract surgery: A population-based analysis. *Ophthalmology*, **114**, 405–410.

16 Benjamin, W.J. (2006) *Borish's Clinical Refraction*, 2nd edn, Butterworth–Heinemann/Elsevier.

Part Two

4
Introduction to Ophthalmic Diagnosis and Imaging

Diagnostic methods and devices are used to detect and/or diagnose a potential disease or functional disorder. They are also used to decide whether a treatment is necessary and if so, which kind of treatment is suitable. Diagnostic methods and devices also allow us to check or monitor the outcome of the treatment and, if necessary, to optimize the therapeutic measures initiated.

Since the eye is, in engineering terms, an optical device, and since most of its parts and structures are easily accessible by optical radiation, the majority of diagnostic devices and methods used in ophthalmology are based on optical principles.[1] Methods and devices can be classified based on either an application- or a technology-oriented approach. In the application-oriented approach, the devices and methods are categorized and described with regard to their relevance for general or specific diagnostic issues (e.g., fundus imaging or glaucoma diagnosis), together with their advantages and drawbacks. This type of representation is highly advantageous from the point of view of the device users, and is the standard approach used in the specialized medical literature. Unfortunately, frequent cross-referencing is unavoidable in this approach, because many methods and devices can be used for a number of different diagnostic tasks. As this book is primarily intended for readers with a technical background (e.g., engineers, scientists, medical technicians, and system device engineers), a more technology-oriented presentation has been chosen in the following chapters. Thus, the principle of classification is no longer a specific clinical application, but rather the overall diagnostic situation to be evaluated.

Classification of methods and devices Taking the essential issues involved in any particular eye examination into account, three primary diagnostic tasks can be determined which will allow classification of different optical methods and devices into the following groups:

1) Other ocular imaging techniques used in clinical practice are high-resolution ultrasound, and magnetic resonance imaging (MRI). These techniques are used if it is impossible to optically obtain the required information. We will not further discuss the nonoptical techniques in this book.

Optical Devices in Ophthalmology and Optometry, First Edition. Michael Kaschke, Karl-Heinz Donnerhacke, and Michael Stefan Rill.

1. Methods and devices for the determination of the eye's refractive status (Chapter 5).
2. Methods and devices for the visualization, imaging, and analysis of ocular structures (two- and three-dimensional; Chapter 6).
3. Methods and devices for the determination of the eye's functional status (Chapter 8).

This type of classification will be applied throughout Part Two of this book, with a separate chapter dedicated to three-dimensional optical coherence tomography (OCT) imaging (Chapter 7). This is justified, as OCT image generation is based on physical principles of optics that fundamentally differ from those applied in other, more traditional imaging methods.

In the following, the application and clinical relevance of the device groups will be summarized. Detailed information can be found in the corresponding sections which describe the individual methods and devices.

4.1 Determination of the Eye's Refractive Status

The most common causes of visual impairment are refractive errors of the eye (Section 3.1). Generally, the purpose of refraction measurement is the determination of optical aberrations of the eye (Section A.1.6). The obtained data are then used to determine a corrective measure (e.g., eyeglass lens, contact lens, or refractive surgery), which allows attainment of the best possible vision (maximum visual acuity) for an ametropic patient.

Approximately 30% of the population over the age of 40 suffer from refractive errors which require correction [1]. Measurements of the refractive status are thus a very common examination method. This, in turn, results in a substantial demand for equipment to determine the refractive status accurately, reliably, and as fast as possible. The methods and devices developed for this purpose are described in Chapter 5.

4.2 Visualization, Imaging, and Structural Analysis

Different optical methods and devices are used to diagnose eye diseases and to collect biometric data such as the eye length, corneal shape, and corneal thickness. However, a number of application challenges have to be overcome to visualize or image ocular structures. The respective issues are listed in Table 4.1 together with the general physical principles of optics for their solution and the related specific ophthalmic diagnostic devices according to the listed reference sections.

Imaging diagnostic methods are based on a structural analysis of the considered object. Structural analyses can be performed subjectively following purely

Table 4.1 Practical challenges and technical solutions for the visualization and diagnosis of ocular structures.

Issue	Solution	Reference sections
Objects are small.	Increase of angular resolution via magnifying glasses and microscopes.	6.1, 6.2, 6.4
Objects (cornea and lens) are highly transparent and thus difficult to visualize.	Usage of scattering properties (optical section) via slit lamps, scanning slit projection techniques (e.g., Scheimpflug imaging).	6.4, 6.5
Objects are not easily accessible via optical devices (e.g., fundus, iridocorneal angle).	Usage of special imaging techniques/devices like ophthalmoscopes, fundus cameras, confocal scanning laser ophthalmoscopes (cSLO), additional optics (e.g., contact lenses) for slit lamps and surgical microscopes.	6.6, 6.7, 6.8.1, 6.4, 6.2
Low contrast of objects makes it difficult to differentiate structures.	Enhancement of contrast via fluorescence methods and multispectral imaging.	6.7, 6.8
Observation/visualization is restricted to near-surface areas (only two-dimensional image of surface).	Generation of virtual cross-sections (three-dimensional visualization) via slit-projection techniques, cSLO, and optical coherence tomography (OCT).	6.5, 6.8.1, 7.6
Monitoring and quantification of the metabolic status at cellular level.	Metabolic mapping.	8.2

qualitative criteria (shape, color, and structure) or by means of measured values of specific diagnostic parameters. The most important quantitative diagnostic parameters, their clinical relevance, and the measurement techniques used for their determination are summarized in Table 4.2.

Quantitative structural parameters are either used directly, for example, in the planning of cataract or refractive surgery, or indirectly, as an indicator for a disease. Challenges regarding disease diagnosis are the large natural variability in most ocular parameters and overlapping intervals of the quantitative structural parameters between "normal" and "diseased" subjects. Hence, for a reliable diagnosis, the use of statistical analysis and normative databases are required.

Table 4.2 Geometric and structural parameters necessary for diagnosis. Their clinical relevance and possible optical measurement techniques are listed as well.

Parameter	Clinical relevance	Optical measuring techniques (section)
Corneal topography (different parameters)	• Diagnosis of corneal diseases • Corneal refractive power determination • Cataract surgery (intraocular lens power calculation) • Refractive laser surgery • Contact lens fitting	• Keratometry (6.3.1) • Topometry (6.3.2) • Scanning slit projection (6.5) • Scheimpflug imaging (6.5.2) • Optical coherence tomography of anterior chamber (7.6.2)
Corneal thickness	• Refractive laser surgery • Glaucoma diagnosis (correction factor for intraocular pressure measurement) • Glaucoma diagnosis (risk factor) • Corneal surgery	• Slit lamp pachymetry (6.4) • Scheimpflug imaging (6.5.2) • Scanning slit projection (6.5.1) • Optical coherence biometry (7.7) • Optical coherence tomography of anterior chamber (7.6.2)
Anterior chamber depth	• Cataract surgery (intraocular lens power calculation) • Refractive lens surgery (phakic intraocular lens) • Glaucoma diagnosis (risk factor)	• Slit lamp pachymetry (6.4) • Scheimpflug imaging (6.5.2) • Optical coherence tomography of anterior chamber (7.6.2) • Optical coherence biometry (7.7)
Anterior chamber angle	• Glaucoma diagnosis (risk factor)	• Scheimpflug imaging (6.5.2) • Optical coherence tomography of anterior chamber (7.6.2) • Slit lamp gonioscopy (6.4.4.1)
Axial eye length	• Cataract surgery (intraocular lens power calculation)	• Optical coherence biometry (7.7)
Retinal nerve fiber layer thickness	• Glaucoma diagnosis	• Scanning laser polarimetry (6.8.3) • Optical coherence tomography (7.6.1.2)
Topography of the optic nerve head	• Glaucoma diagnosis	• Slit lamp biomicroscopy (6.4.4.1) • Stereo fundus imaging (6.7.8) • Confocal scanning laser tomography (6.8.2) • Optical coherence tomography (7.6.1.1)

Table 4.3 Functional parameters required for diagnosis. Their clinical relevance and possible optical measurement techniques are listed as well.

Parameter	Clinical relevance	Optical measuring techniques (section)
Visual acuity	• Most relevant functional parameter of the eye • Quantifies the performance of central vision	• Subjective refraction techniques
Visual field	• Diagnosis of unexplained visual loss or suspected lesions of the visual pathway (e.g., glaucoma diagnosis)	• Perimetry (8.1)
Metabolic status at cellular level	• Diagnosis of retinal diseases	• Metabolic mapping (8.2) • Fundus camera imaging (6.7) • Scanning laser ophthalmoscopy (6.8.1) • Optical coherence tomography (7.6.1)

4.3 Determination of the Eye's Functional Status

The *functional status* of the eye can be described by means of global and/or local functional parameters (Table 4.3).

4.3.1 Global Functional Status

By global functional status we mean the degree of subjective ability to adequately register and/or respond to external visual stimuli. Visual acuity (Section 2.1.5.1) and visual field (Section 2.1) are the most important parameters to describe the global functional status.

Visual acuity Visual acuity quantifies the performance of central vision. It can be impaired by refractive errors of the eye (Section 3.1), pathological changes in one or more ocular segments (e.g., cornea, eye lens, retina), and/or diseases of the retino-cortical visual pathway. The determination of visual acuity is thus the most important functional check in everyday clinical practice. In addition, visual acuity of an optically corrected eye (best corrected visual acuity (BCVA)) is an important measure for the success of any eye treatment. Visual acuity can be quantitatively determined in a rapid and simple manner by means of wall chart tests (optotypes). To determine the best corrected visual acuity, trial lenses in trail frames or phoropters (Chapter 5) are required as well.

Visual field The visual field is the area of the environment out of which an observer can obtain visual information when fixating a particular object. A visual field examination (Section 8.1) delivers relevant information about the functional status of the whole visual system, ranging from the retinal photoreceptors to the neuronal components of the visual cortex. It is particularly essential in the diagnosis and management of glaucoma and can also play an important role in the management of some neurological diseases. The devices and test methods developed for this purpose are described in Section 8.1.

4.3.2 Local Functional Status

The local functional status refers to the metabolism of eye tissue at the cellular level. For example, the retinal metabolism can be characterized, amongst others, by objective parameters related to microcirculation, such as blood flow and oxygen saturation. Dysfunctions in the metabolism can be an early indicator of a disease. The determination of the metabolic status (metabolic mapping) thus may allow early diagnosis of eye diseases. Instruments and testing methods developed for this purpose, many of which are presently in research phase, are described in Section 8.2.

4.4 Light Hazard Protection

We will also address a general requirement that applies to all optical diagnostic and imaging methods, that is, light hazard protection during optical measurements. The optical radiation used for an ophthalmic measurement, imaging, or diagnosis must not cause damage to the eye. The usual radiation safety standards (Section 9.6) apply only to a limited extent, as their maximum permissible exposure (MPE) values are based on assumptions that often do not apply to methods used in the field of ophthalmology. For example, in an ophthalmic examination, the natural defense reactions to very bright light, such as eyelid closure reflex, involuntary eye or head movements, and the pupil reflex, are intentionally restricted or entirely prevented by means of headrests, eye fixation targets, and/or the use of medication to paralyze the pupillary reflex. Many optical methods used in ophthalmology also do not irradiate the surface of the eye, as assumed in standards, but rather use a light source which is imaged into the interior of the eye. Moreover, MPE values apply to individuals with healthy eyes. However, in certain eye diseases, radiation induced injury cannot be ruled out even with values below the MPE limits. As a detailed analysis of potential radiation hazards when using ophthalmic instruments exceeds the scope of this book, we refer the reader to relevant specialized literature [2–4].

References

1 Kempen, J., Mitchell, P., Lee, K., Tielsch, J., Broman, A., Taylor, H., Ikram, M., Congdon, N., and O'Colmain, B. (2006) The prevalence of refractive errors among adults in the United States, Western Europe, and Australia. *Arch. Ophthalmol*, **122** (4), 495–505.

2 DeLori, F.C., Webb, R.H., and Sliney, D.H. (2007) Maximum permissible exposures for ocular safety (ANSI 2000), with emphasis on ophthalmic devices. *J. Opt. Soc. Am. A*, **24** (5), 1250–1265.

3 Sliney, D., Aron-Rosa, D., DeLori, F., Fankhauser, F., Landry, R., Mainster, M., Marshall, J., Rassow, B., Stuck, B., Trokel, S., West, T.M., and Wolffe, M. (2005) Adjustment of guidelines for exposure of the eye to optical radiation from ocular instruments: Statement from a task group of the International Commission on Non-Ionizing Radiation Protection (ICNIRP). *Appl. Opt.*, **44** (11), 2162–2176.

4 International Organisation for Standardization (2007) Fundamental requirements and test methods – Part 2: Light hazard protection (ISO 15004-2:2007).

5
Determination of the Refractive Status of the Eye

The task of the eye's refractive parts is to create an image of the external world on the photoreceptor layer of the retina (Section 1.2). The imaging quality of a real eye is, however, influenced by refractive errors (Section A.1.6), dispersion (Section A.1.9), diffraction effects (Section A.2.1.6), and scattering (9.2).

More than 50% of the global population suffers from refractive errors such as myopia (Section 3.1.1), hyperopia (Section 3.1.1), or astigmatism (Section 3.1.2). We also mentioned in Section 3.8 that uncorrected refractive errors are the main cause of visual impairment worldwide. Accordingly, the determination of the eye's refractive status is the most common examination method in ophthalmology and optometry. In such an examination, data about potentially existing aberrations (Section A.1.6) in the eye's refractive parts is acquired which is then used to find corrective measures to optimize visual acuity (see Section 2.1.5.1).

Defocus and astigmatism (also classified as *lower-order* or *spherocylindric aberrations*; see Table A.3 in Section A.1.8.2) are of particular interest.[1] They are the most dominant ocular aberrations and the only ones which can be handled with "classic correction methods" such as eye glasses and contact lenses (Section 5.5). So, in clinical practice, the determination of the refractive status mainly refers to the correction of spherocylindric visual errors. The instruments and objective methods developed for this purpose are described in Sections 5.1 and 5.2. Higher-order aberrations such as coma (Section A.1.6.2), trefoil, or spherical aberration (Section A.1.6.1) are to date not that significant in standard clinical practice, as they can neither be measured nor corrected with classical methods. However, their importance has increased with the development of more sophisticated correction methods such as customized refractive laser surgery (Sections 10.3 and 10.4). The instruments specially developed for the determination of *all* ocular aberrations and their clinically relevant applications are presented in Section 5.3.

1) The term "spherocylindric aberrations" is derived from the fact that defocus and astigmatism can be corrected by spherical and/or cylindric lenses. In ophthalmology and optometry, these aberrations are often also referred to as *lower-order aberrations* (see also Section A.1.7). We thus use the terms "spherocylindric aberration" and "lower-order aberrations" interchangeably, or directly refer to the distinct types of aberration.

Optical Devices in Ophthalmology and Optometry, First Edition. Michael Kaschke, Karl-Heinz Donnerhacke, and Michael Stefan Rill.

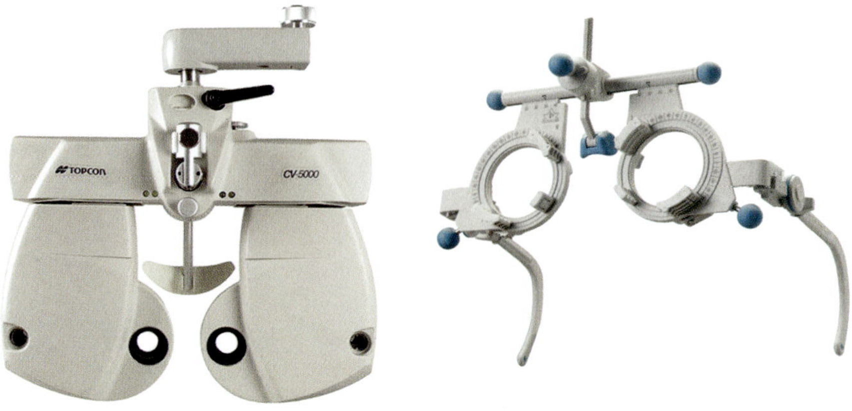

(a) (b)

Figure 5.1 (a) Photograph of Topcon CV-5000 phoropter. Courtesy of Topcon Deutschland GmbH. (b) Photograph of OCULUS New Universal UB4 trial frame. Courtesy of OCULUS Optikgeräte GmbH.

To determine the refractive status of a patient's eye, *subjective* and *objective refraction methods* are used, which we will discuss in the following.

Subjective refraction methods In subjective refraction methods, the refractive status of the eye is indirectly determined by testing the visual acuity (Sections 2.1.5.1). The patient is asked to look at a chart with characters (e.g., Snellen E), numbers, or symbols (e.g., Landolt rings) of different sizes (Section 2.1.5.2). The patient then has to provide direct feedback on which characters or symbols are clearly visible and which are not. The required corrective optics are found by positioning different spherocylindric trial lenses in front of the patient's eye until maximum visual acuity is achieved. After this, the interaction of both eyes is checked and any necessary corrections are performed (binocular balance). To ensure proper centration and corneal vertex distance (Section 5.5), either phoropters (Figure 5.1a) or trial frames (Figure 5.1b) are used. Both tools can be considered as modifiable eye glasses with a set of spherical and cylindrical lenses.

As the patient plays an active role during the subjective refraction measurement, the eye's refractive status as well as the functionality of the entire visual system (including retina and brain) are tested at once. The other important difference between this and most objective methods is that the latter can only provide monocular testing. As a consequence, the measurement result of subjective refraction methods represent the gold standard compared with objective methods.

Objective refraction methods Objective refraction methods can "merely" test the refractive status of the patient's eye. Since the measurements take place without interaction with the patient, disorders in the neuronal area of visual information processing (i.e., from the retinal image to the perceived image) are not considered at all. However, the most common reason for reduced visual acuity is an inadequate refractive status of the eye. Objective refraction methods thus usually serve as a first baseline for a subsequent subjective refraction measurement which defines the final prescription lenses. This approach can substantially simplify the subjective measurement process which is often time-consuming and exhausting. Moreover, objective methods should always be applied whenever active participation on the part of the patient is not possible, for example, in the case of young children.

In the following sections, we will focus on objective refraction methods and corresponding instruments which allow the measurement of the refractive status of the eye. Here, we distinguish between visual–manual devices, for example, retinoscopes (Section 5.1) and automated measuring instruments, for example, autorefractors (Section 5.2) and aberrometers (Section 5.3). In manual devices, the refractive status must be visually evaluated by the physician, whereas automated devices can be used by medical assistants.

The general problem of all objective measurement methods is the fact that the eye's optical system is accessible only from the object side. This means that the test of the imaging properties of a patient's eye is only possible indirectly from the outside, as the image quality on the retina cannot be measured directly. To solve this problem in practice, only a small part of the fundus is illuminated through the pupil by a primary source of radiation in the illumination beam path. The backscattered and reflected light in the illuminated fundus area forms a secondary source of radiation (fundus reflex). Its radiation passes through the eye's refractive parts in the opposite direction and is detected and analyzed by an observation and/or measurement system. For the evaluation, it is assumed that:

1. The secondary source of radiation is located close to the photoreceptor layer. Hence, subjective and objective refraction methods have approximately equal reference planes.
2. The fundus reflex is clearly defined. This means that light scattered or reflected by other ocular structures does *not* affect the measurement.

The secondary source is formed in the retinal pigment epithelium (RPE; Section 1.2), since it is the retinal structure which scatters most in the visible (VIS) and near-infrared (NIR) spectral range. As the photoreceptor layer is adjacent to the RPE, the assumptions made are usually fulfilled (for further details see [1]).

5.1 Retinoscopy

Retinoscopy (which means "visualization of the retina"), also known as *skiascopy* ("visualization of shadows")[2], is a relatively simple but highly accurate objective refraction method used to determine the far point (Section 2.1.4) of the human

2) Although the term "skiascopy" describes the measuring principle correctly, the rather incorrect term "retinoscopy" has established itself in English-speaking countries.

eye. It is derived from Foucault's[3] knife edge test (Figure 5.2), in which the motion of a shadow is used to determine the focal length of an optical system. However, in contrast to Foucault's original knife edge test, the distance between eye (optical system) and knife edge is fixed and the position of the focal point (and not the knife edge) is shifted in retinoscopy. The goal of retinoscopy is to find the far point (flicker or neutrality point) of the patient's eye by placing different trial lenses in front of the eye.

In clinical practice, retinoscopy is the method of choice when the refractive status must be determined in children or patients who are not willing or able to cooperate in the examination. The technique also delivers information about disorders in the anterior segment of the eye (e.g., corneal opacity or cataract). Although the accurate interpretation of the observed light-shadow phenomena requires a certain degree of experience, retinoscopy delivers highly reliable results with minimum equipment requirements. The retinoscope principle has also been used in modern autorefractors (Section 5.2).

5.1.1 Illumination Beam Path

Retinoscopes use an extended white-light source which is imaged on the patient's eye by a movable lens and a beam splitter (i.e., a partially transmitting mirror; Figure 5.3a). Depending on the type of light source, this generates either a circular light spot (*spot retinoscope*) or a narrow, rectangular (slit-shaped) light band (*streak retinoscope*) on the patient's eye. Of these two forms of retinoscopes, the streak retinoscope is clinically more useful, especially for the examination of astigmatic eyes. Therefore, the use of streak retinoscopy has generally replaced the use of spot retinoscopy in ophthalmic practice, and we will restrict our discussion on this illumination design.

In the streak retinoscope, a nonspiraled filament of a halogen bulb is used as a light source. The illumination beam path can become divergent, parallel, or convergent by shifting the lens relative to the light source. For standard eye examinations, a divergent beam path is used[4]. In addition, the long axis of the light source can be rotated around its optical axis, which is helpful for the examination of astigmatic eyes. The angle and the divergence of the light beam are set by means of a shared sleeve or collar ring which can be rotated (to change the angle) and axially displaced (to vary the distance between lens and light source).

The iris (Section 2.1.1) of the patient's eye acts as an aperture stop in the illumination path. The transmitted part of the light bundle is imaged onto the retina by the eye's refractive parts. Depending on the refractive error of the patient's eye, the incident light bundle is more or less well focused on the retina.

3) Léon Foucault (1819–1868).
4) A convergent beam path focuses the light beam before it enters the eye. As a consequence, the motion of the shadow is inverted and thus contrary to the behavior of the original Foucault knife edge test.

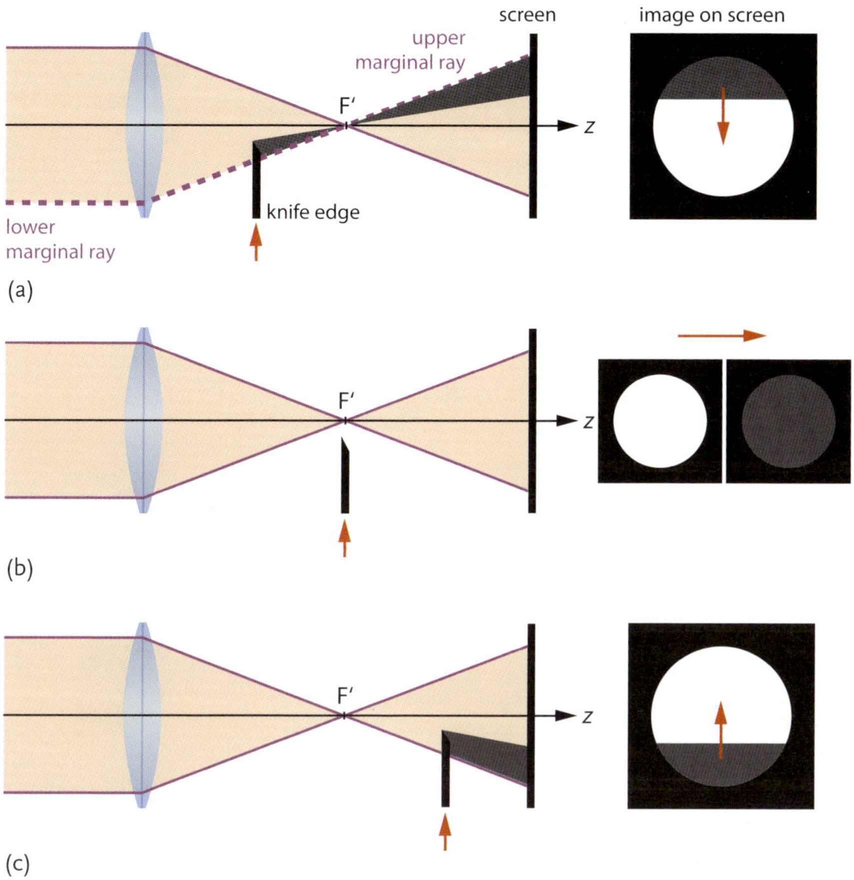

Figure 5.2 Principle of Foucault's knife edge test. A parallel light beam is focused by a lens at point F′ and then projected onto a screen. A knife edge is inserted from below at different positions along the optical axis. (a) The knife edge is located between lens and focal point F′. As we can derive from the ray diagram (left), a shadow is generated at the top of the image (right). When we move the knife edge closer to the optical axis (see red arrow), the area of the shadow expands downwards ("against motion"). (b) The knife edge is aligned with the focal plane. When we move the knife edge upwards, the image does not change as long as we do not cross the optical axis. If the knife edge goes across the optical axis, the image becomes totally dark (neutrality or flicker point). (c) The knife edge is located behind focal point F′. The ray diagram (left) reveals that the shadow expands upwards (right) as the knife edge is moved upwards ("with motion"; see red arrow).

5.1.2
Observation Beam Path

The light beam reflected on the patient's retina, called *retinoscopic reflex* or simply *red reflex*, appears as a reddish light band and acts as a secondary light source. For its part, the reflected light bundle is imaged by the eye into the far point if the

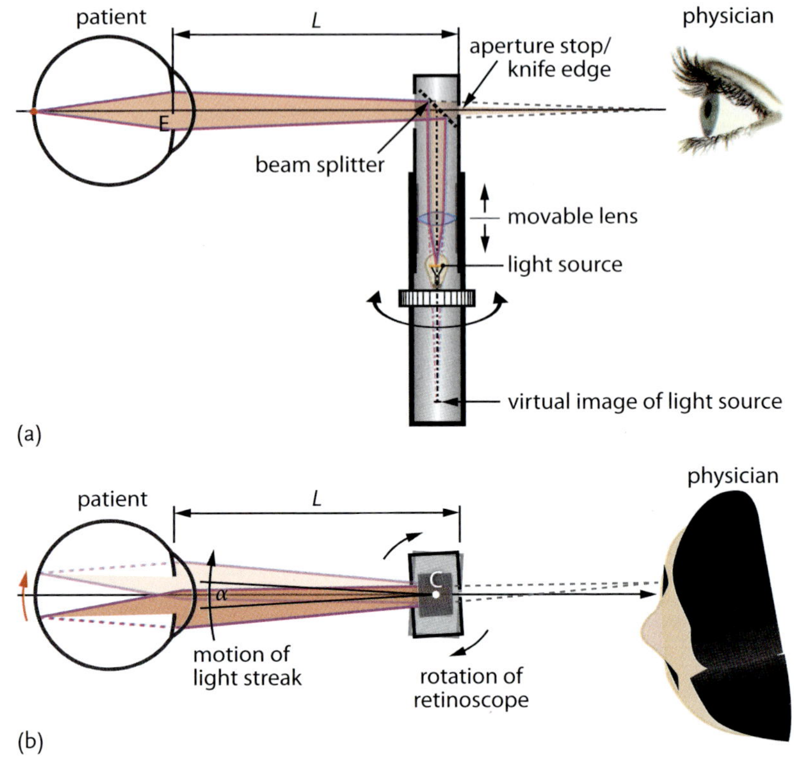

Figure 5.3 Working principle of a streak retinoscope. (a) Side view of the optical setup. The light emitted by a linearly extended light source is collected by a movable lens and then reflected by a beam splitter. The streak light bundle leaves the retinoscope and enters the eye of the patient, where it is reflected at the retina. The physician can now investigate the position of the incident light (on the iris) and the reflected light beam (within the iris aperture) through an aperture stop which acts as a knife edge. The aperture stop is located at a distance L from the patient's entrance pupil. Adapted from [2]. (b) Top view of the measuring arrangement. During the examination, the retinoscope is rotated around its axis C by an angle α and thereby scans the patient's retina (red arrow) with the light band.

patient does not accommodate (Section 2.1.4). The physician observes the red reflex through an aperture stop (retinoscopy aperture) in the observation path arranged at a known distance L from the patient's entrance pupil (see Figure 5.3b). This is, in fact, the smallest aperture in the observation beam path and as such corresponds to a Foucault knife edge.

Figure 5.4 illustrates how the red reflex may behave during an examination. A part of the patient's iris aperture appears as a bright colored reflex when the reflected light bundle enters the aperture stop of the retinoscope. In the case of hyperopic or emmetropic eyes (Figure 5.4a), the reflected ray bundle is focused behind the physician (see right column in Figure 5.4a). As a consequence, the red reflex moves concurrently with the moving direction of the retinoscope ("with motion") and can be thus compared to the situation shown in Figure 5.2c. If the patient has

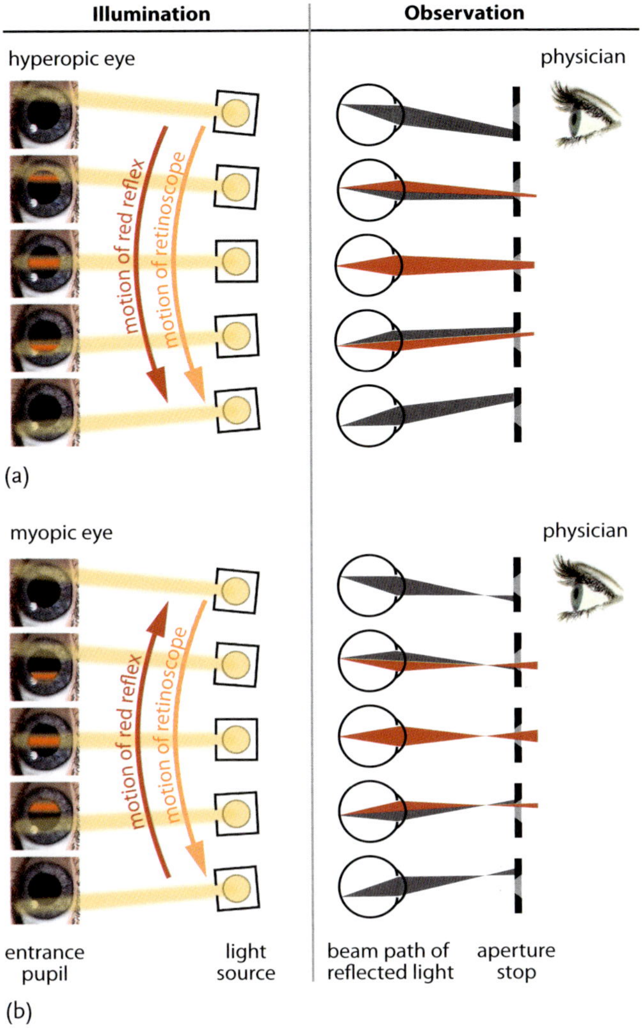

Figure 5.4 Lineup of illumination beam path (left column) and observation beam path (right column) as seen by the physician. The orange arrow represents the moving direction of the light band (retinoscope). The red arrow shows the moving direction of the red reflex within the iris aperture. In the right column, the motion of the reflected light beam is shown from the physician's point of view. The part of the beam which is blocked by the observation aperture is shown in gray. (a) Observation for a hyperopic eye. The light band is scanned over the eye from the right to the left (here: image sequence from top to bottom). The red reflex moves concurrently so that the incident and the reflected light beams overlap. As we have a "with motion", the focal point of the patient's eye lies behind the retinoscope aperture. (b) Observation for a myopic eye. When the retinoscope is scanned over the eye from the right to the left (here: image sequence from top to bottom), the reflected light beam shows a countermovement ("against motion" from left to the right). This happens because the focal point is located in front of the retinoscope's aperture stop (as shown in right column). Adapted from [3].

a myopic eye (Figure 5.4b), the focal point of the emerging rays lies in front of the retinoscope aperture. Hence, the red reflex moves opposite to the moving direction of the light band ("against motion" comparable to Figure 5.2a).

When the far point approaches the aperture stop of the retinoscope, the speed of motion of the red reflex strongly increases while turning the retinoscope, and the reflex becomes bright and wide. If the eye's far point lies exactly in the plane of this aperture stop, the red reflex does not move at all, and we see a short flickering instead. This corresponds to the situation in Figure 5.2b. The short flash at which the red reflex reaches its maximum brightness thus indicates that we found the eye's far point (flicker point).

5.1.3 Measurement Procedure

Measurement of axial-symmetric ametropia (defocus) The far point in emmetropic eyes is located at infinity. As an infinite working distance is not practical, a retinoscopy lens L_{ret} with a refractive power of, for example, $\mathcal{D}_{ret} = +2\,D$ is placed in front of the patient's eye. This shifts the far point of an emmetropic eye to a working distance of $L_{wd} = 0.5\,m$ (see Figure 5.5a).[5] If the physician prefers a different working distance[6], the refractive power of the retinoscopy lens must be chosen in accordance to

$$\mathcal{D}_{ret} = \frac{1}{L_{wd}} . \tag{5.1}$$

To quantify the refractive error, the physician scans the streak illumination over the patient's eye. The relative motion of retinoscope and red reflex reveals whether the eye (including the retinoscopy lens) is myopic ("against motion") or hyperopic ("with motion"). The physician then tries to shift the far point to the working distance by placing additional spherical trial lenses in front of the eye (Figure 5.5b). Once the flicker point is found, L_{ret} can be removed. The far point of the examined ametropic eye with the correcting trial lenses is now located at infinity as for an emmetropic eye (see Figure 5.5c). This means that the far point refraction of the examined eye is given by the refractive power of the trial lenses at the flicker point.

Measurement of astigmatism Until now, we have assumed that the patient's eye has a refractive error which is rotationally symmetric about the measuring axis of the retinoscope. In this case, the red reflex moves perpendicular to the long axis of the light band when the retinoscope is scanned over the eye[7] (Figure 5.6a). In astigmatic eyes, however, the red reflex moves obliquely if the examined princi-

5) We effectively make the combined optical system of eye and retinoscopy lens myopic.
6) The working distance is usually chosen such that it corresponds to the arm length of the physician.
7) The actual moving direction of the retinoscope itself is not relevant during the measurement. Only the motion component perpendicular to the long axis of the light band must be considered.

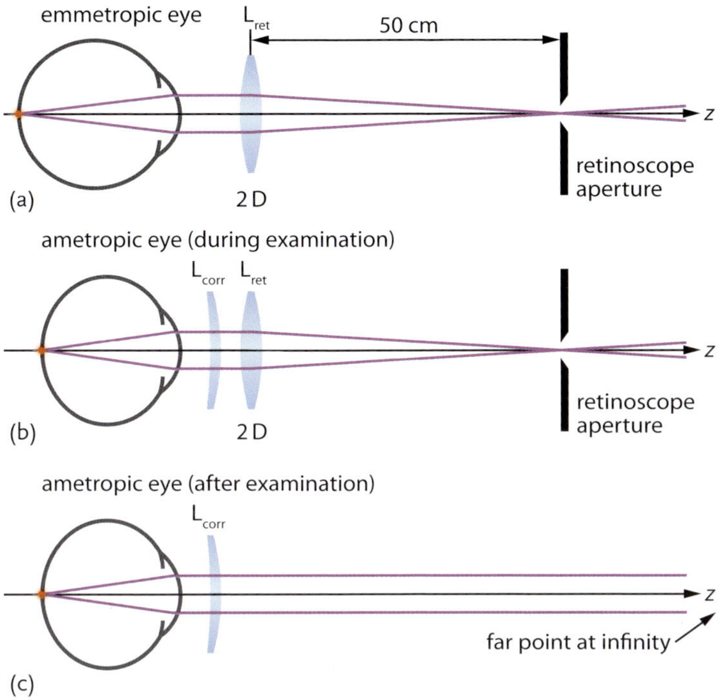

Figure 5.5 Measurement procedure of retinoscopy. (a) A working distance of 50 cm is set by placing a retinoscopy lens L_{ret} with a refractive power of +2 D in front of the patient's eye. In an emmetropic eye, the far point now coincides with the reinoscope aperture and the flicker point is thus found at a distance of 50 cm. (b) The flicker point of an ametropic eye is determined by adding a corrective lens L_{corr} while keeping the working distance constant. (c) Once the flicker point is found at a distance of 50 cm, the ametropic eye is corrected. When we remove the retinoscopy lens L_{ret}, the far point of the patient's eye is shifted to infinity by L_{corr}. Adapted from [24].

pal meridian is tilted against the long axis of the light band (Figure 5.6b). As this is often the case in practice, we first have to find the orientation of the principal meridians (Section 3.1.2) when the far point refraction of an astigmatic eye shall be determined. For this purpose, a collar ring at the retinoscope shaft is used to rotate the light source filament around its optical axis. When the light band is realigned to the examined principal meridian (Figure 5.6c), the red reflex moves again perpendicular to the long axis of the light band. The refractive power along this principal meridian is then determined by adding spherical and/or cylindrical trial lenses. If the patient's eye has a regular astigmatism (Section 3.1.2), the other principal meridian can be examined when the light band is again rotated by 90° and the eye is scanned once again.

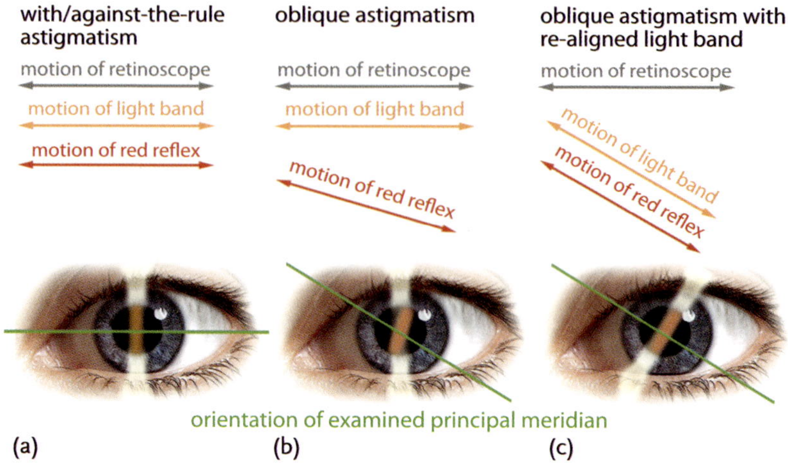

Figure 5.6 Measurement of astigmatism with a retinoscope. (a) Orientation of the fundus (red) reflex for an eye with spherical refractive error (myopia or hyperopia) or in the case of an astigmatic eye (with-the-rule or against-the-rule astigmatism) for which the examined principal meridian (green line) is perpendicular to the long axis of the light band. The red reflex is not tilted against the light band of the retinoscope. (b) Orientation of the red reflex in the case of an astigmatic eye (oblique astigmatism) for which the examined principal meridian (green) is tilted against the long axis of the light band. The directions of motion of the red reflex and the light band now include a nonzero angle. (c) By rotating the light source filament, we can achieve a parallel motion of light band and red reflex. The light band is now oriented perpendicular to the examined principal meridian (green).

5.1.4 Accuracy in Retinoscopy

The theoretical measuring accuracy of retinoscopy is determined by the diffraction-limited depth of field Δz_{dof} (Sections 2.1.8 and 6.2.2) of the red reflex within the pupil plane and is given by [24]

$$\Delta \mathcal{D} = \pm \frac{\lambda}{d_{\text{pupil}}^2} \,. \tag{5.2}$$

d_{pupil} is the diameter of the eye's entrance pupil and λ the wavelength of the used light source (Problem P5.1). For example, with $d_{\text{pupil}} = 3$ mm and $\lambda = 600$ nm, the resulting measuring accuracy is as high as $\Delta \mathcal{D} = \pm 0.07$ D. Hence, retinoscopy is in principle the most accurate objective method to determine the refractive status of the eye. Such a high measurement accuracy is, however, not attained in practice. One reason is that a large pupil diameter increases the effect of spherical aberrations (Section 2.1.5.3) so that it is difficult to find the exact position of the flicker point. In practice, the refraction of the eye can be determined with an accuracy of ± 0.5 D and the orientation of the principal meridians with an accuracy of $\pm 5°$. Additional errors occur if the patient accommodates and/or if the working distance

Figure 5.7 Photograph of HEINE BETA®200 retinoscope. Courtesy of HEINE Optotechnik.

is changed during the measurement. But distance errors become negligible when the working distance is large enough.[8]

In contrast to autorefractors (Section 5.2), retinoscopes use visible (VIS) instead of near-infrared (NIR) light for the measurement. As scattering effects in ocular media increase with decreasing wavelength (Figure 10.3 in Section 10.2.2), retinoscopy is more difficult to perform on eyes with lens or corneal opacities.

5.1.5 Applications

Although the setup of retinoscopes is relatively simple, the hand-held devices are very versatile tools for an accurate determination of the eye's refractive status. However, retinoscopy requires a great deal of experience to ensure correct interpretation of the observed phenomena. Figure 5.7 depicts a commercially available retinoscope by HEINE.

8) For example, if retinoscopy is performed at a distance of $L_{wd} = 50$ cm ($1/L_{wd} = 2$ D) and the actual distance is 55 cm (= 1.82 D), this results in a measuring error of approximately 0.2 D, which is tolerable in practice.

5.2 Automated Objective Refractometers (Autorefractors)

The basic optical setup of all types of automated objective refractometers, shortly called *autorefractors*, consists of an illumination beam path and a detection–observation beam path. The illumination beam path is used to illuminate the fundus of the patient's eye through the iris aperture. The incident light beam is reflected and scattered by the fundus and then exits the eye, which is then detected and analyzed in the detection-observation path of the autorefractor by appropriate methods.

Basically all refractometers available on the market only differ from one another in the implemented measurement methods. In all other respects, the instruments strongly resemble one another in their fundamental design. All of them operate with NIR light sources and are fitted with devices for accommodation control and fixation of the patient's eye. All autorefractors use devices for rapid and secure positioning of the instrument and for reflection suppression. The components which are common for all autorefractors will be addressed in Section 5.2.1. Regarding the measurement methods, autorefractors are based on principles that have long been part of the ophthalmic and general optical measuring technology, that is,

- best focus method (Section 5.2.2.1),
- Scheiner (or coincidence) method (Section 5.2.2.2),
- ray-deflection method (Section 5.2.2.3),
- image size method (Section 5.2.2.4),
- knife edge method (Section 5.2.2.5), and
- retinoscopy method (Section 5.2.2.6).

5.2.1 Common Characteristics of Autorefractors

Light source For the illumination, all autorefractors use NIR light with a wavelength between 800 and 950 nm. Compared to VIS light, it has the following advantages:

- *Higher light yield*: The reflectance of the fundus in the NIR spectral range is approximately $10\times$ higher than in the VIS spectral range. At the same time, the refractive and transparent parts of the eye have the highest transmittance for NIR light (see Section 2.1.10). As a consequence, more light intensity (i.e., a higher signal strength) is available for the measurement.
- *No pupil reflex and accommodation*: The human visual system is insensitive to NIR light so that the "automatic" reflexes of the human eye, for example, pupil reflex (Section 2.1.6) and accommodation (Section 2.1.4), are not activated. The used light does not cause glare to the patient, and the pupil diameter as well as the state of accommodation are preserved during the measurement.

However, the goal of the measurement is to determine the eye's refractive status in the VIS and *not* in the NIR spectral range. For this reason, the following influence factors must be taken into account:

- *Dispersion*: As all optical media of the eye have different refractive indices for NIR and VIS light (Section A.1.1), the measured refractive power for NIR light deviates by 0.7–1.0 D from the actual refractive power in the VIS spectral range.
- *Axial position and expansion of fundus reflex*: Unlike VIS light, NIR light can penetrate deeper into the retinal tissue all the way to the choroid because of lower absorption in the retinal pigment epithelium. As a consequence, the backscattered beam appears more diffuse than VIS light would appear. The axial position of the fundus reflex is thus less well defined for NIR light. In addition, the position of the fundus reflex is axially shifted towards the choroid.

Signal strength and reflection suppression Although the reflectance of the fundus is higher for NIR light than for VIS light, only a small portion of the light intensity can be used for the measurement. Since the reflection of NIR light on the retina is diffuse and the numerical aperture of the eye relatively small (Section 2.1), less than 1% of the reflected light can exit the eye through the iris aperture. As a consequence, high-intensity infrared light sources must be used for the measurement. But we also have to consider the standards of eye safety which determine the maximum permissible exposure (Section 4.4).

In addition, the undesired specular reflections at the optical components of the autorefractor must be suppressed, as they may lead to measurement errors. For this purpose, the illumination and observation beam paths of autorefractors are either separated by a beam splitter, or polarizers are used. To suppress the strong corneal reflections, the entrance pupil of the observation subsystem and the exit pupil of the illumination subsystem are locally separated in the patient's pupil.

Fixation device and accommodation control The major source of errors during the measurement of the far point refraction is the unintended accommodation of the patient. To prevent this, the patient is asked to focus on a fixation target (i.e., a picture of a scene suggesting distance vision). This fixation object is set close to the far point distance by means of lenses and placed on the optical axis of the illumination beam path. The fundus reflex is thus generated in the foveal region.

Adjustment device Before the measurement can be performed, the refractometer has to be centered at a correct distance from the patient's pupil. For this purpose, the physician is typically shown an image of the iris taken by a video camera. All necessary distance and alignment settings can be performed by means of this image. A rough preliminary setting is usually enough. Fine adjustments are often performed automatically by the instrument.

5.2.2
Measuring Methods

The setup of many autorefractors is based on the *optometer principle* (Figure 5.8). An optometer consists of an illumination beam path (optometer subsystem) and an observation beam path (ophthalmoscope subsystem). Both subsystems are separated by a beam splitter so that undesired reflections in the illumination beam path do not impair the detected signal. To suppress the strong corneal reflections, the entrance pupil of the ophthalmoscope subsystem and the exit pupil of the optometer subsystem are locally separated in the patient's pupil (not shown in Figure 5.8).

In the optometer subsystem, a test mire M (i.e., a target pattern) is projected onto the fundus of the patient's eye via optometer lens L_{opt}. The generated fundus reflex is then imaged by ophthalmoscopy lens L_{oph} onto a detection system. For ideal imaging, L_{opt} and L_{oph} must be placed such that their focal points coincide with the eye's nodal point N (Section 2.1.2). The imaging consequences for this optical configuration will be reconsidered in Section 5.2.2.1 (see also Figure 5.10). The physician is, however, not aware of the eye's refractive power so that the exact position of the eye's nodal point is practically unknown. In practice, the lenses are thus placed such that their focal points coincide with the center of the eye's entrance pupil E. This is usually a good approximation, as the magnitude of the resulting measuring errors will be small.

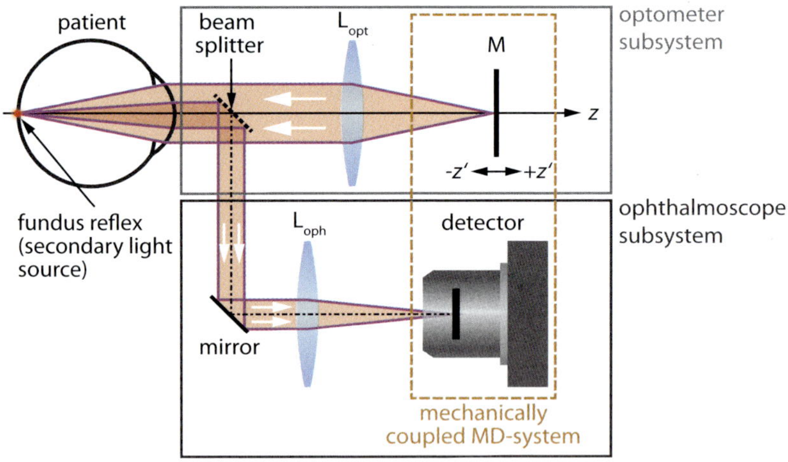

Figure 5.8 General setup of a refractometer based on the optometer principle. The illumination beam path (optometer subsystem) and observation beam path (ophthalmoscope subsystem) are separated by a beam splitter (thick, dashed line). L_{opt} and L_{oph} are the optometer and ophthalmoscopy lenses, respectively. Test mire M and the detector are mechanically coupled (so called MD-system) so that they can be concurrently moved along the *z* axis (optical axis). The white arrows represent the traveling path of light. Adapted from [2].

In an emmetropic eye, test mire M is sharply imaged onto the retina when it is located in the object-side focal plane of optometer lens L_{opt} (situation shown in Figure 5.8). The fundus reflex is, in turn, imaged onto the detector which is located in the image-side focal plane of ophthalmoscopy lens L_{oph}. In an ametropic eye, however, the combined optical system of eye and L_{opt} images M in front of the fundus or behind it. The detector thus acquires a blurred image. To recover a sharp image on the retina, M has to be shifted along the optical axis (z axis in Figure 5.8). With mechanical coupling between M and detector (called "coupled MD-system" in the following), the detected image becomes automatically sharp as soon as M is focused on the retina.

Let us now derive a relation between the required axial shift z' of the coupled MD-system and the far point refraction $A_{far} = 1/s_{far}$. For this purpose, we consider the ray diagrams shown in Figure 5.9, where we restrict ourselves to the imaging of the ophthalmoscope subsystem. Hence, we only look at rays which emerge from the eye after being reflected at the fundus. For the following discussion, we introduce from the geometry of Figure 5.9:

- the object-side and image-side focal lengths f_{oph} and f'_{oph} of ophthalmoscopy lens L_{oph}, respectively,
- the distance $s' = \overline{P_{oph} I}$ from the principal point P_{oph} of L_{oph} to the image plane I,
- the image distance $z' = f'_{oph} - s'$,
- the distance $s = \overline{Q_{far} P} + \overline{PP_{oph}}$ from far point Q_{far} to principal point P_{oph} (via principal point P of the eye), and
- the object distance $z = f_{oph} - s$.

In Figure 5.9a, the imaging properties are shown for an emmetropic eye. Since far point Q_{far} is located at infinity (Section 3.1), the rays emerging from the eye are parallel. Without the ophthalmoscopy lens, the rays (dashed gray lines) would be focused at infinity ($s_{far} = \infty$). With the ophthalmoscopy lens, however, the rays (solid lines) form an image at focal point F'_{oph}, that is, at a distance $s' = f'_{oph}$ from the image-side principal axis of L_{oph}. Thus, we have $z' = 0$.

In Figure 5.9b, the imaging of a myopic eye is illustrated. Rays emerging from the eye aim at far point Q_{far}, which is now located at a finite distance s_{far} (dashed gray lines). When we use the ophthalmoscopy lens, the rays (violet) form an image at a distance $s' = f'_{oph} - z'$ with $z' > 0$.

In Figure 5.9c, we consider a hyperopic eye. The rays which emerge from the eye are divergent and the virtual far point Q_{far} is now located at a finite distance s_{far} behind the eye (Section 3.1). L_{oph} thus forms an image behind focal point F'_{oph} at a distance $s' = f'_{oph} - z'$ with $z' < 0$.

From the lens equation (A14), we may derive a relation between the distances z, z', and the focal lengths of L_{oph} (i.e., the so-called *Newton formula*, see Problem P5.1b) which is given by

$$zz' = -f_{oph} f'_{oph} . \tag{5.3}$$

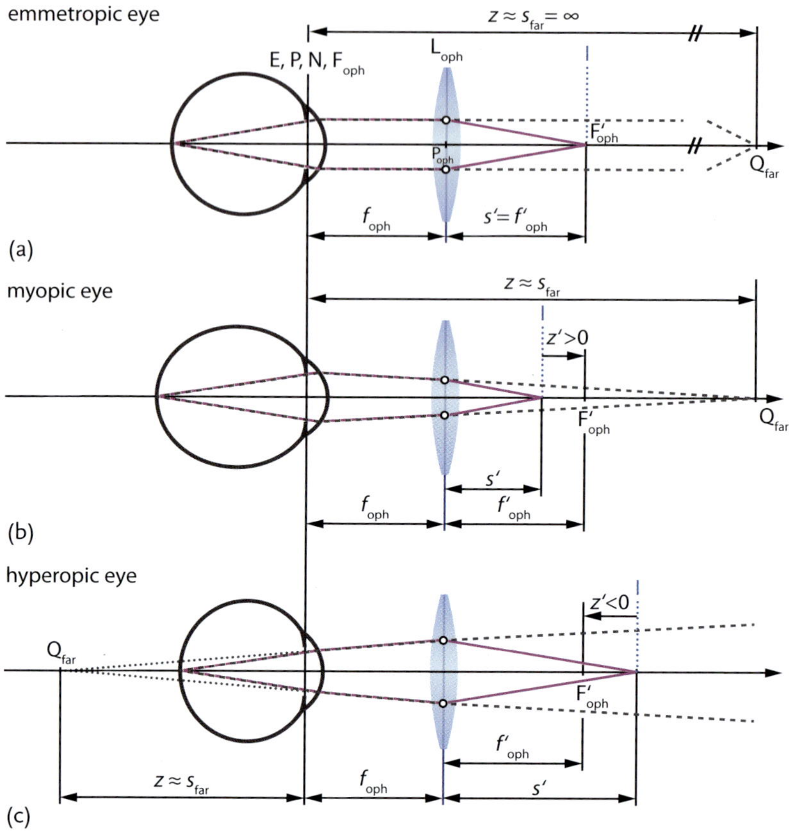

Figure 5.9 Ray diagrams to illustrate the derivation of the optometer formula (5.5). E is the entrance pupil, P the principal point, and N the nodal point of the eye. For the sake of simplicity, E, P, and N are assumed to be located at the same axial position. s_{far} denotes the far point distance. F_{oph} (f_{oph}) and F'_{oph} (f'_{oph}) are the object-side and image-side focal points (lengths) of ophthalmoscopy lens L_{oph}, respectively. P_{oph} is the principal point of L_{oph}. s' is the distance from P_{oph} to the image plane I (dotted blue line). The marginal rays with and without L_{oph} are represented by the violet and dashed gray line, respectively. (a) Ray diagram of the ophthalmoscope subsystem for an emmetropic eye. (b) Ray diagram of the ophthalmoscope subsystem for a myopic eye. The image plane I is located in front of F'_{oph}. The distance between the image point and F'_{oph} is $z' > 0$. (c) Ray diagram of the ophthalmoscope subsystem for a hyperopic eye. The image plane I is located behind F'_{oph}. The distance between the image point and F'_{oph} is thus $z' < 0$.

With the assumptions $z \approx s_{\text{far}}$ [9] and $f'_{\text{oph}} = f_{\text{oph}}$ (same radius of curvature on both sides of the ophthalmoscopy lens), we obtain

$$s_{\text{far}} = -\frac{f_{\text{oph}}^2}{z'} . \tag{5.4}$$

With the refractive power $\mathcal{D}_{\text{oph}}^2 = 1/f_{\text{oph}}^2$ of lens L_{oph}, the inverse of Eq. (5.4) reads

$$A_{\text{far}} = -\mathcal{D}_{\text{oph}}^2 z' , \tag{5.5}$$

which is the so-called *optometer formula* [2].

In practice, the image is captured by a detector and analyzed with regard to its imaging quality. The coupled MD-system is then moved such that the image quality is maximized. From the traveling distance z', the far point refraction A_{far} is calculated by means of Eq. (5.5). If the coupled MD-system must be moved towards the eye ($z' > 0$) to attain a high image quality, the examined eye is myopic. If the coupled MD-system has to be moved away from the eye (i.e., towards the far point; $z' < 0$), the examined eye is hyperopic. To evaluate and/or maximize the imaging quality, different methods are useful which will be discussed in the following Sections 5.2.2.1–5.2.2.6.

5.2.2.1 Best-Focus Method

The *best-focus method* is a "straight-forward" process at which the captured image is analyzed with regard to its contrast and sharpness. A control signal is then derived from the data to adjust the position of test mire M. In principle, this method works like the (passive) autofocus of a photo camera. The choice of the test mire pattern is however critical. On the one hand, the pattern must allow reproducible evaluations of the image sharpness. On the other hand, it must also be adequate to examine astigmatic eyes for which the orientation of both principal meridians and the refractive status along both axes must be determined.

In autorefractors based on the best-focus method, the optometer lens L_{opt} (also called the *Badal lens*) is placed such that its focal point coincides with the eye's nodal point N, which is the standard optometer configuration. As the distance between eye and L_{opt} is constant throughout the measurement, the angle of the incident chief ray is constant as well. When we now move test mire M along the optical axis, the retinal image size, which is defined by the chief ray (Section 2.1.1), basically does not change (Figure 5.10). However, the retinal image appears blurred for an ametropic eye unless the mire M is imaged sharply on the fovea. With such an arrangement, we can thus measure the eye's refractive error by optimizing the image sharpness.

9) According to Section 2.1.4, the far point distance s_{far} is defined as the distance between far point Q_{far} and the eye's object-side principal point P. In practice, it is however impossible to locate the principal point so that focal point F_{oph} of lens L_{oph} is placed such that it coincides with the eye's entrance pupil E. Since the distance between E and P is small (see Table 2.1) compared to the working distance of the autorefractor, this approximation is justified [4].

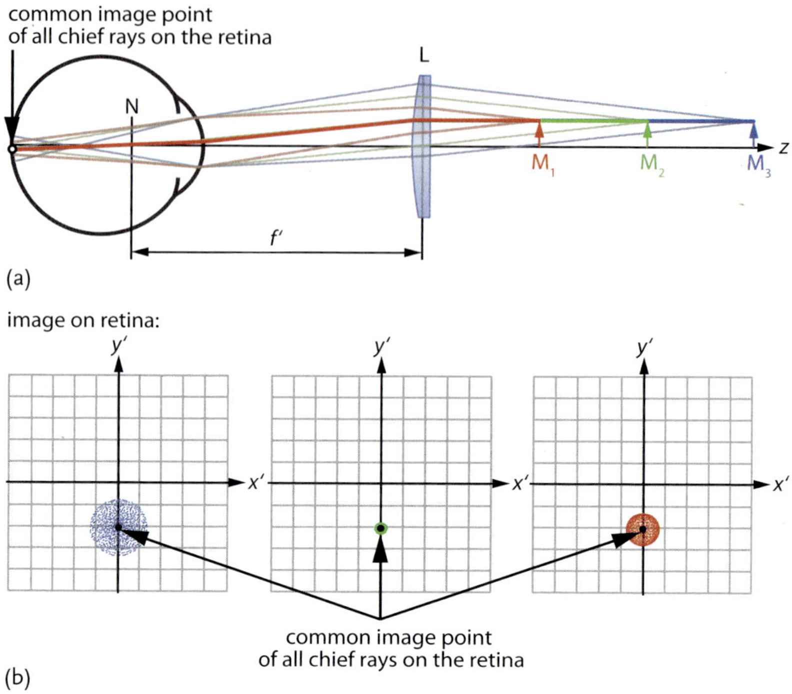

Figure 5.10 Imaging of a lens whose image-side focal point coincides with the eye's nodal point N. (a) The traveling path of rays is shown for three different positions of a test object (M_1, M_2, and M_3). Depending on the distance between object and lens, the resulting ray fan on the retina is either convergent (red), divergent (blue), or focused (green). This means that the retinal image appears blurred for the red and blue rays. But in all cases, the respective chief rays (bold lines) are incident onto the same point on the retina. This property is also illustrated in (b): Each colored dot represents the target position of an incident light ray in the retinal image plane ($x'y'$ plane). The black dots highlight the positions of the corresponding chief rays. Since the retinal image size is determined by the chief ray position on the image plane (Section A.1.4), the image size is equal in all three cases. Courtesy of Dr. Herbert Gross.

The first autorefractors mainly incorporated the best-focus method. Today, however, this measurement principle has been largely replaced by more efficient and cost-effective methods.

5.2.2.2 **Scheiner Method**

Concept The Scheiner[10] method is a method that does *not* analyze the contrast of the test mire image to set the required axial shift z'. Instead, the principle is based on two peripheral partial beams which are masked out from an incident parallel ray bundle. The partial beams are created by a circular disk with two apertures (perforations), called the *Scheiner disk*. After they have passed through the iris aperture,

10) Christoph Scheiner (1573–1650).

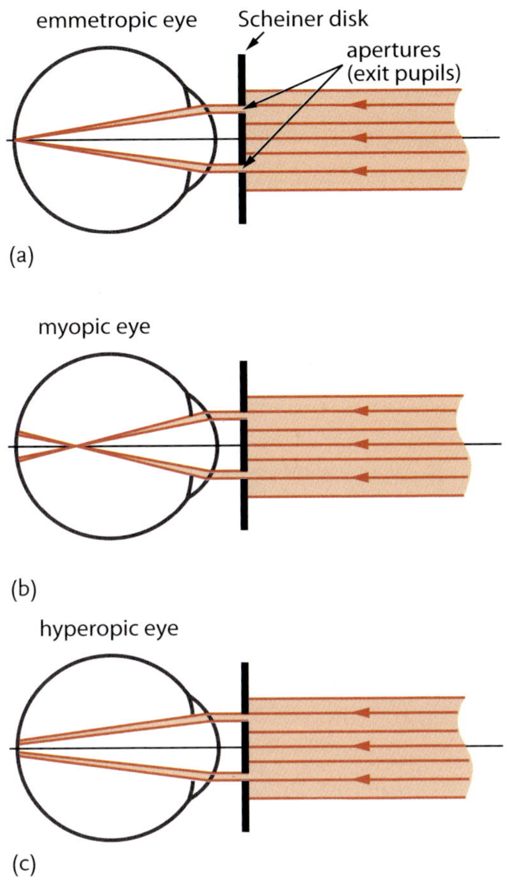

Figure 5.11 Basic principle of the Scheiner method. A bundle of parallel light rays passes through a Scheiner disk with small apertures. Depending on the eye's refractive power, the transmitted rays are focused at different positions. For the sake of simplicity, only the graphs for emmetropia (a), myopia (b), and hyperopia (c) are shown.

the partial beams are refracted by the eye. In an emmetropic eye, the foci of both partial beams coincide on the retina so that just a single circular fundus reflex is generated (Figure 5.11a). For this reason, Scheiner's method is often also called the "coincidence method". In a myopic eye, the partial beams intersect in front of the fundus (Figure 5.11b). In a hyperopic eye, the partial beams intersect behind the fundus (Figure 5.11c). Hence, in the case of ametropia, two separated (blurred) reflexes appear. When the partial beams are alternately blocked, the fundus reflexes will disappear *inversely phased*[11] for myopic eyes and *in-phase* for hyperopic eyes.

To determine the refractive error of astigmatic eyes, the orientation of the apertures must be adapted. The connecting line between both apertures on the

11) This means that the lower light reflex disappears when the upper aperture of the Scheiner disk is blocked and vice versa.

Scheiner disk defines the orientation of the measurement meridian[12]. Thus, the connecting line must be aligned parallel to the principal meridian in which the eye's refractive power shall be determined.

Implementation In autorefractors based on Scheiner's method (Figure 5.12), the classic Scheiner disk is replaced by two NIR light-emitting diodes (LEDs; Section B.5.2) which represent two point sources LED_A and LED_B. The LEDs are vertically displaced from the optical axis and located in the focus of a collimator lens. They are imaged by optometer lens L_{opt} into the pupil plane of the patient's eye (i.e., LED'_A and LED'_B). The images of the LEDs can be considered as a self-luminous Scheiner mask, as they form an exit pupil in front of the eye.

The movable pinhole M, which corresponds to test mire M in Figure 5.8, is arranged in the beam path of both LEDs such that its image forms a sharp light spot on the fundus of an emmetropic eye (zero setting). If we keep the position of M fixed and replace the emmetropic eye with an ametropic eye, *two* blurred spots are formed. The fundus reflexes are then imaged by means of the eye's refractive parts (in reverse direction), the beam splitter, and the ophthalmoscope subsystem onto a four-quadrant photodetector. The quadrants of the detector are oriented such that one of the separation lines is parallel to the connection line of both LEDs.

In the lower graph of Figure 5.13a, the acquired detector image is shown for an ametropic eye at the initial position of M. The fundus reflexes are blurred and vertically displaced relative to each other. When the LEDs are switched on and off in an alternating manner, the detector acquires either an out-of-phase signal[13] (for myopic eyes) or an in-phase signal (for hyperopic eyes). To set the coincidence of both partial beams, aperture stop M is now moved along the optical axis. From Scheiner's principle we know that the image of the test mire is focused on the retina when both fundus reflexes coincide[14]. At this position, the alternating detector signal disappears (Figure 5.13b), and the far point refraction can be calculated from the required displacement z' by means of Eq. (5.5).

If the connecting line of both LEDs does not coincide with one of the principal meridians of the examined eye, the detected fundus reflexes are rotated by an angle α (Figure 5.13c). To realign the optical system, the LEDs, the corneal reflex block, and the detector must be rotated. The signal of the four-quadrant detector is used as a reference. Hence, with autorefractors based on the Scheiner method, we can also determine the angle of astigmatism and the corresponding refractive errors along the principal meridians. Autorefractors by NIDEK work according to the Scheiner principle.

12) The measurement meridian is the curve created by the intersection of corneal surface and a plane which contains the optical axis of the system. A meridian is identified by the angle θ that the plane creating it makes to the horizontal. The value of θ, for a full meridian, takes values from $0°$–$180°$.

13) For example, the upper LED (LED_A) is switched on and a signal is detected in the lower quadrants (or vice versa).

14) As described above, the fundus reflexes already coincide from the very beginning for emmetropic eyes. Hence, an axial shift of M is not required in this case.

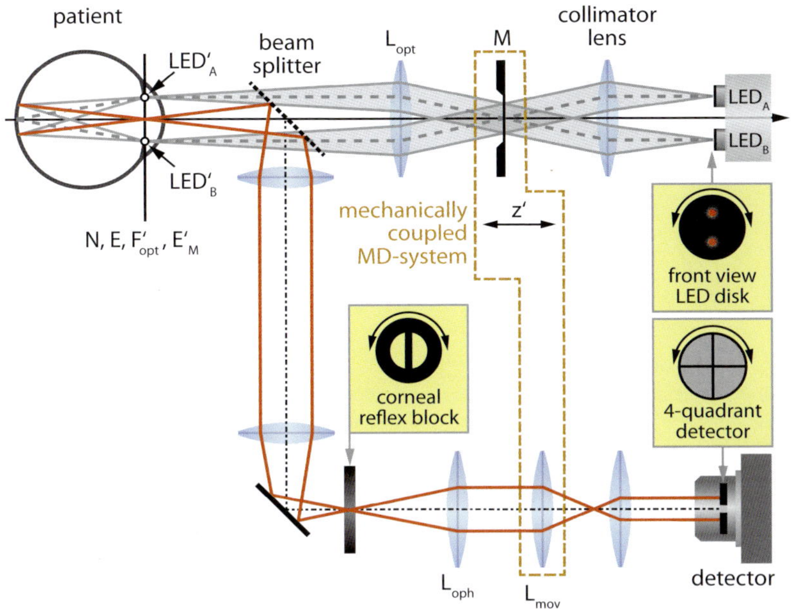

Figure 5.12 Setup of an autorefractor based on Scheiner's method. The two point light sources LED_A and LED_B are located in the focus of a collimator lens which expands the light beam to two bundles of parallel rays. These light bundles pass through pinhole M which acts as a test mire. With optometer lens L_{opt}, they are focused on the entrance pupil E of the patient's eye and form the images LED'_A and LED'_B. These images effectively act as an exit pupil E'_M of a Scheiner disk. If M is located at the image-side focal point of optometer lens L_{opt} (zero setting), a sharp image of M (spot) is formed on the fundus for an emmetropic eye. In the case of an ametropic eye, two blurred spots are formed. The generated fundus reflex is deflected by a beam splitter and imaged by the optical components of the ophthalmoscope subsystem onto a four-quadrant detector (front view of detector is shown in the inset). The corneal reflex block cuts off reflexes from the cornea generated by the incident LED light beams such that they are not detected. Pinhole M and the movable lens L_{mov} in the ophthalmoscope subsystem are mechanically coupled. The light source, the reflex block, and the detector (see insets) can be rotated around the optical axis in order to determine the orientation of the principal meridians in the case of an astigmatic eye. The corneal reflex block is imaged backwards into the entrance pupil of the patient's eye. There, it overlaps with LED'_A and LED'_B and thus cuts off the corneal reflexes generated in the optometer subsystems. Adapted from [2].

5.2.2.3 Ray Deflection Method

Concept In the ray deflection method, the optometer subsystem images a test mire with the greatest possible depth of field (Section 2.1.8) onto the fundus. For the measurement process, the corresponding fundus reflex always has to be a sharp secondary *point* source, independent of the eye's refractive status.

Before the light of the fundus reflex leaves the eye, it is expanded by the eye's refractive parts. Depending on the refractive status, the emerging beam is either parallel (emmetropic eye; Figure 5.14a), divergent (hyperopic eye; Figure 5.14b), or convergent (myopic eye; Figure 5.14c). Right in front of the eye, the pinholes

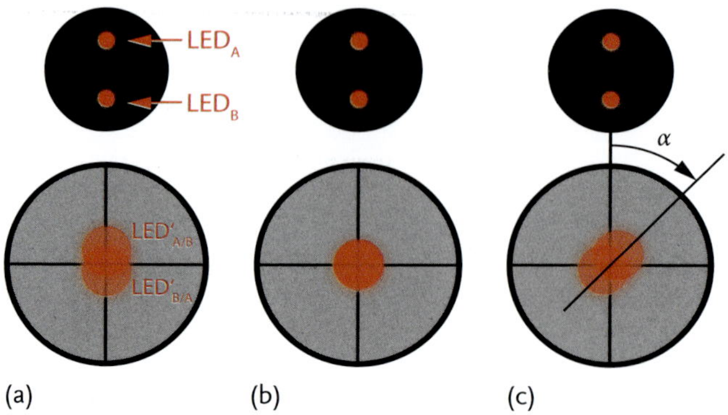

Figure 5.13 Principle of signal evaluation with the four-quadrant detector. Upper image row: position of LEDs. Lower image row: position of detected fundus reflexes in the case of an ametropic eye. (a) The connecting line of both apertures is aligned with the principal meridian of the eye. The reflexes are thus vertically shifted and just partly overlap. (b) Similar situation as in (a), but this time the fundus reflexes are brought into coincidence (i.e., zero setting is attained). (c) The connecting line of both apertures is not oriented along a principal meridian of the eye. Hence, the fundus reflexes are rotated by an angle α. In this case, it is impossible to find a zero setting.

of a Scheiner disk "cut out" two partial beams which are vertically displaced by each other. The ophthalmoscopy lens L_{oph} focuses both beams onto an intermediate imaging plane so that they are finally projected onto a CCD photodetector. In the detector plane, two blurred images of the fundus reflex appear whose relative distance is evaluated. For emmetropic eyes, the reference distance is d_0. If the measured distance is larger than d_0, the examined eye is myopic. If, however, the measured distance is shorter than d_0, the examined eye is hyperopic. The orientation of the measurement meridian is defined by the connecting line of both pinhole apertures of the Scheiner disk.

In practice, the Scheiner disk with two pinholes is replaced by a circular Scheiner aperture. This has the advantage that we can now determine the far point refraction of the patient's eye in all measurement meridians simultaneously. As a consequence, in a non-astigmatic eye, a ring image with a characteristic diameter d is detected instead of two distant points. Accordingly, we can interpret the beam path in Figure 5.14 as a cross-section through the measurement meridian. The term "ray deflection method" is related to the fact that the diameter d of the ring image depends on the deflection of both partial beams that exit the eye through the Scheiner disk.

In contrast to the "original" Scheiner method (Section 5.2.2.2), we do *not* measure the axial (horizontal) shift z' which is required for the coincidence setting of the fundus reflexes, but the shift of both fundus reflex images perpendicular to the optical axis. As no axial shift of optical components is needed to find the coincidence setting, the setup is relatively simple, cost-effective, and allows short measurement times.

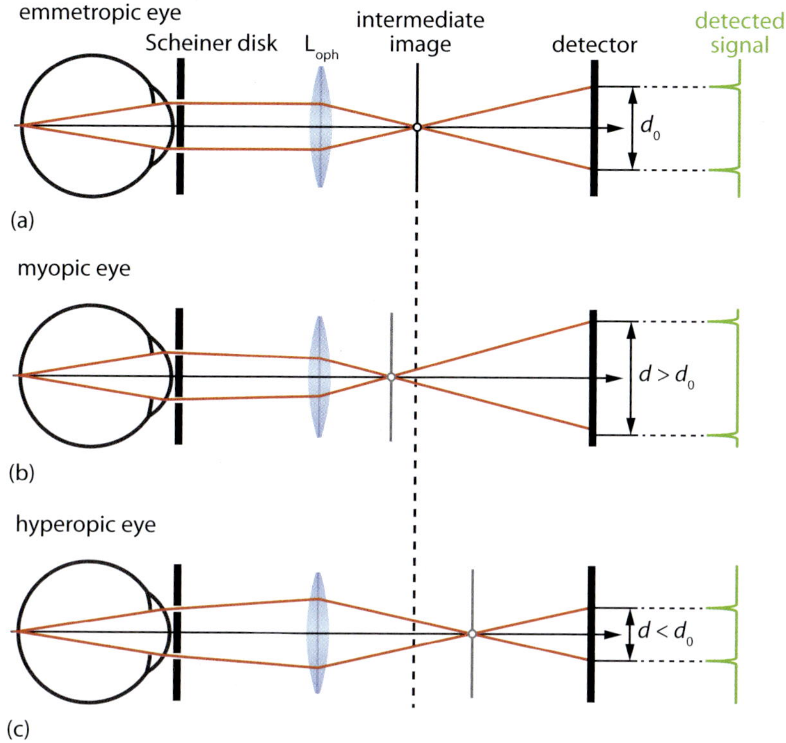

Figure 5.14 Ophthalmoscope subsystem of an autorefractor based on the ray deflection method (optometer beam path not shown). The refractive error of the patient's eye shifts the intermediate image plane (formed by the eye and ophthalmoscopy lens L_{oph}) along the optical axis. In the cross-sectional view, two image points (signal of image is shown in green) are formed on the detector plane whose distance d changes. Actually, the signal peaks are the borders of a circle (see text). d_0 is the reference distance found for an emmetropic eye. (a) Ray diagram for an emmetropic eye. (b) Ray diagram for a myopic eye. The detected distance of the image points is $d > d_0$, as the intermediate image plane (gray) is shifted towards the eye. (c) Ray diagram for a hyperopic eye. The detected distance of the image points is $d < d_0$, as the intermediate image plane (gray) is shifted towards the detector. Adapted from [5].

Implementation In autorefractors based on the ray deflection method (Figure 5.15), an LED with a very small light emitting surface is used as the primary point light source. The emitted light rays are imaged to infinity by a collimator lens and pass through a pinhole mask M. The rays are then focused into the opening of a beam splitter by an auxiliary lens and imaged into the patient's pupil plane by optometer lens L. The image of the point source LED′ acts as a very small exit pupil that ensures a large depth of field. The optical system is arranged such that the image M′ of the pinhole mask is located on the fundus.

The point-shaped fundus reflex is then imaged into an intermediate image plane by the eye's optical system (in reverse direction) and L, which now acts as an oph-

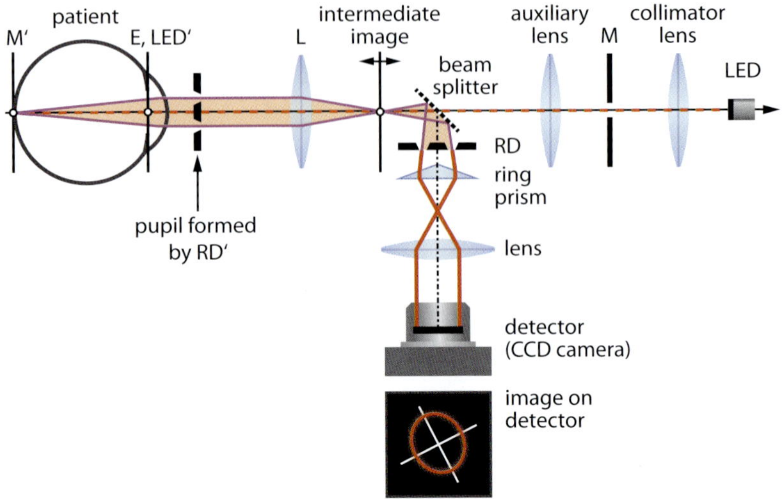

Figure 5.15 Setup of an autorefractor based on the ray deflection method. The point light source LED is expanded by a collimator lens and passes through pinhole M. An auxiliary lens and lens L image M onto the patient's fundus (i.e., M′). For the sake of simplicity, only the central ray is shown for the illumination beam path. LED′ represents the image of the light source which coincides with the pupil plane of the eye. The generated fundus reflex is imaged back into an intermediate image plane, deflected by a beam splitter, and passes through ring diaphragm RD. The image of this ring diaphragm RD′ forms a pupil in front of the eye. After the rays have passed through a ring prism and a lens, they reach a CCD photodetector. In the case of a non-astigmatic eye, the diameter of the ring-shaped image is a measure for the refractive error of the patient's eye. If the image deforms to an ellipse, the examined eye is also astigmatic (as shown in the inset). Adapted from [6].

thalmoscopy lens. After the fundus reflex is deflected by the beam splitter, the rays cross diaphragm RD at which only a ring-shaped light bundle can pass through. RD is actually a Scheiner disk, since it is imaged backwards via L in front of the examined eye, where it forms a pupil. The ring shaped light beam now passes through a ring prism and an additional lens. Finally, a blurred ring image of the fundus reflex is projected onto a CCD camera.

Depending on the refractive error, the position of the intermediate image plane is shifted along the optical axis. This determines the angle at which the rays of the fundus reflex pass through the ring diaphragm as well as shape and size of the image on the CCD camera. In the case of emmetropic eyes, the CCD camera detects a circular ring image whose diameter is used as a reference value. In myopic and hyperopic eyes, the diameter of the ring image changes because of the rotational symmetry of the defocus imaging error (see also Figure A.23 in Section A.1.8.2). In the case of astigmatic eyes, the ring deforms to an ellipse (Figure 5.15). The principal axes of the ellipse characterize the orientation of the principal meridians. The lengths of the minor and major axes of the ellipse are a measure for the refractive errors along the respective principal meridians.

Instead of a continuous ring diaphragm, we can also use a mask with six small circular openings which are arranged on a ring at 60° intervals. In this case, only six points are projected onto the CCD camera, which is generally sufficient for the evaluation of refractive errors. Refractometers by Canon work according to the ray deflection principle.

5.2.2.4 Image Size Method

Concept The major difference of the image size method compared to the other measurement methods is that, in the optometer subsystem, the image-side focus F'_L of optometer lens L does not coincide with the eye's nodal point N (Figure 5.16a), but rather with the eye's object-side focal point F_{eye} (Figure 5.16b). As long as test mire M is located at F_L, it is sharply imaged onto the fundus (i.e., a so-called *4-f imaging*). However, if the refractive power of the eye changes (e.g., in the case of myopia as shown in red in Figure 5.16b), the retinal image size h'_I changes as well. This effect is purposely used in the optometer subsystem of the image size method.

In Figure 5.16a, the focal point of L (Badal lens) coincides with the eye's nodal point N, which is the standard optometer configuration (e.g., used in Sec-

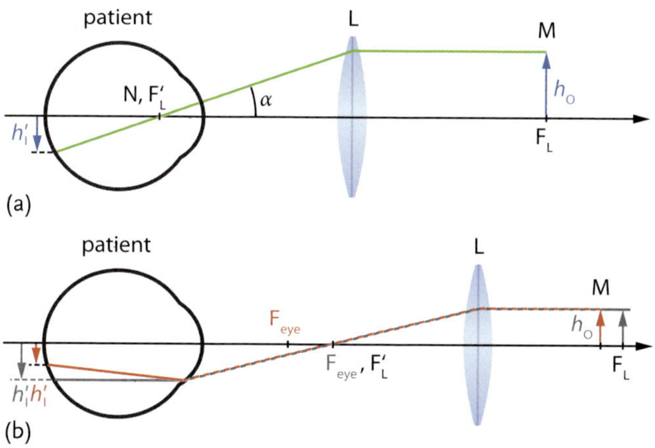

Figure 5.16 Comparison of the optical beam path for a standard optometer setup and for an optometer based on the image size method. (a) In the normal arrangement, the focus of optometer lens L (and the ophthalmoscopy lens) coincides with nodal point N of the eye. In practice, however, the beam is imaged to the entrance pupil of the eye (not shown). As long as the distance between L and the eye is constant, the retinal image size h'_I basically does not change when test mire M is moved along the optical axis. However, the sharpness of the retinal image changes (compare with Figure 5.10). A sharp image is only obtained when M coincides with F_L. (b) For the image size method, the image-side focal point F'_L of the optometer lens coincides with the eye's object-side focal point F_{eye}. As the test mire is always located at F_L, it is always sharply imaged onto the fundus. But, in contrast to the arrangement in (a), the image size changes with the position of the eye's far point. This effect can also be used to evaluate the eye's refractive status. The ray diagrams which lead to the different image sizes are shown for an emmetropic eye (gray beam path) and a myopic eye (red beam path).

tion 5.2.2.1). In this configuration, the retinal image size does not change when test mire M is moved along the optical axis. However, the sharpness of the retinal image changes. This configuration is actually used in the ophthalmoscope subsystem of the image size method for *measuring* of the retinal image size. Therefore, we have an arrangement comparable to Figure 5.16a, where M is replaced by a detector. Here, the detected image size does not depend on the axial position of the detection system.

Implementation In the optometer subsystem, an LED with a small emitting area acts as a primary point light source. The LED is expanded with a collimator lens and then imaged with an optometer lens into the object-side focal plane of the patient's eye. Right after the collimator lens, we have a ring-shaped test mire (annular aperture) which can be moved along the optical axis. The optometer lens images the test mire via a fixed ring aperture onto the fundus. The fixed ring aperture is, in turn, imaged into the eye's pupil plane and acts as a Scheiner aperture. As the image-side focal point of the optometer lens is approximately located at the eye's object-side focal point, the size and shape of the ring image depends on the eye's refractive status. Repositioning the test mire relative to the optometer lens can produce a focused image on the fundus for analysis in the detection system. The focusing only increases the accuracy of the image evaluation, but the sharpness of the image itself is not a measurement criterion.

To measure the retinal image size, the ring-shaped fundus reflex is imaged by an ophthalmoscopy lens onto a CCD camera. As the object-side focal point of the ophthalmoscopy lens is located close to the eye's nodal point, the detected image size does not depend on the axial position of the camera system. Similar to the ray deflection method (Section 5.2.2.3), the change in diameter of the ring image is used as a measure for the refractive error. For astigmatic eyes, the fundus reflex becomes an ellipse whose shape and diameter determine the angle of the principal meridians and the refractive error, respectively.

The image size method is implemented, for example, in autorefractors by Topcon.

5.2.2.5 Knife Edge Method

Concept In the knife edge method, a NIR LED is focused onto the edge of a special diaphragm (Figure 5.17b), where it acts as a primary point light source. The NIR LED is then imaged onto the fundus via lens L. The generated fundus reflex is imaged backwards by L into the plane of the knife edge aperture (Figure 5.17a). In this arrangement, lens L takes on the role of an optometer and an ophthalmoscopy lens. As usual, the image-side focal point of L is located in the plane of the eye's entrance pupil. Depending on the quality (sharpness) of the formed image, some rays may arrive at the detector by passing through the opening in the knife edge diaphragm.

If the luminous knife edge is exactly located at the far point of the patient's eye, it is imaged back into itself so that the amount of detected light is minimal (zero

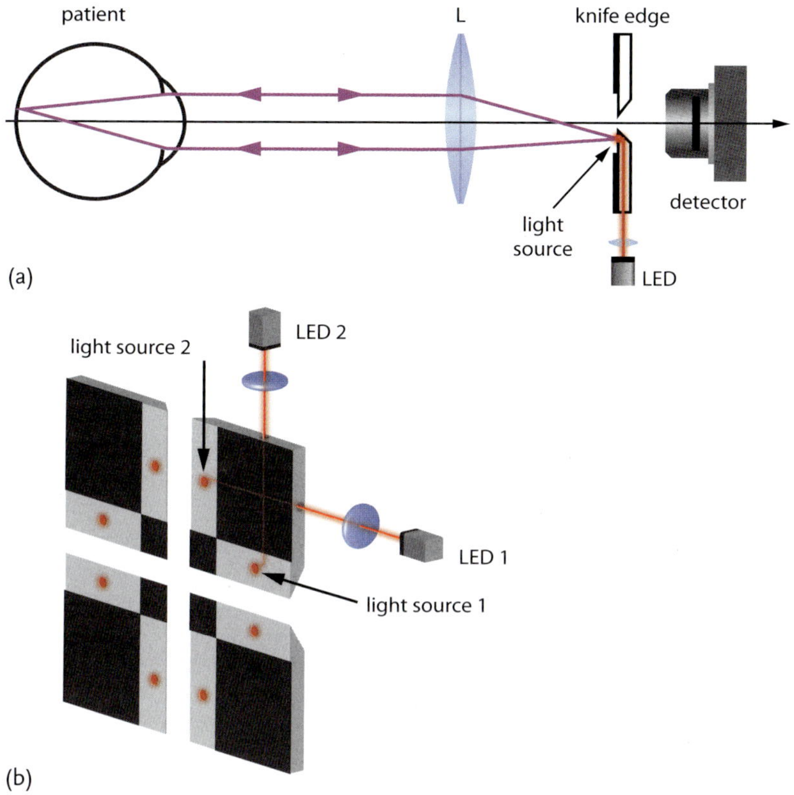

Figure 5.17 (a) Setup of an autorefractor based on the knife edge method. A NIR point light source (formed by an LED) is located at the inner edge of a knife edge diaphragm. It is imaged via lens L and the eye's optical system onto the patient's fundus. The fundus reflex is imaged back into the plane of the knife edge diaphragm. At the far point of the eye, the reflex is projected back into the light source. If, however, the knife edge diaphragm is *not* located at the far point, some rays can pass through the aperture and reach the detector. (b) Oblique front view of the used knife edge diaphragm. The point light sources (fuzzy red dots) are realized by projected LED beams. Adapted from [4].

setting). During the measurement, the knife edge is moved axially via a closed-loop control until the zero setting is attained.

Implementation A typical design of the knife edge diaphragm is depicted in Figure 5.17b. The knife edge consists of four narrow square prisms which are arranged in pairs. Each prism has two slanted and mirrored inner edges (white areas in Figure 5.17b) and is illuminated by two NIR LEDs (LED_1 and LED_2), which are perpendicularly arranged to each other. The LEDs are located right next to the opening of the knife edge diaphragm and represent primary light sources. The eight inner edges thus form a luminous double cross which consists of two horizontal and two vertical pairs of edges. The central opening of the diaphragm is used to detect the reflected rays with a four-quadrant photodetector. The detector is placed such that L forms an image of the eye's entrance pupil in the sensor plane.

The NIR illumination of each knife edge pair is switched on and off in an alternating manner so that the detector signal can be assigned to the respective knife edge pair. The detected signal is used for a controlled approach to the zero setting.

The knife edge diaphragm and the photodetector are only shifted along the optical axis. Hence, only axial-symmetric refractive errors (myopia and hyperopia) can be compensated. For the examination of astigmatic eyes, an assembly composed of four cylindric lenses is used to compensate and simultaneously measure the cylindric error. The corresponding control signal is provided by the four-quadrant detector.

Refractometers like the Humphrey® HARK 599 (Carl Zeiss) are based on the knife edge principle.

5.2.2.6 **Retinoscopy Method**

Concept The working principle of the retinoscopy method used for autorefractors (Figure 5.18) is very similar to hand-held streak retinoscopes (Section 5.1). Retinoscopy allows the determination of the eye's refractive status with high accuracy and a minimum of equipment. Logically, this concept also became the basis for the development of autorefractors. The first instrument of this kind (Bausch+Lomb Safir Ophthalmetron (1971)) was an automated version of the hand-held retinoscopes in that the flicker point was searched for. However, this approach was soon replaced by speed measurements of the fundus reflexes in the pupil plane (e.g., by Nikon). When the speed of the fundus reflexes is known, the refractive error of the patient's eye can be calculated. Compared to the flicker point method, the concept of speed measurement results in simpler and more cost-effective devices. In addition, the measurement times can be significantly decreased. For this reason, the working principle is used today in all autorefractors based on the retinoscopy method.

Implementation To form moving light bands, a rotating slit drum is used. A NIR LED located in the center of the slit drum acts as a light source. The LED is imaged via lens L_{opt} (and a beam splitter) into the object-side focal point of the patient's eye so that a fairly collimated slit of light falls on the retina, thereby creating an extended fundus reflex. The collimated beam is chopped by rotating the slit drum with a constant angular speed. In this way, rectangular light bands are formed moving on the retina (i.e., moving fundus reflexes) at a defined speed v_f in a direction perpendicular to their long axes. The speed v_f of the moving fundus reflexes on the retina is independent of the refractive status of the eye.

When the slit drum rotates counterclockwise as shown in Figure 5.18a, the rectangular light bands projected onto the retina move from the top to the bottom. Speed and direction of the moving fundus reflexes seen in the pupil plane depend on the refractive status of the eye and are recorded by a four-quadrant photodetector. The four detector elements D_i (see inset of Figure 5.18a) are imaged by lens

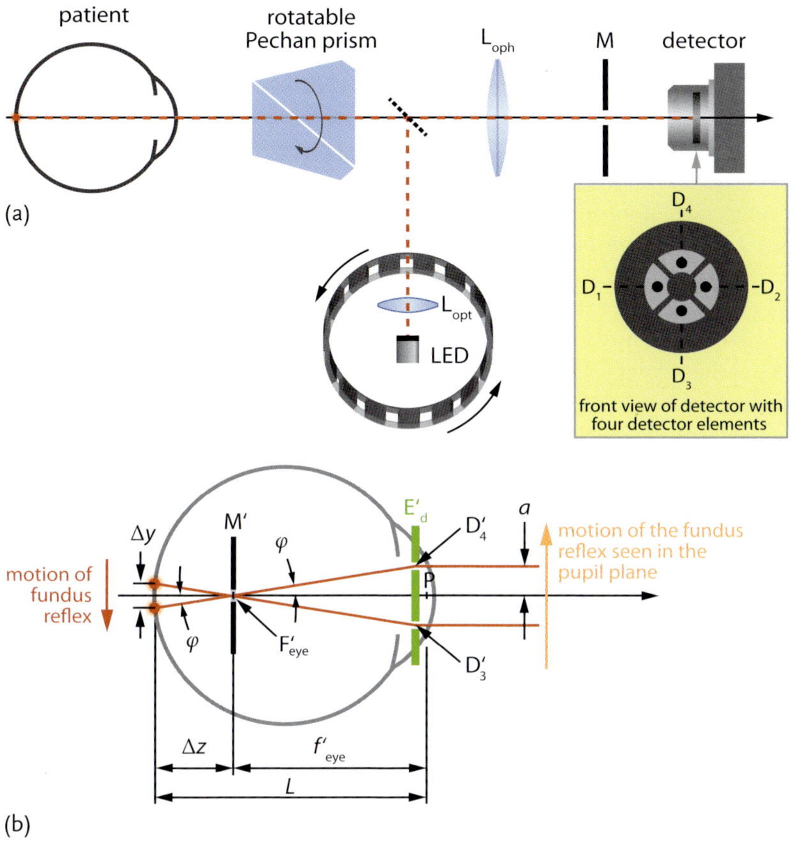

Figure 5.18 (a) Setup of an autorefractor operating according to the retinoscopy principle. For the sake of simplicity, only the path of the central ray (dashed line) is shown. The moving light bands are formed by a rotating slit drum. In the center of rotation of the slit drum, a NIR LED is placed which serves as a light source. The LED is imaged by lens L_{opt} via a beam splitter and a Pechan prism into the object-side focal point of the patient's eye so that a fairly collimated slit of light falls on the retina, thereby creating an extended fundus reflex. The collimated beam is chopped by rotation of the slit drum. In this way, rectangular fundus reflexes are produced moving on the retina at a defined speed. The light beam emerging from the eye is then imaged onto the detector by means of lens L_{oph}. A slit-shaped observation aperture M is placed in the image-side focal point of L_{oph}. A front view of the detector's four-quadrant sensor is shown in the inset. (b) Optical scheme of a myopic eye to explain the measuring principle. M′ represents the image of slit aperture M which is formed in the image-side focal plane of the patient's eye at a distance Δz in front of the retinal surface. E'_d is the exit pupil effectively formed by the image of the detector elements in the pupil plane created by lens L_{oph}. a is the distance between the optical axis and the images of the detector elements (only D'_3 and D'_4 are shown). φ is the angle between the optical axis and rays which are "seen" by the detector elements. L is the distance from the fundus to the eye's principal plane, and f'_{eye} is the image-side focal length of the eye. Adapted from [4].

L_{oph} into the pupil plane of the patient's eye (only D'_3 and D'_4 are shown in E'_d). So, the detector elements effectively form an exit pupil E'_d in which the images of the detector elements represent the corresponding apertures. In this way, the four detector elements analyze four different peripheral areas within the eye's pupil in that they record a signal when the fundus reflexes become visible in the respective area. There are no photo elements in the center of the detector so that the central corneal reflex does not influence the measurement. A slit aperture M is placed in the image-side focal point of L_{oph}. It acts as an observation aperture, and its image M′ is located at the eye's image-side focal plane at F'_{eye} (shown in Figure 5.18b for a myopic eye). The long axes of slit aperture M, detector, and light bands lie all in a common plane. Perpendicular to this common plane we have the meridional plane along which the refractive status is determined.[15] For example, in Figure 5.18, the meridional plane lies in the plane of projection.

Let us now consider the principle of determination of the refractive status for a myopic eye (Figure 5.18b). In this case, image M′ of the slit aperture is *not* located on the retina but shifted by Δz.[16] While the slit drum rotates, the fundus reflexes move on the fundus by a distance Δy within a time period Δt according to the known constant velocity v_f. As mentioned before, speed and moving direction of the fundus reflexes observed in the pupil plane are directly related to the refractive error of the eye. Therefore, a temporal analysis of the light signals from the fundus reflexes recorded by detector elements "3" and "4" allows the determination of refraction of the patient's eye in the meridional plane along the connecting line between both detector elements (plane of projection in Figure 5.18b).

From Figure 5.18b, it follows that only those rays of the moving fundus reflexes are "seen" by the detector elements which pass through M′ and exit the eye parallel to the optical axis (red rays shown in Figure 5.18b). When a fundus reflex moves from the top to bottom, D'_3 will at first detect a signal. While the slit drum further rotates, the rays of the moving fundus reflex cross the central part of E'_d, and signals are neither detected by D'_3 nor by D'_4. After a period Δt, the rays are incident on D'_4. The detected rays include an angle φ with the optical axis. From geometric considerations based on Figure 5.18b, we obtain

$$\varphi \approx \frac{a}{f'_{eye}} = \frac{\Delta y/2}{\Delta z} , \tag{5.6}$$

where a is the distance of the images of the detector elements to the optical axis and f'_{eye} the image-side focal length of the eye. With the distance from the fundus to the eye's principal point L[17] and $\Delta z = L - f'_{eye}$, the refractive power of the examined eye is given by

$$\mathcal{D}'_{eye} = \frac{\Delta y}{2aL} + \frac{1}{L} = \frac{v_f \Delta t}{2aL} + \frac{1}{L} . \tag{5.7}$$

15) As the Pechan prism is used to rotate the meridional plane in the eye, it is not relevant for the discussion of rotational symmetric refractive errors like myopia and hyperopia. We will thus ignore this optical component for a moment.

16) In an emmetropic eye, M′ falls exactly on the retina ($\Delta z = 0$).

17) Distance L approximately equals the distance from the fundus to the pupil E'_d.

With the lens equation (A14) and the sign convention of Section 2.1.4, we may calculate the far point refraction via

$$A_{\text{far}} = \frac{1}{L} - \mathcal{D}'_{\text{eye}} = \frac{v_f \Delta t}{2aL} \,. \tag{5.8}$$

A_{far} is thus proportional to the time period Δt between the detected signals. From Eq. (5.8), it becomes clear that we see flickering if *no* refractive error is present ($A_{\text{far}} = 0$). We have $\Delta t = 0$ in this case, as the detector elements D'_3 and D'_4 record the fundus reflex simultaneously.

Info Box 5.1: Pechan Prism

A Pechan prism is a frequently used image rotator which, when inserted into an optical system, does *not* change the position and orientation of the system's optical axis. It consist of two prisms which are usually held mechanically or bonded to a common mounting plate in order to create a narrow air space between them. This optical component can be used for convergent as well as divergent light beams.

Specifics of the measurement in astigmatic eyes In an astigmatic eye, we first have to determine the orientation of the principal meridians. For this purpose, we use the fact that red reflexes observed in the pupil plane move obliquely with respect to the incident beam as long as the long axes of the light bands are not oriented perpendicular to any of both principal meridians (Section 5.1). In this case, the photoelements "1" and "2" acquire a delayed signal. With a *Pechan prism* (see Info Box 5.1) placed between the beam splitter and the patient's eye, the long axes of the incident light bands are rotated such that the signal delay disappears. When this is achieved, the corresponding refractive powers can be measured. As the Pechan prism also rotates the light beam leaving the eye pupil by the same amount, it is possible to determine astigmatism without rotating the slit diaphragm and the four-quadrant detector.

There are also alternative instrument designs available for measurement of astigmatic eyes which do not require a rotating Pechan prism. In these designs, the slit geometry is modified. For example, slit drums with oblique slits are used that are inclined by 45° (or 135°) to the axis of rotation. In this configuration, the four-quadrant detector has to be rotated by 45° as well, and the slit mask M must be replaced by a pinhole. With such a design, we can measure both principal meridians simultaneously.

Alternative design concepts In the NIDEK OPD-Scan III refractive power/corneal analyzer, the retinoscopy principle is also used for refraction measurements [7]. In this device, the slit drum and a detector with eight linear photoelements[18] are

18) The photoelements are arranged similar to Figure 5.18. In addition, three elements are placed above D_4 and three elements are placed below D_3. The additional elements are vertically aligned.

simultaneously rotated so that no Pechan prism is needed. The linear detector segments are again imaged into the pupil of the patient's eye. With this detector arrangement, at any location within the 6 mm pupil area, the optical path length differences (OPD; Section 5.3.1.3) between retina and aperture image M′ are measured. For this purpose, the time differences of the peak intensities of all sensor elements are compared to a reference sensor. As the scanning slit and the linear detector rotate, the different elements of the detector trace out annular rings in the pupil of the eye. This arrangement allows the measurement of refractive errors in four peripheral ring-shaped pupil areas along each meridian in 1° steps. The results are displayed in a so-called *OPD map* which shows the distribution of refractive errors over the pupil area. Alternatively, the refractive errors are reported as spherocylindric values (sph, cyl, axis; Section 5.4.1.1) within pupil areas with a diameter of 3 or 5 mm.

5.2.3 Measurement Accuracy and Limitations of Automatic Refractometers

Autorefractors deliver reliable data which are a good starting point for subsequent subjective refraction methods. For example, a study [6] showed that the deviation between subjectively and objectively measured refractive errors (spherical equivalent; Section 5.4.1.1) was smaller than 0.5 D in 80% of all cases. Autorefractors deliver incorrect or no measurement results if

- the patient has fixation problems and/or accommodates unintentionally (which shifts the far point that we want to determine),
- the instrument is not adequately centered with regard to the center of the eye pupil,
- the pupil diameter is too small (smaller than approximately 2–3 mm),
- the patient has excessive ametropia (at the limit or outside of the measuring range),
- the anterior eye segment (cornea, lens) shows substantial light scattering (e.g., caused by cataracts (Section 3.2)),
- the corneal surface is irregular (e.g., keratoconus or after refractive surgery), and
- fundus anomalies and inhomogeneities result in a poor fundus reflex.

It should be pointed out that most autorefractors can measure the refractive status of only one eye (monocular measurement). Of these, only some (e.g., ZEISS Humphrey HARK) provide an estimate of (monocular) best corrected visual acuity (BCVA).

Automated objective refractometers are offered by numerous manufacturers, both as standalone instruments and as multifunctional devices in combination with a keratometer (Section 6.3.1). Figure 5.19a shows an example of a standalone autorefractor and Figure 5.19b a combined autorefractor/keratometer.

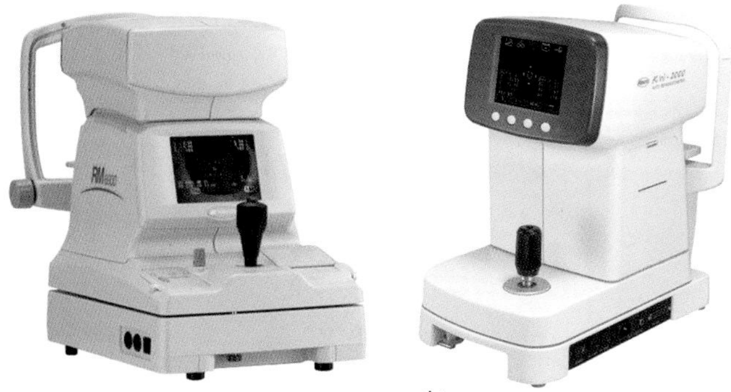

(a) (b)

Figure 5.19 Photographs of two autorefractors. (a) Photograph of Topcon RM-8900 autorefractor. Courtesy of Topcon Deutschland GmbH. (b) Photograph of Kowa KW-2000 autorefractometer/keratometer. Courtesy of Kowa Optimed Deutschland GmbH.

5.3 Aberrometers

With autorefractors, we can solely measure the lower-order aberrations (LOAs; Section A.1.8.2) of a patient's eye, for example, defocus and astigmatism. For classic refractive corrections, this is sufficient, because only these refractive errors can be corrected with eye glasses and contact lenses (Section 5.5). Moreover, the influence of other potential aberrations (e.g., spherical aberration, coma, trefoil) on the quality of vision is relatively low for small pupil diameters ($\leq$ 3 mm). However, laser-supported refractive corneal surgery has gained acceptance during recent years which has created a demand for precise aberration analysis techniques. The corresponding instruments are called *aberrometers* or *wavefront analyzers.*

5.3.1 Fundamentals of Aberrometry

Before we consider the working principles of aberrometers, let us briefly discuss some general facts about aberrations and parameters that are important to understand the measurement processes and the representation of the acquired data. An introduction to the appearance and imaging consequences of (monochromatic) optical aberrations is presented in Sections A.1.5 and A.1.6, respectively. In Sections A.1.7 and A.1.8, we introduce two mathematical frameworks (Taylor and Zernike polynomials) which are used to describe aberrations in a quantitative manner. The origin and imaging consequences of (poly)chromatic aberrations are explained in Section A.1.9.

In the following, we focus on the most relevant concepts related to the measurement of aberrations in the human eye (*aberrometry*). Therefore, we will only consider monochromatic aberrations which can be described by

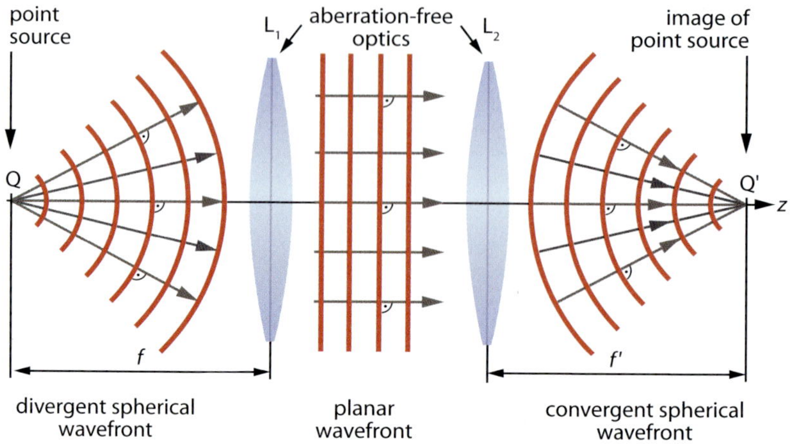

Figure 5.20 Ideal (aberration-free) imaging of a point light source Q by two lenses. The distance between Q and the first lens corresponds to the object-side focal length f of the lens. The light rays are perpendicular to the wavefronts and point in the traveling direction of the light beam (Section A.2.1.3). Thus, both (wave and geometric) descriptions of light can be used in an equivalent manner to characterize the imaging behavior of an optical system. *Geometric optics*: Rays (gray) are emanated radially from Q and are refracted by two lenses such that they converge on a single image point Q′. *Wave optics*: Spherical wavefronts (red) are emanated from Q and are refracted (and/or diffracted) by the first lens such that they become planar. A second lens refracts the wave such that it converges at Q′. The wavefronts on the image side are again spherical and centered at Q′.

- transverse or longitudinal ray aberrations (Section 5.3.1.1),
- wavefront aberrations (Section 5.3.1.2), or
- the optical path difference (Section 5.3.1.3).

All three approaches are equivalent descriptions of imaging errors, as they are different interpretations of the same facts. Which approach is used in the end depends on the specific application. Wavefront aberration is preferably used to characterize aberrations in the eye.

5.3.1.1 Ray Aberrations

In geometric optics (Section A.1), an optical system is considered to be "ideal" (i.e., free of optical errors) if all rays emitted from object point Q converge in a single image point Q′ (Figure 5.20). For optical systems which show aberrations, rays are deflected from the ideal path so that they do not converge in one image point anymore. Consequently, we obtain a blurred image. We differentiate between transverse and longitudinal ray abberations.

Transverse ray aberration Transversal ray aberrations are specified by angular deviations of the ray path from the respective ideal reference path measured (in milliradians, mrad) in all locations of the eye's entrance pupil ($x_p y_p$ plane). In general, the angular ray deviations are described by two components, that is, horizontal

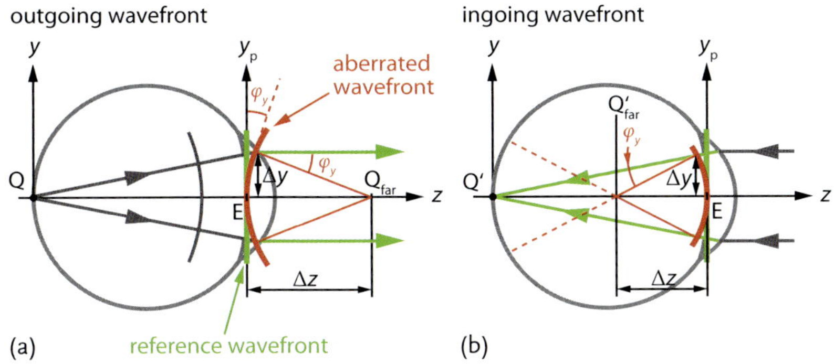

Figure 5.21 Wavefronts and rays of a myopic eye (red) compared to an optical perfect eye without aberrations (green). φ_y describes the transverse angular ray deviation in y direction. The longitudinal aberration is characterized by the distance Δz. Point E denotes the center of the eye's entrance pupil at which the aberration is specified. The arrows of the marginal rays for ideal imaging (green) represent the traveling direction of light. For reasons of clarity, the occurring aberrations are strongly exaggerated. In general, the ray aberration also has a component that runs perpendicular to the drawing plane (in the x direction). E is thus a point in the $x_p y_p$ plane. (a) Outgoing wavefront (rays) emitted from object point Q on the retina. Due to aberrations of the myopic eye, the outgoing light beam (red) is focused in the far point Q_{far} at a finite distance Δz away from E. In the case of ideal imaging, the beam would be focused at infinity (parallel marginal rays, planar wavefronts). (b) Ingoing planar wavefront. The aberrated wavefront (red) forms a blurred image at Q′ on the retina, whereas the wavefront is exactly focused on the retina in the case of ideal imaging (green).

deviations described by $\varphi_x(x_p, y_p)$ and vertical deviation described by $\varphi_y(x_p, y_p)$ (Figure 5.21).[19] The measured local data points within the pupil plane are then illustrated by a *beam deflection map*.

Longitudinal ray aberration The longitudinal ray aberration is specified by the inverse of the distance Δz between the entrance pupil plane and the point of intersection of the aberrated ray with the optical axis. $1/\Delta z$ depends on the angular ray deviation $\varphi_y(x_p, y_p)$ and the vertical distance to the optical axis Δy (Figure 5.21) so that we have

$$\frac{1}{\Delta z} \approx \frac{\varphi_y(x_p, y_p)}{\Delta y} \tag{5.9}$$

with $[\Delta z] = 1/\mathrm{m}$.

If the considered optical system is affected by astigmatism, we have to determine the ray deviation in the planes of both principal meridians. In the case of rotational symmetric errors (e.g., defocus or spherical aberration), the longitudinal ray deviation corresponds to the far point refraction ($A_{far} = 1/\Delta z$).

19) In the following, we use the same nomenclature as in Section A.1.7.

5.3.1.2 Wavefront Aberration

When light travels in space, it can be regarded as a wave with wave crests and troughs which form wavefronts (Sections A.1.7 and A.2.1). Wavefronts are locations at which light waves have exactly the same phase (Section A.2.1) at a given point in time. The shape and propagation of a wavefront is directly related to light rays, since the rays are always perpendicular to the wavefront and point in the traveling direction (Figure 5.20 and Section A.2.1.3).

The wavefronts emitted by a point light source Q (into a homogeneous medium) are spherical. When they pass through an ideal lens L_1 whose focal point coincides with Q, the spherical wavefronts are transformed into planar wavefronts (Figure 5.20). Planar wavefronts are thus an equivalent description for a parallel bundle of rays. When planar wavefronts pass through a second ideal lens L_2, they are transformed into convergent spherical wavefronts. In the focus of L_2 (i.e., the center of the convergent wave), an image Q' of the point light source is formed. For such a spherical lens, ideal imaging is only achieved in the case of paraxial approximation. If we also take those parts of a wavefront into account which pass through the outer zone of the lens, we will see that the wavefront has a distorted form. In this case, the wave does not converge at a single point Q' anymore and thus forms a blurred image.

Let us now apply these general considerations to a relaxed eye (no accommodation). In Figure 5.21a, a fundus reflex acts as a point source (secondary light source) which emits spherical waves. Since the spherical waves cross the eye lens and cornea, their wavefronts are altered by refraction. In the case of ideal imaging, the emerging wavefronts will be planar (green reference wavefront) and the far point will be located at infinity. In real eyes, however, aberrations distort the wavefronts (red line) as compared to the reference wavefront. For example, Figure 5.21a depicts an aberrated wavefront due to defocus of a relaxed myopic eye.

Figure 5.21b shows the imaging of a relaxed eye placed into a (collimated) light beam with planar wavefronts. In the case of ideal imaging, the incident wavefronts would be refracted such that they form centered spheres around point Q' which lies on the retina. In a myopic eye, however, the incident wavefronts are refracted such that they have a stronger curvature. As a consequence, the waves converge in front of the retina, and the image on the retina becomes blurred. If higher-order aberrations (HOAs) are also present, the aberrated wavefront can have a very complex surface shape. In summary, the shape of the aberrated wavefront contains all information about the monochromatic aberrations of the eye. As a first step, the aberrated wavefront has to be compared to a reference wavefront expected from an aberration-free (optically perfect) eye to obtain this information. In the case of an outgoing wavefront (Figure 5.21a), it is common to use the flat wavefront in the plane of the eye's entrance pupil as a reference.[20]

Wavefront aberrations are measured in the axial direction (i.e., parallel to the line of sight) from the pupil plane towards the distorted wavefront. By convention, the

20) To reveal only the effects of HOAs, a spherical wavefront centered on the eye's nominal far point can be used as reference.

wavefront error is set to zero at the pupil center by subtracting the central value from values at all other pupil locations (see Eq. (A34) in Section A.1.7). The local differences between the reference and distorted wavefronts over all locations in the $x_p y_p$ pupil plane are described by the wave aberration function $\mathcal{W}$ (also called *wavefront error function*). Wavefront aberrations can be stated either in micrometers or in fractions of the light wavelength λ. The values of $\mathcal{W}$ are positive if the aberrated wavefront is locally propagating ahead of the reference wavefront (positive phase difference as in the case shown in Figure 5.21) and negative if the aberrated wavefront is lagging behind.

5.3.1.3 Optical Path Difference

The optical path length OPL of a light ray is given by the product of its geometric path length L (from the object to the image) and the refractive index n of the medium through which the ray travels. Hence, we have

$$\mathrm{OPL} = nL \tag{5.10}$$

or, generally, with m partial distances L_i in media with different refractive indices n_i

$$\mathrm{OPL} = \sum_{i=1}^{m} n_i L_i \,. \tag{5.11}$$

The optical path difference OPD of two individual rays is given by the difference of their corresponding optical path lengths (j indicates the individual light rays). So, we have

$$\mathrm{OPD} = \mathrm{OPL}_j - \mathrm{OPL}_{j+1} \,. \tag{5.12}$$

The wavefront is related to the OPL in that it is formed by the endpoints of all light rays that originate from the same light source and have equal optical path lengths.[21] For example, let us consider a point light source which is located in a homogeneous medium with a constant refractive index. In this case, the end points of the emitted rays with equal OPL will form a spherical surface (Figure 5.20).

In an ideal optical system without optical aberrations, the OPL from object point to image point is equal for all light rays entering the entrance pupil of the system. The OPD is thus zero for every point in the pupil plane. If, however, aberrations are present, the OPL varies between the individual rays, which can be expressed by an OPD function in the pupil plane with the OPL of the chief ray as a reference. The OPD function is thus a measure for aberrations. As with the wave aberration function, it is stated either in micrometers or in fractions of the light wavelength. In fact, the absolute values of the OPD function and the wave aberration function $\mathcal{W}$ are equal. Only the algebraic sign is different for both quantities due to the opposed definitions ($\mathrm{OPD}(x_{pi}, y_{pi}) = -\mathcal{W}(x_{pi}, y_{pi})$). This can be understood using the following example: A marginal ray path length that is longer than the chief ray path

21) In fact, this is just another definition of the phase of an optical wave.

length corresponds to a positive value of OPD. Due to the greater OPL, light travels slower. Thus, the phase of the wavefront is delayed with respect to the reference wavefront, which is considered a negative wavefront aberration.

The description of eye aberrations through an OPD function provides immediate baseline data for the required ablation profile in wavefront-guided refractive corneal surgery (Section 10.3.1.1). The OPD function then represents the correction of the OPL needed to correct the wavefront error.

5.3.2
General Measurement Principles for Aberrometers

In all aberrometers, the wavefront is analyzed indirectly via the measurement of the local ray deflection (corresponding to the transverse ray aberration) at the location (x_{pi}, y_{pi}) in the pupil plane. From the local values of $\varphi_x(x_{pi}, y_{pi})$ and $\varphi_y(x_{pi}, y_{pi})$, we may calculate the corresponding slopes of the wave aberration function via

$$\frac{\mathrm{d}\mathcal{W}}{\mathrm{d}x_\mathrm{p}} = \varphi_x(x_{pi}, y_{pi}) , \tag{5.13}$$

$$\frac{\mathrm{d}\mathcal{W}}{\mathrm{d}y_\mathrm{p}} = \varphi_y(x_{pi}, y_{pi}) . \tag{5.14}$$

From these derivatives (wavefront slopes), standard algorithms then allow the reconstruction of the shape of $\mathcal{W}(x_\mathrm{p}, y_\mathrm{p})$ (Section 5.4). In principle, two techniques exist to measure the local ray deviations in the pupil plane caused by the optical system of the eye, which are referred to as *outgoing* and *ingoing light aberrometry*.

Outgoing light aberrometry In outgoing light aberrometry (Figure 5.21a), a NIR light source (not shown) is used to create a fundus reflex. It can be regarded as a secondary point source for a bundle of "test" rays. The deflection of rays in the pupil plane is then analyzed with regard to a nondeflected ray which is parallel to the line of sight.[22] In fact, most commercial aberrometers work according to this principle and use a *Hartmann–Shack wavefront sensor* (Section 5.3.4) as the key measurement component for determination of local transverse ray deviations.

Ingoing light aberrometry In ingoing light aberrometry (Figure 5.21b), a bundle of parallel rays emitted by a NIR light source (not shown) enters the eye through its entrance pupil. The transverse ray deviation is measured from the considered ray (red) to a reference ray (green) that enters the eye parallel to the line of sight at the same pupil location. The reference ray is refracted such that it intersects the retina at the same point as the line of sight. Depending on the degree of eye aberration, the intersection point of the considered ray (fundus reflex) on the retina deviates more or less from the reference ray position on the retina. This deviation

22) In other words, the deflection is measured from the considered ray (red) to a reference ray (green) at the same pupil location but parallel to the line of sight.

can be measured by using a standard ophthalmoscope system (Section 5.2). This is why ingoing light aberrometers are sometimes also referred to as *retinal imaging aberrometers.*

In ingoing light aberrometry, we assume that the incident light is uniformly reflected by the retina. Inhomogeneities and/or an irregular surface shape may thus lead to measurement errors. Outgoing light aberrometry is less sensitive in this respect, because just a small spot illuminates the retina and all measurement points are equally affected by potential inhomogeneities.

5.3.3
General Remarks on Aberrometry

- The line of sight (Section 2.1.3) is used as the reference axis for the measurement of eye aberrations. It is the line joining the fixation point and the center of the entrance pupil.[23] The measured aberrations are represented in a coordinate system whose origin is the center of the entrance pupil (as done in Section A.1.7).
- Since aberrations depend on the size of the eye's entrance pupil, it is essential to measure and specify the pupil diameter.
- All aberrometers use NIR light for their measurements. The advantages and disadvantages of this method have already been listed in Section 5.2.1. Analogous to refractometers, the measurement results must be corrected with regard to the wavelength.

5.3.4
Hartmann–Shack Wavefront Aberrometer (Outgoing Light Aberrometer)

Setup The Hartmann–Shack wavefront aberrometer (Figure 5.22) is a typical outgoing light aberrometer. The incident beam of an infrared LED or infrared laser diode (Section B.5.2) is reflected by the retina. Since the beam diameter is small (approximately 1 mm), the size and location of the spot on the retina are largely independent of potential eye aberrations[24]. The fundus reflex acts as a secondary point light source which emits spherical waves. Depending on the refractive errors of the eye, the outgoing wavefronts are more or less aberrated as compared to the (planar) reference wavefront which would emerge from an ideal eye without any refractive error. The wavefronts pass through a (relay) lens system and are eventually analyzed by a Hartmann–Shack wavefront sensor (HS-WFS).

23) According to the ISO 24157 standard, the line of sight is defined as "... the line from the point of interest in object space to the center of the entrance pupil of the eye and continuing from the center of the exit pupil to the retinal point of fixation".
24) Eye aberrations affect only the quality of the spot, but not its centroid location.

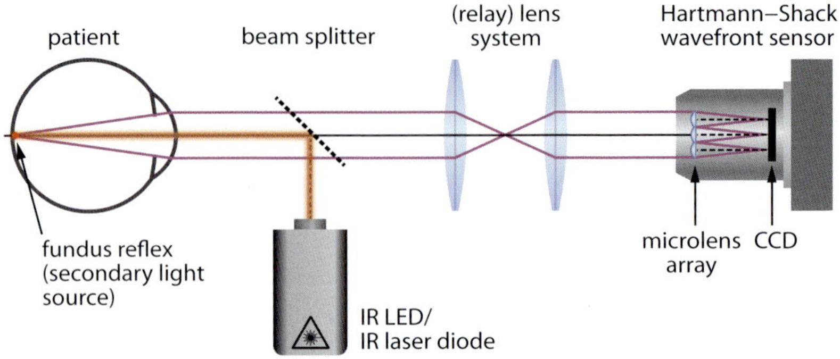

Figure 5.22 Setup of a Hartmann–Shack wavefront aberrometer. The thin beam of an infrared LED or laser is focused onto the retina. The generated reflex acts as a secondary point light source from which spherical waves are emanated. After the refractive parts of the eye have deformed the wavefronts, the beam is imaged into a Hartmann–Shack wavefront sensor. The sensor consists of a microlens array where each lens focuses a part of the wavefront onto a CCD chip. Adapted from [8].

Hartmann–Shack wavefront sensor The HS-WFS[25)] consists of a two-dimensional microlens array and a CCD camera. The surface of the CCD sensor is located such that it coincides with the focal points of the microlenses (Figure 5.23). By a (relay) lens system, the patient's entrance pupil is imaged onto the front surface of the microlens array. In this way, the wavefront of light emerging from the eye's entrance pupil is reproduced at the wavefront sensor's entrance pupil. The microlens array itself consists of several hundred identical, typically plano-convex lenses. The individual lenses usually have a diameter of approximately 100–600 μm and focal lengths in the range of a few mm (maximum 20–30 mm). With this arrangement, we actually subdivide the entire eye pupil into many small areas which all have the same diameter d as the microlenses. As a consequence, each microlens images just a small part of the whole wavefront which is locally considered to be planar (gray lines in Figure 5.23b,c). This small part of the wavefront is then focused onto the CCD camera as a diffraction-limited spot.

Measurement and analysis In an (aberration-free) eye with ideal imaging, the entire outgoing wavefront is planar (i.e., the emitted light rays are parallel). Since each microlens focuses just a part of the wavefront on its "individual optical axis", the CCD camera detects a regular spot pattern (Figure 5.23a). In an aberrated eye, however, the outgoing wavefront is deformed and has a complex surface shape. When this wavefront is incident on the microlens array, it is again virtually divided into many parts (Figure 5.23b). Each individual part of the wavefront can be considered as a planar wavefront which is now tilted by an angle $\varphi(x_{pi}, y_{pi})$ with regard

25) In a way, the HS-WFS is an advanced replacement for a Scheiner disk (Section 5.2.2.2). The first step of development is attributed to Johannes Hartmann (1865–1936) who used a pinhole screen with numerous apertures (referred to as the *Hartmann screen*) instead of a Scheiner disk to test optical systems. Due to recent advances in optical microfabrication, Roland Shack (born 1927) was able to replace the Hartmann screen by a microlens array.

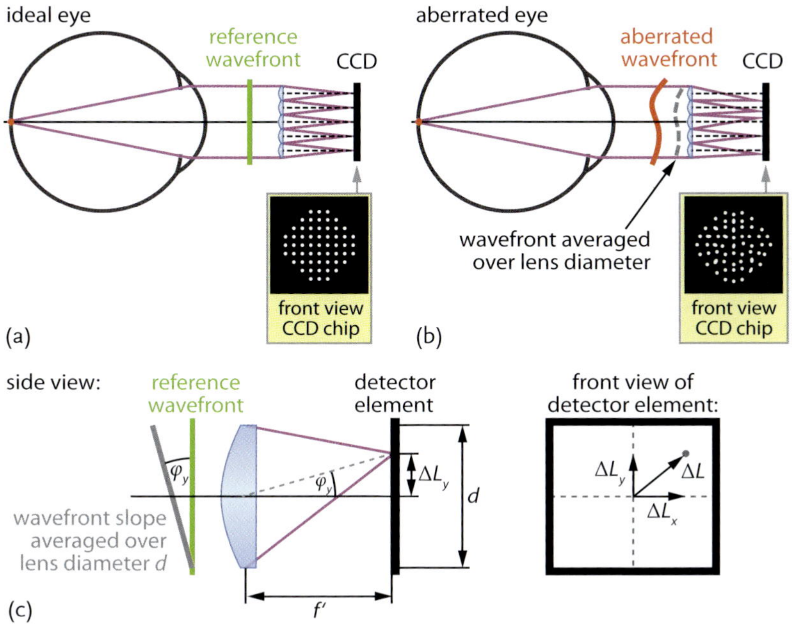

Figure 5.23 Wavefront detection with a Hartmann–Shack wavefront sensor. (a) In an ideal eye without aberrations, the wavefronts emitted by the secondary point light source are deformed to planar wavefronts (parallel rays) by the eye's refractive parts (marginal rays are shown in violet). Each microlens of the microlens array images a part of the whole wavefront into the CCD chip. As no aberrations are present, we obtain a regular pattern of spots (see inset). (b) In an aberrated eye, the spherical wavefronts of the secondary point light source are deformed such that they have a complex surface shape (red). Again, the microlenses image just a part of the wavefront (gray). In this case, the spot pattern becomes irregular (see inset), since each part of the wavefront is differently deflected. (c) Each part of the aberrated wavefront (gray) is tilted by $\varphi(x_{pi}, y_{pi})$ with regard to the reference wavefront (green) so that it is not centrally focused anymore. Instead, the partial wave is focused onto a point which is displaced by $\Delta L(x_{pi}, y_{pi})$. d denotes the diameter of one microlens and the corresponding CCD sensor segment (see Problem P5.2). Adapted from [9].

to the reference wavefront (Figure 5.23c). The microlenses now refract the corresponding wavefront parts to different angles so that the spot pattern on the detector becomes irregular (inset of Figure 5.23b). The CCD camera acquires the local shifts $\Delta L(x_{pi}, y_{pi})$ of the spots relative to the reference spot positions for all microlenses. If we assume only small deviation angles, the shift $\Delta L(x_{pi}, y_{pi})$ of the ith spot is proportional to the respective ray deflection $\varphi(x_{pi}, y_{pi})$ in the pupil plane. So, we have

$$\tan\varphi(x_{pi}, y_{pi}) \approx \varphi(x_{pi}, y_{pi}) = \frac{\Delta L(x_{pi}, y_{pi})}{f'}\,, \tag{5.15}$$

where f' is the image-side focal length of a microlens. The local values of the ray deviation $\varphi_x(x_{pi}, y_{pi})$ and $\varphi_y(x_{pi}, y_{pi})$ represent directly the local slope of the wave

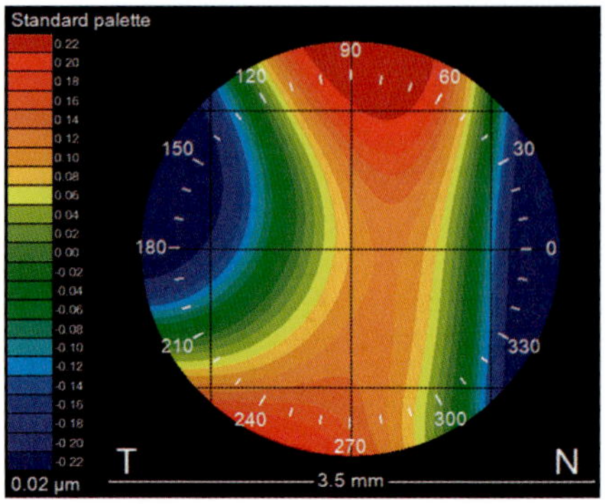

Figure 5.24 Example of a reconstructed wave aberration function determined with a Hartmann–Shack aberrometer (ZEISS i.Profiler®plus). The advancement and retardation of the calculated wavefront compared to a reference wavefront is represented by color codes (values shown in color bar). "T" denotes the temporal and "N" the nasal side of the patient's pupil. The azimuthal scale around the wavefront is used to determine the orientations of the principal planes in the case of astigmatism. Courtesy of Carl Zeiss.

aberration function (Section 5.3.2). From these derivatives, standard algorithms (Section 5.4) then allow reconstruction of the shape of the wave aberration function over the entire pupil plane (Figure 5.24).

Accuracy, sensitivity, and dynamic range The smaller the diameter of the individual microlens, the more accurately the wavefront shape can be reconstructed.[26] If the acquisition quality of the wavefront shape is considered to be the *sampling resolution* or *wavefront sampling accuracy*, this parameter is inversely proportional to the microlens diameter or, alternatively, directly proportional to the number of microlenses in the array. Another important parameter of the HS-WFS *measurement sensitivity* is the smallest detectable angular deviation $\varphi_{\min}$. We may calculate $\varphi_{\min}$ from the smallest detectable spot shift $\Delta L_{\min}$ with

$$\varphi_{\min} \approx \frac{\Delta L_{\min}}{f'} \,. \tag{5.16}$$

The smallest detectable spot shift $\Delta L_{\min}$ depends on the spot diameter on the CCD sensor as well as the pixel size, the sensitivity, and signal-to-noise ratio of the CCD. In addition, the quality of the evaluation algorithms and the scattering of light in the eye play an important role. Special evaluation algorithms are used to determine the center of the blurry spot on the sensor. They may actually fail when the wavefronts are considerably deformed such that the spots of adjacent microlenses overlap (or even cross). To obtain consistent measurement results, the slope of the wavefront must not exceed $\varphi_{\max}$ in any microlens area. If the spot size is assumed

26) Because of potential diffraction effects, we cannot make the diameter of the microlenses too small!

to be very small, then we obtain from Figure 5.23c

$$\varphi_{\max} \approx \frac{\Delta L_{\max}}{f'} = \frac{d/2}{f'} \,. \tag{5.17}$$

Equation (5.17) defines the *dynamic range* of the HS-WFS. Dynamic range, wavefront sampling accuracy, and measurement sensitivity are interconnected and can thus not be optimized independently. If, for example, the dynamic range is increased with microlenses of larger diameter d (or by reducing focal length f), this will reduce the wavefront sampling accuracy. To resolve this issue, the requirement on the dynamic range is reduced by precompensating the defocus of the patient's eye. In practice, this is done by changing the relative distance of the lenses in the (relay) lens system. As defocus usually contributes to the total eye aberration, the required dynamic range decreases in favor of higher wavefront sampling accuracy (Problem P5.2).

5.3.5
Ingoing Light Aberrometers

Let us consider the basic designs for ingoing light aberrometers. The *laser ray tracing aberrometer* (Section 5.3.5.1) and the *Tscherning aberrometer* (Section 5.3.5.2) measure the transversal ray deviations of an external light source.

5.3.5.1 Laser Ray Tracing Aberrometer

In a ray tracing aberrometer, a collimated infrared laser beam with a diameter of approximately 0.3 mm is displaced by an xy scanner over the eye's entrance pupil parallel to the line of sight. The different entry positions of the beam in the pupil plane are recorded (see "ingoing spot diagram" in Figure 5.25).

For an ideal (aberration-free) eye, a test beam always intersects the retina at the same point as the line of sight, no matter where the beam enters the eye pupil. In an aberrated eye, the incident rays are differently refracted depending on their entry positions. Thus, a specific pattern of fundus reflexes is formed which is illustrated as a *retinal spot diagram* (see "retinal spot diagram" in Figure 5.25). The pattern is imaged onto a spatially resolving detector (e.g., a CCD camera) by means of ophthalmoscopy lens L. Note that the aberrations of the examined eye do not contribute twice to the measuring result. Aberrations only affect the imaging quality of the retinal spots in the sensor plane, but not the centroid locations of the spots on the retina. During analysis, the deviation of each position is compared to a reference ray. This data then allows the calculation of the local slope of the wavefront and the total eye aberration.

Laser ray tracing aberrometers have a large dynamic range, as each spot is separately detected and analyzed. However, this step-by-step approach makes the measurement much slower than parallel-operating aberrometers and thus more sensitive to eye motion. To date (in 2013), only the iTrace™ aberrometer by Tracey Technologies is based on the laser ray tracing principle.

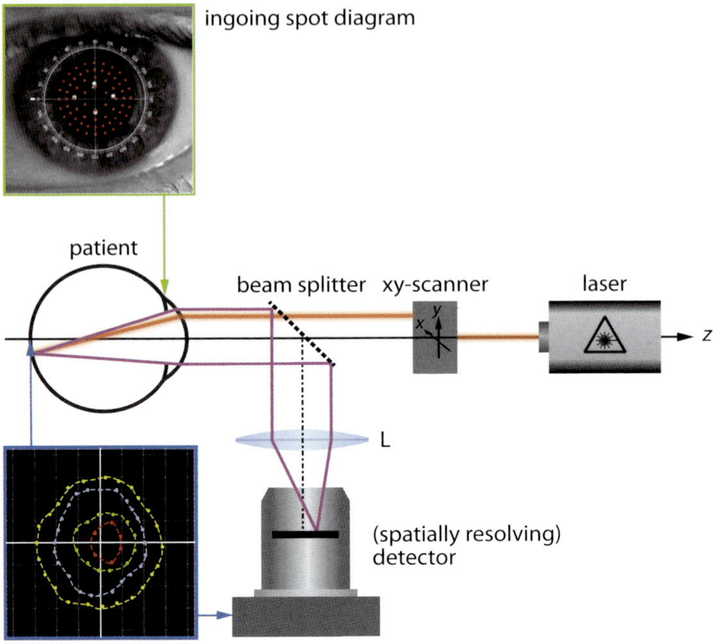

Figure 5.25 Setup of a laser ray tracing aberrometer. A collimated infrared laser beam is displaced by an *xy* scanner over the eye's entrance pupil parallel to the line of sight. The different entry positions of the beam in the pupil plane are recorded and displayed in an ingoing spot diagram (upper inset). For an ideal (aberration-free) eye, the test beam is always imaged to the same position on the retina, no matter where the beam enters the eye pupil. In an aberrated eye, the incident beams are differently refracted depending on their entry positions. The retinal spot diagram (lower inset) illustrates the corresponding retinal spot positions. This pattern is imaged onto a spatially resolving sensor (e.g., a CCD camera) by means of ophthalmoscopy lens L. Adapted from [8].

5.3.5.2 **Tscherning Aberrometer**

Tscherning[27)] aberrometers (Figure 5.26a) use a fixed two-dimensional array of test beams which are simultaneously guided to the patient's eye. The partial beams are generated by means of a pinhole mask (called *Tscherning* or *Hartmann screen*) which is placed in the path of an expanded laser beam. To obtain a sufficiently large retinal spot diagram, the pinhole mask is imaged by an aberroscope lens into a plane right in front of the retina. For this purpose, the optical system must be able to compensate potential defocus of the patient's eye. The spots on the retina are then imaged onto the sensor surface of a spatially resolving detector (e.g., CCD camera) by means of an ophthalmoscopy lens system[28)]. Aperture stop A is placed in the image plane of the eye pupil in order to improve image quality and thus allow

27) Marius Tscherning (1854–1939).

28) In a small area around the eye's pupil center, it is not possible to measure aberrations, as this area is used by the ophthalmoscopy lens system for imaging.

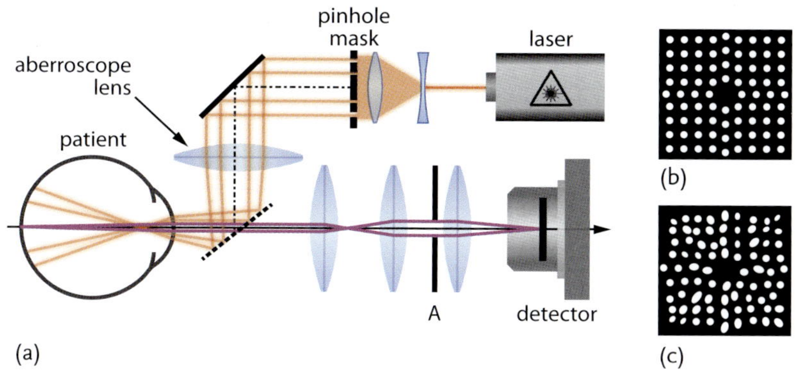

Figure 5.26 (a) Setup of a Tscherning aberrometer. An expanded (collimated) laser beam crosses a pinhole mask so that a two-dimensional array of thin test beams is created (as shown in (b)). To obtain a sufficiently large retinal spot diagram, the pinhole mask is imaged by an aberroscope lens into a plane right in front of the retina (strongly exaggerated in this scheme). The spots on the retina are then imaged onto the sensor surface of a CCD camera by means of an ophthalmoscopy lens system. An aperture stop A is added to the ophthalmoscope (observation) subsystem such that it improves the image quality and thus makes the centroid detection robust. (b) Pinhole mask. (c) Detector image obtained for an eye with aberrations.

a robust detection of the spot centroid. During data analysis, the ray deviations are determined from the positions of the displaced spots on the detector as compared to the reference positions (Figure 5.26b,c).

In contrast to laser tracing aberrometers, the Tscherning aberrometer allows parallel processing of all retinal spots. The measurement is thus faster and less sensitive to eye movements compared to laser ray tracing aberrometers. However, Tscherning aberrometers have a smaller dynamic range due to the spot crossing issue, which we already discussed in the context of Hartmann–Shack aberrometers in Section 5.3.4.

5.3.6 Commercial Aberrometers

In almost all commercially available aberrometers, a HS-WFS is used. Some commercially available systems are combined, multifunctional devices. Figure 5.27 shows photographs of aberrometers by Carl Zeiss and Topcon.

5.4 Wavefront Reconstruction and Wavefront Analysis

The aberrometers discussed in Section 5.3 measure the wavefront slopes $\mathrm{d}\mathcal{W}/\mathrm{d}x_\mathrm{p}$ and $\mathrm{d}\mathcal{W}/\mathrm{d}y_\mathrm{p}$ in the $x_\mathrm{p}y_\mathrm{p}$ pupil plane of the eye (see Eqs. (5.13) and (5.14)). This raw data then has to be analyzed by adequate methods [10] to reconstruct the wave

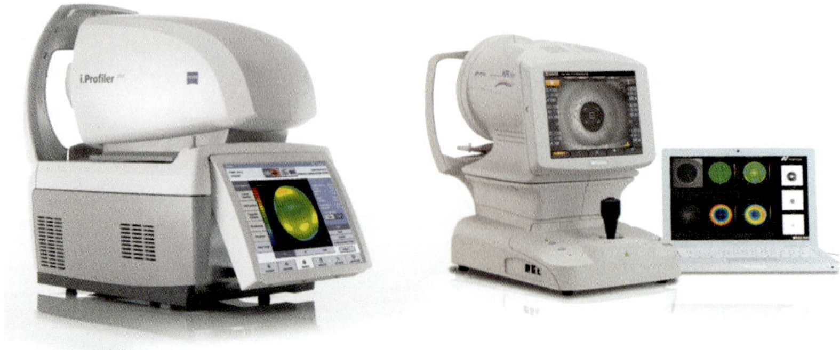

(a) (b)

Figure 5.27 Photographs of commercially available aberrometers which both use a Hartmann–Shack wavefront sensor. (a) Photograph of ZEISS i.Profiler. Courtesy of Carl Zeiss. (b) Photograph of Topcon KR-1W. Courtesy of Topcon Deutschland GmbH.

aberration function $\mathcal{W}$. With the method of least squares (see Info Box 5.2), the measured dataset of wavefront slopes is fitted to a set of suitable basis functions V_j. In the next step, corresponding coefficients c_j are calculated to obtain the best fit for $\mathcal{W}$. Hence, we have

$$\mathcal{W}(x, y, x_{\mathrm{p}}, y_{\mathrm{p}}) = \sum_{j=0}^{N} c_j \, V_j(x, y, x_{\mathrm{p}}, y_{\mathrm{p}}) \, , \tag{5.18}$$

where N denotes the number of used basis functions, and (x, y) is a coordinate in the object plane from which the wavefront has emerged. In ophthalmology, it is common to use Zernike polynomials $\mathcal{Z}_n^m$ (detailed description in Section A.1.8.2) as basis functions. Alternatively, we could also select other basis functions such as Taylor (discussed in Section A.1.8.1) or Seidel[29] polynomials, but Zernike polynomials are more practical for a couple of reasons [10, 11]:

1. The complex wavefront error can be subdivided into the "standard" imaging errors such as defocus, astigmatism, coma, spherical aberration, and so on.[30]
2. Zernike polynomials allow us to classify the different types of monochromatic aberration by different orders.
3. Zernike polynomials are orthonormal on a circular area. This means that the coefficients c_n^m determined for the best fit do not change, regardless of which or how many polynomials are used for the approximation.
4. All terms of the polynomial (with the exception of the term describing piston) have a mean value of zero.

Since Zernike polynomials are best suited to describe aberrations in the *circular* pupil plane, we replace the Cartesian coordinates by the ray's radial distance from the optical axis r and the azimuthal angle α (i.e., polar coordinates). In accordance

29) Philipp von Seidel (1821–1896).
30) In Section A.1.6, other kinds of aberration are explained as well.

with Section A.1.7, the wavefront expansion Eq. (5.18) then reads

$$\mathcal{W}(h, r, \alpha) = \sum_n \sum_{m=-n}^{n} c_n^m(h) \mathcal{Z}_n^m(r, \alpha) . \tag{5.19}$$

According to the ISO 24157 standard for reporting aberrations of the human eye [12], the recommended notations of the Zernike polynomials are $\mathcal{Z}_n^m$ or $\mathcal{Z}(n, m)$ and for the normalized Zernike coefficients c_n^m or $c(n, m)$, respectively.

In addition to the standard double indexing, Zernike coefficients can also be identified with just one index j. We may replace the indices n and m by using the relation

$$j = \frac{n^2 + 2n + m}{2} . \tag{5.20}$$

The single-index notation is simple to use, but the double-index notation is more meaningful. The radial index n indicates the order of the Zernike polynomial. The azimuthal index m tells us the character of a polynomial term. For example, all terms with index $m = \pm 2$ are related to astigmatism and, in general, terms with the same index m have a similar shape (Figure A.23 in Section A.1.8.2). In the following, we will thus primarily use the double-index notation.

From Eq. (5.19), it follows that with the best-fit set of coefficients c_n^m one can describe the wavefront error to any desired accuracy (depending on the number of coefficients used in the fit). But it is often more practical to derive a set of measurable quantities, called *metrics*, which describe eye aberrations in a more illustrative manner.

Info Box 5.2: Method of Least Squares

The method of least squares is a standard mathematical tool for fitting a curve to a set of measured data. This method was first described by Carl Friedrich Gauss (1777–1855) around 1794. The "best fit" is obtained as follows. Firstly, the differences between the measured values a_i and the corresponding values of a suitable model b_i are calculated. The index i counts the N measured points. The differences are called the *residuals* r_i, and we thus have $r_i = a_i - b_i$. The best fit results from the minimized sum of the squared residuals, that is,

$$\min\left(\sum_{i=0}^{N} r_i^2\right) = \min\left(\sum_{i=0}^{N} (a_i - b_i)^2\right) . \tag{5.21}$$

5.4.1
From Wavefront to Refraction (Wavefront Analysis)

The wave aberration function $\mathcal{W}(h, r, \alpha)$ contains all the information about the eye's monochromatic aberrations. However, suitable metrics are more practical for applications of wavefront analysis. The commonly used metrics can be categorized into *pupil plane metrics* (Section 5.4.1.1) and *image plane metrics* (Section 5.4.1.2). As the name suggests, pupil plane metrics describe the wavefront properties in the pupil plane. Image plane metrics refer to the impact of wavefront aberrations on the image quality in the retinal plane. The latter also take interactions between different types of aberrations (within the pupil plane) into account.

In the following, we will briefly discuss a few suitable metrics. Further information about this topic can be found in the literature see [10, 13–16].

5.4.1.1 Pupil Plane Metrics

Root mean square error The most frequently used pupil plane metric is the *root mean square wavefront error* which is defined by Eq. (A35). As the Zernike polynomials are orthonormal, the root mean square wavefront error can simply be calculated from the individual coefficients c_n^m via

$$\mathrm{RMS_{wfe}} = \sqrt{\sum_{n>1,m} \left(c_n^m\right)^2}\,. \tag{5.22}$$

Here, we have neglected the aberrations piston ($n = 0$, $m = 0$) and tilt ($n = 1$ and $m = \pm 1$), as they have no impact on the image quality (aside from an image shift). The maximum value of n is determined by the highest order of the Zernike polynomial used for the wavefront fitting. The total $\mathrm{RMS_{wfe}}$ indicates how strongly the measured wavefront deviates from the reference wavefront.

With Eq. (5.22), we can also identify the individual contribution of selected terms (types of aberration) on the total wavefront error. For example, the contribution of HOAs can be calculated for $n > 2$. In aberrometer reports, the total $\mathrm{RMS_{wfe}}$ as well as the RMS values for LOA and HOA are usually provided additionally to the Zernike coefficients c_n^m (Figure 5.28).

Equivalent defocus Another useful pupil plane metric is the *equivalent defocus* (also called *spherical equivalent error*; see also Section 3.1.3) which reads

$$M_\mathrm{e} = \frac{4\sqrt{3}}{r^2}\sqrt{\sum_{n>1,m}^{n_\mathrm{max}} \left(c_n^m\right)^2}\,. \tag{5.23}$$

r is the radius of the eye's entrance pupil and n_max the highest considered order of the Zernike polynomial. M_e is stated in diopters if $[c_n^m] = \mu\mathrm{m}$ and $[r] = \mathrm{mm}$.

The equivalent defocus can be calculated for total aberrations and/or just for one or more HOAs. In the latter case, it represents the amount of defocus required to obtain the same $\mathrm{RMS_{wfe}}$ as realized by one or more HOAs. The concept of equivalent defocus is clinically useful, because it indicates the order of magnitude of HOAs in the familiar diopter unit.

Spherocylindric refraction values: From the Zernike coefficients (Table A.3), it is possible to calculate the spherocylindric refraction. To retrieve the spherocylindric

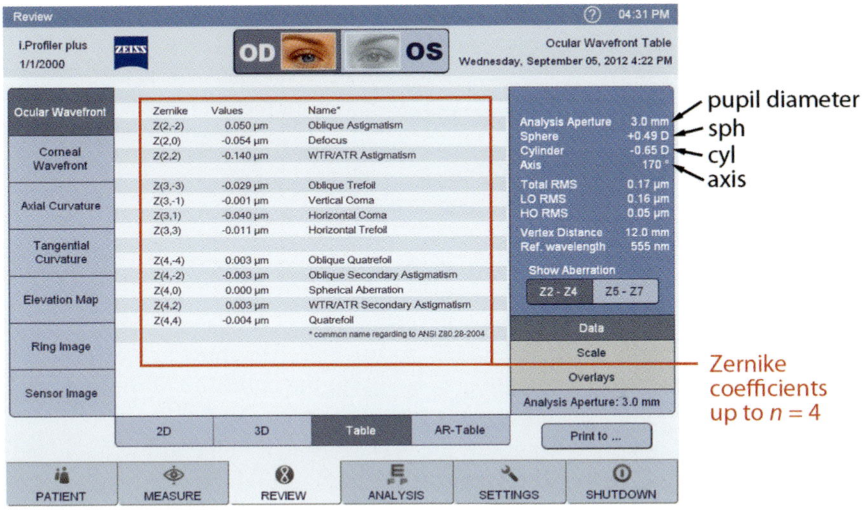

Figure 5.28 Screenshot of the evaluation window for a right test eye (ZEISS i.Profiler). In the red box, the determined Zernike coefficients are listed in micrometers up to the order $n = 4$. The values for sph, cyl, axis (corrected for the given vertex distance), and the pupil diameter are shown as well. "OD" is the abbreviation for *oculus dexter* (right eye) and "OS" the abbreviation for *oculus sinister* (left eye). Courtesy of Carl Zeiss.

refraction from a measured aberration, the aberrated wavefront is approximated by a spherocylindric surface in the entrance pupil plane.[31] For this purpose, the *least-squares fitting* or the *paraxial curvature fitting* method can be used [16].

The goal of least-squares fitting is to minimize the sum of squared deviations between the aberrated wavefront and the spherocylindric surface within the pupil area. By this method, the deviations over the whole pupil are minimized at the expense of possible deviations in the pupil center. Since the method of least squares is used for the Zernike expansion of wavefronts and the Zernike polynomials are orthonormal functions, the best-fit surface is given by second-order Zernike coefficients ($n = 2$) only. These coefficients can be converted to spherocylindric refraction values in power vector notation (Section 3.1.3) which reads

$$M = \frac{-c_2^0 4\sqrt{3}}{r^2}\,, \tag{5.24}$$

$$J_0 = \frac{-c_2^2 2\sqrt{6}}{r^2}\,, \tag{5.25}$$

$$J_{45} = \frac{-c_2^{-2} 2\sqrt{6}}{r^2}\,. \tag{5.26}$$

r is the radius of the entrance pupil analyzed by the aberrometer. The refractive data in power notation can be easily transformed into the commonly used minus

31) The refraction values obtained from Zernike coefficients refer to the entrance pupil plane. If the refractive error is to be corrected at another vertex distance, these values must be converted. The influence of the vertex distance on the refraction values is discussed in Section 5.5.

cylinder or plus cylinder notations. In minus cylinder notation, we have

$$\mathrm{sph} = -\frac{4\sqrt{3}(c_2^0)}{r^2} + \frac{\mathrm{cyl}}{2}\,, \tag{5.27}$$

$$\mathrm{cyl} = -\frac{4\sqrt{6}\sqrt{(c_2^{-2})^2 + (c_2^2)^2}}{r^2}\,, \tag{5.28}$$

$$\mathrm{axis} = \frac{1}{2}\arctan\left(\frac{(c_2^{-2})}{(c_2^2)}\right)\,, \tag{5.29}$$

where [sph] = [cyl] = D in the case of $[c_n^m] = \mu\mathrm{m}$ and $[r] = \mathrm{mm}$.

The goal of *paraxial curvature fitting* is to match the curvature of the aberrated wavefront and the spherocylindric surface in the pupil center so that a common reference point is obtained. Logically, the wavefront deviations in the pupil center are minimized. However, deviations at the pupil margin appear according to the Seidel criterion for emmetropia.[32] The corresponding values in power vector notation read

$$M = \frac{-c_2^0 4\sqrt{3} + c_4^0 12\sqrt{5} + c_6^0 24\sqrt{7} + \ldots}{r^2}\,, \tag{5.30}$$

$$J_0 = \frac{-c_2^2 2\sqrt{6} + c_4^2 6\sqrt{10} - c_6^0 12\sqrt{14} + \ldots}{r^2}\,, \tag{5.31}$$

$$J_{45} = \frac{-c_2^{-2} 2\sqrt{6} + c_4^{-2} 6\sqrt{10} - c_6^{-2} 12\sqrt{14} + \ldots}{r^2}\,. \tag{5.32}$$

Again, refractive data in power notation can be easily transformed into the commonly used minus cylinder or plus cylinder notations. The paraxial curvature fitting method predicts the results of subjective refraction more accurately than the least squares method, since additional terms are included. Furthermore, according to the Stiles–Crawford effect (Section 2.1.7), inner portions of the pupil are preferred to the pupil margin [16].

Figure 5.28 depicts a screenshot of the results of a wavefront measurement. The normalized Zernike coefficients up to the order $n = 4$ are presented, which form the basis for calculated metrics. As the normalized coefficients refer to a unit circle, the pupil diameter must also be stated. On the left-hand side, the corresponding metrics calculated with Eqs. (5.27)–(5.29) are listed, already corrected for the given vertex distance of eye glasses (12 mm; Problem P5.3).

32) The Seidel criterion for emmetropia states that the wavefront must be flat in the paraxial domain near the pupil center and may be warped elsewhere.

5.4.1.2 Image Plane Metrics

Pupil plane metrics are reasonable parameters to quantify the overall wavefront error and the relative share of the individual types of aberration. However, they only provide indirect information about the impact of the individual Zernike terms on the eye's image quality (Section 5.4.2). For example, a corrective lens with the optical parameters determined from the values given by Eqs. (5.27)–(5.29) (and after an appropriate vertex distance correction) will certainly remove LOAs. But this does not automatically mean that the patient's vision is perfectly optimized. HOAs may also have a considerable impact.

To address this issue, a number of image plane metrics have been developed which can be mostly derived from the *point-spread function* PSF (Sections A.1.5 and A.1.7). In short, the PSF describes how aberrations modify the intensity distribution of the image of an object point. It is directly related to the wave aberration function via (A37). If the intensity distribution of an object $O(x, y)$ is convolved (see Info Box 5.3) with $PSF(x', y')$, we obtain the intensity distribution of the image $I(x', y')$. A corresponding simulation is shown in Figure 5.29. Here, we used the letter "E" as a test object and the point-spread functions of an arbitrary imaging error (Figure 5.29a), coma (Figure 5.29b), and spherical aberration (Figure 5.29c). The resulting images are shown in the right column.

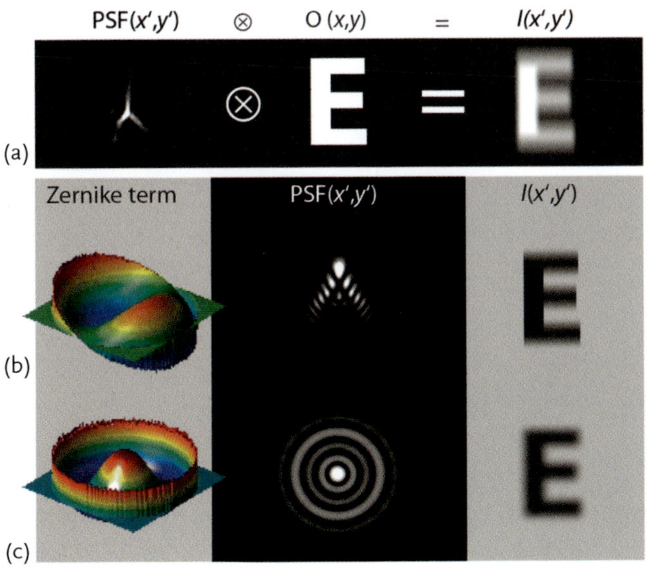

Figure 5.29 (a) When the point-spread function PSF in the $x'y'$ image plane is convolved with the intensity distribution of object $O(x, y)$, we obtain the shape of intensity distribution of the image $I(x', y')$. By example, we simulate $I(x', y')$ for two Zernike terms which represent (b) coma and (c) spherical aberrations. $PSF(x', y')$ and simulated image of the letter "E" (as it would be seen by the patient) are shown. Adapted from [17].

Info Box 5.3: Convolution

Convolution is a mathematical operation which measures the "overlap" between two functions as one ($f(x)$) is shifted across the other ($g(x)$). In mathematical terms, the convolution is defined as

$$(f \otimes g)(x) = \int_{-\infty}^{\infty} f(\tau)g(x-\tau)\mathrm{d}\tau = \int_{-\infty}^{\infty} f(x-\tau)g(\tau)\mathrm{d}\tau \,. \tag{5.33}$$

The *cross-correlation* is similar in nature to the convolution of two functions. A special case is given by the *autocorrelation* (Section A.2.4.2) for which the function $f(x)$ is convolved with a shifted version of itself.

5.4.2 Applications of Wavefront Analysis

Refraction measurements Aberrometers determine eye aberrations with high accuracy. In fact, these instruments are also very accurate autorefractors as standard spherocylindric refraction values (sph, cyl, and axis) can be directly extracted from the measured results via Eqs. (5.19)–(5.29). In all aberrometers, the measured wavefront aberration is quantified with image plane metrics, and the resulting retinal image quality is provided by means of simulations (Figures 5.29). We should, however, keep in mind that the real visual performance also depends on chromatic aberrations (Section A.1.9) and the neuronal processing of the retinal image. Thus, no matter how precise the aberrometer measurements are, they cannot replace subjective refraction methods. But aberrometers certainly improve the existing objective methods. As they also take HOAs into account, optimized correction values can be calculated so that the retinal image quality is improved even under unfavorable conditions (e.g., twilight/mesopic vision). The first commercially available eyeglass lenses optimized on the basis of a wavefront measurement and analysis (ZEISS i.Scription®) were introduced in 2007 by the company Carl Zeiss.

Refractive surgery The most recent generation of aberrometers are multifunctional devices that analyze the imaging properties of the eye as a whole and all its refractive parts individually (to some extent). In addition to a "pure" wavefront analyzer, a corneal topography system is often included to determine the shape of the corneal surface (Sections 6.3.2 and 6.5). Such devices can thus also determine the contribution of the corneal front surface to the overall wavefront aberration. Once the total and the corneal aberrations are known, the error contributions of the corneal back surface and the eye lens follow directly by simple subtraction.

For example, in planning refractive surgeries (Chapter 10), these individual contributions to eye aberrations are essential. The data from the wavefront analysis serves as a baseline for the calculation of the ablation profile (i.e., wavefront-guided or wavefront-supported refractive surgery; Section 10.3).

For lens-based refractive surgeries (e.g., cataract surgery or refractive lens exchange), the data of the wavefront analysis can be used to optimize the intraocular lens to attain the best quality of vision. Once the refractive surgical procedure has been completed, wavefront analysis is used to check the quality of the optical correction achieved (Problem P10.16).

High-resolution fundus imaging When imaging devices such as fundus cameras (Section 6.7), confocal scanning-laser ophthalmoscopes (Section 6.8.1), or optical coherence tomographs (Section 7.6) are used to characterize the patient's eye, we always face the problem that the eye itself is a component of the whole optical system. Potentially existing eye aberrations thus impair the transversal resolution of the imaging devices. If, however, the measured aberrations of the eye are corrected by means of adaptive optical systems, the transversal resolution can be significantly improved up to the μm range.

5.5 Excursus: Refractive Correction with Eye Glasses and Contact Lenses

Refractive corrections aim to compensate low-order aberrations of an ametropic eye such that it effectively becomes emmetropic. In principle, two approaches exist to achieve this aim, that is,

1. adding corrective lenses in front of the eye (e.g., eye glass lenses or contact lenses) or
2. modifying the eye's refractive power by refractive surgery.

For now, we restrict ourselves to method (1); refractive surgery is then discussed in detail in Chapter 10. As mentioned in Section 3.1, an eye is ametropic if the far point Q_{far} is *not* located at infinity for a relaxed eye (no accomodation). In this case, the refractive error can be compensated by a corrective lens which is located at a distance L_c away from the corneal vertex. The corrective lens shall have a refractive power of $\mathcal{D}'_{corr} = 1/f'_c$, such that an object at optical infinity is imaged to the far point of the eye.

In optometry, $\mathcal{D}'_{corr}$ is replaced by the back vertex power $\mathcal{D}'_v = 1/f'_{bv,c}$ ($f'_{bv,c} \neq f'_c$), that is, the inverse distance from the back vertex of the lens to its image-side focus. In fact, this definition has practical advantages. For example, we consider two different corrective lenses with the same back vertex power $\mathcal{D}'_v$ which are fitted at the same distance from the corneal vertex. Both lenses have the same corrective effect even if they differ considerably in form and/or thickness. If the far point of the eye Q_{far} is located at a distance s_{far} from the corneal vertex (Figure 5.30), the required back vertex focal length $f'_{bv,c}$ of the corrective lens is given by

$$f'_{bv,c} = s_{far} + L_c \,, \tag{5.34}$$

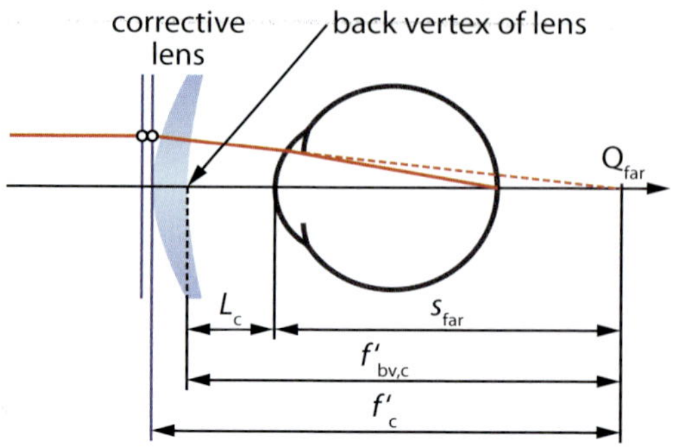

Figure 5.30 Refractive correction of a hyperopic eye with a corrective lens. Q_{far} denotes the far point and s_{far} the far point distance. L_c is the distance from the back vertex of the corrective lens to the corneal vertex. $f'_{bv,c}$ is the image-side back vertex focal length of the corrective lens.

whereas the vertex distance L_c is always positive. As L_c may vary and s_{far} is constant, $f'_{bv,c}$ generally depends on L_c. For the required back vertex power, we thus obtain

$$\mathcal{D}'_v(L_c) = \frac{1}{f'_{bv,c}(L_c)} = \frac{1}{s_{far} + L_c} . \tag{5.35}$$

The required refractive power of the corrective lens is determined by subjective refraction methods at which trial lenses with different $\mathcal{D}'_v$ are placed in front of the eye at a distance L_{c1}. The physician then finds out which back vertex power $\mathcal{D}'_v(L_{c1})$ is needed to achieve the best visual acuity (Section 2.1.5.1). Often, the distance L_{c1} of the trial lenses is not equal to the distance L_{c2} of the corrective lenses used for the actual eye glasses ($L_{c1} \neq L_{c2}$). In this case, the back vertex power must be adapted according to

$$\mathcal{D}'_v(L_{c2}) = \frac{\mathcal{D}'_v(L_{c1})}{1 + \mathcal{D}'_v(L_{c1}) \cdot (L_{c2} - L_{c1})} . \tag{5.36}$$

Example 5.1

Adaption of the back vertex power From an examination with trial lenses, we obtain a back vertex power of $\mathcal{D}'_v(L_{c1}) = -5$ D, whereas the lenses are located at a distance of $L_{c1} = 20$ mm. The patient gets a spectacle frame at which the distance between the back vertex of the corrective lens and corneal vertex is now $L_{c2} = 12$ mm. As a consequence, the back vertex power must be adapted to

$$\mathcal{D}'_v(L_{c2}) = \frac{-5\,\text{m}^{-1}}{1 + (-5\,\text{m}^{-1}) \cdot (0.012\,\text{m} - 0.020\,\text{m})} = -4.8\ \text{D} . \tag{5.37}$$

For the eye glasses, the required back vertex power thus decreases by 0.2 D.

The refractive power of eye glass lenses is usually available in interval steps of 0.25 D. Thus, only refractive power differences of $|\mathcal{D}'_{\mathrm{v}}(L_{\mathrm{c2}}) - \mathcal{D}'_{\mathrm{v}}(L_{\mathrm{c1}})| \geq 0.125$ D are relevant. Such differences, however, only appear if the absolute value of $\mathcal{D}'_{\mathrm{v}}(L_{\mathrm{c1}})$ is high and/or the distance between lens and corneal vertex changes considerably, for example, if contact lenses are used. In the case of contact lenses, we have $L_{\mathrm{c2}} = 0$ so that the back vertex power would be reduced by 0.5 D in the above mentioned Example 5.1.

For cylindric eye glass lenses, the refractive correction must be processed for both principal meridians individually.

5.6 Recommended Reading

For further information about the determination of the eye's refractive status, please refer to the following references:

- Retinoscopy (Section 5.1): [1, 2, 18]
- Automated Objective Refractometers (Section 5.2): [2, 19]
- Aberrometers/Wavefront Analyzers (Section 5.3): [9, 10, 20, 21]
- Wavefront Reconstruction and Wavefront Analysis (Section 5.4): [12, 14, 15, 22, 23]

5.7 Problems

P5.1. Retinoscope

1. Show that the accuracy of a retinoscope can be determined by the equation

$$\Delta\mathcal{D} = \pm\frac{\lambda}{d_{\mathrm{pupil}}^2} . \tag{5.38}$$

2. Derive the Newton formula

$$zz' = f_{\mathrm{L}} f'_{\mathrm{L}} \tag{5.39}$$

from the thin lens equation (A12).

P5.2. Hartmann–Shack wavefront sensor

1. The wavefront aberrations for defocus and spherical aberration at a wavelength of $\lambda = 550$ nm of a collimated beam with a diameter of 3 mm are to be measured with a Hartmann–Shack wavefront sensor. The CCD detector has a pixel size of 7 µm. With suitable algorithms, the centroid can be determined to within 1/20th of a pixel. Which focal length should a microlens have to allow

the determination of defocus as a Zernike coefficient down to $\lambda/20$? How accurately can the Zernike coefficient of the aperture error be determined for this focal length? What causes this difference?

2. If the finite size of the detector elements is taken into consideration, then in the case of standard evaluation algorithms for the centroids, the dynamic range of the sensor is given by a spot leaving the surface assigned to the detector element on the sensor. In a simple geometric image, what is the extent of the maximal measurable defocus of the above sensor if $N = 30$ elements are assumed across the beam diameter and the fill factor is set to 1?
   If you assume that the lenses are diffraction-limited in the small elements, then finite-sized spots are obtained. What is the above-calculated maximal defocus when taking diffraction into consideration? Remember that the detector elements are squared (Figure 5.23).
3. Discuss the influence of different coherence states of the incident signal wave on the measuring result. What happens if the CCD sensor is positioned exactly in the focal plane of the microlens array? What is the effect of using various wavelengths? What happens at the edge of a sharply limited wave to the signal of partially illuminated detector elements? How can this problem be solved in practical applications if the intensity of the waves is constant?

P5.3. Aberrometry Figure 5.28 shows the ocular wavefront table measurement as obtained by a ZEISS i.Profiler. In the red box, the determined Zernike coefficients for an eye with a pupil diameter of 3.4 mm (analysis aperture) are listed in micrometers up to the order $n = 4$. On the left, the calculated metric values are shown:

1. Calculate the spherocylindric refraction values (polar notation) for a vertex distance of 12 mm.
2. Calculate the root mean square wavefront error RMS_{wfe} for the lower-order aberrations ($n = 2$), higher-order aberrations ($n > 2$), and the total RMS_{wfe}.

References

1 Campbell, C.E., Benjamin, W.J., and Howland, H.C. (2006) Objective refraction: Retinoscopy, autorefraction, and photorefraction, in *Borish's Clinical Refraction*, Butterworth–Heinemann/Elsevier, pp. 682–752, 2nd edn.

2 Diepes, H. (2004) *Refraktionsbestimmung*, DOZ Verlag Optische Fachveröffentlichungen GmbH, 3rd edn.

3 Grehn, F. (2008) *Augenheilkunde*, Springer.

4 Benjamin, W.J. (2006) *Borish's Clinical Refraction*, Butterworth–Heinemann/Elsevier, 2nd edn.

5 Rassow, B. (ed.) (1987) *Ophthalmologisch-optische Instrumente*, Enke, Stuttgart.

6 Wesemann, W. (2004) Funktionsprinzipien und Messgenauigkeit moderner Autorefraktometer. *Dtsch. Opt. Z.*, **10**, 38–44.

7 Buscemi, P.M. (2003) The Nidek OPD-Scan – Basic principles, technical features, data analysis, in *The Aberrometers*

– *Theory, Clinical and Surgical Applications*, Fabiano SRL, Canelli, pp. 269–275.
8 Rozema, J.J., Dyck, D.E.V., and Tassignon, M.J. (2005) Clinical comparison of 6 aberrometers. Part 1: Technical specifications. *J. Cataract Refract. Surg.*, **31**, 1114–1127.
9 Yoon, G. (2006) Wavefront sensing and diagnostic uses, in *Adaptive Optics for Vision Science: Principles, Practices, Design and Applications*, John Wiley & Sons, pp. 63–82.
10 Dai, G. (2008) *Wavefront Optics for Vision Correction*, SPIE Bellingham.
11 Gross, H., Zügge, H., Peschka, M., and Blechinger, F. (2007) *Handbook of Optical Systems – Correction of Optical Systems*, vol. 3, Wiley-VCH Verlag GmbH.
12 International Organization for Standardization (2008) Reporting aberrations of the human eye (ISO 24157:2008(e)).
13 Williams, D., Applegate, R., and Thibos, L. (2004) Metrics to predict the subjective impact of the eye's wave aberration, in *Wavefront Customized Visual Correction: The Quest for Super Vision II*, Slack Inc., Thorofare, pp. 78–84.
14 Lombardo, M. and Lombardo, G. (2010) Wave aberration of human eyes and new descriptors of image optical quality and visual performance. *J. Cataract Refract. Surg.*, **36**, 313–331.
15 Thibos, L. (2006) From wavefronts to refractions, in *Adaptive Optics for Vision Science: Principles, Practices, Design and Applications*, John Wiley & Sons, Inc, pp. 331–365.
16 Thibos, L.N., Hong, X., Bradley, A., and Applegate, R.A. (2004) Accuracy and precision of objective refraction from wavefront aberrations. *J. Vis.*, **4**, 329.
17 Roorda, A. (2004) A review of basic wavefront optics, *Wavefront Customized Visual Correction: The Quest for Super Vision II*, Slack Inc., Thorofare, pp. 9–17.
18 International Organization for Standardization (2006) Retinoscopes (ISO 12865:2006).
19 International Organization for Standardization (2010) Eye refractometers (ISO 10342:2010).
20 Atchison, D.A. (2005) Recent advances in measurement of monochromatic aberrations of human eyes. *Clin. Exp. Optom.*, **88**, 5–27.
21 Maeda, N. (2009) Clinical applications of wavefront aberrometry – A review. *Clin. Exp. Ophthalmol.*, **37**, 118–129.
22 Thibos, L.N., Himebaugh, N.L., and Coe, C.D. (2006) Wavefront refraction, in *Borish's Clinical Refraction*, Butterworth–Heinemann/Elsevier, pp. 765–789, 2nd edn.
23 American National Standards Institute (2004) Methods for reporting optical aberrations of eyes (ANSI Z 80.28-2004).
24 Grimm, W., Ucke, Ch., and Friedberg, D. (eds) (1978) *Strichskiaskopie*, Enke, Stuttgart.

6
Optical Visualization, Imaging, and Structural Analysis

In the previous chapter, we discussed various methods and devices of measuring the refractive status of the human eye. Equally important are methods and devices for the visualization, imaging and structural analysis of fine structural details of the anterior and posterior eye segments. This is required both as part of a pre- and/or postoperative diagnosis and during surgery. The challenge is to make small structures visible, sometimes under difficult illumination conditions. Hence, one has to find an optimal balance between resolution (down to some tens of a micrometer), image quality, field of view, and speed of data/image acquisition.

The scope of this chapter ranges from the more versatile visualization and imaging devices, such as medical magnifying systems (Section 6.1), surgical microscopes (Section 6.2), ophthalmoscopes (Section 6.6), slit lamps (Section 6.4), to the more dedicated and specialized instruments like keratometers (Section 6.3.1), corneal topographers (Section 6.3.2), scanning-slit projection devices (Section 6.5), fundus cameras (Section 6.7), and scanning-laser devices (Section 6.8).

6.1
Medical Magnifying Systems

Because of the fine and detailed eye structures, there is a need for magnified visualization in the ophthalmic surgical environment. The simplest form of corresponding magnifying systems are known as *visual aids* or *magnifier loupes*. In principle, a simple loupe (Section 6.1.1) can also be considered as a visual aid. However, as we will see, the optical demands in ophthalmic surgery require more complex optical systems, which can be considered as "head-worn microscopes". In fact, these are combinations of eye glasses with a Galilei[1)] or Kepler[2)] telescope system (Section 6.1.2).

1) Galileo Galiei (1564–1642).
2) Johannes Kepler (1571–1630).

Optical Devices in Ophthalmology and Optometry, First Edition. Michael Kaschke, Karl-Heinz Donnerhacke, and Michael Stefan Rill.

6.1.1
Optics of a Single Loupe

Loupes are single lenses or lens systems which form a virtual image in the object space. They are used to magnify the viewing angle and thus increase the viewing resolution of nearby objects. The loupe and the eye of the observer together form an optical system. The loupe magnification thus depends on the relative positions of object (patient), loupe, and eye (observer). In general, the magnification of the loupe L is defined by the ratio of the two retinal image sizes with and without loupe $h''_{I,L}$ and h'_I, respectively. We thus have (Figure 6.1; Problem P6.1)

$$\beta_L = \frac{\tan \gamma'}{\tan \gamma} = -\mathcal{D}_L s_{ref} + A_{set} s_{ref}(1 - \mathcal{D}_L L_{LP}) \tag{6.1}$$

with the refractive power of the loupe $\mathcal{D}_L$ ($[\mathcal{D}_L] = D$), the reference viewing distance s_{ref}, the set refraction (degree of accommodation) of the observer's eye A_{set}, and the distance L_{LP} between the eye's principal point P and the principal point of the loupe P_L. γ' and γ denote the visual angles with and without a loupe, respectively. The

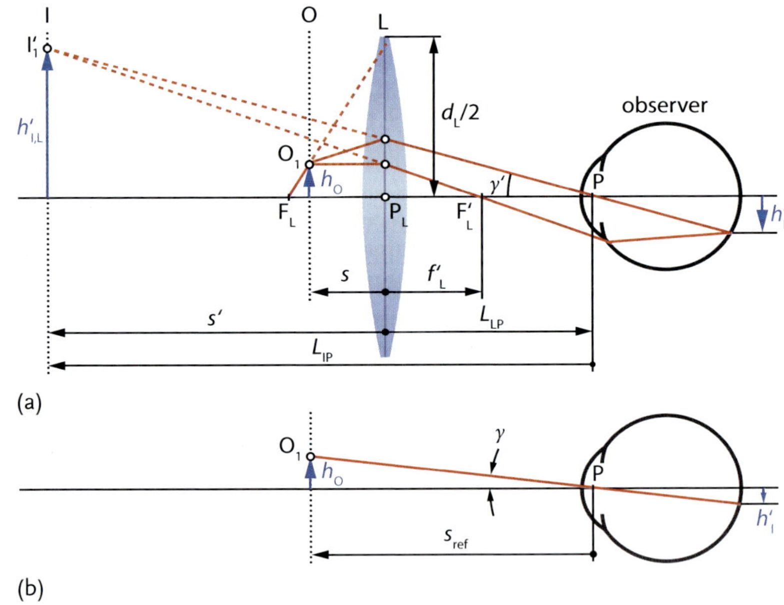

Figure 6.1 (a) Ray diagram illustrating the magnification of a simple magnifier loupe which consists of a single lens L. s and s' are the object and image distances, respectively. L_{LP} denotes the distance between the principal point of the eye (P) and principal point of the lens (P_L), f'_L the image-side focal length of the lens, and L_{IP} the distance between the principal point of the eye P and image plane I. $h'_{I,L}$ is the size of the virtual image and $h''_{I,L}$ the retinal image size. (b) Object observed without a lens. s_{ref} is the reference viewing distance.

Table 6.1 Parameter sets of aspheric magnifier loupes (see also Problem P6.1). $\mathcal{D}_L$ denotes the refractive power of the loupe, β_L the loupe magnification, $\beta_{L,nominal}$ the nominal loupe magnification, s the object distance, L_{LP} the distance between the principal points of the eye and loupe, d_L the lens diameter, and d_{fov} the field of view. Courtesy of Carl Zeiss.

		Optimum working condition				
$\mathcal{D}_L$ (D)	$\beta_{L,nominal}$	s (mm)	L_{LP} (mm)	β_L	d_L (mm)	d_{fov} (mm)
6	1.5	145	185	2.1	100	90
8	2.0	110	220	2.7	85	50
12	3.0	70	210	3.5	70	30
16	4.0	55	130	3.25	60	30
20	5.0	40	90	3.1	55	28

often specified nominal magnification of a loupe

$$\beta_{L,nominal} = \frac{\mathcal{D}_L}{4D} \tag{6.2}$$

is only valid if the reference viewing distance s_{ref} equals the typical near viewing distance $s_{nv} = 25$ cm (Section 2.1.9) and if either the object lies in the focal plane of the loupe or $f_L' = L_{LP}$. From Eq. (6.1) and Figure 6.1, another useful formula for the magnification can be derived in which only two distances (L_{LP}, s) have to be measured. In this case, we have

$$\beta_L = \frac{s_{ref}}{s - L_{LP}(1 + \mathcal{D}_L s)}\,, \tag{6.3}$$

where s is the distance between object and loupe. From Eqs. (6.1) and (6.3), we can now calculate the parameters for optimal use of loupes. Note that the above formulas apply only to thin spherical lenses (Section A.1.2.1). Table 6.1, which shows data of commercial aspheric magnifier loupes, indicates that it is difficult to achieve magnifications β_L significantly larger than 3. Even more critical is the field of view, that is, the diameter of the visible area[3] d_{fov} (Problem P6.1) and the object distance s, which becomes excessively small for any practical use in surgery. As a consequence, telescope systems (Section 6.1.2) are used to form a magnified image of distant objects with an object-side loupe.

6.1.2
Medical Loupes

A visual aid for surgery with a separate illumination system must fulfill the following requirements:

1. The magnified image (2–8 ×) must be upright and nonreversed to relieve the surgeon from any reorientation of microsurgical instruments.

3) Aspheric lenses or aplanatic/achromatic (Section 6.2.3.2) lens systems are used to obtain a higher imaging quality.

2. The imaging must be three-dimensional (3D) in that it must allow stereoscopic viewing (Section 2.1.9).
3. The working distance between the front surface of the visual aid and the operating field must be at least 150 mm so that microsurgical tools can be used.
4. The color reproduction of the imaging system must be adequate so that all details can be clearly recognized by the physician.

These requirements can be fulfilled by surgical microscopes (Section 6.2) and medical telescope loupes. According to the optical principle of the telescope, we differentiate between Galilei and Kepler telescopes.

Galilei telescope The Galilei telescope consists of a positive objective lens and an eyepiece with a negative lens (framed parts in Figure 6.2). Medical loupes based on this arrangement (e.g., Figure 6.3a) have a good image quality up to a magnification of 2.5 ×. For higher magnifications, the image quality rapidly deteriorates.

The exit pupil is located between objective lens and eyepiece and is virtual so that the field of view is not sharply defined. This allows certain freedom to change the viewing direction during surgery. In a medical loupe which uses a Galilei telescope, the object (with height h_O) lies in the focal plane of the loupe and is thus imaged to infinity. The formed image (with image size $h'_{I,L}$) is viewed through the telescope with magnification β_{tele}. The total magnification then follows from

$$\beta_{tot} = \beta_L \beta_{tele} \,. \tag{6.4}$$

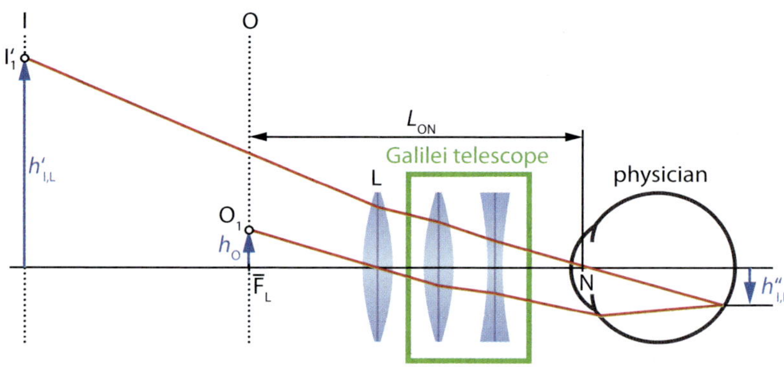

Figure 6.2 Ray diagram of a medical loupe system based on the Galilei telescope (green frame) with two positive lenses and one negative lens. Two possible configurations exist to image an object at a finite distance with an afocal Galilei telescope. One solution is to use a positive lens L, as shown in the scheme, to image the object to optical infinity. Alternatively, we may simply increase the distance between the positive lens and negative lens of the Galilei telescope such that a parallel bundle of rays is formed in front of the physician's eye (not shown). The second solution does not require the additional lens L, but is less instructive for the following discussions. L_{ON} is the distance between the nodal point of the physician's eye N and the object plane O. $h'_{I,L}$ is the size of the formed virtual image, h''_I the retinal image size, and $\bar{F}_L$ the focal point of the entire optical system.

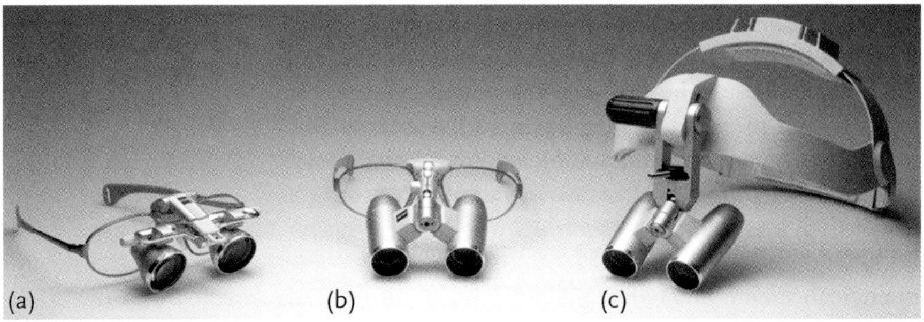

Figure 6.3 Various forms of medical loupes by ZEISS. The product in (a) is based on a Galilei telescope. The products in (b) and (c) are based on a Kepler telescope. Courtesy of Carl Zeiss.

The individual magnifications of the lens β_L and the telescope β_{tele} also determine the working distance of Galilei telescope-based loupes (see also Eq. (A15) in Section A.1.2.1).

Kepler telescope In a Kepler telescope, both objective and eyepiece lenses are positive lenses. The image of such a telescope is inverted. To form an upright image, an additional inverting prism (e.g., a Pechan prism; see Info Box 5.1 in Section 5.2.2.6) is introduced in medical loupes based on this type of telescope (e.g., Figure 6.3b,c). The image quality of a Kepler telescope is very good even for magnifications of up to $6\times$. However, as the exit pupil is real and lies behind the eyepiece, the entrance pupil of the eye has to be placed at the same position. Consequently, the viewing direction can only be minimally changed during surgery.[4] In addition, Kepler telescopes must be centered very precisely. The magnification formula (6.4) for Galilei telescopes can also be applied to loupes based on a Kepler telescope.

Medical loupes are often used in ophthalmic surgery for extraocular procedures (e.g., correction of strabimus[5].) For intraocular procedures, only a surgical microscope can provide the required magnification, illumination, and ergonomics, which will be discussed in the following.

6.2 Surgical Microscopes

Surgical or *operating microscopes* have become an indispensable tool for ophthalmic surgeons as well as for microsurgeons in many other medical fields such as neurosurgery, ear, nose, and throat surgery, gynecology, plastic surgery, and recently in dentistry. The use of the surgical microscope in microsurgery has led to a radical change in how surgical procedures are performed. The ability to view small

4) More precisely, the extent to which the viewing direction can be changed depends on the relative size of the loupe exit pupil and the entrance pupil of the eye.
5) Strabismus is a condition in which the eyes are not properly aligned with each other so that normal binocular vision is prevented.

but well-illuminated objects coupled with the high mobility of a surgical microscope (Section 6.2.3.6) has expanded the potential of microsurgery. In ophthalmic surgery, surgical microscopes are used for all types of intraocular surgery.

Info Box 6.1: Microsurgery

Microsurgery is an operation technique that uses special visual aids such as head-worn loupe systems or surgical microscopes to perform surgery on tiny structures like vessels, nerves, and organs.

History Over a hundred years ago, magnifying visual aids were used for medical surgery. Even the use of head-mounted telescope systems was in place around 1880 [1]. The limitations of such systems were soon realized, which led to the development of binocular microscopic systems. In 1921, Carl Nylén (1892–1978) proposed the use of such binocular microscopes for surgery [2]. This idea was popularized by Gunnar Holmgren (1875–1903) who used a binocular instrument by ZEISS in ear, nose, and throat surgery. However, mechanical problems in stability, focusing, and positioning have long delayed the wider acceptance of microscopy in surgery. The surgical microscope as we know it today thus has a surprisingly short history considering that the first instrument fully suitable for surgical purposes was introduced almost 100 years after Carl Zeiss (1816–1888) and Ernst Abbe (1840–1905) presented their scientifically designed microscopes in Jena. In 1953, the introduction of the ZEISS OPMI® 1 surgical microscope was a major milestone in the advancement of surgical visualization. Already in the same year, Heinrich Harms (1908–2003) reported the first use of a surgical microscope in eye surgery [3]. From this moment on, the development as forecasted and promoted by Hans Littmann (1908–1991) [4] and José Barraquer (1916–1998) [5] led to increasingly advanced surgical procedures.

6.2.1 Requirements for Surgical Microscopes

In a surgical environment, a surgical microscope has to meet the following requirements:

- Resolution and magnification must be sufficiently high (typically between 3 and 40 ×) so that fine tissue structures and details can be visualized (Section 6.2.2). Higher magnification would be impractical due to the lack of eye-to-hand coordination.
- The surgical microscope must have a variable magnification (Section 6.2.3.4) as well as a variable field of view (Section 6.2.2).
- The stereoscopic imaging must allow sufficiently high depth resolution (Section 6.2.2). At the same time, the surgeon requires a depth of field (Section 2.1.8) as large as possible.

- The optical image quality must not cause any ambiguity in the interpretation of details. All relevant aberrations must therefore be largely compensated by appropriate optical design solutions (Section 6.2.3.2).
- The working distance between microscope and patient must be large enough so that the surgeon can easily use surgical instruments without blocking the view of the surgical area. Typically, a working distance between 200 and 300 mm is required in surgical ophthalmology. In other microsurgical disciplines such as neurosurgery, the working distance must also be variable during surgery (Section 6.2.3.3).
- The illumination of the surgical area must be optimized for appropriate brightness (Section 6.2.3.5), high contrast, and realistic color fidelity. In ophthalmology, the microscope must provide *red reflex illumination* (Section 6.2.3.5), which is a kind of transillumination (as opposed to incident illumination) that can make structures in the transparent anterior media of the eye visible.
- Vignetting (Section A.1.4.1) and reflexes must be minimized to allow an undisturbed view.
- The microscope must be installed such that it allows stable visualization even in cases of high magnification. At the same time, a suitable mount must allow easy and rapid adjustments and repositioning of the field of view (Section 6.2.3.6).
- The surgical microscope, or at least all components which are used for repositioning and change of magnification (or other adjustments), must be able to become sterilized.
- Often, a co-observation microscope is needed for training, instruction, and operational assistance.

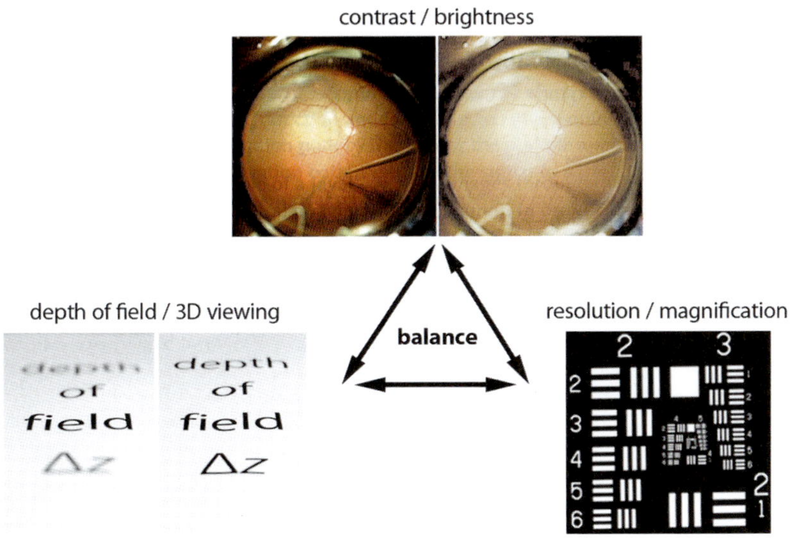

Figure 6.4 The design of a surgical microscope has to provide an optimal balance of several parameters, which can be a challenge in practice. The optimization space is spanned by the brightness/contrast, depth of field, and magnification/resolution.

Some requirements concerning the imaging are summarized in a schematic diagram (Figure 6.4), which reveals that it can be quite challenging to find an optimal balance between all relevant parameters.

6.2.2
Functional Principle

All requirements mentioned in Section 6.2.1 can be fulfilled by a binocular optical system (for 3D viewing) with a relatively low total magnification,[6] a relatively small numerical aperture of $NA < 0.05$ (and thus a sufficiently large working distance), a field of view adapted to the surgical area, and an adequately large depth of field. Such an optical system is ideally realized by a *stereo microscope*.

A stereo microscope consists of a binocular telescope with a magnifying loupe mounted on the object side (Figure 6.5). The examined object is located in the object-side focal plane of a magnifying objective lens O so that it is imaged to optical infinity with a magnification according to Eq. (6.3) and viewed with a telescope. Between objective lens O and tube lenses T, two separate parallel beam paths are created by means of so-called *pupil splitting*. In this way, two separate images can be formed for the left and right eye of the physician which provides a 3D impression of the observed object. In general, a magnification changer or zoom system with variable magnification[7] Γ is also placed between O and T. The intermediate images formed by the tube lenses are viewed with eyepieces EP. The distance between eyepieces and tube lenses is approximately $f_{\text{tub}} + f_{\text{ep}}$, where f_{tub} and f_{ep} represent the focal lengths of the individual tube lenses and eyepieces, respectively.

Magnification From Figure 6.5 it follows that T and EP form a binocular telescope (green frame) with a total magnification of

$$\beta_{\text{tot}} = \frac{f_{\text{tub}}}{f_{\text{ep}}} \Gamma \frac{s_{\text{nv}}}{f_{\text{obj}}} = \frac{f_{\text{tub}}}{f_{\text{obj}}} \Gamma \beta_{\text{ep}} \,. \tag{6.5}$$

$s_{\text{nv}} = 250\,\text{mm}$ is the typical near viewing distance (Section 2.1.9), f_{obj} the focal length of objective lens O, and $\beta_{\text{ep}} = s_{\text{nv}}/f_{\text{ep}}$ the magnification of the eyepiece EP. Equation (6.5) can be easily understood in terms of a telescope magnifier. The factor $(s_{\text{nv}}/f_{\text{obj}})$ describes the magnification of the objective, and $(f_{\text{tub}}/f_{\text{ep}})$ is the telescope magnification of the binocular tube.

Depending on the application in ophthalmology, focal lengths of 175–300 mm are selected for the objective lens, and 125–175 mm for the tube lenses. f_{ep} typically

6) At first sight, the low total magnification seems to be a discrepancy of what we usually expect from a microscope. However, the low magnification allows a sufficiently large working distance with a relatively large depth of field and a high brightness, which is usually most relevant in clinical practice.

7) In order to be consistent with standard literature, we denote the magnification of a zoom system by Γ. Please note that, in principle, this quantity refers to the "common" magnification β.

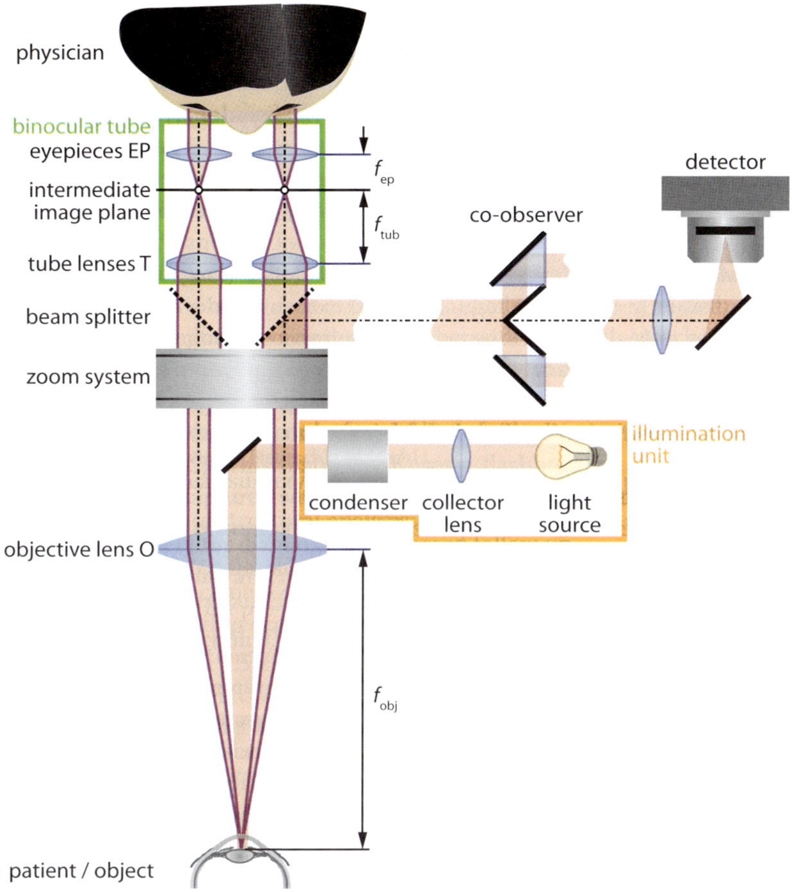

Figure 6.5 Schematic setup of a surgical stereo microscope (compare with Figure 6.9). In the plane of the beam splitters, possible attachments are shown, that is, optics for the co-observer (e.g., a medical assistant) and optics for a camera detector typically used for documentation. The binocular tube and illumination unit are grouped with green and orange frames, respectively.

ranges between 12.5 and 25.0 mm. This results in an overall working distance[8] of 150–300 mm and a total magnification of 2–30 × (Table 6.2).

Field of view The field of view d_{fov} can be calculated from the diameter of the field stop in the eyepiece d_{fs} (typically 10 mm) and Eq. (6.5) as

$$d_{\text{fov}} = \frac{d_{\text{fs}} s_{\text{nv}}}{\beta_{\text{tot}} f_{\text{ep}}} = \frac{d_{\text{fs}} \beta_{\text{ep}}}{\beta_{\text{tot}}} . \tag{6.6}$$

8) The working distance L_{wd} corresponds to the free space between object and objective lens so that we have $L_{\text{wd}} \leq f_{\text{obj}}$. In practice, L_{wd} is in the range of 200 mm.

Table 6.2 Typical parameter set of surgical microscopes.

f_{obj} (mm)	f_{tub} (mm)	Γ	f_{ep} (mm)	β_{tot}	d_{fov} (mm)
200	125	2.5	12.5	19.5	10.2
300	125	2.5	12.5	13.0	15.4
200	125	0.4	12.5	3.1	64.0
300	125	0.4	12.5	2.1	96.0
200	175	2.5	12.5	27.3	7.3
300	175	2.5	12.5	18.2	11.0
200	175	0.4	12.5	4.4	45.7
300	175	0.4	12.5	2.9	68.6

Typical values for the magnification and field of view are listed in Table 6.2 for reference.

Numerical aperture and lateral resolution The numerical aperture (Section A.1.4) of a surgical microscope $\mathrm{NA}_{\mathrm{mic}}$ is determined by the effective object-side aperture diameter of the zoom system d_{zoom}. In contrast to standard microscopy, it is *not* the effective aperture of the objective lens which must be taken into account. This can be easily understood when we consider the pupil splitting into a left and right imaging path (Figure 6.6). The numerical aperture of the surgical microscope is then given by

$$\mathrm{NA}_{\mathrm{mic}} = \frac{d_{\mathrm{zoom}}}{2 f_{\mathrm{obj}}} , \tag{6.7}$$

from which we obtain the lateral spatial resolution (compare with Eq. (A78) in Section A.2.1.6)

$$\Delta(x, y)_{\mathrm{mic}} = \frac{1.22\lambda}{\mathrm{NA}_{\mathrm{mic}} + \mathrm{NA}_{\mathrm{c}}} . \tag{6.8}$$

λ denotes the wavelength of the illuminating light and NA_{c} the numerical aperture of the condenser (Figure 6.5). If the aperture of the condenser equals that of each imaging channel, we have $\mathrm{NA}_{\mathrm{c}} = \mathrm{NA}_{\mathrm{mic}}$. From Eq. (6.8), we find spatial resolutions in the range of 10–20 μm. Compared to the optical resolution of the human eye (typically 30–40 μm; Section 2.1.5), we thus obtain a resolution enhancement by a factor of 2–4.

Usable magnification As for standard microscopes, it makes sense to define a range of *usable magnifications* β_{um} for which details resolvable by the eye are correctly imaged by the microscope. Higher magnifications ($\beta_{\mathrm{tot}} > \beta_{\mathrm{um,max}}$) require exit pupils that are too small and render the image dark. In this case, the image

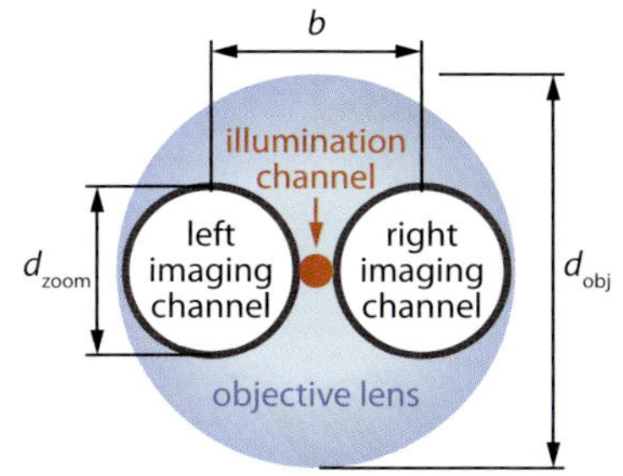

Figure 6.6 Cross-section of the stereoscopic zoom system of a surgical microscope. b denotes the center-to-center distance between both observation channels, and d_{zoom} is the aperture diameter of one individual observation channel in the zoom system. d_{obj} is the diameter of the objective lens.

is still magnified but shows no additional details[9]. Below the minimum usable magnification ($\beta_{tot} < \beta_{um,min}$), the resolution of the microscope is not used to full capacity.

We can determine the usable magnification starting from Eq. (6.5). With the typical near viewing distance $s_{nv} = 250\,\text{mm}$ and the factor $\kappa = f_{ep}/(f_{tub}\Gamma)$, we obtain

$$\beta_{tot} = \frac{250\,\text{mm}}{\kappa f_{obj}} = \frac{250\,\text{mm}}{f_{obj,eq}}\,, \tag{6.9}$$

where $f_{obj,eq} = \kappa f_{obj}$ can be regarded as an "equivalent focal length" of the microscope system for the purpose of determining the optical resolution. According to Figure A.33 in Section A.2.1.6, the diffraction pattern of the Airy disk is seen under an viewing angle of

$$\tan\left(\frac{\theta}{2}\right) = \frac{\Delta(x,y)_{mic}}{2f_{obj,eq}} = \frac{0.61\lambda}{2\text{NA}_{mic}\,f_{obj,eq}}\,, \tag{6.10}$$

in which NA_{mic} is the numerical aperture of the surgical microscope and $\Delta(x,y)_{mic}$ its lateral resolution. Hence, we have

$$\beta_{tot} = \tan\left(\frac{\theta}{2}\right)\frac{2\cdot 250\,\text{mm}}{0.61\lambda}\text{NA}_{mic}\,. \tag{6.11}$$

If we assume a typical angular eye resolution θ_{eye} between $2'$ and $4'$, the range of usable magnification for a wavelength of 500 nm is

$$500\,\text{NA}_{mic} \le \beta_{um} \le 1000\,\text{NA}_{mic}\,. \tag{6.12}$$

Table 6.3 shows that the actual magnification for typical configurations of surgical microscopes normally agrees with the usable magnification.

9) Magnifications beyond the maximum usable magnification $\beta_{um,max}$ are sometimes called *empty magnifications.*

Table 6.3 Typical resolutions and usable magnifications for surgical microscopes.

f_{obj} (mm)	d_{zoom} (mm)	NA_{mic}	Γ	β_{tot}	$\Delta(x,y)_{mic}$ (μm)	$\beta_{um,min}$	$\beta_{um,max}$
200	10	0.025	2.5	19.5	13.4	12.5	25.0
300	10	0.017	2.5	13.0	20.1	8.3	16.7
200	15	0.038	0.4	3.1	8.9	18.8	37.5
300	15	0.025	0.4	2.1	13.4	12.5	25.0

Depth of field The *depth of field* Δz_{dof} (Section 2.1.8) is defined as the distance in object space within which the image of an object has an acceptable sharpness. To determine this parameter, the diffraction, angular resolution, and the ocular range of accommodation have to be taken into account [6]. According to [7, 8], we find

$$\Delta z_{dof} = \underbrace{\frac{\lambda}{2\,NA_{mic}^2}}_{1} + \underbrace{\frac{0.34\,\text{mm}}{\beta_{tot} NA_{mic}}}_{2} + \underbrace{\frac{s_{nv}^2}{\beta_{tot}^2}\left(\frac{1}{s_{near}} - \frac{1}{s_{far}}\right)}_{3} , \tag{6.13}$$

where s_{near} and s_{far} are the near and far point distances of the eye (Section 2.1.4), respectively. In this equation, term 1 describes the "wave-optical" (diffraction-limited) depth of field (Section A.2.1.6), term 2 describes the "geometric-optical" depth of field, and term 3 takes the accommodation deduced depth of field into consideration. Term 3 is not an instantaneous depth of field in the classical optical sense, because it takes into account different states of the overall microscope–eye interaction (see also Problem P6.8). For emmetropic eyes, the far point is located at infinity so that we have $1/s_{far} = 0\,\text{D}$. In the case of near vision, the near point has a distance of 25 cm from the eye so that $1/s_{near} = 4\,\text{D}$. As can be seen from Eq. (6.13) and Figure 6.7, Δz_{dof} decreases with increasing magnification, but can be compensated to some extent by accommodation (gray area).

Stereoscopic depth perception In addition to the depth of field, the minimum stereoscopic depth perception (Section 2.1.9) is an important parameter for appropriate image perception. For a human eye without visual aids and the typical near viewing distance $s_{nv} = 250\,\text{mm}$, it is given by (see Eq. (2.19))

$$\Delta L_{eye} = \frac{\varepsilon_{min} s_{nv}}{PD} = 45\,\mu\text{m} , \tag{6.14}$$

where $\varepsilon_{min} = 10''$ is the minimum stereo angle for photopic vision and PD the interpupillary distance. For a stereo microscope (Figure 6.8; Problems P6.2 and P6.3), we have [8]

$$\Delta L_{mic} = \frac{\varepsilon_{min} f_{obj}^2}{b\Gamma(f_{tub}/f_{ep}) \pm \varepsilon_{min} f_{obj}} , \tag{6.15}$$

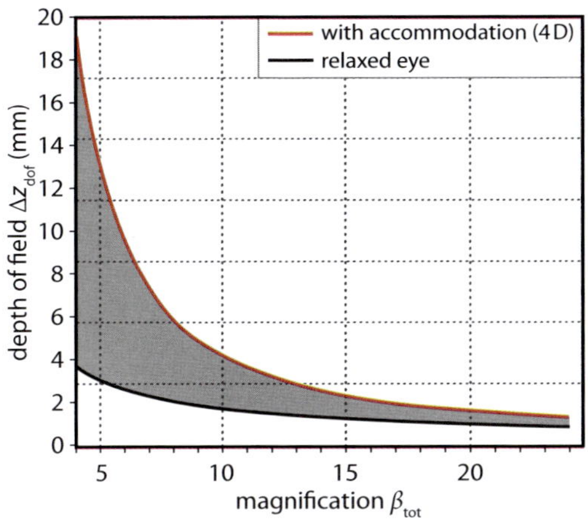

Figure 6.7 Depth of field Δz_{dof} (mm) versus total magnification β_{tot} for the relaxed eye (black) and an accommodation of 4 D (red). The values of Δz_{dof} are calculated according to Eq. (6.13) with $\lambda = 550$ nm and $NA_{mic} = 0.025$. Since an accommodation of 4 D corresponds to the inverse typical near viewing distance $1/s_{nv}$, the gray area represents the depth of field range for the standard range of accommodation.

in which the stereo base is given by the distance b between both observation channels (Figures 6.6 and 6.8). Since a smaller value of ΔL_{mic} means a better stereoscopic resolution, it would be logical to increase b in order to obtain a high stereoscopic resolution. This quantity is, however, typically limited to 20–30 mm due to the objective lens diameter, which largely determines the overall diameter size of the microscope. As Eq. (6.15) and Table 6.4 reveal, an improved stereo effect is possible for given microscope magnifications as compared to the unaided eye.

It is also important that the stereoscopic depth perception is a good representation of the spatial reality in the object field at typical magnifications used for microsurgery. A certain distance in the z direction must seem the same to the surgeon as a distance in x and y.

Table 6.4 Stereoscopic depth perception ΔL_{mic} with a stereo microscope for typical parameter sets. The minimum stereo angle ε_{min} is taken as $10'' = 4.85 \times 10^{-5}$ rad (Section 2.1.9) and $f_{tub} = 125$ mm.

f_{obj} (mm)	d_{zoom} (mm)	f_{ep} (mm)	b (mm)	β_{tot}	ΔL_{mic} (μm)
200	0.4	20	22	3.1	35.3
300	0.4	20	22	2.1	79.4
200	1.0	20	22	7.8	14.1
200	2.5	20	22	19.5	5.6
200	1.0	20	26	7.8	11.9
200	2.5	20	26	19.5	4.8

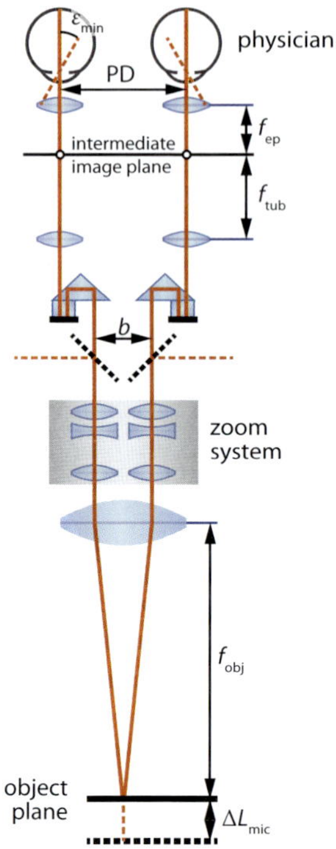

Figure 6.8 Ray diagram of a stereo microscope used to illustrate the stereoscopic depth perception. ΔL_{mic} is the minimum stereoscopic depth perception, *b* the stereo base of the optical system, PD the interpupillary distance, and ε_{min} the minimum stereo angle. Please compare this diagram with Figure 2.11 in Section 2.1.9, where we considered binocular vision without visual aids.

6.2.3 Modular Structure of Surgical Microscopes

In the following, we will take a closer look at some components and modules of standard surgical microscopes (Figure 6.9). We will also discuss some aspects (e.g., chromatic aberration; Section A.1.9) which are relevant for a good image quality.

6.2.3.1 Objective Lens

The image quality of an optical system depends on the extent to which all optical aberrations (Section A.1.6) can be eliminated in the design of its components and modules. In particular, the objective lens plays an important role for imaging and must thus fulfill the following properties:

- low chromatic aberration (Section A.1.9),
- absence of distortion (Section A.1.6.5),
- low field curvature (Section A.1.6.4), and
- low reflection by means of anti-reflection coatings.

The working distance of a surgical microscope is defined by the distance between the front (objective) lens vertex and the object plane. In most cases, this distance can be approximated by the focal length of the objective lens f_{obj}. A notable exception can be *varioscopes* (Section 6.2.3.3).

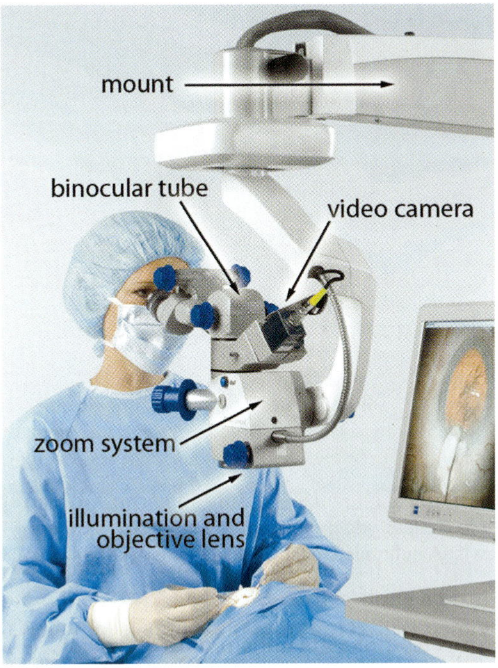

Figure 6.9 Use of a surgical microscope (ZEISS OPMI LUMERA®) in ophthalmic surgery. Courtesy of Carl Zeiss.

6.2.3.2 **Technical Optics: Chromatic Aberration of Objective Lenses**

Let us now address the question of chromatic correction of objective lenses in somewhat more detail. At first, we derive the longitudinal chromatic aberration (Section A.1.9) for paraxial rays (Section A.1.2) for the case of a thin lens with focal length f, magnification β, and Abbe number ν. With regard to dispersion of optical media (Section A.1.1), the Abbe number ν is a useful parameter which allows us to switch from differential variations to difference quantities of the refractive index. To calculate f and β, the refractive index for the spectral emission line of mercury ($\lambda_e = 546\,\text{nm}$) is used as a reference. The longitudinal chromatic aberration can be derived for paraxial rays directly from the lens maker's equation (A12) for a thin lens with radii r_1 and r_2. For this purpose, we assume the object position to be fixed and calculate the change in the image distance $\Delta s'$ when the wavelength λ (and hence the refractive index n) is altered. If we differentiate with respect to λ, we obtain

$$\Delta\frac{1}{s'} = -\Delta n\left(\frac{1}{r_2} - \frac{1}{r_1}\right) = \frac{\Delta n}{n-1}\frac{1}{f} \tag{6.16}$$

with $\Delta(1/s') = -(1/s'^2)\Delta s'$ and $\Delta n \approx n_{F'} - n_{C'}$. $n_{F'}$ and $n_{C'}$ are the refractive indices for light wavelengths of $\lambda_{F'} = 480\,\text{nm}$ and $\lambda_{C'} = 644\,\text{nm}$, respectively.

With the imaging equation of a thin lens equation (A14) and $s' = f(1-\beta)$, we find

$$\Delta s' = \frac{-\Delta n}{n-1} f(1-\beta)^2 \,. \tag{6.17}$$

To determine the local range of focus positions $\Delta s'_{F'..C'}$ on the optical axis for wavelengths between $\lambda_{F'}$ and $\lambda_{C'}$, we may use the Abbe number ν, which yields

$$\Delta s'_{F'..C'} = -f\frac{(1-\beta)^2}{\nu} \,. \tag{6.18}$$

The longitudinal aberration is thus proportional to the focal length f, inversely proportional to the Abbe number[10] ν, and squared to the lateral magnification β. Abbe numbers range between 67.0 (low dispersion glasses like N-BK10) and 20.9 (high dispersion glasses like N-SF66) [9].

But how can chromatic aberration be corrected? It is evident from Figure 6.10 that a positive lens generates a longitudinal chromatic aberration that is opposite to that of a negative lens, and that both positive and negative chromatic aberrations increase with the Abbe number ν. Thus, the idea is to combine lenses with positive and negative optical powers. To ensure that some refractive power remains, different materials (with different dispersion or Abbe numbers) must be used. Let us calculate the color correction for a pair of attached lenses. The total refractive power of the lens doublet is given by

$$\begin{aligned} \mathcal{D}_{tot} &= \frac{1}{f_{tot}} = \frac{1}{f_{L1}} + \frac{1}{f_{L2}} \\ &= (n_{L1}-1)\left(\frac{1}{r_{L1,2}} - \frac{1}{r_{L1,1}}\right) + (n_{L2}-1)\left(\frac{1}{r_{L2,2}} - \frac{1}{r_{L2,1}}\right) \,, \end{aligned} \tag{6.19}$$

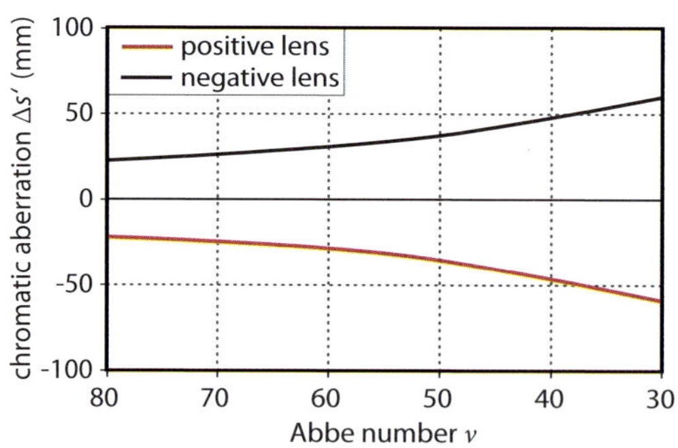

Figure 6.10 Longitudinal chromatic aberrations in the image of a thin positive (red) and a thin negative (black) lens versus the Abbe number. The values are calculated according to Eq. (6.17) for an object distance of $1.5f$.

We can see that $\Delta s'$ develops to different "directions". Hence, we can compensate chromatic aberration by means of a lens doublet which consist of one positive and one negative lens.

10) The smaller the Abbe number ν, the higher the dispersion of the medium.

where the parameters of lens 1 and lens 2 are labeled by the indices L1 and L2, respectively. We can simplify Eq. (6.19) with $K_{L1} = 1/r_{L1,2} - 1/r_{L1,1}$ and $K_{L2} = 1/r_{L2,2} - 1/r_{L2,1}$ to

$$\mathcal{D}_{tot} = (n_{L1} - 1)\,K_{L1} + (n_{L2} - 1)\,K_{L2}\,. \tag{6.20}$$

The condition for fully compensated chromatic aberration then reads

$$\frac{d}{d\lambda}\mathcal{D}_{tot} = \frac{d}{d\lambda}\left[(n_{L1} - 1)\,K_{L1} + (n_{L2} - 1)\,K_{L2}\right] = 0 \tag{6.21}$$

which, in turn, simplifies to

$$\frac{dn_{L1}}{d\lambda}K_{L1} + \frac{dn_{L2}}{d\lambda}K_{L2} = 0\,. \tag{6.22}$$

By setting $dn_{L1} = n_{L1}(\text{blue}) - n_{L1}(\text{red})$ and, similarly, $dn_{L2} = n_{L2}(\text{blue}) - n_{L2}(\text{red})$, we obtain

$$\frac{K_{L1}}{K_{L2}} = -\frac{n_{L1}(\text{blue}) - n_{L1}(\text{red})}{n_{L2}(\text{blue}) - n_{L2}(\text{red})}\,. \tag{6.23}$$

When we insert the medium refractive indices for green light $\lambda_e = 546\,\text{nm}$ $(= n_e)$, for blue light $\lambda_{F'} = 480\,\text{nm}$ $(= n_{F'})$, and for red light $\lambda_{C'} = 644\,\text{nm}$ $(= n_{C'})$, the Abbe number ν of the respective material can be calculated. The ratio of the focal distances of both lenses is then related to the Abbe numbers via

$$\frac{f_{L2}}{f_{L1}} = -\frac{\nu_{L1}}{\nu_{L2}}\,, \tag{6.24}$$

where we used $K_{Li} = 1/[f_{Li}(n_{Li} - 1)]$ with the lens indicator i. Due to the minus sign and $\nu > 0$, the lens doublet of an objective lens has to consist of a negative *and* a positive lens. As shown in Figure 6.11 and illustrated by the dashed curve in Figure 6.12, the foci for the red and blue wavelengths coincide in the case of an achromatic lens system (Problem P6.4). But we still have a minor deviation for the central color (green) which is referred to as the *secondary spectrum*. In an *apochromatic* lens (Problem P6.5), the focal points coincide for all three or even more wavelengths (dash-dotted curve in Figure 6.12).

6.2.3.3 Varioscope System

A *varioscope* (Figure 6.13) is an optical system which allows rapid change of the working distance while keeping the object in focus (Problem P6.6). In other surgical areas than ophthalmology, this is a desired feature for many applications. In ophthalmology, we only find this solution in a few specialized applications (Figure 6.14). The principle of a varioscope is based on the variation of the focal length of the objective lens, which is often called the *internal focus*. This can be achieved with a system of a positive and a negative lens element in which the lenses are displaced relatively to each other.

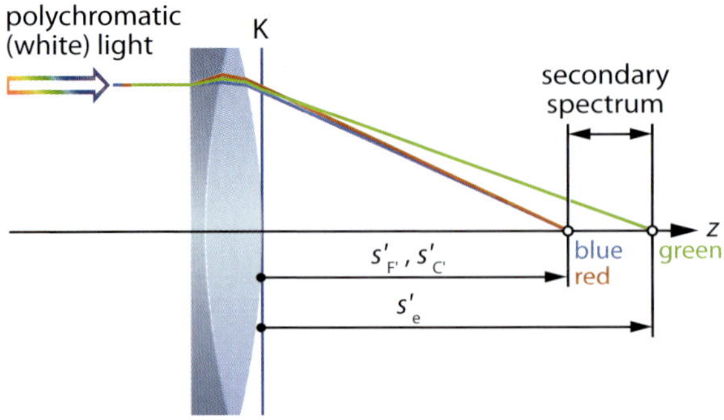

Figure 6.11 Ray diagram for the blue ($\lambda_{F'} = 480$ nm), red ($\lambda_{C'} = 644$ nm), and green ($\lambda_e = 546$ nm) spectral components of incident white light imaged by an achromatic lens doublet. The blue and red light components are focused onto the same point, whereas the focus of the green light component is shifted along the optical axis (secondary spectrum). K denotes the principal plane of the lens doublet.

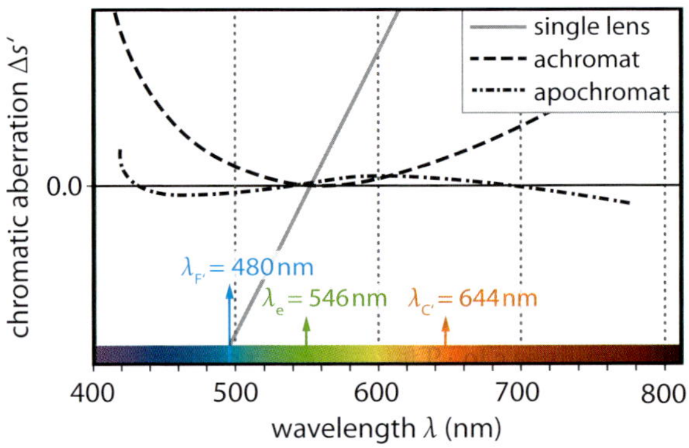

Figure 6.12 Comparison of the longitudinal chromatic aberrations for a single lens (gray curve), an achromatic lens (dashed curve), and an apochromatic lens (dash-dotted curve).

The negative lens element of the varioscope is the "final" optical component of the surgical microscope and is fixed in position. The positive lens element is displaced relative to the negative element by a distance s_{var} such that the working distance L_{wd} and the focal length f are changed simultaneously.

In an objective lens with a fixed focal length, the focal length and the working distance are almost identical. The characteristic feature of the varioscope is, however, that f is always larger than L_{wd} (Figure 6.13). As, according to Eq. (6.5), the total magnification of a surgical microscope is inversely proportional to the focal length of the objective lens, the magnification in a surgical microscope with a varioscope is always smaller than in a surgical microscope with fixed focal length for the same working distance.

As mentioned, varioscopes are rarely used in ophthalmic surgery. In most procedures, surgeons rather use microscopes with a fixed focal length objective lens. One noticeable exception are surgical microscopes for vitreo-retinal surgery (Fig-

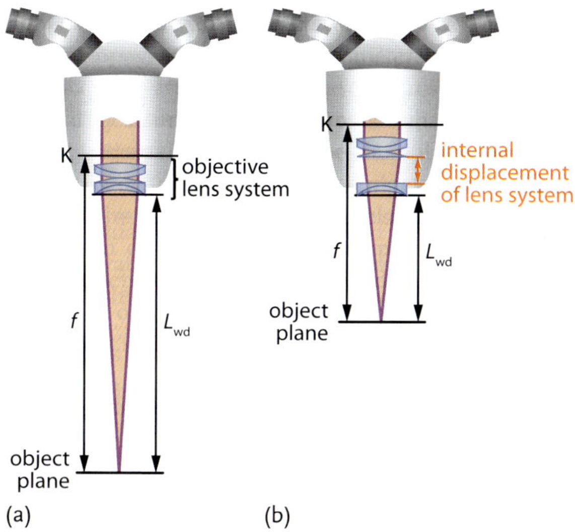

Figure 6.13 Principle and optical design of a varioscope which is able to continuously vary the working distance L_{wd} by changing the focal length f of the objective lens system. (a) Configuration for a large working distance. (b) Configuration for a small working distance at which the objective lens system is internally displaced. A typical characteristic of a varioscope is that $L_{wd} < f$, as the principal plane K lies in front (upstream) of the objective lens system. A typical value for a small working distance $L_{wd} = 200$ mm is $f = 270$ mm. For a larger working distance of $L_{wd} = 400$ mm, we have $f = 450$ mm.

ure 6.14; Problem P6.6) which are combined with a wide-angle fundus imaging lens. A wide-angle aspheric lens, typically with a refractive power between 60 and 130 D, and the optical system of the eye create a real image of the fundus in an intermediate plane. This plane then becomes the focal plane of the varioscope objective and is viewed through the surgical microscope (inset of Figure 6.14c). The system is designed such that when the fundus viewing system (including the varioscope optics) is moved out of the beam path, the microscope is focused on the cornea of the patient's eye (Figure 6.14b). The optical system of the varioscope ensures that the microscope can always stay in a fixed position even if the intermediate image plane is shifted. This is indeed an important safety aspect during surgery, as in particular the distance between wide-angle lens and cornea remains constant. As the intermediate image is inverted, another image inversion must happen in the microscope, which is achieved by a so-called *inverter tube* (Figure 6.14c).

6.2.3.4 **Variable Magnification**

The most convenient way to change the overall magnification of a surgical microscope is a variable magnification system in the parallel beam path between the objective lens and the binocular tube (Figure 6.5). Two principal designs can be found in commercial systems, that is, the *step magnification changer* [8] and the either manual or motorized *zoom telescope system* [8, 10–12].

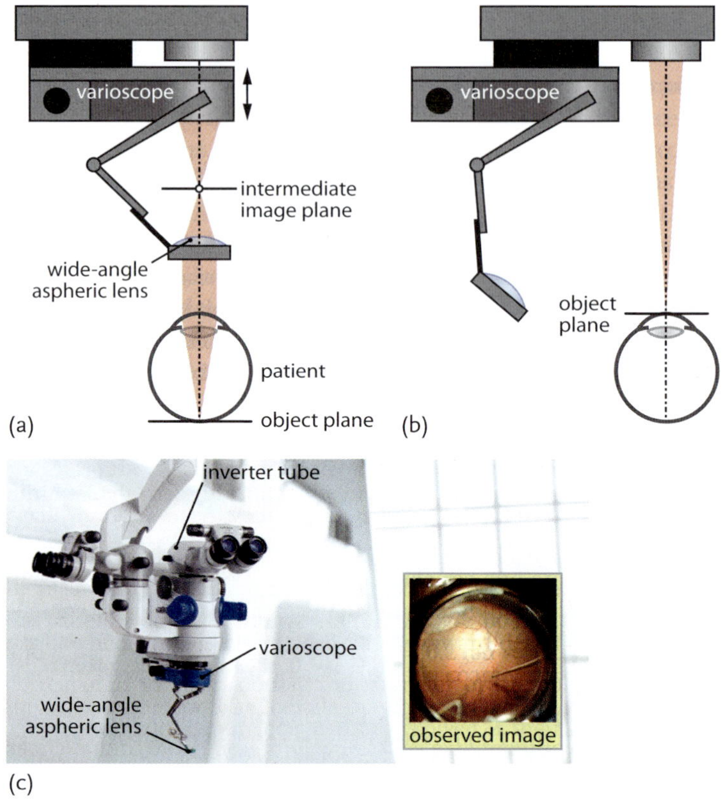

Figure 6.14 (a) Setup of a fundus-imaging microscope. With an additional wide-angle aspheric fundus lens, the surgical microscope can also be used to visualize the patient's eye fundus. For this purpose, the additional lens is placed such that its focal point coincides with the focus of the varioscope. As a consequence, an intermediate image plane is formed. The distance between the additional wide-angle lens and the cornea is kept fixed during an intervention by dynamic readjustment of the varioscope focal length (see arrow). (b) When the fundus viewing system (wide-angle spherical lens including varioscope optics) is "moved out" of the beam path, the microscope is optically focused on the cornea of the patient's eye for nonintraocular (external) surgery without having to mechanically reposition the system. (c) Surgical microscope for vitreo-retinal surgery (ZEISS OPMI LUMERA with ZEISS RESIGHT® unit). Courtesy of Carl Zeiss. Inset: Observed fundus image through surgical microscope ZEISS OPMI LUMERA with ZEISS RESIGHT unit. Courtesy of Prof. Dr. Peter Esser.

Galilean step magnification changer The design of a *Galilean step magnification changer* is based on a Galilei telescope (Section 6.1.2) which consists of a positive objective lens with focal length f_{obj} and a negative eyepiece lens with focal length f_{ep} separated by a distance of about $f_{\text{obj}} + f_{\text{ep}}$. The Galilei telescope forms an upright magnified image (magnification is given by $\beta = f_{\text{obj}}/f_{\text{ep}}$) and has a very compact design. When we look inversely into a Galilean telescope, the formed image will be reduced in size by $1/\beta$. This concept is now used by arranging two Galilean tele-

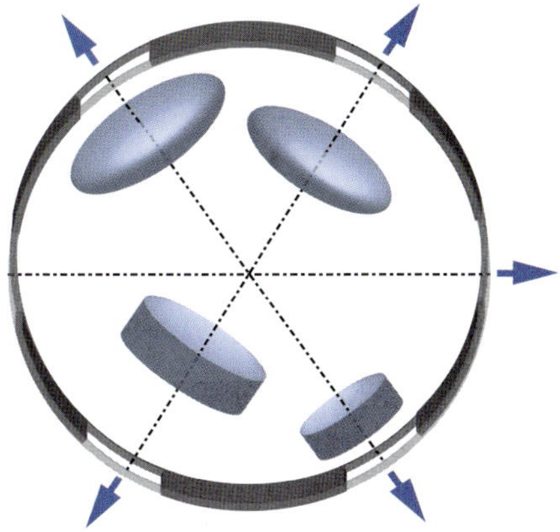

Figure 6.15 Arrangement of a five-step magnification changer based on the optical principle of a Galilei telescope. In a rotating drum, four lenses (i.e., two telescopes) are placed such that they can be used in four viewing directions (blue arrows). In addition, an empty (clear) path is included which can be used if no magnification is required (horizontal blue arrow).

scopes with magnifications $\beta = 2.5\times$ and $\beta = 1.6\times$ as well as a clear path ($\beta = 1$) in a drum-shaped wheel, as depicted in Figure 6.15. Depending on the viewing direction, the system also provides the magnifications $\beta = 0.4\times = (1/2.5)\times$ and $\beta = 0.63\times = (1/1.6)\times$. Such an arrangement is thus called a *five-step magnification changer.*

Zoom telescope magnification changer To obtain a continuously changeable magnification, we have to use a zoom magnification system. Such an optical system consists of at least three lenses with variable relative distances. Zoom lenses are commonly known in photography, where an object at optically infinite or close to infinite distance is usually imaged into a nearby image plane (film or optoelectronic sensor). In a surgical microscope, however, we need an *afocal zoom telescope* (Problems P6.7 and P6.8), since the zoom magnification changer has to be placed between the objective loupe imaging "to infinity" and the binocular tube imaging "from infinity". This requires a design in which the image plane, that is, the exit pupil of the zoom system, is kept in a fixed position. A corresponding zoom system with three lenses is shown in Figure 6.16. L_1 is the positive *compensator lens* (front lens), L_2 a negative *variator lens,* and L_3 a locally fixed (relay) lens. The whole optical zoom system is afocal, and the zoom function is achieved by moving two lenses (L_1 and L_2) in a complex, nonlinear dependence on each other. Typical requirements for zoom designs for surgical microscopes are:

- a zoom range between 1:6 and 1:10,
- a magnification of 0.4 to $3\times$,
- an entrance pupil diameter of approximately 16 to 20 mm,
- a compact design (length preferably between 80 and 100 mm),

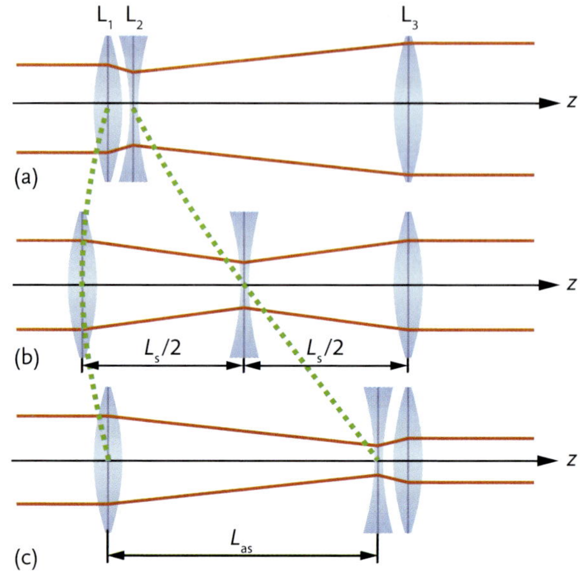

Figure 6.16 Arrangement of lenses in an afocal zoom system (Problem P6.7). To stabilize the position of the image plane, the lenses L_1 (compensator lens) and L_2 (variator lens) have to be moved on a nonlinear curve (green dashed line). The focal lengths of L_1 and L_2 are chosen like $-f_{L2} < f_{L1}$. Lens L_3 is fixed in position. (a) Lens configuration for a magnified image. (b) Symmetry configuration of the zoom system. The (negative) variator lens L_2 is placed in the middle between compensator lens L_1 and lens L_3 (symmetric configuration). L_s is the distance from L_1 to L_3. In this case, the magnification of the zoom system is $\Gamma = 1$. (c) Asymmetric configuration. Compared to the arrangement in (a), the variator and compensator lenses are now shifted along the optical axis such that the distance between L_1 and L_2 is now L_{as}. As a consequence, the resulting image magnification is < 1.

- a precise motion along the ideal nonlinear curve,
- a binocular balance, and
- a stable image position.

The simplest design of a zoom system with three lenses (*three-member zoom*) comparable to the arrangement in Figure 6.16 was introduced by ZEISS in the early 1970s. A drawback of this design is that it is not possible to find a suitable place for an aperture stop, as the two moving lenses (or lens groups) travel through the area where the device pupil of the instrument is located. However, an aperture stop is advantageous for the zoom system. It allows the balancing of aberrations (Section A.1.6) and the diameters of the lens elements, and it makes the zoom system highly flexible for the adaption of video systems.

With optical zoom designs based on four lens elements (*four-member zoom*), the first and last (relay) lens can be kept fixed and only the inner two lenses are moved (Figure 6.17) [11]. With such a design, we can achieve a good aberration correction

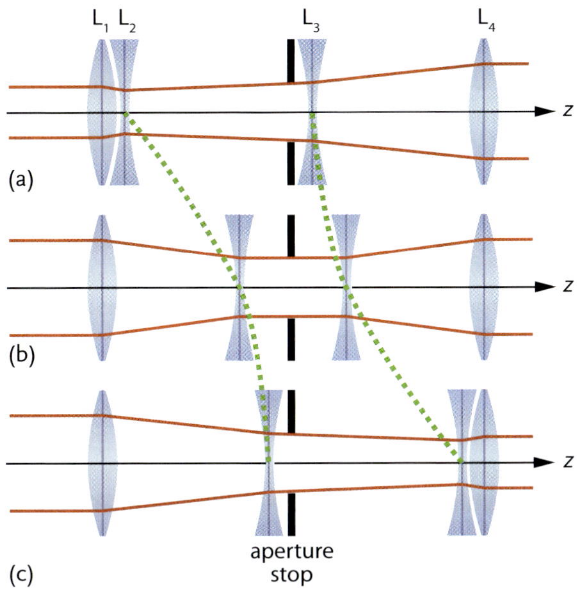

Figure 6.17 Afocal zoom system with four lens elements which allows implementation of a fixed aperture stop. Lenses L_1 and L_4 are fixed in position. Lenses L_2 and L_3 are moved on a nonlinear curve (green dashed line). (a) Lens configuration for a magnified image. (b) Symmetric lens configuration for 1:1-scale imaging. (c) Lens configuration which forms an image with a reduced size.

at high zoom range. At the same time, we may use lenses with compact size[11], the aperture stop is easily accessible, the mechanical design is affordable, and the zoom system can be adapted to different use cases.

Matrix calculation of an afocal zoom system To illustrate the functional principle in more detail, let us use the matrix formalism (Section A.1.3) and consider a three lens symmetric zoom. In principle, the same algorithm can also be used for symmetrical zoom systems with four lenses. In the symmetric configuration (index "s"; Figure 6.16b), the negative variator lens is located at a distance $L_s/2$ from the positive compensator lens. Quantities which refer to lenses L_1 or L_2 are labeled with the indices 1 and 2, respectively. The ABCD matrix for the symmetric position of L_2 is thus given by (Table A.1 in Section A.1.3)

$$\begin{pmatrix} A_s & B_s \\ C_s & D_s \end{pmatrix} = \begin{pmatrix} 1 & 0 \\ -\mathcal{D}_1 & 1 \end{pmatrix} \begin{pmatrix} 1 & \frac{L_s}{2} \\ 0 & 1 \end{pmatrix} \begin{pmatrix} 1 & 0 \\ -\mathcal{D}_2 & 1 \end{pmatrix} \begin{pmatrix} 1 & \frac{L_s}{2} \\ 0 & 1 \end{pmatrix} \begin{pmatrix} 1 & 0 \\ -\mathcal{D}_1 & 1 \end{pmatrix} \tag{6.25}$$

11) Because of the symmetry, we have two pairs of lenses which lowers the cost of manufacturing.

with a refractive power $\mathcal{D}_i$ of the respective lens. The resulting equations in the symmetric case read

$$A_s = 1 - L_s\mathcal{D}_1 - \frac{1}{2}L_s\mathcal{D}_2 + \frac{1}{4}L_s^2\mathcal{D}_1\mathcal{D}_2 , \tag{6.26}$$

$$B_s = L_s - \frac{1}{4}L_s^2\mathcal{D}_2 , \tag{6.27}$$

$$C_s = -2\mathcal{D}_1 - \mathcal{D}_2 + \frac{1}{2}L_s(2\mathcal{D}_1^2 + 2\mathcal{D}_1\mathcal{D}_2) - \frac{1}{4}L_s^2\mathcal{D}_1^2\mathcal{D}_2 , \tag{6.28}$$

$$D_s = 1 - L_s\mathcal{D}_1 - \frac{1}{2}L_s\mathcal{D}_2 + \frac{1}{4}L_s^2\mathcal{D}_1\mathcal{D}_2 . \tag{6.29}$$

For the angle behind the zoom system, the general matrix equation yields (see Eq. (A18))

$$\gamma' = Ch + D\gamma . \tag{6.30}$$

A system is afocal, if $\gamma' = 0$ is valid for all object heights h with $\gamma = 0$. This condition is only fulfilled for $C = 0$. From $C_s = 0$, a quadratic equation results for the symmetric lens distance which is given by

$$\frac{1}{2}L_s = \frac{\mathcal{D}_1^2 + \mathcal{D}_1\mathcal{D}_2 \pm \mathcal{D}_1^2}{\mathcal{D}_1^2\mathcal{D}_2} \tag{6.31}$$

with the solution[12)]

$$\frac{1}{2}L_{s,1} = \frac{2\mathcal{D}_1 + \mathcal{D}_2}{\mathcal{D}_1\mathcal{D}_2} = f_1 + 2f_2 . \tag{6.32}$$

The overall length of the zoom system is thus

$$L_{tot} = L_s = 2f_1 + 4f_2 \tag{6.33}$$

and the magnification $\Gamma_s = 1$.

If the variator lens L_2 is displaced as shown in Figure 6.16a and c, we have an asymmetric case (index "as"), and the matrix changes to approximately[13)]

$$\begin{pmatrix} A_{as} & B_{as} \\ C_{as} & D_{as} \end{pmatrix} = \begin{pmatrix} 1 & 0 \\ -\mathcal{D}_1 & 1 \end{pmatrix} \begin{pmatrix} 1 & L_{as} \\ 0 & 1 \end{pmatrix} \begin{pmatrix} 1 & 0 \\ -\mathcal{D}_1 - \mathcal{D}_2 & 1 \end{pmatrix} . \tag{6.34}$$

So, we have

$$A_{as} = 1 - L_{as}(\mathcal{D}_1 + \mathcal{D}_2) , \tag{6.35}$$

12) A second solution yields $L_{s,2} = 2f_1$. In this case, the negative lens in the middle is located in the focus of the first lens element and, as a field lens, does not act on the marginal rays. This results in a zoom system with image inversion and a long overall length which is not desired for our application.

13) This is an approximation, since we assume that the distance between the lenses L_2 and L_3 is zero.

$$B_{as} = L_{as} , \tag{6.36}$$

$$C_{as} = -2\mathcal{D}_1 - \mathcal{D}_2 + L_{as}\mathcal{D}_1(\mathcal{D}_1 + \mathcal{D}_2) , \tag{6.37}$$

$$D_{as} = 1 - L_{as}\mathcal{D}_1 . \tag{6.38}$$

The afocal case is again obtained for $C_{as} = 0$ which yields the length of the optical system for the asymmetric case

$$L_{as} = \frac{2 f_1 f_2 + f_1^2}{f_1 + f_2} . \tag{6.39}$$

In the asymmetric case, the length of the optical system is thus shorter than in the symmetric case ($L_{as} < L_s = L_{tot}$). In this position, the telescope magnification is

$$\Gamma_{as} = \Gamma_{min} = D_{as} = -\frac{f_2}{f_1 + f_2} . \tag{6.40}$$

With $\Gamma_{max} = 1/\Gamma_{min}$, the zoom factor follows as

$$M = \frac{\Gamma_{max}}{\Gamma_{min}} = \left(\frac{f_1 + f_2}{f_2}\right)^2 . \tag{6.41}$$

From this equation and Eq. (6.33), we obtain the focal lengths as a function of the zoom factor M and the overall length of the zoom system L_{tot}, that is,

$$f_1 = \frac{L_{tot}(\sqrt{M} + 1)}{2(\sqrt{M} - 1)} , \tag{6.42}$$

$$f_2 = -\frac{L_{tot}}{2(\sqrt{M} - 1)} . \tag{6.43}$$

Hence, we can calculate from a given length and a zoom factor the focal lengths of a zoom system with three lens elements.

If, in practice, a new telecentric[14] zoom system shall be designed (Figure 6.18), we proceed according to the following steps:

1. Define the minimum magnification Γ_{min} for the asymmetric position L_{as} according to

$$\frac{1}{A(L_{as})} = \Gamma_{min} < 1 . \tag{6.44}$$

2. Define the refractive power of the compensator lens $\mathcal{D}_1 = 1/f_1$.

14) In a telecentric optical system, the entrance or exit pupil is located at optical infinity. As a consequence, the chief rays are parallel to the optical axis in front of or behind the optical system, respectively. If the entrance pupil is at infinity, the image magnification does *not* depend on the object distance or position in the field of view.

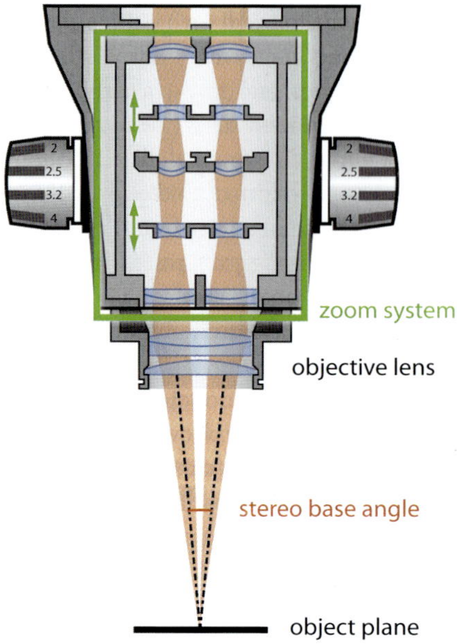

Figure 6.18 Schematic cross-section through the telecentric zoom system and the objective lens system of a surgical microscope (inside green frame). Courtesy of Carl Zeiss.

3. Attend the requirement that the variation of the image position should disappear in the case of an afocal system, that is, $C_{as} = C_s = 0$.
4. Calculate the distance L between L_1 and L_2 (from Eq. (6.38)) as well as the scaling factor ε with

$$L_{as} = \frac{1 - \Gamma_{min}}{\mathcal{D}_1} , \tag{6.45}$$

$$\varepsilon = -\frac{1 + \Gamma_{min}}{\Gamma_{min} + 3/16(1 - \Gamma_{min})^2} . \tag{6.46}$$

5. The scaling factor ε yields, in turn, the refractive power of the second lens, because we have $\mathcal{D}_2 = \varepsilon \mathcal{D}_1$.
6. The magnification of the inverse asymmetric case is then given by

$$\Gamma_{max} = 1 - L_{as}\mathcal{D}_{min}(1 + \varepsilon) . \tag{6.47}$$

7. Finally, the zoom factor is obtained with $M = \Gamma_{max}/\Gamma_{min} = d_E/d_{E'}$, where d_E and $d_{E'}$ represent the diameters of the entrance and exit pupils of the afocal zoom system, respectively.

6.2.3.5 **Illumination**

Besides the quality of optics, the illumination of the observed object is essential for the imaging performance of a surgical microscope. In surgical microscopes for ophthalmology, halogen bulbs and light-emitting diodes (LEDs; see Info Box B.4 in Section B.5.2) are normally used as light sources. The spectral ranges of the light sources (also referred to as the *color temperature*) must be chosen such that the illuminated object appears as "natural" as possible. In addition, we have to pay attention that the illuminated tissue is not damaged due to high light intensity (Section 4.4).[15] For the various segments of the eye, different phototoxic effects have to be considered (see Table 9.2 in Section 9.6).

Köhler illumination *Köhler illumination* with halogen or xenon bulbs [8, 14] has been the standard illumination technique of surgical microscopes for a long time. However, it is now also complemented by other configurations which make use of optical fibers and LED illumination techniques. Köhler illumination has the advantage that the image of the light source is perfectly defocused in the object plane and in the intermediate image planes. In the ray diagram of Figure 6.19a, this can be seen by the rays which originate from any point of the light source. All of these rays pass parallel through the object plane.

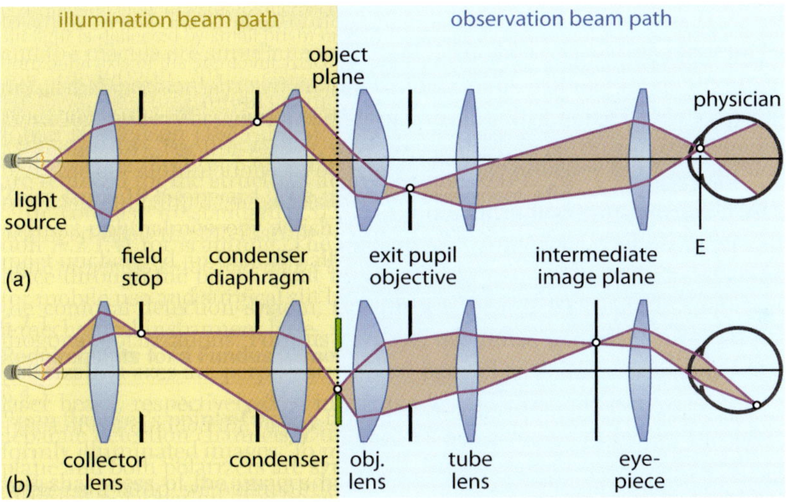

Figure 6.19 Principle of the Köhler illumination. (a) Imaging of the light source. A collector lens forms a magnified image of a light source in the condenser diaphragm, which is located in the image-side focal plane of the collector lens. In combination with a microscope objective lens, we thus have a telecentric beam path. Another image of the light source is formed in the entrance pupil E of the patient's eye so that the fundus is uniformly illuminated. The condenser diaphragm is used to control the brightness for a constant object field. (b) Imaging of the field stop. The field stop is an iris diaphragm and lies in the object-side focal plane of the condenser lens. The condenser lens thus forms an image of the field stop with a reduced image size (green) in the object plane. The field stop determines the size of the illuminated object field for constant light intensity.

15) To derive threshold data for maximum irradiation levels before damage, extensive animal studies have been carried out [13].

The setup of a Köhler illumination consists of a collector lens (field lens), a field stop, a condenser diaphragm, and a condenser lens (system). Figure 6.19a shows that the collector lens forms a magnified image of the light source in the condenser diaphragm. For this purpose, the condenser diaphragm must be located in the image-side focal plane of the collector lens. With the condenser lens and the objective lens, another image of the light source is formed in the exit pupil of the objective. The condenser diaphragm is used to control the brightness of the Köhler illumination for a constant object field. Behind the collector lens, a field stop is placed in the object-side focal plane of the condenser lens (Figure 6.19b). The condenser lens thus forms an image of the field stop with a reduced image size in the object plane (dotted line). As the field stop is an iris diaphragm, it can be used to set the size of the illuminated object field for constant light intensity.

Retro-illumination from the fundus (red reflex) For certain surgical conditions, it is particularly important to optimize the illumination in such a way that the image becomes bright and has a high contrast. In such cases, a *collinear* (e.g., the Stereo Coaxial Illumination SCI™) or *quasi-collinear illumination* is used in which the observation and illumination paths either coincide or form a small angle, respectively. A special arrangement, particularly used for cataract surgery, is referred to as the *red reflex* illumination (Figure 6.20) [15, 16]. Here, the reflected reddish light from the fundus (see also Section 5.1.2) illuminates the observed object (i.e., the anterior eye chamber) uniformly. For this purpose, a *stereo-coaxial* design is used which means that observation and illumination beam paths are overlapped with a beam splitter. Each channel of this illumination is, in fact, a true Köhler-type illumination with a defined illuminated area. The imaging of the red reflex illumination is designed such that the patient's eye itself is taken into account by using the Gullstrand Eye model (Section 2.2.1).

6.2.3.6 Stands and Mounts

In surgical microscopes, not only optical but also mechanical requirements must be taken into account. In particular, the following issues are essential:

- *Usability and ergonomic comfort:* In contrast to laboratory microscopes, surgical microscopes must be freely movable and must provide sufficient free space for the surgeon during an intervention. In addition, the user should be able to comfortably handle the system while sitting or standing.
- *Mobility:* The surgical microscope must allow precise and effortless (re-)positioning before and during a surgery.
- *Stability:* Once a suitable visual field has been found, the system must allow fast position locking such that it remains fixed in space in any arbitrary location. In the case of slight vibrations, the image observed with the surgical microscope should neither move nor drift.

In general, surgical microscopes are mounted to the suspension system of a floor stand or ceiling mount ([17, 18], Figure 6.21). The weight of the surgical microscope

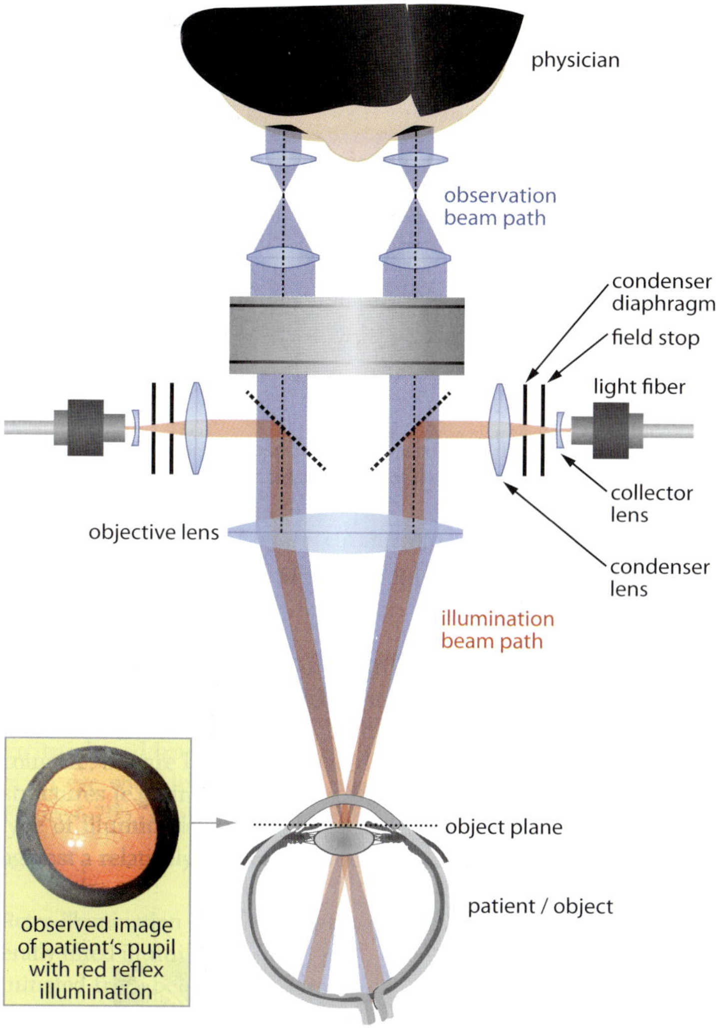

Figure 6.20 Stereo coaxial illumination (SCI) which uses light reflections on the fundus as a secondary light source. In this way, the eye structures can be back-illuminated from inside. As the illumination beam path (red) and the observation beam path (blue) are co-aligned, it is possible to view and efficiently illuminate hollow objects (eye ball) with a relatively small aperture stop (see inset). The SCI is realized by means of a light fiber which is imaged according to the Köhler principle (compare with Figure 6.19) into the patient's eye.

is dynamically balanced with a counter weight or a spring system. As long as the microscope is freely movable, it thus seems to "float" and can be pushed to all directions without effort. Once the desired position has been found, the suspension system can be locked with electromagnetic or mechanical brakes.

Figure 6.21 Photograph of two ZEISS OPMI LUMERA microscopes with floor stand and ceiling mount. Courtesy of Carl Zeiss.

Floor stands stabilize the microscope system because of their relatively high weight (up to 400 kg). Ceiling mounts are tightly bolted to the ceiling of the operating room and thus allow outstanding flexibility with regard to positioning. In addition, ceiling mounts can be easily handled by means of a smooth lift function and magnetic brakes and have sufficient headroom. Surgical microscope stands are also suitable for integrating power lines, data cables, and a computer system used for hardware control, data analysis, and image processing.

6.2.4 **Prospects**

Surgical microscopes are continuously optimized with regard to ergonomic comfort. Current trends in digital projection and optoelectronics also play an important role. Modern surgical microscopes thus do not only provide raw image data, but also help to provide an overview during an intervention. For this purpose, all relevant information about the patient and the examined object is directly visible when looking into the eyepiece so that the surgeon does not have to turn away from the examined object. Furthermore, all modern surgical microscopes are equipped with state-of-the-art video documentation capabilities (often in 3D and high definition).

Another important development is the separation of the so-called *boom system* (consisting of binocular tube and eyepiece) from the detector unit (consisting of objective lens, zoom system, and camera). The digital image data acquired by the detector unit is sent via wireless communication to the mechanically decoupled boom system. As a consequence, the surgeon is no longer tied to a distinct working distance and/or the limits of optical imaging. As the boom system can be freely positioned, this solution allows optimal ergonomic handling. Moreover, the digital image can be easily duplicated such that all medical assistants are able to see exactly the same image as the surgeon.

6.3 Reflection Methods for Topographic Measurements

In ophthalmic diagnosis, the term "topography" refers to the measurement and characterization of the corneal surface shape. The front surface of the cornea has the highest refractive power in the eye (Section 2.1). Thus, it is very important to know the corneal topography for many applications in optometry and ophthalmology, such as contact lens fitting, cataract surgery, corneal surgery, and in particular for refractive surgeries of the cornea. Relevant parameters of an arbitrary point P_j on the corneal surface that can be measured by topographic methods are the elevation from a reference surface (Figure 6.32) and the radius of curvature (Figure 6.33). In addition, other parameters which follow from those (e.g., curvature, refractive power) are relevant as well. We will consider these parameters in detail in Sections 6.3.2.5 to 6.3.2.7. For a short overview, we assign the corneal topography parameters to their particular application in Table 6.5.

To determine the topography of the cornea, *reflection methods* (discussed in this section) and *slit projection techniques* (Section 6.5) are used. Reflection methods use the imaging properties of the reflective surface of the cornea[16] which acts as a convex mirror with a reflectance of approximately 2%. When a test mire is pro-

Table 6.5 Topography parameters and corresponding applications which are necessary for the characterization of the cornea.

Parameter	Application
Radius of curvature	Contact lens fitting
Refractive power	Cataract surgery (determination of the refractive power of the intraocular lens to be implanted)
Radius of curvature, refractive power, elevation profile	Corneal surgery (diagnosis and therapy follow-up)
Elevation profile	Refractive corneal surgery for the determination of the ablation profile (Sections 10.3 and 10.5)

16) In strict terms, reflection methods examine the smooth tear film on top of the corneal surface (Figure 1.7) and *not* the corneal surface itself! However, in most cases, the resulting measuring error is not significant. Special care must only be taken with abnormal tear films and severe corneal irregularities. In the latter case, other methods like slit projection techniques (Section 6.5.1) should be used instead, as they directly characterize the corneal surface.

jected onto the cornea, the reflected image appears as a virtual, upright image (Section A.1.2.1) behind the cornea, which is referred to as the first Purkinje[17] image. The detected size of the Purkinje image depends on the corneal radius of curvature, and its specific shape provides information about present aberrations (Sections 5.3.1 and A.1.6). In keratometers (Section 6.3.1), crosshairs or staircase test mires are quite common, while in reflection-based corneal topographers (Section 6.3.2) circular rings are used.

Info Box 6.2: Glossary of Commonly Used Terms

Keratometer (*synonym*: Ophthalmometer) an instrument which measures the radii of curvature in the central part of the anterior corneal surface in a noncontact manner. Historically, keratometers were the first devices to characterize the corneal surface shape.

Keratometry a method of examining of the patient's cornea with a keratometer.

Corneal topographer an instrument or system which measures the shape of the cornea of the human eye over the entire anterior surface in a noncontact manner.

Placido ring corneal topographer a special type of corneal topographer which measures the corneal surface by analyzing the reflected image of a Placido ring system created by the corneal surface (Section 6.3.2). As a video camera system is used to capture the reflected image, it is sometimes also referred to as a *Placido ring video keratoscope* or *videokeratograph*.

Optical sectioning corneal topographer a special type of corneal topographer which measures the corneal surface by analyzing multiple optical sections of that surface created by scanning-slit projection techniques (Section 6.5).

Topometry/Topography a method of examining the cornea with a corneal topographer.

6.3.1 Keratometer

Keratometers are ophthalmic instruments used to determine the radius of curvature r_C in the central part of the patient's cornea. For this purpose, reflections from the (spherical or toroidal) corneal surface are measured and analyzed. Keratometers are frequently used for contact lens fitting and preoperative diagnosis for cataract surgery. In the case of contact lens fitting, the central corneal radius and the astigmatic error of the cornea are still the most important parameters for selecting the first trial lens. In preoperative cataract surgery diagnosis, the central corneal radius

17) Jan Purkyně (1787–1869).

is the most critical parameter for calculating the refractive power of the implanted intraocular lens (IOL).[18]

History In 1854, the first keratometer was built by Hermann von Helmholtz (1821–1894).[19] This device was developed for research purposes and allowed measurement of the corneal radii of curvature with high precision for the very first time. The respective measuring conditions laid the foundation for the subsequent development of a highly accurate keratometer by ZEISS scientist Hans Littmann in 1950. Importantly, the measuring result of the Helmholtz-type keratometer (details in Section 6.3.1.2) does not depend on the distance between the patient's eye and the instrument.

In 1881, Émile Javal (1839–1907) and Hjalmar Schiötz (1850–1927) designed a conveniently-sized keratometer that rapidly achieved wide acceptance as a diagnostic device. It was mainly used to determine corneal astigmatism. In the early days of optometric refraction, the measured corneal astigmatism served as a basis to calculate the total astigmatism of the eye. The precision of the Javal-type keratometer (details in Section 6.3.1.2) is lower than that of the Helmholtz model, because the measuring result depends on the distance between the patient's eye and the instrument.

The Sutcliffe[20] keratometer was designed in 1907. Instruments produced by the company Bausch+Lomb from 1932 onwards are based on this concept. The Sutcliffe keratometer allows the physician to measure the radii of curvature of an astigmatic cornea in both principal meridians (Section 3.1.2) simultaneously. Although the Sutcliffe keratometer is a distance-dependent, the differences between the radii of curvature are still correctly measured.

6.3.1.1 Functional Principle

The measuring principle of all keratometers is based on (at least) two luminous test mires M_1 and M_2 (Figure 6.22) which are projected onto the corneal surface. M_1 and M_2 are vertically separated by a known distance y and are located at a given distance s in front of the corneal vertex V. The angle included by the mires and the instrument's optical axis is thus predefined by the setup. At the surface of the cornea, M_1 and M_2 form two images M_1' and M_2' which are separated by distance y' ($y' \neq y$). With a suitable observation or detection system, we measure y' in order to determine the desired radius of curvature[21] r_C. Specifically, we start from the lens equation in paraxial approximation (Section A.1.2.1)

$$\frac{1}{s'} = \frac{1}{s} + \frac{1}{f'}, \tag{6.48}$$

18) A measuring error of the corneal radius of 0.1 mm can lead to an IOL fitting error of 0.8 D.
19) Hermann von Helmholtz called his device an "Ophthalmometer". In German-speaking countries, this term is usually used instead of "keratometer".
20) John Sutcliffe (1867–1941).
21) Here, the corneal surface is assumed to be spherical at least in the examined meridian section.

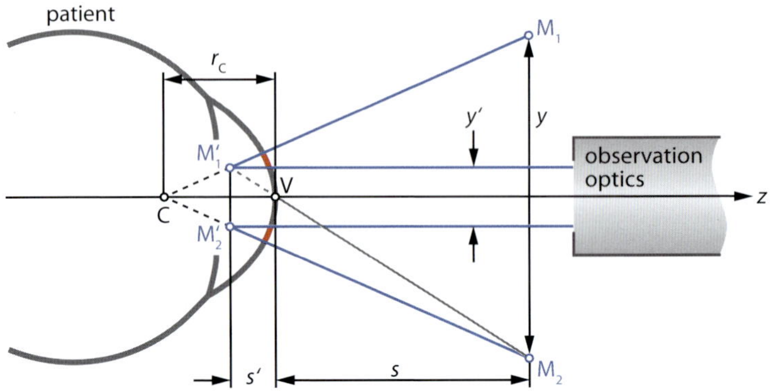

Figure 6.22 Measuring principle of a keratometer. M_1 and M_2 are test mires separated by a distance of y. M_1' and M_2' represent their corresponding images which are separated by a distance of y'. s is the distance between the corneal vertex V and the plane which contains both test mires, and s' is the distance from the corneal vertex to the plane which contains the test mire images. r_C and point C denote the corneal radius of curvature and the center of curvature, respectively. For the measurement, only two small, separated, usually circular segments (red) of the corneal surface are used. Adapted from [21].

$$\frac{y'}{y} = \frac{s'}{s} , \tag{6.49}$$

where f' is the focal length of the cornea. With $f' = r_C/2 = \overline{CS}/2$ and the assumption $y \gg y'$, we obtain (see Problem P6.9) the radius of curvature

$$r_C = \frac{2sy'}{y} . \tag{6.50}$$

In practice, it is critical to keep testing distance s fixed during an examination. Since the calibration of keratometers is based on a fixed projection distance, measuring errors occur if s changes. These errors depend on the *relative* change in distance so that keratometers with small values of s show a considerable distance-dependence concerning the imaging of mires. In this case, additional tools such as fixation aids are needed to minimize the calibration-induced measuring errors. If, however, the test mires are placed in the focal plane of a collimator optic, their apparent distance s is infinite. Keratometers with such an arrangement are thus distance-independent with regard to the imaging of mires. In this case, we may use the simplified relation

$$r_C = Cy' \tag{6.51}$$

with the instrument constant C.

For the measurement, only two small, separated, usually circular segments of the corneal surface are used (red segments in Figure 6.22). Their diameter usually ranges between 0.1 and 0.5 mm. When the test mires are rotated around the optical axis of the instrument, both surface segments trace an annulus on the corneal surface. The outer diameter d of this annulus is approximately equal to y'. d generally ranges between 1.5 and 4.0 mm. In optoelectronic keratometers at least six test mires are used, which are arranged in the form of a circle (Figure 6.26a).

Manual keratometers (Section 6.3.1.2) and optoelectronic keratometers (Section 6.3.1.3) use different measuring principles to determine the desired image distance y'. Let us now discuss them in brief.

6.3.1.2 Manual Keratometer

With manual keratometers, it is impossible to measure y' directly because of involuntary eye movements. As a consequence, these instruments use the *image doubling method* in which a "copy" of the original mire images serves as a measuring scale (see Info Box 6.3). In Helmholtz-type keratometers, image doubling is achieved by parallel plates which can be rotated with respect to each other. Littmann-type keratometers realize image doubling through prisms with a partially transmitting beam splitter. In Javal-type keratometers, the image intensity is split by a Wollaston prism [20], and Sutcliffe-type keratometers separate the geometric image through Scheiner disks (Section 5.2.2.2). In all designs of manual keratometers, the coincidence setting is performed with an observation telescope (Figure 6.23).

Info Box 6.3: Image Doubling Method

In practice, it is quite difficult to measure the distance y' of two points, as shown in Figure 6.23a, if they are continuously moving in a random manner (as in the case of involuntary eye movements). A solution for this issue is the image doubling method in which a "copy" of the original image is overlaid. The image copy now acts as a measuring scale. For this purpose, it is placed next to the original image so that the right point of the original image overlaps with the left point of the copied image (or vice versa; see, for example, coincidence of M_2'' and M_1' in Figure 6.23c). The unknown distance y' equals the known displacement Δy (i.e., $y' = \Delta y$).

Figure 6.23a shows the test mire images M_1' and M_2' in the visual field of the observation system without image doubling. When M_1' and M_2' are doubled, we additionally see the copies M_1'' and M_2'' which are generally displaced by a distance Δy (Figure 6.23b). Since the copied images M_1'' and M_2'' can be moved independently from M_1' and M_2', we can bring them into coincidence such that M_1' overlaps exactly with M_2'' (Figure 6.23c). The shift Δy, which does *not* depend on eye movements, now equals the required distance y'. Examples of frequently used test mire pattern are depicted in Figure 6.23d.

For illustration, let us consider the coincidence setting performed with test mires used in Javal–Schiötz keratometers (Figure 6.23d, middle image). We assume that the examined corneal surface has an astigmatic shape and that the measuring plane formed by M_1 and M_2 does not coincide with one of the principal meridi-

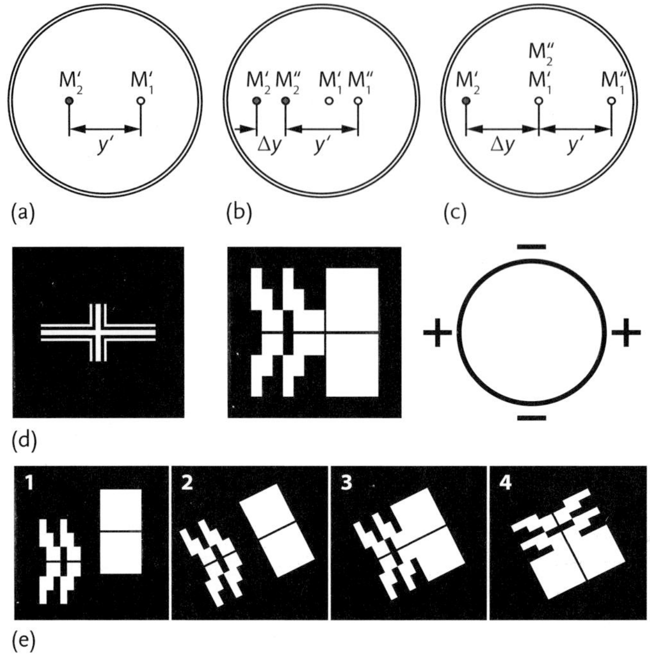

Figure 6.23 Test mires of a manual keratometer. (a) Image of the test mires M_1' and M_2' as seen with the observation system without image doubling. (b) Generated double images which are shifted by a distance Δy relative to each other. (c) The mire images M_1' and M_2'' are brought into coincidence so that we have $y' = \Delta y$. (d) Left: Test mire design used in Littmann keratometers (e.g., instruments by ZEISS). It consists of a hollow cross and crosslines. Middle: Square and staircase mires used in Javal-type keratometers (e.g., by Haag–Streit). Right: +/—-sign used for Sutcliffe-type keratometers (e.g., instruments by Bausch+Lomb). (e) Setting of coincidence with a Javal-type keratometer in the case of an astigmatic cornea. Adapted from [19].

ans. As a consequence, the double images are not only horizontally shifted as in Figure 6.23a–c, but also have a vertical offset (Figure 6.23e, inset 1). We can get rid of the vertical offset by rotating the projection system around the optical axis so that the projection plane is aligned to the principal meridian (Figure 6.23e, inset 2). After coincidence is completed (Figure 6.23e, inset 3), the projection system is turned by 90° and the measurement is performed once again along the other principal meridian (Figure 6.23e, inset 4). In manual keratometers with only two mires, two measurements are required to characterize the corneal astigmatism. This is why they are also called *two-position keratometers*. In contrast, Sutcliffe-type keratometers do not require step 4, since they measure simultaneously in both orthogonal principal planes (*one-position keratometer*).

To set the coincidence, instruments with a variable image doubling and constant mire distance y can be used (Helmholtz–Littmann or Sutcliffe keratometers). Alternatively, the image doubling may be fixed and y is variable (Javal–Schiötz keratometers). The image distance y' can either be determined by measuring the linear distance or the angular separation at which the mire images appear from the viewpoint of the observation optics. Keratometers which measure the linear distance through parallel displacement of the principal rays are distance-independent with regard to observation. One example of such a device is the Helmholtz keratometer, in which the image shift is achieved by plane-parallel plates in the entrance pupil of the observation system. Javal–Schiötz and Sutcliffe keratometers measure the angular separation. They are thus distance-dependent, as the angle depends on the distance between the entrance pupil of the observation optics and the corneal image plane.

We will now consider the Helmholtz and Javal–Schiötz keratometer designs in detail to better understand the specific applications of both principles.

Helmholtz–Littmann keratometers The optical principle of the classic Helmholtz keratometer (Figure 6.24a) is based on two test mires M_1 and M_2 which are fixed in location, usually at a distance of 5 m away from the eye. After reflection at the cornea, the mire images are doubled by means of plane-parallel plates. Hans Littmann slightly modified this design by replacing the plane-parallel plates by an arrangement of prisms (I and II in Figure 6.24b). With regard to the test mire projection and observation, both setups are distance-independent. In the Littmann design, the mire images M_1' and M_2' are located in the object-side focal plane of collimator lenses K_1 and K_2. In addition, M_1' and M_2' lie in the focal plane of objective lens O_1. The reflected beam is then incident on a beam splitter BS_1 which doubles the mire images (intensity splitting). Both partial beams travel along prism arrangements I and II and pass through the movable lenses L_1 and L_2, respectively. A second beam splitter BS_2 recombines the partial beams, and objective lens O_2 eventually generates an image which is viewed through the eyepiece. The double images can be brought into coincidence by shifting lenses L_1 and L_2 in opposite directions (see arrows in Figure 6.24b).

The keratometer design by Hans Littmann meets all requirements made on the original Helmholtz keratometer. The test mires are imaged at infinity, the beam intensity is split in the observation beam path, and we have a telecentric optical system to measure the linear distance between the mire images. This means that the measuring result of a Helmholtz-type keratometer is neither influenced by imperfect focusing nor by ametropia or unintended accommodation of the physician. To date, the Littmann keratometer is still considered by many as the "gold standard" in keratometry. Figure 6.24c shows a photograph of a corresponding, formally commercially available system by ZEISS.

Javal–Schiötz keratometer In contrast to a Helmholtz-type keratometer, the Javal–Schiötz keratometer keeps the distance y' of the double images fixed and measures the variation of mire distance y. As shown in Figure 6.25a, two test mires M_1 and M_2 are arranged on an arc-shaped measuring scale. Both mires can be moved on the arc by the same amounts in opposite directions. During this, the angular separation of the mires α is continuously changed. The objective lens system of the observation system consists of two sub-systems O_1 and O_2. The mire images M_1' and M_2' reflected at the cornea lie in the focal plane of O_1. A birefringent Wollaston

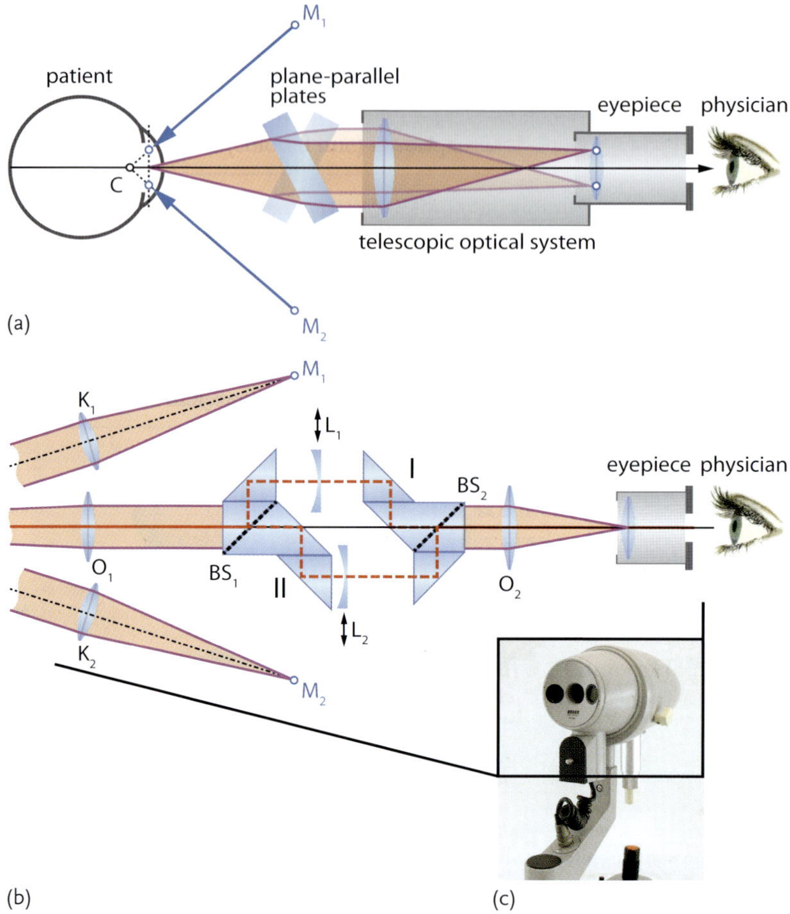

Figure 6.24 Setup of a Helmholtz-type keratometer. (a) Original optical setup of the Helmholtz keratometer. Test mires M_1 and M_2 are projected onto the patient's eye. The reflected image passes through plane-parallel plates for image doubling. An optical (telescope) system then images the mires onto the physician's eye. Adapted from [21]. (b) Modified imaging principle based on Hans Littmann's design. Test mires M_1 and M_2 are projected onto the patient's eye such that their images appear to be located at optical infinity. The reflected beam passes through objective lens O_1 and is divided by beam splitter BS_1 (image doubling). The partial beams (dashed red lines) then travel through prism arrangements I and II. For simplification, only the beam center (red) is shown for the passage through I and II. Along the way, the partial beams pass through movable lenses L_1 and L_2 which are used for the coincidence setting. Beam splitter BS_2 recombines both partial beams, and lens O_2 eventually generates an image which can be viewed through the eyepiece. (c) Front view of a Littmann keratometer by ZEISS. Courtesy of Carl Zeiss.

prism P, which is placed between O_1 and O_2, divides the incident bundle of parallel rays into two partial beams. The partial beams are polarized perpendicularly to each other (Section A.2.1.4) and enclose a fixed angle. Objective lens O_2 then images the shifted test mire pattern onto an eyepiece through which they are ob-

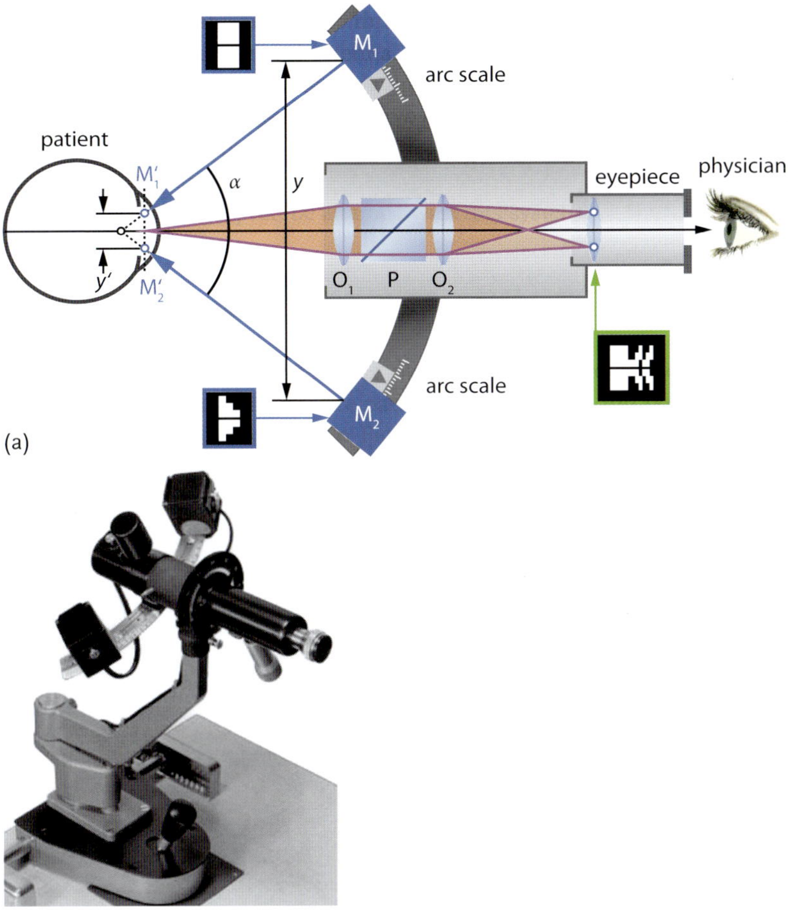

(a)

(b)

Figure 6.25 Design principle of a Javal–Schiötz keratometer. (a) Optical imaging principle. Test mires M_1 and M_2 are arranged on an arc-shaped measuring scale. The projected mire patterns are shown in the insets with blue frames. The reflected mire images M_1' and M_2' are imaged by objective lens system O_1, pass through a Wollaston prism P, and are focused by lens system O_2 onto an eyepiece. The physician then observes both mire images through the eyepiece (inset with green frame). In order to bring the components of the mire images into coincidence, M_1 and M_2 are moved along the arc in opposite direction. The radius of curvature of the patient's eye can be read from a calibrated arc scale. Adapted from [21]. (b) Photograph of a Javal–Schiötz keratometer by Haag–Streit. Courtesy of Haag–Streit Deutschland GmbH.

served. The deflection angle and the average mire distance y are chosen such that both test mire images coincide in the case of a normal corneal curvature. If the corneal radius of curvature is abnormal, we have to change mire distance y until the images overlap again. Once the mire images are brought into coincidence, the radius of curvature and the refractive power of the cornea can be read from the arc scales.

Javal-type keratometers are simple and very robust instruments, as the observation system does not contain any moving optical components. However, a major disadvantage of this design is the distance-dependence with regard to mire projection and observation. Since both errors add up, the instrument must be set very carefully to a defined distance from the patient's eye to prevent potential measuring errors. Figure 6.25b shows a photograph of a formerly commercially available Javal–Schiötz keratometer by Haag–Streit which is still widely used today.

6.3.1.3 **Optoelectronic Keratometer**

In optoelectronic keratometers, near-infrared (usually imaged to optical infinity) LEDs are used as test mires. They are arranged in a circle around the instrument axis (Figure 6.26a). Two opposite diode pairs span a projection plane (meridional plane) in which the corneal radius of curvature is measured. At least three diode pairs are required to determine the corneal astigmatism. However, in modern devices, up to 16 diodes are used simultaneously to minimize measuring uncertainties. The LEDs can be arranged in one ring or two concentric rings. The corneal reflections M_1' and M_2' are then imaged by a telecentric optical system, as shown in Figure 6.26b. Hence, the measuring result of mire image distance y' does not depend on the distance s between the patient's eye and the objective lens. In the image-side focal plane F' of the objective lens, an aperture stop is placed which limits the marginal ray bundles. Eventually, a CCD camera captures the mire images, and mire distance y' is calculated directly from this image. Eye movements during the measuring process have no effect on the measurement, because only short exposure times are needed to take the photo.

As optoelectronic keratometers do not need any components for image doubling and coincidence setting, they are simply constructed, easy-to-use instruments which can be produced very cost-effectively. Due to these advantages compared to manual systems, they are generally integrated as additional components into autorefractors (Section 5.2) and other measuring systems (e.g., optical biometry sys-

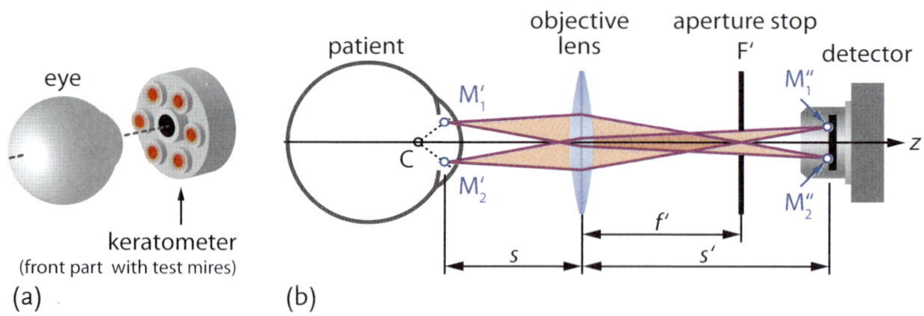

Figure 6.26 (a) Arrangement of illuminating light sources in the optoelectronic keratometer ZEISS IOLMaster® from the patient's point of view. (b) Ray diagram of the observation path in an optoelectronic keratometer. s is the distance between the test mire images M_1' and M_2' and the objective lens. s' is the distance between the objective lens and the mire images M_1'' and M_2'' projected onto the detector (CCD camera). An aperture stop is located at the image-side focal point F' of the objective lens (with focal length f'). Adapted from [21].

tems; Section 7.7). In this case, many mechanical and electronic components of the autorefractors or biometers can be used as well which, in turn, reduces production costs. Consequently, users not only benefit from the enhanced measuring functions of the combined devices, but also from the lower price.

6.3.2
Placido Ring Corneal Topographer

Keratometers are only able to characterize the central part of the anterior corneal surface. If the peripheral areas of the cornea shall be examined as well, instruments such as reflection-based corneal topographers are required. Most of them use concentric rings as a test mire pattern. Today, more than a dozen companies offer a broad range of Placido ring-based corneal topographers equipped with different features. The current devices are used for many applications, for example, in

- refractive corneal surgery (Sections 10.3 and 10.5) for pre- and postoperative diagnostics,
- corneal surgery for diagnosis and therapy follow-up,
- contact lens fitting and measurement,
- orthokeratology for pre- and postprocedural diagnostics, and
- keratoconus (Section 3.1.6) screening.

In addition, the systems can be used to analyze the tear film quality (measurement of the tear breakup time (TBUT)) and for cataract surgery (measurement of the corneal refractive power to determine the required refractive power of the IOL).

History Today's Placido ring corneal topographers use concentric rings as a test mire pattern. In fact, this kind of pattern was already used by António Plácido (1848–1916) in 1880 for qualitative characterization of the corneal surface. At that time, a back-illuminated, opaque disk with alternating black and white rings (Figure 6.27a) was held before the patient's eye. The image of the ring pattern formed by reflection on the cornea was observed through a small diaphragm in the center of the disk. If the corneal surface was deformed, the physician saw a distorted image of the ring pattern. The image could thus be used for a qualitative diagnosis of the corneal shape.

The first successful *quantitative* evaluation was performed by Alvar Gullstrand in 1896. He measured the distances between the individual rings with a microscope and developed suitable algorithms to reconstruct the corneal shape. To date, the principles of Gullstrand's arc-step algorithm (Section 6.3.2.4) are still used in modern corneal topographers. The first commercial instrument which could photograph the deformed ring structures was the photo topographer developed by ZEISS in 1930 together with Marc Amsler (1891–1961) (Figure 6.27b). In the same year, Henri Dekking (1902–1966) designed an instrument in which the Placido rings were arranged inside a conical body instead of on a disk. This device made it easier to evaluate the peripheral areas of the cornea.

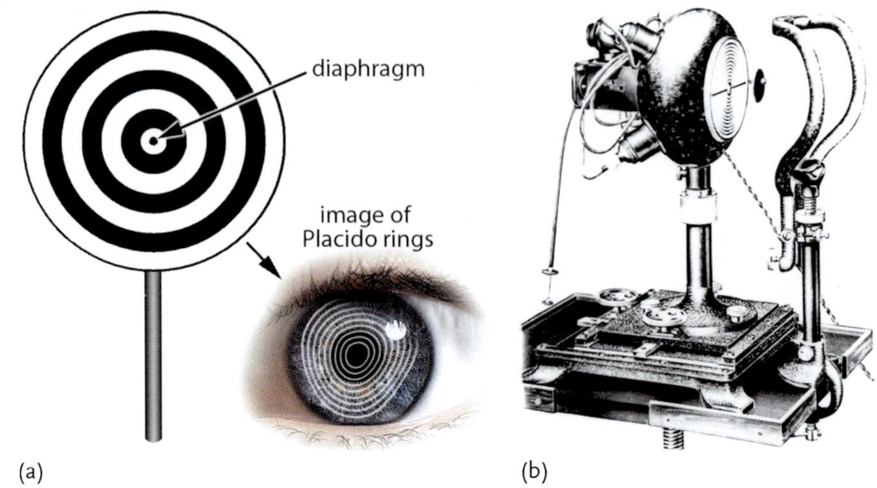

Figure 6.27 (a) Early Placido disk topography. Concentric black and white rings of a Placido disk are projected onto the patient's cornea. The image of the ring structure formed by reflection on the cornea is then viewed through a small diaphragm in the center by a physician or detection system. Inset: Placido ring image on the cornea. The regular ring structure of the Placido disk is deformed by irregularities of the corneal shape. (b) First commercial Placido disk topographer (by the company Carl Zeiss, 1930). A photograph is taken from the image of a back-illuminated Placido disk. Courtesy of Carl Zeiss.

Although a number of companies started to produce photo topographers from the 1950s onwards, the market acceptance was low. The reason for this was mainly the difficult manual evaluation of the recorded Placido ring images and the time-consuming reconstruction of the corneal shape. The situation changed towards the late 1980s with the advent of video technology and the development of powerful computers. The first computer-aided automatic video-based Placido ring topographer was the Corneal Modeling System (CMS-1) by Computed Anatomy introduced in 1987. The CMS-1 was the first system which presented the measured data in the form of color-coded corneal maps. In the early 1990s, the increasing acceptance of laser-assisted refractive surgery of the cornea became an additional driver for innovation. This surgical treatment required novel techniques to determine the corneal shape with high accuracy. However, the spherical approach-based reconstruction algorithms implemented in the early computer-aided corneal topographers were not suitable for this purpose. Only the development of nonspherically-based arc-step algorithms delivered a sufficient accuracy and reproducibility. In 1993, the company Optikon 2000 launched the first corneal topographer with an embedded arc-step algorithm (Keratron) on the market. This system allowed very precise determination of the local corneal curvature.

6.3.2.1 **Functional Principle**

A corneal topographer is based on a projection system (Section 6.3.2.2) which consists of a ring pattern arranged within a conical housing. For the measurement, the

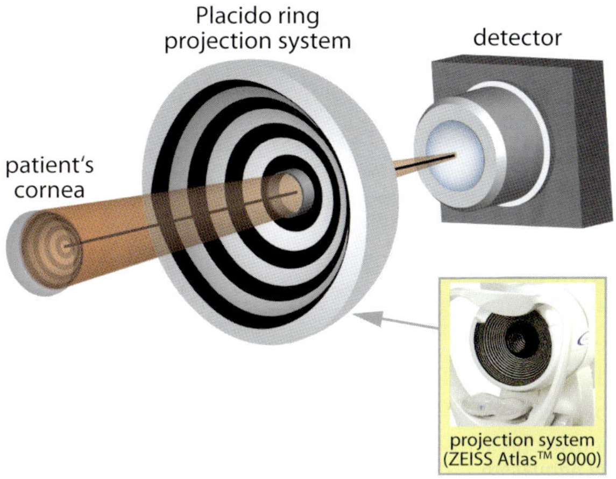

Figure 6.28 In a Placido ring corneal topographer, luminous rings on a concave, aspheric surface are projected onto the patient's cornea, and their images formed by reflection on the cornea are recorded by a detector (e.g., CCD camera) through a central hole in the surface. If the cornea has an irregular curvature, the reflection image of the rings is deformed. A computer algorithm then reconstructs the corneal surface shape from the detected ring shapes. Inset: Commercial corneal topographer which uses Placido ring structures, that is, the ZEISS Atlas™ 9000. Courtesy of Carl Zeiss.

ring pattern is projected onto the patient's cornea, as shown in Figure 6.28. The virtual image behind the corneal vertex is then recorded by means of a CCD camera. As the projection system is usually located between eye and CCD camera, it has a hole in the center for recording. During postprocessing, the image data is digitized and evaluated (Section 6.3.2.3) by suitable computer algorithms (Section 6.3.2.4).

6.3.2.2 **Projection System**

Most corneal topographers use patterns which consist of about 20–30 back-illuminated concentric rings. Usually, the rings are alternating black and white, but some devices also use colored rings. For patients with highly irregular corneal surfaces, colored rings may simplify the identification of overlapping rings. It is common to use a visible (white) light source for back-illumination, whereas some devices have infrared LEDs.

In contrast to the classic Placido disk shown in Figure 6.27a, the rings of a corneal topographer are not arranged in a plane. They are rather inside a spherical or rotationally-symmetric aspherical surface. With such an arrangement, the peripheral areas of the cornea can also be captured. For this purpose, the ring mires must be projected onto the cornea at the widest possible angles. In addition, the non-planar arrangement allows capturing completely sharp images on the CCD chip if the corneal reflections of all rings are imaged into the same plane. The desired flattening of the whole image field is attained by a suitable shape of the interior surface of the projection system. An adequate surface shape avoids blurring of the ring images on the cornea, which would render accurate edge detection[22] more difficult. The wide projection angle can be either generated with a strongly curved projection surface which has a short working distance[23] L_{wd} (*small-target devices*; Figure 6.36b), or with a less curved projection surface which has a longer working distance (*large-target devices*; Figure 6.36a).

For exact and reproducible measurements, the projection system must be oriented as accurately as possible in the lateral directions (see Figure 6.30 for reference) relative to the corneal vertex. In addition, L_{wd} must be set exactly by focusing. For this purpose, most corneal topographers have automatic focusing and adjustment tools for the fine setting after a coarse manual adjustment. As the ring mires are projected onto the cornea from a finite distance, their image sizes depend on the working distance (in accordance with conditional equation (6.50)). The measuring error, in turn, depends on the relative change in distance so that an adjustment error in the working distance ΔL_{wd} leads to a greater measuring error when the working distance is short. As a consequence, the demands made on the adjustment accuracy and focusing are greater for small-target devices than for large-target devices. In addition, the angular error, which arises from an incorrect lateral adjustment, is inversely proportional to the working distance. However, small-target keratoscopes are able to cover a large area of the corneal surface with small instrument footprints. In these devices, the risk of a shadowing effect of the peripheral ring mires by eyelids, eyebrows, and the nose is also reduced. In an ideal measurement, almost the entire corneal surface (maximum diameter of 12 mm) is detected.

6.3.2.3 **Image Acquisition and Analysis**

The reflection image of the ring pattern is acquired with a CCD camera through a central opening in the projection system. The digital data is then stored as an image frame. Usually, several images are recorded, and the most suitable is chosen based on particular selection criteria such as the adjustment status. To allocate the raw data to a polar coordinates system, a central reference point must be defined in the image plane. Either the center of the innermost (smallest) ring mire or, alternatively, the corneal reflex of the fixation light[24] is used for this purpose. In a second initialization step, the exact position of the Placido rings is identified in the digital image by means of edge detection algorithms. Then, data points (raw data) are acquired along defined corneal semi-meridians[25] starting from the

22) Edge detection is a digital image processing algorithm which is used to identify the position and shape of the images of Placido rings formed by reflection on the patient's cornea.

23) We define the working distance as the distance between the examined cornea (corneal vertex) and the entrance pupil of the corneal topographer.

24) The fixation light is aimed at the patient from the center of the projection system.

25) The corneal meridian is the curve formed by the intersection of the *reconstruction plane* (plane which contains the optical axis) with the corneal surface. It is identified by the angle φ between the reconstruction plane and the 0° horizontal plane. The angle φ for a meridian takes values from 0° to 180°. The corneal semi-meridian is the part of a meridian extending from the optical axis towards the periphery in *one* direction. The corresponding angle φ_{sm} for a semi-meridian takes values from 0° to 360°.

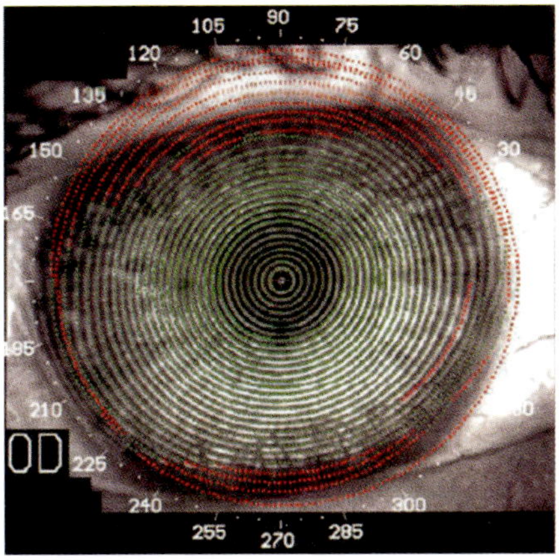

Figure 6.29 Image of Placido rings reflected at a patient's cornea. The picture was taken with the ZEISS Atlas 9000 corneal topographer. Missing data points are interpolated by the data analysis software and shown as red dots. Courtesy of Carl Zeiss.

center. The locations of the data points are determined by the intersections of the examined semi-meridian with the Placido rings. For example, if we assume 360 semi-meridians rotated in 1° steps and 30 Placido rings, 10 800 data points would be acquired. In real examinations, however, some data is lost due to shadowing effects or ring areas that cannot be evaluated for various reasons. In such cases, the missing data points are automatically completed by interpolation, and the affected areas are highlighted (Figure 6.29). Manual corrections by the physician are often allowed as well.

6.3.2.4 **Algorithms to Reconstruct the Corneal Shape**

Once data acquisition is completed, a software algorithm is used to reconstruct the corneal shape. For this purpose, the data points of each examined semi-meridian are processed individually. Each examined semi-meridian lies in a corresponding two-dimensional (2D) *reconstruction plane*[26] which is tilted by an angle of φ with regard to the 0° reference plane (Figure 6.30a). Figure 6.30b shows the ray diagram within such a reconstruction plane. The whole 3D corneal surface is then "put together" step-by-step from the calculated 2D profiles of each semi-meridian.

Let us now briefly consider how the corneal shape is determined from the measured data points along each semi-meridian created by the corresponding reconstruction plane. Usually, so-called *arc step algorithms* are used for this purpose. The

26) The orientation of the reconstruction plane in space is defined by the optical axis of the corneal topographer and by the chief ray (Figure 6.30a). So, the reconstruction algorithm assumes that all rays which are emitted by the rings lie in the same plane as the rays reflected from the cornea. In other words, it is assumed that all incident and reflected rays travel in a common plane. In fact, this condition is only fulfilled if the corneal surface is exactly rotationally symmetric, which is often not the case in practice. However, a more sophisticated analysis [22] reveals that potential errors can be neglected.

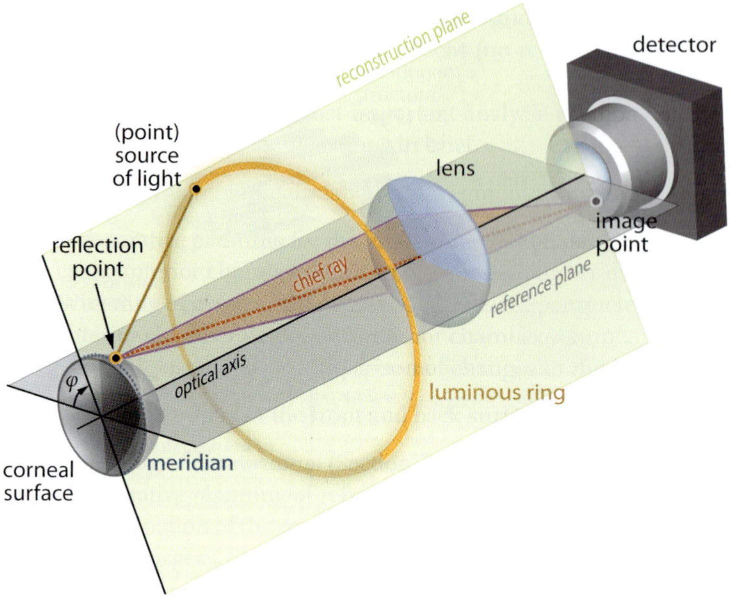

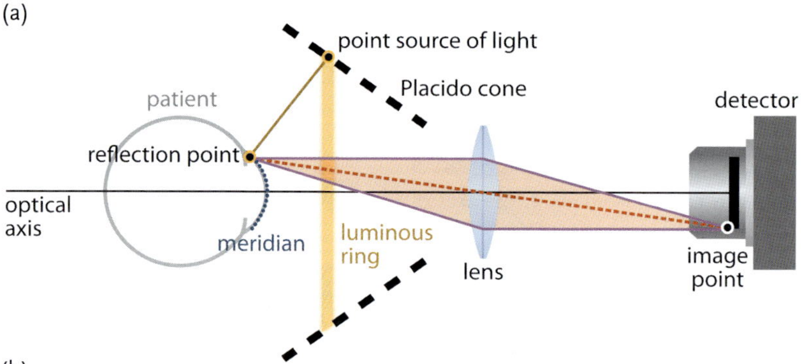

Figure 6.30 (a) Geometric arrangement of a corneal topographer. We consider the point of a luminous Placido ring which intersects with the reconstruction plane (green). The reconstruction plane includes a defined angle φ with the 0° reference plane. The point of intersection is considered as a point source which is reflected on the surface of the patient's cornea. The image of the reflected point is then imaged by a lens into a defined point of the detector plane. In this way, the corneal shape is reconstructed for each meridian (blue dotted intersection of corneal surface with reconstruction plane). (b) Optical imaging of a corneal topographer within the reconstruction plane. The light ray emitted from each Placido ring (i.e., pair of lines in two dimensions) is incident on the cornea. From the reflection point, the ray travels through the optics, and an image is formed on the CCD chip of a detector. Adapted from [22].

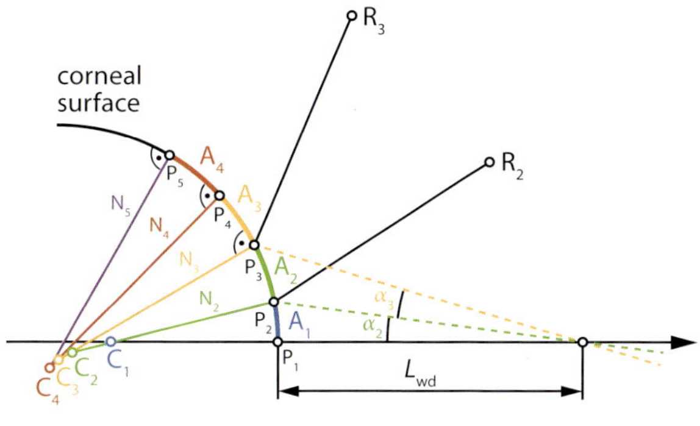

Figure 6.31 Schematic representation of the arc-step algorithm according to van Saarloos. P_j represent reflection points and R_j Placido ring sections (light sources). Between two adjacent points P_j and P_{j+1}, a circle with center C_j is fitted to the corneal surface. The corresponding arc between these points is denoted by A_j. α_j are the angles which define the image size $h'_{\mathrm{I},j}$ (see text) and L_{wd} the working distance. N_j denotes the surface normal at point P_j.

van Saarloos[27] *arc step algorithm* [23] is a typical example of this iterative method. Here, the 2D profile of the cornea along a semi-meridian is approximated by small circular arc segments (Figure 6.31). The center of curvature C_j of an arc A_j is determined by the intersection of the surface normals N_j and N_{j+1} at P_j and P_{j+1}, respectively. To accurately reconstruct the corneal profile, the distances between adjacent Placido rings must be short enough. At first, the exact working distance L_{wd} between camera system and corneal vertex must be found by means of calibration. From this, the image height $h'_{\mathrm{I},j}$ of the jth reflection point P_j (i.e., the object point) can be determined. A possible way to calibrate the system is to intentionally place one ring outside the Placido disk plane. Due to the resulting parallax to the adjacent rings, we may calculate the working distance. In the next step, the local curvature at the corneal vertex is calculated by means of the radius of the innermost ring. The obtained data from the vertex then serve as the initial points for an iterative reconstruction process. As a result, we obtain the local coordinates (x_j, y_j) relative to the corneal vertex and the slope (angle between surface normal and optical axis) for every measured point on the corneal surface.

Once the procedure has been performed for all data points of a certain semi-meridian, it is sequentially applied to all remaining semi-meridians. Missing data from the areas between the rings and between the examined meridians are completed via interpolation. The large amounts of data generated in corneal topographers are summarized in the form of color-coded maps. This kind of representation allows the experienced user to comprehend the whole dataset at a glance. Most frequently used maps are the surface elevation map (Section 6.3.2.5), the curvature map (Section 6.3.2.6), and the power map (Section 6.3.2.7). Curvature and power maps correspond to the traditional keratometry-based description in clinical practice.

27) Wim van Saarloos (born 1955).

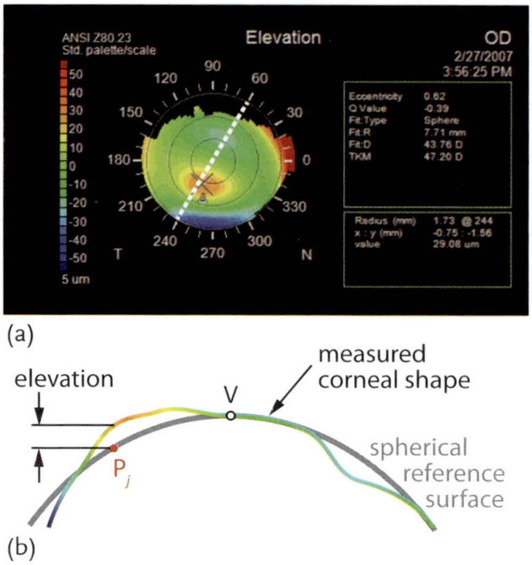

Figure 6.32 (a) Screenshot of the corneal elevation map of a patient's (right) eye with suspect keratoconus analyzed with the ZEISS Atlas 9000. The colors of the elevation map indicate the surface deviation of the examined cornea from a best-fit reference sphere with a radius of 7.71 mm. "Warm" colors (red) represent positive and "cool" colors (blue) negative elevation values. The map shows a maximum elevation (protrusion) of 29 μm above a best-fit reference sphere. Courtesy of Carl Zeiss. (b) Scheme of a corneal cross-section along an intersecting line from 240° to 60° (white dashed line in (a)) compared to the best-fit reference sphere (gray line).

6.3.2.5 **Surface Elevation Map**

The reconstructed corneal topography can be represented by an elevation profile relative to a reference surface (Figure 6.32). Showing the global shape of the cornea relative to a flat reference surface is quite illustrative for the user. However, in this case, surface details get lost. If small deviations of the local shape shall be visualized as well, spherical, aspherical, or toroidal "best-fit" reference surfaces are applied. Depending on the respective application, other types of reference surface can be used as well. For example, the back surface of a contact lens is a suitable reference for contact lens fitting, or the preoperative surface of the cornea is useful to visualize the treatment outcome of a refractive corneal surgery. The elevation map of a patient's cornea with suspect keratoconus is shown in Figure 6.32a as an example.

6.3.2.6 **Curvature Map**

The curvature of the corneal surface at a point P_j of a certain semi-meridian can be described by means of the axial radius of curvature r_a (Figure 6.33a) or the meridional radius of curvature[28] r_m (Figure 6.33b). To describe the corneal shape, the inverse radii are often used, which are referred to as the *axial curvature* $K_a = 1/r_a$ and *meridional curvature* $K_m = 1/r_m$, respectively. The curvature values are usu-

28) Instead of the term "meridional", "tangential" or "instantaneous" are often used.

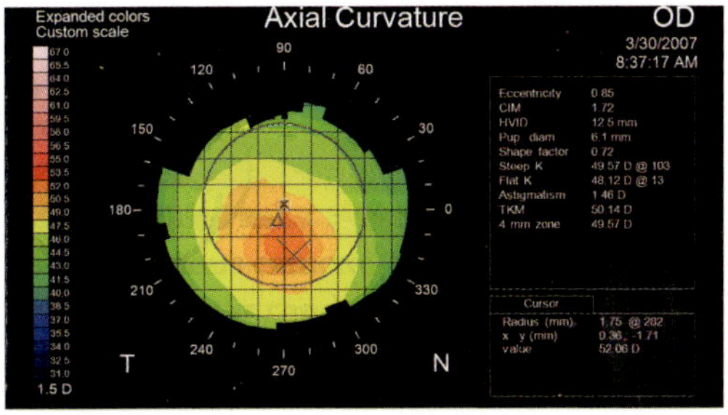

(a)

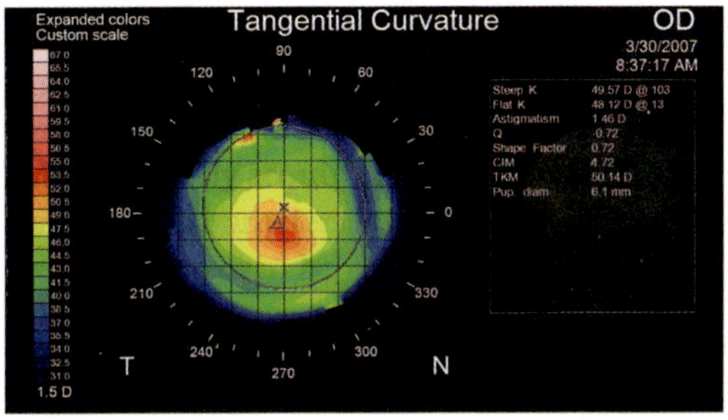

(b)

Figure 6.33 Screenshots of curvature maps of a patient's eye with suspect keratoconus analyzed with the ZEISS Atlas 9000. (a) Axial curvature map. (b) Meridional (tangential) curvature map. The curvature values are specified in keratometric diopters. "Warm" colors (red) are used for large curvature values (small radii of curvature). "Cool" colors (blue) are used for small curvature values (large radii of curvature). The meridional curvature map is better suited to display suspect keratoconus and the peripheral areas of the cornea. Courtesy of Carl Zeiss.

ally *not* specified in inverse millimeters (1/mm) but multiplied by the *keratometric index* $n^* = 337.5$ in *keratometric diopters* (see also Section 6.3.2.7).

Axial radius (axial curvature) In accordance with Figure 6.34a, the axial radii of curvature are given by

$$r_{a,j} = \frac{y_{P,j}}{\sin \alpha_j} , \tag{6.52}$$

and the corresponding axial curvatures by

$$K_{a,j} = \frac{1}{r_{a,j}} = \frac{\sin \alpha_j}{y_{P,j}} . \tag{6.53}$$

The axial radius of curvature $r_{a,j}$ at point P_j is the distance from the surface normal to the optical axis (green arrow in Figure 6.34a). In contrast to meridional radii, axial radii always originate from the symmetry axis of the considered surface.

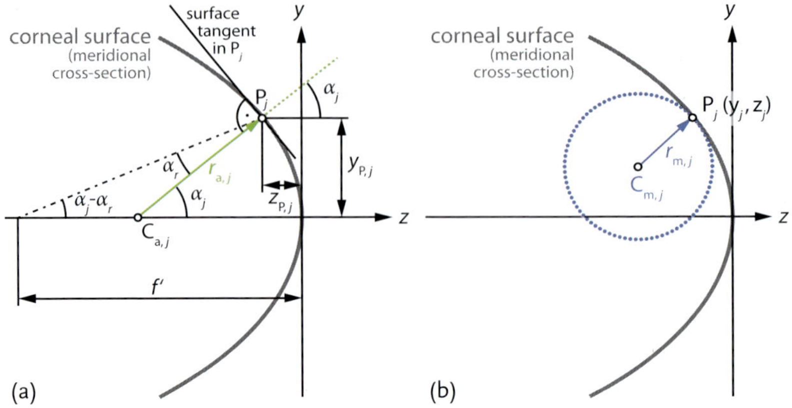

Figure 6.34 (a) The axial radius of curvature is given by $r_{a,j} = \overline{C_{a,j} P_j}$. $y_{P,j}$ is the distance from the optical axis to the considered point P_j on the patient's cornea, f' the image-side focal length of the cornea, and $z_{P,j}$ the distance of the corneal vertex plane (y axis) to P_j. (b) The meridional radius of curvature is determined by $r_{m,j} = \overline{C_{m,j} P_j}$.

Strictly speaking, axial radii can thus only be defined for surfaces which have certain symmetry features, for example, spherical or toroidal surfaces. Thus, with axial radii, only the central portion of the cornea which is approximately spherical can be adequately described.

Meridional radius (meridional curvature) The meridional radii are better suited to characterize the peripheral areas of the cornea. In the simplest case, the meridional radius of curvature $r_{m,j}$ at a point P_j of a semi-meridian is determined by the mean value of the radii of curvature of two adjacent arcs which pass through P_j. However, the shape of the cornea is usually approximated in the semi-meridian by means of a polynomial function $y(z)$. The meridional radius is thus given by

$$r_{m,j} = \frac{\left(\sqrt{1 + (\mathrm{d}y_j/\mathrm{d}z_j)^2}\right)^3}{\mathrm{d}^2 y_j/\mathrm{d}z_j^2} \tag{6.54}$$

($[r_{m,j}] = \mathrm{mm}$) and the corresponding meridional curvature specified in keratometric diopters by

$$K_{m,j} = \frac{n^*}{r_{m,j}} = \frac{337.5}{r_{m,j}} \,. \tag{6.55}$$

The meridional radius of curvature at a point P_j is the radius of a circle that fits the curve best at P_j (see tangent/osculating circle at P_j in Figure 6.34b).

6.3.2.7 Power Map

An alternative way to describe the corneal surface curvature at a distinct point P_j is the (refractive) power map. To calculate the corresponding data, we could in principle use the paraxial equation (2.22) for the total refractive power of the cornea $\mathcal{D}'_c$ or

Eq. (2.21) for the corneal back vertex power[29] $\mathcal{D}'_{\mathrm{cv}}$ from Section 2.2.1.1. However, in corneal topographers (and keratometers; Section 6.3.1), we can only determine the radius of curvature of the corneal front surface r_{C}. The radius of curvature of the corneal back surface is unknown though. For this reason, $(n' - n)$ must be replaced by $(n^* - 1)$ so that we obtain the general *keratometer equation*

$$\mathcal{D}'_{\mathrm{c}} = \frac{n^* - 1}{r_{\mathrm{C}}} , \tag{6.56}$$

$$\mathcal{D}'_{\mathrm{cv}} = \frac{n^* - 1}{r_{\mathrm{C}}} . \tag{6.57}$$

n^* denotes again the keratometric index which is not a refractive index in the common sense but rather a calculated effective value. Depending on which refractive power we consider ($\mathcal{D}'_{\mathrm{c}}$ or $\mathcal{D}'_{\mathrm{cv}}$), different values for n^* must be used. If Eq. (6.56) is inserted into Eq. (2.20) or Eq. (2.21), we obtain the conditional equation for n^*. With the values of the Exact Gullstrand Eye (Table 2.1 in Section 2.2.1), the ratio of the radii of curvature is $7.7\,\mathrm{mm}/6.8\,\mathrm{mm} = 1.132$.[30] As a result, we have $n^* = 1.3315$ when Eq. (2.20) is used and $n^* = 1.3377$ when Eq. (2.21) is used. If the radius of curvature of the corneal front surface is given in millimeters, the keratometer equations can be approximated by

$$\mathcal{D}'_{\mathrm{c}} \approx \frac{331.5}{r_{\mathrm{C}}} , \tag{6.58}$$

$$\mathcal{D}'_{\mathrm{cv}} \approx \frac{337.7}{r_{\mathrm{C}}} . \tag{6.59}$$

Some manufacturers of keratometers and corneal topographers use $n^* = 1.3320$ (e.g., Carl Zeiss and Topcon) or $n^* = 1.3375$ instead. Thus, if we want to compare the refractive powers measured with different instruments, the respective keratometer indices and the measured "raw" data must be available.

Equations (6.58) and (6.59) are only valid in paraxial approximation. This means that they can only be applied within a small area (approximately 3 mm) around the corneal vertex, which actually corresponds to the typically acquired geometric range of keratometers (Section 6.3.1). In this region, the refractive power values obtained with a corneal topographer are thus equivalent to the values measured with a keratometer. This is the reason why these measurements are referred to as *simulated keratometer readings* (abbrev., "simulated K" or simply "Sim K"). If we

29) In optometry, the back vertex power $\mathcal{D}'_{\mathrm{cv}}$ is often considered instead of the total corneal refractive power $\mathcal{D}'_{\mathrm{c}}$.$\mathcal{D}'_{\mathrm{cv}}$ is determined by the inverse distance from the corneal back vertex to its image side focus.

30) In corneal refractive surgery (e.g., photoablation; Section 10.3), the radius of curvature of the corneal front surface is modified by intention. As a consequence, the ratio of the radii of curvature of the corneal front and back surfaces deviates from the (supposed) constant relation so that we cannot use Eq. (6.56). To calculate the refractive power of a modified corneal surface, the exact Eqs. (2.20) and (2.21) must be used instead. As they also contain the refractive power of the corneal back surface and the corneal thickness, we have to measure these quantities with alternative methods, such as the scanning-slit techniques (Section 6.5).

use the equations for the characterization of the peripheral areas of the cornea, the deviations of the measured from the real refractive power values will increase with the distance from the corneal vertex. The deviations are mainly caused by the increasing influence of spherical aberrations. As a consequence, if "power" maps calculated with Eqs. (6.58) and (6.59) are applied to the entire cornea, they should be rather called (dioptric) curvature maps (Section 6.3.2.6), as they indicate the surface shape, but in general not the true refractive power.

A more suitable description of the true distribution of the corneal refractive power results when we apply Snell's law to each point of the calculated surface shape and assume a refractive index of $n_k = 1.3357$ for the keratometer. In this case, we may plot the so-called *ray-tracing refractive power map* whose values are given by (Figure 6.34a)

$$\mathcal{D}_c' = \frac{n_k}{f'} = \frac{1.3357}{y_P / \tan(\alpha_j - \alpha_r) + z_P} . \tag{6.60}$$

Again, for the user-friendly representation of the measurement results, a color code has been established. Green and yellow represent the value range for normal corneal surfaces. "Warm" colors like red and orange are used for upward deviations, for example, stronger curvature and higher refractive powers. "Cool" colors like blue and violet are used for downward deviations. However, the scaling is sometimes critical. If the gradation is too coarse, small deviations may get lost. If the gradation is too fine, details may be emphasized which are not of clinical relevance. Usually, gradations with an interval of 1.5 D are a good compromise with regard to sensitivity, specificity, and the measurement range.

6.3.2.8 Precision and Measurement Range

Placido disk-based corneal topographers allow a very accurate determination of the refractive power distribution and the radii of curvature of the patient's cornea. The accuracy and reproducibility of the refractive power lies in the range of ±0.10 D and the corresponding radii of curvature have a tolerance of ±0.02 mm. Such narrow tolerances cannot be attained by any other commercially available topography measurement method. Only keratometers are even more accurate. However, they can only characterize the central part of the cornea and do not deliver reliable results in the periphery. Placido disk-based measurement systems are thus often integrated into other diagnostic devices which require precise data about the corneal surface shape. In commercial systems, the measurement range of the refractive power varies between 10 and 100 D. The range of radii of curvature varies between 33.75 and 3.375 mm.

6.3.2.9 Application Packages for User-Specific Software Modules

In addition to the standard measuring procedures, other evaluation routines are usually implemented:

- Software modules have been developed as tools to assist with refractive surgery screening and to help to identify abnormal corneal conditions, for example, keratoconus.

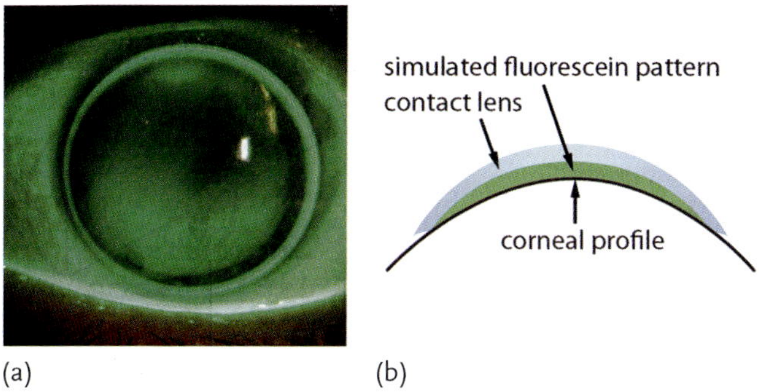

Figure 6.35 (a) Fluorescein imaging of the cornea used for contact lens fitting. Courtesy of Carl Zeiss. (b) Scheme of corresponding corneal cross-section which shows the layer of the contact lens and the fluorescein pattern.

- Fluorescein imaging of the cornea for contact lens fitting (Figure 6.35) can be simulated so that no fluorescein dye is required.
- Other application packages allow determination of the tear film quality by measuring the tear break-up time (TBUT). During this procedure, the image quality of the ring mires serves as a measure so that no fluorescein dye is required.
- Software modules exist which characterize the corneal surface shape via Fourier (see Info Box A.2 in Section A.2.4) and/or Zernike (Section A.1.8.2) analysis.

6.3.2.10 **Commercial Devices**

A broad range of Placido disk-based corneal topographers is commercially available. Figure 6.36 shows two typical examples of the two conceptual designs, that is, large-target and small-target corneal topographers.

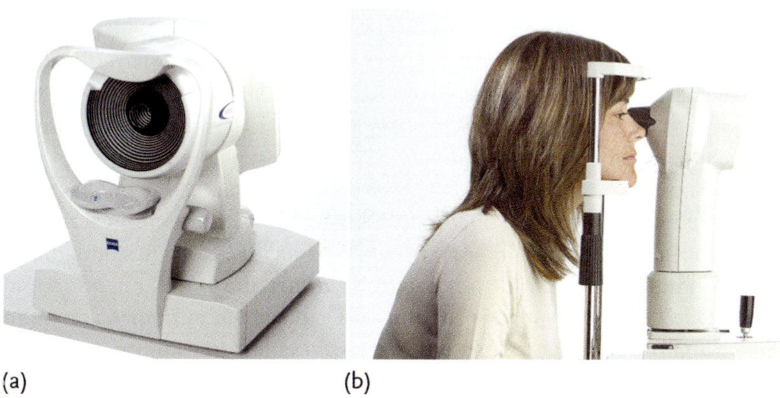

Figure 6.36 (a) Photograph of the large-target system ZEISS ATLAS 9000 with a working distance of 70 mm. Courtesy of Carl Zeiss. (b) Example for a small-target device. The Medmont E300 has a working distance of 30 mm. Courtesy of Medmont Pty Ltd.

6.3.2.11 Prospects

Placido disk-based corneal topographers are increasingly integrated into multifunctional hybrid devices which allow characterization of the corneal topography and/or the imaging properties of the cornea in a comprehensive manner. For example, Placido disk-based subsystems are included in scanning-slit devices (Section 6.5.2) and aberrometers (Section 5.3).

6.4
Slit Lamp

Today, the slit lamp is the most flexible and widely-used instrument for ophthalmic diagnosis. It is mainly used for the visual examination of the anterior eye segment, that is, cornea, eye lens, anterior and posterior chamber, and the anterior part of the vitreous. With additional optics (auxiliary lenses; Section 6.4.4.1), it is also possible to examine the posterior eye segment (fundus) and the iridocorneal angle (Figure 1.4 in Section 1.1), which are usually not directly accessible. The slit lamp also serves as a mechanical and optical support for many accessories, for example, systems to measure the intraocular pressure, systems to measure the lengths and angles of ocular parts, and video and photo cameras for documentation. In addition, slit lamps are often used as beam delivery devices for visually controlled laser treatment (Sections 10.1.2 and 10.2.4.3).

History The first slit lamp was presented by Alvar Gullstrand in 1911 (Figure 6.37a). The company Carl Zeiss produced commercial systems based on this concept from 1912 onward. Strictly speaking, Gullstrand's slit lamp was "merely" a slit light projector. The scattering images of the anterior eye segment had to be observed with external devices, for example, with a handheld monocular or binocular loupe (see also Section 6.1). In 1915, Leonard Koeppe (1884–1969) proposed combining Gullstrand's slit light projector with the ZEISS binocular corneal microscope[31] into a single unit (Figure 6.37b).

In 1933, the handiness and compactness of the first slit lamp was considerably improved according to a proposal by Wilhelm Comberg (1885–1958). The slit light projector was arranged in a vertically downward position, as schematically shown in Figure 6.40a. A small prism deflected the slit light onto the horizontal axis of a microscope. Both slit light projector and microscope could be swiveled independently of each other around a common rotary axis which lies in the focal planes of the corneal microscope and the slit light projector. This design principle was further enhanced by Hans Littmann in 1953 in many ways and is to date still used for all ZEISS slit lamps.

A similar design solution with a common rotary axis was proposed by Hans Goldmann (1899–1991). In contrast to the design concept of Wilhelm Comberg

31) The binocular corneal microscope was developed by the ZEISS scientist Siegfried Czapski (1861–1907) in 1889.

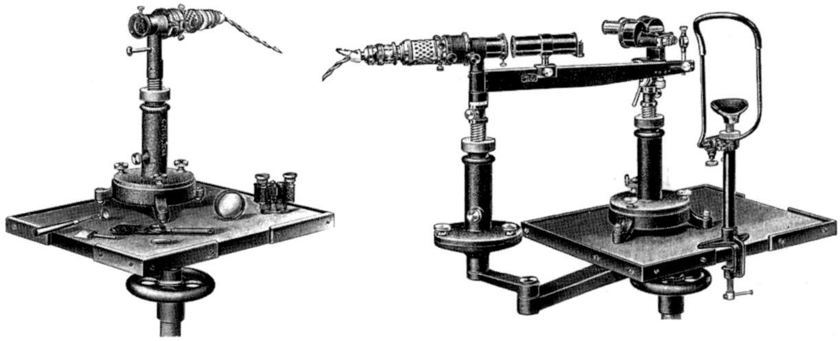

(a) (b)

Figure 6.37 (a) Drawing of Gullstrand slit lamp (1912). (b) Drawing of Gullstrand slit lamp with corneal microscope (1915). Courtesy of Carl Zeiss.

and Hans Littmann, the slit light projector was horizontally arranged in relation to the microscope. From 1933 onward, Goldmann's concept has been used for slit lamps by Haag–Streit. The special feature of these instruments was the mechanical instrument base that could be moved horizontally with a single control element (joystick) and vertically with a thumb wheel. In this way, the slit lamp could be adjusted relative to the patient's eye with only one hand. This concept not only simplified the handling, but also extended the functionality of the slit lamp. In 1958, Haag–Streit introduced the slit lamp series 900 with the characteristic vertical upward positioning of the tiltable slit light projector (Figure 6.41a). In 1969, a 3D joystick was added for simple adjustment of the instrument base in all three spatial directions. To date, this design principle is still used in all Haag–Streit slit lamps.

6.4.1 Functional Principle

Both the cornea and the lens of a healthy eye are highly transparent for visible light (Section 2.1.10), as they consist of an ordered tissue structure and do not contain any strongly absorbing chromophores. It is well-known that such structures are difficult to image in transmitted and reflected light, since the relative amplitude modulation (variation of the light wave amplitude) of light is too weak and the phase modulation (variation of the light wave phase) cannot be perceived by the human eye. However, such objects can be observed well in scattered light (Section 9.2) if a bright, narrow slit of light is projected onto the transparent object with otherwise low ambient brightness. The situation can be easily understood when we consider an intense light beam entering a dark room through a tight window slit. In this beam, dust particles become visible which could never be seen if the room were brightly illuminated. With a narrow slit and a sufficiently small aperture angle, the illumination beam has a shape defined by two knife edges placed end to end. When the beam passes through transparent structures in the anterior part of the eye, it

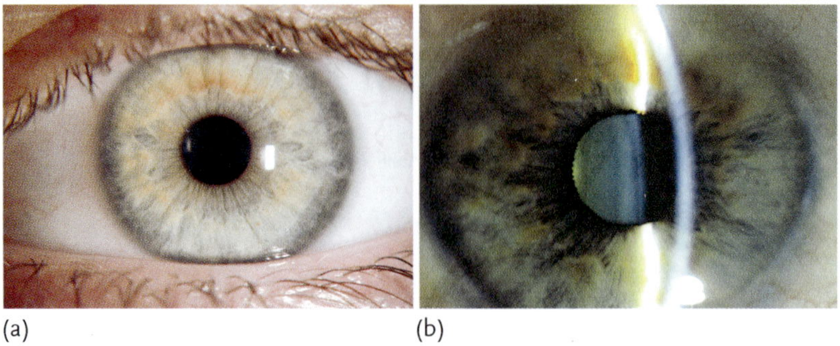

Figure 6.38 (a) Front view of an eye in normal (diffuse) illumination. The cornea can only be recognized via reflections (Purkinje image). (b) Front view of the eye with slit illumination in which the scattering images provide an optical sectioning. With such an illumination technique, we can also see the cross-sections of cornea and eye lens. Courtesy of Carl Zeiss.

is scattered at microscopically small inhomogeneities[32]. The slit-shaped scattering image of the structures is referred to as the *optical section* (Figure 6.38b). The intensity of the scattering light within the optical section increases with increasing slit illuminance $\mathcal{E}_v$ (Section A.2.1.5) and increasing portion of blue light in the spectrum of the light source used for the slit illumination (high color temperature).

6.4.2 Modular Structure

As shown in Figures 6.40a and 6.41a, a slit lamp essentially consists of a slit light projector and a stereo microscope which allows a detailed observation of the scattering image. The two units are attached to a mechanical base to position the measuring head relative to the patient's eye. Let us now consider the individual assemblies and modules.

6.4.2.1 Slit Illumination Device

The slit light projector of a slit lamp is intended to form a bright slit image with maximum homogeneity on the patient's eye. In addition, the length, width, and angle of the slit image must be precisely adjustable. With regard to user-friendliness, the working distance between the patient's eye and the instrument must be convenient, and the depth of field (see also Sections 6.2.2 and 6.5.2.2) of the slit image must be as large as possible. Unfortunately, the requirements in terms of brightness, working distance, and depth of field cannot be met at the same time, since these parameters depend on the illumination aperture to different extents.

32) Unfortunately, very little information is available on the wavelength dependence of the scattering intensity I of cornea and lens. Roughly speaking, the cornea tends to be a Rayleigh scatterer ($I \propto \lambda^{-4}$; Section 9.2.1), and the eye lens is more a Rayleigh-Gans scatterer [24].

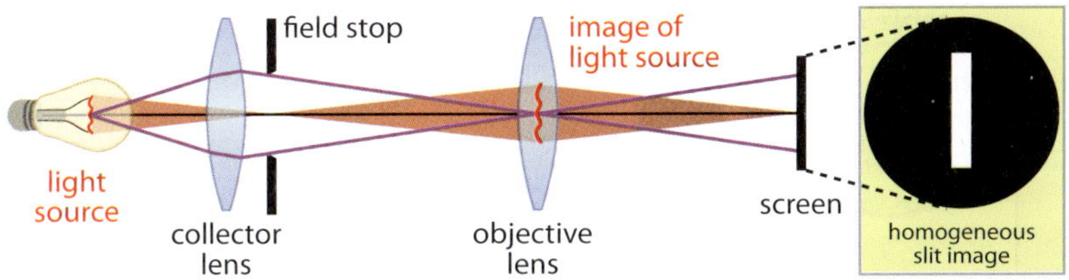

Figure 6.39 Köhler-type illumination. The light source (glow filament) is imaged by a collector lens in the objective lens. The objective lens images the field stop (mechanical slit) in front of the collector lens onto a screen. As a consequence, a homogeneously illuminated slit image appears on the screen despite the structured filament of the light source.

Formation of the slit image In the slit lamp projector, a Köhler-type illumination (see also Section 6.2.3) is used to provide a homogeneously illuminated slit image (Figure 6.39). The light source is imaged by a collector lens in the objective lens. The objective lens, in turn, images the field stop near the collector lens onto the object to be examined. As a consequence, a very homogeneous slit image is obtained even with a structured light source (inset in Figure 6.39). Form and position of the slit illumination can be adapted to the different observation requirements. The width of the slit image can usually be adjusted continuously in the range from 0–14 mm and the slit image length in steps or continuously in the same range. Typically, the slit illumination can be rotated horizontally by $\pm 90°$ and vertically from $0°$ to $-20°$ relative to the common rotary axis of illumination and microscope (Figure 6.40a). The decentration of the horizontal slit image relative to the focal point of the microscope lies usually in the range of $\pm 4°$.

The brightness of the slit lamp illumination is characterized by the illuminance (Section A.2.1.5) in the slit image given by

$$\mathcal{E}_\mathrm{v} = \frac{S_\mathrm{ep}}{L^2} T \mathcal{L}_\mathrm{v} \,. \tag{6.61}$$

$\mathcal{L}_\mathrm{v}$ denotes the luminance of the light source, T the transmittance of the imaging optics, S_ep the area of the exit pupil[33], and L the distance between exit pupil and slit image.

Light sources Due to the strong wavelength dependence of the scattering behavior of cornea and eye lens, the required light sources should have a considerable blue spectral portion. Thus, tungsten and halogen filament bulbs are usually integrated, but LEDs (see Info Box B.4 in Section B.5.2) are increasingly replacing the classic light sources.

Optical accessories for special applications For certain examination methods (Section 6.4.4), a large-area and diffuse illumination of the eye is more suitable than a bright, thin light slit. For diffuse illumination, a diffusing disk or ground glass

33) This corresponds to the area of the deflecting prism/mirror in the slit light projector (Figures 6.40a and 6.41a).

screen is placed in front of the light source image at the exit pupil. These components "interrupt" the beam path and act as a secondary light source.

For other applications, a modified spectrum of the light source is needed. This can be achieved with additional filters in the illumination beam path. Gray filters reduce the illuminance $\mathcal{E}_v$ while keeping the spectrum unchanged. Green filters are mainly used to enhance the image contrast, and blue filters act as excitation filters for fluorescence methods.

6.4.2.2 **Slit Lamp Microscope**

The stereo microscope included in the slit lamp allows observation of the patient's eye with selectable magnification and adequate resolution (Problem P6.11). The field of view (Section 6.2.2)) and depth of field should be as large as possible. A sufficiently large working distance is additionally required in order to provide enough space for manipulation of the patient's eye and/or to place certain accessories (e.g., contact lenses). These properties are fulfilled by telescope systems which are also common for surgical microscopes (Section 6.2). However, in contrast to surgical microscopes, slit lamp microscopes usually have a number of fixed magnification settings in the range between 10 and 50 ×.[34] The magnification changers are located in the parallel beam path between the main objective lens and the tube lenses. In this part of the microscope, further (plane-parallel) optical components such as filters and beam splitters can be easily added without decreasing the image quality.

Typical optical parameters for stereo microscopes used in slit lamps are an objective lens focal length of 140 mm, back-focal distances from 90–120 mm, working distances between 60 and 70 mm, and stereo base angles (Section 2.1.9) between 10° and 12°. All other optical parameters of a slit lamp, that is, resolution, brightness, focal length, stereoscopic depth perception, and so on, must be chosen (and balanced) as described in Section 6.2.1.

6.4.2.3 **Mechanical Components**

Handling and functionality of a slit lamp are mainly determined by the mechanical components. On the one hand, the mechanical base of a slit lamp must provide a simple linking of slit light projector and stereo microscope. On the other hand, the positioning relative to the patient's eye must be precise and easy to handle. Typically, the slit light projector and stereo microscope can be rotated independently around a common axis (Figure 6.40a). The rotary axis is realized by mechanical bearing shafts located below the patient's eye. Since the axis lies in the focal planes of the microscope and the slit light projector, the rotation of both components does not change the position of the adjusted focus.

The mechanical base also features a cross-slide guideway for the horizontal movement and a tool for vertical adjustments. The entire illumination and observation system of the slit lamp thus can be positioned relative to the patient's eye with a three-coordinate joystick. A headrest attached to the instrument table is used to lock the position of the patient's eye during examination.

34) The magnifications 10 ×, 16 × and 25 × are most frequently used.

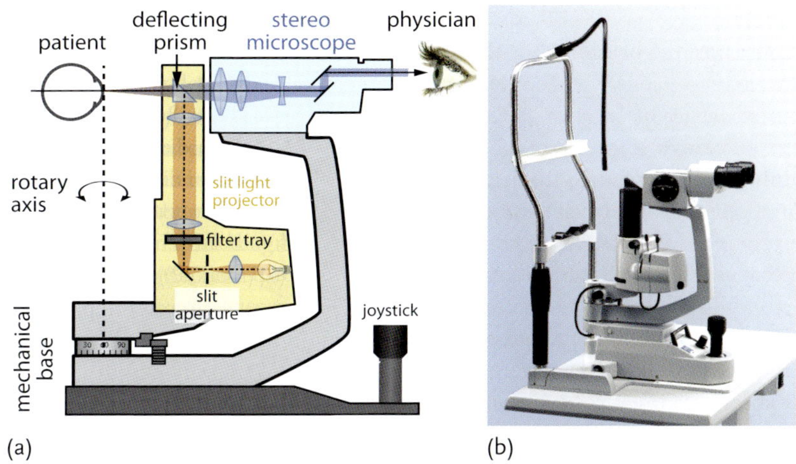

Figure 6.40 (a) Schematic optical setup of the ZEISS SL 120 slit lamp. The slit light projector is arranged below the stereo microscope. The slit light is deflected by small prism onto the horizontal axis of the stereo microscope. Both the slit light projector and the stereo microscope are attached to a common mechanical base and can be swiveled around a common axis (dashed line) independently of each other. Normally, the common rotary axis lies in the focal planes of microscope and slit light projector. (b) Photograph of the ZEISS SL 120 slit lamp. Courtesy of Carl Zeiss.

As the slit illumination can be arranged above or below the corneal microscope, two basic slit lamp designs exist. In ZEISS-type slit lamps, the slit light projector is arranged *below* the microscope (Figure 6.40). In Haag-Streit-type slit lamps, the slit light projector is located *above* the microscope (Figure 6.41). Handheld slit lamps for mobile use and surgical slit lamps attached to surgical microscopes do not have a mechanical instrument base.

6.4.3
Types of Illumination for Various Applications

With a slit lamp, we obtain different information about eye structures in the anterior segment depending on the chosen type of illumination.

Diffuse illumination Diffuse illumination (Figure 6.42a) is used to obtain a general overview of the entire anterior eye segment. For generation of diffuse illumination, the slit is adjusted to the full aperture and a diffusing disk is swung into the illumination beam path. The anterior segment is then usually examined with a microscope magnification of 5–10 ×.

Direct focal illumination For standard observations, direct focal illumination (Figure 6.42b) is most often used (see also Figure 6.52). With a narrow slit image (width of about 0.1–0.2 mm) and a sufficiently small angular aperture, the illuminating beam takes the form of two knife blades placed edge to edge (see also Figure 6.52a).

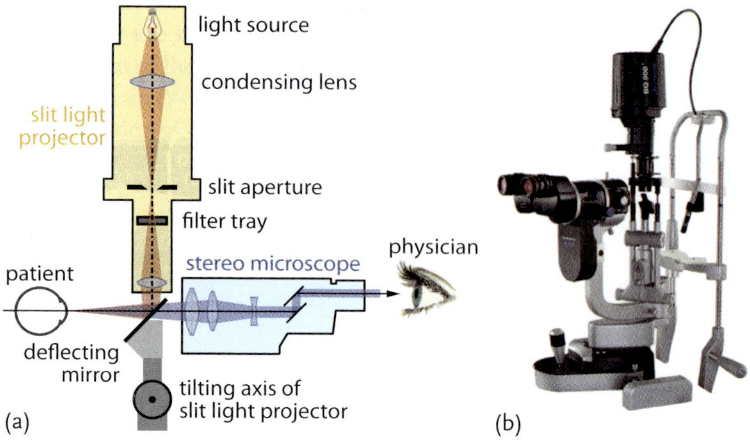

Figure 6.41 (a) Schematic optical setup of the Haag–Streit 900 BQ slit lamp. In contrast to the arrangement in Figure 6.40a, the slit light projector is located above the stereo microscope. Note that the mechanical base is not shown. (b) Photograph of the Haag–Streit 900 BQ LED slit lamp. Courtesy of Haag–Streit Deutschland GmbH.

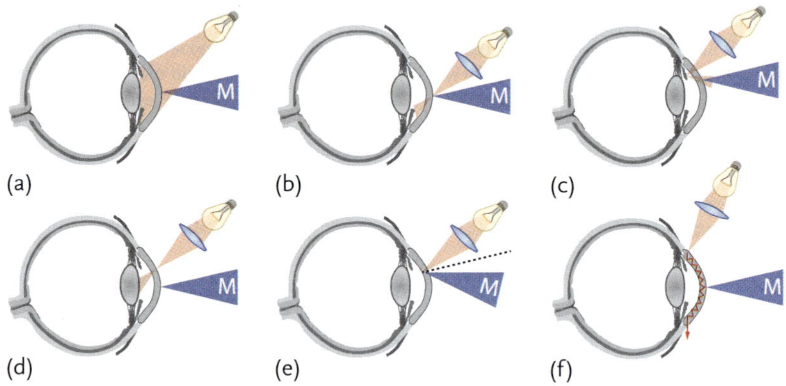

Figure 6.42 Different settings for slit lamp illumination (red) of the eye. M (blue) represents the viewing angle of the stereo microscope (i.e., the observation path). (a) Direct diffuse illumination (incident light observation). (b) Direct focal illumination (observation in the optical section). (c) Indirect focal illumination. (d) Retro-illumination. (e) Specular illumination for specular microscopy of the corneal endothelium. (f) Scattering sclerocorneal illumination which uses the effect of total internal reflection inside the cornea. Courtesy of Carl Zeiss.

As scattering appears only in the illuminated region, the observed scattering image (optical section) clearly reveals an optical cross-section (Figure 6.43a) which corresponds to a mechanical section, as shown in Figure 6.43b. To observe the optical cross-section, the angle between illuminating and observation path should be as large as possible. Regardless of the selected magnification, the slit image is always located in the focal plane of the microscope objective and in the center of the

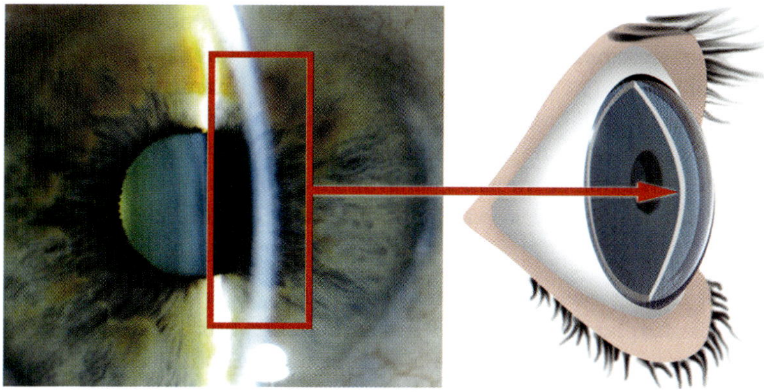

Figure 6.43 (a) Observed cross-section of the cornea (framed) with direct focal illumination and (b) corresponding schematic cross-section of the (sliced) cornea.

microscope's field of view. In combination with the stereo microscope, the optical sectioning provides precise depth information of structures within the anterior segment.

Indirect focal illumination For indirect focal illumination (Figure 6.42c), the axis of the slit light projector is tilted by a small angle ($\pm 4°$ horizontally) away from the normal position. In this configuration, the illuminated and observed areas do not coincide, but are laterally displaced. In this way, the slit light reflected by internal structures (e.g., iris) illuminates the object to be examined (e.g., the cornea). This type of illumination allows the examination of the cornea and anterior chamber against a relatively dark background.

Retro-illumination Retro-illumination (Figure 6.42d) is also performed with a laterally tilted illumination beam path. In this case, the examined eye structure is illuminated from behind and viewed with the directly or diffusely back-reflected light from iris, lens, or fundus. In analogy with standard transmitted light microscopy, abnormalities and defects in the examined eye structure can be observed, as the reflected light is either refracted or absorbed by the defect.

Specular illumination Specular illumination (Figure 6.42e) is a special form of direct focal illumination. It is mainly used to examine the corneal endothelium (Figure 1.7 in Section 1.1). While corneal reflections are usually disturbing in the case of standard direct focal illumination, we now intentionally localize them. For the analysis, we set the angle between the illumination and observation beam paths as large as possible. Hence, a bright reflection image is obtained in the focal plane. But instead of this bright image, we focus on the weaker reflection image of the endothelium behind it. With sufficiently large magnification (40–240 $\times$ with a special endothelial microscope objective lens), the cell structure of the endothelium can be examined.

Scattering, sclero-corneal illumination In the scattering, sclero-corneal illumination (Figure 6.42e) we use the effect of total reflection (Section A.1.1) between the front and back surfaces of the cornea. For this purpose, we tilt the slit light projector such that the emitted light is aimed at the *limbus* region[35]. The angle of incidence must be as flat as possible so that the light is guided inside the corneal tissue. As structural changes in the cornea distort the propagation of the totally reflected light beam, they become visible as bright or dark areas.

6.4.4 Accessories for Other Examinations and Measurements

Slit lamps are mainly used to examine the anterior segment of the eye including the lens and adjacent areas of the vitreous. However, with additional optics (auxiliary lenses; Section 6.4.4.1), other parts of the eye (e.g., fundus and iridocorneal angle) can be examined which are usually not directly accessible. In addition, various accessories extend the slit lamp from a pure observation system to a versatile measuring system. For example, it can be used to determine the intraocular pressure (Section 6.4.4.3) or the lengths and angles of the anterior segment (Section 6.4.4.1).

6.4.4.1 Auxiliary Lenses

Auxiliary lenses are used to make regions of the eye visible which are not directly accessible. In the following, we will discuss the typical applications of these lenses, such as fundus observation, gonioscopy, and laser therapy.

Fundus observation With a standard slit lamp, a direct observation of the posterior eye segment (fundus) is not possible due to the refractive power of the eye's optical system. The fundus will be imaged sharply into the far point plane by the optical system of the eye. The far point of an emmetropic eye in the case of relaxed accommodation lies at infinity, or at least more than 6 m in front of the corneal vertex (Section 3.1). For an ametropic eye, the far point is located in front (myopia) or behind (hyperopia) the eye. Therefore, the fundus image normally lies outside the focal (object) plane of the corneal microscope.

However, we can observe the fundus with an auxiliary lens placed in front of the eye which shifts the far point, and thus the intermediate image of the fundus, to the focal plane of the slit lamp microscope. For this purpose, positive or negative lenses (Section A.1.2.3) can be used (Figure 6.44). Positive lenses provide an inverted, real intermediate image. In principle, fundus observation with positive lenses corresponds to indirect ophthalmoscopy (Section 6.6.3). The only difference is that the intermediate image is not viewed with the "naked eye", but with the slit lamp microscope. Compared to negative lenses, the field of view of the fundus is

35) The corneal limbus is the border between cornea and sclera.

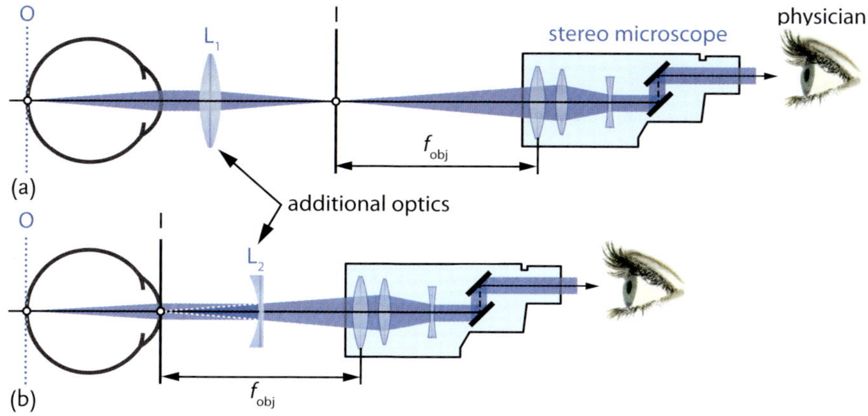

Figure 6.44 Fundus observation with a slit lamp in the case of an emmetropic eye. (a) To observe the fundus, a positive lens L_1 is placed between the patient's eye and the stereo microscope to form an intermediate image of the fundus I in the focal plane of the microscope. (b) Alternatively, a negative lens L_2 can be placed in front of the eye which forms an upright virtual intermediate image I of the fundus in the focal of the microscope.

generally larger with positive lenses, as positive lenses image the entrance pupil of the microscope with reduced size into the patient's pupil. In this case, the pupil does not act as field stop (for details see Section 6.6.3). Hence, with an additional positive lens, the fundus can be viewed stereoscopically and with high magnification. In practice, handheld aspheric lenses with high optical power (60 D or 90 D) are often used.

Negative lenses form an upright, virtual intermediate image. They are mainly placed directly onto the cornea in the form of a contact lens with an optical power of about −64 D. In this case, the image of the fundus lies in the anterior part of the eye (Figure 6.45b). With such a contact lens, the normal working distance of the slit lamp must be changed only slightly.

An important sub-group of auxiliary optics are mirror contact glasses. As they have built-in reflective surfaces, the accessible field of view reaches beyond the central fundus area. This optical arrangement thus allows viewing of the peripheral parts of the eye right up to the iridocorneal angle (Figure 6.45e). Commercially available mirror contact glasses usually contain three mirror surfaces which are arranged at angles of 59°, 66°, and 73°.

Gonioscopy The examination of the iridocorneal angle in the anterior chamber is referred to as *gonioscopy*. Gonioscopy is an important method to diagnose angle-closure glaucoma (Section 3.3) and to decide which therapy is most suitable. If the iridocorneal angle is too small, the risk of glaucoma increases. With a standard slit lamp setup, this part of the eye cannot be viewed, since the sclera is opaque due to strong scattering. A lateral observation at an oblique angle through the iris

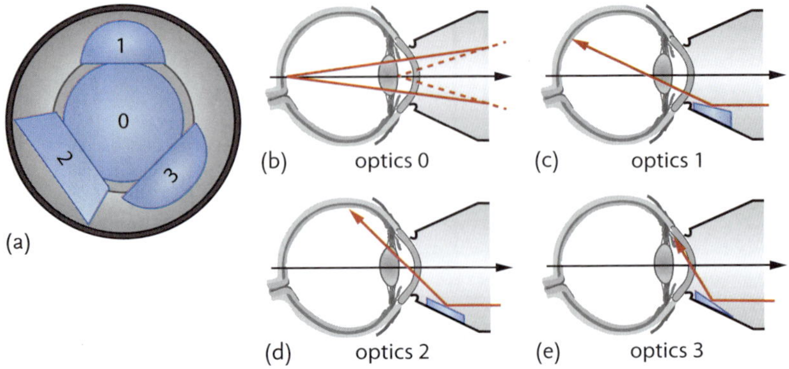

Figure 6.45 (a) Scheme of three-mirror contact lens for different applications. (b) Observation of central areas of the fundus without a mirror. (c) Observation of off-center areas of the fundus with mirror 1. (d) Observation of peripherial areas of the fundus with mirror 2. (e) Observation of the iridocorneal angle using mirror 3. Courtesy of Carl Zeiss.

is also not possible because of total reflection at the cornea.[36] However, the total reflection can be suppressed if we place a mirror contact glass onto the cornea which has approximately the same refractive index as the cornea. For gonioscopy, special *gonio-lenses* exist which contain up to six reflective surfaces arranged at an angle of 64°. They allow an almost perfect 360° visibility of the iridocorneal angle.

Laser therapy Ophthalmic laser therapies are usually performed with a slit lamp so that the treatment outcome is directly observable (Sections 10.1.2.2, 10.2.4.3, and 10.4.3). To apply the laser beam to the fundus of the eye, special laser contact lenses are required which have higher radiation resistance than auxiliary lenses for diagnostic purposes. Diagnostic lenses are typically made from acrylic glass, whereas laser lenses consist of mineral glass with suitable anti-reflection coatings.

6.4.4.2 Filters for Spectral and Fluorescence Imaging

The spectral composition of the light source of a slit lamp can be changed by filters which are swung into the illumination beam path.

Contrast enhancement Green filters (i.e., red-free illumination) enhance the contrast of red eye structures (e.g., the fundus). In addition to contrast enhancement, another barrier filter (yellow filter) can be inserted to the beam path.

Fluorescent excitation Blue filters act as excitation filters in examinations which use fluorescent effects (see also Section 6.7.5). Fluorescence imaging with exogenous fluorescence dyes plays an important role in ophthalmic diagnosis. Corresponding methods allow visualization of eye structures which can normally only be observed with difficulty or not at all.

36) The critical angle is approximately 46°.

For contact lens fitting, fluorescence is used to determine the layer thickness distribution of the tear film between the front surface of the cornea and the back surface of the contact lens (Figure 6.35a). In addition, the tear film quality can be assessed by measuring the tear film breakup time (TBUT). In applanation tonometry (Section 6.4.4.3), the fluorescence image is used to correctly set the diameter of the flattened (applanated) corneal surface.

For fluorescence imaging with a slit lamp, the dye *sodium fluorescein* is typically used. With this dye, we may visualize corneal defects (e.g., damage of the epithelium) and the tear film. In practice, a small amount of the fluid is inserted in the patient's conjunctival sac[37] either by eye drops or via a fluorescein paper strip. Maximum fluorescence is obtained with a dye concentration of 0.2–0.4% in the tear fluid. The most effective excitation is achieved with blue light, that is, wavelengths between 450 and 500 nm. The emitted fluorescent light then has a maximum intensity in the yellow-green spectral range ($\approx$ 530–535 nm wavelength; Section 6.7). For optimal excitation, a cobalt blue filter is added to the beam path of the slit light projector. An additional yellow filter ($\geq$ 530 nm) in the observation beam path of microscope considerably enhances the contrast of the observed structures and blocks the blue excitation light.

6.4.4.3 Goldmann Applanation Tonometer

The most frequently used accessory for slit lamps is the Goldmann[38] applanation tonometer (GAT) used for measuring the intraocular pressure. A transparent measuring body with an integrated image doubling prism is gently pressed with variable force against the cornea. The quotient of applied force and resulting flattening of the cornea (applanation) is a measure of the intraocular pressure. In principle, the measurement is influenced by the natural rigidity of the cornea and the adhesive force of the tear film between the measuring body and the corneal surface. However, both effects compensate for each other if the applanated surface has a diameter between 2.5 and 3.5 mm. For practical reasons, in the Goldmann applanation tonometer, a value of 3.06 mm is selected for the measurement, as an applied force of 9.81 mN (weight of 1 g) then corresponds to an intraocular pressure of 1.33 kPa $\approx$ 10 mmHg.

During the measurement, we want to determine the force required to flatten the corneal surface as specified. To do this, the tear film is dyed with sodium fluorescein and illuminated through a cobalt blue filter (Section 6.4.4.2). In the area of the contact surface between cornea and measuring body, the tear film is displaced. This can be easily examined, as the boundary between applanated and curved cornea appears as a fine green-yellow ring. An integrated image doubling system (see Info Box 6.3 in Section 6.3.1.2) splits the original image and shifts both halves of the ring by 3.06 mm relative to each other. The applied pressure is now increased until both parts of the ring merge. In this case, the applanated surface has the desired specified diameter of 3.06 mm. Because of the image-doubling principle, the mea-

37) The conjunctival sac is a "bag"-like tissue beneath the eyelid.
38) Hans Goldmann (1899–1991).

surement of the ring diameter is very exact. The calculated measuring error of the interocular pressure is below 1 mmHg.

6.4.5
Prospects

Slit lamps are relatively low-cost instruments that belong to the standard equipment of any eye care professional. Thus, other diagnostic devices such as keratometers (Section 6.3.1) or even optical coherence tomographs (Section 7.6) are increasingly attached to the slit lamp as accessory units. As the existing infrastructure of the slit lamp can be used, cost-effective hybrid systems can be fabricated which extend the application spectrum.

6.5
Scanning-Slit Projection Devices

Scanning-slit projection techniques are a logical extension of the optical sectioning principle used in slit lamp biomicroscopy (Section 6.4). In these methods, the examined tissue structures are scanned with a slit light illumination in a sequential manner. The scattering light images are then captured by a camera and evaluated.

The cornea, lens, and iris are the structures of the anterior eye segment which generate the highest degree of scattering. In comparison, scattering effects in the tear film and the aqueous humor are negligible, aside from any disease-related changes. The scanning-slit projection techniques use these scattering properties for a contact-free, quantitative, 3D structural analysis (optical tomography) of the entire anterior eye segment. Specifically, the topographic data (elevation) of the corneal front and back surfaces as well as of the front surfaces of iris and lens can be measured. Further clinically relevant dimensions such as the corneal thickness, the depth of the anterior chamber, and the iridocorneal angle can then be calculated from the measured data. In fact, a precise geometric analysis of the anterior segment is of great clinical importance. In particular, the increasing acceptance of refractive surgical methods has substantially driven the development of new instrument systems for this purpose.

The designs of the optical topography and/or tomography instruments based on scanning-slit projection techniques can essentially be classified by

1. the type of scanning system (lateral or azimuthal scan), and
2. the imaging technique for scattering light images (e.g., Scheimpflug principle or others).

The most relevant representatives of these two instrument designs will be described in the following. A summary of the most important measurement parameters, their user-friendly presentation, and medical applications are listed in Section 6.5.3 together with some examples.

6.5.1 Lateral Scanning-Slit Projection Techniques

In 1995, the company Orbtek, Inc. (later acquired by Bausch+Lomb) launched the first topography system (Orbscan) based on the lateral scanning-slit projection technique. From 1999 onwards, a Placido ring topography system (Section 6.3.2) was additionally integrated to combine the advantages of both methods in a single device. In this instrument, the radii of curvature are measured with a Placido-ring system, and the elevation data is taken from the scanning-slit projection system (Orbscan IIz; Figure 6.46b). Here, we will address only the lateral scanning-slit projection technique [25].

6.5.1.1 Functional Principle

In the lateral scanning-slit technique (Figure 6.46a), two slit beam projectors are arranged to the right and the left of a centrally positioned Placido ring system. Each slit projector images 20 slit-shaped beams (approximately 12 mm high and 0.3 mm wide) from the right and the left at an angle of 45° relative to the instrument axis onto the surface of the cornea in a sequential manner. This ensures that practically the entire surface of the cornea is covered by the measurement. The scattering images of the two slit-shaped beams, which move in opposite directions over the surface of the cornea, are recorded with a video camera through a central opening in the Placido disk.

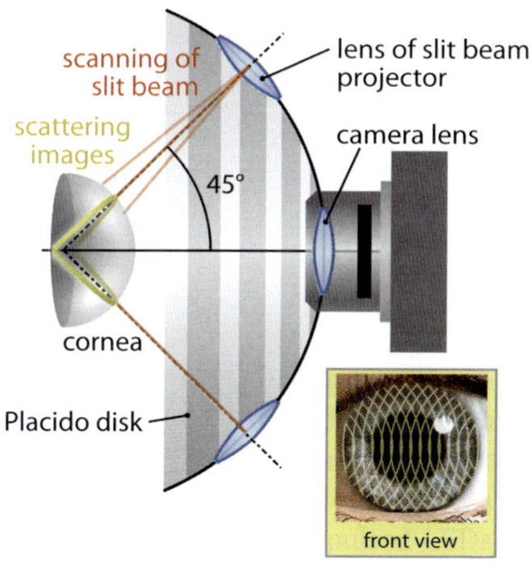

(a)

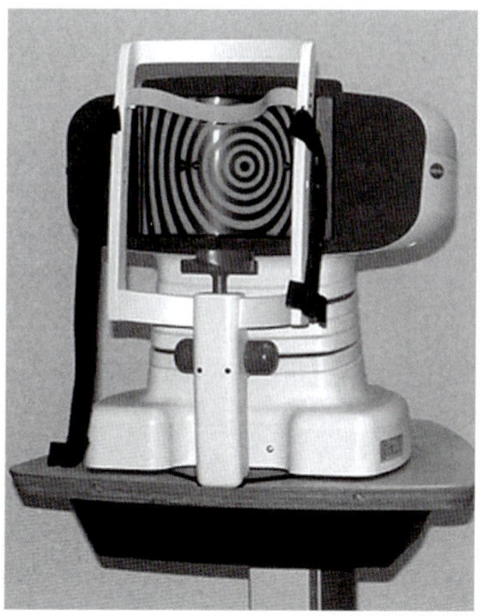

(b)

Figure 6.46 (a) Top view of the arrangement used for the lateral scanning-slit projection technique. The yellow lines correspond to the bent scattering images of the light slits projected onto the cornea. The Placido-ring projection system is used to measure the radii of curvature of the corneal front surface. Inset: Animated front view of the slit beam scattering images as seen by the camera. (c) Photograph of the Orbscan IIz which works according to the lateral scanning-slit projection technique. Courtesy of TECHNOLAS Perfect Vision GmbH – A Bausch+Lomb Company.

The optical axes of projector and camera intersect in the reference plane Z_0, which is perpendicular to the camera axis and imaged by the camera lens onto the camera sensor plane (Figure 6.47a). When a straight light slit is projected onto a plane scattering surface in the reference plane, the formed scattering image is straight as well. If, however, the slit-shaped beam is projected onto the curved cornea placed in front of the reference plane, the video camera acquires a *curved* scattering light image in its sensor plane (Figure 6.47b). In this case, the lateral shift $\Delta s(x, y)$ of a point $P_i(x, y)$ in the scattering light image is proportional to its distance $\Delta h_i(x, y)$ from reference plane Z_0. Hence, we have

$$\Delta s_i(x, y) = \Delta h_i(x, y) \tan \gamma_i \,, \tag{6.62}$$

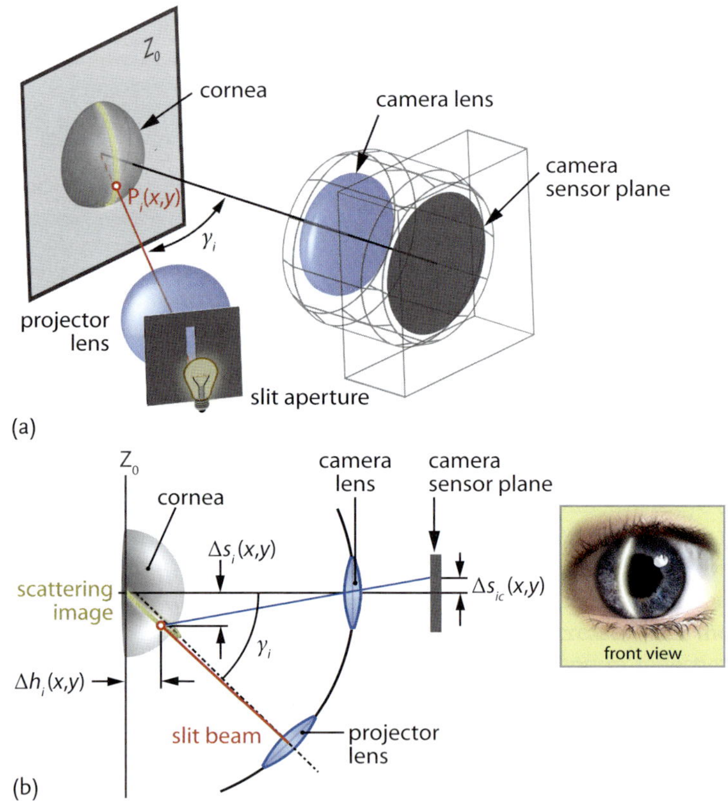

Figure 6.47 Geometric arrangement of the slit lamp projector, video camera, and examined cornea. The optical axes of the slit lamp projectors and the video camera intersect in the object plane Z_0 which is imaged by the camera lens onto the camera sensor plane. For the sake of simplicity, the second beam projector is *not* depicted. When a straight light slit is projected onto the cornea, the scattering light image (yellow) appears curved. The distance of the curved image from the corneal center $\Delta s_i(x, y)$ is a measure for the height $\Delta h_i(x, y)$ of the corneal surface in point $P_i(x, y)$ with reference to Z_0. (a) Oblique view. (b) Top view. Inset: Front view of the cornea illuminated with the slit-shaped beam.

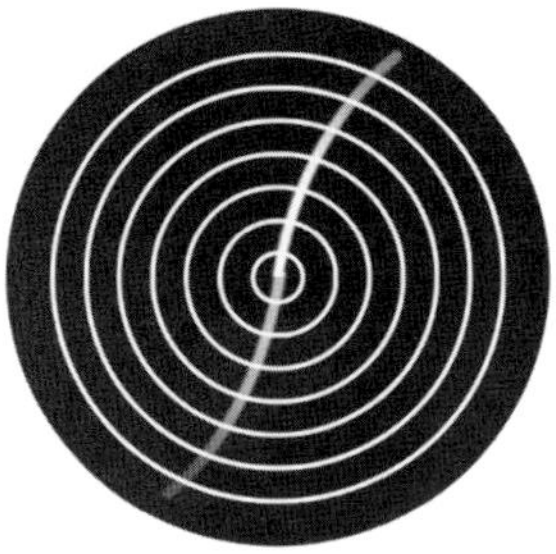

Figure 6.48 The corneal reflection image of Placido-ring mires is used to set the position of the instrument correctly. In the case of optimal alignment, the two half slit images are exactly merged in the center such that they form an "S".

where γ_i is the angle of incidence of the slit light illumination relative to Z_0. With an edge detection algorithm applied to the scattering light images captured in the camera plane, we may determine the image shift $\Delta s_i(x, y)$ for all points in all scattering light images which, in turn, provides the raw data for the elevation profile of the corneal surface (Figure 6.49).

In the factory calibration of the measurement system, a flat, isotropically scattering plate (called *calibration surface*) is arranged perpendicular to the camera axis at a known distance from the camera lens. The scattering plate is then sequentially illuminated with slit images, and the exact position of the corresponding scattering light image is detected for every slit light beam. A sequence of measurements with different distances between the calibration surface and the camera lens delivers the required baseline data which is used to calculate the exact corneal topography $h(x, y)$ for a known camera magnification.

To determine the true geometry of the eye structures behind the cornea, ray tracing is used to simulate the deflections of a light beam which travels through the refractive media of the cornea. The more accurately the curvature of the corneal front surface is known, the more accurately the true position of these structures can be determined. For this reason, the Orbscan system uses the Placido-ring topography system to measure the radii of curvature of the corneal front surface.

To set the position of the instrument correctly, two half slit images are projected onto the cornea. An optimal alignment is attained when both slit image halves exactly merge to an S-curve in the center of the Placido-ring images (Figure 6.48). A tracking system compensates for potential involuntary eye movements during the measurement process.

6.5.1.2 **Applications and Representation of Data**

With the implemented scanning-slit technique, it is possible to measure the topographic data of the corneal front and back surfaces as well as of the front surfaces of the iris and eye lens. Clinically relevant parameters such as the thickness profile of the entire cornea (wide-field pachymetry), the depth of the anterior chamber, and the size of the iridocorneal angle can be derived from this data. The corresponding results are then presented in the form of color-coded maps.

Surface elevation map and best-fit sphere Unlike a curvature map, the surface elevation map displays the true shape of the cornea. To also make small local shape deviations visible, spherical, aspherical, or toroidal "best-fit" reference surfaces are used (see also Section 6.3.2.5). In Figure 6.49, the shape deviation of an astigmatic corneal front surface (colored line) is compared to a best-fit sphere (gray line).

Wide-field pachymetry maps The thickness distribution of the entire cornea (Figure 6.50) can be derived from the elevation data of the corneal front and back surfaces. This representation is of great clinical value, in particular for the planning of corneal surgeries.

Other applications The scanning-slit projection technique is also used in the Retinal Thickness Analyzer (RTA) by Talia Technologies [26] to generate a topographic

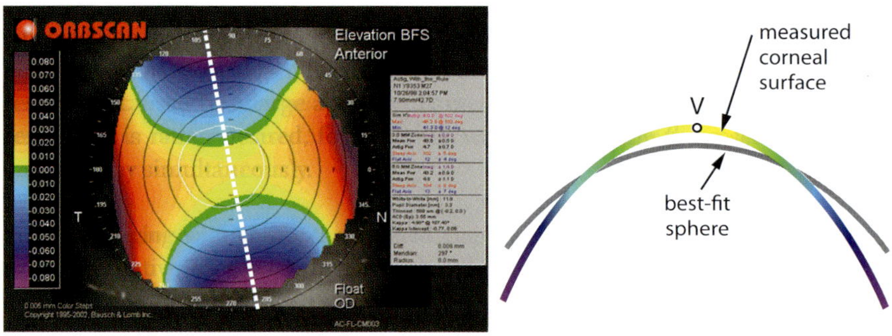

Figure 6.49 (a) Screenshot of the elevation map of the corneal front surface generated with the TECHNOLAS Orbscan IIz. The values are referenced to a best-fit sphere. Courtesy of TECHNOLAS Perfect Vision GmbH – A Bausch+Lomb Company. (b) Scheme which shows how the elevation profile is determined from the deviation of the corneal height profile from a best-fit curve (gray line). The color codes are directly related to the elevation map in (a) along the white dashed line from 100° to 280°.

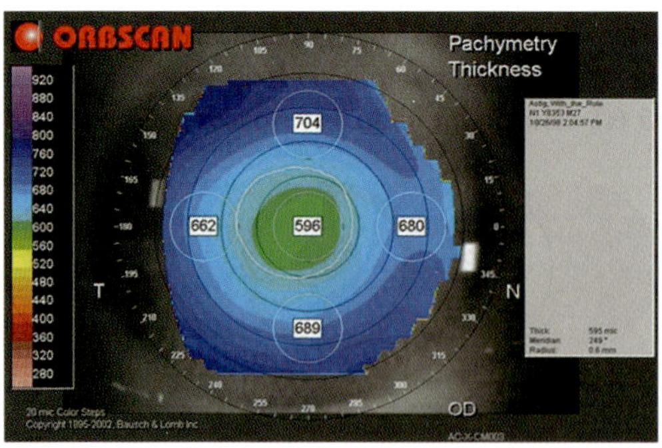

Figure 6.50 Screenshot of a wide-field pachymetry map generated with the TECHNOLAS Orbscan IIz. Courtesy of TECHNOLAS Perfect Vision GmbH – A Bausch+Lomb Company.

map of the retinal thickness. For this purpose, a slit image of a green laser beam is projected obliquely onto the retina and is sequentially scanned across the fundus. The reflected and backscattered light is captured at a certain angle by a black-and-white CCD camera. A special curve-fitting tool is used to identify the distance between the inner limiting membrane and the retinal pigment epithelium, which corresponds to the total retinal thickness (Section 1.2). An axial resolution of approximately 50 μm [27] can be attained with this system.

6.5.2
Scheimpflug Imaging of Rotating-Slit Projections

In conventional photo cameras, the objective lens plane L, the image (detector) plane I, and the focal plane F are parallel to each other and perpendicular to the optical axis (black dotted lines in Figure 6.51a). Objects located within the depth of field range (Section 2.1.8) of the camera focal plane (gray area in Figure 6.51a) reveal an acceptable sharpness in the image plane. Hence, if a planar object O is tilted relative to the focal plane of the camera, only a narrow region within the depth of field range is sharply imaged (Figure 6.51b). However, a large number of applications exist for which the entire tilted object must be imaged sharply.

A typical application in ophthalmology is the visualization of the cross-section of the anterior eye segment created by optical sectioning with a slit lamp (Section 6.4). During a slit lamp examination of the anterior segment, the slit-shaped beam used for optical sectioning is usually tilted relative to the optical axis of the slit lamp microscope (Figure 6.52a). The optical cross-section (plane of the scattering images of objects within the anterior segment) is thus tilted relative to the focal plane of the microscope. Given a finite depth of field of the microscope, only a limited area of the visualized cross-section in front of and behind the focal plane appears sharp (Figure 6.52b). As a consequence, simultaneously sharp imaging of the optical sec-

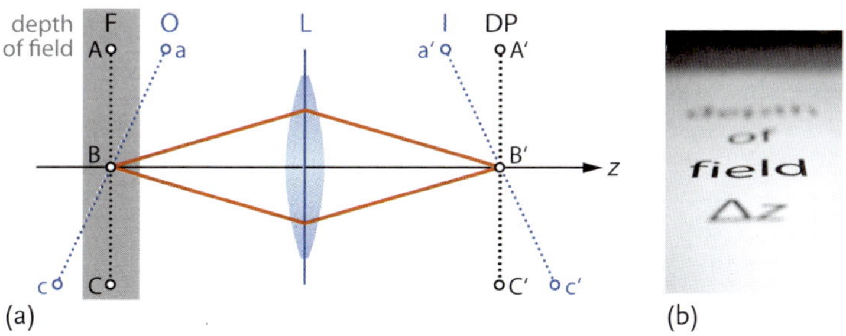

Figure 6.51 (a) Geometry of a conventional camera with a planar object O tilted relative to the focal plane of the imaging lens L. A, B, and C are distinct points in the vertical focal plane F (dotted black line). A′, B′, and C′ are images of these points in the detector plane DP (dotted black line). a and c are object points in the *tilted* object plane O (dotted blue line). a′ and c′ are points of the respective *tilted* image I of the object (dotted blue line). (b) On the detector, the image of the tilted, planar object only appears sharp within the depth of field.

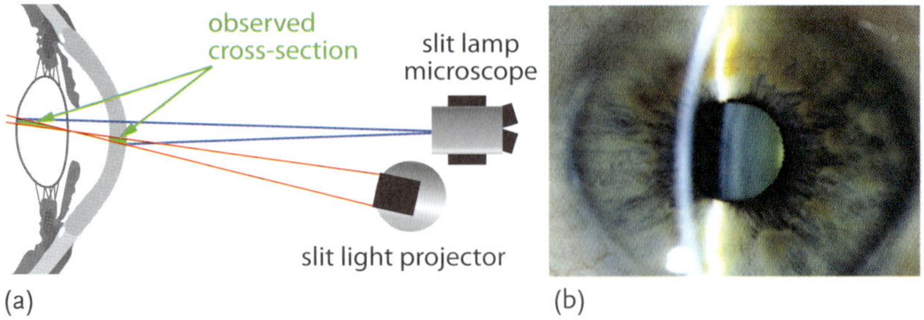

Figure 6.52 (a) Imaging of the anterior eye segment (green) with a standard slit lamp arrangement. (b) Photograph of a patient's eye taken with a standard slit lamp. When focusing on the eye pupil, we can see that only the inner rim of the iris (pupil) appears sharp, whereas the outer part of the iris and the scattering images of cornea and eye lens (generated by the slit beam) are already blurred. This happens due to the limited depth of field of the slit lamp microscope. Courtesy of Carl Zeiss.

tioning in the cornea *and* lens is not possible with the standard optical arrangement used for slit lamp examination.

Interestingly, a tilted, planar object can be imaged entirely sharply with a so-called *Scheimpflug arrangement* (Figure 6.53). Here, the lens plane L and/or the image plane I are tilted such that they intersect object plane O in a single *Scheimpflug line* SL (see also *Scheimpflug principle* in Section 6.5.2.2). In this sense, we can actually consider "normal" imaging setups with parallel lens and image planes such that the Scheimpflug line lies at optical infinity.

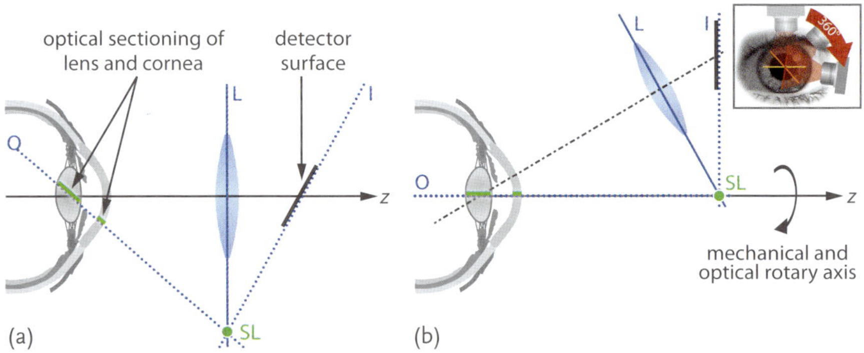

Figure 6.53 Principle of Scheimpflug photography. (a) In the static arrangement, the slit illumination is incident from the side (along object plane O) and the image is captured centrally in the tilted image plane I. Lens plane L is perpendicular to the optical axis of the eye. O, L, and I intersect in the Scheimpflug line SL (green). (b) Arrangement with a laterally rotating Scheimpflug camera and a central slit illumination (along object plane O). Lens plane L is tilted with regard to the optical axis of the eye. Again, O, L, and I intersect in the Scheimpflug line SL (green). Inset: View along the rotary (z) axis. The cornea is scanned by simultaneously turning the slit beam orientation and the detector around the optical axis.

6.5.2.1 History

The imaging principle that allows entirely sharp imaging of obliquely tilted objects with minimal image distortion was first described by Theodor Scheimpflug (1865–1911) in 1904. It was originally used, and is still used today, in large format photography (Problem P6.12). In ophthalmology, the Scheimpflug principle was first used by Drews, Niesel, and Brown, as well as Dragomirescu and Hockwin to photograph the optical sectioning of the anterior eye segment [28].

Dragomirescu and Hockwin also developed the principle of a rotating Scheimpflug camera (Figure 6.53b). In contrast to the standard (static) Scheimpflug arrangement, shown in Figure 6.53a, the slit light projection is performed centrally, and the formed optical sectioning is imaged with a laterally attached Scheimpflug camera. With this configuration, it is possible to visualize cross-sections of the anterior chamber and eye lens with a high degree of reproducibility. Based on this design principle, the first commercial Scheimpflug camera was developed by Topcon (Topcon SL-45). This instrument uses a flashbulb for slit light illumination, and the image is captured with a film camera. The first rotating Scheimpflug video system with an electronic image analysis system was developed by the company Carl Zeiss (ZEISS SLC). In this instrument, prisms are arranged in the optical path such that the camera does not need to be rotated.

At first, Scheimpflug cameras were primarily used for research purposes, for example, for examinations of the eye lens (classification, monitoring, and evaluation of cataracts) and the cornea by means of densitometry. But due to the growing importance of cornea and lens-based refractive surgery, Scheimpflug cameras also became a versatile tool to acquire biometric data of the entire anterior eye segment. The first instrument of this kind was the NIDEK EAS-1000 Eye Analysis System. It is a rotating Scheimpflug camera which uses a video camera for electronic image acquisition. The recorded images are then analyzed by a digital system. Today, ophthalmic Scheimpflug cameras are offered by various companies, either as stand-alone systems or in combination with other diagnostic instruments.

Before we discuss the design and function of two commercial Scheimpflug cameras in Sections 6.5.2.3 and 6.5.2.4, let us first consider the geometry of a typical Scheimpflug arrangement to understand the underlying optical principle.

6.5.2.2 Scheimpflug Principle

Figure 6.54 shows a schematic side view of a typical Scheimpflug setup at which the object plane O intersects with the lens plane L and image plane I in the Scheimpflug line SL. As it is sufficient to consider a 2D cross-section of the arrangement, let us derive that O and L intersect I at point $(x, y, z) = (0, -y_{\mathrm{SL}}, 0)$. We may describe O by the linear function

$$y_{\mathrm{O}} = -a z_{\mathrm{O}} - y_{\mathrm{SL}} , \tag{6.63}$$

where a represents the slope. According to the imaging equation of a thin lens equation (A14), it follows that

$$\frac{1}{z'_{\mathrm{I}}} - \frac{1}{z_{\mathrm{O}}} = \frac{1}{f} . \tag{6.64}$$

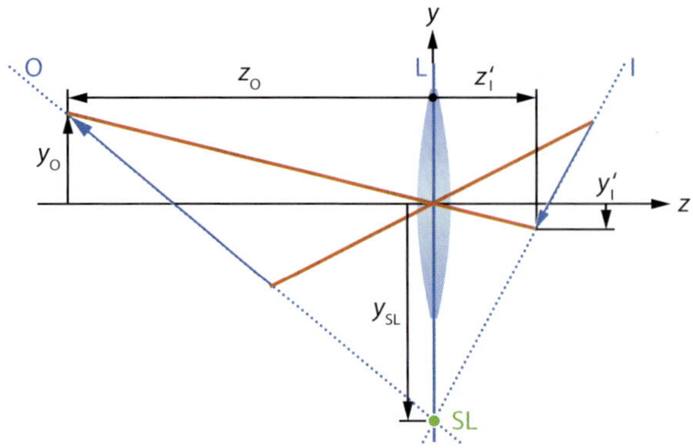

Figure 6.54 Arrangement of lens plane L, image plane I, and object plane O used to illustrate the Scheimpflug principle. All of these planes intersect in the Scheimpflug line SL (green). y_O and z_O are the coordinates of any point in O, and y'_I and z'_I are the coordinates of the corresponding image point in I. y_{SL} denotes the distance of the Scheimpflug line from the lateral center of the lens.

With the magnification factor (according to Eq. (A15))

$$\frac{y'_I}{y_O} = \frac{z'_I}{z_O}, \tag{6.65}$$

we thus obtain a linear function for the image plane I which reads

$$y'_I = -\left(a - \frac{y_{SL}}{f}\right) z'_I - y_{SL}. \tag{6.66}$$

When we compare (6.66) with (6.63), it becomes obvious that image plane I intersects with lens plane L at a common point $(0, -y_{SL}, 0)$. In this way, the Scheimpflug line SL is formed.

6.5.2.3 Device Description: OCULUS Pentacam

The basic optical-mechanical setup of the rotating OCULUS Pentacam® Scheimpflug camera (Figure 6.55) corresponds to that of the Topcon SL-45 (as shown in Figure 6.35b). Additionally, state-of-the-art technical upgrades are implemented. The radial arrangement of the Scheimpflug images leads to a maximum density of measured points in the center of the rotation, which actually corresponds to the center of the cornea. A blue LED (with a wavelength of 475 nm) is used for the slit illumination, and the Scheimpflug images are acquired by a CCD camera. The entire anterior segment of the eye is measured in less than two seconds with an automatic scanning program. With this solution, up to 100 Scheimpflug images (Pentacam HR®) are recorded in a sequential manner. The boundary surfaces of the eye structures are then identified in the captured cross-sectional images, and the resulting point clouds of the individual layers (front and back surfaces of the cornea, iris, eye lens, and so on.) are saved. The involuntary motion of the patient's eye during image acquisition is recorded with a second camera in order to correct

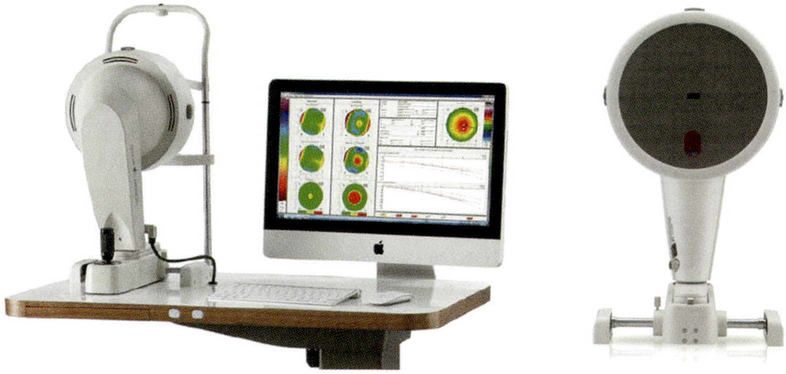

(a) (b)

Figure 6.55 (a) Photograph of rotating OCULUS Pentacam Scheimpflug camera system used for the characterization of the anterior eye segment. (b) Front view of the Pentacam measuring head with one lateral opening for Scheimpflug imaging. Courtesy of OCULUS Optikgeräte GmbH.

the axial position of the radially arranged cross-sectional images to one reference point. As the sectional images are captured at an oblique angle, they are slightly distorted. Additional image distortion of internal eye structures occurs due to light refraction at the boundary surfaces of cornea and lens. To correct these imaging errors, ray tracing from the raw data is used to determine the true positions of all key edge points in the anterior eye segment. The processed data set is finally merged to create a 3D model of the entire anterior chamber which can be used to generate different kinds of tomographic images and to extract biometric parameters, such as elevation data, radii of curvature, thickness (pachymetry) data of the cornea, depth and volume of the anterior chamber, and size of the chamber angle.

The different levels of brightness in the eye structures provide additional information about their specific scattering properties. A densitometric analysis allows determination of the lens opacity in the case of cataract (Section 3.2). For this purpose, multiple Scheimpflug images are captured from a fixed camera position to improve the signal-to-noise ratio.

6.5.2.4 **Device Description: Galilei G2**

The Ziemer Galilei™ Dual Scheimpflug Analyzer differs from the OCULUS Pentacam in two points:

1. The images are captured simultaneously by two rotating Scheimpflug systems.
2. A Placido-ring topography system is additionally integrated.

The two optically identical Scheimpflug channels are arranged symmetrically to the mechanical and optical rotary axis (Figure 6.56a). The slit lamp illumination is aligned to the common rotary axis and uses a blue LED (wavelength of 470 nm). The simultaneous image acquisition with two camera systems offers the following advantages:

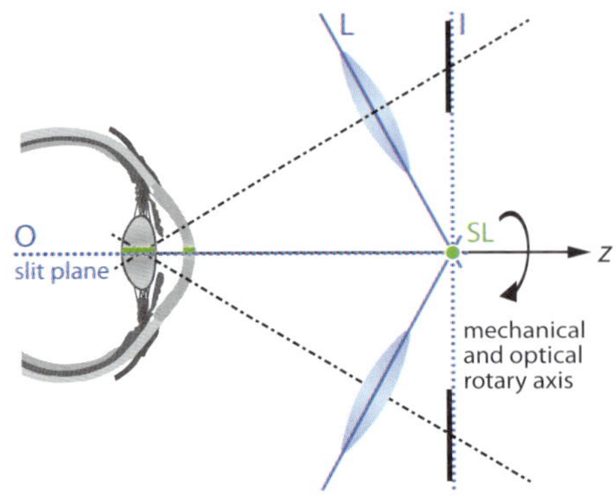

(a)

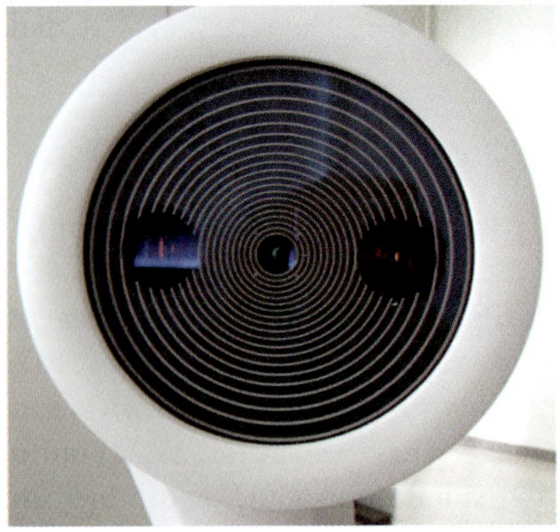

(b)

Figure 6.56 (a) Schematic top view of lens plane L, image plane I, and object plane (slit plane) O used for the Dual Scheimpflug Analyzer by Ziemer. (b) Photograph of the measuring head (detector) of the Ziemer Galilei G2 Dual Scheimpflug Analyzer with integrated Placido disk and two lateral openings for dual Scheimpflug imaging. Courtesy of Ziemer Ophthalmic Systems AG.

- The measurement of corneal thickness is not affected by potential decentrations of the optical rotary axis relative to the corneal vertex. This allows exact pachymetry even if decentrations, which may appear during the scanning process, are not known (Figure 6.57).
- The whole anterior eye segment can be recorded in three dimensions with a camera rotation of just 180°. This reduces the measurement time by a factor of 2.

Figure 6.57 illustrates the influence of decentration when the corneal thickness is measured with two camera systems (L/R) simultaneously. The instrument configuration is optimally centered if the optical axis (i.e., the slit plane) is perpendicular to the corneal vertex V (Figure 6.57a). As the viewing angles of both cameras are equal, the cross-sections of the cornea have the same width. If, however, the camera systems are displaced by Δs relative to the corneal vertex (Figure 6.57b), the corneal cross-sections appear at different widths. This is caused by the fact that the slit light no longer penetrates the corneal surface perpendicularly. As the measuring errors caused by such a decentration have opposite algebraic signs, it is possible to correct them by averaging both measured values.

The Galilei Scheimpflug analyzer also contains a Placido ring-based corneal topography system. For this reason, the central curvature of the corneal front surface can be more precisely determined than with the Scheimpflug approach. To achieve a comparable measurement precision with the Scheimpflug technique, the pixel resolution would have to be extremely high. But compared to the Placido disk method, the Scheimpflug technology is much better suited for the analysis of the corneal elevation data. Hence, both measuring techniques complement each other. In the Galilei G2, the Placido disk and Scheimpflug data is acquired simultaneously and merged into a single data set.

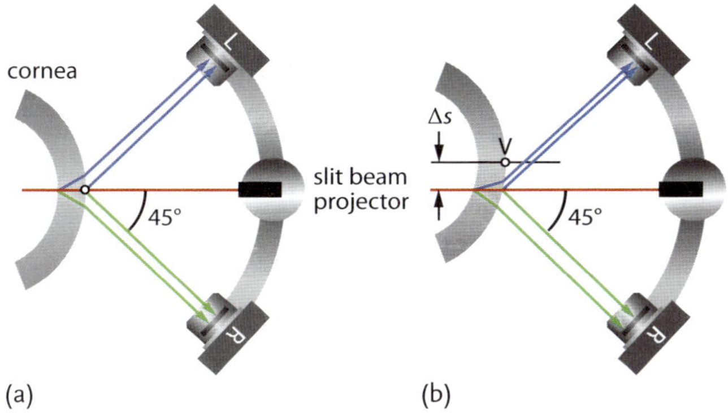

Figure 6.57 Centration sensitivity of a two-camera arrangement. (a) Optimally centered instrument configuration, that is, the slit beam intersects the cornea perpendicularly at the corneal vertex V. (b) Decentration of the setup by Δs. As a consequence, the apparent corneal cross-section appears thicker for the right (R) and thinner in the left (L) camera (or vice versa). Adapted from [29].

6.5.3 Clinical Relevance and Applications

In contrast to the Placido ring-based techniques which measure the radii of curvature, scanning-slit techniques directly acquire the topography elevation data of the front and back surfaces of the cornea. The elevation map can then be compared to a best-fit reference surface. The corneal thickness, the global or local radii of curvature, and the corresponding refractive powers can be derived from the elevation data without any additional assumptions.

For cataract surgery and lens-based refractive surgery procedures (Section 10.5), the refractive power data of the cornea is crucial for the determination of the required refractive power of the IOL. In scanning slit-based devices, the total refractive power of the cornea can be calculated directly from the refractive powers of the front and back surfaces, and the corneal thickness by means of the Gullstrand formula (2.20). The scanning-slit techniques thus also allow an *exact* determination of the refractive power for a cornea whose surface has been modified by refractive surgery, for example, following a LASIK treatment (Section 10.3.3.2). This is, however, not possible with keratometers (Section 6.3.1) or Placido corneal topographers (Section 6.3.2), as these devices measure only the radii of curvature of the corneal *front* surface[39]. In addition, the rotating-slit technique attains its highest measured point density in the center of the cornea, which is of particular optical interest. In contrast, keratometers and Placido ring-based devices cannot measure in the corneal center at all, or just with low resolution. In summary, slit projection

39) To a certain extent, the advantage of being able to measure the back surface of the cornea is offset by the disadvantage of having to obtain curvatures as the second derivative of the surface height. This may explain why, for normal corneas (i.e., not modified by refractive surgery), reflection-based methods (keratometry and corneal topography) are still considered as the standard.

systems are more versatile than Placido-based and reflection-based systems. They are even applicable if the tear film is absent (no reflection) or if the cornea has a very irregular surface shape.

Let us now consider the most important analysis methods of the scanning-slit devices and their clinical applications in brief.

- 3D anterior chamber analysis/anterior segment tomography (Figures 6.58a,b):
  - Preoperative planning for the implantation of phakic[40] IOLs. Required parameters: anterior chamber depth and anterior chamber diameter.
  - Screening of glaucoma (Section 3.3). Required parameters: iridocorneal angle, anterior chamber depth, and anterior chamber volume.
  - Pre- and postoperative comparison of changes in the anterior chamber.
- Corneal topography of the front and back surfaces (Figure 6.59):
  - Screening of keratoconus (Section 3.1.6).
  - Preoperative planning of refractive corneal surgery (Section 10.3).
  - Determination of the corneal power of patients who have undergone refractive corneal surgery.

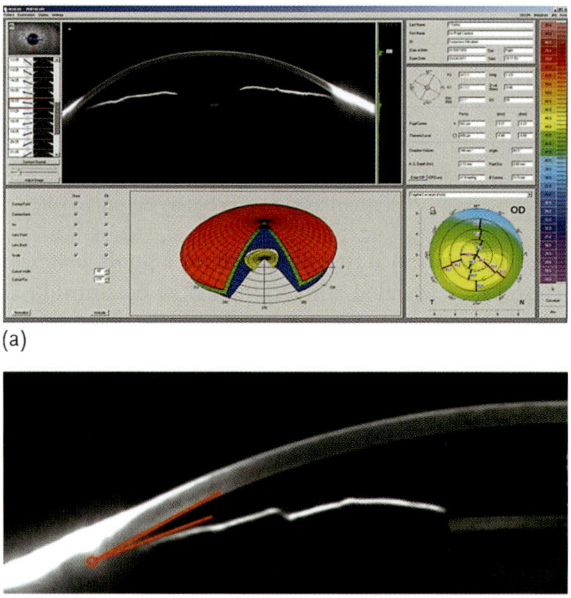

(a)

(b)

Figure 6.58 (a) Screenshot of the anterior segment tomography window of the OCULUS Pentacam. (b) Measurement of the iridocorneal angle for glaucoma screening. Courtesy of OCULUS Optikgeräte GmbH.

40) An eye which has a natural eye lens is called *phakic*. Thus, a *phakic intraocular lens* (phakic IOL) is an artificial lens placed into the anterior chamber in addition to the natural eye lens.

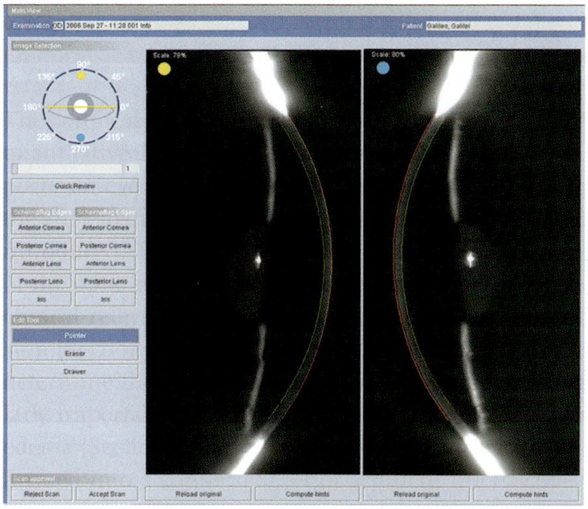

Figure 6.59 Screenshot of the topography analysis of the corneal front and back surfaces measured with the Ziemer Galilei Dual Scheimpflug Analyzer. Courtesy of Ziemer Ophthalmic Systems AG.

  - Follow-up examination after corneal surgery or a refractive procedure.
  - Corneal wavefront analysis, for example, to specify customized IOLs used to correct spherical aberration.

- Pachymetry (corneal thickness measurement):
  - Preoperative planning for refractive corneal surgery. The corneal thickness is a key exclusion criterion and an important parameter for treatment planning (Section 10.3).
  - Screening of glaucoma (low corneal thickness is a risk factor).
  - Keratoconus diagnosis and quantification.
  - Measurement of the intraocular pressure. The central corneal thickness is used as a correction factor for applanation tonometry (Section 6.4.4.3).

- Densitometry: Visualization and objective quantification of the eye lens and corneal opacity in 2D and 3D (Figure 6.60).

6.6 Ophthalmoscope

The visual inspection of the posterior eye segment is referred to as *ophthalmoscopy* or *funduscopy* and represents one of the most important techniques in the basic diagnostics of the human eye. Ophthalmoscopes are relatively simple (hand-held) optical instruments with which ophthalmoscopy can be performed in a fast and uncomplicated manner. For this reason, ophthalmoscopes have become part of the basic equipment used by any eye care professional.

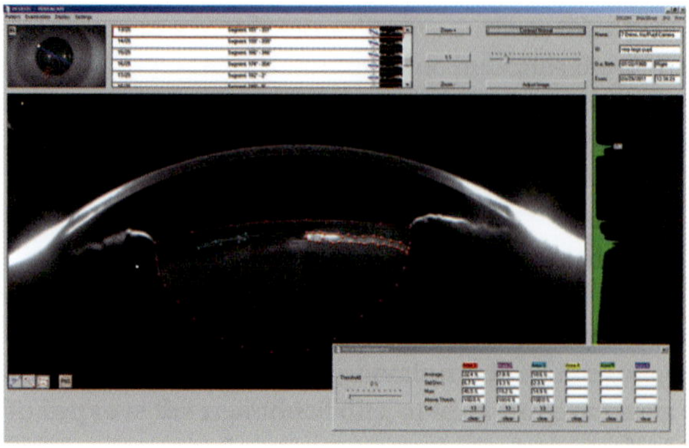

Figure 6.60 Screenshot of cataract analysis (3D densitometry) window of the OCULUS Pentacam. Courtesy of OCULUS Optikgeräte GmbH.

History The first (direct) ophthalmoscope was developed by Hermann von Helmholtz in 1850. This instrument allowed viewing of the interior of the living eye through the eye pupil for the very first time. Helmholtz called it an "Augenspiegel" (eye mirror), since he illuminated the interior of the eye in the direction of observation with a partially transmitting mirror. For this purpose, he used a stack of microscope cover slips. After several improvements, the Helmholtz ophthalmoscope quickly found its way into clinical practice. A second model allowed adding positive or negative lenses to the observation beam path in order to correct the patient's and/or physician's ametropia and to obtain a sharp image of the fundus.[41]

The Helmholtz ophthalmoscope heralded a new era in diagnosis and therapy of diseases in the posterior eye segment. Until then, the diagnostic possibilities of ophthalmologists had been very limited. Although the eye is, in principle, accessible for optical examination methods, the lack of an appropriate combination of illumination and observation beam paths prevented the observation of the posterior eye segment. Hence, the invention of the ophthalmoscope is considered the advent of modern ophthalmology and the birth of ophthalmic instrumentation.

6.6.1
Functional Principle

The fundus can only be observed if it is illuminated by an additional light source. The illumination and the observation beam paths must be arranged such that the illuminated and observed areas overlap on the fundus. If this condition is not fulfilled (Figure 6.61a), a fundus examination is *not* possible, as the part of the fundus observable through the pupil appears totally dark. Consequently, a beam splitter (Figure 6.61b), pinhole mirror (Figure 6.61c), or half mirror (Figure 6.61d) must be added to the observation beam path to attain the required overlap. If a pinhole mir-

41) In 1848, Charles Babbage (1791–1871) developed an ophthalmoscope without dioptric correction that was thus not accepted for clinical practice.

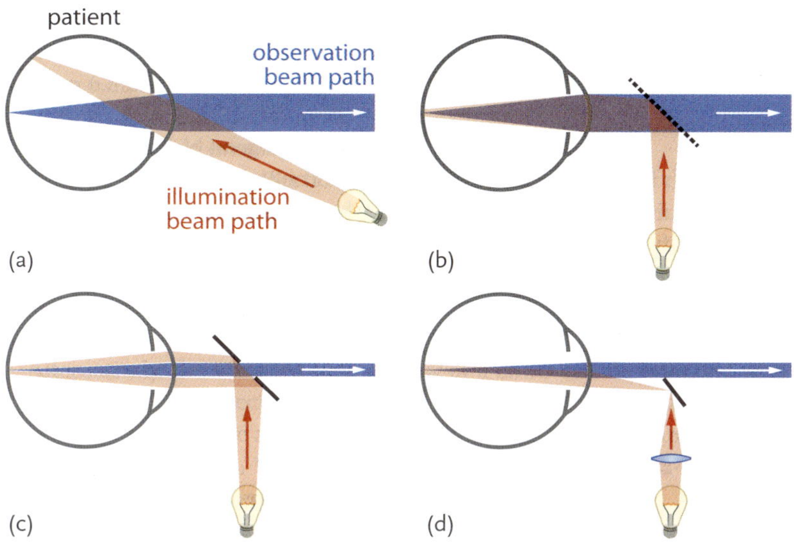

Figure 6.61 General arrangement of illumination and observation beam paths for fundus observation. (a) The part of the fundus observable through the pupil is not illuminated and is thus not visible. Illumination and observation beam paths must be arranged such that the illuminated and observed areas overlap on the fundus. To attain the required overlap (b) a beam splitter, (c) a pinhole mirror, or (d) a half mirror (or prism) can be used.

ror is used (as in the case of fundus cameras; Section 6.7), the fundus is illuminated via the mirror surface and observed through the pinhole. In ophthalmoscopes, half mirrors (or small prisms) are implemented instead. The lower part of the shared beam path is used for illumination, and the upper part is used for observation. Based on the configuration of the observation system, we distinguish between

1. *direct ophthalmoscopes* (Section 6.6.2) in which the patient's fundus is directly observed, and
2. *indirect ophthalmoscopes* (Section 6.6.3) in which a real intermediate image is used for evaluation.

6.6.2 Direct Ophthalmoscope

The working principle of a direct ophthalmoscope is illustrated in Figure 6.62. Figure 6.62a shows the basic setup and Figure 6.62b an oblique-view scheme of the arrangement during an examination. The illumination and observation beam paths are usually tilted relative to each other to minimize disturbing corneal reflections (details in Section 6.6.3.1).

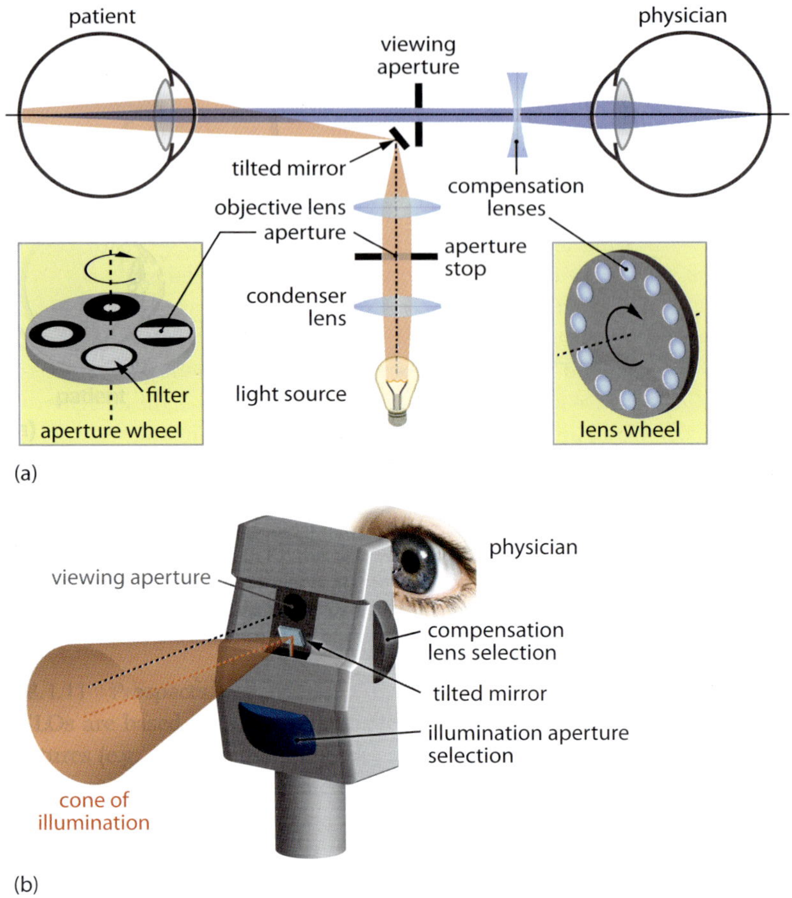

Figure 6.62 Principle of a direct ophthalmoscope. (a) Optical design of a direct ophthalmoscope. The light of a xenon–halogen bulb or LED is collected by a condenser lens and passes through an aperture wheel (left inset) with which different aperture shapes can be selected. With an objective lens, the light is then focused onto a tilted mirror and projected onto the patient's eye. There, it illuminates the eye fundus. With a suitable compensation lens (chosen from a lens wheel; right inset), the fundus is then imaged into the physician's eye. Adapted from [30]. (b) Scheme of a typical direct ophthalmoscope which shows the cone of illumination and the selection wheels.

6.6.2.1 **Illumination Beam Path**

For the illumination of the fundus, a xenon–halogen bulb or an LED (see Info Box B.4 in Section B.5.2) light source is used which is imaged by a condenser and an objective lens onto a small, tilted (45°) mirror or a deflecting prism. The image of the light source on the mirror acts as a secondary light source and illuminates the fundus through the pupil. As during examination the ophthalmoscope, and thus the tilted mirror (or prism), is located relatively close to the patient's eye, a bright circular light spot (luminous field) is formed on the retina instead of a sharp light source image. The diameter of the circular light spot is smaller than the pupil diameter (about 50% in emmetropic eyes). The aperture wheel between condenser and objective lens allows selection of different shapes or colors of illumination. It typically contains slit- and circular-shaped apertures as well as blue and green filters. In some applications, the filters are used to enhance the visibility of certain structures on the fundus. For example, a green filter (*red-free filter*) is suitable to improve the visibility of blood vessels. Blood vessels appear black due to the strong absorption of blood in the green spectral range (Figure 9.3 in Section 9.3). The distance between aperture wheel and objective lens is chosen such that the aperture is focused on the retina if the patient is emmetropic. In this way, the light spot on the patient's retina is fairly sharp.

6.6.2.2 **Observation Beam Path**

If the physician and the patient are both emmetropic and do not accommodate (relaxed eye; Section 2.1.4), a sharp, inverted image of the patient's fundus is formed on the physician's fundus which is of equal size (magnification $\beta = -1$) (Figure 6.62a). Because of the image inversion in the visual cortex of the brain (Section 2.1), the physician perceives an upright image. Direct ophthalmoscopy is thus often called "ophthalmoscopy with an upright image". If the patient and/or the physician are ametropic, the refractive errors must be compensated by appropriate corrective lenses (typical refractive power range: -35 D to $+40$ D) so that the fundus or other segments of the eye are imaged sharply.

Magnification and the field of view play an important role for the applicability of direct ophthalmoscopes. Let us discuss them in more detail.

Magnification Although the lateral magnification of the overall optical system comprising both the patient's and the physician's eye is -1, the physician sees a highly enlarged image of the patient's fundus. The image magnification is actually caused by the optical system of the patient's eye that acts as a powerful loupe with the fundus in its object-side focal plane. Analogous to Eq. (6.2), we have a nominal magnification of

$$\beta_{\text{nominal}} = \frac{\mathcal{D}_{\text{eye}}}{4\,\text{D}} \tag{6.67}$$

for a direct ophthalmoscope, where $\mathcal{D}_{\text{eye}}$ denotes the refractive power of the patient's eye. If the standard value $\mathcal{D}_{\text{eye}} \approx 60$ D (Table 2.1 in Section 2.2.1) is used in Eq. (6.67), we obtain a nominal magnification of $\beta_{\text{nominal}} \approx 15$[42]. Due to this relatively high magnification, very fine details on the fundus can be identified with a direct ophthalmoscope. As the magnification depends on the refractive status of the patient's eye, the actual magnification is higher (lower) than β_{nominal} in the case of myopia (hyperopia).

42) In accordance with the definition of the nominal magnification of a loupe (Section 6.1.1), the physician sees an object on the patient's fundus 15 × larger than if the same (virtually cropped) object were viewed from the typical near viewing distance of 25 cm without optical aids.

Field of view The *field of view* is the retinal area that the physician can immediately see without moving his eyes and/or the ophthalmoscope relative to the optical axis of the patient's eye. For an emmetropic patient, the maximum angle of the field of view is given by ([31]; Problem P6.13)

$$\alpha_{\text{fov}} = \frac{d_{\text{pupil}} + d_{\text{phys}}}{L_{\text{pp}}}, \tag{6.68}$$

with $[\alpha_{\text{fov}}] = \text{rad}$. d_{pupil} is the diameter of the patient's pupil and d_{phys} the diameter of the physician's pupil. d_{phys} can also represent the diameter of the observation aperture of the ophthalmoscope, depending on which value is smaller. L_{pp} denotes the distance between the patient's and the physician's pupil. With Eqs. (2.6) and (2.7) in Section 2.1, the diameter of the field of view on the retina results as[43]

$$d_{\text{fov}} \approx \frac{\alpha_{\text{fov}}}{\mathcal{D}_{\text{eye}}}. \tag{6.69}$$

Similar to the magnification, the field of view depends on the refractive status of the patient's eye. Compared to emmetropic eyes, the field of view becomes smaller in the case of myopia and larger for hyperopia. To obtain a large field of view, L_{pp} must be as small as possible. In addition, dilating eye drops are typically used to enlarge the patient's pupil. When we assume $\mathcal{D}_{\text{eye}} \approx 60\,\text{D}$, $L_{\text{pp}} = 35\,\text{mm}$, $d_{\text{pupil}} = 5\,\text{mm}$, and $d_{\text{phys}} = 2\,\text{mm}$, it follows that $\alpha_{\text{fov}} = 0.2\,\text{rad} \approx 11.5°$ and $d_{\text{fov}} = 3.3\,\text{mm}$, which corresponds to about double the diameter of the optic nerve head (Section 1.2). However, the results calculated with Eqs. (6.68) and (6.69) can only be considered as maximum theoretical values. Because of vignetting (Section A.1.4.1), the brightness of the fundus image gradually decreases towards the periphery, even if the patient's fundus is uniformly illuminated. As a consequence, the effective field of view is practically reduced so that the diameter d_{fov} is only about 2 mm for a standard direct ophthalmoscope.[44] As the field of view is quite small, the patient's fundus must be screened section by section. This is usually done according to the keyhole principle in that the direction of illumination and observation is changed relative to the optical axis of the patient's eye. For this purpose, the instrument is moved along an arc-shaped path around the patient's eye and/or the patient is asked to look in different directions, which means a change of the line of sight (Section 2.1.3).

6.6.3 Indirect Ophthalmoscope

The main reason for the limited field of view of the direct ophthalmoscope is the relatively large distance between the patient's and physician's pupils which cannot

43) For an emmetropic eye, the distance between the image-side nodal point and the image-side focal point of the eye $\overline{N'F'}$ is equal to the distance between the image-side nodal point and the image point on the retina $\overline{N'I'_0}$.

44) The diameter of the field of view corresponds to the diameter of the optic nerve head with a horizontal diameter of 5° (1.5 mm) and a vertical diameter of 7° (2.1 mm).

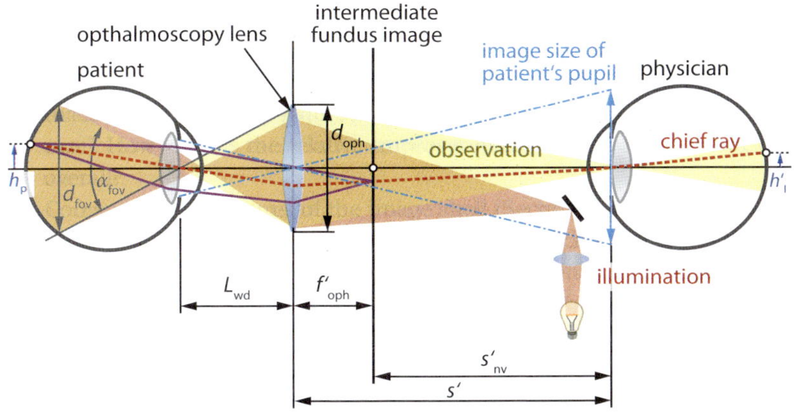

Figure 6.63 Ray diagram of an indirect ophthalmoscope which shows the illumination beam path (red) and the observation path (yellow). L_{wd} is the working distance of the ophthalmoscopy lens, s' the distance between the ophthalmoscopy lens and the physician, f'_{oph} the image-side focal length of the ophthalmoscopy lens, and s'_{nv} the typical near viewing distance between the intermediate image plane and the physician. h_p is the object size on the patient's fundus and h'_I the corresponding image size formed on the physician's fundus. α_{fov} denotes the angle of the retinal field of view, d_{fov} the corresponding diameter of the field of view, and d_{oph} the free diameter of the ophthalmoscopy lens.

be further reduced. However, this limitation can be overcome when we image both pupils into one common plane with a positive (field) lens. This optical principle is also referred to as *pupil matching* and was first introduced by Christian Ruete (1810–1867) in 1852.

Let us consider the optical system of an indirect ophthalmoscope, for which we initially assume that the physician and the patient are both emmetropic. In contrast to direct ophthalmoscopy, an *ophthalmoscopy lens* is added between physician and patient (Figure 6.63). It is arranged close to the patient's eye and forms a magnified image of the patient's pupil in the physician's pupil plane (see blue dash-dotted line). Alternatively, we may consider the optical imaging the other way round in that the ophthalmoscopy lens forms a smaller image of the physician's pupil and the nearby light source in the pupil plane of the patient's eye. In addition to this, the ophthalmoscopy lens together with the optical system of the patient's eye form an enlarged, inverted, and reversed intermediate image of the patient's fundus. If the patient is emmetropic, the intermediate image is formed in the focal plane of the ophthalmoscopy lens. It can be viewed by the physician from a near viewing distance of $s'_{nv} = 25–40$ cm. For this purpose, the physician must either accommodate or insert an appropriate positive lens into the viewing aperture. On the retina of the physician, an upright fundus image of the patient is formed (Figure 6.63). As an image inversion happens in the visual cortex of the brain, the physician perceives an inverted image. Indirect ophthalmoscopy is thus often called “ophthalmoscopy with an inverted image”.

When the refractive power of the ophthalmoscopy lens is given, we may calculate the distance between the two pupils from the ophthalmoscopy lens (according to Eq. (A14)) and the lateral magnification β (according to Eq. (A15)). For an ophthalmoscopy lens with a refractive power of $\mathcal{D}'_{\text{oph}} = 20\,\text{D}$ ($f'_{\text{oph}} = 5\,\text{cm}$) and a viewing distance of $s'_{\text{nv}} = 25\,\text{cm}$, we obtain a working distance of the ophthalmoscopy lens of $L_{\text{wd}} = -5.6\,\text{cm}$, $s' = 45\,\text{cm}$, and a magnification of $\beta = s'/L_{\text{wd}} \approx -8$ (used parameters specified in Figure 6.63). Under these conditions, the image of a 4 mm-large pupil of the patient in the plane of the physician's pupil has a diameter of 32 mm. A fundus observation is only possible if the physician's pupil and the image of the secondary light source are located within this circle. Conversely, the image of the physician's pupil and the nearby secondary light source are reduced in size by a factor of 8 in the patient's pupil so that they do not have a beam-limiting effect. Hence, the free diameter[45] of the ophthalmoscopy lens d_{oph} is the effective aperture stop for the illumination and observation beam paths.

Field of view With Figure 6.63, we can now relate the angle of the retinal field of view α_{fov} to the respective diameter of the field of view d_{fov}. We have [31]

$$\frac{\alpha_{\text{fov}}}{2} = \arctan\left(\frac{d_{\text{oph}}/2}{L_{\text{wd}}}\right), \tag{6.70}$$

$$\Rightarrow \alpha_{\text{fov}} \approx \frac{d_{\text{oph}}}{L_{\text{wd}}} \tag{6.71}$$

and

$$d_{\text{fov}} \approx \frac{\alpha_{\text{fov}}}{\mathcal{D}_{\text{eye}}} \tag{6.72}$$

with the refractive power of the patient's eye $\mathcal{D}_{\text{eye}} = 1/f_{\text{eye}}$. For the above sample data, we thus obtain $d_{\text{fov}} \approx 11.6\,\text{mm}$ if $d_{\text{oph}} = 40\,\text{mm}$. This value is approximately $3.5\times$ larger than the maximum theoretical value or nearly $6\times$ larger than the free diameter of the field of view in direct ophthalmoscopy (Problem P6.14). The same factors apply to α_{fov}. However, the price paid for the considerably larger field of view is a reduction in magnification.

Magnification The angular magnification is defined as the ratio of the angular size of the intermediate image at the near viewing distance s'_{nv} to the angular size of the retinal field of view at the same viewing distance. Hence, the magnification of an indirect ophthalmoscope is given by

$$\beta = -\frac{\mathcal{D}_{\text{eye}}}{\mathcal{D}'_{\text{oph}}}, \tag{6.73}$$

where the negative sign takes into account the image inversion of the real intermediate image. For the typical values of $\mathcal{D}'_{\text{oph}} = 20\,\text{D}$ and $\mathcal{D}_{\text{eye}} \approx 60\,\text{D}$, it follows

45) In practice, the border of a lens is framed by an opaque lens mount. With "free diameter" we mean the optically transparent area through which light can travel.

that $\beta = -3$. Thus β is about $5\times$ smaller than for direct ophthalmoscopy (Problem P6.15).

In indirect ophthalmoscopy, the field of view and the magnification both depend on the patient's ametropia. But in contrast to direct ophthalmoscopy, the influence of the refractive error is negligible in indirect ophthalmoscopy, as the resulting displacements of the intermediate image due to ametropia can be easily compensated for by the physician. In fact, a slight accommodation and/or a minimal change of the visual distance is sufficient.

Besides the larger field of view, pupil matching has two further advantages. First, it is easier to realize a reflection-free observation (Section 6.6.3.1). Second, it is possible to implement stereoscopic viewing of the fundus (Section 6.6.3.2).

6.6.3.1 Reflection-Free Observation

When the illumination and observation beams travel through the same optical system, the optical components reflect/backscatter light from the light source which may enter the observation path and thus impair the visibility of the fundus image. As the reflectance of the retina is very low in the visible spectral range, even relatively low intensities of reflected or scattered light may outshine the faint fundus image. For this reason, the image quality is particularly reduced in the center of the fundus. The portion of reflected light originating from the ophthalmoscopy lens is relatively low, because of anti-reflection coatings on the lens surface. The remaining scattered light can be further reduced when the ophthalmoscopy lens is slightly tilted. To remove the reflections of cornea and eye lens, the observation and illumination beam paths must be separated (see also Section 5.2.1). In this case, we use the principle of *geometric pupil separation* which states that the illumination and observation beam paths must be geometrically separated in all locations where scattering and reflections may occur. This principle was formulated by Alvar Gullstrand in 1910 and was successfully implemented into the ZEISS ophthalmoscopes (Large Ophthalmoscope) just one year later.

As the ophthalmoscopy lens delivers an image of the physician's pupil and the light source which is reduced in size in the patient's pupil plane, it is much easier to fulfill the principle of geometric pupil separation in indirect ophthalmoscopes than in direct ophthalmoscopes. Figure 6.64 shows the images of the physician's pupil (blue) and the light source (red) in the patient's pupil. In addition, the illumination and observation beam paths are traced through the anterior segment of the patient's eye (right column). In Figure 6.64a, the situation is shown for which a reflection-free observation is possible. The geometric separation is, however, not sufficient in Figure 6.64b so that reflected and scattered light can enter the observation beam path.

6.6.3.2 Binocular Indirect Ophthalmoscope

The ophthalmoscope setups which we have considered so far only allow monocular viewing of the fundus. This means that clinically relevant depth information (e.g., the shape of the optic nerve head for glaucoma diagnosis) cannot be gathered.

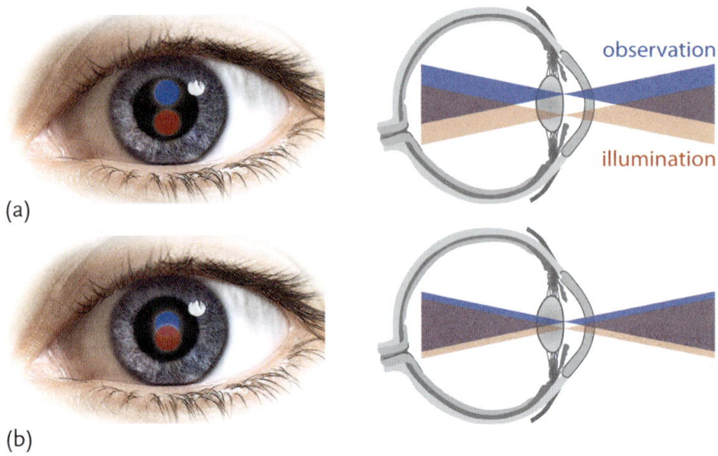

Figure 6.64 Principle of geometric pupil separation of the observation and illumination beam paths for reflex-free examination of the posterior eye segment. (a) Illumination (red) and observation (blue) beam paths are separated from the corneal front surface to the rear surface of the eye lens. In this case, no interfering reflection or scattering light generated in the anterior segment can be observed in the fundus image. (b) Illumination and observation beam paths are not sufficiently separated so that corneal reflections and eye lens scattering will impair the observed image.

However, with some design modifications of an indirect ophthalmoscope, it is possible to provide a 3D impression of the patient's fundus (stereopsis; Section 2.1.9). As the image of the physician's pupil formed by the ophthalmoscopy lens is reduced in size, two laterally separated observation beam paths can be fitted into the patient's pupil for binocular viewing, as shown in Figure 6.65. However, binocular visualization of the patient's fundus is only possible if both observation pupils lie completely *inside* the enlarged image of the patient's pupil (Problem P6.16). In most cases, the interpupillary distance PD is much larger than the image diameter of patient's eye pupil in the physician's pupil plane so that an adjustable arrangement of mirrors or prisms is required to reduce the effective PD.

A typical binocular indirect ophthalmoscope (BIO) is either head-mounted or attached to an eyeglass frame so that both hands of the physician are left free to hold the ophthalmoscopy lens and to manipulate the patient's eye. It features a light source and contains a setting device for the PD. The patient's pupil is usually enlarged with dilating (mydriadic) eye drops. This not only simplifies the geometric pupil separation, but also creates the free space which is needed to increase the stereopsis. As discussed in Section 6.7.8, the stereo base b is given by the distance between the physician's pupils (both observation channels) in the plane of the patient's pupil, which in turn determines the maximum depth resolution. For optimal adaption to a given pupil diameter of the patient's eye, BIOs allow setting the distance between the physician's pupils and the illumination beam path in the pupil plane of the patient (Figure 6.65b). As a consequence, small pupils can be examined as well, logically at the expense of stereopsis.

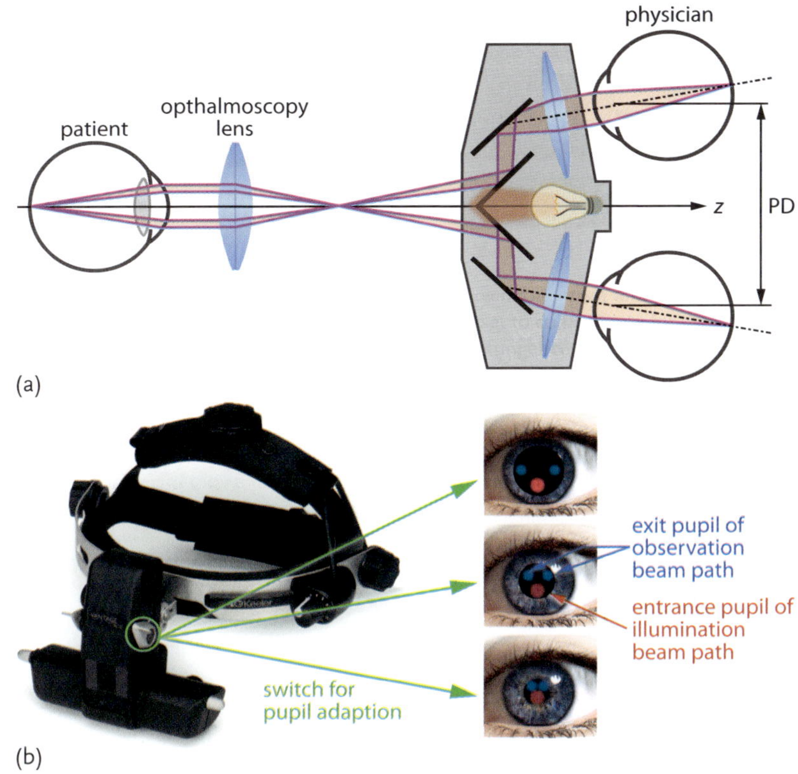

Figure 6.65 (a) Ray diagram of a binocular indirect ophthalmoscope. The eyepieces in front of the physician's eyes are arranged in a convergence angle with respect to the optical axis to adapt the natural convergence position of the pair of eyes during accommodation (see also Problem P6.16). (b) Photograph of the Keeler Vantage *Plus* binocular indirect ophthalmoscope. With a switch, the entrance pupil of the illumination beam path and the exit pupils of the observation beam paths can be adapted to the actual diameter of the patient's eye pupil. Courtesy of Keeler Limited.

Other features of standard BIOs are:

- Green (red-free) and blue filters can be added to the illumination beam path.
- Additional optics can be swung into the observation beam paths for additional magnification and to invert the intermediate image.
- An integrated video camera can be used for co-observation or image documentation.

BIOs typically utilize ophthalmoscopy lenses with refractive powers in the range of 15–30 D, where the 20 D lens is commonly accepted as a standard. The lenses have a free diameter of 30–50 mm and generally have an aspheric design. In this way, the required image quality for stereoscopic viewing can also be achieved in the peripheral areas.

History The first binocular indirect ophthalmoscope was presented by Félix Giraud-Teulon (1816–1887) in 1861. The stereoscopic and reflection-free observation was, however, not implemented until ZEISS introduced the Gullstrand-type Large Ophthalmoscope in 1911. In contrast to the hand-held instrument by Giraud–Teulon, this was a floor-mounted device which became the standard instrument in ophthalmoscopy for decades. From the 1930s onwards, the instrument was produced under license by Bausch+Lomb. The breakthrough of binocular indirect ophthalmoscopy into clinical routine was then achieved with the binocular indirect ophthalmoscope of Charles Schepens (1912–2006) in 1946. Schepens' design was used for commercial devices from 1953 onwards by the American Optical Company and is currently offered by a large number of companies in different configurations.

Prospects Due to the relatively large field of view and the possibility of stereoscopic viewing, the BIO became a standard instrument for visual inspection of the posterior eye segment. Compared to other competing methods such as slit lamp microscope with auxiliary lenses (so-called *slit lamp biomicroscope*; Section 6.4.4.1), the BIO is flexible in use (e.g., for disabled or bedridden patients) and thus an essential tool for ambulant diagnosis. It is also used as a cost-effective beam delivery device in laser photocoagulators (Section 10.2.4.4).

6.7 Fundus Camera

Fundus cameras are instruments which are primarily used for photographic documentation of the eye fundus.[46] The acquired images are used to perform structural analyses of the visualized fundus structures. In this way, it is possible to examine potential deviations from the normal condition and/or to detect any morphological changes of specific diseases such as age-related macular degeneration (Section 3.4), glaucoma (Section 3.3), diabetic retinopathy (Section 3.5), and retinal vein occlusions (Section 3.6). As the findings are usually available in the form of a digital image, the (often computer-aided) diagnosis can be rapidly completed. The fundus camera also simplifies the clinical documentation and progression control of diseases.

Fundus imaging is usually performed with white light so that color images of the fundus are obtained just like in ophthalmoscopy. In addition, the fundus camera can be operated in different imaging modes which substantially expand the diagnostic options:

- *Monochrome images* with filtered light of different spectral ranges (Section 6.7.5) highlight specific fundus structures. For example, when blue light is used, the retinal nerve fiber layer becomes clearly visible.

46) Fundus cameras can also be used for documentation of the anterior eye segment.

- *Fluorescence angiography imaging* (Section 6.7.6) allows visualization of blood vessels of the retina and choroid with high contrast. Thus, leakages and blockages can be easily detected, and the flow dynamics (*hemodynamics*) can be analyzed.
- *Fundus autofluorescence imaging* (Section 6.7.7) is used to visualize the topographic distribution of substances which are relevant for the metabolism of the eye. With this imaging mode, the functional status of the retina can thus be diagnosed at the cellular level (metabolic mapping; see also Section 8.2). A typical application is the diagnosis of disease-related changes in the retinal pigment epithelium (Section 1.2) by means of the autofluorescence of the metabolic pigment lipofuscin.
- *Stereo imaging:* Usually, fundus cameras produce monocular images. However, stereo image pairs (Section 6.7.8) can be generated as well to provide a stereoscopic view of the fundus. For example, the stereo imaging mode is used to evaluate the shape and structure of the optic nerve head in glaucoma diagnosis.
- *Wide-field images* (Section 6.7.4) are generated from individual fundus images which are captured from different points of view. The fundus images are then automatically stitched together by a image processing software to get an overview of a large part of the posterior eye segment.

History The first fundus camera suitable for clinical practice was developed by Johan Nordenson (1883–1965) based on Gullstrand's reflection-free indirect ophthalmoscope (Section 6.6.3). The instrument was fabricated by the company Carl Zeiss from 1926 onward. The underlying concept of optical image acquisition by indirect ophthalmoscopy which was developed at that time is still used today, almost unchanged, in basically all fundus cameras.

6.7.1 Requirements for a Fundus Camera

From the user's point of view, a fundus camera must deliver reflection-free and uniformly illuminated images. To make details clearly visible, the resolution, contrast, and sharpness of the images have to be sufficiently high. In addition, a fundus camera must provide the largest possible field of view, avoid excessive light stress for the patient, and should be easy to use. The implementation of all these features requires a complex optical system, because the human eye with all its aberrations (Sections 5.3 and 5.4) is an integral part of the illumination and imaging system. In addition, the curved surface of the fundus must be imaged sharply onto a flat detector surface. To obtain a good image quality, all these influence factors must be taken into consideration for the optical design.

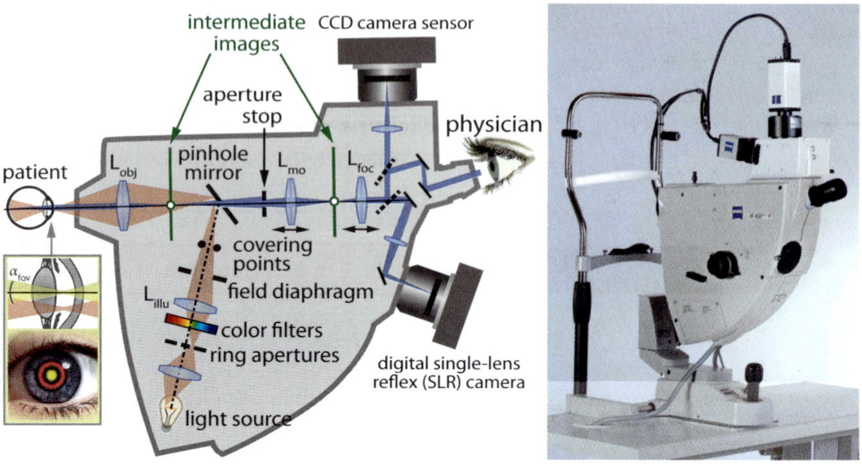

(a) (b)

Figure 6.66 (a) Simplified schematic beam path of the ZEISS FF 450^{plus} fundus camera. L_{obj} denotes the objective lens which corresponds to the ophthalmoscopy lens of an indirect ophthalmoscope. L_{foc} and L_{mo} are the focusing and main objective lenses, respectively. L_{illu} denotes the illumination lens. The individual optical components and their function are explained in the text. Top inset: Beam path separation in the anterior eye segment to avoid disturbing reflections in the image. Between the corneal front surface and the anterior lens surface, the illumination beam path (red) is separated from the observation path represented by the field of view (yellow). Bottom inset: Front view of the examined eye in which the ring-shaped entrace pupil of the illumination path (red) and the exit pupil of the observation path (yellow) are marked. (b) Photograph of the ZEISS FF 450^{plus} fundus camera. Courtesy of Carl Zeiss.

6.7.2
Functional Principle

The main optical components of a fundus camera are the illumination system (red beam path in Figure 6.66a) and the observation system (blue beam path in Figure 6.66a). The beam paths of both systems are merged by a pinhole mirror and pass through an aspheric objective lens L_{obj} as well as the refractive parts of the patient's eye. Here, the objective lens plays the same role as the ophthalmoscopy lens of an indirect ophthalmoscope (Section 6.6.3).

6.7.2.1 Illumination Beam Path

To obtain reflection-free images, a geometric pupil separation is required (Section 6.6.3.1). That means the illumination and the observation beam paths must be separated within the anterior segment of the patient's eye,[47] that is, from the

47) Reflection-free observation can also be achieved by means of *spectral pupil separation*. In fluorescence imaging, the illumination and observation take place at different wavelengths. For this purpose, suitable filters are used to "select" different spectral ranges.

corneal front surface to the back surface of the eye lens. To fulfill this condition, the separation zone between the two beams in the pupil plane must be large enough. In fundus cameras, the patient's eye pupil is thus usually illuminated with an annular light beam (ring illumination), while the exit pupil of the observation beam path passes through the pupil center (inset of Figure 6.66a), where the optical imaging quality of the eye is highest.

The ring illumination of the fundus camera is formed by a system of ring apertures which is imaged by the illumination lens L_{illu} onto a pinhole mirror. The objective lens L_{obj} then images the annular light beam into the patient's pupil so that the fundus is uniformly illuminated. The ring illumination must be designed such that the efficiency of the illumination is maximal, that is, a large ring area is required. However, for a sufficiently large separation zone between the illumination and observation beam paths, a large inner diameter of the illuminating light bundle is necessary. The diameter of the patient's pupil, which limits the outer diameter of the ring illumination, must also be taken into account [32].

If the diameter of the eye pupil is too small, the outer rim of the ring illumination is cut off which, in turn, leads to a nonuniform illumination of the fundus and reduced brightness. In order to capture fundus images with good quality, the diameter of the patient's pupil must be large enough. For example, for a field angle of 45°, the pupil diameter must be at least 4 mm. In addition to this, fundus cameras have an exchangeable field diaphragm which allows gradual adjustment of the diameter of the ring illumination to the pupil diameter. To enlarge the pupil diameter, dilating eye drops containing mydriatic agents are applied or the natural adaption of the eye to ambient light conditions is used. Depending on which method is used to increase the pupil diameter for fundus imaging, we distinguish between *mydriatic* and *nonmydriatic* fundus cameras. In mydriatic fundus cameras, visible light (e.g., from a tungsten halogen lamp) can be used for the adjustment and focusing of the instrument, as dilating eye drops disable the pupil reflex (Section 2.1.6). However, mydriatic agents can only be applied by qualified personnel due to potential adverse effects (Section 3.3). Hence, so-called *nonmydriatic fundus cameras* have also been established on the market. They are based on the natural pupil enlargement in a dark environment so that the ambient light level must be low before and during image acquisition. For adjustment, the patient's eye is illuminated with infrared light which does not stimulate the pupil reflex. The observation is then performed by means of a suitable infrared camera. For image acquisition with both types of fundus cameras, high-power pulsed light sources such as xenon flash lamps or pulsed LEDs[48] are used because of the low reflectance of the fundus in the visible spectral range. Within the pupil response time, several high-quality color images can be taken with a nonmydriatic fundus camera before the pupil diameter appreciably decreases. If, however, the image acquisition takes a few minutes, for example, in the case of fluorescence angiography (Section 6.7.6), the pupil reflex must also be disabled with mydriatic eye drops.

48) The pulsed operation mode is essential, as eye motion during image acquisition might impair the image quality.

So far, we have not considered the reflection of illuminating light on both surfaces of the objective lens, which might be present even if the lens has an anti-reflection coating. The reflected light may enter the observation beam path and lead to undesirable reflexes on the fundus image. To avoid such reflexes, so-called *covering points* (or *anti-reflection points*; Figure 6.66a) are added to the illumination beam path. As every covering point impairs the uniformity of the fundus illumination, the position and size of the corresponding apertures must be carefully designed.

The illumination for the above mentioned imaging modes is realized by distinct color filters which are inserted into the illumination beam path as shown in Figure 6.66a.

6.7.2.2 **Observation Beam Path**

In analogy with an indirect ophthalmoscope, the objective lens forms a real intermediate image of the illuminated fundus in front of the pinhole mirror. Behind the pinhole mirror, a second intermediate image is formed by the main objective lens L_{mo}. With a movable focusing lens L_{foc}, the rays are then parallelized (focus at infinity). In this way, we get a flexible optical interface which can be used to attach

- high-resolution cameras for image acquisition,
- observer cameras (e.g., infrared CCDs for adjustment purposes in nonmydriatic fundus cameras), and
- projection systems (e.g., for fixation targets and/or image-assisted microperimetry; Section 8.1.3.4).

Behind the pinhole mirror, an aperture stop is located which is imaged by L_{obj} into the patient's pupil plane with a decreased image size. It defines the diameter of the exit pupil of the observation beam path and thus the optical resolution of the fundus image. The highest resolution (approximately 6 μm) is achieved with an exit pupil diameter of 3 mm, as up to this diameter the emmetropic eye can be considered as a diffraction-limited optical system (Sections 2.1.5.3 and A.2.1.6). However, the maximum resolution can only be obtained for a small field of view and if the pupil is dilated. This follows from the fact that the separation zone required for reflection-free observation (Gullstrand condition) increases with the angle of the field of view α_{fov} (inset of Figure 6.66a). Thus, to capture reflection-free fundus images with a large field of view, we need a small aperture stop which, in turn, reduces the resolution (e.g., approximately 10 μm resolution for $\alpha_{fov} = 50°$).

L_{mo} is used to "flatten" the intermediate fundus image in the case of a large field of view. It is advantageously arranged in a telecentric optical system so that the image size of the intermediate image does not depend on potential refractive errors of the patient's eye. In addition, L_{mo} can be axially moved to correct the refraction-dependent position of the intermediate image. In this regard, a main objective lens with a short focal length is able to compensate for a large range of refractive errors by slight displacements. With these measures, it is possible to keep the observed image size constant for a given field of view so that the fundus image is well-suited for reproducible geometric measurements in the posterior part of the eye.

6.7.3
Field of View and Magnification

According to the indirect ophthalmoscope, the maximum angular field of view of a fundus camera α_{fov} depends on the ratio between the objective lens' diameter d_{obj} and the working distance L_{wd} (see Eq. (6.71)). Since the working distance is constant in practice, the *usable* field of view is, however, determined by the diameter of the illumination beam at the objective lens. Normally, the maximum field of view of a fundus camera is $50°$. Only with special mydriatic wide-field fundus cameras, a larger field of view of up to $60°$ can be realized. Many commercially available systems allow imaging of the patient's fundus with a reduced field of view and increased resolution. Typical gradations for α_{fov} are $50°$, $30°$, and $20°$, which can be set with an exchangeable field diaphragm.

Focusing lenses with different focal lengths ensure that the fundus can be imaged with a reduced field of view such that the whole camera sensor is filled. In this case, an aperture stop with a larger diameter can be introduced behind the pinhole mirror. As mentioned, the exit pupil then becomes larger, and a higher image resolution results. The achievable image magnification lies in the range from $1.5\times$ (for $\alpha_{\text{fov}} = 45°$) to $5\times$ (for $\alpha_{\text{fov}} = 20°$). The total observation magnification then ranges between 10 and $30\times$.

6.7.4
Wide-Field Imaging

Peripheral areas of the retina which lie outside the central field of view are also accessible when the patient looks in different directions, which means a change of the line of sight (see also Section 6.6.2.2). With a so-called *Auto Mosaic* (or *Montage*) *Software*, the individual images are then stitched together to form a panorama which spans an angular range of up to $110°$.

Other than creating montages, two additional methods exist with which wide-field images of the retina can be obtained. We can either use a special contact or noncontact wide-field lens in combination with a conventional fundus camera or a dedicated wide-angle camera system. In these special fundus imaging systems, the objective lens is directly attached to the cornea with a contact gel, and the fundus is illuminated either directly through the pupil (transpupillary illumination) or from the side through the sclera (transscleral illumination).

6.7.5
Color and Monochrome Imaging

Fundus imaging is commonly performed with white light and thus delivers color images as depicted in Figure 6.67a. However, certain retinal structures can be visualized with better contrast, if the white light is spectrally filtered [33]. In the following, the standard imaging modes are listed:

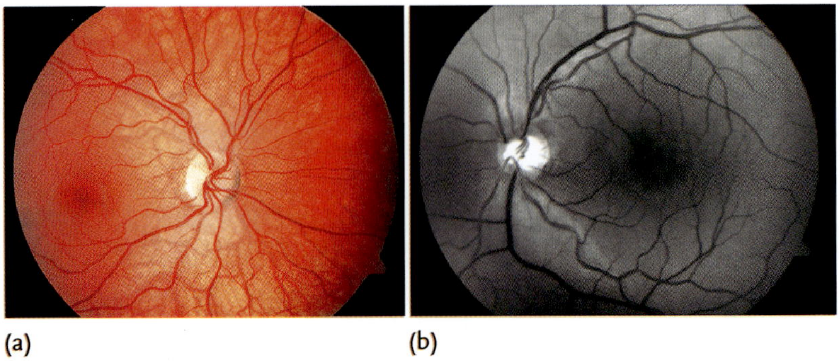

Figure 6.67 (a) Color image of the fundus. (b) Red-free monochrome image of the fundus. Courtesy of Carl Zeiss.

- *Red-free images with green filter*: Monochrome imaging in the green spectral range between 540 and 580 nm provides high-contrast pictures of the retinal blood vessels. As shown in Figure 6.67b, the vessels appear dark with this type of illumination, since the green light is strongly absorbed by hemoglobin (Figure 9.3 in Section 9.3).
- *Blue filter images*: With blue light of wavelengths between 490 and 530 nm, the stripe pattern of the retinal nerve fiber layer becomes clearly visible due to its scattering properties (Section 9.2).
- *Red filter images*: Red light with wavelengths between 620 and 650 nm is only weakly absorbed by the retinal pigment epithelium so that it penetrates deeply into the tissue. For this reason, we can use red light to take pictures of the choroid.
- *Multispectral imaging*: This imaging mode is an expansion of the established monochrome imaging techniques. Here, the fundus is simultaneously or sequentially illuminated with light of several spectral ranges in order to detect the local distribution and concentration of specific, metabolically relevant substances. The substances are then identified by means of their spectral signature. Multispectral methods, which are still in the research or early clinical application phase, are described in Section 8.2.2.

6.7.6
Fluorescence Angiography

Fluorescence angiography (FAG) imaging is a method to visualize the blood vessels of the retina and choroid. It can also be used to analyze the blood flow dynamics for finding leakages or blockages. For this purpose, a fluorescent dye is directly injected into the vein of the patient's arm. As soon as the dye has been transported to the blood vessels of the eye, it is excited with monochromatic light at a wavelength for which the dye's absorption coefficient (Section 9.1) is high. The required monochromatic light is generated by adding an *excitation band-pass filter* to the illumination beam path. The light-excited dye emits fluorescence radiation of a longer wavelength (so-called *Stokes shift*) which is then detected via a *barrier band-pass filter* in the observation beam path. As the intensity of the fluorescent light is substantially weaker than that of the excitation light, a sufficiently strong spectral separation of both filter characteristics is required. Otherwise, the intensity of the nonblocked excitation light is also recorded by the detector and thus cannot be distinguished from the pure fluorescent light (pseudo-fluorescence).

In fluorescence imaging, a black-and-white image is evaluated in which bright areas with a strong fluorescence signal indicate a high local dye concentration. We can use this property to detect pathological changes in eye structures. For example, we observe strong fluorescence (*hyperfluorescence*) in the case of leakages of the blood vessels, defects of the retinal pigment epithelium, and edema. But if the blood vessels of the patient's eye are partially or totally occluded, weak or no fluorescence (*hypofluorescence*) is observed instead.

Two dye substances are commonly used in FAG, that is, sodium fluorescein (NaF) and, less frequently, indocyanine green (ICG). FAG with NaF is referred to as *fluorescein angiography (FA)* and FAG with ICG as *ICGA*, respectively.

Fluorescein angiography (FA) NaF is a suitable substance to highlight retinal blood vessels. It is a water-soluble yellow-red dye with a high fluorescence efficiency which means that a high fluorescence intensity, and thus a high image contrast, is provided. As NaF is not able to penetrate intact retinal and choroidal blood vessels, a visible leakage of the dye directly means leakages in the blood vessels or abnormal vessels (e.g., in the case of neovascularization; Section 3.4).

In FA, at first red-free fundus images are taken which serve as a zero reference (signal background). Approximately 8–15 s after injection into the arm vein, the dye enters the choroidal and retinal blood vessels. Starting during the inflow phase (*early-transit phase*), FAG images are acquired every second for some 20 s. These pictures allow examination of the inflow process of the dye into the vascular system of the retina and identification of potential passage disorders. During the mid-transit (3–5 min after injection) and late-transit (10–15 min after injection) phases, the time intervals between successive image acquisitions are usually longer, depending on the disease to be diagnosed. These pictures reveal potential leakages or dye accumulation in tissue (*staining*) or in an anatomic space (*pooling*).

Figure 6.68a shows the absorption and emission spectra of NaF as well as the characteristics of the relevant excitation and barrier band-pass filters. A typical NaF angiogram during the early-transit phase is depicted in Figure 6.68b. NaF cannot be used for the angiography of the choroid, as the required blue-green excitation light is strongly absorbed by the retinal pigment epithelium.

Indocyanine green angiography (ICGA) ICG has a relatively high absorption in the near-infrared spectral range (Figure 6.69a), where the retinal pigment epithelium has a rather low absorption (approximately 10%). Consequently, this substance is appropriate for angiography of the choroid. However, the fluorescence efficiency of ICG is substantially lower than for NaF. For this reason, ICG angiography requires

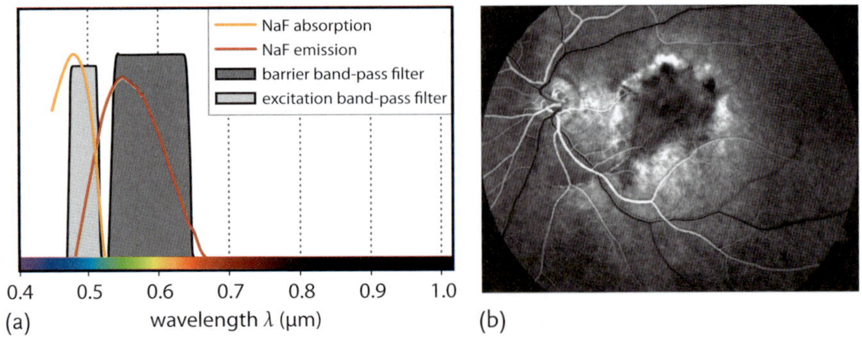

Figure 6.68 Fluorescein angiography (FA). (a) Absorption (orange) and emission (red) spectra of sodium fluorescein (NaF) and spectral characteristics of the excitation and barrier band-pass filters used for FA. (b) Fluorescein angiogram of the fundus during the early-transit phase. The arteries (light) are already filled with NaF, while the veins (dark) are still not filled with the dye. Courtesy of Carl Zeiss.

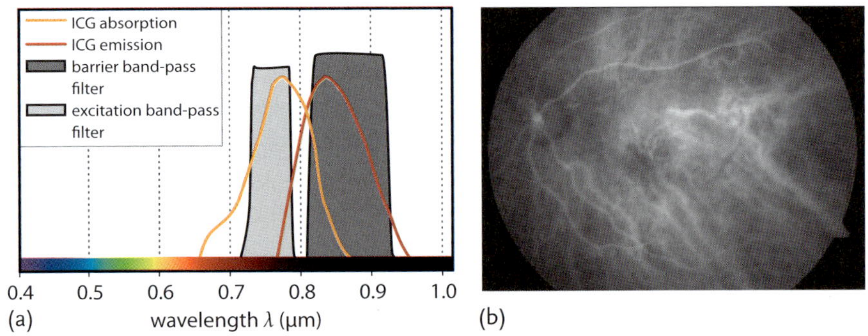

Figure 6.69 Indocyanine green angiography (ICGA). (a) Absorption (orange) and emission (red) spectra of indocyanine green (ICG) and characteristics of the excitation and barrier band-pass filters used for ICG angiography. (b) Indocyanine green angiogram of the choroid. The retinal vessels which arise from the optic nerve head are visible as well. Courtesy of Carl Zeiss.

higher excitation intensities and more sensitive image sensors. Compared to the NaF angiogram (Figure 6.68b), the ICG angiogram appears more diffuse and is more difficult to interpret (Figure 6.69b). This is mainly caused by the complex 3D mesh of the choroidal blood vessels and the light scattering in the tissue structures in front of the choroid.

6.7.7 Fundus Autofluorescence

Fundus autofluorescence (FAF) imaging is a relatively new method to visualize the topographic distribution of pigments or metabolically relevant substances. FAF thus allows examination of the functional status of the retina at the cellular level [34]. For example, the distribution of the macular pigment xanthophyll can be measured [35], or the progression of age- and disease-related changes in the retinal pigment epithelium can be examined via the autofluorescence of the metabolic end product lipofuscin (lipofuscin granules; Section 3.4). A detailed description of other techniques which also utilize FAF can be found in Section 8.2.2.

In contrast to FAG, FAF uses the fluorescence characteristics of *endogenous fluorophores*[49]. This approach has the advantage that we do not have to inject substances into the bloodstream. However, autofluorescence imaging is much more difficult due to the following reasons:

- It is quite difficult to distinguish the individual endogenous fluorophores, as their excitation and emission spectra are very broad and partially overlap. Thus, multiple excitations may occur simultaneously.
- The autofluorescence in the eye lens may overlay the fluorescence signal of the fundus.
- The fluorescence efficiency of endogenous fluorophores is relatively low. For safety reasons (Section 4.4), it is also not possible to substantially raise the intensity of the excitation light.
- The absorption of endogeneous dyes is maximal in the ultraviolet spectral range. But this spectral range cannot be used for excitation light, as most of the ultraviolet light is absorbed by cornea and lens (Figure 2.12 in Section 2.1.10).

All of these factors must be taken into account when appropriate excitation and barrier band-pass filters are selected. In Figure 6.70a, the absorption and emission spectra of the main retinal fluorophore lipofuscin and the characteristics of relevant excitation and barrier band-pass filters are shown. A typical lipofuscin FAF image of an eye with dry age-related macular degeneration (Section 3.4) is depicted in Figure 6.70b. The areas of geographic atrophy appear in black, as in these regions

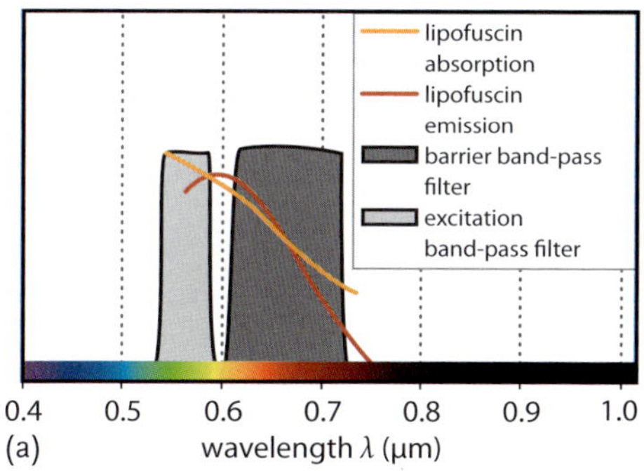

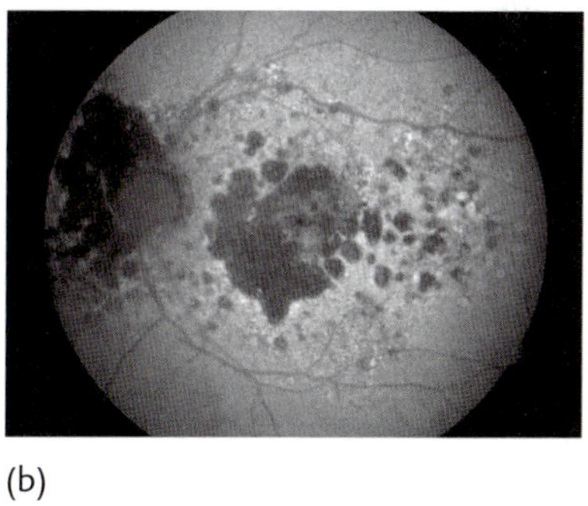

Figure 6.70 Lipofuscin autofluorescence imaging. (a) Absorption (orange) and emission (red) spectra of lipofuscin and characteristics of the excitation and barrier band-pass filters. Data taken from [36]. (b) Lipofuscin autofluorescence image of an eye with dry age-related macular degeneration. The areas of geographic atrophy appear in black, because in these regions the retinal pigment epithelium is destroyed, and lipofuscin is lacking. Courtesy of Dr. Thomas Behling.

49) Endogenous fluorophores are substances which originate or are produced by certain (tissue) cells.

the retinal pigment epithelium is destroyed, and lipofuscin is lacking. For some fundus cameras, a special set of lipofuscin FAF filters is offered as an optional accessory.

6.7.8
Stereoscopic Imaging and Analysis

With a set of properly acquired 2D pictures, it is possible to render a 3D picture of a 3D object. For this purpose, a *stereo image pair* of the considered object must be captured from different perspectives. In this way, the image details of the object are laterally shifted depending on their corresponding distance from the camera used for capturing (image disparity; see also Problem PI.11). When someone looks at the stereo image pair with a special stereo-viewing setup[50], he or she has a 3D impression of the object. As for "normal" binocular vision (Section 2.1.9), the brain is able to combine the perceived image disparity and "forms" a 3D picture. With this technique, it is thus possible to distinguish between any object details which are located in different planes along the viewing direction (see also Figure 2.11 in Section 2.1.9).

We can now use this principle to generate virtual 3D stereo images with a fundus camera, for example, to analyze the topography of functional areas on the fundus (Section 1.2.2), which is relevant for the diagnosis of certain eye diseases. For this purpose, two fundus images are captured which are laterally displaced with regard to the pupil center (Figure 6.71a). The lateral shift is then "translated" by the optical system of the patient's eye to a corresponding change of the viewing angle. The stereoscopic parallax generated by this change is saved in the stereo image pair as an image disparity. The virtual 3D image is then observed with a special stereo viewer.

In the following, we estimate the minimal differences in depth which can be perceived on the fundus. For normal binocular vision, the minimum stereoscopic depth perception ΔL_{min} depends on the minimum stereo angle ε_{min} of the human eye, the stereo base given by the interpupillary distance PD, and the distance L between observer and object. According to Eq. (2.18), we have

$$\Delta L_{\text{min}} = \frac{L s_{\text{p,min}}}{\text{PD}} = \frac{\varepsilon_{\text{min}} L^2}{\text{PD}} , \qquad (6.74)$$

where $s_{\text{p,min}} = \varepsilon_{\text{min}} L$ is the minimum stereoscopic parallax distance. For the stereo image pair of a fundus camera, the stereo base is given by the distance between both observation channels b in the plane of the exit pupil, and L is now determined by the distance between the nodal point N′ of the patient's eye and the fundus. As the distance $\overline{\text{N}'\text{I}_0}$ is equal to the image-side focal length of the eye f'_{eye} (see Eqs. (2.3) and (2.6)), we have $L = f'_{\text{eye}}$. For $s_{\text{p,min}}$, the smallest distinguishable structure in

50) With special optical separation methods (e.g., red/green color filters or polarization filters), these devices ensure that the left (right) eye of the physician looks only at the left (right) image. This technique is also used in modern 3D televisions and consumer electronics for 3D movie presentation.

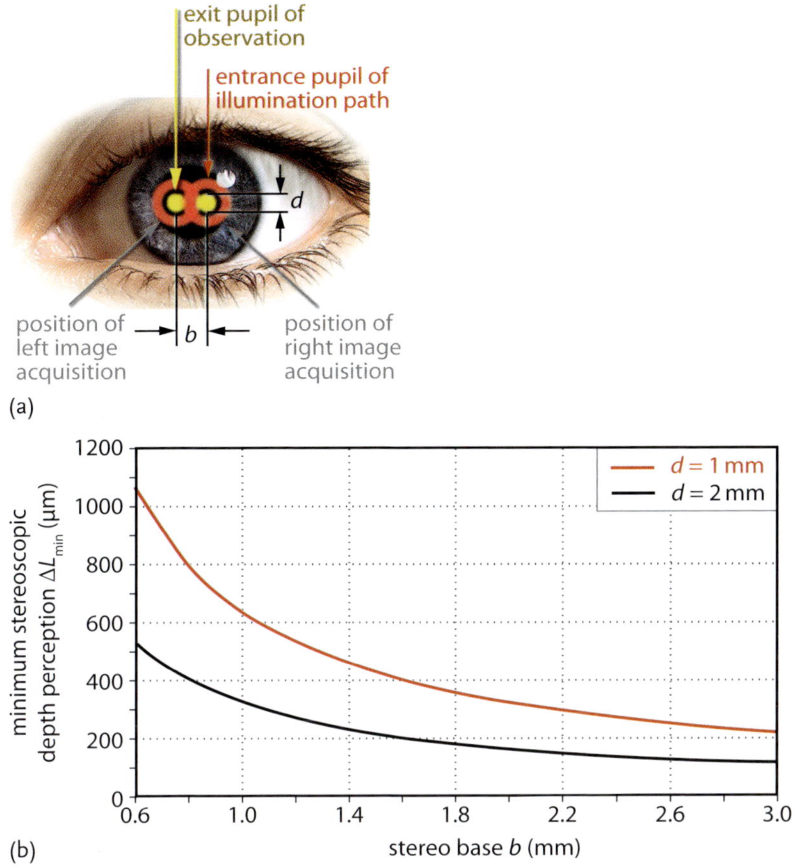

Figure 6.71 (a) Location of the entrance pupils of the illumination beam path (red circle) and of the exit pupils of the observation path (yellow area) used for stereo photography of the fundus. b is the stereo base, that is, the distance between both observation channels, and d the diameter of both. (b) Minimum stereoscopic depth resolution ΔL_{min} in stereoscopic fundus imaging as a function of the stereo base b. Two curves are shown for different exit pupil diameters d of the observation beam path.

the fundus image must be used. As $s_{p,min}$ depends on the resolution limits of the eye (Section 2.1.5) and the image sensor, it is given by

$$s_{p,min} = \frac{4\lambda f'_{eye}}{d} , \qquad (6.75)$$

where d denotes the diameter of the exit pupil of the observation beam path. The factor 4 is an experimentally determined correction factor[51] which takes the reduced contrast sensitivity of the detector (compared to the eye) and the reduced imaging quality of the eye (due to aberrations in the case of off-axis image acquisi-

51) Private communication by Detlef Biernat.

tion) into account. From Eq. (6.74) it follows that

$$\Delta L_{\min} = \frac{4(f'_{\text{eye}})^2 \lambda}{b d} . \tag{6.76}$$

This is a conditional equation which allows us to calculate $\Delta L_{\min}$ for virtual 3D stereo images of the fundus.

In Figure 6.71b, $\Delta L_{\min}$ is plotted versus the stereo base b for a wavelength of $\lambda = 550$ nm and for two different diameters of the exit pupil d. Structures can only be perceived as a 3D object if the depth difference is greater than the corresponding value of the curve. To increase the stereo resolution $(1/\Delta L_{\min})$ as much as possible, the patient's pupil must be dilated by mydriatic eye drops. Under optimal conditions, the maximum achievable stereo base b is approximately 3 mm.

The stereo image pair can either be acquired simultaneously with special stereo fundus cameras or sequentially with standard monocular cameras. When the images are taken sequentially, the two images are recorded shortly one after the other with shifted camera positions. Although product solutions exists which ensure a correct lateral shift, involuntary eye movements between both photographs can lead to errors. Hence, simultaneous imaging with stereo fundus cameras is certainly the better approach, as both images are acquired under identical conditions and with a noninfluenced stereo base. However, stereo fundus cameras are more expensive.

With the generated 3D stereo images, the physician may characterize the topography of the optic nerve head in the case of glaucoma (Section 3.3) or macular edema (Section 3.5). To date, such findings are mainly obtained by visual inspection in a purely qualitative manner. But software modules are also available which allow a quantitative evaluation [37].

6.7.9
Equipment Solutions

Some ten companies provide fundus cameras with different accessories which have more or less comparable standard specifications (field of view, resolution, etc.). However, with regard to functional features, such as the number of available imaging modes and user-friendliness (e.g., autofocus, autoshoot functions), the available products differ. In Figure 6.72a,b, commercial mydriatic and nonmydriatic fundus cameras by Topcon and Kowa are shown, respectively.

6.7.10
Prospects

Fundus imaging with fundus cameras is an established method whose possible applications are steadily expanded by integration of novel optical technologies. Some important development trends are:

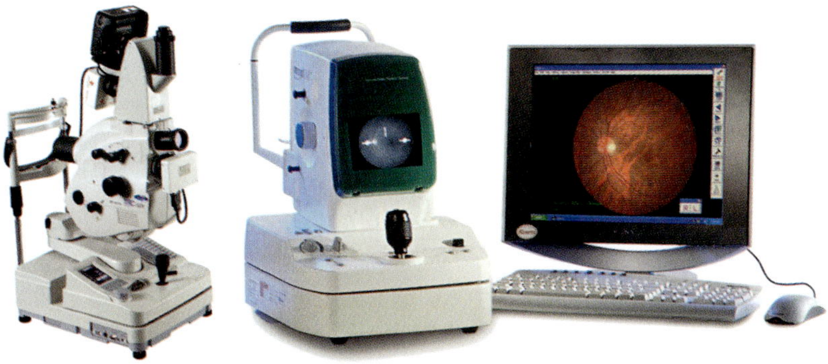

(a) (b)

Figure 6.72 (a) Photograph of mydriatic fundus camera Topcon TRC-50DX. Courtesy of Topcon Deutschand GmbH. (b) Photograph of nonmydriatic fundus camera Kowa nonmyd α-D III. Courtesy of Kowa Optimed Deutschland GmbH.

1. Improvement of structural analysis:
   - Combination with alternative imaging methods, for example, the combination of fundus cameras and optical coherence tomographs complements 2D imaging with 3D cross-sectional images (as realized by the ZEISS Cirrus™ photo; Section 7.5.2).
   - Use of adaptive optical systems to increase the image resolution up to the cellular level, that is, wavefront-corrected fundus imaging (Section 5.3).
2. Simultaneous structural and functional diagnostics:
   - Combination with functional diagnostic methods, for example, fundus-controlled microperimetry (Section 8.1.3.4).
3. Cost-effective and easy-to-operate nonmydriatic cameras:
   - Integration of novel optical technologies, for example, LED illumination.
   - Extensive automation of image acquisition.
   - Integration of intelligent image evaluation algorithms for diagnostic support.

6.8 Scanning-Laser Devices

Scanning-laser devices are a group of imaging diagnostic instruments all of which are based on the same functional principle. During the examination, a focused laser beam (Appendix B) scans the tissue pointwise in a sequential manner by following a preset pattern. At each scanning position, the signal of the backscattered laser light is acquired by a detector. For this purpose, the laser focus is imaged by an optical system into a confocal pinhole aperture which is located in front of the detector. While scanning the patient's fundus, the detector records a specific intensity distribution which is then translated to a video image.

The scanning-laser technology is commonly used for the following purposes:

- Imaging of the posterior and, to a certain extent, the anterior segments of the patient's eye (Section 6.8.1).
- Structural analysis, in particular, analysis of the optic nerve head topography by capturing optical cross-section images (Section 6.8.2).
- Structural analysis including thickness determination of the retinal nerve fiber layer (RNFL; Section 1.2) by polarimetry (Section 6.8.3).

History In 1980, the first scanning-laser device was developed to image the eye fundus (flying-spot TV ophthalmoscope) [38]. Just a few years later, the method was considerably improved in that a confocal stop was used for signal detection [39]. The first commercial devices were fabricated by the companies Rodenstock and Carl Zeiss from 1989 onwards.

6.8.1 Confocal Scanning-Laser Ophthalmoscope

Complementary to fundus cameras, confocal scanning-laser ophthalmoscopes (cSLO) are widely used for fundus imaging. With additional optics, they can also be applied to anterior segment imaging. The acquired images allow evaluation of deviations from the normal condition and/or detection of morphological changes which refer to certain eye diseases. Concerning the clinical relevance and applications, cSLOs are quite similar to fundus cameras (Section 6.7). The major difference between cSLOs and fundus cameras is, however, the imaging process itself. In fundus cameras, a complete image is captured over a certain field of view in a single snapshot. In cSLOs, the examined object is sequentially screened point by point. The corresponding advantages and drawbacks are discussed in Section 6.8.1.2.

6.8.1.1 Illumination Beam Path

In the illumination beam path of a cSLO (Figure 6.73a), a parallel bundle of laser light is coupled into an xy scanner through a pinhole mirror. The xy scanner consists of two synchronized mirrors which deflect the laser beam perpendicular to the optical axis (z axis). Specifically, a rapidly rotating polygonal mirror or a resonant scanner is used for rapid deflection in the x direction (image line). In the y direction, the beam is synchronously deflected by a galvanometric scanner (line-feed). The laser beam is then projected into the patient's eye with a telecentric optical system. For an emmetropic eye, the beam is directly focused onto the retina. The mirror surfaces of the xy scanner are imaged by the telecentric optical system to the entrance pupil of the eye. For this reason, the pupil center acts as a center of rotation for the incident laser beam, and the maximum range of the xy scanner determines the effective field of view.

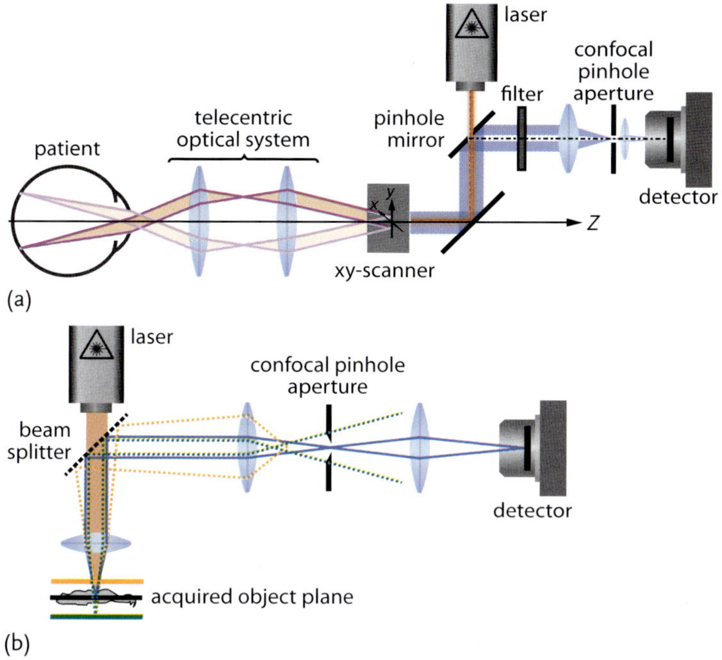

Figure 6.73 Working principle of a confocal scanning-laser ophthalmoscope (cSLO). (a) Ray diagram of a cSLO. A laser beam (red) passes through a pinhole mirror and is deflected by an xy scanner. A telecentric optical system focuses the laser beam onto the patient's fundus. For observation, the reflected light (blue) passes through the same optical components in the opposite direction and is coupled out into the observation path by the pinhole mirror. In the observation path, the light signal is imaged into a confocal pinhole aperture. For fluorescence imaging, an additional filter is used. (b) General principle of confocal detection. A laser beam is focused onto the plane of the examined object (bold black line). The reflected light is then imaged into a confocal pinhole aperture. Only light reflected in the laser focus plane (bold black line) will reach the detector. Laser light reflected at the orange or green plane is blocked by the pinhole aperture. As a consequence, the signal-to-noise ratio of the image is considerably enhanced.

The divergence of the laser beam and, thus, the axial position of the focus plane can be changed by a tunable telescope system in front of the xy scanner (not shown in Figure 6.73a). By stepwise changing the axial position, we can scan through different tissue layers and capture a series of cross-sectional images. From this data, a 3D image of the examined part of the eye can be reconstructed (Section 6.8.2).

6.8.1.2 **Observation Beam Path**

Laser light reflected or backscattered by the illuminated eye structure travels through the telescope system and the xy scanner in opposite directions. At the pinhole mirror, the light is guided to the detection branch, where it is focused on the plane of a confocal pinhole aperture and acquired by a detector. Eventually, the detected signals are amplified and compiled pixel by pixel into a fundus image.

For fluorescence imaging, an additional barrier band-pass filter is used to suppress the excitation laser light. Compared to a fundus camera, the image acquisition of a cSLO has a few advantages which shall be discussed in the following.

Confocal detection A confocal pinhole aperture is situated in front of the detector in which an image of the laser focus is formed. This diaphragm effectively cuts off the portion of light which is backscattered by eye structures outside of the laser focal plane (Figure 6.73b). In this way, if a sufficiently small inner diameter of the pinhole aperture is used, the image quality can be considerably enhanced. This technique offers the following benefits compared to a fundus camera:

- No geometric pupil separation is required for reflection-free observation. Thus, the images can also be captured with a small pupil size.
- In general, the contrast of the captured images is higher.
- Light scattering in the eye lens (e.g., in the case of cataracts (Section 3.2)) is less critical, as only the brightness but not the contrast of the image is reduced.
- Since only the backscattered signal in the focal plane of the laser beam is detected, autofluorescence of optical media in front of the examined structure does not overlay the desired fluorescence signal.
- Confocal detection allows us to capture optical cross-section images, that is, parallel images at different depths along the z axis. This feature can be used for 3D imaging and quantitative 3D structure analysis, for example, to determine the topography of the optic nerve head (Section 6.8.2).

Detection sensitivity As only light intensities are detected with a cSLO, more light-sensitive detectors, such as photo multiplier tubes (PMT) or avalanche photo diodes (APD), can be used. These devices have a higher signal-to-noise ratio than image sensors (CCD chips) so that the images can be recorded with a light intensity which is lower by approximately a factor of 200 compared to a fundus camera. As a consequence, the exposure of the patient's eye is considerably reduced.

Acquisition With cSLOs, videos with a maximum frame rate of 20–30 Hz may be recorded without rapidly repeating flash illumination. Dynamic processes (e.g., during the early-transit phase in fluorescein angiography; Section 6.7.6) can thus be examined with reduced light exposure in real-time.

Compared to a fundus camera, the imaging technique of a cSLO has the following drawbacks:

- Different laser sources must be used depending on the desired acquisition mode (monochromatic images, fluorescence angiography, fundus autofluorescence), while merely the filter has to be exchanged in fundus cameras.
- When color images are captured with a cSLO, the scans are carried out with laser wavelengths in maximal three different (narrow-band) spectral ranges. Hence, the composite color image captured in this way does not provide the familiar

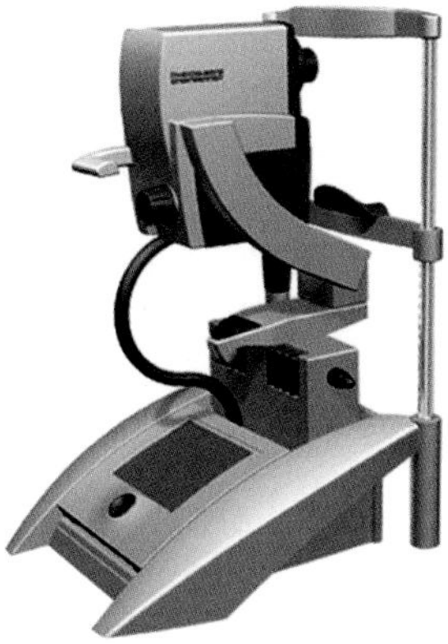

Figure 6.74 Photograph of Heidelberg SPECTRALIS® HRA. Courtesy of Heidelberg Engineering GmbH.

color fidelity of an ophthalmoscope or fundus camera. In addition, cSLO images are slightly pixelated due to the self-interference of laser light (Section A.2.3.1).

- If we use a confocal pinhole aperture with a small inner diameter, fundus structures which lie outside the focal area can hardly be seen. In contrast, such structures are also visible in images taken by a fundus camera, even though they appear blurred.

6.8.1.3 Equipment Solutions

Confocal scanning-laser ophthalmoscopes are commercially available in different configurations (Figure 6.74) but mostly use the discussed imaging principle. The ultra-wide-field cSLO by Optos has, however, another optical design and will be considered separately in Section 6.8.1.10.

6.8.1.4 Imaging Modes

In principle, all imaging modes of a fundus camera (Sections 6.7.5–6.7.8) can be implemented in a cSLO. Table 6.6 summarizes the imaging modes and their respective laser wavelengths as they are offered in two commercial instruments by the companies Heidelberg Engineering and Nidek. Laser diodes and frequency-doubled semiconductor lasers (for the blue spectral range) are commonly used as light sources (Section B.5.2).

6.8.1.5 Imaging Methods

The imaging of a cSLO can either be performed with a confocal pinhole aperture (bright-field mode) or a confocal dark-field diaphragm (indirect imaging/retro mode).

Direct imaging with confocal pinhole aperture Confocal bright-field imaging with a confocal pinhole aperture (Figure 6.75a) is the primarily used standard method for cSLOs. Here, the reflected and/or backscattered light from the laser focus plane is directly acquired by the detector (reflectance imaging). Confocal bright-field imaging is also the preferred method used in fluorescence imaging. As the light inten-

Table 6.6 Imaging modes and their corresponding laser wavelengths as they are currently offered by Heidelberg SPECTRALIS HRA and NIDEK F-10.

Imaging modes	Heidelberg SPECTRALIS HRA	NIDEK F-10
Green (red-free) reflectance imaging	488 nm	532 nm
Blue reflectance imaging	488 nm	490 nm
Red reflectance imaging	–	660 nm
Near-infrared reflectance imaging	820 nm	790 nm
Fluorescein angiography (FA)	488 nm	490 nm
Indocyanine green angiography (ICGA)	790 nm	790 nm
Autofluorescence imaging (AF)	488 nm	490 nm

sity reaching the detector increases with the diameter of the aperture stop, it is also possible to image weakly reflecting or fluorescent structures. The reduced depth resolution caused by a large aperture diameter does not impair the measurement result, but is rather desired, for example, when performing fluorescein and ICG angiography simultaneously.

Indirect imaging with confocal dark-field diaphragm Confocal dark-field imaging is an alternative method which is only used for some special applications. For example, with a ring-shaped aperture stop (Figure 6.75b), we may visualize eye structures outside the laser focal plane. The imaging is based on the acquisition of infrared laser light reflected or scattered by deeper retinal layers (e.g., retinal pigment epithelium) and partially from the choroid and sclera. The backscattered light is then collected through a ring-shaped aperture, while the directly reflected light from the laser focus is blocked by a central stop.

Retro-mode imaging is a modification of the dark-field mode. The retro-mode aperture (Figure 6.75c) consists of a central stop which blocks the direct light reflection, and an off-axis oval-shaped opening through which scattered light of only one direction may pass through. As the detector receives light from a limited solid angle, it casts shadows as if viewing the transparent retinal tissue with obliquely incident light. For this reason, we obtain a pseudo-3D image of the retinal tissue layers. In

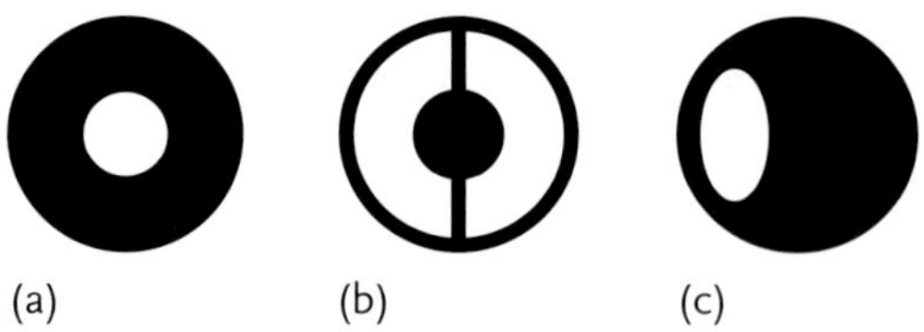

Figure 6.75 Possible aperture stop designs in confocal scanning-laser ophthalmoscopy. (a) Aperture with central opening for standard bright-field confocal imaging. (b) Ring-shaped aperture for indirect dark-field imaging, that is, the central reflected light is blocked. (c) Laterally shifted oval-shaped opening for retro mode imaging.

ophthalmology, the retro mode is often used to visualize drusen, defects in the retinal pigment epithelium (Section 3.4), and edema (Section 3.5). It can also be used to visualize sub-threshold laser spots (Section 10.2.3.2).

6.8.1.6 Image Acquisition Modes

The following modes of image acquisition are possible with a cSLO:

- Individual images are generated by means of overlaid image frames (typical frame rates: 10–20 frames/s). For this purpose, special software algorithms are used which are able to detect and correct eye movements in real-time. In this way, the signal-to-noise ratio of individual images can be substantially improved without motion artifacts by averaging over a longer period of time. The same image processing can also be used to generate color pictures from individual monochrome images which have been acquired with laser light of different wavelengths.
- Time sequences of images. Depending on the featured field of view and digital image size, videos can be captured with a frame rate in the range of 5–30 Hz (Table 6.7).
- In simultaneous (dual display) mode, two imaging modes are recorded at the same time (e.g., fluorescein and ICG angiography or near-infrared reflectance imaging and angiography).
- Virtual 3D images can be generated from optical cross-section images along the z axis (Section 6.8.2).
- Analogous to fundus cameras, cSLOs are also able to generate stereo image pairs (Section 6.7.8) and wide-field images (Section 6.7.4). With fast image processing routines, potential eye movements can be recognized and eliminated.

6.8.1.7 Geometric and Digital Image Resolution

The geometric image resolution per pixel can be calculated from the field of view and the digital image size (Table 6.7). Assuming an emmetropic eye, the field of view d_{fov} can be obtained from the angle α_{fov} according to Eq. (6.69). For the Hei-

Table 6.7 Summary of essential technical imaging parameters of cSLOs for Heidelberg SPECTRALIS HRA (above separation line) and NIDEK F-10 (below separation line).

Field of view	High resolution mode		High speed mode	
	Digital image size (pixel)	Maximum frame rate (Hz)	Digital image size (pixel)	Maximum frame rate (Hz)
$30^\circ \times 30^\circ$	1536×1536	5	768×768	9
$20^\circ \times 20^\circ$	1024×1024	7	512×512	13
$15^\circ \times 15^\circ$	768×768	9	384×384	16
$32^\circ \times 24^\circ$	1600×1200	10	800×600	20
$32^\circ \times 24^\circ$	1280×960	10	640×480	26

delberg SPECTRALIS HRA, we have maximum resolutions of approximately 6 and 11 μm/pixel in high resolution and high speed mode, respectively. For the NIDEK F-10, maximum resolutions of approximately 6 and 12 μm/pixel result, respectively.

6.8.1.8 **Optical Resolution**

In principle, the cSLO is a confocal laser scanning microscope (LSM) [40] for which the patient's eye takes on the role of the objective lens and the retina represents the examined object. We can thus use the imaging equations of an LSM to estimate the optical resolution of a cSLO. In an ideal diffraction-limited scenario (i.e., no aberrations and uniform illumination), the point-spread function PSF (Sections 5.4.1.2 and A.1.5) of an LSM is shaped like an axially stretched ellipsoid (Figure 6.76a), which can be described by the axial and lateral *full widths at half maximum* (FWHM). With a numerical aperture (Section A.1.4) of NA < 0.5 and a small confocal pinhole aperture, the axial resolution of a cSLO then follows as [41]

$$\Delta z_{\mathrm{cSLO,FWHM}} = 1.67\frac{\lambda}{\mathrm{NA}^2} \tag{6.77}$$

and the lateral resolution as

$$\Delta(x, y)_{\mathrm{cSLO,FWHM}} = 0.51\frac{\lambda}{\mathrm{NA}} \,. \tag{6.78}$$

This means that two objects can be resolved if their distance is greater than the respective value (compare Section A.2.1.6). Under ideal conditions, the axial and lateral optical resolution of a cSLO at a wavelength of $\lambda = 550\,\mathrm{nm}$ and for a pupil diameter of 2 mm ($\mathrm{NA} = 0.058\,65$) yields $\Delta z_{\mathrm{cSLO,FWHM}} = 270\,\mu\mathrm{m}$ and $\Delta(x, y)_{\mathrm{cSLO,FWHM}} = 5\,\mu\mathrm{m}$, respectively. If, however, aberrations in the patient's eye are taken into account, especially in the case of larger pupil diameters, the cSLO has practically an axial resolution of 300–500 μm and a lateral resolution of 5–20 μm [42]. Hence, the lateral optical resolution well matches the digital resolution (Section 6.8.1.7).

When many cross-sectional images are acquired along the z axis and evaluated with a small confocal pinhole aperture, we can determine the z position of a struc-

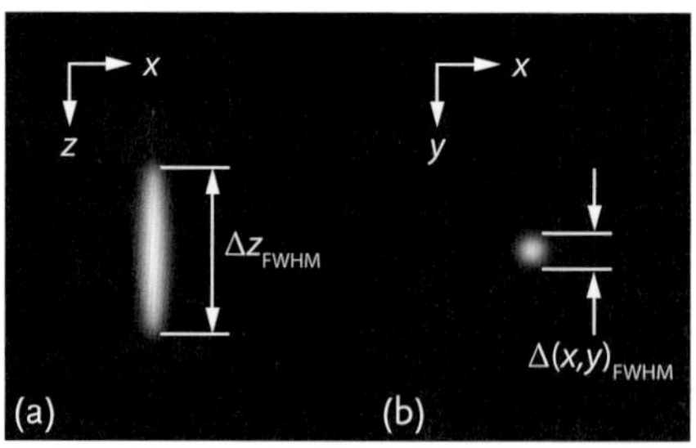

Figure 6.76 Cross-section of the 3D point-spread function of a laser scanning microscope in the (a) axial and (b) lateral directions.

ture with a reflecting surface much more precisely than the axial resolution in principle allows. This fact is used by confocal scanning-laser tomographs described in Section 6.8.2.

6.8.1.9 **Wide-Field Imaging**

As for a fundus camera (Section 6.7.4), the peripheral parts of the fundus outside the field of view can be accessed either by changing the viewing direction of the patient's eye or by moving the measuring head along an arc-shaped path around the patient's eye. The individual images are then stitched together by the Auto Mosaic (or Montage) Software so that a composite image (panorama) of the fundus is obtained which shows an angular range of up to 125°.

Alternatively, the standard field of view of a cSLO can be extended with a noncontact wide-angle objective lens by a factor of 1.5–1.8. In this way, the detector can simultaneously acquire an angular range of 55° without changing the scanning or frame rate parameters. With special contact lenses [43] or ultra-wide-field cSLOs, the range can even be extended to 150°.

6.8.1.10 **Alternative Design Concept: Ultra-Wide-Field Scanning-Laser Ophthalmoscope**

In 2000, the company Optos launched the ultra-wide-field scanning-laser ophthalmoscope P200 which is based on a standard cSLO setup. The P200 was the first cSLO which could generate a panoramic image of the retina with an angular range of 200° in a noncontact manner with a single snapshot. The large field of view is achieved by a large ellipsoidal mirror which forms two foci F_1 and F_2 (Figure 6.77a) [44]. The ellipsoidal mirror is located such that F_2 lies in the entrance pupil of the patient's eye. At the position of F_1, one mirror of the xy scanner is placed which is responsible for the slow vertical line-feed scans. The rotational axis of this scanning mirror is perpendicular to the connecting line between F_1 and F_2. The other mirror of the xy scanner performs fast horizontal line scans. F_1 is thus the common point of rotation of the scanning laser beam.

The extreme fields of view lead to aberrations which must be precompensated. Nonlinearities of the scanning process lead to distortions in the composite image and are thus corrected by software. With the ultra-wide-field scanning-laser ophthalmoscope, approximately 80% of the entire fundus area can be acquired with a single scan. This technique allows examination of lesions in the periphery which would be overlooked with a standard cSLO. For a full field of view, the optical resolution is approximately 20 μm. For a reduced field of view, the maximum resolution is approximately 11 μm. Depending on the instrument configuration, all standard cSLO imaging modes can be implemented. Color images are generated by digital combination of images taken in the red (633 nm wavelength) and green (532 nm wavelength) spectral range. Fluorescein angiography is performed with blue laser light (488 nm wavelength) for excitation.

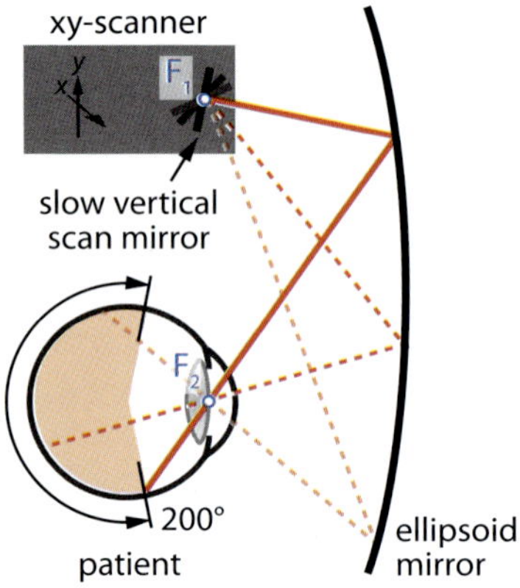

(a)

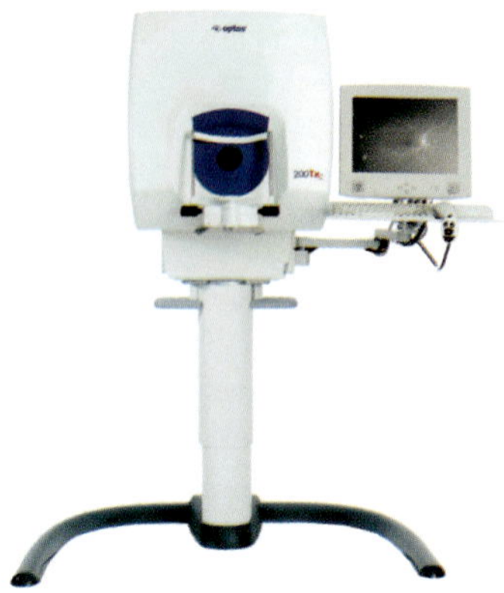

(b)

Figure 6.77 (a) Optical principle of an ultra-wide-field scanning-laser ophthalmoscope (side view). With a large ellipsoidal mirror, two foci F_1 and F_2 are formed. In F_1, the slow vertical scan mirror of an xy scanner is placed. In F_2, an image is formed via the ellipsoidal mirror. F_2 lies in the entrance pupil of the patient's eye. (b) Photograph of the Optos 200Tx™ which is a technically advanced version of the Optos P200. The 200Tx can also be used for fluorescein angiography and autofluorescence imaging. Courtesy of Optos GmbH, Germany.

6.8.1.11 **Prospects**

cSLOs are based on a well-established imaging method. Because of their special features (e.g., 2D live imaging), they are often combined with other diagnostic and therapeutic devices or integrated into them. Some of the recent development goals are:

1. The improvement of the structural analysis by means of
   - combination with alternative imaging methods (simultaneous multimodal imaging, for example, combination of cSLO and optical coherence tomographs (Section 7.6)), and
   - adaptive optical systems to increase the resolution (Section 5.4.2).
2. Image-supported functional diagnostics by combining the cSLO with functional diagnostic devices[52] (e.g., fundus-controlled perimetry (Section 8.1.3.4) and fixation monitoring).
3. Image-supported laser therapy (Section 10.2.6).

52) Any scan patterns can be overlaid on the retina by means of modulation of the scanning laser beam. The scanned patterns can be used as stimuli for image-supported functional diagnostics.

6.8.2 Confocal Scanning-Laser Tomograph

The confocal scanning-laser tomograph is a special cSLO used for 3D imaging and quantitative 3D structure analysis of the human eye. It is primarily used in the posterior eye segment to measure the topography of the optic nerve head. In 1993, the first instrument of this kind, the Heidelberg Retina Tomograph (HRT), was introduced. The shape parameters, also referred to as the *stereometric parameters*, of the optic nerve head which can be derived from the measurement are of great clinical relevance, as they allow diagnosis and monitoring of glaucoma progression (Section 3.3) [45]. The latest instrument release (HRT-3) can also be used to determine the retinal thickness and to characterize retinal edema. This is particularly important for the diagnosis and progression monitoring of diabetic macular edema (Section 3.5). A very interesting feature is that the HRT-3 can be easily transformed into a confocal anterior segment microscope with some additional optics (Rostock Cornea Module). With such an upgrade, the HRT-3 allows examination of corneal and scleral diseases at the cellular level.

6.8.2.1 Functional Principle

The confocal scanning-laser tomograph uses a small confocal pinhole aperture to acquire optical cross-sectional images. For this purpose, a series of equidistant 2D cross-section images of the examined object are taken by means of a step-by-step shift of the focal plane of the scanning laser beam along the optical axis (z axis; Figure 6.78a). After image acquisition, the individual pictures are automatically aligned by software-based correlation algorithms to eliminate potential image shifts and rotations that can happen due to involuntary eye movements.

For each coordinate (x, y) on the image, a set of reflectance values, called *z profile*, is obtained depending on the axial position of the cross-section. If the examined

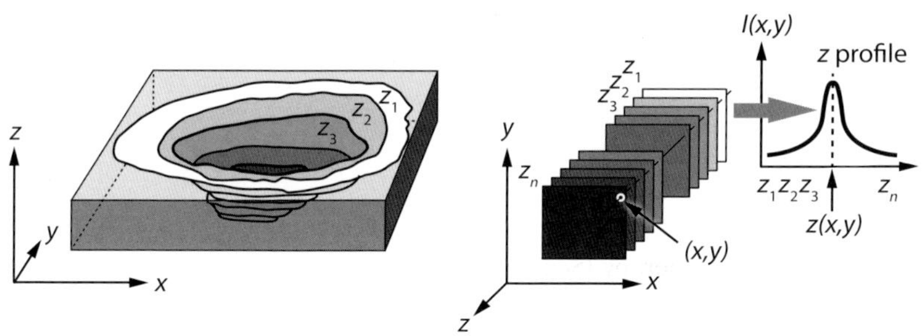

Figure 6.78 (a) Schematic slicing of a 3D object (e.g., optic nerve head) with a confocal scanning-laser tomograph. (b) Each image point $z_n(x, y)$ of every focal plane z_n is represented by a bell-shaped signal $I(x, y)$ (i.e., the so-called z profile). The maximum of $I(x, y)$ corresponds to the actual z position of the examined object. From the information of all focused layers, it is possible to reconstruct a 3D image of the examined object. Adapted from [46].

eye structure has only one reflective surface, the z profile is bell shaped (diagram in Figure 6.78b). The maximum value indicates the actual z position of the reflecting surface at point (x, y). The desired surface profile of the structure is thus determined by the maxima of all z profiles of the individual images.

6.8.2.2 Representation of Data and Data Analysis

The obtained surface profile can be represented as a 2D topography image, where the z values are characterized by graduated red tones (Figure 6.79a) or in the form of a pseudo-3D image. The sum of all intensity values for each image point (x, y) along the z axis is used to generate a reflectance image of the retina (Figure 6.79b). To improve the reproducibility, at least three datasets are recorded and a *mean reflectance image* and a *mean topography image* are calculated from them. Typically, 16 cross-sectional images are produced for each millimeter of a depth scan in the z direction, whereas the maximum scan depth is 4 mm. The necessary scan depth and the optimal focal plane are determined by means of a preliminary scan. From initialization data, the imaging conditions are automatically set. Table 6.8 shows the technical specification of the HRT-3 (Figure 6.79c).

Depending on the specific application (glaucoma diagnosis, 3D retina imaging), several software modules exist which allow derivation of characteristic stereoscopic parameters from a quantitative analysis of topographic data. Important parameters in the analysis of the optic nerve head are, for example, the excavation characteristics (cup-to-disk ratio), rim area, rim volume, and height variation contour (Section 7.6.1.1). For a detailed description of the evaluation procedure (data acquisition, analysis, comparison with normative data, trend analysis, and data presentation), please refer to specialized literature such as [45, 47, 48].

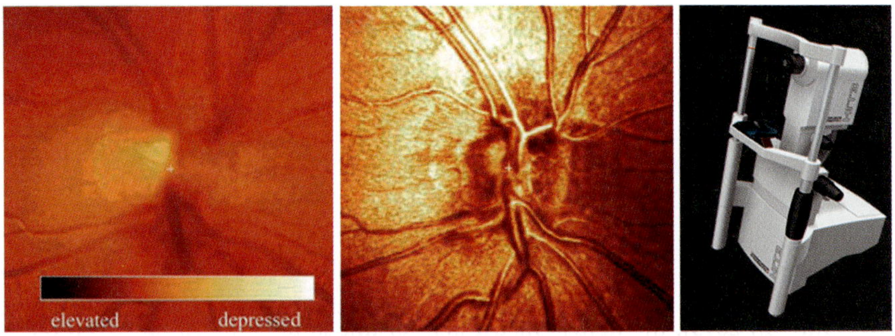

(a) (b) (c)

Figure 6.79 (a) 2D topography image of the optic nerve head (height profile color-coded). (b) Reflectance of the optic nerve head (reflectance color-coded). (c) Photograph of the Heidelberg Retina Tomograph 3 (HRT-3). Courtesy of Heidelberg Engineering GmbH.

Table 6.8 Important functional parameters of the Heidelberg Retina Tomograph HRT-3.

Parameter	Value
Field of view	15° × 15°
Scan depth	1.0–4.0 mm (automatic)
Digital image size	2D image: 384 × 384 pixels
	3D image: up to 384 × 384 × 64 pixels
Optical resolution	10 μm (transverse); 300 μm (axial)
Digital resolution	10 μm/pixel (transverse); 62 μm/pixel (axial)
Repeatability (axial)	±20 μm
Scan time	2D image: 24 ms
	3D image: 1 s typical (2 mm scan depth)
Minimum pupil diameter	≥ 1 mm
Light source	670 nm diode laser

6.8.3 Scanning-Laser Polarimeter

The scanning-laser polarimeter (SLP) is a cSLO with an integrated polarization angle (Section A.2.1.4) measurement system, that is, a so-called *polarimeter*. It is used for the quantitative characterization of the retinal nerve fiber layer (RNFL). The RNFL consists of bundles of retinal ganglion cell axons[53] which build up the optic nerve. In a healthy eye, the RNFL around the optic nerve head (*circumpapillary* region) is thicker in the upper (superior) and lower (inferior) regions and thinner in the temporal and nasal regions, as shown for the left eye (OS) in Figure 6.82.

At the cellular level, glaucoma (Section 3.3) is characterized by an increasing loss of ganglion cells and their axons. This leads, in turn, to structural anomalies such as the reduction of the RNFL thickness and/or changes in the shape of the optic nerve head which manifest as a visual field loss [49]. Since structural changes in the RNFL can be detected at an earlier stage than visual field losses, the RNFL thickness is an important indicator for early diagnosis and progression monitoring of glaucoma.

In 1993, Laser Diagnostic Technologies (LDT) introduced the first SLP named Nerve Fiber Analyzer (NFA). LDT was later acquired by the company Carl Zeiss. Currently (in 2013), the newest instrument of this kind is the ZEISS GDxPRO™ (Figure 6.80b).

6.8.3.1 Functional Principle

The RNFL consists of cylindrical microtubules which are contained in the parallel axons of the retinal ganglion cells. The bundles of cylindrical microtubules are

53) The axons of ganglion cells are fiber-like extensions that transmit the received visual information to the brain (Section 1.1).

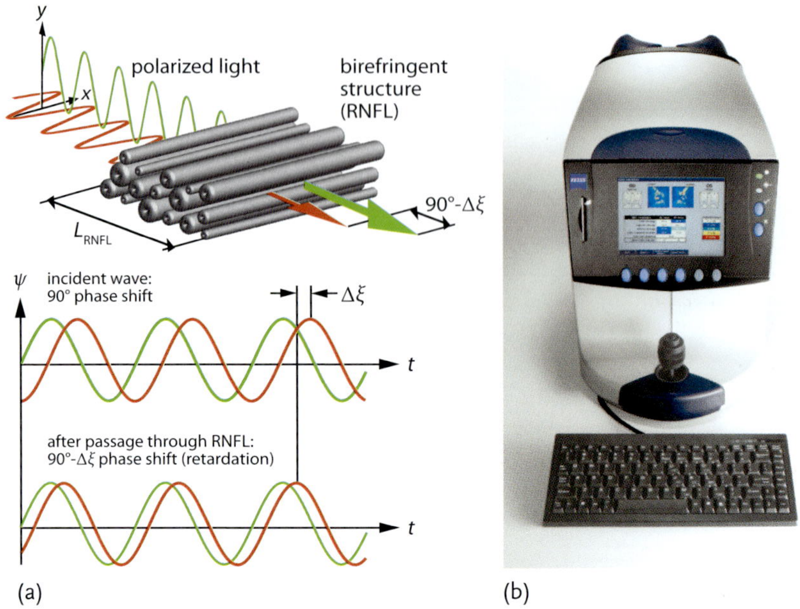

Figure 6.80 (a) Due to its ordered structure, the retinal nerve fiber layer (RNFL) is a birefringent medium. This means that linearly polarized light passes through the medium at different speeds depending on the polarization direction. If the incident light is polarized parallel to the elongated nerve fibers (red), it undergoes a phase retardation $\Delta\xi$, as shown for the wavefunctions $\psi(t)$ of both light polarizations. $\Delta\xi$ is determined by a polarimeter and then translated into a thickness value of the RNFL. (b) Photograph of the scanning-laser polarimeter ZEISS GDxPRO. Courtesy of Carl Zeiss.

oriented along a preferred direction (x direction; Figure 6.80a). When the RNFL is illuminated with light, it thus behaves as a birefringent optical medium (Section A.2.1.4). This means that the refractive index of the RNFL depends on the polarization and traveling direction of the incident light. Specifically, the refractive index is higher for light polarized parallel to the axons than perpendicular to them ($n_x > n_y$). The refractive index is a measure for the speed of light inside a medium (Section A.1.1). Thus, we can also say that a light wave whose polarization plane is oriented parallel to the axons travels slower through the RNFL than a light wave which is polarized perpendicular to them ($c_{\mathrm{m},x} < c_{\mathrm{m},y}$). As the optical anisotropy is caused by the cumulative degree of organization of the cylindrical microtubules (i.e., density and orientation), this property is referred to as *form birefringence*. The SLP now uses the form birefringence of the RNFL to analyze its structure in a quantitative manner. For this purpose, light is used with a wavelength shorter than the diameter of the microtubules. Hence, the RNFL shows relatively low scattering and effectively behaves like a solid (crystalline) medium.

When light passes through the RNFL, partial waves with different polarizations travel with different speeds which leads to a phase difference $\Delta\xi$ (*retardation*; Section A.2.3.1). $\Delta\xi$ is, in turn, proportional to the magnitude of the birefringence

$\Delta n = |n_y - n_x|$ and to the layer thickness of the RNFL L_{RNFL} (Figure 6.80a). In the case of glaucoma, both L_{RNFL} and Δn are reduced [50]. Consequently, we just have to determine $\Delta\xi$ to evaluate these structural changes by means of an integrated polarimeter.

In the latest release of the instrument (ZEISS GDxPRO shown in Figure 6.80b), the measured phase difference is displayed as the RNFL Integrity™ (RNFL-I™) parameter. In previous models, the measured values were converted into a corresponding layer thickness by means of a histologically determined constant conversion factor (0.67 nm/μm). In this case, the converted layer thickness is termed *polarimetric layer thicknesses* (Polarimetric Thickness™ or $P - \mu m$) so that it cannot be confused with the anatomic layer thickness (e.g., measured with optical coherence tomography (Section 7.6)).

6.8.3.2 **Measurement Procedure**

In the SLP, the fundus of the patient's eye is scanned with the focused laser beam of a GaAlAs laser diode ($\lambda = 785$ nm; Section B.5.2). Infrared radiation is used due to its better penetration into retinal layers. The scanned area of the laser corresponds to a field of view of 40° × 20° (horizontal × vertical) so that both the optic nerve head and the macula are simultaneously analyzed. The digital image resolution is 256 × 128 pixels (horizontal × vertical), and each image is acquired within approximately 0.8 s. The optical design used for a SLP is, in principle, identical to that of a cSLO (Section 6.8.1). For this reason, we will only discuss the special components which are required for the structural analysis of the RNFL.

In contrast to standard cSLOs, the SLP uses linearly polarized light (Section A.2.1.4) for scanning. The light beam backscattered from the retina passes twice through the birefringent RNFL and, thus, becomes elliptically polarized. In the confocal detection system, the beam is split into two partial beams with orthogonal polarizations. For this purpose, two linear polarizers are used whose set polarization axes are perpendicular and parallel to the polarization of the incident laser beam, respectively. The intensity of both partial beams is then recorded by separate detection channels. During the measurement procedure, the polarization plane and both polarizers are synchronously rotated so that a series of images with different polarization directions are generated. The two detector channels register sinusoidal signals for each image pixel as a function of the rotation angle. These signals have a relative phase difference of 90° and provide information about the total retardation as well as the orientation of the birefringence axis in the respective image point [51].

Analogous to the RNFL, the cornea and eye lens have birefringent properties. Their contribution to the change of the light polarization must therefore be predetermined and appropriately compensated by a variable corneal compensator (VCC) in the measurement beam path. The VCC consists of two (linear) retarders (see Info Box 6.4) which can be rotated relative to each other. In this way, we are able to set the amount and orientation axis of the desired retardation. With the VCC, the RNFL retardation is measured in two steps:

1. *Corneal scan*: The cornea is scanned with the zero setting of the VCC to determine the retardation and orientation axis of birefringence in the anterior segment of the patient's eye.
2. *RNFL scan* (VCC method): The measured reference value is then used to set the retardation of the VCC such that the birefringent effect of the anterior segment is compensated for the RNFL scan.

Let us now consider the individual steps of the measurement procedure in detail.

Info Box 6.4: Retarder

A *retarder* or *wave plate* is an optical device which is used to change the polarization (Section A.2.1.4) of a light wave which travels through it. A common type of retarder is the *half-wave plate* ($\lambda/2$ plate) which shifts the polarization direction of linearly polarized light. Another frequently used retarder is the *quarter-wave plate* ($\lambda/4$ plate). It converts linearly polarized light into circularly polarized light and vice versa.

Corneal scan To determine the birefringent properties of the anterior eye segment, the VCC is set to zero retardation, and a complete retardation image is acquired. With this data, the retardation profile in the macular region is evaluated (Figure 6.81a). As the birefringence of the Henle fiber layer (HFL) in the macular region is regular and symmetric (Section 1.2), any irregularity in the acquired retardation profile of this region can be referred to the birefringence of cornea and eye lens. The obtained values are then used as a signal background for the subsequent measurement for determination of the RNFL retardation (RNFL scan).

Info Box 6.5: Henle Fiber Layer

The Henle fiber layer (HFL) is located in the macular region (Section 1.2.2). It consists of elongated cone cell axons (Section 1.2) which extend radially from the foveal center. In the macular region, the HFL is spatially uniform so that its radial birefringence can be used as an (intraocular) reference polarimeter for the measurements of birefringence of the cornea.

RNFL scan with VCC method In the VCC method, the contribution of the anterior segment to the total birefringence is compensated with a VCC. A full compensation is achieved when the set value of the VCC is equal to the measured retardation, and when the VCC axis is perpendicular to the birefringence axis of the anterior segment. An RNFL scan image with a properly set VCC compensation (Figure 6.81b) shows the expected regular birefringence profile in the macular region and the local distribution of the RNFL retardation values in the optic nerve head region (RNFL Integrity).

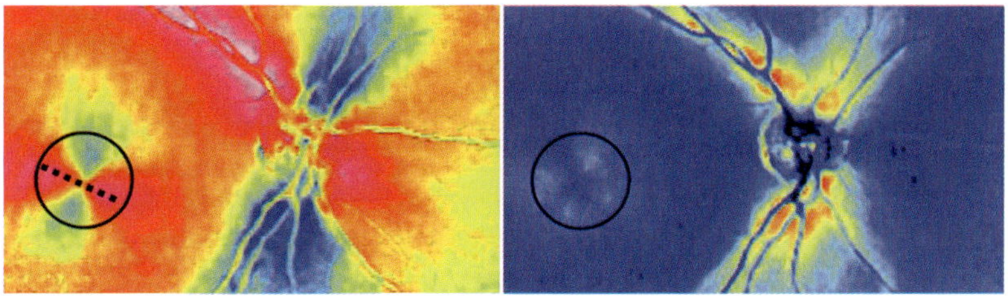
(a) (b)

Figure 6.81 (a) The retardation image for an uncompensated corneal scan contains phase retardations caused by the cornea, eye lens, and retinal nerve fiber layer (RNFL). In the macular region (black circle), the retardation profile is altered by the cornea, lens, and the Henle fiber layer (HFL; see Info Box 6.5). As the birefringence of the HFL in the macular region is regular and symmetric, any irregularity in the retardation profile of this region can be referred to the birefringence of cornea and eye lens. The axis of birefringence can be clearly seen (dashed line). Once the axis and the magnitude values of birefringence are known for cornea and eye lens, the VCC can be set to compensate for the anterior segment birefringence (VCC method). (b) RNFL scan. In the macular region (circle), the resulting retardation image for the compensated scan shows the expected uniform profile. In the region around the optic nerve head shows the undistorted RNFL retardation profile. The retardation value (or corresponding RNFL thickness) is color-coded with smaller (thinner) regions displayed in blue or green and higher (thicker) regions displayed in yellow and red. Courtesy of Carl Zeiss.

RNFL scan with ECC method The *enhanced corneal compensation* (ECC) method is based on a software algorithm which was developed to improve the signal-to-noise ratio. In contrast to the VCC method, it does *not* compensate for the birefringence of the anterior eye segment. Instead, a *bias retarder* with a known significant birefringence is added to the measurement beam path. The bias retarder shifts the retardation measurement into a more sensitive region and is then mathematically subtracted point-by-point from the measured values to yield the actual RNFL retardation.

Fundus image In addition to retardation images (corneal scan, RNFL scan), a fundus image is presented after each image acquisition in the form of a color-coded reflectance image. Its primary purpose is to estimate the quality of the acquisition parameters (e.g., alignment, fixation, refraction) during the scan. If the quality is too low, the scan is repeated.

6.8.3.3 **Data Analysis and Presentation**

The retardation profile determined by an RNFL scan can be interpreted and represented in many ways. The following items are listed in a typical RNFL report:

- The *Nerve Fiber Layer Map* (RNFL map) is a color-coded retardation image which depicts the different RNFL-I values in the $20° \times 20°$ area around the optic nerve head. Dark blue and bright red areas represent smaller and larger retardation (i.e., smaller and larger RNFL-I values), respectively.

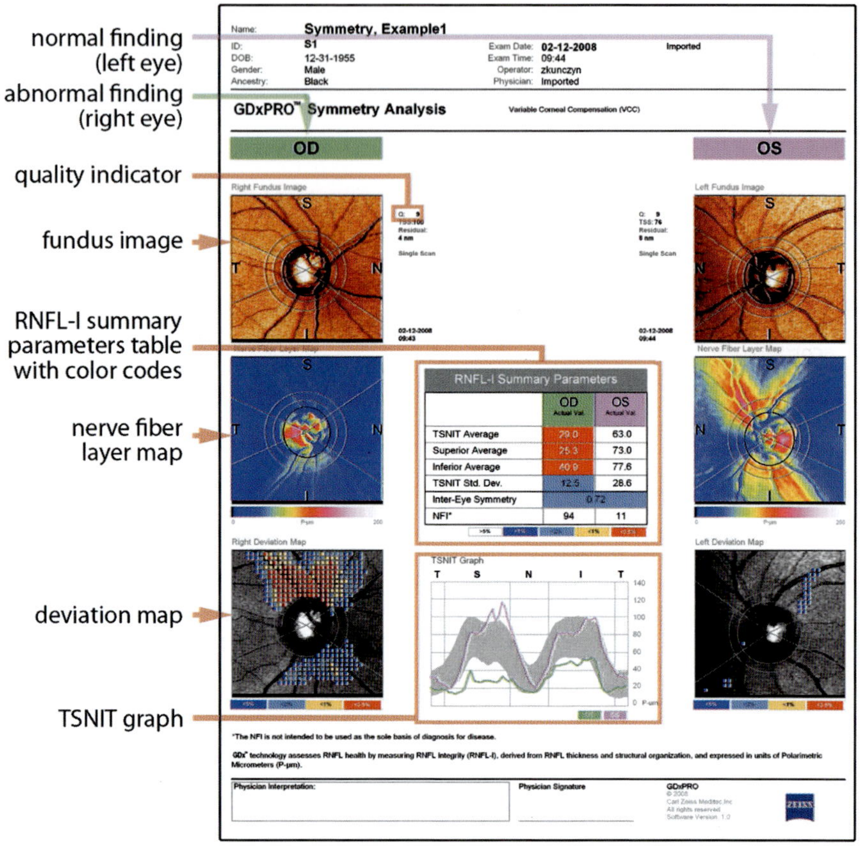

Figure 6.82 Sample Symmetry Analysis report for an abnormal right (OD) and normal left (OS) eye. Courtesy of Carl Zeiss.

- The *Deviation Map* compares the RNFL-I values with corresponding values derived from a normative database. Small color-coded squares indicate the amount of deviation from normal values at each location. The squares are distributed over a black-and-white fundus image for visual reference.
- The *TSNIT graph* shows the RNFL retardation profile along a circle around the optic nerve head. For this purpose, the data is plotted versus temporal, superior (upper), nasal, and inferior (lower) directions into one diagram.
- *RNFL-I Summary Parameters* are specially derived parameters to support the diagnosis. The values are highlighted in colors to illustrate the degree of deviation from the corresponding normative values. For example, the *nerve fiber indicator* (NFI) is an important parameter.
- The *Guided Progression Analysis Report* (GPA) is a proprietary statistical tool which compares all measurements over time and determines whether the differences are statistically significant. As the RNFLs get slowly lost in the case of

glaucoma, the loss rate allows diagnosis of glaucoma in an early stage and helps to measure the treatment effectiveness.

Figure 6.82 shows a typical RNFL analysis report for a patient with a normal finding for the left eye (OS) and an abnormal finding for the right eye (OD). The discussed parameters and data representations are referenced as well. For a more detailed description of RNFL data analysis and presentation, please refer to specialized literature, for example, [52].

6.9 Recommended Reading

For further information about optical visualization, imaging, and structural analysis, please refer to the following references:

- Medical Magnifying Systems (Section 6.1): [10, 31]
- Surgical Microscopes (Section 6.2): [8, 10, 31, 53, 54]
- Reflection Methods for Topographic Measurements (Section 6.3): [55–59]
- Slit Lamp (Section 6.4): [60–63]
- Scanning-Slit Projection Devices (Section 6.5): [25, 28, 55–57, 59]
- Ophthalmoscope (Section 6.6): [31, 62, 64, 65]
- Fundus Camera (Section 6.7): [32, 36, 66, 67]
- Scanning-Laser Devices (Section 6.8): [36, 43, 48, 51, 52, 67, 68]

6.10 Problems

P6.1. Loupes

1. Derive Eq. (6.1) using the geometry shown in Figure 6.1.
2. Calculate the magnification of a loupe which consists of a thin lens with a focal length of $f = 25$ mm. Please consider the following cases
   - Alter the distance L_{LP} between the observer's eye and the loupe from $L_{LP} = 0$ to 100 mm.
   - Place an object in the focus of the lens. How does the resolution of the eye change? At first, determine the magnification and resolution of an eye without visual aids for the near point at the typical near viewing distance $s_{nv} = 250$ mm. At this distance from the loupe, the eye has a focal length of 17 mm. The density of cones in the fovea is approximately 100/mm.
   - Why are loupes with a magnification of 100 × not available? Where is the limit?

3. The field of view d_{fov} of a simple magnifier lens can be calculated via

$$d_{\text{fov}} = \frac{d_{\text{L}}}{L_{\text{LP}}\mathcal{D}_{\text{L}}^*}\,, \tag{6.79}$$

in which $\mathcal{D}_{\text{L}}^*$ is the effective optical power of a magnifier lens given by

$$\mathcal{D}_{\text{L}}^* = \frac{\mathcal{D}_{\text{L}} - 1/s'}{1 - L_{\text{LP}}/s'}\,. \tag{6.80}$$

Derive Eq. (6.79) and determine an approximation for d_{fov} as a function of only the power of the lens.
4. Calculate the parameters of Table 6.1 for a bi-convex spherical lens with a refractive index of $n = 1.7$. Is the assumption of a thin lens justified? What differences can you observe?
5. How can β_{L} and d_{fov} be derived for each magnifier lens with $\mathcal{D}_{\text{L}}$ and d_{L} given an optimum working condition?

P6.2. Stereoscopic vision An observer has an interpupillary distance of PD = 62 mm. He uses a telescope with $8\times$ magnification whose objective lens is located at a distance of 115 mm from the eye. How much is the stereoscopic vision improved compared to the case of a "naked" eye without visual aids? Compare the minimum stereo angles ε_{min} for both cases.

P6.3. Stereoscopic depth perception Why is the stereoscopic depth perception an important quantity in microsurgery? Please refer to the diameter of the nerve fibers, blood vessels, and so on. Compare the stereoscopic depth perception of eyes without visual aids and surgical microscopes. Typical parameters of surgical microscopes are $\varepsilon_{\text{min}} = 48.5 \times 10^{-5}$ rad, $b = 26$ mm, $f_{\text{obj}} = 200$–400 mm, $f_{\text{tub}} = 125$ mm, $f_{\text{ep}} = 20$ mm, and $\Gamma = 0.4$ to 1.

P6.4. Achromat design

1. An achromatic lens system shall be designed for the focal length $f = 300$ mm with crown glass N-FK51A for the (front) positive element and flint glass N-SF6 for the negative element (Table 6.9). The cemented surface should be planar. Calculate the focal lengths and the radii of curvature of the two partial lens elements.

Table 6.9 Data of optical glasses. The refractive indices are given for the wavelengths 486, 587, and 656 nm. ν denotes the Abbe number. Data taken from [9].

Type of glass	n(486 nm)	n(587 nm)	n(656 nm)	ν
N-FK51A	1.490 56	1.486 56	1.484 80	84.47
N-SF6	1.827 83	1.805 18	1.796 08	25.36
K-7	1.517	1.511 12	1.508 54	60.41
N-KZFS11	1.648 28	1.637 75	1.633 24	42.41

2. How large is the distance of the green ($\lambda = 587$ nm) from the blue-red ($\lambda = 486$ nm; $\lambda = 656$ nm) image plane (secondary chromatic aberration)? How large is the distance between the red and green image plane for a simple lens made of the glass K7? Where does the image for the blue wavelength lie in this case? How large is the relative improvement in the axial chromatic difference overall?
3. Compare the curvatures of the image shell (Section A.1.6.4) for the achromatic lens system and the individual lens element. Which system is more favorable? Why can the achromatic lens system not be flattened with regard to its image shell? Let us use the Rayleigh length $z_R = \lambda/\mathrm{NA}^2$ as a measure for the depth of field. How large may the field angle of the achromatic lens system be in order to obtain a sharply defined image, if the incident bundle is collimated to a diameter of 8 mm and only the field curvature is considered?

P6.5. Apochromat design Design a three-element apochromat with a focal length of $f = 300$ mm. As a third glass N-KZFS11 is available. Compare the resulting performance (regarding chromatic aberration) with the achromat in Problem P6.4 and a single lens made of K-7. For calculation of the secondary spectrum, you can use the $n(\lambda)$ dependence given by the Sellmeier equation

$$n^2(\lambda) - 1 = \frac{B_1\lambda^2}{\lambda^2 - C_1} + \frac{B_2\lambda^2}{\lambda^2 - C_2} + \frac{B_3\lambda^2}{\lambda^2 - C_3} \tag{6.81}$$

with the Sellmeier coefficients B_i and C_i given in Table 6.10.

P6.6. Varioscope

1. Calculate a simple varioscope (with a negative and a positive lens element) for a working distance range from 200–400 mm. Also calculate the related focal lengths. Why should the negative lens component be the front lens?
2. How large is the working distance? What focal lengths do you suggest for the individual lens elements?
3. Can chromatic correction be simultaneously achieved? Select two suitable materials.
4. Calculate the magnification of a surgical microscope for vitro-retinal surgery which uses a wide-angle lens with 125 D and an objective lens with $f_{\mathrm{obj}} =$

Table 6.10 Sellmeier coefficients of optical glasses. Data taken from [9].

Type of glass	B_1	B_2	B_3	C_1	C_2	C_3
N-FK51A	0.971 247	0.216 901	0.904 652	0.004 723	0.015 358	168.681 33
N-SF6	1.779 317	0.338 150	2.087 345	0.013 371	0.061 753	174.017 59
K-7	1.127 355	0.124 412	0.827 101	0.007 203	0.026 984	100.384 59
N-KZFS11	1.332 224	0.289 242	1.151 617	0.008 403	0.034 424	88.4310 53

200 mm. All other parameters shall be standard. Assume the lens to be located 20 mm above the eye.

- Where is the intermediate image plane?
- In order to keep the microscope in a fixed position after insertion of the wide-angle lens, what is the required change of the focal length of the varioscope?
- How does the situation change, if the lens with 125 D is replaced by a 60 D lens?

5. Calculate for the case of a fundus-imaging microscope the necessary varioscope focal length change (Figure 6.14) when the wide-angle aspheric lens ($\mathcal{D} = +60$ D or $+125$ D) for fundus viewing is placed into the observation path. The working distance of the system should be 200 mm.

P6.7. Afocal zoom systems Let us consider an afocal zoom system, as shown in Figure 6.16, with the focal lengths $f_1 = f_3 = +100$ mm and $f_2 = -20$ mm.

1. What is the condition for afocality in the ABCD matrix notation? What is the matrix of the overall system for a symmetric setup with equal distances? Use this arrangement to calculate the distances for the afocal case. What does the second solution of the equation look like, and why can it not be used here?
2. The zoom system is longest in the symmetrical case. Calculate the system length as a function of the focal lengths. What is the system matrix for the asymmetrical case when decreasing the first distance to $L_1 = 0$? How large is the second distance L_2 and the telescope magnification for this case? What is the general expression used for the overall length L_{tot} here?
3. The zoom factor is generally determined by the ratio of the telescope magnifications from the two end positions. What is the zoom factor M for a symmetrical zoom? What focal lengths should be selected if, for example, a zoom with overall length of $L_{\text{tot}} = 100$ mm and zoom factor $M = 9$ is required?
4. Let us now look at the above zoom system with optical compensation. In this case, only the negative lens L_2 is displaced by z from its center position with distances $L_s/2$. Without the exact compensation, the image plane does not remain exactly constant. To obtain a finite image position, an objective lens with $f_4 = 100$ mm is mounted on the zoom system. Compile the paraxial matrix of the system up to the image plane for arbitrary z values. To simplify the formulas, $f_1 = f_3 = f_4 = f$ should be used here.

- What is the condition for the image plane in the matrix calculation?
- What is the expression for the defocussing s' of the image plane?
- What minimum and maximum values can z technically assume?
- Plot s' as a function of z.
- At what interval can the negative element be displaced, if the defocussing is not to be greater than 2 mm?

P6.8. Video documentation In camera and video documentation the magnification of the surgical microscope is given by

$$\beta_F = \frac{f_{adp}\Gamma}{f_{obj}} , \tag{6.82}$$

where f_{adp} is the focal length of a camera adapter which is generally arranged directly after the zoom system. Typical values lie at around 100 mm.

1. Discuss the meaningfulness of using an HDTV video recorder. The two HDTV-standard image resolutions are 1280×720 pixels and 1920×1080 pixels, in the full format. The width-to-height ratio of the image is $16 : 9$. The size of CCD image sensors is often specified in inches. Common sizes for professional video cameras are $2/3''$ and $1/2''$. The description of the chip sizes was derived from the external diameter of the old picture tubes. A $1''$ CCD chip has, by definition, the same image diagonal as a $1''$ tube. The length of the image diagonals (8 mm for $1/2''$ CCDs) determines the size of the photosensitive surface.
2. How can you increase the depth of field in video documentation? What drawback does this entail?

P6.9. Keratometer

1. Please derive the keratometer equation for the corneal radius of curvature given in Eq. (6.50).
2. The cornea of a patient with a radius of curvature of $r_C = 7.8$ mm is analyzed with a keratometer. The test mires of the keratometer are located at a distance of 30 cm in front of the corneal vertex. At which distance behind the cornea do we find the reflex images of the mires?
3. We now look at the reflex images with a telescope. The sizes of the reflex images and the original size of the mires shall be equal. How must the telescope magnification be chosen to realize this?

P6.10. Principle of topometry Approximate an optical system as a spherical lens with a refractive index of 1.5 and a radius of 25 mm. Then, superimpose on this system a cylindrical optical system with an axis inclined by 30° with respect to the vertical line and an additional radius of 85 mm.

1. Calculate (numerical-graphical) a topometric image for an exactly axial alignment. Assume reasonable radii and reasonable stripe distances for the rings of the Placido disk.
2. How does this image change if the optical system is tilted by an angle α?
3. Replace the spherical lens with a parabolic aspherical lens, whereby the radius of curvature of one axis is equal to the radius of the spherical lens. How does the topometric image change?

P6.11. Slit lamp

1. As an alternative to a telescopic system, we can also implement a Greenough miscroscope to the observation path of a slit lamp. The total magnification of the Greenough miscroscope is given by the product of the magnification of the objective lens β_{obj} and the magnification of the eyepieces β_{ep}. Show that the magnification of the objective lens is given by $\beta_{\text{obj}} = L_{\text{tub}}/f_{\text{obj}}$, where L_{tub} denotes the optical tube length.
2. Slit lamp microscopes usually have a numerical aperture of NA $= 0.05$. Let us assume that imaging aperture and illumination aperture are equal. Calculate the microscope resolution for a wavelength of $\lambda = 550$ nm. Calculate the minimum and maximum usable magnifications $\beta_{\text{um,min}}$ and $\beta_{\text{um,max}}$, respectively (Section 6.2.2). Discuss the result with reference to typical magnifications used in slit lamp microscopes.

P6.12. Scheimpflug principle Imagine yourself standing in front of the highest church tower in Europe with your medium-format camera (6 cm × 6 cm) in hand. The market square in front of the tower allows you to be at a distance of 100 m from the foot of the tower. You have two objective lenses with focal lengths of $f_1 = 50$ mm and $f_2 = 85$ mm to choose from. The F-number of the objective lenses is $f_i/d_{\text{pupil}} = 1.4$ in each case; with the diameter of the entrance pupil d_{pupil}. You also have a bellow device which allows you to tilt the camera objective lens with respect to the film plane. Select the optimal arrangement allowing the church tower to be image-filling and in sharp focus, if possible, along its entire height.

P6.13. Direct ophthalmoscopes In direct ophthalmoscopy, the maximum angular field of view for an emmetropic eye is given by

$$\alpha_{\text{fov}} = \frac{d_{\text{pupil}} + d_{\text{phys}}}{L_{\text{pp}}}, \tag{6.83}$$

where d_{pupil} denotes the pupil diameter of the patient, d_{phys} the pupil diameter of the physician, and L_{pp} the distance between the patient's and the physician's pupil. Verify this relation by means of a drawing.

P6.14. Indirect ophthalmoscopes I A patient with emmetropic eyes (refractive power of eye $\mathcal{D}_{\text{eye}} \approx 60$ D) is examined with an indirect ophthalmoscope which uses an ophthalmoscopy lens with a refractive power of 13 D. The free diameter of the ophthalmoscopy lens is $d_{\text{oph}} = 40$ mm. The physician observes the intermediate image of the fundus from a distance of 40 cm.

1. At which distance from the patient's eye (pupil plane) must the ophthalmoscopy lens be placed?
2. How large is the field of view d_{fov} on the retina?
3. How large would the field of view be for a direct ophthalmoscope (same distance between patient and physician; pupil diameter of patient is 7 mm; pupil diameter of physician is 3 mm)?

P6.15. Indirect ophthalmoscopes II An eye is examined with an indirect ophthalmoscope. The distance between patient and physician is 50 cm and the ophthalmoscopy lens is placed at a distance of 5 cm in front of the patient's eye.

1. Determine the required refractive power of the ophthalmoscopy lens.
2. Calculate the diameter of the image of the patient's pupil (originally 4 mm) in the plane of the physician's pupil.
3. At which magnification does the physician observe the fundus (refractive power of patient's eye $\mathcal{D}_{\text{eye}} \approx 60\,\text{D}$)?

P6.16. Indirect ophthalmoscopes III The fundus of a patient shall be examined with an indirect ophthalmoscope. The pupil diameter of the patient has been enlarged to 7 mm. The ophthalmoscopy lens has a refractive power of 30 D and is placed at a distance of 50 cm in front of the physician's eye. The physician has an interpupillary distance of PD = 65 mm. Is it possible for the physician to perceive a virtual stereoscopic image of the patient's fundus or does he require an optical device which effectively reduces PD?

References

1 Roper-Hall, M.J. (1967) Microsurgery in ophthalmology. *Br. J. Ophthalmol.*, **51**, 408–414.

2 Nylean, C.O. (1954) The microscope in aural surgery, its first use and later development. *Acta Oto-laryng.*, **43** (s116), 226–240.

3 Harms, H. (1953) Augenoperationen unter dem binokularen Mikroskop. *Ber. Dtsch. Ophthalmol. Ges.*, **58**, 119–122.

4 Littmann, H. (1954) Ein neues Operations-Mikroskop. *Klin. Monatsbl. Augenheilkd.*, **24**, 473.

5 Barraquer, J.I. (1956) The microscope in ocular surgery. *Am. J. Ophthalmol.*, **42**, 916–918.

6 Atchison, D.A. and Smith, G. (2006) *Optics of the Human Eye*, Butterworth–Heinemann.

7 Berek, M. (1927) Grundlagen der Tiefenwahrnehmung im Mikroskop. *Sitzungsber. Ges. Naturwiss. Marbg.*, **62**, 189–223.

8 Lang, W.H. and Muchel, F. (1981) *ZEISS Microscopes for Microsurgery*, Springer.

9 Schott AG (2010) Optical glass – all datasheets, Online Product Specification. http://www.schott.com/advanced_optics/english/download/index.html (accessed 26 August 2013).

10 Gross, H., Blechinger, F., and Achtner, B. (2008) *Handbook of Optical Systems – Survey of Optical Instruments*, vol. 4, John Wiley & Sons, Inc.

11 Zimmer, K.P. (1998) Optical designs for stereomicroscopes, in *Proc. SPIE – Int. Opt. Des. Conf. 1998*, vol. 3482.

12 Murty, A.S., Aravinda, K., Ramanaiah, T.V., Petluri, R.L.V., Murthy, V.V.R., and Reddy, G.R.C. (1997) Design of a high-resolution stereo zoom microscope. *Opt. Eng.*, **36**, 201–209.

13 Stiller, H. and Rassow, B. (1991) Light hazards to the patient's retina from ophthalmic instruments. *Appl. Opt.*, **30**, 2187–2196.

14 Köhler, A. (1893) Gedanken zu einem neuen Beleuchtungsverfahren für mikrophotographische Zwecke. *Z. Wiss. Mikrosk.*, **10**, 433–440.

15 Chang, D.F. (2009) Improving the red reflex and surgical outcomes with the LUMERA microscope. *Cataract Refract. Surg. Today*, March, 27–30.

16 Reimer, P., Strähle, F., and Kolster, D. (2005) Beleuchtungs- und Beobachtungseinrichtung, Patent DE102006013761 A1.

17 Hutten, H. (1991) *Biomedizinische Technik – Medizinische Sondergebiete*, Springer.

18 Brenner, R. (2008) Stativanordnung und Stativ für ein medizinisch-optisches Instrument, Patent EU2016921 A3.

19 Rassow, B. (ed.) (1987) *Ophthalmologisch-optische Instrumente*, Enke, Stuttgart.

20 Saleh, B.E.A. and Teich, M.C. (2007) *Fundamentals of Photonics*, John Wiley & Sons, Inc.

21 Diepes, H. (2004) *Refraktionsbestimmung*, DOZ Verlag Optische Fachveröffentlichungen GmbH, 3rd edn.

22 Campbell, C.E. (1997) Reconstruction of the corneal shape with the MasterVue corneal topography system. *Optom. Vis. Sci.*, **74** (11), 899–905.

23 van Saarloos, P. and Constable, I. (1991) Improved method for calculation of corneal topography for any photokeratoscope geometry. *Optom. Vis. Sci.*, **68**, 960–965.

24 Bohren, C.F. and Huffmann, D.R. (2010) *Absorption and Scattering of Light by Small Particles*, John Wiley & Sons, Inc.

25 Snook, R. (1995) Pachymetry and true topography using the Orbscan system, in *Corneal Topography: the State of the Art*, Slack Inc., Thorofare, pp. 89–103.

26 Asrani, S. and Zeimer, R. (2005) Retinal thickness analyzer, in *Retinal Imaging*, Mosby, pp. 111–124.

27 Zeimer, R., Shahidi, M., Mori, M., Zou, S., and Asrani, S. (1996) A new method for rapid mapping of the retinal thickness at the posterior pole. *Invest. Ophthalmol. Vis. Sci.*, **37**, 1994–2001.

28 Wegener, A. and Laser-Junga, H. (2009) Photography of the anterior eye segment according to Scheimpflug's principle: options and limitations – a review. *Clin. Exp. Ophthalmol.*, **37**, 144–154.

29 Roberts, C.J. and Züger, B.J. (2006), The advantage and principle of dual Scheimpflug imaging for analyzing the anterior segment of the human eye. http://www.ziemergroup.com/fileadmin/media/products/GALILEI/Roberts__Galilei_2006.pdf (accessed 26 August 2013).

30 Timberlake, G.T. and Kennedy, M. (2005), The direct ophthalmoscope – how it works and how to use it. www.learningace.com/doc/2747505/73c066cea46572e4803fda9a25e159a3/thedirectophthalmoscope (accessed 26 August 2013).

31 Smith, G. and Atchison, D.A. (1997) *The Eye and Visual Optical Instruments*, Cambridge University Press.

32 DeHoog, E. and Schwiegerling, J. (2008) Optimal parameters for retinal illumination and imaging in fundus cameras. *Appl. Opt.*, **47**, 6769–6777.

33 DeLori, F.C., Gragoudas, E.S., Francisco, R., and Pruett, R.C. (1977) Monochromatic ophthalmoscopy and fundus photography. the normal fundus. *Arch. Ophthalmol.*, **95**, 861–868.

34 Schmitz-Valckenberg, S., Holz, F.G., AC, A.C.B., and Spaide, R.F. (2008) Fundus autofluorescence imaging: Review and perspectives. *Retina*, **28**, 385–409.

35 Schweitzer, D., Jentsch, S., Dawczynski, J., Hammer, M., Wolf-Schnurrbusch, U.E., and Wolf, S. (2010) Simple and objective method for routine detection of the macular pigment xanthophyll. *J. Biomed. Opt.*, **15**, 061714.

36 Holz, F.G., Schmitz-Valckenberg, S., Spaide, R.F., and Bird, A.C. (eds) (2007) *Atlas of Fundus Autofluorescence Imaging*, Springer.

37 Nakagawa, T., Suzuki, T., Hayashi, Y., Mizukusa, Y., Hatanaka, Y., Ishida, K., Hara, T., Fujita, H., and Yamamoto, T. (2008) Quantitative depth analysis of optic nerve head using stereo retinal fundus image pair. *J. Biomed. Opt.*, **13**, 064026.

38 Webb, R.H., Hughes, G.W., and Pomerantzeff, O. (1980) Flying spot TV ophthalmoscope. *Appl. Opt.*, **19**, 2991–2997.

39 Webb, R.H., Hughes, G.W., and DeLori, F.C. (1987) Confocal scanning laser ophthalmoscope. *Appl. Opt.*, **26**, 1492–1499.

40 Pawley, J.B. (ed.) (2006) *Handbook of Biological Confocal Microscopy*, Springer.

41 Wilhelm, S., Gröbler, B., Gulch, M., and Heinz, H. (2001), Confocal laser scanning microscopy: Principles. http://zeiss-campus.magnet.fsu.edu/referencelibrary/pdfs/ZeissConfocalPrinciples.pdf (accessed 26 August 2013).

42 Donnelly III, W.J. and Roorda, A. (2003) Optimal pupil size in the human eye for axial resolution. *J. Opt. Soc. Am. A*, **20**, 2010–2015.

43 Staurenghi, G., Viola, F., Mainster, M.A., Graham, R.D., and Harrington, P.G. (2005) Scanning laser ophthalmoscopy and angiography with a wide-field contact lens system. *Arch. Ophthalmol.*, **123**, 244–252.

44 David, J. and Barnet, R. (2008) Improvements in or relating to scanning ophthalmoscope, Patent WO/2008/009877 A1.

45 Strouthidis, N.G. and Garway-Heath, D.F. (2009) *Detecting Glaucoma Progression by Imaging in Glaucoma Progress III (Essentials in Ophthalmology)*, Springer.

46 Offenhäuser, F. (1992) Fundustopometrie mit dem Laserscanophthalmoskop, in *Jahrbuch der Augenheilkunde – Laser*, Biermann Verlag, Zülpich.

47 Balasubramanian, M., Bowd, C., and Zangwill, L.M. (2008) Algorithms for detecting glaucomatous structural changes in the optic nerve head, in *Image Modeling of the Human Eye*, Artech House, pp. 147–188.

48 Fingeret, M., Flanagan, J.G., and Liebmann, J.M. (2011), The essential HRT primer. http://www.heidelbergengineering.com/us/academy-education/quick-guides (accessed 26 August 2013).

49 Wollstein, G., Kagemann, L., Bilonick, R.A., Ishikawa, H., Folio, L.S., Gabriele, M.L., Ungar, A.K., Duker, J.S., Fujimoto, J.G., and Schuman, J.S. (2012) Retinal nerve fibre layer and visual function loss in glaucoma: The tipping point. *Br. J. Ophthalmol.*, **96**, 47–52.

50 Huang, X.R., Bagga, H., Greenfield, D.S., and Knighton, R.W. (2004) Variation of peripapillary retinal nerve fiber layer birefringence in normal human subjects. *Invest. Ophthalmol. Vis. Sci.*, **45**, 3073–3080.

51 Zhou, Q. (2006) Retinal scanning laser polarimetry and methods to compensate for corneal birefringence. *Bull. Soc. Belge Ophthalmol.*, **302**, 89–106.

52 Carl Zeiss Meditec AG (2009) GDx-PRO user manual. http://www.zeiss.de/C125679E00525939/EmbedTitelIntern/GDxPRO_User_Manual.pdf/$File/GDxPRO_User_Manual.pdf (accessed 26 August 2013).

53 International Organization for Standardization (2000) Optics and optical instruments - Operation microscopes – Part 1: Requirements and test methods, ISO 10936-1:2000.

54 International Organization for Standardization (2010) Optics and photonics – Operation microscopes – Part 2: Light hazard from operation microscopes used in ocular surgery, ISO 10936-2:2010.

55 Wang M. and Swartz, T.S. (2011) *Corneal Topography: A Guide For Clinical Application in the Wavefront Era*, 2nd edn, Slack Inc. Thorofare.

56 Horner, D.G., Salmon, Th.O., and Soni, P.S. (2006) *Corneal Topography, in Borish's Clinical Refraction*, (ed. W.J. Benjamin), 2nd edn, Butterworth-Heinemann/Elsevier, pp. 645–681.

57 Corbett, M.C., Rosen, E.S., and O'Brart, D.P.S. (eds) (1999) *Corneal Topography Principles and Applications*, London BMJ Books.

58 International Organization for Standardization (2009) Ophthalmic instruments – Ophthalmometers, ISO 10343:2009.

59 International Organization for Standardization (2012) Ophthalmic instruments – Corneal topographers, ISO 19980:2012.

60 Müller, O. (Second revision by Wagner, B. 2001) Eye Examination with the Slit Lamp, http://www.meditec.zeiss.com/88256DE3007B916B/0/506FBA0E8FCB598E882571D8007D4B40/$file/spaltlampen_eye_exam_en.pdf (11 September 2013).

61 Martonyi, C.L., Bahn, C.F., and Meyer, R.F. (2007) *Slit Lamp: Examination and Photography*, Time One Ink, Ltd., Sedona.

62 Rabbetts, R.B. (2003) *Clinical Visual Optics*, 3rd edn, Butterworth-Heinemann.

63 International Organization for Standardization (2007) Ophthalmic instruments – Slit-lamp microscopes, ISO 10939:2007.

64 International Organization for Standardization (2006) Ophthalmic instruments – Direct ophthalmoscopes, ISO 10942:2006.

65 International Organization for Standardization (2011) Ophthalmic instruments – Indirect ophthalmoscopes, ISO 10943:2011.

66 International Organization for Standardization (2009) Ophthalmic instruments – Fundus cameras, ISO 10940:2009.

67 Dua, S., Rajendra Acharya, U., and Ng, E.Y.K. (eds) (2011) *Computational Analysis of the Human Eye with Applications*, World Scientific Publishing Co. Pte. Ltd., Singapore.

68 Dithmar, S. and Holz, F.G. (2008) *Fluorescence Angiography in Ophthalmology*, Springer, Berlin, Heidelberg, New York.

7
Optical Coherence Methods for Three-Dimensional Visualization and Structural Analysis

In the previous chapter, we discussed a number of optical methods for visualization and structural analysis for the anterior and posterior segments of the eye. We will now focus on devices based on *low-coherence interferometry* (LCI) or *optical coherence tomography* (OCT). OCT enjoys a special position among the diagnostic devices, as it has revolutionized ophthalmology. It is estimated that in 2012 a total of more than 30 million OCT exams were conducted. In other words, every second an ophthalmic OCT scan or other LCI-based measurement on the eye is performed somewhere in the world. This widespread clinical adoption of a relatively recent technology has emerged from a convergence of modern optical technology, clinical application needs, and software capabilities, resulting in a superior clinical solution with demonstrable benefits for physicians and patients alike. This success, achieved over the last 20 years, is an example of what can be made possible through close collaboration between scientific and clinical research on the one hand, and through product development in the medical technology industry on the other. A large number of excellent review papers [1–4] and outstanding books [5–10] have been published on almost every particular aspect of OCT. We refer the interested reader to this literature.

In the following, in line with the objective of this book, we will focus on the engineering aspects of OCT in the medical device development process. We will give a brief introduction to the technology and its genesis (Section 7.1), followed by a more comprehensive description of the fundamentals of OCT (Section 7.3) and elements of the underlying theory (Section 7.4). Then, we will deal in detail with some design and layout aspects and their implementation in commercial systems (Section 7.5).

Last but not least, we will provide an overview of applications in the anterior and posterior segments of the eye (Section 7.6) as well as optical biometry (Section 7.7) in order to give a feel for the current clinical OCT activities. Prospects for the still dynamically evolving OCT development in ophthalmology will be presented in Section 7.8.

Optical Devices in Ophthalmology and Optometry, First Edition. Michael Kaschke, Karl-Heinz Donnerhacke, and Michael Stefan Rill.

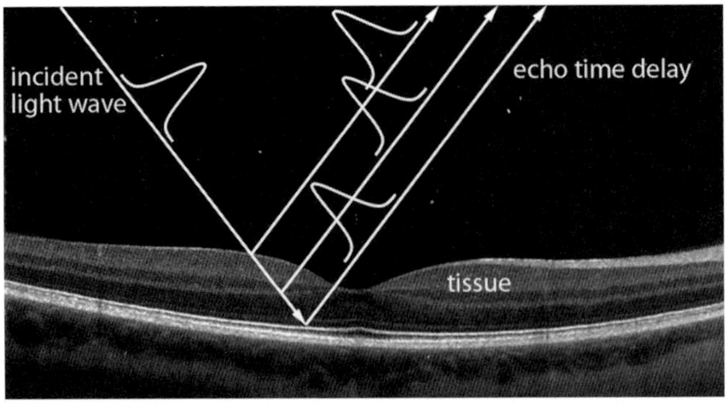

Figure 7.1 OCT as an "optical biopsy" technique. The time delay of the reflected light "echo" is used to reconstruct the axial position of different tissue layers. Courtesy of Carl Zeiss.

7.1 Introduction to Optical Coherence Tomography

Optical coherence tomography[1] (OCT) is a relatively new medical imaging modality with resolution in the μm range and an in-tissue penetration depth of up to 5 mm. The modality is thus capable of producing high-resolution cross-sectional images of inhomogeneous and scattering tissue (Figure 7.1). An OCT system is based on low-coherence interferometry (LCI) which will be explained in detail in Section 7.3.

OCT is analogous to *ultrasound B-mode imaging*. Similar to the acoustical echo in ultrasound, in OCT backreflected or backscattered echo signals from various tissue layers are acquired, with intensity being measured over the time delay (Figure 7.1). This measurement of backscattered light yields a so-called *A-scan* or *1D-scan* (Figure 7.2a). When the optical beam is scanned across the tissue and the signal is plotted as a function of depth (z direction) and transverse position (x direction), a two-dimensional (2D) cross-section (*B-scan*) of the tissue will result (Figure 7.2b). Finally, 3D imaging can be obtained by stacking the B-scans at different (orthogonal) transverse positions (y direction; Figure 7.2c).[2]

OCT has a number of characteristics that make it attractive for applications in ophthalmic imaging and structural analysis. A fundamental property is the independence of transverse and longitudinal resolution. In laser scanning light microscopy ([11]; Section 6.8.1), both axial (depth) resolution Δz_{cSLO} and transverse (lateral) resolution $\Delta(x, y)_{\mathrm{cSLO}}$ depend on the numerical aperture (NA) of the entire optical system involved in the measurement (compare also Eqs. (6.77) and (6.78)

1) The term "tomography" is typically used in radiology when two-dimensional cross-sections are derived from a set of projection data. In OCT, two-dimensional cross-sections are obtained from a series of z scans.

2) Alternatively, with modern OCT technologies and 2D sensors, it is possible to obtain a series of xy scans (*en-face imaging*) as a function of the axial depth z. In this way, a 3D image can be composed. "*En-face* imaging" means that OCT data is axially integrated over a given slab of tissue to create a 2D xy image.

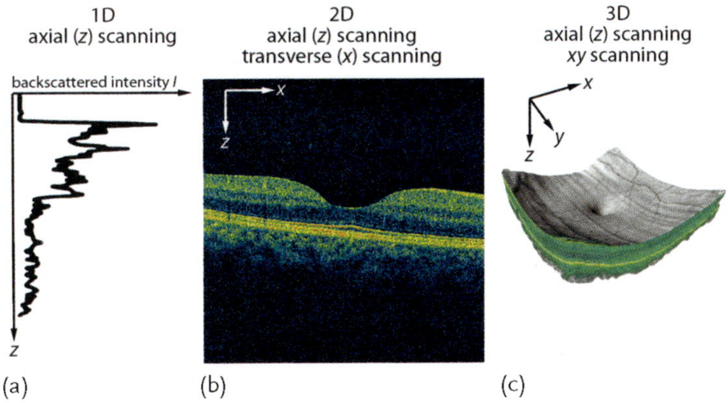

Figure 7.2 OCT imaging scans. (a) Backscattered light signal of an axial scan in the z direction. (b) Two-dimensional (2D) OCT scan of a healthy human retina. The different retinal layers (Section 1.2.1) are clearly visible. The depression corresponds to the fovea. (c) Three-dimensional (3D) OCT scan of a healthy human retina with overlaid fundus image. Courtesy of Carl Zeiss.

in Section 6.8.1.8). In concrete terms, we have

$$\Delta z_{\mathrm{cSLO}} \propto \frac{\lambda}{\mathrm{NA}^2}, \tag{7.1}$$

$$\Delta(x, y)_{\mathrm{cSLO}} \propto \frac{\lambda}{\mathrm{NA}}, \tag{7.2}$$

where λ is the wavelength of light. Scanning laser ophthalmoscopes (Section 6.8.1) thus have limited depth (axial) resolution, since the human eye has a relatively small pupil, and therefore a small NA, typically below 0.1. In contrast to laser scanners, the axial resolution in OCT is only limited by the bandwidth of the incident light $\Delta\lambda$ and is given by ([5], Section 7.4)

$$\Delta z_{\mathrm{OCT}} \propto \frac{\lambda_0^2}{\Delta\lambda} \tag{7.3}$$

with the center wavelength of the spectrum λ_0. As a consequence, a high axial resolution can be attained even at relatively low numerical apertures, for example, for large working distances or a small pupil. The axial and transverse resolutions become decoupled. Besides the enhanced axial resolution at low NA values, OCT is also a powerful technology for the following reasons:

- OCT is a contact-free, noninvasive measurement with relatively large penetration depth into biological tissue. The light level is relatively low and no special preparation of the specimen is required, as in, for example, a biopsy).
- OCT allows a high imaging speed with up to real-time measurement capabilities.
- OCT is quantitative in that volumetric and biometric measurements are possible (Sections 7.6 and 7.7).

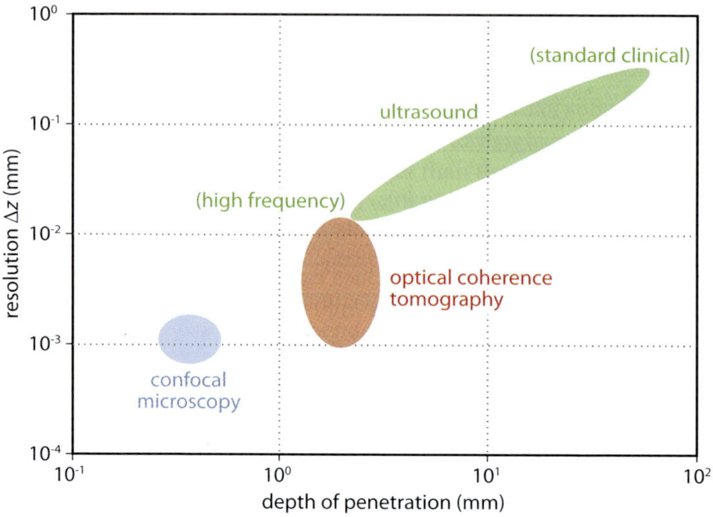

Figure 7.3 Depth of penetration in typical human tissue of optical coherence tomography (red) in comparison to laser scanning ophthalmoscopes (cSLO; blue) and ultrasound scanners (green). Courtesy of J.G. Fujimoto.

- OCT allows integrated structural and functional measurements. For example, Doppler[3], polarization, and spectroscopic measurements can be performed simultaneously (Section 7.8).
- OCT fills the "imaging gap" between conventional ultrasound and confocal microscopy imaging in terms of resolution and depth of penetration in human tissue (Figure 7.3).

7.2
Development of OCT and LCI as an Example of Modern Medical Technology Innovation

LCI and OCT technologies have, without a doubt, revolutionized ophthalmology within the last two decades. Cross-sectional scans of the eye and quantitative biometric analysis have now become the standard of care in the diagnosis and monitoring of many eye diseases. However, the adoption of OCT took time. In retrospect, ophthalmic OCT evolved through a series of steps from invention to adoption and proliferation. It is instructive to take a historical perspective on the development of LCI and OCT, although, unlike some other optical ophthalmic techniques, its history spans a period of only twenty years [12].

3) Christian Doppler (1803–1853).

7.2.1
Academic Research – Conception of OCT (until 1993)

By the late 1980s, low-coherence interferometry with short-pulse lasers had already been used for optical ranging (Optical Coherence Domain Reflectometry; OCDR) and in optical communication [13, 14]. The interferometric detection used to measure the echo time delay (Figure 7.1) is based on LCI and is also called *coherence gating*. The basic principle is that interference between broadband light reflected from a sample and a reference arm of an interferometer will generate a signal only if the path lengths match. By varying the path length of the reference arm, the scattering intensity of the sample can be measured as a function of depth. The technique will be described in detail in Section 7.3.

The first application of LCI in biology was described by Adolf Fercher *et al.* for the measurement of the axial eye length [15] in 1988. The first *in vivo* measurement of an axial eye length by means of dual-beam interferometry was reported in 1991 [16], which eventually led to the development of a new class of optical biometry devices (Section 7.7). Meanwhile, further work on using LCI for imaging by Naohiro Tanno [17] in Japan and, in particular, by James Fujimoto's group (working together with Carmen Puliafito) had led to the adaptation of OCDR to tissue imaging by laterally scanning light from a superluminescent diode across tissue [18]. Backscattered light was coupled into an interferometer to capture depth resolved scattering OCT profiles which, in turn, were used to create B-scans of an object. Among the first OCT images published were cross-sectional *in vitro* images of the retina and optic nerve head. These achievements clearly suggested that OCT could provide a revolutionary method for visualizing the living retina. First *in vivo* OCT images were published as early as 1993 [19, 20].

7.2.2
First Generation of Commercial OCTs (1993–2002)

As mentioned, the evolution of OCT into a clinical standard passed through several distinct phases (Table 7.1). The invention of OCT in academic research led to a startup phase during which industrial partnerships were established. OCT was not patented until 1991 and, as soon as 1992, was then transferred from Fujimoto's group to the company Advanced Ophthalmic Devices (AOD). AOD was formed to further develop the technology, incorporate it into prototypes, and explore first clinical applications. Although the "academic" generation OCT setup was far from being a practical clinical instrument, the inventors, the medical device industry, and the early clinical adopters recognized the potential of this technology. The company Humphrey Instruments, which became part of Carl Zeiss in 1993, had purchased AOD in the same year to commercialize OCT. The first commercially available instrument, the ZEISS OCT 1, was launched in 1996. This instrument gave researchers around the world the opportunity to investigate clinical applications of OCT. By today's standards, the performance of the OCT 1 was modest, and there was a steep learning curve for interpreting images of retinal pathology. Neverthe-

Table 7.1 Milestones of OCT development. Data taken from [12, 21–23].

Year	Milestone
1991	Demonstration of OCT *in vitro*
1993	First *in vivo* images of the retina
1994	Carl Zeiss (Humphrey Instruments) acquired OCT technology
1995	Clinical studies with first prototypes
1996	First commercial TD-OCT system introduced (ZEISS OCT 1)
1999	Approximately 200 units sold
2000	Improved commercial system ZEISS OCT 2
2001	Approximately 400 units sold
2002	Second generation TD-OCT system introduced (ZEISS Stratus OCT™)
2005	OCT becomes a standard of care
2006	Approximately 6000 units of ZEISS Stratus OCT sold
2006	Multiple companies enter ophthalmic market with FD-OCT devices
2009	Worldwide OCT revenues exceed US$ 250 million
2010	US Medicare reimbursed OCT procedures exceed 8 million
2010	Worldwide reimbursement payments exceed US$ 1 billion
2012	ZEISS Cirrus™ HD-OCT reaches 10 000 installations
2012	Inventors of OCT receive Antonio Champalimaud Vision Award

less, clinical researchers embraced the technology, investigating both a wide range of diseases and measurement methodologies. Analysis methods and manufacturing expertise quickly accumulated with the expanding base of OCT instruments. Within a few years, the instrument design was modified to take advantage of this knowledge. The instrument was consolidated into a single desk-type instrument, the user interface was improved, and (very importantly) a quantitative glaucoma analysis package including a normative database was added.

Early users of OCTs, mostly specialists in clinical research, guided OCT development and also educated other ophthalmologists in OCT use. OCT provided in particular retinal specialists with dramatically new visualization capabilities, and glaucoma specialists with new metrics for diagnosis. During this time, the first edition of the textbook “Optical Coherence Tomography of Ocular Diseases” [8] was published which illustrated the use of OCT in ophthalmology. As researchers continued to gain access to OCT technology, the number of clinical publications mushroomed, and by the end of 2001, about 400 OCT units had been sold. Knowledge gained in this phase was consolidated in the second OCT generation, which became a clinical “workhorse”.

7.2.3 Second Generation of OCTs – ZEISS Stratus OCT (2002–2006)

Broader market demand required a new platform with better scanning technology and detection electronics which incorporated clinical know-how and which was engineered for routine clinical use. Successful clinical research had led to better understanding of clinical applications, but technology had advanced as well. In particular, a higher axial scan speed was achieved by a so-called *rapid scanning optical delay* (RSOD) line [24, 25] which was a key component of a complete redesign. The RSOD line made the device platform practical and ready for widespread and routine clinical use. When the ZEISS Stratus OCT was launched in 2002, ophthalmic applications of OCT had advanced dramatically. Quantitative analyses of the data improved as well, with inclusion of normative data for both glaucoma and retinal analysis. As more ophthalmologists adopted OCT, evidence mounted for the use of OCT for common diseases like glaucoma (Section 3.3), age-related macular degeneration (Section 3.4), diabetic retinopathy (Section 3.5), as well as retinal artery and vein occlusions (Section 3.6). Validation of ZEISS Stratus OCT analysis tools was a key element for achieving clinical reimbursement for OCT scans, which was an important milestone in the commercial development of ophthalmic OCTs.

7.2.4 Third Generation of OCTs – Frequency-Domain OCT (2007–current)

Expanding adoption drove further research and commercial success of "generation 2" OCT medical devices and provided further funding to support ongoing OCT research. The first two generations of OCT were so-called *time-domain* (TD-OCT; Section 7.3.3) implementations as in the original invention, in which the low-coherence (broadband) light is detected by a single detector and the interferometer's reference arm is scanned for axial resolution (Section 7.3.3). The implementation of new higher speed and resolution OCT technology led to today's third generation frequency-domain OCTs (FD-OCT; Section 7.3.4), which are now in widespread clinical use. A first demonstration of FD-OCT in an ophthalmic application was reported in 2003 [26]. In FD-OCT devices, which dramatically increased efficiency, all backscattering information for all depths of the sample is acquired simultaneously (Sections 7.3.4 and 7.4.2) [27–29]. Development of FD-OCT instruments has also been profoundly different from previous OCT generations. Product development of first and second generation OCTs occurred in an era of early adopters, but the third generation was only developed after the ZEISS Stratus OCT had firmly established the clinical importance of OCT, which allowed multiple companies to enter the ophthalmic market having been primed by the installation of about 6000 ZEISS Stratus OCT. The capability of this generation of FD-OCT devices to visualize the retina and to provide automated analysis of complex retinal structures is illustrated in Figure 7.4. Here, scans from OCT 1 (1996) are compared to the current generation of FD-OCTs (2010). Improved axial resolution and higher

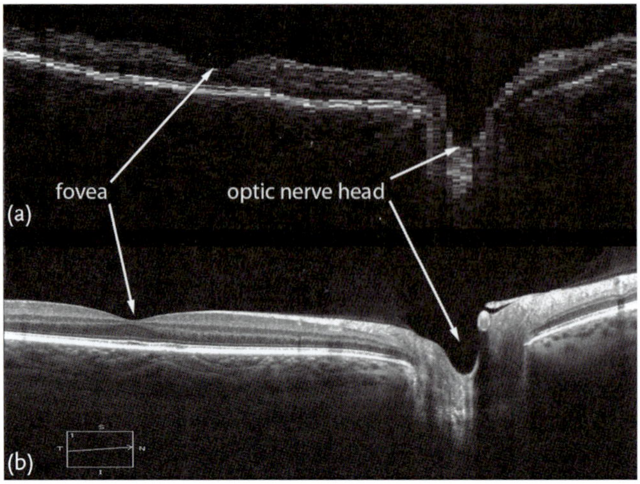

Figure 7.4 OCT sample images of first generation and FD-OCT generation instruments showing the foveal depression and optic nerve head. (a) OCT image scanned with ZEISS OCT 1. (b) OCT image scanned with ZEISS Cirrus HD-OCT. Courtesy of Carl Zeiss.

scanning speed show new details in the photoreceptor layer and retinal tissue layers compromised by age-related macular degeneration.

Currently, about ten companies in the diagnostic ophthalmology OCT market exist. Four more are using OCT to guide laser surgical procedures in ophthalmic applications (Section 7.6 and Chapter 10). As we have pointed out, a remarkable technology like OCT is not immediately embraced by the medical community on its own merit, but follows a fairly common pathway for medical device innovations (Figure 7.5). On this pathway, close partnerships between researchers and industry are crucial. In particular, these partnerships help to transfer technology from the university lab to the industry. They are also key to demonstrating tangible medical advantages, which improve physicans' workflows and expand their diagnostic

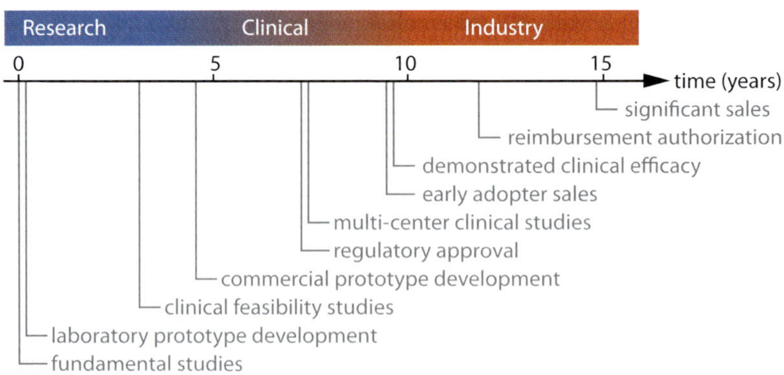

Figure 7.5 Development pathway of medical innovations. Adapted from [12].

abilities. At the same time, they generate patient benefit, such as noninvasive, non-contact, and fast exams. Today, a new medical product is rarely introduced into the market solely on the basis of improved patient outcomes. It is often essential to show that a new product cannot only improve outcomes but also has economic benefits. For the clinical practice manager as well as the insurer, a product to be purchased will need to pay back its investment cost within a reasonable period through cost savings and/or increased revenue in addition to improving the work-flow and quality of care. Regulatory as well as reimbursement authorizations are further reasons for the long innovation pathway of medical devices, of which OCT is an excellent but not unique example [12].

7.3 Principles of Low-Coherence Interferometry and Optical Coherence Tomography

To understand the principles of LCI and OCT, it is most instructive to start from standard Michelson interferometry with coherent and incoherent light.

7.3.1 Michelson Interferometry with Coherent Light

In Section A.2.3, we describe the superposition of coherent monochromatic light waves which leads to the total intensity (see also Eq. (A98))

$$I = I_1 + I_2 + 2\sqrt{I_1 I_2}\cos(\Delta\xi)\ . \tag{7.4}$$

I_1 and I_2 are the light intensities of each wave, and $\Delta\xi$ is the relative phase difference. This interference signal can be measured at the detector of a Michelson interferometer in which the (monochromatic) beam is directed to a sample ("S") and a reference ("R") arm (Figure 7.6a). The lengths of these arms are z_S and z_R, respectively. The mirror of the reference arm is movable and causes a phase difference $\Delta\xi = \xi_2(t) - \xi_1(t)$ for the path length difference $2n\Delta z = z_S - z_R$ between both arms, which is given by

$$\Delta\xi(\Delta z) = 2\pi\left(\frac{2n\Delta z(t)}{\lambda_0}\right)\ . \tag{7.5}$$

n denotes the refractive index, which is assumed to be equal in both arms. An exemplary time-dependent detector signal for the case of a completely coherent radiation with wavelength λ_0 (or frequency ω_0) is shown in Figure 7.6b. In the case of monochromatic light, the signal is characterized by a sinusoidal modulation that follows the displacement Δz of the reference mirror with constant speed.

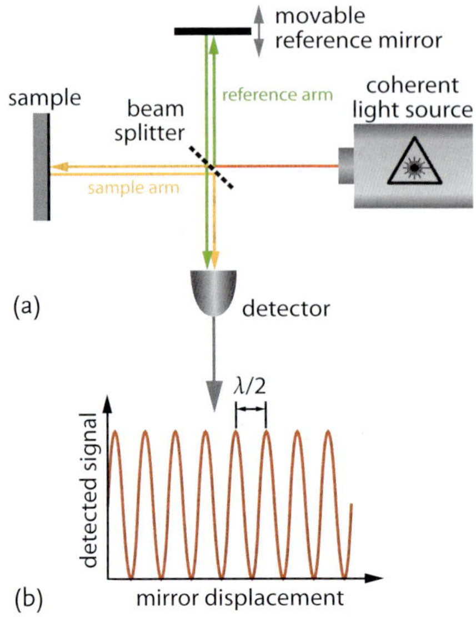

Figure 7.6 (a) Setup of a Michelson interferometer with a coherent light source (e.g., laser). (b) The diagram shows the detected signal versus the displacement of the reference mirror.

7.3.2
Michelson Interferometry with Low-Coherence Light

Let us now move from coherent light sources to partially or low-coherence light sources. In Section A.2.3.2, we define the coherence time t_c as the time interval during which the light field is correlated. The coherence time is related to the coherence length via (Section A.2.3.2)

$$L_c = ct_c \,, \tag{7.6}$$

where c denotes the speed of light. With the autocorrelation function defined in Eq. (A111) and the angular frequency $\omega = 2\pi c/\lambda$, the power spectral density is related to the autocorrelation function by

$$\sigma(\omega) = \int_{-\infty}^{\infty} \mathcal{G}(\Delta t) e^{i\omega\Delta t} d(\Delta t) \,. \tag{7.7}$$

From Eq. (7.7), we can derive the well-known interpretation according to which a spectrally broadened light source has a shorter coherence length and vice versa (Figure A.41 in Section A.2.4.2). A good approximation for the coherence length is given by (Section A.2.4.2; Problem P7.1)

$$L_c = ct_c = \text{const} \cdot \frac{c}{\Delta\omega} = \text{const} \cdot \frac{\lambda_0^2}{\Delta\lambda} \tag{7.8}$$

in which $\Delta\omega$ is some measure of the spectral bandwidth, and the constant is a value between 1 and 5 depending on the exact definition of the coherence length

Table 7.2 Approximate coherence lengths and spectral widths of a number of light sources. Data taken from [30–32].

Light source	$\Delta\nu$ (Hz)	t_c	L_c
Filtered visible sunlight	3.7×10^{14}	2.7 fs	0.8 µm
fs Ti:Al_2O_3 (titanium:sapphire) laser pulse at $\lambda_0 = 780$ nm	1.8×10^{13}	50 fs	15 µm
Superluminescent diode at $\lambda_0 = 1325$ nm	1.3×10^{13}	50 fs	15 µm
LED at 1050 nm	1.5×10^{13}	67 fs	20 µm
Low-pressure sodium lamp	5.0×10^{11}	2 ps	600 µm
Multimode helium-neon laser	1.5×10^{9}	670 ps	20 cm
Single-mode helium-neon laser	1.0×10^{6}	1 µs	300 m

and the spectral shape of light (Problem P7.2). For the special case of a light source with a Gaussian-shaped spectrum, the coherence length can be calculated with bandwidth (defined as the full width at half maximum (FWHM)) $\Delta\lambda$ or $\Delta\omega$, respectively, and center wavelength λ_0 as

$$L_c = \frac{4\ln 2}{\pi}\frac{\lambda_0^2}{\Delta\lambda} = \frac{8c\ln 2}{\Delta\omega} . \tag{7.9}$$

Representative examples of spectral bandwidths and coherence lengths for several light sources are listed in Table 7.2.

Let us now consider a Michelson interferometer with spectrally broadened (low-coherence) light (Figure 7.7). An interference signal can only be expected if the optical path lengths in both interferometer arms coincide at least within the coherence length.[4)] A short light pulse is one form of low-coherence light, where the pulse length correlates (but does not have to be identical) with the coherence time. In order to generate an interference signal in the Michelson interferometer, the individual pulses from both arms of the interferometer must overlap in time at the detector (see signals in schematic setup of Figure 7.7).

In analogy with Eq. (7.4), the intensity signal measured by the photodiode can be expressed as

$$I(z) = I_S + I_R + 2\sqrt{I_S I_R}\cos(\xi_S - \xi_R(\Delta z)) , \tag{7.10}$$

where Δz is the displacement coordinate in the reference arm which can be varied over time. For illustration purposes, let us consider a Gaussian pulse with a short pulse duration as the broadband light source. The pulse shall have a temporal evolution of

$$I(t) = \underbrace{I_0 \exp\left(-\frac{2t^2}{t_p^2}\right)}_{\text{pulse envelope}} \exp(-i\omega_0 t) \tag{7.11}$$

4) In addition to the requirement of temporal coherence, spatial coherence is needed in OCT [33].

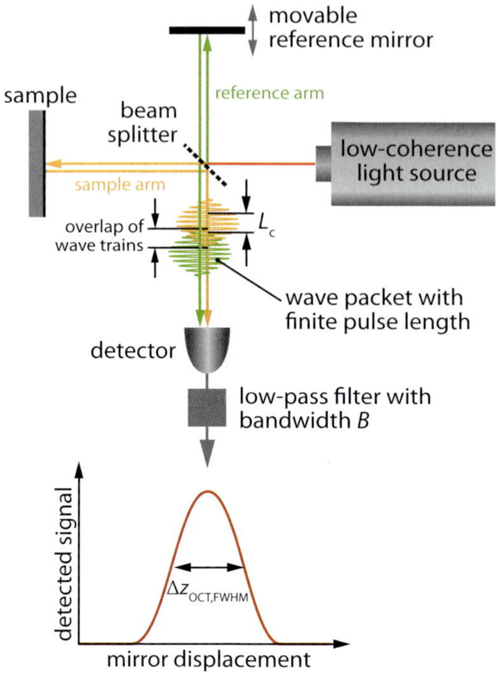

Figure 7.7 Setup of Michelson interferometer with a low-coherence light source. Here, as one form of low-coherence light, we consider ultrashort laser pulses. The pulses from the reference and the sample arm only interfere if the arm lengths are nearly equal. In this case, we have an overlap of both pulses, and their interference can be detected (signal in the diagram). A light pulse can also be understood as a wave packet with limited axial (= temporal) coherence. In fact, it is this limited temporal coherence that is relevant here.

in which t_p is the pulse parameter[5]. From Eqs. (A111) and (A112), the (energy) spectral density can be derived as

$$\sigma(\omega) = \frac{I_0 t_p}{4\pi} \exp\left(-\frac{t_p^2}{2}(\omega - \omega_0)^2\right) \tag{7.12}$$

$$= \frac{I_0 t_p}{4\pi} \exp\left(-2\pi^2 t_p^2 (\nu - \nu_0)^2\right) . \tag{7.13}$$

This is again a Gaussian profile, from which the relationship for the full spectral width $\Delta\nu = \Delta\omega/(2\pi)$ (FWHM) follows as

$$\Delta\nu = \frac{\sqrt{2\ln 2}}{\pi t_p} = \frac{0.375}{t_p} = \frac{2\ln 2}{\pi \tau_{FWHM}} . \tag{7.14}$$

5) The pulse parameter t_p should not be confused with the pulse duration τ. Different definitions for the pulse duration are possible. We use in this book the full width at half maximum (FWHM) pulse duration which is determined by $\tau_{FWHM} = \sqrt{2\ln 2}t_p = 1.18t_p$. In contrast, the 1/e-pulse width $\tau_{1/e}$ is defined via $I(\tau_{1/e}) = I_0/e \approx 0.37\, I_0$ (see also Problem P7.2).

From these results, we can derive the coherence length (compare with Eq. (7.8); see Problem P7.2)

$$L_c = \frac{\pi c \tau_{\mathrm{FWHM}}}{2 \ln 2} . \tag{7.15}$$

An interference signal is only obtained in the range of approximately the duration of τ_{FWHM} where the pulses overlap.

7.3.3 Time-Domain OCT

All time-domain OCT (TD-OCT) systems are essentially based on the original invention [18]. Thus far, we have assumed the interferometer has reflecting mirrors in each arm. Let us now replace the sample mirror with a backscattering sample. For the sake of simplicity, we shall assume that the sample has distinct layers with individual backscattering properties. By scanning the mirror in the reference arm, the reflectance (i.e., backscattering) depth profile of the sample can be detected. The backscattered signal of each layer is essentially "coherence-gated". This means that the depth resolution is determined by the coherence length of the light source. Any backscattered light from the sample that arrives at the photodiode outside the coherence gate is rejected.

Figure 7.8 shows an example of recorded intensities for a sample with only two reflecting surfaces. For a Gaussian-shaped spectrum, the axial resolution in TD-OCT is given by (Section 7.4.1)

$$\Delta z = \frac{2 \ln 2}{\pi} \frac{\lambda_0^2}{\Delta \lambda} \tag{7.16}$$

which is half the coherence length. The reference mirror scan thus generates an axial depth profile which is very analogous to ultrasound A-mode measurements. 2D and 3D OCT images can be obtained by stacking several A-scans. Adjacent A-scans are obtained by orthogonal displacement of the beam with an xy scanner (Figure 7.9). The transverse resolution[6] is determined by the beam waist which, in turn, depends on the effective numerical aperture given by the optical system or the sample.[7]

In TD-OCT, the path length of the reference arm is varied. This changing path length creates a Doppler shift[8]. As the optical frequency ν_0 is shifted by a small amount $\Delta \nu_{\mathrm{Doppler}}(t)$, heterodyne detection proves advantageous to detect the envelope signal resulting from the coherence gating.

6) We also have to consider the transverse sampling density. If the sampling density is less than the optical transverse resolution, the effective transverse resolution will be determined by the sampling density.

7) In ocular measurements, the numerical aperture of the eye (Section 2.1) represents the limiting numerical aperture.

8) The Doppler shift is the change in frequency of a wave for a detector moving relative to the source, or vice versa.

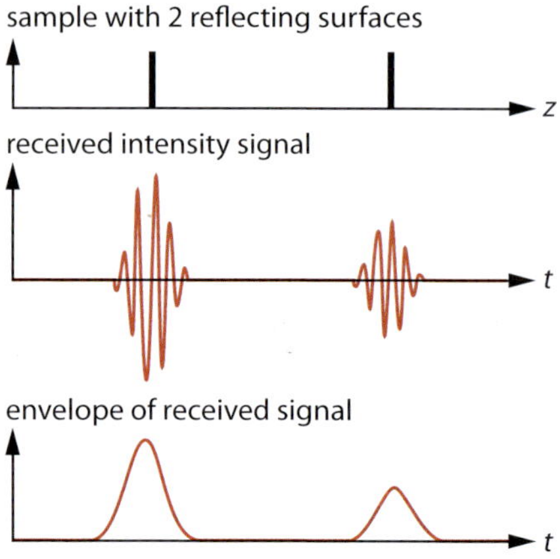

Figure 7.8 TD-OCT intensity and envelope signal for a sample consisting of two reflecting surfaces.

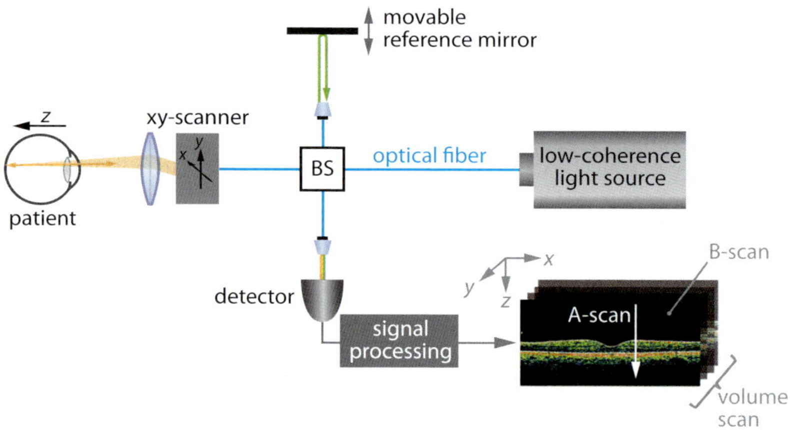

Figure 7.9 Setup of a time-domain OCT. Light from a low-coherence light source is fed into a fiber (blue) and is split by a fiber beam splitter (BS). One part of the wave travels along the reference arm and is reflected at the reference mirror. The other partial wave travels along the sample arm, is deflected by an *xy* scanner, and then reflected at the object (here: eye of a patient). Both partial waves are then combined by the fiber beam splitter and detected. A vertical line in the *z* direction (A-scan) is obtained by one linear scan of the movable reference mirror. By stacking multiple A-scans, we obtain a cross-sectional image (*xz* plot). Stacking multiple B-scans results in a 3D volume scan (*xyz* plot). Adapted from [34].

A number of elegant ways to achieve rapid reference arm length variations have been developed. The most common is a moving retro-reflector, but also fiber-optic Michelson interferometers with RSOD lines have been used. In the latter case, fiber stretchers [35] and grating- or prism-based delay lines have been used [10, 24].

7.3.4 Frequency-Domain OCT

In TD-OCT, the optical length of the reference arm is continuously changed, and the intensity of the interference signal is measured without analyzing the spectrum of the low-coherence radiation. In contrast, frequency-domain OCT (FD-OCT) detects the interference of the individual spectral components of the low-coherence light (Figure 7.10). In this case, the reference mirror is in a fixed position, and the photodiode detector is replaced by a spectrometer. The spectrum is recorded by a CCD, CMOS, or a photodiode line-array.

Low-coherence backscattered (backreflected) light from a sample is recombined in a Michelson interferometer with the light from the reference arm and spectrally

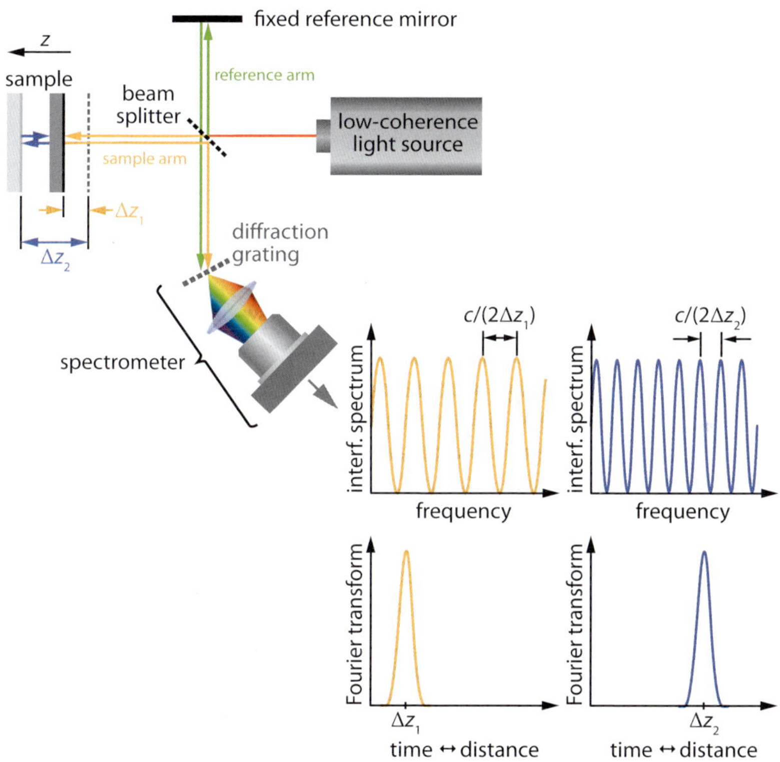

Figure 7.10 Setup of a frequency-domain OCT. In contrast to time-domain OCT, the reference mirror is fixed, and the detected signal is split into its spectral components by a diffraction grating. The information about the object's depth is obtained from the Fourier transform of the detected interference spectrum. Here, the detected signals for two different object distances (orange and blue) are shown. In the sample arm, one object has a distance of Δz_2 (blue) from the reference plane (dashed gray line). The other object has a shorter distance of Δz_1 (orange) from the reference plane. When we compare both Fourier transform diagrams, we see that the different positions can be retrieved from the detected signal. Adapted from [7].

dispersed by a spectrometer. The signal is then imaged onto a CCD camera. Path length differences Δz between sample and reference arm produce a periodic modulation in the interference spectrum. This can be easily understood if one recalls Eq. (7.5) in Section 7.3.1. This equation states that for a fixed Δz, constructive and destructive interference[9] will occur for different wavelength components λ. The Michelson interferometer acts in effect as a "wavelength modulator" for broadband light. The modulation is performed across the spectrum of the recorded intensity on the camera detector. The modulation frequency in the spectrum depends on Δz. Small Δz values will produce a low frequency modulation, while for bigger delays the modulation frequency will increase. By inverse Fourier transformation (see Info Box A.2 in Section A.2.4), the actual value of Δz can be retrieved, which corresponds to the position of the backscattering surface. Backscattering from a number of planes along the z axis will result in a superposition of modulations in the interference spectrum. As the Fourier transformation is a linear operation, it allows the retrieval of the entire A-scan from the measurement of the spectrum (see also simulation in Section 7.4.2)[10].

Advantages of FD-OCT

- The main advantage of FD-OCT is the simultaneous measurement of all backscattering locations along the A-scan within the sample. Accordingly, all depth information is collected without the need for any moving parts.
- The simultaneous measurement of all backscattering locations in FD-OCT has a considerable sensitivity advantage compared to TD-OCT (Section 7.4.4). This results from the parallel signal measurement with a CCD featuring a large number N of pixels. At first sight, this may seem counterintuitive, as each CCD pixel in the spectrometer has about the same signal-to-noise ratio (SNR) as the individual photodiode used in a TD-OCT. Still, the FD system shows a significant sensitivity increase. This is because, in the Fourier transformation the signals from each pixel add up coherently, whereas the noise components add up incoherently. As a consequence, the SNR of the FD-OCT is increased by a factor of N (Section 7.4.4). Systems which have a sensitivity advantage compared to TD-OCT higher than 20 dB have been demonstrated which allows a much higher A-scan rate (100 ×). A higher speed translates to a higher scan density, larger scan volumes, and less sensitivity to eye motion artifacts.

Drawbacks of FD-OCT

- *Maximum scan depth:* Typically, the number of CCD pixels N is taken as a power of 2 (i.e., $N = 2^i$) which allows the use of the Fast Fourier Transformation (FFT) algorithm [36]. For a (symmetrically assumed) spectrum resolved in N frequency

9) Constructive interference occurs for $\Delta\xi(\Delta z) = 2m\pi$, and destructive interference occurs for $\Delta\xi(\Delta z) = (2m+1)\pi$, with m being an integer.

10) From this rationalization, it follows directly that two completely identical arms do not yield any interference fringes on the spectrum.

channels of width $\delta\omega$ around a center frequency ω_0, the FFT algorithm results in two symmetrical series of $N/2$ data points in the time domain (Figure 7.11) with a "temporal range" of

$$\Delta t_R = \frac{\pi}{\delta\omega} = \frac{\lambda_0^2}{2c\delta\lambda} . \tag{7.17}$$

This range in the time domain corresponds to a (spatial) range along the z axis. The maximum scan depth is then given by the highest channel number in the time domain ($N/2$). The time axis has to be scaled with c/n_S, where n_S is the average refractive index of the sample. Here, we have taken into account the round-trip factor of 2. In this way, the maximum scan depth follows from

$$2z_{max} = \frac{c\Delta t_R}{n_S} = \frac{1}{2n_S}\frac{\lambda_0^2}{\delta\lambda} = \frac{N}{2n_S}\frac{\lambda_0^2}{\Delta\lambda} , \tag{7.18}$$

where we assumed $\Delta\lambda \approx N\delta\lambda$. $\Delta\lambda$ is the maximum spectral width the spectrometer can measure and $\delta\lambda$ the channel width. The price we pay for the increased sensitivity is a maximum scan depth for a given spectral width. As an example, for $N = 2048$, $\lambda_0 = 1300\,\text{nm}$, $\Delta\lambda = 80\,\text{nm}$, and $n_S = 1.35$, we find a maximum scan depth of $z_{max} \approx 8\,\text{mm}$, which is just sufficient to image the anterior eye segment. However, it is not enough to image the entire eye. Any signal that comes from depths beyond z_{max} will appear as an aliased signal.

- *Signal drop-off:* Another drawback of FD-OCT is an inherent signal drop-off with depth [27]. This can be understood if we take the spectrometer resolution into account. Lower spectral modulation (orange curve in Figure 7.10a) stemming from backscatter planes close to $\Delta z = 0$ can be better resolved than higher frequency modulations coming from planes with larger positive or negative Δz. This results in a reduced sensitivity, and thus image points along the A-scan close to $\Delta z = 0$ will appear brightest with highest SNR. The problem can be overcome to some extent by properly selecting the zero delay position close to the area of interest.
- *Mirror image artifacts:* In FD-OCT, mirror image artifacts exist, which can be understood from the inherent characteristics of Fourier transformation. As can be seen from Figure 7.11, the spectrum cannot distinguish between positive and negative depths Δz of the same absolute value. Optical signals from $+|\Delta z|$ and $-|\Delta z|$ will generate the same modulation frequency on the interference spectrum and thus cannot be distinguished[11]. The resulting mirror images can, however, be avoided if the reference arm is aligned in an experimental setup (Figure 7.12) such that the entire sample to be imaged is either positive or negative in Δz. In retinal imaging, the $\Delta z = 0$ position is chosen to be in the vitreous or above the retinal surface so that all retinal structures can be imaged without mirror image artifacts. To specifically image the sub-retinal layers at

11) The optical signals can be distinguished by different dispersion characteristics if, like in the ZEISS Cirrus HD-OCT, dispersion is not matched between reference and sample arm (Section 7.4.3).

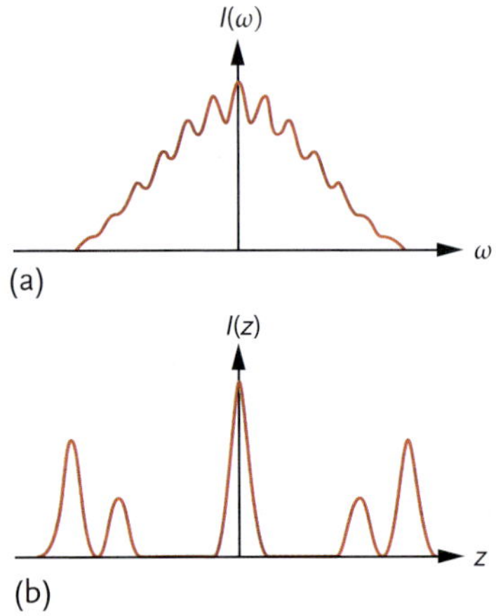

Figure 7.11 Typical detected signal of an FD-OCT. (a) Measured intensity spectrum $I(\omega)$. (b) Fast Fourier Transformation of the spectrum corresponds to the spatial intensity distribution.

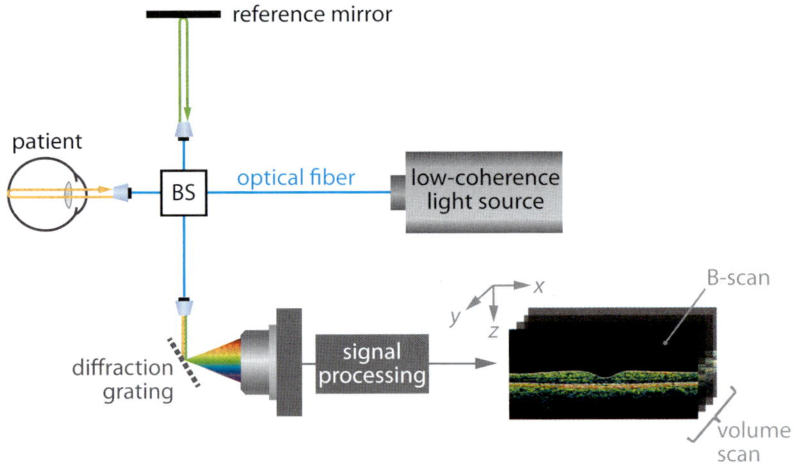

Figure 7.12 Schematic setup of an FD-OCT. In contrast to TD-OCT, it is possible to directly retrieve an A-scan (1D cross-sectional image) from one measurement. To obtain a B-scan, multiple A-scans are combined. A 3D volume eventually consists of a sequence of adjacent B-scans.

higher sensitivity (see comment on signal drop-off above), the zero-delay position is sometimes moved deep into the choroid, which results in so-called *enhanced depth imaging* (EDI).

7.3.5
Swept-Source OCT

Swept-source OCT (SS-OCT), also referred to as *wavelength-tuning OCT* or *time-encoded frequency-domain OCT*, is a variant of FD-OCT. However, in contrast to the FD-OCT described in Section 7.3.4, the signal is not generated by a broadband light source and a spectrometer, but by a single photodetector and a light source with rapidly tunable wavelength. The functional principle of SS-OCT can be understood from a similar picture which we used to illustrate FD-OCT. We have redrawn Figure 7.10 into Figure 7.13 and have replaced the spectrometer and CCD with a standard photodiode. Instead of a broadband light source, a tunable laser with a frequency sweep indicated by a "saw-tooth" frequency profile over time is used. In Figure 7.13, the reference (green) and backscattering signal (orange/blue) are in-

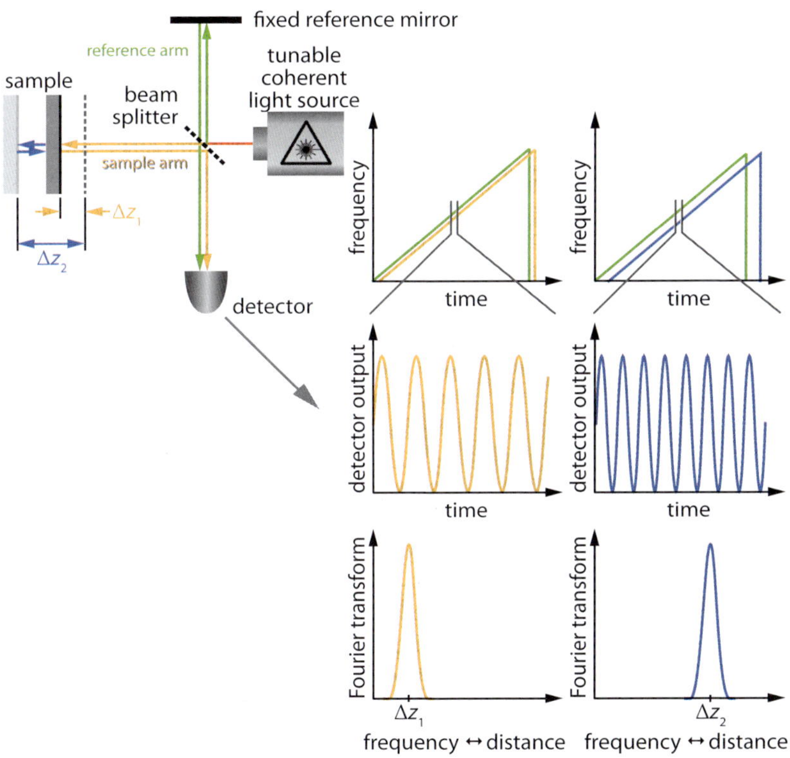

Figure 7.13 Functional principle of SS-OCT. For SS-OCT, a tunable coherent light source is used. Hence, in contrast to FD-OCT, no array detector is required. Similar to the FD-OCT setup, the reference mirror is spatially fixed. The depth information about the sample is retrieved by rapidly sweeping the wavelength of a coherent light source. Note that the time-scales of the frequency-time and detector output-time diagrams are different and not to scale. As an example, for a total 30 THz bandwidth sweep at 100 kHz (corresponding to 100 nm at 1 μm wavelength), a delay of $\Delta z = 1$ mm results in a 20 MHz frequency shift only. Adapted from [5].

dicated by the time shifted "saw-tooth" curve representations. The signals in the sample and reference arms have a delay determined by the path length difference (depth of tissue structure). Due to swept light frequency, the echoes in the sample beam will have a different frequency offset than in the reference beam. The interference of both signals leads to a modulation (beat) in intensity at a frequency given by this frequency offset. The modulated frequency spectrum, where the modulation frequency depends on Δz, is measured by the detector as a function of time. Via Fourier transform this time-dependent modulation then relates to spatial (axial) sample information. As in FD-OCT, it is the change of the phase difference as a function of light frequency (wavelength) that encodes the spatial (delay) information.

Equations (7.4) and (7.5) in Section 7.3.1 can be rewritten to describe the signal measured by the photodiode according to the arrangement shown in Figure 7.13. For this purpose, we take into account that Δz is fixed and $\lambda(t)$ varies in time. We obtain

$$I(t) = I_S + I_R + 2\sqrt{I_S I_R}\cos(\Delta\xi(t)) , \tag{7.19}$$

$$\Delta\xi(t) = 2\pi\left(\frac{2n_S\Delta z}{\lambda(t)}\right) . \tag{7.20}$$

The frequency of the photodiode signal is given by the phase change over time

$$f = \frac{\mathrm{d}(\Delta\xi(t))}{\mathrm{d}t} = 4\pi n_S \Delta z\left(\frac{\mathrm{d}}{\mathrm{d}t}\frac{1}{\lambda}\right) = \frac{4\pi n_S \Delta z}{c}\frac{\mathrm{d}\nu}{\mathrm{d}t} . \tag{7.21}$$

Thus, the frequency of the photodiode signal is directly proportional to the path difference Δz and the sweep rate of the frequency $\mathrm{d}\nu/\mathrm{d}t$. When we assume a constant sweep rate (as for example, in a "saw-tooth" profile), Δz can easily be obtained by Fourier transformation of the recorded signal.[12)]

Advantages of SS-OCT:

- The main advantage of the SS-OCT relative to TD-OCT is the simultaneous measurement of the entire A-scan of the sample without any moving parts, as in FD-OCT.
- In contrast to FD-OCT, no spectrometer or CCD/CMOS camera is required, which also eliminates the problem of signal drop-off due to limited spectral resolution discussed in Section 7.3.4. The use of high speed A/D converters allows another speed and sensitivity improvement (up to a factor of 5 has been reported). In SS-OCT, sensitivity and measurement speed are in principle not higher than for FD-OCT. However, for technical reasons, the scan speed is currently higher, since the sweep rate of state-of-the-art SS-OCT laser sources is higher than the readout speed of used camera detectors.

12) The easiest way to visualize the analogy is to think of the measurement of SS-OCT as a digitization of the photodiode signal by an analog-digital (A/D) converter. The digitized recording is completely equivalent to the spectrum of Figure 7.10.

- From Eq. (7.21), we can see that the maximum scan depth only depends on the bandwidth of the A/D converter and the inverse of the sweep rate. An SS-OCT system can thus be tuned to deliver either high-speed/high-resolution images (e.g., of the retina) or ultrawide range imaging (e.g., full eye imaging of anterior and posterior segments in one scan; Problem P7.3).

Current challenges of SS-OCT: The complexity and cost of the tunable laser source are currently the major challenges of SS-OCT. Although SS-OCT was already demonstrated in 1997 [37], the tunable laser sources in those days were mostly external-cavity semiconductor lasers (Section B.5.2) with relatively low sweep rates. In 2010, this limitation was overcome by frequency-domain mode-locked lasers [38, 39] and, more recently, by microelectromechanical systems (MEMS) in conjunction with vertical cavity surface emitting lasers (VCSEL) [40]. As the description of such lasers for SS-OCT is beyond the scope of the book, please refer to specialized literature such as [40, 41] and some comments on prospects in Section 7.8. Currently, operation at 1060 and 1300 nm has been reported [41].

7.3.6 Overview and Comparison of OCT Systems

In the literature, there has been somewhat inconsistent use of the terminology for "frequency-domain OCT". Some authors prefer to use the term "spectral-domain OCT", which can be confusing if we speak about spectroscopic OCT (Section 7.8). Another part of the literature uses the term "Fourier-domain OCT". Again, we believe that this is somewhat confusing, as both the time and the frequency are Fourier domains with respect to each other. We have thus consistently employed the terms "time domain" and "frequency domain" throughout the book, depending on whether raw data is acquired as a function of time (axial scan) or frequency (wavelength) of low-coherence light. In this sense, any SS-OCT is in fact an FD-OCT, because the signal, although varying with time, is acquired as a function of the swept light frequency.

Another useful criterion for categorization of the currently available OCT systems is by the way that 2D scanning and 3D volume data acquisition is achieved [42]. Here, in addition to the time- and frequency-domain differentiation, the information encoding principle is taken into account. Either temporal or spatial encoding can be used to derive the structural information in the three dimensions x, y (transversal), and z (axial) from a sample. For example, in FD-OCT the optical frequency components are captured either in a sequence, that is, time encoded (te) ("swept-source OCT"), or simultaneously (parallel) by spatial encoding (se) with a dispersive element ("frequency-domain OCT", "Fourier-transform OCT", "spectral-domain OCT"). TD-OCTs use either sequential scanning (te) or parallel imaging based on linear superposition of light fields along the z direction (se). For both TD- and FD/SS-OCT systems, one can have either parallel or point-

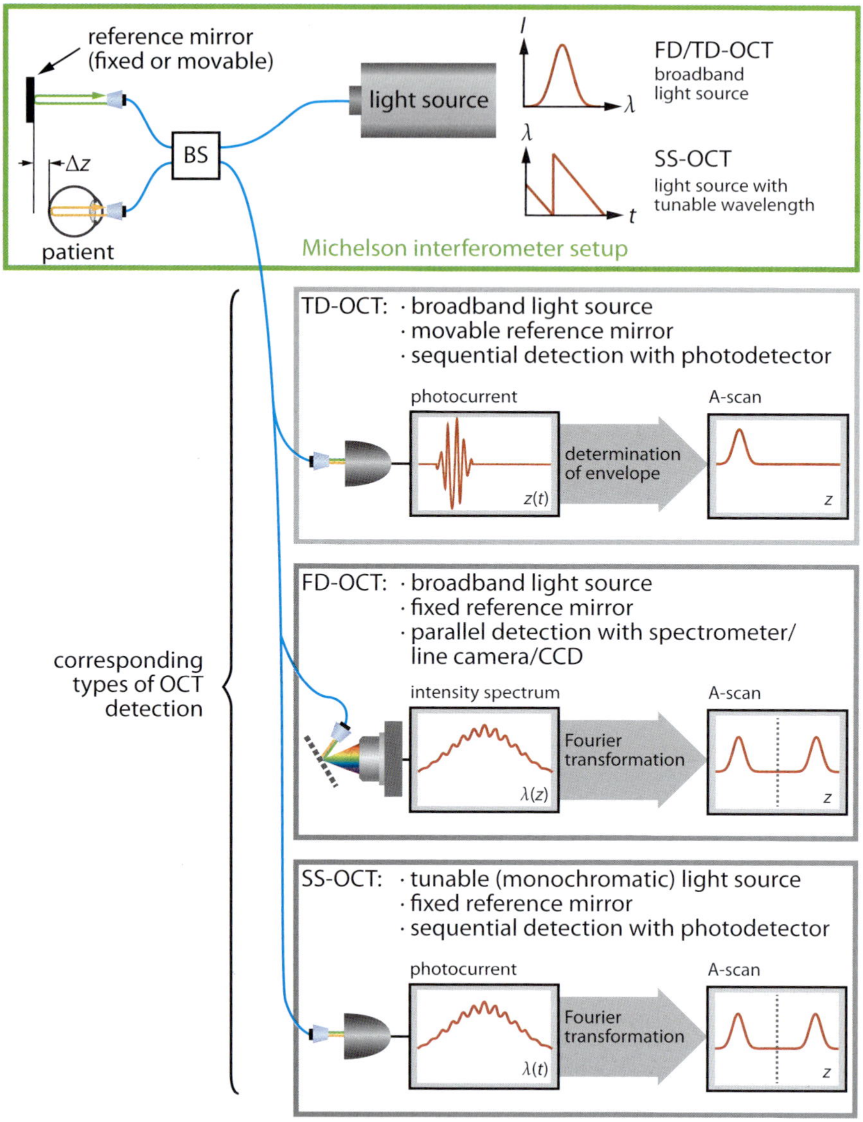

Figure 7.14 Overview of available OCT techniques.

scanning systems to retrieve the transverse/lateral information. Further details can be found in Table 7.3.

We can thus distinguish four variants of OCT (Figures 7.14 and 7.15, Table 7.3):

Table 7.3 Overview on current OCT systems. Adapted from [42].

	Time domain		**Frequency Domain**	
Used terminology	Scanning TD-OCT		Spectral-domain OCT, Fourier-domain OCT	swept-source OCT
Encoding	time	spatial	spatial	time
Alternative name (Figure 7.14)	time-encoded TD-OCT	spatially-encoded TD-OCT	spatially-encoded FD-OCT	time-encoded FD-OCT
z scan	sequential in time	parallel	parallel	sequential in time
Type of *z* scan	moving reference mirror, rapid scanning optical delay line (RSOD)	no moving part	no moving part	no moving part
Light source	superluminescent diode, low-coherence laser	superluminescent diode, low-coherence laser	superluminescent diode, low-coherence laser	tunable wavelength light source
Detector	diode, avalanche photodiode	CCD, CMOS, diode line array	prism or grating and line detector	diode, avalanche photodiode
Interferometer	beam splitter	expanded beam	beam splitter	beam splitter
A-scans/s	4000	4000	25 000–300 000	100 000–1 000 000
Voxels/s	10 000–100 000	n.a.	20–40 million	100–2500 million
Complexity of full-field 3D volume scanning realization	relatively easy through *en-face* OCT, but slow	n.a.	difficult through parallel spectrometers (problematic matrix detection)	relatively easy through *en-face* OCT, fast
Advantages	simple components	complex	extremely fast, high SNR, no moving parts	extremely fast, high SNR, no moving parts
Challenges	slow, low SNR	low SNR, complex	signal processing, dispersive elements, and detector	signal processing, laser source
Section	7.3.3		7.3.4	7.3.5

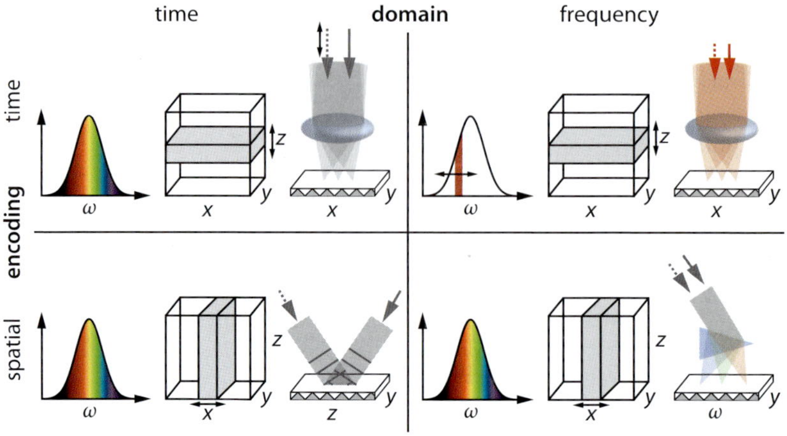

Figure 7.15 Comparison of "full field"-OCT technologies using various 2D arrays for acquisition of a 3D volume by accumulating 2D images. Insets from left to right depict the characteristics of used light, the sample volume from which information is extracted (*z* corresponds to axial depth coordinate), and the measuring arrangement of the reference arm (represented by dashed arrows) and sample arm (represented by solid arrows) incident on the detection unit. Gray arrows indicate scanning with a broadband light beam, and red arrows indicate scanning with (frequency-swept) monochromatic light. Adapted from [42].

1. teTD (time-encoded time-domain) OCT,
2. seTD (spatially-encoded time-domain) OCT based on linear superposition of sample and reference fields,[13)]
3. teFD (time-encoded frequency-domain) imaging,[14)] and
4. seFD (spatially-encoded frequency-domain) OCT.

In ophthalmic applications, we find different versions of teTD, teFD, and seFD imaging.

7.4 Elements of OCT Theory

Let us consider the elements of the theoretical formalism of OCT in some more detail. For a more thorough theoretical treatment, please refer to [1, 8, 9]. In the following, the intention is to gain a physical understanding of the OCT signal in both TD and FD-OCT by providing the cornerstones of OCT theory. Here, we will largely follow the approach taken in [10]. In addition, we will also derive important conclusions with regard to axial resolution, signal processing, SNR, and measuring artifacts.

13) The depth is encoded by spatial distribution of different delays. This method has only been demonstrated with a linear array [25].

14) The full volume is acquired successively at different optical frequencies.

7.4.1
Theory of Time-Domain OCT – Axial Resolution

We express the sample and reference light fields as functions of frequency after their passage through the respective arms of the interferometer as

$$\psi_{\mathrm{R}}(\omega) = \psi_{\mathrm{R},0}(\omega)\mathrm{e}^{2\mathrm{i}k_{\mathrm{R}}(\omega)z_{\mathrm{R}}-\mathrm{i}\omega t} , \tag{7.22}$$

$$\psi_{\mathrm{S}}(\omega) = \psi_{\mathrm{S},0}(\omega)\mathrm{e}^{2\mathrm{i}k_{\mathrm{S}}(\omega)z_{\mathrm{S}}-\mathrm{i}\omega t} . \tag{7.23}$$

The factor 2 in the exponent accounts for the round-trip passage through each arm. k_{R} and k_{S} are wave numbers as defined in Eq. (A53). z_{R} and z_{S} denote the lengths of the reference and sample arm, respectively. $\psi_{\mathrm{R},0}(\omega)$ and $\psi_{\mathrm{S},0}(\omega)$ represent the amplitude spectral distribution of the low-coherence light (see also Eq. (A108)). For the sake of simplicity, let us first assume that there is only a single 100% reflecting surface in the sample beam path.[15] According to Eq. (A98), the total backscattered intensity signal for the backscattering surface at the detector can then be obtained by linear superposition as[16]

$$\begin{aligned} I(\omega) &= (\psi_{\mathrm{R}}(\omega) + \psi_{\mathrm{S}}(\omega))\,(\psi_{\mathrm{R}}(\omega) + \psi_{\mathrm{S}}(\omega))^* \\ &= |\psi_{\mathrm{R}}(\omega)|^2 + |\psi_{\mathrm{S}}(\omega)|^2 + 2\mathrm{Re}\left(\psi_{\mathrm{R}}(\omega)\psi_{\mathrm{S}}^*(\omega)\right) . \end{aligned} \tag{7.24}$$

If we now integrate over all spectral components of the interference signal, both $\int |\psi_{\mathrm{R}}(\omega)|^2 \mathrm{d}\omega$ and $\int |\psi_{\mathrm{S}}(\omega)|^2 \mathrm{d}\omega$ yield time-independent terms which do not depend on the mirror position. The terms thus only contain amplitude information about each interferometer arm. We are, however, interested in the coherence-gated signal formed by the backscattered sample light and "gated" by the reference light. We thus only have to consider the interference term which represents the measurable signal I_{TD} of the detector in the Michelson interferometer for TD-OCT, that is,

$$I_{\mathrm{TD}} = 2\mathrm{Re}\left(\int_{-\infty}^{\infty} \psi_{\mathrm{R}}(\omega)\psi_{\mathrm{S}}^*(\omega)\mathrm{d}\omega\right) . \tag{7.25}$$

Inserting Eqs. (7.22) and (7.23) yields

$$I_{\mathrm{TD}}(\Delta\xi) = 2\mathrm{Re}\left(\int_{-\infty}^{\infty} \psi_{\mathrm{R},0}(\omega)\psi_{\mathrm{S},0}^*(\omega)\mathrm{e}^{-\mathrm{i}\Delta\xi(\omega)}\mathrm{d}\omega\right) \tag{7.26}$$

in which the phase difference is given by

$$\Delta\xi(\omega) = 2\left(k_{\mathrm{S}}(\omega)z_{\mathrm{S}} - k_{\mathrm{R}}(\omega)z_{\mathrm{R}}\right) . \tag{7.27}$$

15) In Problem P7.7, a straightforward extension to multiple scattering surfaces with different backscattering coefficients will be considered.

16) The stars in the following equations denote the complex conjugate of the corresponding function.

Equation (7.26) reveals that the different optical path lengths of the Michelson interferometer arms determine the interference signal.

Next, we define the spectral cross-correlation signal $S(\omega)$ of the sample and reference beams by $\psi_{R,0}(\omega)\psi^*_{S,0}(\omega) = R_S S(\omega)$ in which R_S is the average frequency-independent reflection/backscatter coefficient of the sample arm. This term contains all information about the backscattering. The interference term now reads

$$I_{TD}(\Delta\xi) = R_S \mathrm{Re}\left(\int_{-\infty}^{\infty} S(\omega) e^{-i\Delta\xi(\omega)} d\omega\right) . \tag{7.28}$$

We can expand the wave numbers $k_S(\omega)$ and $k_R(\omega)$ as functions of ω_0 around the center frequency ω_0 into a Taylor series which yields

$$k_S(\omega) = k_R(\omega) = k(\omega) \approx k(\omega_0) + \left.\frac{dk}{d\omega}\right|_{\omega_0} (\omega - \omega_0) . \tag{7.29}$$

We also consider the group velocity[17] of a wave packet (see also Eq. (A101)), which is, because of the duality in time and frequency, both a packet in time and a packet in frequency space. It is given by

$$c_g = \left.\frac{d\omega}{dk}\right|_{\omega_0} . \tag{7.30}$$

Similarly, we can write the phase velocity as

$$c_p = \frac{\omega}{k(\omega_0)} = \left.\frac{\omega}{k}\right|_{\omega_0} . \tag{7.31}$$

With Eqs. (7.30) and (7.31), the phase difference can be written as

$$\Delta\xi \approx 2\Delta z \left(\frac{\omega_0}{c_p} + \frac{\omega - \omega_0}{c_g(\omega_0)}\right) = \omega_0 \Delta t_p + (\omega - \omega_0)\Delta t_g . \tag{7.32}$$

$\Delta z = z_R - z_S$ is the actual geometric path length difference between the two arms. $\Delta t_p = 2\Delta z / c_p$ can be understood as the round-trip *phase* delay between the two interferometer arms, and analogously $\Delta t_g = 2\Delta z / c_g$ as the corresponding round-trip *group* delay (Problem P7.4).

When we insert Eq. (7.32) into Eq. (7.28), it follows that

$$I_{TD}(\Delta z) = R_S \mathrm{Re}\left(e^{-i\omega_0 \Delta t_p} \int_{-\infty}^{\infty} S(\omega) e^{-i(\omega-\omega_0)\Delta t_g} d\omega\right) . \tag{7.33}$$

17) It is not surprising that we encounter the group velocity here. Whenever we deal with polychromatic light (short pulses, wave packets, broadband, or any other low-coherence light), we have to take the group velocity into account. In Section 7.4.3, we will see that the effect of dispersion on the group velocity (Section 9.5.3) has a significant influence on the axial resolution of OCT devices.

The functional dependence $I_{TD}(\Delta z)$ has to be understood as a functional dependence $I_{TD}(\Delta t_g(\Delta z), \Delta t_p(\Delta z))$. For further simplification, we assume that the spectrum $S(\omega)$ is symmetrical around the center frequency ω_0, as is the case for a Gaussian spectrum. The integral is then real and the prefactor becomes a cosine term (Problem P7.5) so that the interference term now reads

$$I_{TD}(\Delta z) = R_S \cos(\omega_0 \Delta t_p) \int_{-\infty}^{\infty} S(\omega) e^{-i(\omega-\omega_0)\Delta t_g} d\omega \ . \tag{7.34}$$

The prefactor $\cos(\omega_0 \Delta t_p)$ corresponds to the oscillation of a carrier wave, while the second term with the integral describes the envelope and thus the width of the coherence gate. As we can see, the envelope depends on both the spectral distribution and the round-trip *group* delay of light.

For a Gaussian spectral distribution

$$S(\omega) = \frac{1}{\sqrt{2\pi \Delta\omega}} \exp\left(-\frac{(\omega-\omega_0)^2}{2(\Delta\omega)^2}\right) \tag{7.35}$$

with bandwidth $\Delta\omega$ (FWHM) related to the spectral bandwidth $\Delta\lambda$ (FWHM) via

$$\Delta\omega = \frac{2\pi c}{2\sqrt{2\ln 2}} \frac{\Delta\lambda}{\lambda_0^2} \ , \tag{7.36}$$

the integral in Eq. (7.34) can be calculated exactly (Problem P7.5). In this case, we obtain for the OCT time-domain signal[18)]

$$I_{TD}(\Delta z) \propto \cos(\omega_0 \Delta t_p) \exp\left(-\frac{1}{2}(\Delta t_g)^2 (\Delta\omega)^2\right) \ . \tag{7.37}$$

Equation (7.37) is very instructive and graphically displayed in Figure 7.16, which shows the signal at the detector prior to demodulation for a wavelength of $\lambda = 1\ \mu m$ and a coherence length of $L_c = 20\ \mu m$. As expected, the TD-OCT signal is a wave packet with a carrier wave ω_0 (which experiences some phase delay Δt_p) and an envelope whose width is determined by the bandwidth $\Delta\omega$.

So far, we have assumed that the group velocity is equal in both arms. In fact, this is only valid in the case of a free space interferometer and one backscattering surface in the sample arm. However, in the case of more complex samples and if other optical materials and components are present in one of the arms, this assumption no longer holds. The resulting effects of such a group velocity mismatch in both arms shall be discussed in more detail in Section 7.4.3.

For now, we will restrict ourselves to the case of free space (i.e., a classical Michelson interferometer) and can simplify Eq. (7.37) further, which allows us to create a link between resolution and the coherence length of the used light. In free space,

18) For the sake of simplicity, we have dropped all the constant prefactors in Eq. (7.37) to focus on the relevant parameters of the signal.

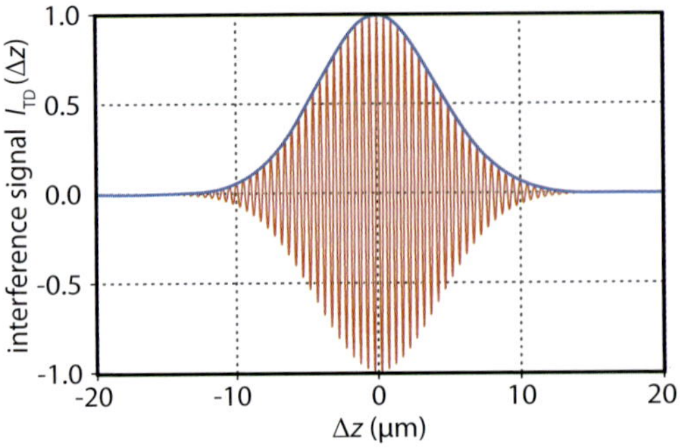

Figure 7.16 Detector signal (before low band pass) for a wavelength of 1 μm and a bandwidth of 44 nm corresponding to a coherence length of 20 μm. Please note that the detector is a low-pass filter and will only record the envelope (blue curve) and not the carrier wave (red curve).

we have $\Delta t_p = 2\Delta z/c$ and we also would like to express Eq. (7.37) in terms of the coherence length Eq. (7.9). We thus obtain

$$I_{TD}(\Delta z) \propto \cos\left(4\pi \frac{\Delta z}{\lambda_0}\right) \exp\left(-\frac{(\Delta z)^2}{2\sigma_z^2}\right) \tag{7.38}$$

in which $\sigma_z = c/(2\Delta\omega)$.

The axial resolution of TD-OCT can be defined as the FWHM of the $I_{TD}(\Delta z)$ envelope (Figure 7.16). With Eq. (7.15) and the expression for the coherence length Eq. (7.9), we directly obtain from Eq. (7.38) and with $\Delta\omega$ from Eq. (7.36)

$$\frac{1}{2} = \exp\left(-\frac{(\Delta z_{OCT,FWHM}/2)^2}{2\sigma_z^2}\right), \tag{7.39}$$

$$\Delta z_{OCT,FWHM} = 2\sqrt{2\ln 2}\sigma_z = \frac{\sqrt{2\ln 2}c}{\Delta\omega} = \frac{2\ln 2}{\pi}\frac{\lambda_0^2}{\Delta\lambda} = \frac{L_c}{2}. \tag{7.40}$$

The axial resolution of OCT (in free space) corresponds to about half the coherence length of the light source employed, which coincides with Eq. (7.16) and our discussion in Section 7.3.2. For a material with a refractive index n, the axial resolution is reduced by a factor of $1/n$. In Figure 7.17, the dependence of $\Delta z_{OCT,FWHM}$ on the bandwidth $\Delta\lambda$ is plotted for various center wavelengths λ_0.

7.4.2
Theory of Frequency-Domain OCT

In general, an equivalent frequency-domain method exists for each time-domain measurement method. As the autocorrelation function of a signal and its frequency spectrum are linked via the Wiener–Khinchin theorem (Section A.2.4.2), an analogous relationship also applies in optics for the optical spectrum and the wave function. This means that we can expand our TD-OCT theory to an appropriate FD-OCT description, which follows the one given in [10] starting from Eqs. (7.22)

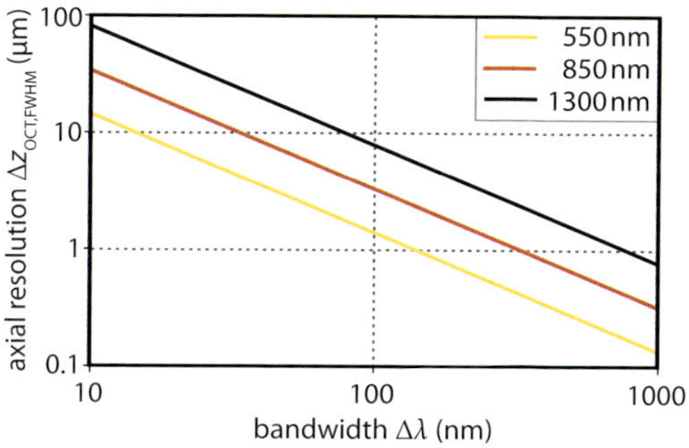

Figure 7.17 Axial resolution $\Delta z_{OCT,FWHM}$ of a TD-OCT for various bandwidths $\Delta\lambda$ and center wavelengths λ_0, that is, 550 nm (yellow), 850 nm (red), and 1300 nm (black).

and (7.23). As the FD-OCT setup has the advantage of no moving parts, the reference path length is fixed, and the diode detector is replaced by a spectrometer. In order to extract the structural depth information, the measured interference spectrum must be Fourier transformed. Since all depth information along z_S is now gathered in one spectral measurement, we must include in our theoretical model all "backscatterers" (reflectors) in the sample arm by means of integration. In FD-OCT, the wave functions in both arms thus read

$$\psi_R(\omega) = \psi_{R,0}(\omega) R_R e^{2ik_R(\omega)z_R - i\omega t} , \tag{7.41}$$

$$\psi_S(\omega) = \psi_{S,0}(\omega) \int_{-\infty}^{\infty} R_S(z_S) e^{2ik_S(\omega)z_S - i\omega t} dz_S . \tag{7.42}$$

R_R denotes the reflectance in the reference arm (assumed to be independent of frequency), and $R_S(z_S)$ is the reflectance in the sample arm (again assumed to be independent of frequency, but now dependent on the position z_S of the backscatterer). We describe the backscattering planes (indicated by i) in the sample by a sequence of planes with backscattering coefficients $R_S(z_{Si})$ at position z_{Si}, that is, $R_S(z_S) = \sum_i R_S(z_{Si})\delta(z_S - z_{Si})$. For the sake of simplicity, we will disregard dispersion and write

$$k = \frac{2\pi}{\lambda} = \frac{\omega}{c} = \frac{k_R}{n_R} = \frac{k_S}{n_S} . \tag{7.43}$$

We also assume the reference arm to be free of optical media ($n_R = 1$) and rewrite the expression for the interference signal this time as a function of the wave number $k = 2\pi/\lambda$, as we want to work in the frequency domain, which means with the wave number, wavelength, or frequency. It follows that

$$I_{FD}(k) = |\psi_R(kc) + \psi_S(kc)|^2 . \tag{7.44}$$

From this, we obtain an expression for the spectral density function of the interference signal at the spectrometer as a function of the wave number by inserting

Eqs. (7.41) and (7.42) into Eq. (7.44), that is,

$$I_{\mathrm{FD}}(k) = \underbrace{S(k)\,R_{\mathrm{R}}^2}_{1} + \underbrace{2S(k)\,R_{\mathrm{R}} \int_{-\infty}^{\infty} R_{\mathrm{S}}(z_{\mathrm{S}}) \cos\left(2k(n_{\mathrm{S}} z_{\mathrm{S}} - z_{\mathrm{R}})\right) \mathrm{d}z_{\mathrm{S}}}_{2}$$

$$+ \underbrace{S(k) \left| \int_{-\infty}^{\infty} R_{\mathrm{S}}(z_{\mathrm{S}}) \mathrm{e}^{2ik(n_{\mathrm{S}} z_{\mathrm{S}})} \mathrm{d}z_{\mathrm{S}} \right|^2}_{3} . \tag{7.45}$$

Here, we have used the spectral density function of the light source $S(k) = |\psi_{\mathrm{S},0}(kc)|^2$. Equation (7.45) is the fundamental starting equation for the theoretical treatment of FD-OCT. Obviously, it is more complex than the corresponding equation in TD Eq. (7.33). This is not surprising if we take into account that the signal I_{FD} contains the entire depth information of the sample. The three terms have the following physical interpretation:

1. Signal from the reference arm which is also measured if the sample arm is blocked.
2. Cross-interference signal between sample and reference arms, which is only measured if light from both arms reach the detector (spectrometer).
3. Interference between signals from backscatterers at various depths z_{S}. This signal is also measured even when the reference arm is blocked.

The objective of the measurement (tomography) is to obtain the distribution function $R_{\mathrm{S}}(z_{\mathrm{S}})$ of the sample. This is most easily extracted from the cross-interference term 2 by means of an inverse Fourier transformation.[19] We can simplify the calculation by shifting the coordinate system to $z_{\mathrm{R}} = 0$ which leads to the new (shifted) distribution of backscatterers

$$\hat{R}_{\mathrm{S}}(z_{\mathrm{S}}) = \begin{cases} R_{\mathrm{S}}(z_{\mathrm{S}}) & \text{if } z_{\mathrm{S}} \geq 0\,, \\ R_{\mathrm{S}}(-z_{\mathrm{S}}) & \text{if } z_{\mathrm{S}} \leq 0\,. \end{cases} \tag{7.46}$$

$\hat{R}_{\mathrm{S}}$ is an even function of z_{S} which will later simplify the integration considerably. We can now rewrite Eq. (7.45) (see also Problem P7.6) as

$$I_{\mathrm{FD}} = S(k)\,R_{\mathrm{R}}^2 + S(k)\,R_{\mathrm{R}} \int_{-\infty}^{\infty} \hat{R}_{\mathrm{S}}(z_{\mathrm{S}}) \mathrm{e}^{2ik(n_{\mathrm{S}} z_{\mathrm{S}})} \mathrm{d}z_{\mathrm{S}}$$

$$+ \frac{1}{4} S(k) \left| \int_{-\infty}^{\infty} \hat{R}_{\mathrm{S}}(z_{\mathrm{S}}) \mathrm{e}^{2ik(n_{\mathrm{S}} z_{\mathrm{S}})} \mathrm{d}z_{\mathrm{S}} \right|^2 . \tag{7.47}$$

19) In this context, it is important to take the Nyquist theorem [43] into account. In our context, this requires that the length of the "imaged reference arm" is situated outside the sample (see also Section 7.3.4, Figure 7.12, [9, 10], and Problem P7.6).

With refractive index n_S, normalized coordinates $\hat{z}_S = z_S/(2n_S)$, and the Fourier transform (A104), we find

$$I_{FD}(k) = S(k)\left[R_R^2 + \frac{R_R}{2n_S}\mathcal{F}_k\left\{\hat{R}_S\left(\frac{\hat{z}_S}{2n_S}\right)\right\} + \frac{1}{16n_S^2}\left|\mathcal{F}_k\left\{\hat{R}_S\left(\frac{\hat{z}_S}{2n_S}\right)\right\}\right|^2\right] \tag{7.48}$$

in which $\mathcal{F}$ is the Fourier transform between the $\hat{z}_S$ and k domain. $\mathcal{F}_k$ denotes the Fourier transform direction into the k domain.[20)]

Again, we have three terms, as explained in the discussion of Eq. (7.45). As a reminder, $I_{FD}(k)$ or, equivalently, $I_{FD}(\lambda)$ correspond to the *measured* signal in an FD-OCT system. From this measurement, we then have to retrieve the structural information $\hat{R}_S(\hat{z}_S/(2n_S))$ or $R_S(z_S)$, respectively. Applying the inverse Fourier transform $\mathcal{F}^{-1}$ (A105) to the measured spectrum (7.48) leads to

$$\begin{aligned}\mathcal{F}_{\hat{z}_S}^{-1}\{I_{FD}(k)\} = \mathcal{F}_{\hat{z}_S}^{-1}\{S(k)\} \otimes &\left\{\frac{R_R^2}{2n_S}\delta\left(\frac{\hat{z}_S}{2n_S}\right) + \frac{R_R}{2n_S}\hat{R}_S\left(\frac{\hat{z}_S}{2n_S}\right)\right.\\ &\left.+\frac{1}{16n_S^2}\mathcal{G}\left[\hat{R}_S\left(\frac{\hat{z}_S}{2n_S}\right)\right]\right\} .\end{aligned} \tag{7.49}$$

The symbol "⊗" describes the convolution (see Info Box 5.3 in Section 5.4.1.2) between two functions and $\mathcal{G}$ the autocorrelation function according to Eq. (A111). The Wiener–Khinchin theorem (A112) was used in the derivation of Eq. (7.49). In order to return to the measurable quantity z_S, we replace $\hat{z}_S = z_S 2n_S$ and obtain

$$\begin{aligned}\mathcal{F}_{\hat{z}_S}^{-1}\{I_{FD}(k)\} &= \mathcal{F}_{\hat{z}_S}^{-1}\{S(k)\} \otimes \left\{\frac{R_R^2}{2n_S}\delta(z_S) + \frac{R_R}{2n_S}\hat{R}_S(z_S) + \frac{1}{16n_S^2}\mathcal{G}\left[\hat{R}_S(z_S)\right]\right\}\\ &= \mathcal{F}_{\hat{z}_S}^{-1}\{S(k)\} \otimes \left\{\mathcal{T}_1 + \mathcal{T}_2\left[\hat{R}_S(z_S)\right] + \mathcal{T}_3\left[\hat{R}_S(z_S)\right]\right\} .\end{aligned} \tag{7.50}$$

After the somewhat lengthy calculation, we have finally found an equation which can be used as a basis for the evaluation of the measured spectral data $I_{FD}(k)$ in FD-OCT.

We still have to extract $\hat{R}_S$ from Eq. (7.50), which is not straightforward. The first term $\mathcal{T}_1$ in the bracket is equal to zero if, as required, the reference mirror image is situated outside the sample. Then, the evaluation problem resides with the third term and its convolution with the spectral function $S(k)$. This can be solved by performing and analyzing another measurement at which the value of $k(z_S n_S - z_R)$ is shifted by 180°. This can be achieved by varying the delay between sample and reference light. This phase shifted spectral measurement is described by a modified version of Eq. (7.48) in which the second term in the bracket now

20) The $\hat{z}_S$ and k Fourier transformation is equivalent to the time-frequency transformation, as $\hat{z}_S$ and k are linked with time and frequency via the speed of light.

has a minus sign, that is,

$$I_{\mathrm{FD}}(k) = S(k)\left[R_{\mathrm{R}}^2 - \frac{R_{\mathrm{R}}}{2n_{\mathrm{S}}}\mathcal{F}_k\left\{\hat{R}_{\mathrm{S}}\left(\frac{\hat{z}_{\mathrm{S}}}{2n_{\mathrm{S}}}\right)\right\} + \frac{1}{16n_{\mathrm{S}}^2}\left|\mathcal{F}_k\left\{\hat{R}_{\mathrm{S}}\left(\frac{\hat{z}_{\mathrm{S}}}{2n_{\mathrm{S}}}\right)\right\}\right|^2\right]. \tag{7.51}$$

For evaluation, we thus have to measure two spectra and calculate the difference of Eqs. (7.48) and (7.49), which yields

$$\Delta I_{\mathrm{FD}}(k) = S(k)\frac{R_{\mathrm{R}}}{n_{\mathrm{S}}}\mathcal{F}_k\left\{\hat{R}_{\mathrm{S}}\left(\frac{\hat{z}_{\mathrm{S}}}{2n_{\mathrm{S}}}\right)\right\}. \tag{7.52}$$

The A-scan $\hat{R}_{\mathrm{S}}$, that is, the depth profile of the backscattering sample, can be obtained from Eq. (7.52) simply by means of an inverse Fourier transform of the difference of the two spectra, divided by the excitation spectrum. We thus obtain

$$\hat{R}_{\mathrm{S}}(z) = \frac{n_{\mathrm{S}}}{R_{\mathrm{R}}}\mathcal{F}_{z_{\mathrm{S}}}^{-1}\left\{\frac{\Delta I_{\mathrm{FD}}(k)}{S(k)}\right\}, \tag{7.53}$$

which is the main expression for evaluation of interference spectra in FD-OCT. Three spectral measurements are necessary to obtain the entire A-scan, that is, two 180° phase shifted interference spectra and, for normalization, the spectrum of the low-coherence light.

In an alternative measurement approach, $\mathcal{T}_1$ and $\mathcal{T}_3$ in Eq. (7.50) are determined individually such that only $\mathcal{T}_2$ remains after calculating the difference. As shown in the discussion of Eq. (7.45), terms $\mathcal{T}_1$ and $\mathcal{T}_2$ in the measured spectrum (7.48) vanish if the reference arm is blocked. Terms $\mathcal{T}_2$ and $\mathcal{T}_3$ in Eq. (7.48) vanish if the sample arm is blocked. This allows us to determine the cross-interference term $\mathcal{T}_2$. As a consequence, we may again use Eq. (7.53) for the calculation of the depth profile. With three spectral measurements and a digital FFT, we can indeed recover the entire A-scan. The speed and sensitivity advantages compared to TD-OCT are substantial and will be considered in Section 7.4.4.

Here again as a summary, the measurement and analysis process in FD-OCT:

Method A:

1. Measurement M1 of the spectrum $I_{\mathrm{FD}}(k)$ according to Eq. (7.48).
2. Measurement M2 of the spectrum $I_{\mathrm{FD}}(k)$ according to Eq. (7.51) with 180° phase shift of sample and reference.
3. Subtraction M1 − M2 leads to the cross-correlation spectrum $\Delta I_{\mathrm{FD}}(k)$ in Eq. (7.52).
4. Measurement of the spectrum $S(k)$ and normalization.[21)]
5. Inverse Fourier transformation (7.53) which delivers the desired quantity $\hat{R}_{\mathrm{S}}(z)$.

21) If the spectrum of low-coherence light is sufficiently broad and uniform over the spectral measurement range, it may be taken as a constant.

Method B:

1. Measurement M1 of just the light from the reference arm (sample arm blocked) yields term 3 in Eq. (7.48).
2. Measurement M2 of only the light from the sample arm (reference arm blocked) yields term 1 in Eq. (7.48).
3. Measurement M3 of the entire spectrum yields Eq. (7.48).
4. Subtraction M3 – M2 – M1 leads to the cross-correlation spectrum $\Delta I_{FD}(k)$.
5. Measurement of the spectrum $S(k)$ and normalization.
6. Inverse Fourier transformation (7.53) which delivers the desired quantity $\hat{R}_S(z)$.

In Figures 7.18b,c, simulated FD-OCT spectra are shown for a simplified structure of the anterior eye segment (Figure 7.18a). We again assumed a Gaussian spectrum with $\Delta\lambda = 200$ nm (FWHM) and a center wavelength of $\lambda = 1300$ nm. The simulated spectrum shows modulations of varying depth which are caused by the different reflectances at the layer boundaries. In the enlarged view (Figure 7.18c), one can distinguish at least two different frequencies, which is consistent with the phenomenological picture we discussed in Section 7.3.4. The low frequency (long period, 3.14/mm) with high modulation depth can be attributed to the front layer reflection (air–cornea interface). The higher frequencies (shorter periods, approximately 0.25/mm) with lower modulation can be related to the deeper lying lens–vitreous interfaces. It is possible to ascertain that the described data evaluation method is very robust by superimposing a noise signal onto the synthetic spectrum (Problem P7.7).

7.4.3
Effect of Group Velocity Dispersion in OCT Systems

In the previous section, we assumed that the group velocity is equal in both arms and that there is also no group velocity dispersion (GVD). However, this will generally only hold for the free space Michelson interferometer[22]. If a sample is placed in one of the arms, the dispersive properties of the sample will have an effect on the group velocity and thus also on the resolution. This effect appears for TD- and FD-OCT systems.

The effect of GVD broadens the interferogram and consequently decreases resolution.[23] In the experimental realization of OCT, the effect of unbalanced GVD in both arms is minimized by keeping the GVD mismatch in the arms as small as

22) Even in the case of a free space Michelson interferometer, the use of a thick glass beam splitter creates different dispersive path lengths for both beams. This effect has been known since the early days of the Michelson interferometer. Compensation has been achieved by placing additional dispersive elements in the respective arm. When using beam splitter cubes, both beams always travel the same distance through the beam splitter, as it is self-compensating.

23) To be more precise, unbalanced GVD in both arms results in a larger cross-correlation time than the coherence time of the light. Please remember that the cross-correlation term contains the information and determines the resolution of the system (see Eq. (7.45)).

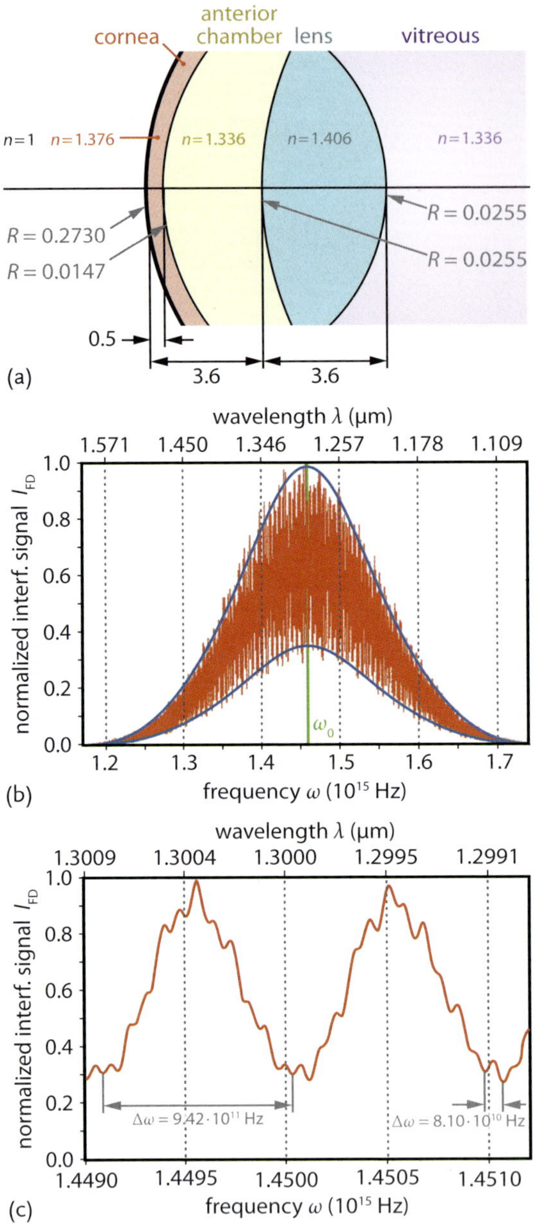

Figure 7.18 (a) Simplified cross-sectional model of the anterior eye segment which is used for the analysis of detected FD-OCT signals. The distances are given in millimeters. *n* denotes the refractive index of the corresponding ocular medium, and *R* is the reflectance at each surface. (b) Simulated interference spectrum (coarse view). (c) Simulated interference spectrum. Detailed view around the center frequency ω_0 (center wavelength λ_0). See text as well as Problems P7.7 and P7.8 for further details.

possible. This can be done by introducing a medium with a similar GVD as the sample to the reference arm.

For Gaussian spectra, the effect of GVD can be estimated relatively well (Problem P7.9) according the following procedure:

1. We expand the wave number $k(\omega)$ into a Taylor series up to the quadratic term[24)]

$$k(\omega) \approx k(\omega_0) + \left.\frac{dk}{d\omega}\right|_{\omega_0} (\omega - \omega_0) + \frac{1}{2}\left.\frac{d^2k}{d\omega^2}\right|_{\omega_0} (\omega - \omega_0)^2 . \qquad (7.54)$$

   The coefficient

$$k'' = \left.\frac{d^2k}{d\omega^2}\right|_{\omega_0} = \frac{d}{d\omega}\left(\frac{1}{c_g}\right) \qquad (7.55)$$

   describes the GVD and has the unit s^2/m.
2. We calculate the phase difference $\Delta\xi = 2k_S(\omega)z_S - 2k_R(\omega)z_R$ between sample and reference arms according to Eq. (7.27) with the expansion to the quadratic term.
3. We replace the obtained phase difference and the Gaussian spectrum of the source

$$S(\omega) = \frac{1}{\sqrt{2\pi}\Delta\omega} \exp\left(-\frac{(\omega - \omega_0)^2}{2(\Delta\omega)^2}\right) \qquad (7.56)$$

   in Eq. (7.28).
4. We follow the same steps from Eqs. (7.28)–(7.38) to determine a modified expression for the axial resolution which shows the broadening of the axial point-spread function (PSF)[25)] (Section A.1.5). The broadening for a dispersive element of thickness L_g and with a GVD coefficient k'' is given by (Problem P7.9)

$$\begin{aligned} \delta z_{\text{OCT,FWHM}} &= \Delta z'_{\text{OCT,FWHM}} - \Delta z_{\text{OCT,FWHM}} \\ &= 4c\frac{\sqrt{2\ln 2}}{\Delta\omega}\left(\sqrt{1 + \left(\frac{k''(\Delta\omega)^2 L_g}{4\ln 2}\right)^2} - 1\right) . \end{aligned} \qquad (7.57)$$

As an example, Figure 7.19 shows the calculated broadening of the axial PSF for a block of glass (BK7) with a thickness of 10 cm (or a glass fiber equivalent of the same length) in the sample arm. The low-coherence light is assumed to have a fixed coherence length $L_c \approx 7.5$ μm. Around a center wavelength of 1.3 μm, where the GVD vanishes, the broadening $\delta z_{\text{OCT,FWHM}}$ is negligible. In the visible spectral range, however, it can assume significant values.

7.4.4 Sensitivity and Signal-To-Noise Ratio in TD-OCT and FD-OCT

The most striking advantage of FD-OCT over TD-OCT is that in one measurement all backscattered light is collected to obtain (through Fourier transformation) an

24) In the approximation of Eq. (7.29) in Section 7.4.1, we have only done this for the linear term.
25) Equation (7.38) is often called the conditional equation for the axial point-spread function.

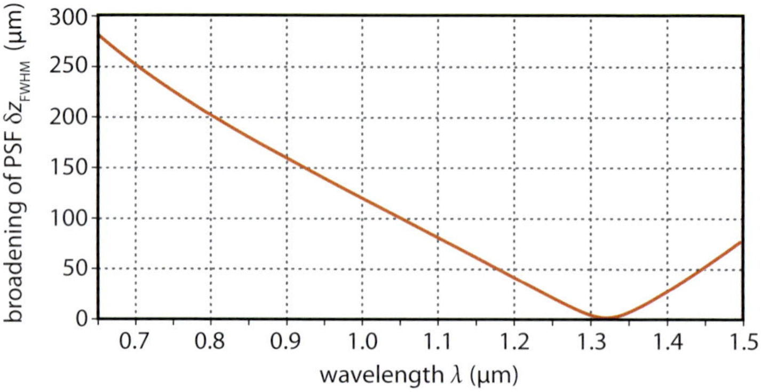

Figure 7.19 Broadening of the axial PSF for various center wavelengths according to Eq. (7.57) in an OCT setup with constant coherence length of $L_c \approx 7.5\,\mu m$ for which in one of the interferometer arms a BK7 glass block has been introduced.

entire axial scan. In TD-OCT, the backscattered light is coherence-gated and only a small fraction of light is used. This makes TD-OCT inherently slower than FD-OCT, or a loss in sensitivity must be accepted if one increases the scan speed. In contrast, in FD-OCT, we may integrate over a longer measuring time per CCD/CMOS pixel of the spectrometer detector, which yields a significantly higher sensitivity compared to TD-OCT.

The SNR for TD-OCT can be simply derived from the SNR of a photodiode in heterodyne mode (if the detection is shot-noise limited) which is given by

$$\mathrm{SNR}_{\mathrm{TD}} = \frac{\eta \gamma_{\mathrm{S}} P_0}{h \nu_0 B} \, . \tag{7.58}$$

η denotes the quantum efficiency of the detector, $\nu_0 = c/\lambda_0$ the center frequency, P_0 the input power of light (in watts), and B the electronic bandwidth of the detector. h is Planck's constant (see Info Box B.1 in Appendix B) and γ_{S} the effective beamsplitting and backscattering coefficient in the sample arm. For FD-OCT, the SNR can be derived as [1]

$$\mathrm{SNR}_{\mathrm{FD}} = \frac{\eta \gamma_{\mathrm{S}} N \tau_{\mathrm{int}} P_0}{h \nu_0} \tag{7.59}$$

with the integration time in the measurement τ_{int} and the number of channels in the spectrometer N.[26] When we take the ratio of both, the SNR advantage amounts to $N \tau_{\mathrm{int}} B$. If we want to compare both OCT modalities under equal conditions, we have to set the electronic bandwidth of the TD photodiode detector as the inverse of the integration time in the FD measurement, that is, $B = 1/\tau_{\mathrm{int}}$. As a consequence,

26) Equations (7.58) and (7.59) suggest an improvement of sensitivity by a factor of N. Due to the nature of the FFT, FD-OCT produces redundant data for positive and negative frequencies. Thus, the real improvement is $N/2$. However, even a factor of $N/2$ is too optimistic, as this would assume that the power is equal in each spectral channel. A more realistic Gaussian spectrum would reduce the sensitivity advantage by approximately another factor of 2.

the sensitivity advantage between both OCT methods scales with the number of channels N. Hence, the more wavelength channels N we have in the spectrometer, the higher the sensitivity advantage of FD-OCT.[27] In [3], a detailed analysis of the sensitivity of signal and noise contributions for the various OCT modalities can be found. Please also refer to [28, 29] and Problem P7.10.

7.5 Device Design of OCTs

7.5.1 Light Sources

As the light source bandwidth determines the axial resolution in TD- and FD-OCT, the selection of a suitable light source is of paramount importance. For clinical application, the most commonly used light sources are superluminescent diodes (SLDs), because they are compact and relatively inexpensive. SLDs are semiconductor diodes (Section B.5.2) which emit broadband optical radiation based on *amplified spontaneous emission* (Section B.1.2). They are very similar to laser diodes, but do not have any optical cavity. Most SLDs work in one of the wavelength regions around 850, 1300, and 1550 nm, which have particular advantages for different ophthalmic applications (Section 7.6). SLDs have output powers in the mW range, high spatial coherence, good beam quality, and can easily be coupled into optical fibers. The spectral bandwidth of an SLD ranges, depending on the wavelength, from some tens of nanometers to above 100 nm, corresponding to a coherence length of a few tens of micrometers. Increasing the output power will generally reduce the bandwidth due to the so-called *gain-narrowing* [44]. SLDs should not be exposed to any optical feedback from surface reflections, as this will automatically reduce the spectral bandwidth. For example, the Fresnel reflection from a cleaved fiber end is already nontolerable (Problem P7.12). Another important design consideration when SLDs are used is the wavelength dependence on temperature, which is typically below one nanometer per Kelvin temperature change. More details can be found on the manufacturers' web pages [30, 32] and in [44, 45]. A typical example for an SLD specification is shown in Table 7.4.

For research applications, femtosecond (fs) laser sources such as the titanium:sapphire ($Ti:Al_2O_3$) laser or fiber-based lasers can be used [46]. They provide a large bandwidth and high output power, but are currently too expensive for commercial applications. In Figure 7.20, we show a comparison of an SLD and a very broadband fs $Ti:Al_2O_3$ laser both centered around a wavelength of 830 nm which allows a good comparison (Problem P7.11).

27) Obviously, it is not useful to increase the number of spectrometer channels by any amount. As the physical spectral resolution of any spectrometer $\delta\lambda_{res}$ is limited, any increase in N beyond $\Delta\lambda/N < \delta\lambda_{res}$ will just lead to oversampling. In SS-OCT, the minimal $\delta\lambda_{res}$ is determined by the spectral bandwidth of the tunable laser source.

Table 7.4 Typical specification of the Thorlabs SLD1325 superluminescent diode (SLD) (see also Problem P7.12). Data taken from [30].

Parameter	Value
Center wavelength	1325 nm
Bandwidth (FWHM)	> 100 nm
Fiber-coupled power	> 10 mW
Maximum SLD injection current	780 mA
Maximum voltage	4 V
Operating temperature range	0–40 °C
Intensity spectrum	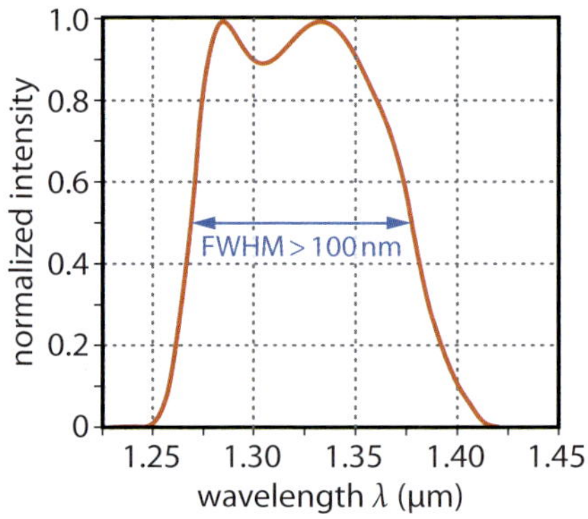

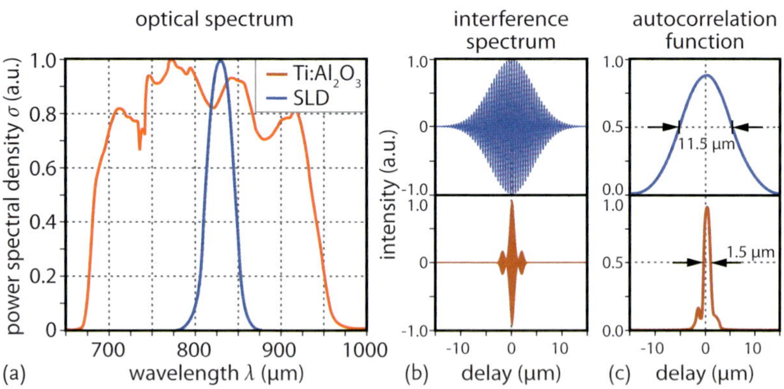

Figure 7.20 Optical characteristics of a fs titanium:sapphire laser (red) and a superluminescent diode (blue). (a) Optical spectra. (b) Interference spectra. (c) Autocorrelation functions. Adapted from [46].

Tunable laser sources for SS-OCT systems have for long time also suffered from prohibitively high cost. But recently, they have gained significant momentum and are experiencing cost reductions, especially the all-semiconductor variants [41, 47, 48]. There is substantial research activity which makes us believe that tunable laser sources and SS-OCT have significant upside potential. Hence, the SS-OCT system is likely to become the gold standard in OCT, at least for the wavelength range $> 1\ \mu m$.

7.5.2 Commercial Systems

As outlined in Section 7.2, Carl Zeiss pioneered the commercial application of OCT in ophthalmology. For more than a decade, ZEISS OCT systems were the only ones on the ophthalmic market. With the onset of FD-OCT, a number of companies (e.g., Optovue, Heidelberg Engineering, Topcon, Nidek, Tomey, Bioptigen, etc.) have also introduced commercial OCT systems for ophthalmic applications. It would go beyond the scope of the book to describe all these systems in detail. They have largely similar technical specifications (resolution, acquisition time, SNR, etc.), but differ in their software algorithms, user interface, and workflow concept. The integration and connectivity into an image archiving and data management network as well as the combination of the setup with other imaging and diagnostic modalities is also solved differently for each individual system. We will thus only highlight a few examples for illustrative purposes.

Heidelberg SPECTRALIS The Heidelberg SPECTRALIS (by Heidelberg Instruments; Figure 7.21a) is a multimodality diagnostic imaging device consisting of an integrated FD-OCT and confocal scanning laser ophthalmoscope (cSLO; Section 6.8.1). The combination of these two technologies allows new imaging capabilities, such as TruTrack™ active eye tracking, and BluePeak™ blue laser autofluorescence, providing physicians augmented views of the structure of the eye.

ZEISS Cirrus photo The ZEISS Cirrus photo is one integrated system for fundus imaging and OCT (Figure 7.21b). It combines a full mydriatic/nonmydriatic fundus camera (Section 6.7) with an FD-OCT. Specifically, the system allows correlation of data from high-density OCT cubes, thickness and layer maps with results from color fundus images, as well as fluorescence angiography (Section 6.7.6) and fundus autofluorescence (Section 6.7.7) images.

Topcon DRI OCT-1 Atlantis The Topcon DRI OCT-1 Atlantis is the first commercial retinal SS-OCT system (Figure 7.21c). It has an axial scan range of 3–12 mm, a scan speed of 100 000 A-scans per second with an axial resolution of 8 μm. The wavelength is in the 1-μm region. 3D volumetric measurements of 512×128 A-scans can be acquired in less than a second.

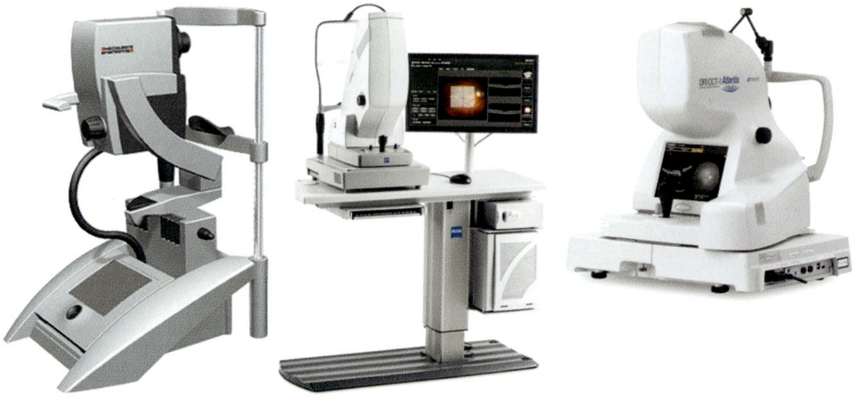

(a) (b) (c)

Figure 7.21 Photographs of commercial OCT systems. (a) Heidelberg SPECTRALIS OCT system. Courtesy of Heidelberg Engineering GmbH. (b) ZEISS Cirrus photo OCT system. Courtesy of Carl Zeiss. (c) Topcon DRI OCT-1 Atlantis OCT system. Courtesy of Topcon Deutschland GmbH.

7.6 Ophthalmic Applications of OCT

Since the first clinical tests in the early 1990s, ophthalmic applications have expanded rapidly. This was mainly driven by a number of characteristics of OCT:

1. OCT is a high-resolution imaging modality approaching the resolution and information content of histology (compare Figures 7.22b,c).
2. OCT is inherently 3D. This makes OCT images intuitive and relatively easy to interpret as structural information. OCT also allows quantitative structural assessments. It thus also supports a standardized interpretation of structural information.
3. OCT is a noninvasive *in vivo* technique.
4. OCT instruments (at least the more recent FD- and SS-OCTs) are real-time or quasi-real-time imaging devices.
5. OCT measurements can be performed in a repetitive manner, which allows early diagnosis as well as disease progression monitoring.
6. OCT allows the correlation of the eye structure with the results of a functional diagnostic.
7. OCT devices can be made compact, portable, robust, and can be easily integrated into other ophthalmic devices due to their fiber-optic core components.

All of these features have contributed to establish OCT as a powerful, even invaluable diagnostic modality for retinal diseases (Sections 3.4 and 3.5) and glaucoma (Section 3.3). In addition, it is a superior and versatile imaging and measuring device for the anterior segment of the eye. In many areas, OCT has become the clinical standard of care. A detailed treatment of these applications is beyond the

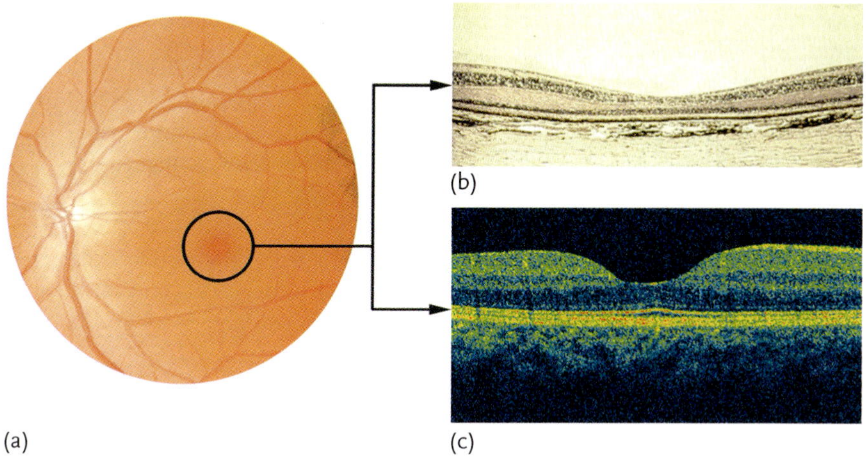

Figure 7.22 Comparison of (a) a fundus image, (b) a histology cross-section, and (c) an OCT scan in the macular area. Adapted from [6].

scope of the book, and we refer to specialized literature, for example, [7, 8, 49]. We will only give an outline to make the reader familiar with the variety of applications and the modifications in the OCT design to realize a particular application. In the following, we will focus on imaging applications. In Section 7.7, we will then cover in somewhat more detail the biometric applications of OCT and LCI technology.

7.6.1 Posterior Segment of the Eye

Traditional examination methods (e.g., fundus photography; Section 6.7) allow only the surface of the retina to be imaged. Morphological changes (i.e., changes of color and/or shape) that are evident in the fundus images provide important information about retinal diseases. However, the diagnostic options improve significantly if it is possible to image retinal areas below the surface at high lateral and axial resolution (Figure 7.24). It is also advantageous to visualize the 3D structure of the retina so that structural information about the early manifestation or progression of a retinal disease can be obtained. OCT offers a variety of new diagnostic options for retinal ophthalmology, the most important of which are:

- Diagnosis of macular diseases (including macular holes, edema, retinal thickening).
- Early diagnosis and progression control of age-related macular degeneration (AMD; Section 3.4).
- Glaucoma (Section 3.3) diagnosis by measuring the thickness of the retinal nerve fiber layer (RNFL) and/or ganglion cell layer (GCL).
- Glaucoma diagnosis by means of optical nerve head analysis.

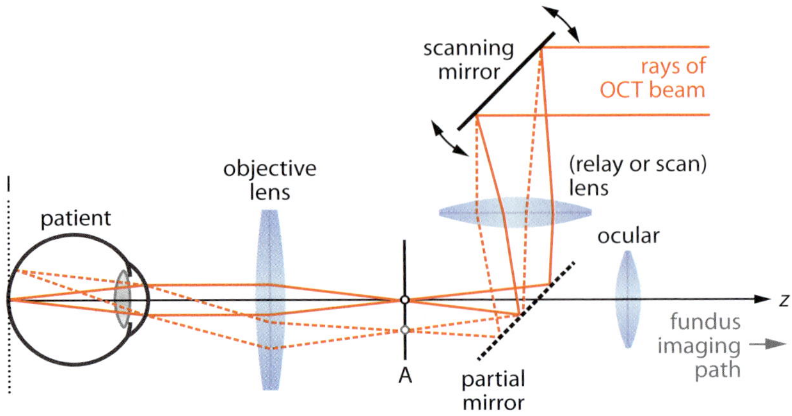

Figure 7.23 Ray diagram of an OCT device setup for posterior segment (retinal) imaging. The instrument setup is similar to a fundus camera (Section 6.7) or a laser scanning ophthalmoscope (Section 6.8.1). By example, the dashed red line represents the path of rays when the scan mirror is rotated. Plane A denotes the intermediate image plane formed between partial mirror and objective lens. Adapted from [8].

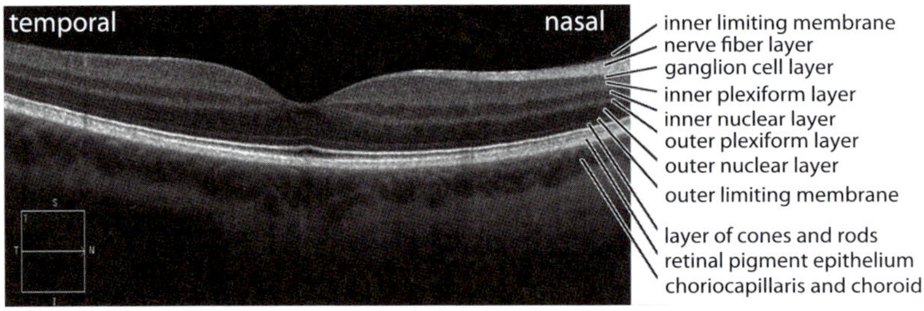

Figure 7.24 Retinal layers as imaged by means of OCT. Adapted from [8]. Courtesy of Carl Zeiss.

For retinal diagnostic applications, the OCT is configured essentially as a fundus imaging device or a confocal retinal scanner (Figure 7.23). An *xy*-scanning mirror in combination with a (relay) lens focuses the OCT beam (with a wavelength of typically 850 nm) onto the intermediate image plane from where it is imaged by the objective lens through the eye pupil onto the retina. The maximum theoretically achievable transverse resolution is thus limited by the numerical aperture of the eye. Vignetting (Section A.1.4.1) is minimized by rotating the OCT beam around the pupil. The position of the OCT image on the fundus can be acquired by either a video camera system or, as in some combination devices, by a fundus camera or a scanning laser ophthalmoscope. In this way, it is possible to correlate the OCT image to pathology landmarks and to make follow-up measurements at a later stage at exactly the same location on the retina. In particular, this feature is crucial for progression analysis of retinal diseases. OCT scans are often performed according to specific scan patterns which are advantageous for a certain disease diagnostic.

Let us consider some illustrative examples and procedural descriptions for OCT-based retinal disease diagnostics. Here, we follow a historical approach. In Sec-

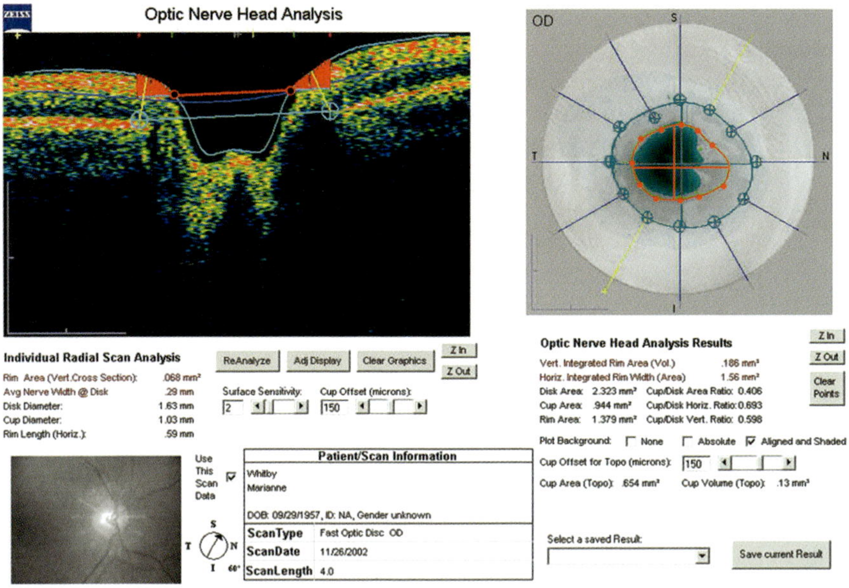

Figure 7.25 Diagnosis based on optic nerve head analysis. Left: Individual radial scan analysis. Right: Composit diagram of ONH analysis results. The yellow line indicates the selected individual scan also displayed on the left hand-side. Courtesy of Carl Zeiss.

tion 7.6.1.1, glaucoma diagnosis by means of optical nerve head (ONH) analysis with TD-OCT is described. Then, in Sections 7.6.1.2 and 7.6.1.3, we discuss glaucoma diagnosis based on analysis of the retinal nerve fiber layer (RNFL) and AMD diagnosis with FD-OCT, respectively.

7.6.1.1 **Glaucoma Diagnosis Based on Optic Nerve Head Measurements**

An optic nerve head analysis report is shown in Figure 7.25. It involves measurements on the topography of the optic nerve head (ONH). An important parameter is the disk diameter (light blue line in left part of Figure 7.25) measured by the line through the anatomical endpoints of the RPE layer. Another parameter is the cup diameter (red line in left part of Figure 7.25) which is determined by the endpoints of line that is shifted upward by 150 μm parallel to the disk diameter. The rim area (red in Figure 7.25) is determined by the cup diameter line and the lines from the RPE layer endpoints (which are perpendicular to the disk diameter line) to the anterior retinal surface. It has been found that the ratio of cup area to disk area is an important indicator in glaucoma diagnosis. Healthy eyes show a ratio between 0.2 and 0.5. Any value above 0.5 may indicate early or manifest glaucoma. Again, these data are derived from large-scale dedicated clinical studies with broad representations of ethnic groups and other demographic parameters including gender and geographic location. They are also compiled in a normative database.

7.6.1.2 Glaucoma Diagnosis Based on Thickness Analysis of the Retinal Nerve Fiber Layer

As we will see in Chapter 8, one of the gold standards for glaucoma diagnosis is functional imaging by means of visual field perimetry. In Section 6.8.3, we also discussed that structural changes in the RNFL can be measured which may provide complementary data and early diagnostic hints about the onset of glaucoma. Logically, using OCT as a high resolution measurement device to determine the RNFL thickness has proven to be a viable method to complement functional diagnostics.

The measurement is based on the concept that, due the high axial resolution of OCT, a reduction of the thickness of the RNFL can be detected very reproducibly before impairments of the visual field begin. In the OCT measurement, a certain optimized scan pattern is performed around the ONH to determine the RNFL thickness (Figure 7.26). In healthy eyes (Figure 7.26a), the RNFL thickness has a certain normative value (see highlighted green arrow in the RNFL diagram of Figure 7.26a), whereas in a glaucomatous eye (Figure 7.26b), it is significantly reduced in the temporal, superior, and inferior regions around the ONH (see highlighted red arrow in the RNFL diagram of Figure 7.26b). The green and red bands represent statistical comparisons, which are referred to as *normative data*. They represent age-matched reference values for the RNFL thickness and are derived from clinical studies. They are typically approved for clinical use by regulatory agencies, such as the Food and Drug Administration (FDA) in the United States.

7.6.1.3 Age-Related Macular Degeneration

Age-related macular degeneration (AMD; Section 3.4) manifests itself in a number of forms. OCT has been proven to not only detect early structural changes (e.g., drusen, integrity of the retinal pigment epithelium), but more importantly to be able to monitor the progression of the disease very reliably due to its registration and quantitative analysis capabilities. OCT thus has become, together with other imaging methods like fluorescence angiography (Section 6.7.6), an important tool to monitor the onset and optimize the treatment of wet AMD, in particular in optimizing frequency and timing of anti-VEGF injections (Section 3.4). Figure 7.27 illustrates the formation of drusen by means of an OCT image, and Figure 7.28 shows a report of RPE analysis by which the integrity of the RPE layer can be quantified and monitored.

7.6.2 Anterior Part of the Eye

OCT as an imaging modality was initially developed for retinal imaging, although a first anterior segment OCT was already demonstrated in 1994 [50]. However, the retinal OCT wavelengths around 830 nm proved to be sub-optimal for anterior imaging. In particular, for the characterization of the iridocorneal angle (anterior chamber angle), light with a wavelength of 850 nm exhibits strong scattering,

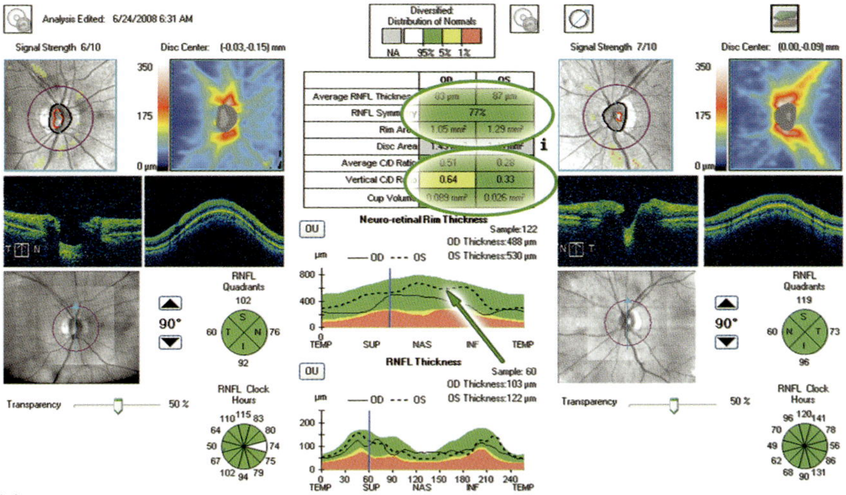

(a)

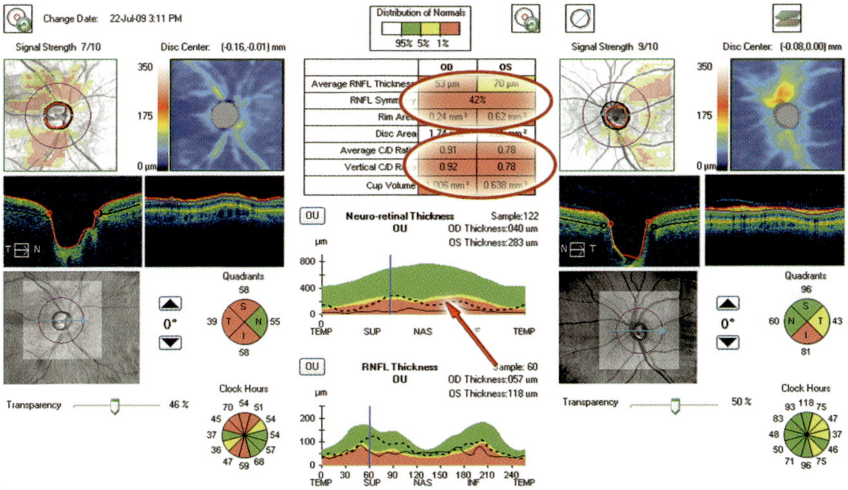

(b)

Figure 7.26 Screenshots of glaucoma diagnosis reports based on a thickness analysis of the retinal nerve fiber layer. (a) Diagnostic results for a healthy eye. Green circles and arrows show measured data well within the normative "healthy" range of parameters. (b) Diagnostic report for an eye with a reduced retinal nerve fiber layer, which is an indicator for the onset of glaucoma. Red circles and arrows indicate measurement results of a glaucomatous eye falling into the "disease" cluster of the normative database. Courtesy of Carl Zeiss.

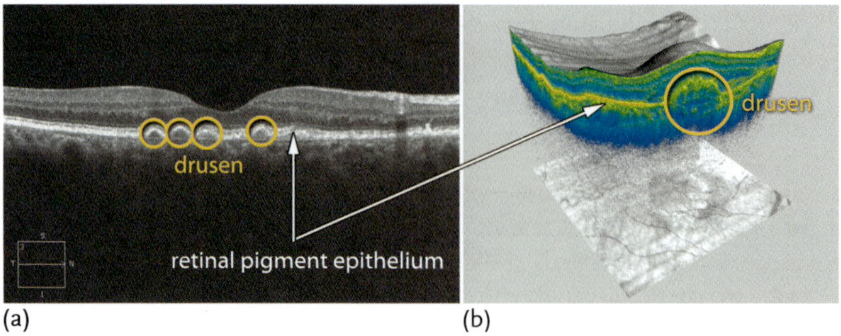

Figure 7.27 Drusen formation in AMD. (a) Cross-sectional image of the retina. (b) Three-dimensional cube scan. The corresponding fundus image is shown as a layer below the colored cross-section. Courtesy of Carl Zeiss.

especially in the sclera. Thus, for anterior segment imaging with OCT, a longer wavelength around 1050 nm or 1300 nm is favorable.

Currently, three dedicated anterior-segment products are available on the market, that is, Tomey SS-1000 CASIA, Heidelberg SL-OCT™ (Figure 7.29a), and ZEISS Visante® OCT (Figure 7.29b). Due to the improved SNR relative to TD-OCT, standard retinal FD-OCT systems with center wavelengths around 830 nm allow anterior segment imaging in real-time. But these commercial retinal systems employing FD-OCT still have inherent disadvantages in imaging the iridocorneal angle. SS-OCT at a wavelength of 1050 nm may provide a good compromise for imaging both anterior and posterior structures with high resolution and speed.

The ZEISS Visante OCT essentially operates according to the principles shown in Figure 7.30. An observation microscope consisting of objective and ocular lenses generates an image of the anterior eye segment, which is then detected by a video camera. An *xy*-scanning mirror is used to position the OCT measurement beam. The beam is focused by a (relay) lens onto the image plane. In this way, the OCT scan generates (quasi-)real-time, cross-sectional images of selected structures in the anterior eye segment. The OCT system operates at a wavelength of 1310 nm. As this wavelength is more strongly absorbed in the vitreous than 850 nm, higher incident powers can be used. This allows real-time imaging at several B-scans per second.

Anterior OCT can be used for

- full biometry and cross-sectional imaging of the anterior chamber of the eye (Figure 7.31a and Section 7.7),
- visualization and measurement of the anterior chamber angle and diagnosis of angle-closure glaucoma (Figure 7.31b),
- measurement of LASIK flap and stromal bed thickness (Figure 7.31c) before and after LASIK (Section 10.5.4.1),
- visualization and measurement of the results of corneal implants and lamellar procedures,

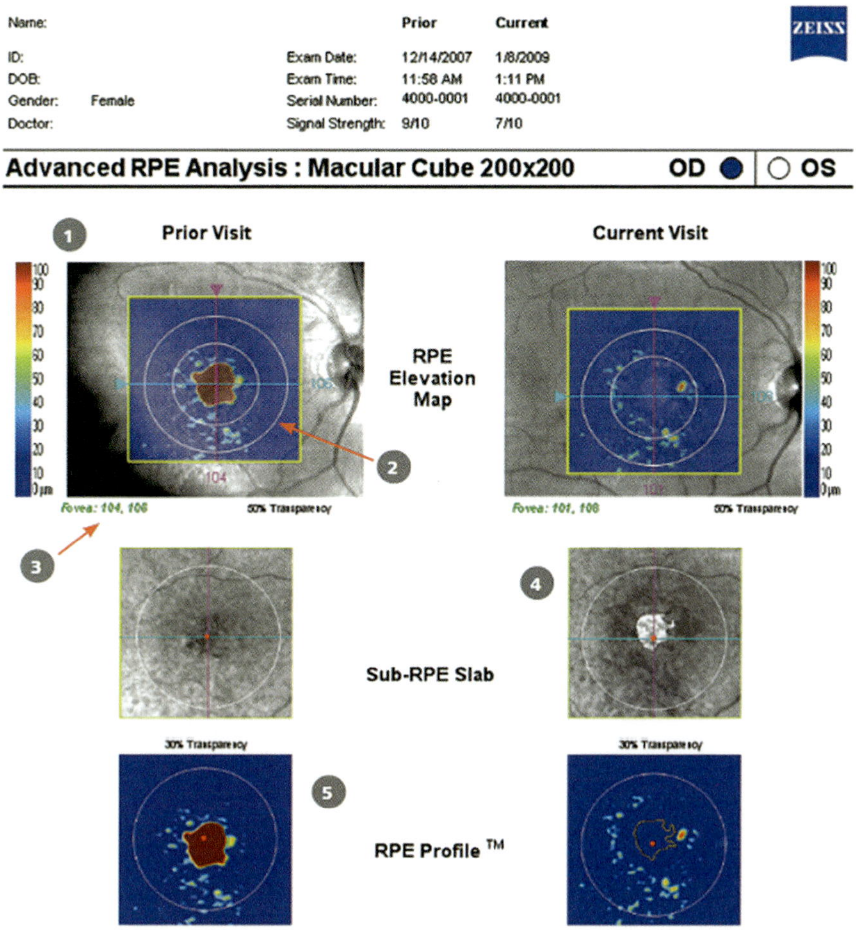

*The calculated difference does not consider test-retest variability.

6

RPE Elevations	Prior	Current	Difference*	% Change
Area in 3 mm Circle (mm²)	2.9	0.7	-2.2	-75.9%
Area in 5 mm Circle (mm²)	3.5	1.7	-1.8	-51.4%
Volume in 3 mm Circle (mm³)	0.41	0.03	-0.38	-92.7%
Volume in 5 mm Circle (mm³)	0.43	0.06	-0.37	-86.0%
Sub-RPE Illumination	**Prior**	**Current**	**Difference***	**% Change**
Area in 5 mm Circle (mm²)	0.0	1.3	1.3	xxx
Closest distance to Fovea (mm)	xxx	0.0	xxx	xxx

Figure 7.28 Advanced RPE Analysis report for the examination of the macular region (right eye). (1) Elevation map of the retinal pigment epithelium (RPE) overlaid on fundus image. (2) Circles on the RPE elevation map centered on the fovea location. (3) Fovea location coordinates. (4) Sub-RPE slab, that is, an *en-face* projection image of the reflectance of tissue beneath Bruch's membrane. (5) RPE Profile™ which combines the RPE elevation map and the areas of the sub-RPE illumination identified by the software. (6) Table of values. Courtesy of Carl Zeiss.

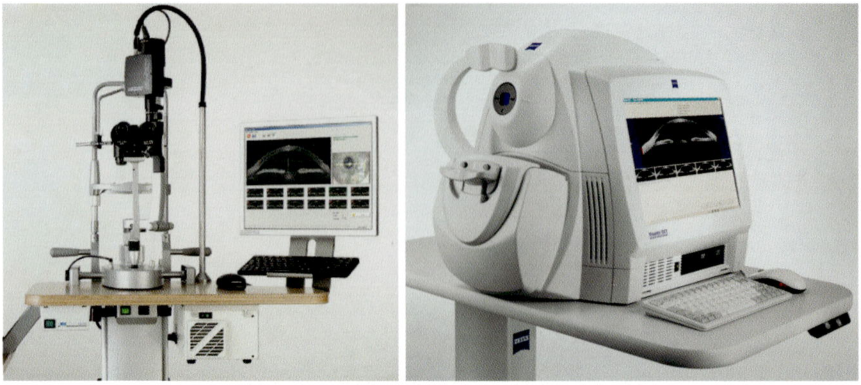

(a) (b)

Figure 7.29 (a) Photograph of Heidelberg SL-OCT. Courtesy of Heidelberg Engineering GmbH. (b) Photograph of ZEISS Visante OCT for anterior segment imaging. Courtesy of Carl Zeiss.

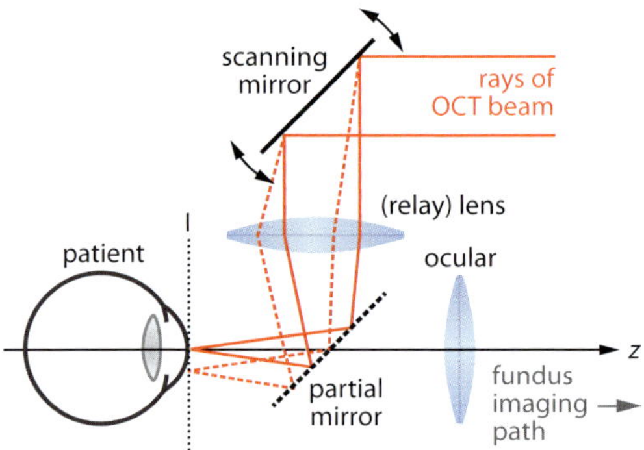

Figure 7.30 Schematics of an OCT device setup for anterior segment imaging. The imaging path is analogous to a slit-lamp microscope (Section 6.4.2.2). Adapted from [8].

- measuring the dimensions of the anterior chamber and assessing the fit and positioning of an anterior chamber intraocular lens (IOL) (Section 7.7.2),
- mapping of corneal thickness and keratoconus evaluation (Section 3.1.6), and
- imaging through corneal opacity to access internal eye structures.

7.7
Optical Biometry by Low-Coherence Interferometry

Low-coherence interferometry (LCI) is a valuable tool for the diagnosis of the anterior chamber of the eye, in particular for biometric measurements related to cataract surgery. Prior to any cataract (or refractive lens exchange) surgery, both eyes must be characterized to obtain relevant data regarding the required refractive

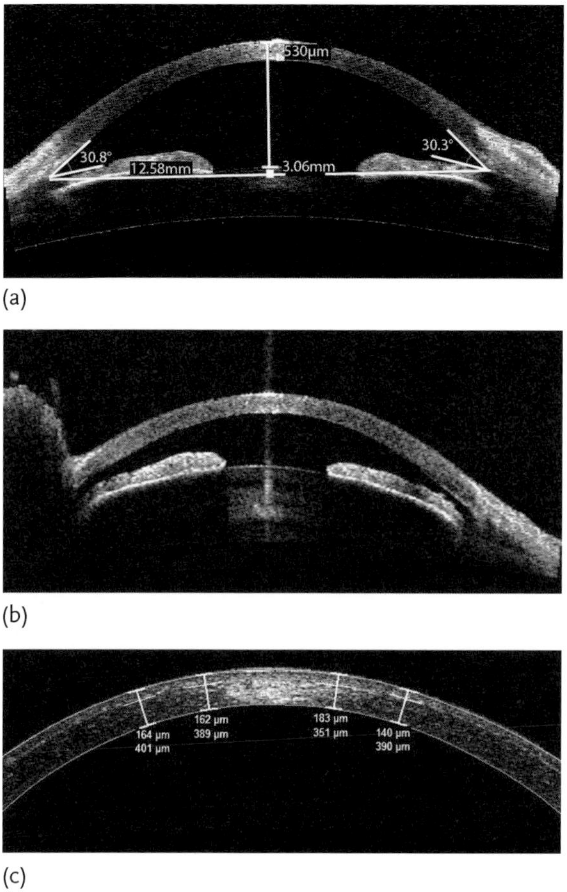

Figure 7.31 Cross-sectional images (screenshots) of the anterior chamber obtained with the ZEISS Visante OCT. (a) Analysis of angles and dimensions in the anterior segment of the patient's eye. (b) Manifestation of angle-closure glaucoma. (c) Cross-sectional image of cornea obtained seven days after a LASIK procedure. Courtesy of Carl Zeiss.

power of the IOL to be implanted. The relevant parameters for the IOL determination are (see also Section 2.2)

- the (optically relevant) axial eye length L_{eye}, that is, the length of the eye along the visual axis from the corneal vertex to the photoreceptor layer in the fovea,
- the anterior chamber depth L_{ac}, that is, the distance between the back surface of the cornea and front surface of the eye lens, and
- the corneal radius of curvature.

Until the year 2000, in clinical practice, two separate devices were used for measuring eyes. The axial length and the anterior chamber depth were measured by means of an ultrasound pulse echo procedure ([51]; ultrasound biometer; Figure 7.32a). The corneal radius of curvature was measured with a keratometer (Section 6.3.1).

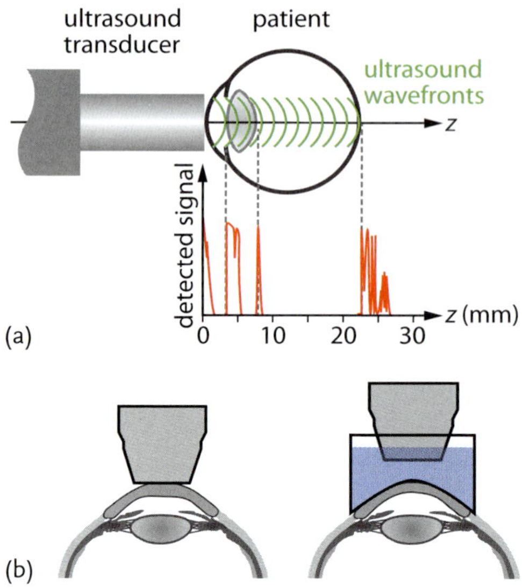

Figure 7.32 Ultrasound biometry. (a) Measurement principle. (b) Different modes of ultrasound power transfer. Left: Classical direct-contact mode. Right: Immersion mode. The ultrasound power is transferred via a liquid immersion layer.

Ultrasound biometry is based on a pulse-echo time-delay measurement which provides the axial length of the eye from the corneal vertex to the inner limiting membran (anatomic eye length or ultrasound eye length $L_{\text{eye,ac}}$). Over the years, a number of standard formulas for the calculation of the refractive power of the IOL to be implanted have been developed which depend on the used eye model and the given actual characteristics of the measured eye. As input parameters, the formulas use the ultrasound-measured axial length and the corneal refractive power. For efficient ultrasound power transfer into the eye, acoustic biometers typically require direct contact with the cornea[28] (left image in Figure 7.32b), which is certainly a discomfort for the patient. More importantly, contact biometry lacks in reproducibility because a varying (operator dependent) contact pressure directly influences the eye length measurements. Immersion biometry uses a liquid interface between the ultrasound probe and the cornea for efficient ultrasound power transfer. It is virtually a noncontact technique, as the cornea is not compressed. Therefore, it is more precise, but also more time-consuming. In addition, in all cases, an experienced ultrasound technician or ophthalmologist is needed to conduct the measurement.

If one compares the setup scheme in Figure 7.32 with the principle of LCI/OCT in Figure 7.1, the obvious analogy between optical and acoustic delay time measurements becomes apparent. Interestingly, although the basic principles of applying LCI to eye length measurements had already been laid out by Fercher *et al.* in 1986 ([53], Figure 7.33), quite some time before OCT, optical biometry (which

28) Noncontact ultrasound biometers have also been reported [52]. However, they have not found wide clinical acceptance yet.

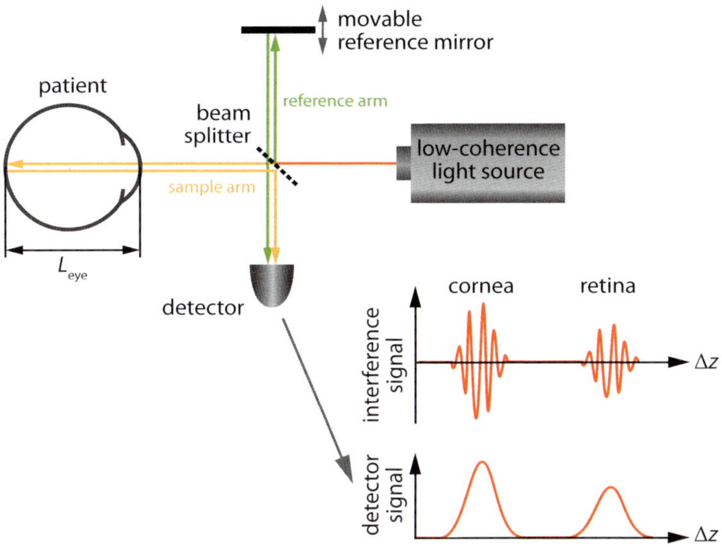

Figure 7.33 Basic principle of optical biometry based on LCI. All boundary layers that are present in the eye and capable of reflecting or backscattering radiation can be measured in this way with high precision. For the sake of clarity, only two layer reflections are shown (cornea, retina). The length of the eye can be obtained from the signal distance of reflections from these two layers.

is essentially a dedicated LCI A-scan) was only introduced into clinical practice in 1999.

7.7.1
Dual-Beam Low-Coherence Interferometry

The reasons why it took optical biometry some more years to find its way into clinical practice may be understood if one visualizes the particular challenges for optical biometry in measuring cataract eyes (Section 3.2):

1. As an eye lens affected by cataract is fairly opaque and the light has to travel twice through it, the signal is strongly attenuated. In addition, there is a large dynamic range between the signal from the cornea and the one from the retina.
2. Eye safety requirements limit the applicable laser power.
3. To measure the entire eye length, a relatively large A-scan range of up to 40 mm is required, which limits the achievable scanning time. During the scanning time, the eye may move in the z direction, which would automatically reduce the desired accuracy of the axial length measurement. The desired precision is $\pm 30\ \mu m$ in order to be comparable to or better than acoustic biometry.

One possible solution for these requirements is an extremely fast and, at the same time, high SNR scan, which only became possible with the advent of FD- and SS-OCT (Sections 7.3.4 and 7.3.5, respectively). Another solution is a special dedicated

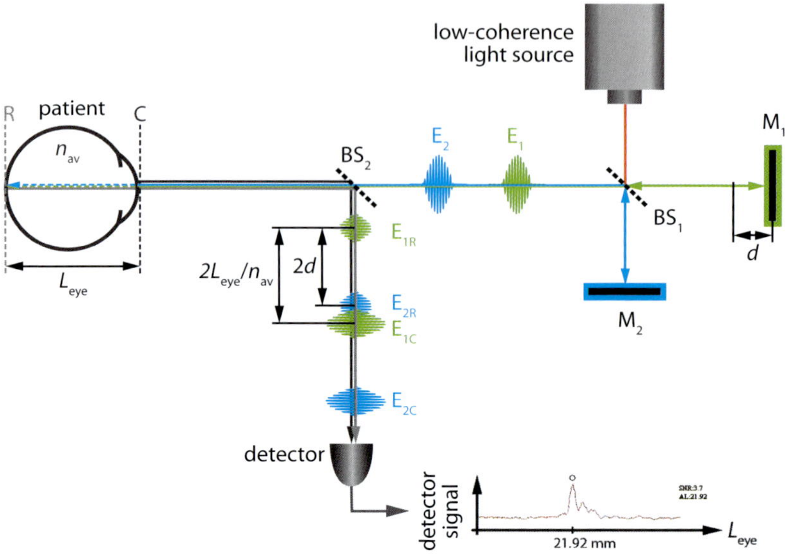

Figure 7.34 Principle of dual-beam low-coherence interferometry (dual-beam LCI). E_1 is the signal from mirror M_1 (green), and E_2 the signal from mirror M_2 (blue). A portion of the light signal is backscattered/reflected by the corneal front surface C, and the other portion by the retinal surface R. The backscattered/reflected signals are then guided to the detector arm by means of beam splitter BS_2, where they interfere. In the diagram, a typical detection signal of the ZEISS IOLMaster® is shown (measured eye length $L_{eye} = 21.92$ mm).

measuring setup of LCI, which is insensitive to axial changes in z position of the eye during the scan time. The latter can be achieved by *dual-beam LCI* (Figure 7.34), which is sometimes also referred to as *dual-beam PCI* (partial-coherence interferometry). The ZEISS IOLMaster was the first commercial device based on this principle.

In dual-beam LCI, a multimode laser diode ($\lambda = 780$ nm) with high spatial but relatively low temporal coherence (coherence length $L_c \approx 100$ μm) is used as a light source. The low-coherence radiation is split up in a Michelson interferometer comprising a beam-splitter BS_1 and reflecting mirrors (M_1, M_2) to produce two coaxial partial beams E_1 and E_2. E_1 is reflected on the fixed interferometer mirror M_1, while E_2 is reflected on the movable interferometer mirror M_2. The phases of both partial beams are shifted relatively by $4\pi d/\lambda$, where d is the path length difference between the two arms with M_1 and M_2. The beams are imaged into the eye along the visual axis and pass through another beam splitter BS_2. The portions of radiation that are reflected and/or backscattered from the corneal surface (E_{1C} and E_{2C}) and from the retina (E_{1R} and E_{2R})[29] interfere and are finally detected by a photodiode. If the path difference d in the Michelson interferometer is equal to the optical path length of the eye L_{eye}/n_{av}, constructive interference appears between

29) The distance between cornea and retina corresponds to the axial eye length.

E_{2R} and E_{1C} which, in turn, leads to a strong signal at the photodiode. As the plane C through the corneal vertex is used as the reference surface, relative movements of the patient's eye and the measuring device while scanning M_2 do not affect the eye length measurement result. This makes the dual-beam LCI suitable for applications in clinical practice.

The precision of the procedure is mainly determined by the coherence length of the radiation source. For multimode laser diodes typically used, the length of the eye can be determined with a precision (standard deviation) of $\pm 30\ \mu$m. Again, as in the case of TD-OCT, the reference mirror (M_2) is moved at constant speed v which leads to a Doppler shift in the interference signal at a frequency of $2v/\lambda$. This allows highly sensitive signal detection by means of heterodyne procedures.

7.7.2
Applications of Optical Biometry

With the dual-beam LCI-based biometry principle, an entirely new technique was found for the measurement of the axial eye length. Since the introduction in 1999, optical biometry has continuously replaced ultrasound biometry and has achieved a penetration of more than 80% [54] in all biometries performed in the United States. Approximately 20 000 units have been installed worldwide. The success of optical biometry originates from a number of advantages:

- The measurement is noncontact. There is no infection or injury risk, and no need to apply any local anesthesia.
- The measurement with the ZEISS IOLMaster is performed along the visual axis and requires a minimum of patient compliance.
- The measurement can be performed by a medical assistant (Figure 7.35). Ease of handling means time and cost savings for the ophthalmic practice, although the initial device cost is higher than in the case of ultrasound biometry.
- Measurements produce highly reproducible and accurate results. This leads to improved postoperative results.

A typical ZEISS IOLMaster measurement for an eye with an axial length of 21.92 mm is shown in the diagram of Figure 7.34. When we speak about an axial length, we need to distinguish between the axial length $L_{\text{eye,opt}}$ as measured by the IOLMaster and $L_{\text{eye,ac}}$ measured via ultrasound biometry. $L_{\text{eye,opt}}$ is the distance from the corneal vertex to the RPE measured along the visual axis (red line in Figure 7.36a). As mentioned before, $L_{\text{eye,ac}}$ denotes the distance from the vertex to the inner retinal membrane usually measured along the optical axis[30] (green line in Figure 7.36a). In the analysis of the correlation between $L_{\text{eye,opt}}$ and $L_{\text{eye,ac}}$, a constant correction factor for $L_{\text{eye,opt}}$ has been determined. Figure 7.36b demonstrates the excellent agreement between acoustical and optical eye length measurements after consideration of this correction factor [55] (Problem P7.12).

30) The different definitions of both cardinal axes are considered in Section 2.1.3.

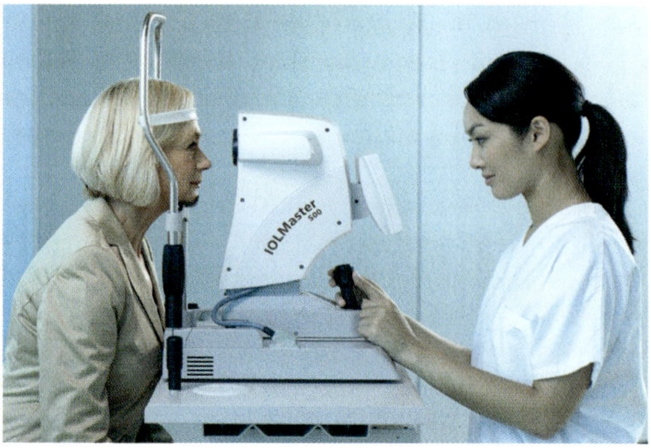

Figure 7.35 ZEISS IOLMaster 500 in clinical application. Courtesy of Carl Zeiss.

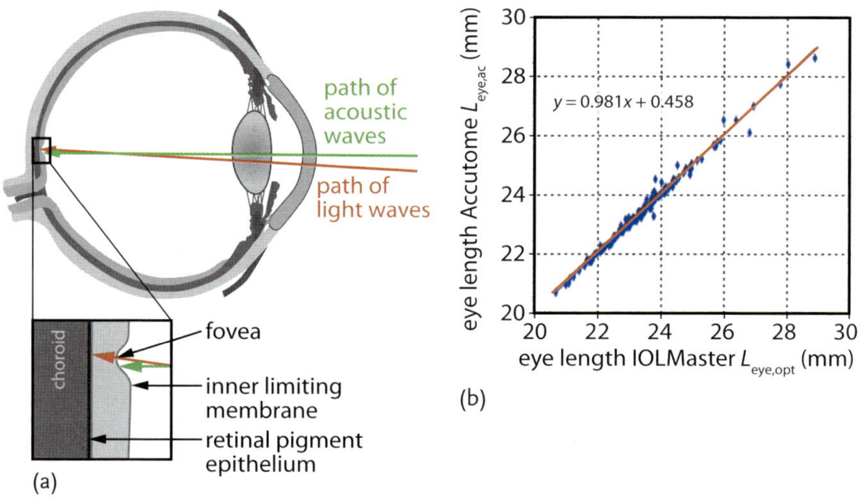

Figure 7.36 (a) Axial eye lengths as measured by optical and acoustical biometry and (b) their correlation. The ZEISS IOLMaster measures the distance from the corneal vertex to the retinal pigment epithelium but displays a corrected value that represents the distance from the corneal vertex to the inner limiting membrane, as measured in acoustical biometry. Adapted from [56].

The ZEISS IOLMaster is a good example of the benefits of medical instrumentation technology. Although a capital cost is associated with the diagnostic instrument, the recurring cost is relatively low, patient throughput can be high, and significant treatment outcome improvements can be verified. As a consequence, significant cost savings in patient care with a socio-economic impact (see also Section 3.10) result.

Meanwhile, a number of devices from various manufacturers are available which are either based on dual-beam LCI or an ultrafast A-scan, often referred to as *opti-*

Table 7.5 Overview of specifications of available biometry systems by Carl Zeiss, Haag–Streit, Tomey, Nidek, and Topcon.

	ZEISS IOLMaster	Haag–Streit Lenstar LS 900	Tomey OA 1000	NIDEK AL-Scan	Topcon ALADDIN
Technology	dual-beam LCI (PCI)	OLCR	OLCR	OLCR	LCI (PCI)
Axial length	14–38 mm	14–32 mm	14–40 mm	14–40 mm	15–38 mm
Anterior chamber depth	1.5–6.5 mm	1.5–6.5 mm	1.5–7 mm	1.5–6.5 mm	1–7 mm
Corneal thickness	n.a.	300–800 μm	300–1200 μm	250–1300 μm	n.a.
Lens thickness	n.a.	0.5–6.5 mm	n.a.	n.a.	n.a.
Retinal thickness	n.a.	1 μm	n.a.	n.a.	n.a.
Keratometry	5–10 mm	5–10 mm	n.a.	5–13 mm	5–10.5 mm
White-to-white distance	8 – 16 mm	7–16 mm	n.a.	7–14 mm	8–16 mm
Pupil Diameter	5–10 mm	2–13 mm	n.a.	1–10 mm	2–9 mm
Other features	Connectivity to ultrasound biometer, data management system		Connectivity to topography, ultrasound biometer, Scheimpflug imaging	Optional built-in ultrasound biometer; 3D auto tracking and auto shot	Placido topography system (built-in); Scheimpflug imaging for L_{ac} measurement
Internal IOL power calculation	yes	yes	no	yes	yes

cal low-coherence reflectometry (OLCR) (Table 7.5). The Haag–Streit Lenstar LS 900® (Figure 7.37a) uses OLCR and achieves fast A-scans by means of a rotating prism cube. The advantage of the method is that all intraocular eye distances (L_{eye}, corneal

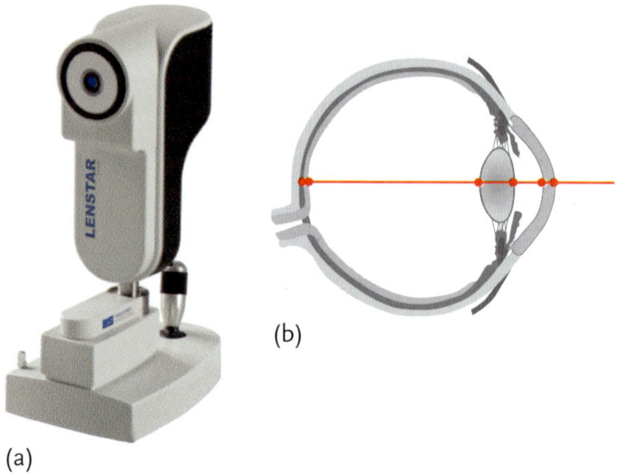

Figure 7.37 (a) Photograph of Haag–Streit Lenstar LS 900. (b) Intraocular eye distances measured by Lenstar LS 900. Courtesy of Haag–Streit Deutschland GmbH.

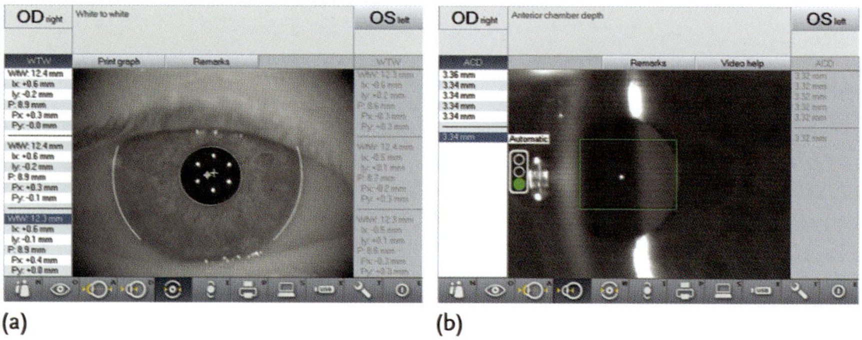

Figure 7.38 User Interface screens of ZEISS IOLMaster for (a) white-to-white measurements and (b) the measurement of the anterior chamber depth L_{ac}. Courtesy of Carl Zeiss.

thickness, L_{ac}, lens thickness, retinal thickness) can be measured by means of interferometry (Figure 7.37b).

Typically, these devices measure, in addition to the axial eye length, the corneal radius, the anterior chamber depth, and the white-to-white distance (Figure 7.38). The white-to-white distance is the horizontal iris diameter, which is an important parameter for IOLs and refractive surgery with lasers. A number of devices have been developed into combination instruments. For example, Scheimpflug devices (Section 6.5.2) can be included to measure L_{ac}, a built-in ultrasound biometer can be used for extremely dense cataracts, or a Placido topography system (Section 6.3.2) can be implemented instead of the optoelectronic keratometer (Section 6.3.1.3) for corneal radius measurements.

Let us now consider in brief how the measured biometric data is used to calculate the optical power of the eye lens and the optical power of the IOL to achieve a

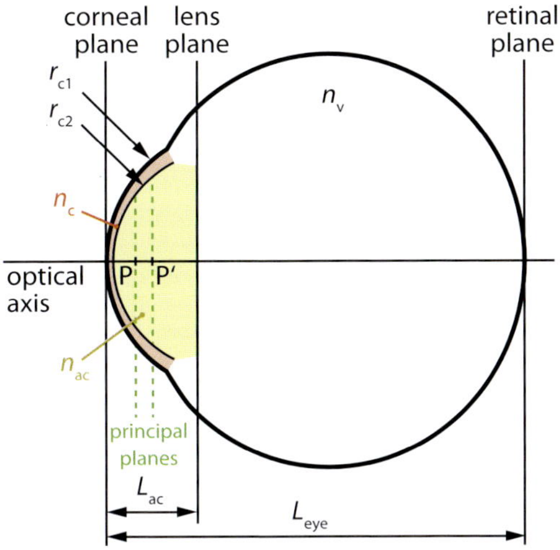

Figure 7.39 Simplified eye model for IOL calculations based on biometric data (see also Section 2.2). Adapted from [56].

certain desired refraction. A detailed treatment is beyond the scope of the book. Please refer to [56–58] as well as Problems P7.13 and P7.14 for further information.

We use a simplified model of an emmetropic eye, as shown in Figure 7.39 in which both cornea and lens are treated as thin lenses. L_{ac} is the anterior chamber depth and L_{eye} the eye length. With the refractive index of the aqueous humor n_{ac}, the total refractive power of the eye is then given by (see also Gullstrand formula (2.20))

$$\mathcal{D}'_{eye} = \mathcal{D}'_c + \mathcal{D}'_l \tag{7.60}$$

$$= \mathcal{D}'_{c1} - \frac{L_{ac}}{n_{ac}}(\mathcal{D}'_{c1})^2 + \mathcal{D}'_l \,. \tag{7.61}$$

$\mathcal{D}'_c$ is the total refractive power of the cornea (refraction at the front *and* back surface). $\mathcal{D}'_{c1}$ and $\mathcal{D}'_l$ are the refractive power values of the air–cornea interface and the eye lens, respectively. We further make the assumption that the anterior (r_{c1}) and posterior (r_{c2}) radii of the cornea are proportional to each other so that the corneal refractive power can be written as (Exact Gullstrand Eye #1; Table 2.1)

$$\mathcal{D}'_{c1} = \frac{0.3315}{r_{c1}} \,. \tag{7.62}$$

The refractive power of the nonaccommodated eye lens $\mathcal{D}'_l$ can then be written as

$$\mathcal{D}'_l = \frac{n_v}{L_{eye} - L_{ac}} - \frac{n_{ac}}{\frac{n_{ac}}{\mathcal{D}'_{c1}} - L_{ac}} \,, \tag{7.63}$$

in which n_v is the refractive index of the vitreous. From Eq. (7.63), we follow that the three biometric parameters of interest are L_{ac}, L_{eye}, and the corneal refractive power $\mathcal{D}'_{c1}$ (derived from the corneal radius). Equation (7.63) can now be used before cataract surgery to calculate the refractive power of the IOL to achieve emmetropia. In certain cases, it is desired to "undercorrect" or "overcorrect" the eye. For a desired refractive power A_x, it is straightforward to show (Problem P7.13) that the required IOL power is determined by

$$\mathcal{D}'_1 = \frac{n}{L_{\text{eye}} - L_{\text{ac}}} - \frac{n}{\frac{n}{z(A_x)} - L_{\text{ac}}} \tag{7.64}$$

with

$$z(A_x) = \mathcal{D}'_{c1} + \frac{A_x}{1 - A_x L_{\text{ac}}} \tag{7.65}$$

and $n \approx n_{\text{ac}} \approx n_{\text{v}} = 1.336$.

Based on the principles which lead to Eq. (7.64), there are more sophisticated optical model-based formulas in use. On the other hand, based on statistical correlations, a number of purely empirical formulas have been derived which link the measured biometric data to the IOL power. We refer to [56–58] and references given therein for further details. It is, however, already instructive to use Eq. (7.63) in order to estimate how measuring uncertainties influence the IOL power (Problem P7.14).

7.8 Prospects

OCT technology is still a very dynamically developing technique which attracts a significant amount of research interest and funding. We see a number of trends in OCT application which we would like to highlight:

- OCT instrumentation will become more and more designed to facilitate the clinical practice, which means that ergonomics and workflow will play a more important role.
- Software and image processing will become increasingly more important.
- Combination with other imaging and diagnostic modalities will lead to image and data fusion for a better diagnosis, for example, cSLO (Section 6.8.1) and OCT (Heidelberg Instruments), fundus camera and OCT (Carl Zeiss), or perimeter and OCT (Carl Zeiss). The latter is an example of so-called *structure–function diagnostics* for glaucoma.
- High-speed (FD-OCT) and ultrahigh-speed (presumably SS-OCT) devices allow real-time sampling of relatively densely packed ocular volume images and thus pave the way for OCT to become an intraoperative measurement, control, and imaging modality (see Section 10.5.4.2 and [41]).
- OCT has already proven its diagnostic imaging potential in a few other medical disciplines (e.g., cardiology, dermatology, oncology). A coverage of these very promising and interesting fields is beyond the scope of this book, and we refer the interested reader to [5, 6, 9, 46].

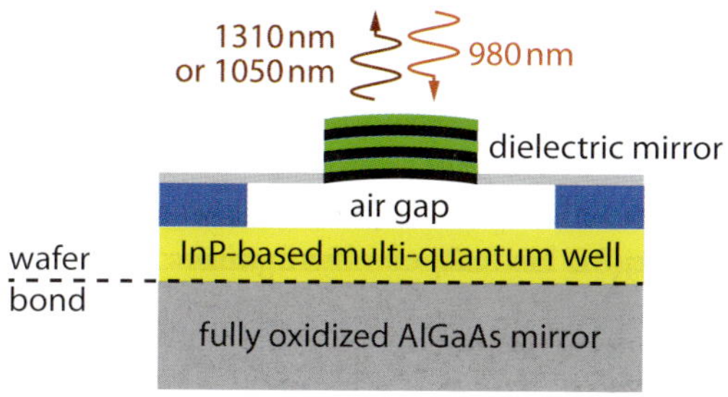

Figure 7.40 Schematic layer system of a vertical-cavity surface emitting laser (VCSEL). Adapted from [59].

In technology development, we see the following promising trends.[31]

Ultrahigh speed OCT Development activities referring to new OCT instrumentation will have a strong focus on ultrahigh imaging speeds. In particular with SS-OCT, scan speeds have already been achieved above 1 GHz. Here, the recent development of new semiconductor laser sources (e.g., vertical-cavity surface emitting lasers (VCSEL)) has opened up a dynamically developing field (Figure 7.40) [59].

Ultrahigh resolution OCT With the advancement of more sophisticated fs and wavelength-tunable laser sources, ultrahigh resolution OCT will become available, although the costs and complexity of suitable laser sources currently prevent their use in clinical practice. In the lab, however, the use of these laser sources together with a careful balancing of the Michelson interferometer with regard to dispersion, spectral transmittance, and polarization mismatch has lead to axial resolutions in the range between 1 and 3 μm [60, 61].

Polarization-sensitive OCT (PS-OCT) The OCT interference signal depends on the relative polarization of the two interfering beams. OCT has thus the inherent potential to detect polarization-dependent or polarization-changing properties of the sample tissue (see also Section A.2.1.4) [1]. Three different mechanisms can change the light polarization in tissue:

1. *Birefringence:* The speed of light depends on the refractive index of the medium through which the light wave travels. If the refractive index of a medium depends on its orientation, the polarization state varies with the direction of the incident light beam.
2. *Dichroism:* In the case of *dichroism*, one polarization direction is filtered out by the medium, whereas the others pass through.
3. *Optical rotation:* In the case of optical rotation, the polarization axis is rotated.

Polarization-sensitive OCT (PS-OCT) may be attractive for medical applications, as it provides an additional image contrast which depends on certain pathologies al-

31) An excellent overview on modern technology trends in OCT can be found in [3].

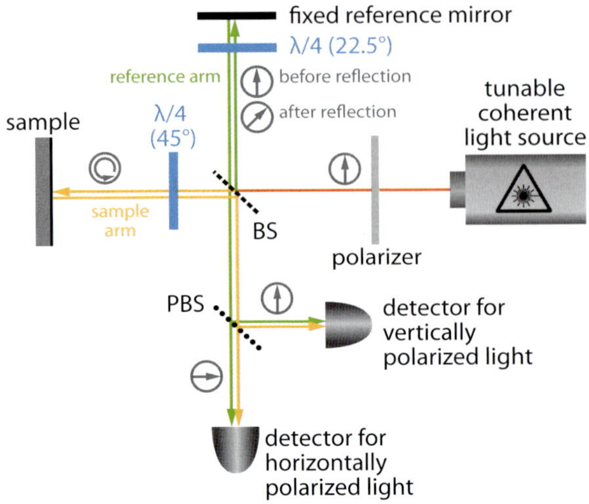

Figure 7.41 Schematic setup of a polarization-sensitive OCT (PS-OCT) based on SS-OCT. "$\lambda/4$" refers to quarter-wave plates with their corresponding settings in brackets. BS denotes the beam splitter and PBS represents a polarizing beam splitter. The gray circles with the arrow show the polarization direction of light in the respective arm. Adapted from [62].

tering the polarization state of low-coherence light. First clinical results have been reported which correlate well with the findings of scanning laser polarimetry (Section 6.8.3) [1, 3].

A typical setup of a PS-OCT based on SS-OCT is shown in Figure 7.41. The data processing for both acquired, polarization-resolved signals is similar to "standard" FD-OCT. Once the Fourier transformation has been performed, the amplitudes and phases of the orthogonal polarization states are obtained as a function of depth z. The phase retardation ξ_{pol} of the horizontal and vertical polarization states (ψ_{h} and ψ_{v}, respectively) is a measure of birefringence of the examined tissue and determined by

$$\xi_{\text{pol}} = \arctan\left(\frac{\psi_{\text{v}}(z)}{\psi_{\text{h}}(z)}\right) . \tag{7.66}$$

Figure 7.42 shows an example of a PS-OCT intensity image of the human retina together with the phase retardation plot (see also Problem P7.15).

Doppler OCT (DOCT) Doppler OCT (DOCT) [1, 3], sometimes also called *optical Doppler tomography* (ODT), is essentially a combination of coherent detection and laser Doppler flowmetry (Section 8.2.1.1). Laser Doppler flowmetry allows a determination of the fluid (blood) flow by measuring the Doppler-shifted frequency from moving objects. Object velocity v and Doppler-shifted frequency $\Delta\nu_{\text{Doppler}}$ are related (for small velocities) via

$$v = \frac{\lambda_0 \Delta\nu_{\text{Doppler}}}{2\cos\theta} , \tag{7.67}$$

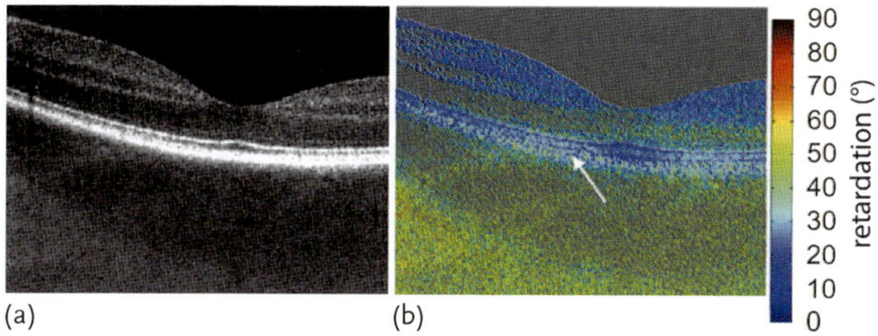

Figure 7.42 (a) Polarization-sensitive tomogram of intensity. (b) Measured phase retardation. The arrow indicates the depolarizing structure of the retinal pigment epithelium. Reprinted from [3] with permission from Elsevier.

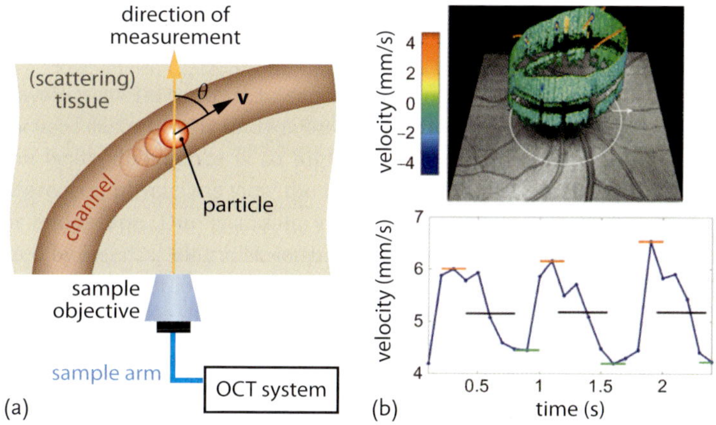

Figure 7.43 Principle of Doppler OCT. (a) The sample arm is aligned to the scattering tissue such that the direction of measurement includes an angle of θ with the direction of particle flow in the channel. The optical signal falls on a particle which moves with a velocity v. The frequency of the light backscattered to the sample objective is then Doppler shifted by $\Delta\nu_{\text{Doppler}}$. Adapted from [1]. (b) Example of quantitative Doppler analysis. Doppler OCT data of a circular scan around the optical nerve head. In the diagram, the extracted average flow velocity versus time is shown for one cross-section. Typical velocities are a few ten mm/s. Ultrahigh-speed OCT systems yield a resolution in the μm/s range. Reprinted from [3] with permission from Elsevier.

where λ_0 is the wavelength of the incident light. In Eq. (7.67), the relative angle θ between the direction of the fluid flow and the direction of measurement (i.e., the traveling direction of light) has also been taken into account (Figure 7.43a). When we apply this measuring principle to the OCT technique, the Doppler shifted frequency $\Delta\nu_{\text{Doppler}}$ has to be acquired in addition to the structural data. The differences in the phases of subsequent z scans (in time) at the same xy position can be used to calculate $\Delta\nu_{\text{Doppler}}$. Alternate algorithms, similar to the ones known from ultrasound Doppler flowmetry, have also been used (see, e.g., [63]; Problem P7.16).

With DOCT, it is possible to determine the blood flow in turbid, scattering media, even below the surface.[32] DOCT imaging can be carried out in any type of OCT. However, FD- and SS-OCT are advantageous, since they provide an easier way to derive the phase of the signal from the measurement.

Spectroscopic OCT As OCT uses broadband light sources (or light sources with a tunable wavelength), it should be possible to analyze the backscattered light with respect to any depth-dependent absorption or other spectroscopic properties of the sample (e.g., density of contrast agents in a sample). In principle, any spectroscopic OCT[33] should be able to deliver quantitative volumetric data about absorption and scattering properties of the examined tissue. A treatment of the methodology is beyond the scope of the book. We thus refer to [3] and give an outline on the methodology in Problem P7.17.

The methods described (in particular Doppler and spectroscopic OCT) can be potentially used for the diagnosis of other human diseases, as the measured data can be functionally dependent on the metabolism of the body (Section 8.2) for instance. In this case, the eye serves as a diagnostic window to the body and OCT as the corresponding noninvasive measurement "tool".

7.9 Recommended Reading

A large number of review papers and books have been published also on particular aspect and applications of OCT. The interested reader is referred to these references [1–10].

7.10 Problems

P7.1. Coherence time Derive the coherence length of a Gaussian pulse of spectral bandwidth $\Delta\omega$. Express it also in terms of λ_0 and $\Delta\lambda$. Compare this to the pulse duration or the product of pulse duration and speed of light. Use the coherence length of a superluminescence diode (see e.g., [30, 32]). What differences do you notice?

P7.2. Autocorrelation function, spectral density, and coherence length Calculate the autocorrelation function $\mathcal{G}$, the spectral density $\sigma(\omega)$, and coherence length L_c for various pulse forms and spectral distributions:

32) This is again similar to ultrasound Doppler flowmetry.

33) Do not confuse spectroscopic OCT with spectral-domain OCT. For this reason, we introduced the term "frequency-domain" instead of "spectral-domain".

1. Gaussian pulse
2. Rectangle pulse
3. Lorentz spectrum
4. sech^2 pulse

Use the definitions for $\mathcal{G}(\Delta t)$ and $\sigma(\omega)$ as given in Eqs. (A111) and (A109), respectively. For the coherence length as well as the spectral density use appropriate definitions such as the full width at half maximum (FWHM), second momentum of a normalized function, or second momentum of a squared normalized function.

P7.3. Swept-source OCT How would you design a 1050 nm swept-source OCT with variable sweep rates up to 500 MHz and a tunable range of 100 nm for the various tasks?:

1. High speed/high resolution/short "scan" depth (e.g., used for retinal scans)
2. Medium speed/medium resolution/medium "scan" depth (e.g., used for scans in the anterior chamber)
3. Low speed/medium resolution/ultrawide "scan" depth (e.g., used for full-field scans)

Determine relevant parameters for sweep rate, bandwidth, SNR, measuring time per axial scan, and so on.

P7.4. Group velocity delay Calculate the group velocity delay in a Michelson interferometer for various broadband light sources:

1. SLD with $\lambda_0 = 850$ nm and 50 nm bandwidth
2. Titanium–sapphire laser with $\lambda_0 = 800$ nm and 70 nm bandwidth
3. SLD with $\lambda_0 = 1050$ nm and 100 nm bandwidth

Assume the anterior chamber of the human eye to be the object (sample). Use as material the data from the Gullstrand Eye (Section 2.2.1) and treat the chamber as one slab.

P7.5. Theory of TD-OCT Derive in detail Eqs. (7.34) and (7.37) from Eq. (7.28). Why is the resolution in TD-OCT half the coherence length and not the coherence length itself (Figure 7.6)?

P7.6. Theory of FD-OCT Derive in detail Eq. (7.49) from Eq. (7.47).

P7.7. Theory of FD-OCT

1. Use a mathematical software tool (e.g., Matlab or MathCad) to simulate an FD-OCT spectrum resulting from the reflections of the 4 major interfaces of the anterior segment of the eye (assumed to be δ functions in space). Assume a Gaussian pulse of spectral bandwidth for various broadband light sources, for example, SLD with $\lambda_0 = 850$ nm and 50 nm bandwidth, titanium:sapphire laser with $\lambda_0 = 800$ nm and 70 nm bandwidth, and/or SLD with $\lambda_0 = 1050$ nm

and 100 nm bandwidth. What can you say about the required resolution of the spectrometer and the dynamic range?
2. Overlay on the simulated spectra a simulated white noise spectrum. Then, Fourier transform the simulated results to obtain an A-scan of the anterior eye segment.

P7.8. **Theory of SS-OCT** Let us consider an SS-OCT with 1050 nm center wavelength and a sawtooth-like sweep.

1. Simulate the detector signal for a typical sweep rate of 200 kHz and various arm length mismatches (100 μm, 1 mm, 10 mm).
2. How does the result change when the direction of the sweep is reversed? How, if the sweep rate is reduced to 50 kHz?
3. Use a mathematical software tool (e.g., Matlab or MathCad) to simulate the SS-OCT spectrum resulting from the reflections of the 4 major interfaces of the anterior eye segment (assumed to be δ functions in space). Use 1050 nm as a center wavelength and a sawtooth-like sweep with 200 kHz. Then, Fourier transform the simulated results to obtain an A-scan of the anterior segment of the eye.

Compare with the results of Problem P7.7.

P7.9. **Group velocity dispersion in OCT** We assume Gaussian pulses of spectral bandwidths for various broadband light sources (e.g., those considered in Problem P7.4).

1. One of the arms of the OCT Michelson interferometer contains glass (type BK7) with a (geometric) length of 10 cm. What axial loss in resolution results from the group velocity dispersion (GVD) as compared to an "empty" OCT?
2. Use the results of 1.) to estimate how much BK7 glass you must place into the reference arm of the OCT to compensate for the GVD of the eye in the signal arm. What would be even better than BK7? As an approximate model of the human eye, we assume as slab with a length of 24.4 mm having the same dispersion as water (aqueous humor).

P7.10. **Signal-to-noise ratio** Calculate the SNR for TD-OCT, FD-OCT, and SS-OCT for shot-noise limited detection (see Eqs. (7.58) and (7.59)).

P7.11. **Light sources for OCT**

1. What can be done to avoid any optical feedback from the Fresnel reflection of the fiber into which the SLD light should be coupled in?
2. Calculate from the spectrum in Table 7.4 the autocorrelation function and the OCT interference signal. Do a best fit to the spectrum by assuming a sum of two spectrally shifted Gaussian beams. What is the effect of the trough? How do the autocorrelation function and interference signal change if the trough becomes deeper? What can you say about the coherence length? Alternatively

you may use numerical methods (FFT) to simulate the influence of the shape of the spectrum on autocorrelation function and interference signal.
3. Calculate numerically and assume as Gaussian fit from the spectra in Figure 7.20 the autocorrelation functions and the OCT interference signals. What is the difference between the numerical result and the Gaussian approximation? Why does the autocorrelation function of the SLD exhibit side lobes?

P7.12. **Optical and acoustic biometry** Show that the correlation displayed in Figure 7.36b follows directly from Figure 7.36a. Estimate the slope of the correlation curve.

P7.13. **Intraocular lens formula** Derive the IOL formula (7.64) from the model shown in Figure 7.39. How does the calculation change if we use a more elaborated model, in particular, if both cornea and lens are treated as thick lenses.

P7.14. **Intraocular lens power determination with biometry** Estimate the uncertainties in the determination of the IOL power which result from uncertainties in the measurement of the biometric parameters:

1. eye length uncertainty: 10 and 50 μm
2. uncertainty of the anterior chamber depth: 10 and 50 μm
3. corneal radius

Which other factors influence the calculated power of an IOL?

P7.15. **Polarization-sensitive OCT** Describe the measurement concept of PS-OCT and calculate (simulate) the FD-signal as a function of the depolarization of a sample. Assume a simple model for the retinal pigment epithelium consisting of two layers with different depolarization behavior.

P7.16. **Doppler OCT** Describe the measurement concept of DOCT and calculate (simulate) the FD-signal as a function of the blood flow of a sample. Assume a simple model consisting of a layer in which blood flows at an angle of 45° with a velocity of v below a layer of scattering tissue.

P7.17. **Spectroscopic OCT** Describe the measurement concept of spectroscopic OCT and calculate (simulate) the FD-signal as a function of the diameter of a strongly absorbing pigment sphere located inside scattering tissue of approximately the same refractive index and with equal backscattering properties.

References

1 Tomlins, P. and Wang, R. (2005) Theory, developments and applications of optical coherence tomography. *J. Phys. D: Appl. Phys.*, **38**, 2519–2535.

2 Fercher, A.F., Drexler, W., Hitzenberger, C.K., and Lasser, T. (2003) Optical coherence tomography – principles and applications. *Rep. Prog. Phys.*, **66**, 239–303.

3 Leitgeb, R.A. (2011) Current technologies for high-speed and functional imaging with optical coherence tomography.

Adv. Imaging Electron Phys., **168**, 109–192.
4 Schmitt, J.M. (1999) Optical coherence tomography (OCT): A review. *IEEE J. Sel. Top. Quantum Electron.*, **5**, 1205–1215.
5 Chen, Y., Bousi, E., Pitris, C., and Fujimoto, J.G. (2011) Optical coherence tomography: Introduction and theory, in *Handbook of Biomedical Optics*, CRC Press, pp. 255–279.
6 Bouma, B.E. and Tearney, G.J. (eds) (2002) *Handbook of Optical Coherence Tomography*, Informa Healthcare.
7 Schuman, J.S., Puliafito, C.A., Fujimoto, J.G., and Duker, J.S. (2012) *Optical Coherence Tomography of Ocular Diseases*, Slack Inc., Thorofare, 3rd edn.
8 Schumann, J.S., Puliafito, C.A., and Fujimoto, J.G. (2007) *Everyday OCT: A Handbook for Clinicians and Technicians*, Slack Inc., Thorofare.
9 Drexler, W. and Fujimoto, J.G. (2008) *Optical Coherence Tomography – Technology and Applications*, Springer.
10 Wang, L.V. and Wu, H.I. (2007) *Biomedical Optics*, John Wiley & Sons, Inc.
11 Liang, R. (2010) *Optical Design for Biomedical Imaging*, SPIE Press.
12 Kaschke, M., Meyer, S., Everett, M., and Grahl, M. (2009) Optical coherence tomography – a case study in medical imaging adoption. *Opt. Photonik*, **4**, 24–28.
13 Takada, K., Yokohama, I., Chida, K., and Noda, J. (1987) New measurement system for fault location in optical waveguide devices based on an interferometric technique. *Appl. Opt.*, **26**, 1603–1608.
14 Youngquist, R.C., Carr, S., and Davies, D.E.N. (1987) Optical coherence domain reflectometry: a new optical evaluation technique. *Opt. Lett.*, **12**, 158–160.
15 Fercher, A.F., Mengedoht, K., and Werner, W. (1988) Eye-length measurement by interferometry with partially coherent light. *Opt. Lett.*, **13**, 1867–1869.
16 Hitzenberger, C.K. (1991) Optical measurement of the axial eye length by laser Doppler interferometry. *Invest. Ophthalmol. Vis. Sci.*, **32**, 616–624.
17 Tanno, N., Ichikawa, T., and Saeki, A. (1990) Lightwave Reflection Measurement. Japanese Patent #2010042.
18 Huang, D., Swanson, E.A., Lin, C.P., Schuman, J.S., Stinson, W.G., Chang, W., Hee, M.R., Flotte, T., Gregory, K., A.Puliafito, C., and Fujimoto, J.G. (1991) Optical coherence tomography. *Science*, **254**, 1178–1181.
19 Fercher, A.F., Hitzenberger, C.K., Drexler, W., Kamp, G., and Sattmann, H. (1993) In vivo optical coherence tomography. *Am. J. Ophthalmol.*, **116**, 113–114.
20 Swanson, E.A., Izatt, J.A., Hee, M.R., Huang, D., Lin, C.P., Schuman, J.S., Puliafito, C.A., and Fujimoto, J.G. (1993) In vivo retinal imaging by optical coherence tomography. *Opt. Lett.*, **18**, 1864–1866.
21 Swanson, E.A. (2012), Estimates of ophthalmic OCT market size and the dramatic reduction in reimbursement payments. http://www.octnews.org/articles/4176266/estimates-of-ophthalmic-oct-market-/size-and-the-dr/ (accessed 26 August 2013).
22 Carl Zeiss Meditec AG (2012) Cirrus HD-OCT record number of 10 000 installations and prestigious award for OCT inventors. http://www.businesswire.com/news/home/20121106006719/en/Cirrus-HD-OCT-Record-Number-10000-10000-Installations-Prestigious (accessed 26 August 2013).
23 Swanson, E.A. and Huang, D. (2011) Ophthalmic OCT reaches $1 billion per year: But reimbursement clampdown clouds future innovation. http://www.octnews.org/articles/2844561/ophthalmic-oct-reaches-1-billion-per-year-but-reim/ (accessed 26 August 2013).
24 Tearney, G.J., Bouma, B.E., and Fujimoto, J.G. (1997) High-speed phase- and group-delay scanning with a grating-based phase control delay line. *Opt. Lett.*, **22**, 1811–1812.
25 Hauger, C., Worz, M., T., and Hellmuth (2003) Interferometer for optical coherence tomography. *Appl. Opt.*, **42**, 3896–3902.
26 Wojtkowski, M., Leitgeb, R., Kowalczyk, A., Bajraszewski, T., and Fercher, A.F. (2002) In vivo human retinal imaging

by Fourier domain optical coherence tomography. *J. Biomed. Opt.*, **7**, 457–463.

27 Leitgeb, R., Hitzenberger, C.K., and Fercher, A.F. (2003) Performance of Fourier domain vs. time domain optical coherence tomography. *Opt. Express*, **11**, 889–894.

28 de Boer, J.F., Cense, B., Park, B.H., Pierce, M.C., Tearney, G.J., and Bouma, B.E. (2003) Improved signal-to-noise ratio in spectral-domain compared with time-domain optical coherence tomography. *Opt. Lett.*, **28**, 2067–2069.

29 Choma, M., Sarunic, M., Yang, C., and Izatt, J. (2003) Sensitivity advantage of swept source and Fourier domain optical coherence tomography. *Opt. Express*, **11**, 2183–2189.

30 Thorlabs (2013) Superluminescent Diode (SLD) Light Source for OCT Systems, http://www.thorlabs.de/newgrouppage9.cfm?objectgroup_id=3753 (accessed 26 August 2013).

31 Saleh, B.E.A. and Teich, M.C. (2007) *Fundamentals of Photonics*, John Wiley & Sons, Inc.

32 Superlum (2013) Superlum Website, http://www.superlumdiodes.com (accessed 26 August 2013).

33 Gross, H., Singer, W., and Totzeck, M. (2005) *Handbook of Optical Systems – Physical Image Formation*, vol. 2, John Wiley & Sons, Inc., Hoboken.

34 Hacker, M. and Kempe, M. (2009) Tief in die Augen geschaut. *Phys. J.*, **8**, 31–36.

35 Yan, L., Yao, X.S., Lin, L., and Chen, X. (2008) Improved fast scanning delay line in optical coherence tomography applications utilizing fiber stretcher. *Opt. Eng.*, **47**, 5011–5013.

36 Gross, H. (2005) *Handbook of Optical Systems – Fundamentals of Technical Optics*, vol. 1, John Wiley & Sons.

37 Chinn, S.R., Swanson, E.A., and Fujimoto, J.G. (1997) Optical coherence tomography using a frequency tunable source. *Opt. Lett.*, **22**, 340–342.

38 Huber, R., Wojtkowski, M., and Fujimoto, J.G. (2006) Fourier Domain Mode Locking (FDML): A new laser operating regime and applications for optical coherence tomography. *Opt. Express*, **14**, 3225–3237.

39 Wieser, W., Biedermann, B.R., Klein, T., Eigenwillig, C.M., and Huber, R. (2012) Multi-megahertz OCT: High quality 3D imaging at 20 million A-scans and 4.5 GVoxels per second. *Opt. Express*, **18**, 14685–14704.

40 Liu, A.Q. and Zhang, X.M. (2007) A review of MEMS external-cavity tunable lasers. *J. Micromech. Microeng.*, **17**, R1–R13.

41 Grulkowski, I., Liu, J.J., Potsaid, B., Jayaraman, V., Lu, C.D., Jiang, J., Cable, A.E., Duker, J.S., and Fujimoto, J.G. (2012) Retinal, anterior segment and full eye imaging using ultrahigh speed swept source OCT with vertical-cavity surface emitting lasers. *Opt. Express*, **3**, 2733–2751.

42 Považay, B., Unterhuber, A., Hermann, B., Sattmann, H., Arthaber, H., and Drexler, W. (2006) Full-field time-encoded frequency-domain optical coherence tomography. *Opt. Express*, **14**, 7661–7669.

43 Marks, R.J. (2011) *Introduction to Shannon Sampling and Interpolation Theory*, Springer.

44 Paschotta, R. (2008) *Encyclopedia of Laser Physics and Technology*, Wiley-VCH Verlag GmbH, Weinheim.

45 Paschotta, R. (2013) Superluminescent diodes. http://www.rp-photonics.com/superluminescent_diodes.html (accessed 26 August 2013).

46 Fujimoto, J.G. and Farkas, D.L. (2009) *Biomedical Optical Imaging*, Oxford University Press.

47 Gora, M., Karnowski, K., Szkulmowski, M., Kaluzny, B.J., Huber, R., Kowalczyk, A., and Wojtkowski, M. (2009) Ultra high-speed swept source OCT imaging of the anterior segment of human eye at 200 kHz with adjustable imaging range. *Opt. Express*, **17**, 14880–14894.

48 OCTNews (2013) Tunable sources. http://www.octnews.org/category/37/frequency-tunable-sources/ (accessed 26 August 2013).

49 Steiner, R. and Huang, D. (2008) *Anterior Segment Optical Coherence Tomography*, Slack Inc., Thorofare.

50 Izatt, J.A., Hee, M.R., Swanson, E.A., Lin, C.P., Huang, D., Schuman, J.S.,

Puliafito, C.A., and Fujimoto, J.G. (1994) Micrometer-scale resolution imaging of the anterior eye in vivo with optical coherence tomography. *Arch. Ophthalmol.*, **112**, 1584–1589.

51 Mundt, G.H. and Hughes, W.F. (1956) Ultrasonics in ocular diagnosis. *Am. J. Ophthalmol.*, **41**, 488–498.

52 Santodomingo-Rubido, J., Mallen, E.A.H., Gilmartin, B., and Wolffsohn, J.S. (2002) A new non-contact optical device for ocular biometry. *Br. J. Ophthalmol.*, **86**, 458–462.

53 Fercher, A. and Roth, E. (1986) Ophthalmic laser interferometry. *Proc. SPIE*, **658**, 48–51.

54 Haigis, W. (2012) Challenges and approaches in modern biometry and IOL calculation. *Saudi J. Ophthalmol.*, **26**, 7–12.

55 Haigis, W. and Lege, B.A.M. (2000) Optical biometry as alternative to acoustical biometry. *Ophthalmo-Chirurgie*, **3**, 75–80.

56 Haigis, W. (2008) Biometrie, *Augenärztliche Untersuchungsmethoden*, Thieme, pp. 358–404.

57 Shammas, H.J. (2003) *Intraocular Lens Power Calculations*, Slack Inc., Thorofare.

58 Haigis, W., Dick, B., Schwenn, O., and Weidler, S. (2000) Optimierung von Linsenkonstanten für mono- und multifokale Silikon-Linsen, in *13. Kongress der Deutschsprachigen Gesellschaft für Intraokularlinsen Implantation und Refraktive Chirurgie (DGII)*, (eds T. Kohnen, C. Ohrloff, and M. Wenzel), Biermann, Frankfurt, pp. 194–199.

59 Jayaraman, V., Jiang, J., Li, H., Heim, P.J.S., Cole, G.D., Potsaid, B., Fujimoto, J.G., and Cable, A. (2011) OCT imaging up to 760 kHz axial scan rate using single-mode 1310 nm MEMS-tunable VCSELs with > 100 nm tuning range. 2011 Conference on Lasers and Electro-Optics (CLEO), 1–2.

60 Baumann, B., Potsaid, B., Kraus, M.F., Liu, J.J., Huang, D., Hornegger, J., Cable, A.E., Duker, J.S., and Fujimoto, J.G. (2011) Total retinal blood flow measurement with ultrahigh speed swept source/Fourier domain OCT. *Biomed. Opt. Express*, **2**, 1539–1552.

61 Potsaid, B., Baumann, B., Huang, D., Barry, S., Cable, A.E., Schuman, J.S., Duker, J.S., and Fujimoto, J.G. (2010) Ultrahigh speed 1050 nm swept source/Fourier domain OCT retinal and anterior segment imaging at 100 000 to 400 000 axial scans per second. *Opt. Express*, **18**, 20029–20048.

62 Pircher, M., Hitzenberger, C.K., and Schmidt-Erfurth, U. (2011) Polarization sensitive optical coherence tomography in the human eye. *Prog. Retin. Eye Res.*, **30**, 431–451.

63 Kim, D.Y., Fingler, J., Werner, J.S., Schwartz, D.M., Fraser, S.E., and Zawadzki, R.J. (2011) In vivo volumetric imaging of human retinal circulation with phase-variance optical coherence tomography. *Biomed. Opt. Express*, **2**, 1504–1513.

8
Functional Diagnostics

In addition to the imaging and structural diagnostic devices and methods described in Chapters 5 to 7, functional diagnostics are complementary and, at the same time, essential in the diagnosis and monitoring of diseases of the human eye. Functional status can be described by *globally* defined functional parameters as measured with subjective methods. In the following, by "global functional status" we mean the degree of the patient's subjective ability to adequately recognize and/or respond to external visual stimuli. Alternatively, functional status can be described by *locally* defined parameters which characterize the metabolic status of the eye tissue (Table 4.3 in Section 4.3) at the cellular level.

Global (visual) functional status The most important parameters to describe the eye's global visual functional status are visual acuity (Section 2.1.5.1) and visual field (Section 2.1). Visual acuity quantifies the performance of central vision and can be impaired by refractive errors (Section 3.1), pathological changes in one or more ocular segments (e.g., cornea, eye lens, retina), and/or diseases of the retino-cortical visual pathway. For this reason, determination of visual acuity is the most important functional check in everyday clinical practice. In addition, the visual acuity of an optically corrected eye (best corrected visual acuity (BCVA)) is a very important criterion for assessment of the success of any eye treatment. Visual acuity can be quantitatively determined by means of simple wall chart tests. To determine the best corrected visual acuity, trial lenses in trial frames or phoropters (Chapter 5) are additionally required.

The *visual field* is the area of the environment out of which an observer can obtain visual information when fixating on a particular object. Visual field examination allows identification of functional disorders in the whole visual system, ranging from the photoreceptors to the neuronal components of the visual cortex (Sections 1.1 and 1.2.1). It is particularly essential in the diagnosis and management of glaucoma (Section 3.3) and also can play an important role in the management of some neurological diseases. We will discuss the concepts and applications of the visual field examination in Section 8.1.

Local (visual) functional status By the term "local functional status", we mean the metabolic status of the retina at the cellular level. Objective parameters for the

Optical Devices in Ophthalmology and Optometry, First Edition. Michael Kaschke, Karl-Heinz Donnerhacke, and Michael Stefan Rill.

characterization of the functional status are, amongst others, parameters related to microcirculation (e.g., blood flow and oxygen saturation). Metabolic mapping may allow early diagnosis, since dysfunctions in the metabolism may be an early indicator of eye diseases. Instruments and testing methods developed for this purpose, many of which are currently in the research phase, are described in Section 8.2.

8.1 Visual Field Examination

Visual field examination is also referred to as *perimetry*. Corresponding instruments, called *perimeters*, are used to determine the size of the visual field and to analyze the functionality of the visual system within the visual field. Today, most perimeters are fully automated, allowing for standardized testing methods and complex methods for data analysis. At present, perimetry is the most relevant functional method for diagnosing glaucoma-related visual defects and for tracking their progression in a quantitative manner. With this technique, loss of visual function can usually be detected and quantified before the patient becomes aware of it. As perimetry methods have a broad dynamic range, a substantially wider damage spectrum of glaucoma can be diagnosed and monitored over longer periods of time than with current structural analysis methods such as fundus imaging (Section 6.7), confocal scanning laser ophthalmoscopy (Section 6.8.1), scanning laser polarimetry (Section 6.8.3), or optical coherence tomography (Section 7.6). In particular, for the management of advanced stages of glaucoma, perimetry is the method of choice. Other applications of visual field examination include:

- Certification of the existing visual performance, for example, for driver license testing and for legal blindness tests.
- Diagnosis of retinal diseases (e.g., of macular diseases) as a complementary tool to the above-mentioned analysis methods.
- Diagnosis and management of neurological disorders of the visual pathway, as a complementary method to established imaging methods such as computer tomography (CT) and magnetic resonance imaging (MRI).

8.1.1 Physiological Aspects and Functional Principles

To analyze the visual field, different response functions of the eye's visual system can be used such as

- *differential light sensitivity*,
- *color discrimination*, and
- *peripheral motion vision* or *motion detection*.

In most commonly used perimeters, the goal of examination is to quantify the differential light sensitivity (DLS). Hence, in the following, we will focus on the physiological foundations and functional principles central to the determination of DLS. Selected alternative perimetry methods that utilize other vision response functions are described in Section 8.1.3.

8.1.1.1 **Differential Light Sensitivity**

In perimetry (Figure 8.1), a patient focuses on a fixation target which is located in the center of a uniformly illuminated bowl with constant background luminance $\mathcal{L}_b$ (Section A.2.1.5). For visual field examination, a light stimulus with a luminance of $\mathcal{L}$ is projected onto the bowl surface outside of the central fixation target.[1] The patient is now requested to push a button as soon as he or she can see the light spot on the background. The minimum luminance difference which can be perceived as a light stimulus with a 50% probability when presented many times is the so-called *threshold differential luminance* $\Delta\mathcal{L}_{th} = \mathcal{L}_{th} - \mathcal{L}_b$.[2] According to the patient's response, the location on the retina and the corresponding threshold value $\Delta\mathcal{L}_{th}$ are recorded. This procedure is then repeated several times for different locations within the patient's visual field. Due to the inverse relationship between thresh-

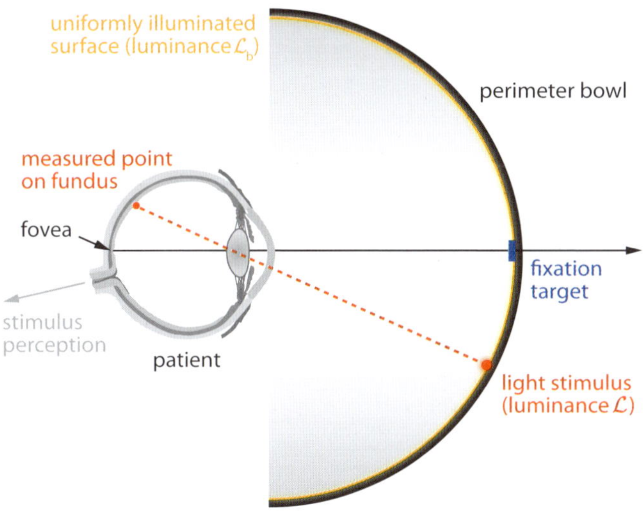

Figure 8.1 General principle of a bowl perimeter. The patient is asked to fixate steadily on a fixation target (blue) which is located in the center of a uniformly illuminated bowl with constant background luminance $\mathcal{L}_b$. Light stimuli with a luminance of $\mathcal{L}$ are presented at various locations in the peripheral visual field (red). The patient pushes a button as soon as he or she perceives the light spot. If a stimulus is seen, brightness is decreased until not seen. If not seen, brightness is increased, thus determining the visual sensitivity at each tested point.

1) In perimetry, the luminance $\mathcal{L}$ is often stated in the non-SI unit "apostilb" (asb), where $1\,\text{asb} = 1/\pi\,\text{cd/m}^2 = 0.3181\,\text{cd/m}^2$.
2) Most perimeters use a luminance of $\mathcal{L}_b = 10\text{cd/m}^2$ for background illumination, which is high enough for photopic (daylight) vision (Section 2.1.6). Hence, only the cones' ability of contrast distinction is tested with a perimeter.

old differential luminance and retinal sensitivity (lower values of $\Delta\mathcal{L}_{th}$ correspond to higher values of the retinal sensitivity). The luminance values (in cd/m^2) are expressed by an inverted decibel (dB) scale, that is,

$$\mathcal{L}\text{ (in dB)} = 10\log\left(\frac{\mathcal{L}_{max}\text{ (in cd/m}^2\text{)}}{\mathcal{L}\text{ (in cd/m}^2\text{)}}\right) . \tag{8.1}$$

$\mathcal{L}_{max}$ denotes the brightest stimulus that the perimeter can produce (Problem P8.2). The higher the dB value of the stimulus luminance, the dimmer the stimulus. In the case of $\mathcal{L} = \mathcal{L}_{max}$, the dB value equals zero.

The threshold differential luminance $\Delta\mathcal{L}_{th}$ in the inverted dB scale is referred to as the *differential light sensitivity* and is given by

$$\text{DLS (in dB)} = 10\log\left(\frac{\Delta\mathcal{L}_{max}\text{ (in cd/m}^2\text{)}}{\Delta\mathcal{L}_{th}\text{ (in cd/m}^2\text{)}}\right) . \tag{8.2}$$

$\Delta\mathcal{L}_{max}$ represents the maximum luminance difference that can be achieved with the perimeter. With this inverted dB scale, we obtain the desired direct relationship between threshold luminance and light sensitivity of the retina. Here, < 0 dB means that no light stimulus can be perceived by the patient even for maximum luminance difference. Note that the decibel scales are not standardized, because the maximum values of $\mathcal{L}_{max}$ and $\Delta\mathcal{L}_{max}$ vary between individual instruments (Problem P8.1).

The definition of the DLS as a dB value has the advantage that the broad range of luminance values are "compressed" by the logarithmic scale and transformed into an equidistantly subdivided dB scale. Additionally, the DLS is an intuitive quantity, since areas of the visual field with a high sensitivity have high dB values and vice versa.

Local DLS values can be represented by a three-dimensional *hill of vision* (Figure 8.2a). If all locations having equal DLS values are connected by a line, so-called *isopters* are formed, which are comparable to contour lines in cartography. Under photopic lighting conditions (Section 2.1.6), the DLS is normally maximal in the fovea and declines with increasing distance from the fovea.

The normal shape of the hill of vision depends on testing conditions such as size, duration, and color of the light stimulus, as well as on the brightness of the background illumination. To obtain comparable data, testing conditions are specified in standard examination programs. Since the typical shape of the hill of vision changes with age, it is necessary to determine the changes from a representative group of normal subjects with standard testing conditions. Visual field defects are identified as deviations from the normal age-matched hill of vision. Some manifestations of the visual field defects are *relative scotomas* (local defects), generalized depressions of the whole visual field, and combinations of both. Areas in which the stimuli cannot be perceived by the patient even at the maximum stimulus brightness (i.e., DLS < 0 dB) are termed as *absolute defects* or *scotomas*. At the location of the optic nerve head, we always have an absolute scotoma in the visual field, since no photoreceptors are present in this area. Uncorrected refractive errors and/or

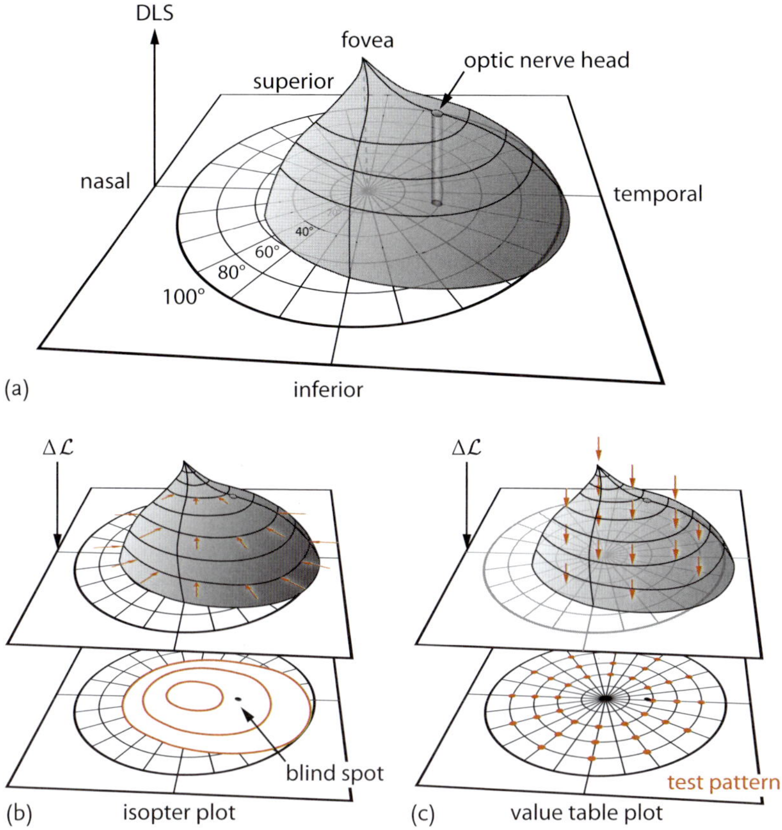

Figure 8.2 (a) Schematic three-dimensional representation of a hill of vision for a normal right eye. The top of the hill is usually the fovea (point of fixation). On the optic nerve head, no photoreceptors are present so that this point is effectively blind (blind spot). (b) Schematic representation of kinetic perimetry and corresponding isopter plot. (c) Schematic representation of static perimetry and corresponding value table plot. Adapted from [1].

media opacities such as cataracts (Section 3.2) may lead to a general decrease in DLS over the entire visual field.

The 0 dB isopter marks the outer rim of the hill of vision and thus represents the border of the peripheral visual field that can be perimetrically measured. Typical limits for the size of the visual field (Figure 8.2a), measured in degrees from the fovea, are:

- approximately 50°–60° in superior direction,
- approximately 70°–75° in inferior direction,
- approximately 50°–60° in nasal direction, and
- approximately 100°–110° in temporal direction.

Medical visual field examinations are usually carried out in the central field only, that is, within 30° of the fovea, since the diagnostic information obtained there is most important to patients. As the visual fields of both eyes overlap, each eye of the patient is almost always examined separately. In contrast, testing for disability certification or for driving eligibility may be done binocularly, in order to better reflect the patient's every day visual performance. This is usually done in both the central and peripheral fields, again because of the importance of peripheral vision in daily tasks.

8.1.1.2 Examination Methods

Two methods are used for visual field examination, that is, *kinetic perimetry* (Figure 8.2b) and *static perimetry* (Figure 8.2c). Let us now consider these methods in brief.

Kinetic perimetry In kinetic perimetry, a stimulus of defined brightness and size is placed in an area of the visual field where it cannot be seen, and moved at constant velocity until it enters an area where it is seen. In the simplest case, the stimulus is placed out of sight in the far periphery, and moved towards the fovea until seen. If this process is repeated step by step from multiple directions around the hill of vision, we will obtain a set of points of equal differential light sensitivity. We can then connect these points to form an isopter. The luminance or size of the light stimulus is then decreased and the process repeated. In this way, a sequence of isopters may be determined, from which the elevation profile of the patient's hill of vision can be represented in two dimensions (in analogy to topographic contour lines). In practice, 3–5 isopters are usually determined and presented on a single report, along with a map of the physiological blind spot. Manual kinetic perimetry is a precise and globally-accepted method for determining the borders of the peripheral visual field, as well as any localized defects therein. However, because this method is time-consuming and requires a highly skilled technician, it is mostly used in academic settings and for patients with limited cooperative ability.

Kinetic perimetry may also be used to certify the remaining visual performance, for example, in driver license testing or legally required blindness tests. Kinetic perimetry is often performed manually with Goldmann manual perimeters by specially trained examiners (Section 8.1.2). However, computer-aided kinetic perimetry can also be performed by means of modern static perimeters with special software.

Static perimetry The goal of static perimetry is the determination of DLS at a set of predefined locations in the visual field (test pattern). Stimulus brightness is systematically varied in order to determine DLS at each tested location. This form of visual field testing is commonly termed *static threshold perimetry*.

In clinical practice, computerized static threshold perimetry is normally performed under standardized testing conditions, in which circular white light stimuli are projected using fixed and well-documented testing protocols on a white background of fixed luminance. This method is called *standard automated perimetry* (SAP) and is the most widely used perimetric method worldwide.

SAP is most commonly used for diagnosis and management of glaucoma, and often also for neurological diseases. Perimetric testing in these medical conditions usually is restricted to the central visual field, within 30° of the center of vision. The most frequently used tests determine DLS at either 54 or 76 test locations in the central field. These numbers of test points have proven to provide reasonable trade-offs between testing speed, patient fatigue, and diagnostic completeness.

8.1.2 Basic Perimeter Design

A perimeter is an instrument which is able to present a light stimulus of known size and brightness for a defined amount of time in a known location of the visual field [2]. Efficient visual field testing can only be achieved if all of these requirements are fulfilled.

History The first perimeter which met all these requirements was developed by Hans Goldmann (1899–1991) in the early 1940s and fabricated by the company Haag–Streit from 1945 onwards. Goldmann used the uniformly illuminated inner surface of a hemisphere as a projection surface for reproducible presentation of light stimuli. Light stimuli were moved on the inner surface of the hemisphere, and their location coordinates were simultaneously documented by means of a pantograph. The size and luminance of the test points, as well as the brightness and size of the hemisphere introduced by Goldmann are still used almost unchanged in modern perimeters. Even today, the Goldmann perimeter serves as a reference instrument in kinetic perimetry. Other important development steps in perimetry are [3, 4]:

- Development of static perimetry by Elfriede Aulhorn (1923–1991) and Heinrich Harms (1908–2003) in the late 1950s. Based on their research, the company OCULUS produced the Tübinger Perimeter from 1959 onwards. The Tübinger Perimeter was the first instrument which allowed perimetrists to perform manual static perimetry under standardized testing conditions.
- Development of automated static perimetry by Franz Fankhauser and colleagues. This work began in 1956 in collaboration with Jean Marie Parel in Berne and culminated in the commercial introduction of the automated perimeter Octopus® 201 in 1974 by the company Interzeag (later acquired by Haag–Streit).
- Improvement of SAP test and analysis methods by Anders Heijl and colleagues from the 1970s onwards. These developments were then incorporated into the Humphrey® Field Analyzer by the company Humphrey Instruments (later Carl Zeiss).

Currently, several companies offer perimeters with comparable basic features such as size of the visual field, size of the light stimuli, and luminance. However, the

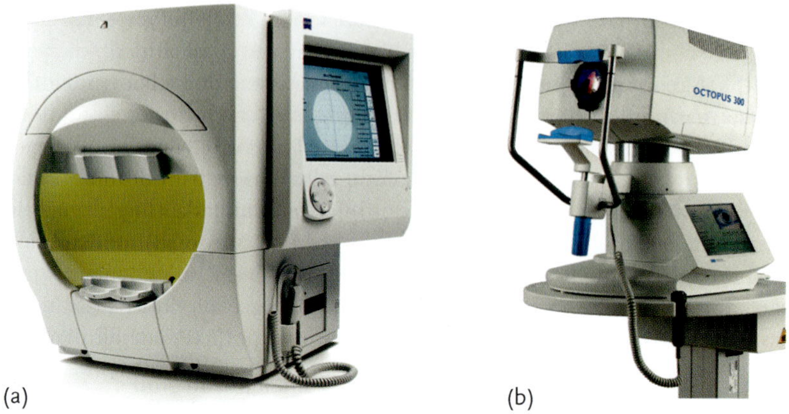

Figure 8.3 (a) Photograph of the Humphrey Field Analyzer HFA II-i by the company Carl Zeiss. Courtesy of Carl Zeiss (b) Photograph of the direct projection perimeter Haag–Streit Octopus® 300. Courtesy of Haag–Streit Deutschland GmbH.

efficiency of available testing algorithms, data management strategies, and user-friendliness differ significantly among commercial products.

8.1.2.1 **Projection System**

Depending on the type of presentation of the light stimuli, perimeters can be classified into bowl perimeters (e.g., shown in Figure 8.3a) and compact perimeters (e.g., shown in Figure 8.3b).

Bowl perimeters The bowl perimeter is the most common perimeter design and generally follows the design principles originally laid down by Goldmann. Light stimuli are projected with a mirror system onto a uniformly illuminated and diffusely scattering inner surface of a hemisphere or slightly aspherical shell (Figure 8.1). The patient's eye is positioned in the center of the bowl by means of a headrest. By pressing a button, the patient can record the perception of a light stimulus. Advantages of the bowl perimeter design:

- Constant visual field testing distance (usually 30 cm).
- Defined examination conditions, as the hemisphere substantially shields the testing surface from scattered ambient light.
- Large spatial range of the visual field measurement (maximum range of 90°).
- Flexible test target patterns.

Disadvantages of the bowl perimeter design:

- Large footprint.
- Slightly dimmed room required for examination.
- Because of the finite object distance of the light spots (30 cm), older patients may frequently require a near correction with a positive lens in addition to their refractive correction for distance vision to sharply see the light stimuli.

Compact perimeters In compact perimeters, the fixation target, the light stimuli, and the background illumination are virtually presented at optical infinity. For this purpose, the image is either directly projected onto the examined retina (direct projection perimeters) or presented on a cathode ray tube (monitor-based perimeters). Compared to bowl perimeters, compact perimeters have the following advantages:

- The examination is less dependent upon ambient light conditions.
- The visual field testing distance is infinite due to the chosen projection setup. Thus, no further correction is required in addition to the patient's distance refraction.
- Substantially smaller instrument footprint.

The major drawback of compact perimeters, as compared to bowl perimeters, is the limited range of the visual field measurement (maximum range of 30°).

8.1.2.2 Test Conditions

The result of a visual field examination depends on the following testing conditions:

- background luminance (and color),
- size (and color) of light stimuli, and
- exposure time of light stimuli.

To obtain comparable data, certain testing conditions are specified in standard examination programs. Examination under standardized testing conditions simplifies data exchange and is of particular importance for progression monitoring of diseases. In the following, standard values are listed for SAP:

- *Background illumination:* The luminance of the inner surface of the perimeter bowl is usually 10 cd/cm^2. In this regime, vision is primarily mediated by cone photoreceptors. Cone vision depends more on the image contrast than on absolute image brightness. Modern perimeters use the principle, first elucidated by Goldmann, of making small adjustments in stimulus brightness in order to keep the stimulus contrast consistent and relatively independent of unintended variations in background luminance.
- *Size of light stimuli:* While perimeters usually offer all standard stimulus sizes introduced by Goldmann (Figure 8.4), automated static perimetry has generally been standardized around Goldmann's Size III stimulus. Stimulus size V is sometimes used in advanced visual field loss. Compact perimeters are generally limited to the stimulus sizes III and V.
- *Duration of light stimuli:* For light stimuli which are shown longer than approximately 100 ms, visibility is largely independent of the duration of presentation, thus reducing the effects of small variations in stimulus duration. If, however, the stimulus duration is longer than the patient's reaction time, typically on the order of 200 ms, patients may be tempted to try to look in the direction of the stimulus instead of gazing steadily at the central fixation target. Thus, durations of 100 or 200 ms are typically used.

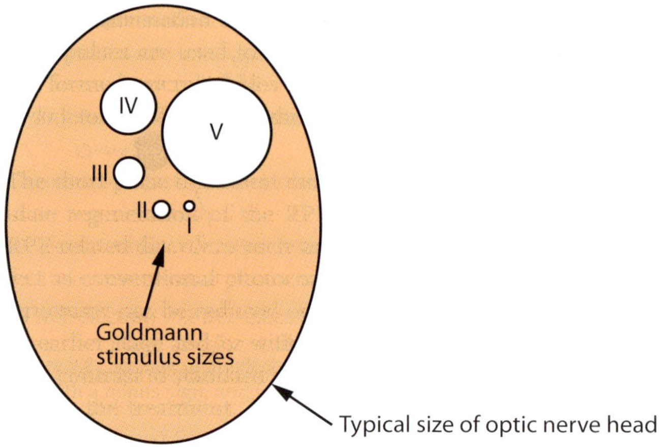

Goldmann notation	I	II	III	IV	V
Size of stimulus (arc min.)	6.45′	12.9′	25.7′	51.5′	103.0′

Figure 8.4 Standard Goldmann stimulus sizes as compared to the typical size (5° × 7°) of the physiological blind spot (red). Bottom: Allocation of Goldmann notation and size of stimuli in arc minutes. Adapted from [2].

- *Fixation monitoring:* To obtain reliable visual field examinations, it is important to identify patients who are not able to steadily gaze at the fixation target during testing. Fixation monitoring is often accomplished by means of gaze tracking or related methods. In this way, we can detect fixation fluctuations and prevent measuring errors.

8.1.2.3 **Test Patterns and Strategies**

In recent years, testing with automated perimetry has converged upon the use of just a few standard testing patterns and strategies. In the diagnosis and management of glaucoma and neurological disease, the manufacturer of the most commonly used perimeter recommends the use of a single specific testing pattern and a single testing strategy, except in rather specific situations. While other manufacturers may make alternative recommendations, based on the specific capabilities of their products, the tendency toward simplification and standardization has been quite clear in recent years.

Exceptions to these recommendations include disability certification and driver's license testing, where the goal is identification of profound loss, rather than subtle early disease. Here, stimuli are presented which are so bright that they should only be missed by patients having profound loss. Another exception is testing for retinal toxicity in patients chronically taking hydroxychloroquine and similar medications. In these patients, the goal is to detect subtle early loss in the very central vision, within 10° of the fovea.

Testing strategies have evolved significantly over the years to the point where, for example, the most commonly used threshold test for glaucoma and neurological loss now takes only about 5–7 min (i.e., less than a quarter of the time required by the original commercial devices) with no loss in diagnostic performance. Even faster strategies are now available, although with perhaps some compromise in data quality. In general, threshold testing takes longer than, for example, the simple strategies used in driver's license testing. In the latter case, a single very bright stimulus is presented at each tested point, which is either seen or not seen. In the former case, the goal is to present at each tested point at least one stimulus which was seen and at least one that was not seen. Then, the threshold of vision is empirically determined at each test point location. Quite obviously, efficient threshold testing is anything but simple, and requires careful use of everything that can be known about the patient's visual function status. Please refer to the specialized literature for a detailed description of these test patterns and strategies, for example, [1, 2, 5].

8.1.2.4 **Data Analysis and Reporting Visual Field Data**

After the visual field examination, perimeters use software packages which store and analyze the local DLS values (raw data) and prepare the results for presentation. Due to the variability of visual field measurements, the significant overlap between normal and pathological visual field findings, and the slow progression of defects in glaucoma, statistical analysis of the data is needed to help answer important diagnosis related questions such as:

- Is the measured visual field within the normal age-related range or not?
- Are changes observed over time statistically significant or not?
- Is the observed rate of change over time rapid enough to pose a threat of visual disability to the patient? Or is it necessary to change the therapy?

The results of a visual field examination are presented in the form of instrument-specific reports. In addition to the graphic representations of the local DLS distribution (in the form of numeric values or interpolated grayscales), reports also contain so-called *deviation plots*. The latter highlight areas of the visual field that fall outside the normal range after correction of the DLS values for the patient's age (total deviation) or for any generalized loss, for example, as the result of cataract or a small pupil (pattern deviation). To monitor the changes over longer periods of time (progression analysis), characteristic global indices are used, such as the *mean deviation* and *visual field index* (VFI). In Figure 8.5, a typical visual field report is shown in which relevant parameters are highlighted.

With software programs like ZEISS FORUM® or Haag–Streit EyeSuite, it is possible to import ophthalmic images and/or structural data which were previously acquired with suitable imaging methods (e.g., fundus photography (Section 6.7), confocal scanning laser ophthalmoscopy (Section 6.8.1), and optical coherence tomography (Section 7.4)). The structural data can then be presented together with visual field data in combined diagrams and/or plots. Suitable combinations of func-

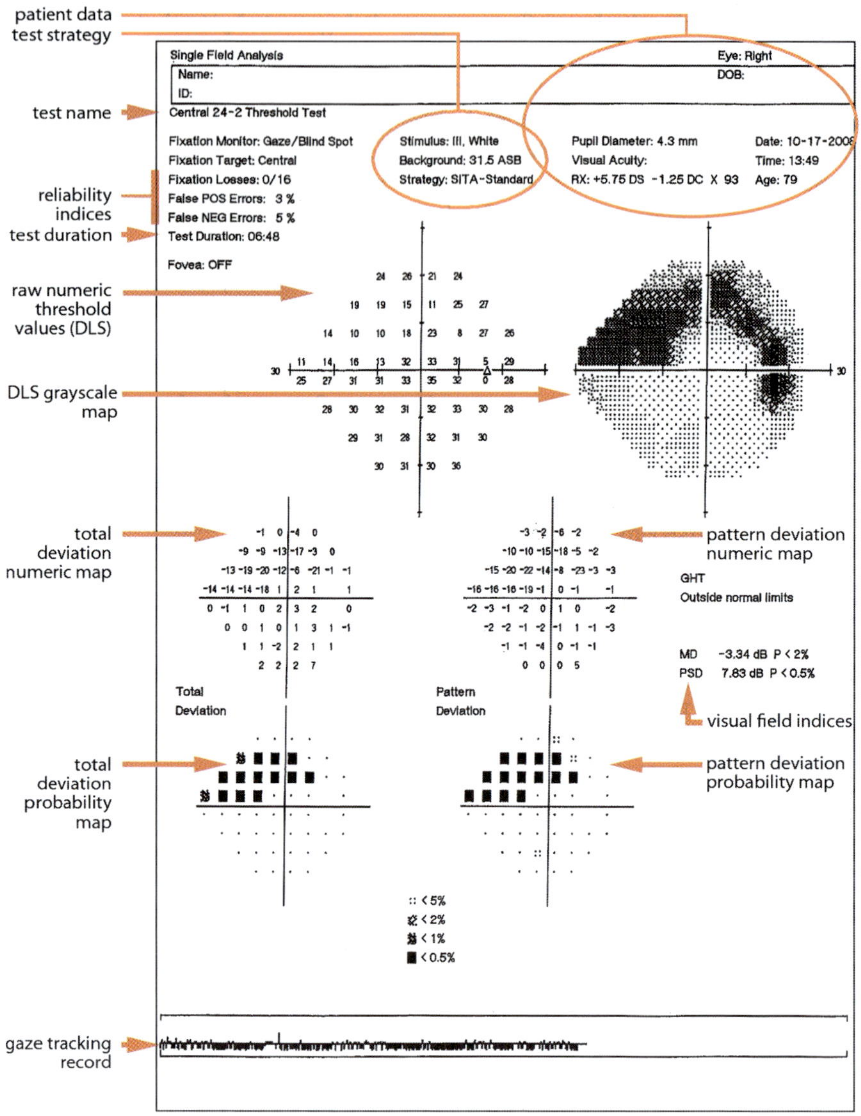

Figure 8.5 Typical STATPAC single field analysis for an eye which suffers from visual field loss caused by glaucoma. For example, probability symbols in the total deviation probability map and the pattern deviation probability map for $p < 0.5\%$ indicate that fewer than 0.5% of normal, tested eyes are expected to have such a low sensitivity. Courtesy of Carl Zeiss.

tional and structural data have been shown to be more valuable than either functional or structural data taken alone [6, 7]. Please refer to the specialized literature for a detailed description of the data analysis and presentation programs, for example, [1, 2, 5].

8.1.3
Alternative Perimetric Concepts

In addition to the previously described traditional SAP, other perimetry methods have been developed for visual field examination. The most important alternative functional tests are

- *short-wavelength automated perimetry* (SWAP),
- *frequency-doubling technology* (FDT),
- *flicker-defined form (edge) perimetry* (FDF), and
- *fundus-controlled perimetry*.

The SWAP, FDT, and FDF methods were designed with the goal of detecting functional losses of the visual system earlier than with SAP. These methods seek to stimulate a subgroup of the retinal ganglion cells, and to thereby reduce the effects of redundancy which may occur in SAP. While these methods have not been shown to provide compellingly earlier diagnoses than SAP [8], they do offer alternatives to SAP, each with their own strengths and limitations.

Fundus-controlled perimetry is used to examine the functional status of the retina in selected areas, that is, under the perimetrist's direct visual control, with retinal imaging methods. Let us now consider the alternative function tests in brief.

8.1.3.1 Short Wavelength Automated Perimetry

In short wavelength automated perimetry (SWAP)[3], blue light stimuli (λ = 440 nm ± 35 nm) are presented on a yellow background ($\lambda \geq 530$ nm). Because of the intense yellow background luminance ($\mathcal{L}_b \approx 100\,\text{cd/m}^2$), the responsiveness of the green-sensitive *M* cones and red-sensitive *L* cones (Section 1.2) is reduced. The blue stimuli are thus primarily detected by the blue-sensitive *S* cones which are only connected with *koniocellular ganglion cells* (*K* cells). As koniocellular ganglion cells represent only 10% of the total ganglion cell population, just a small, highly specialized subgroup of the ganglion cells is stimulated.

Compared to SAP, SWAP has a lower dynamic range (approximately 15 dB). In addition, SWAP results are strongly affected by the transmittance of the eye lens in the blue spectral range. Hence, it is also largely affected by light scattering in ocular media, which restricts its usefulness particularly in elderly patients.

SWAP has been implemented in a number of commercially available perimeters as a standard examination program which complements SAP. Data analysis and presentation are performed as described in Section 8.1.2. Please refer to [9] for a detailed description of the SWAP technology.

8.1.3.2 Frequency Doubling Technology

The proportion of *magnocellular ganglion cells* (*M* cells) in the total ganglion cell population is approximately 10%. In frequency doubling technology (FDT), these *M*

3) This method is also referred to as *blue-yellow perimetry*.

cells are preferentially stimulated by rapidly changing grating structures. The square-like stimuli consist of sine gratings with a spatial frequency of $\leq$ 0.5 cycles per degree which are counterphased flickered with a frequency of $\geq$ 12 Hz (Figure 8.6a). The stimuli are presented on a white background with the goal of determining the minimum stimulus contrast that can be detected at each tested location.

FDT is based on an effect which was first described by Kelly *et al.* [10] in 1966. With sufficiently high flicker frequencies and from a certain contrast threshold onwards, the grid is perceived with twice the spatial frequency (*frequency doubling illusion*). When the visual field is increasingly impaired by glaucoma, the contrast required to see this illusion rises [11].

In 1997, the company Welch Allyn developed the first FDT perimeter. Today, FDT perimeters are offered by Carl Zeiss in two versions, that is, the Humphrey FDT and, since 2005, the Humphrey Matrix (Figure 8.6b). Both instruments are able to analyze the central visual field (30°) and provide the following standardized testing conditions:

- duration of stimulus: 300 ms,
- size of stimulus: 10° (Humphrey FDT); 10°, 5°, 2° (Humphrey Matrix),
- spatial frequency of stimulus: 0.25 cycles per degree (10° stimulus), 0.5 cycles per degree (5° and 2° stimuli),
- temporal frequency of stimulus: 25 Hz (10°), 15 Hz (5°), 12 Hz (2°), and
- background luminance: 100 cd/m^2.

In both instruments, stimuli are presented on a cathode ray tube that is imaged at optical infinity. The measurement parameter is the minimum contrast of the sine grid pattern at which the patient can still detect the stimulus structures.

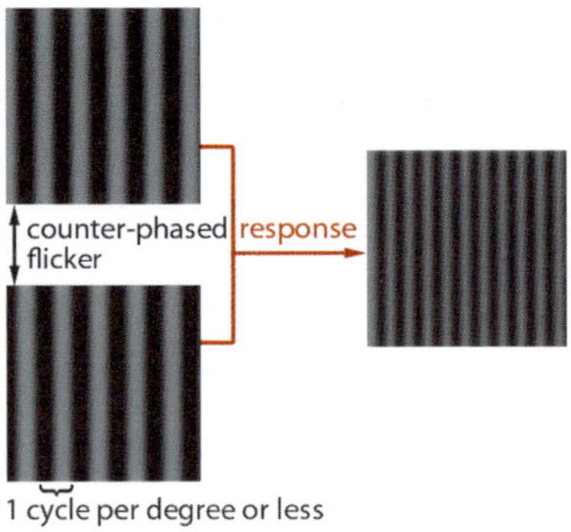

(a)

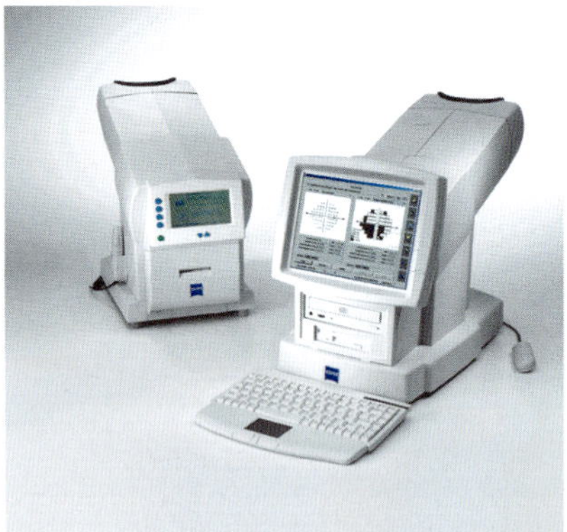

(b)

Figure 8.6 (a) The principle of frequency doubling technology is based on flickering of a sine grid with a frequency greater than 15 Hz. As a consequence, the eye perceives a grid image with twice the spatial frequency (or half the period), that is, the so-called *frequency doubling illusion.* (b) Photograph of Humphrey FDT (left) and Humphrey Matrix (right) by ZEISS which are based on the frequency doubling technology. Courtesy of Carl Zeiss.

The Humphrey FDT by ZEISS is particularly well-suited for suprathreshold screening testing, in which the primary goal is to identify patients having early to moderate visual field loss. Because of the relatively large size of the stimulus ($10^\circ \times 10^\circ$), the examination is less dependent upon exact correction of refractive errors. Testing time for suprathreshold screening tests usually is less than one minute. The smaller stimuli of the Humphrey Matrix by ZEISS allow the examination of the central visual field with a spatial resolution comparable to that of SAP.

FDT perimeters are compact, portable, and relatively low-cost instruments for visual field examination. Studies have shown that these instruments are able to detect visual field losses due to glaucoma and other ophthalmo-neurological disorders with high sensitivity and specificity. Please refer to the specialized literature for a detailed description of FDT perimetry, for example, [12, 13].

8.1.3.3 **Flicker-Defined Form Perimetry**

Similar to FDT, *M* cells are preferentially stimulated in flicker-defined form (FDF) perimetry. On an evenly illuminated background, multiple randomly arranged background dots rapidly alternate between black and white (Figure 8.7). The actual stimulus, called *flicker defined form (FDF)*, is superimposed upon this randomly changing background and appears as a 5° (diameter) circular area in which points flicker in a counter-phased manner relative to the background dots. Above a certain contrast level between the light and dark dots, the phase difference creates a gray circular pop-up illusory stimulus against the uniformly illuminated background.

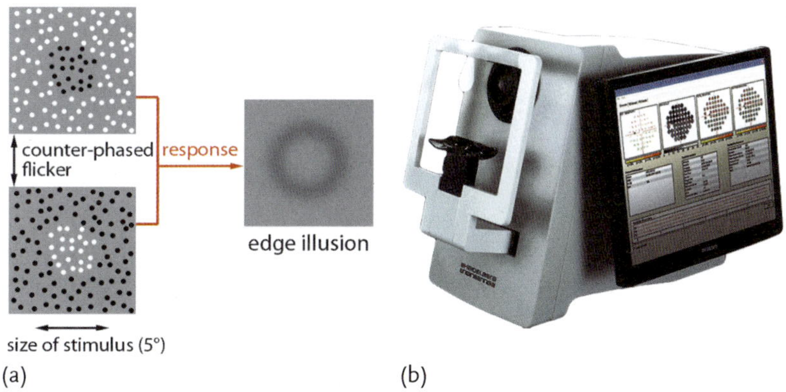

Figure 8.7 (a) The principle of flicker-defined form perimetry is based on the presentation of multiple test dots which are shown in a rapid alternating sequence. The background random dots subtend an angle of 0.34°. The actual stimulus is a 5° (diameter) circular area in which dots flicker in a counter-phased manner relative to the background dots. Above a certain contrast level between the light and dark points, the eye perceives a gray ring-shaped structure in this area. This visual effect is called *edge illusion*. (b) Photograph of the Heidelberg Edge Perimeter (HEP) which is based on flicker-defined form perimetry. Courtesy of Heidelberg Engineering GmbH.

This visual effect is called *edge illusion*. It was discovered in 1991 by Ramachandran *et al.* [14] and investigated in detail by Quaid *et al.* [15].

In 2007, the FDF stimulus was applied to the Heidelberg Edge Perimeter (HEP). In this instrument, the stimuli are presented on a cathode ray tube at optical infinity to the patient's eye. The HEP allows examination of the central visual field in the ranges of 10°, 24°, and 30°. The range can also be extended to 60° by peripheral refixation in which the patient looks at different fixation targets. With the HEP-FDF stimulus, the minimum contrast of the inversely phased flickering points is determined at which the patient perceives the edge illusion. Standard test conditions for the visual field examination are as follows:

- duration of stimulus: 400 ms,
- size of stimulus: 5° (optional 2°),
- size of the random dots: 0.34°,
- flicker frequency: 15 Hz, and
- average background luminance: 50 cd/m^2.

In addition to FDF, a modified version of standard automated perimetry (SAP) is implemented in the HEP.

8.1.3.4 Fundus-Controlled Perimetry (Microperimetry)

In fundus-controlled perimetry or *microperimetry*, light stimuli are projected onto selected areas on the retina. The positions of the light stimuli are visually controlled by the physician with a fundus camera (Section 6.7) or a confocal laser-scanning ophthalmoscope (Section 6.8.1). In this way, the correlation of structural and functional changes of the patient's fundus can immediately be examined. This method allows us to check in a direct and objective manner whether, and to what extent, visible morphological (structural) changes of the fundus lead to functional impairments. Fundus-controlled perimetry can also provide quantitative information about the preferred retinal location of fixation (foveal or extrafoveal) and the stability of fixation. In fact, these are important parameters for the management of macular diseases (e.g., age-related macular degeneration; Section 3.4) and the rehabilitation of low-vision patients.

Fundus-controlled perimeters essentially require

- a basic fundus-imaging system,
- a rapid (high frequency) eye tracker,
- a stimulus projection system,
- fixation targets,
- background illumination, and
- a highly trained technician.

Basic fundus imaging system The primary benefit of fundus-controlled microperimetry is its ability to relate visible retinal changes to functional loss. For this purpose, a basic fundus imaging system such as a scanning laser ophthalmoscope (Section 6.8.1) or a nonmydriatic fundus camera (see Section 6.7) is used which acquires a live near-infrared image of the fundus.

High frequency eye tracker With a high frequency eye tracker ($\geq$ 25 Hz), eye motion is recorded and compensated in real-time so that the stimuli can be exactly projected onto the desired location on the retina. Eye tracking is achieved by automatic recognition of suitable retinal structures or landmarks in the live image. The eye tracker also delivers quantitative information about fixation stability and the patient's preferred location of fixation.

Projection system White light stimuli, colored fixation targets, and the white background illumination are coupled into the beam path of the basic fundus imaging system by means of a projection system (e.g., internal LCD display). The light is then projected onto the desired location on the retina while taking the eye motion into account.

Examination process Local DLS is determined with the standard procedure described in Section 8.1.1. Similar to SAP, the patient has to push a button as soon as he or she recognizes a light stimulus. The software for the automatic control of the measurements as well as the analysis and presentation of the data will not be discussed in further detail, as there are no fundamental differences compared to standard perimeters described in Section 8.1.2.

In fundus-controlled perimetry, the measured data are usually presented as an overlay of the visual field data and the respective fundus image acquired at the end of examination. For this purpose, the local DLS values are added to the fundus image in the form of color-coded dots. Compared to SAP, fundus-controlled perimetry allows a more precise point-to-point correlation of the visual field findings to the fundus image. Depending on the used technique, the basic fundus imaging sys-

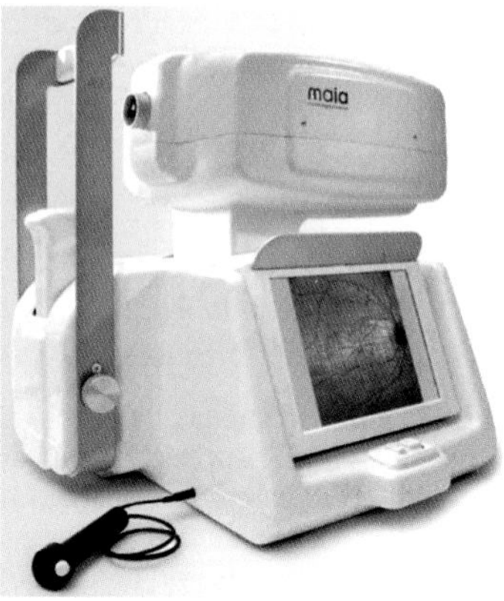

Figure 8.8 Photograph of the Topcon Macular Integrity Assessment (MAIA™). Courtesy of Topcon Deutschland GmbH.

Table 8.1 Technical specifications of two commercial devices for fundus-controlled perimetry, that is, the NIDEK MP1 microperimeter and the Topcon MAIA.

Parameter	NIDEK MP1	Topcon MAIA
Fundus-imaging system	Nonmydriatic fundus camera (IR video and color flash imaging)	Scanning laser ophthalmoscope ($\lambda \approx 850$ nm)
Field of view for fundus imaging	45°	36° × 36°
Field of view for perimetry	22.5°	20° × 20°
Background luminance	1.27 cd/m^2	1.27 cd/m^2
Size of stimulus	I, II, III, IV, V	III
Duration of stimulus	100–2000 ms	n.a.
Maximum stimulus luminance	127 cd/m^2	127 cd/m^2

tems can only cover the central visual field area. Hence, this method is preferably used for functional diagnosis in the macula.

Commercial devices For the two instruments currently offered commercially, that is, the NIDEK MP1 microperimeter and the Topcon Macular Integrity Assessment (MAIA) (Figure 8.8), we summarize the most relevant technical specifications in Table 8.1.

8.1.4 Prospects

Visual field examination (perimetry) is an established functional diagnosis method to detect and monitor visual field loss. Current development trends:

- Increased diagnostic accuracy and reliability by means of improved test or analysis strategies. In this way, potential damage might be detected earlier, and progression might be quantified with greater certainty than is possible with currently available methods.
- Cost-effective, space saving, and user-friendly systems by integration of modern optical technologies (e.g., virtual reality displays), extensive automation, and integration of intelligent analytic algorithms.
- Combination of functional and structural data to improve the diagnostic accuracy.

8.2 Metabolic Mapping

The first symptoms of many diseases may become manifest as disordered metabolism, long before morphological anomalies and functional losses occur. Metabolic mapping thus may allow diagnosis of diseases and monitoring of therapeutic measures at the cellular level [16]. In the analysis of retinal metabolism, extracellular and intracellular aspects must be taken into consideration [17].

Extracellular aspects Extracellular aspects are described by the functional parameters of the blood flow in *arterioles*[4] and *venules*[5] (Section 8.2.1.1) as well as the concentration of *metabolites* in the blood. In the latter case, oxygen saturation plays an important role (Section 8.2.1.2). If blood flow parameters (amount of blood flowing through the vessels per unit time) and oxygen saturation in the retinal arterioles and venules are known, oxygen consumption can be calculated from the *arterio-venous difference* [17]. This parameter may provide important information about the metabolic status of the retina.

Blood flow and oxygen saturation are influenced by complex physiological control and steering mechanisms. Unfortunately, no procedure for a simultaneous, quantitative determination of both parameters is currently available.

Intracellular aspects The autofluorescence of fluorophores can be used to evaluate metabolism at the cellular level (Section 8.2.2). The fluorophores of interest are those that are directly or indirectly involved in the cellular metabolic processes.

8.2.1 Microcirculation Mapping

The functional status of the retina may be characterized by blood flow and oxygen saturation in the retinal arterioles and venules, which is also called *microcirculation*. If the oxygen supply to the retinal tissue is not sufficient, functional losses may appear. In the case of low oxygen consumption, there may be a metabolic disturbance in the retina.

8.2.1.1 Blood Flow Mapping

Measurement of blood flow in retinal vessels requires simultaneous determination of blood flow speed and vessel diameter. Usually, flow speed is measured with *laser Doppler velocimetry* (also related to laser Doppler flowmetry) for selected vessel segments [18]. The vessel diameter can be determined from fundus reflectance images (Section 6.8.1.5). For basic imaging, fundus cameras (Section 6.7) or confocal scan-

4) An arteriole is a blood vessel with a small diameter which extends from an artery and leads to capillaries.

5) A venule is a blood vessel with a small diameter which allows blood to return from the capillaries to the veins.

ning laser ophthalmoscopes (Section 6.8.1) are used. Since currently no clinically accepted methods or systems are available to determine retinal blood flow, we skip further considerations and refer to [19] which discusses the benefits and disadvantages of the used approaches. Optical coherence methods like Doppler FD-OCT (Section 7.8) seem to be promising candidates for blood flow mapping, since they allow simultaneous determination of blood flow in many different vessel segments within a region of interest [20].

8.2.1.2 **Oxygen Saturation Mapping**

Oxygen (O_2) is transported in blood cells via hemoglobin. Hemoglobin is an iron-containing protein situated inside red blood cells which has four oxygen binding sites. In its oxygenated state in which all four bindings are occupied, hemoglobin is referred to as *oxyhemoglobin* (HbO_2). If no binding site is occupied by oxygen molecules (deoxygenated state), it is referred to as *deoxyhemoglobin* (Hb).

Oxygen saturation (OS) is defined as the percentage of HbO_2 molecules relative to the total amount of hemoglobin molecules. We have

$$\text{OS} = \frac{C_{\text{HbO}_2}}{C_{\text{HbO}_2} + C_{\text{Hb}}} \tag{8.3}$$

with the oxyhemoglobin and deoxyhemoglobin concentrations C_{HbO_2} and C_{Hb}, respectively. To measure OS in retinal vessels, we use the differing absorbance (Section 9) of Hb and HbO_2 in the visual spectral range (Figure 9.3 in Section 9.3) [21]. For the absorbance A (see Eq. (9.1)), Lambert–Beer's law (Section 9.1) states

$$A(\lambda) = \ln\left(\frac{I_0}{I}\right) = \mu_{\text{a,mol}}(\lambda) C L \,, \tag{8.4}$$

where I_0 is the intensity of the incident light, I the intensity of the transmitted light, $\mu_{\text{a,mol}}(\lambda)$ the molar absorption coefficient ($[\mu_{\text{a,mol}}] = 1/(\text{cm mol})$), C the concentration of the light-absorbing substance ($[C] = \text{mol/l}$), and L the layer thickness ($[L] = \text{cm}$). For a mixture of Hb and HbO_2, the absorbance is given by

$$A(\lambda) = L(\mu_{\text{a,mol,HbO}_2} C_{\text{HbO}_2} + \mu_{\text{a,mol,Hb}} C_{\text{Hb}}) \,. \tag{8.5}$$

With $C = C_{\text{HbO}_2} + C_{\text{Hb}}$ and Eq. (8.3), we obtain

$$A(\lambda) = C L[\mu_{\text{a,mol,Hb}} + \text{OS}(\mu_{\text{a,mol,HbO}_2} - \mu_{\text{a,mol,Hb}})] \,. \tag{8.6}$$

In this equation, the oxygen saturation OS and the product CL have to be determined. As the molar absorption coefficients $\mu_{\text{a,mol,Hb}}$ and $\mu_{\text{a,mol,HbO}_2}$ are known [22, 23], we obtain OS by measuring $A(\lambda)$ for two different wavelengths. To quantify the product CL, we have to measure at a wavelength for which the absorbances of Hb and HbO_2 are equal, that is, at the so-called *isobestic point*. In this case, we have

$$A(\lambda_{\text{iso}}) = C L \mu_{\text{a,mol}}(\lambda_{\text{iso}}) \,. \tag{8.7}$$

For the second measurement, it is convenient to use a wavelength for which $\mu_{\text{a,mol,Hb}}$ and $\mu_{\text{a,mol,HbO}_2}$ exhibit as large a difference as possible. In a pure hemoglobin solution in which all red blood cells have been removed (*hemolyzed* blood), oxygen saturation mapping can be used to determine the oxygen saturation with high accuracy. However, in practical applications with a living eye, we face a number of problems two of which are:

1. The measurement on the eye must be performed in reflection. When calculating the absorbance, the fraction of the incident light reflected back by the tissues surrounding the vessels is considered to be equal to the incident light. In the blood vessels, however, the incident light interacts not only with hemoglobin, but also with other blood components and the vessel wall itself. This effect must be taken into consideration via suitable evaluation algorithms.
2. Hemoglobin molecules are located inside the red blood cells. Light scattering caused by red blood cells must be taken into account in Eq. (8.6) with a wavelength-dependent scattering light factor.

Thus, when we select the measuring wavelengths, a suitable compromise must be found between the absorbance properties of hemoglobin, scattering in the eye structures, and the wavelength-dependent reflection properties of the fundus.

Various methods have been developed for determining OS in retinal vessels. A fundus camera (Section 6.7) or a confocal scanning laser ophthalmoscope (Section 6.8.1) is typically used as a base instrument, with which the absorbance can be recorded simultaneously or consecutively in selected spectral ranges. For a detailed description of the applied measuring techniques and calculation models, please refer to [21].

8.2.1.3 **Commercial Devices for Microcirculation Mapping**

Currently (in 2013), the Retinal Functional Imager (RFI) by the company Optical Imaging [24] is the only commercial instrument to record the metabolic status of the retina. It consists of a fundus camera, a special stroboscopic illumination system (100 Hz maximum flash frequency), and fast switching filters (30 ms filter switching frequency). With these components, it is possible to capture image series in a fast sequence in different spectral ranges. With red-free illumination (green exciting light), the flow speed in the blood vessels is indirectly measured via motion detection of the red blood cells. Due to their absorption properties, red blood cells or clusters of red blood cells can be identified as mobile dark spots in a sequence of red-free images. However, it is not yet possible to measure the blood flow in a direct manner (volume per time).

The RFI can also be used to qualitatively determine oxygen saturation. For this purpose, fundus images are captured in a rapid sequence for two different spectral ranges. Relative oxygen saturation in the retinal vessels is then presented in a color-coded *oximetric retinal map*. Although the instrument cannot provide quantitative data, the oximetric retinal map delivers relevant information about the relative local oxygen concentration or consumption in the examined retinal region. The map also allows visualization of changes in oxygen concentration or consumption after light stimulation.

8.2.2 Fluorophore Mapping

Autofluorescence of fluorophores (Sections 6.7.7 and 6.8.1.2) directly or indirectly involved in metabolic processes may be used to assess metabolism at the cellular level. Here, the *redox pairs* of the coenzymes NAD^+/NAD (oxidized and reduced nicotinamide adenine dinucleotide) and $FAD/FADH_2$ (oxidized and reduced flavin adenide dinucleotide) are of particular interest. The autofluorescence of these redox pairs depends on the partial pressure of oxygen and the cellular metabolic rate [25]. This property allows estimation of the oxygen supply to cells required for energy production. Additional information about the metabolism inside the cells is provided by the autofluorescence of metabolic end products such as lipofuscin (see also Section 6.7.7) or advanced glycation end products (AGE).

Lipofuscin Lipofuscin (or its autofluorescing component A2E) has been reported to be the most intensively investigated natural fluorophore in the retina. It is the metabolic end product of the continuously regenerating outer segments of photoreceptors (Section 1.2.1). Normally, lipofuscin is steadily catabolized by retinal pigment epithelial (RPE) cells and partially deposited in the retina as lipofuscin granules so that a normal distribution is formed. Any changes in this distribution are thought to be an indicator of functional disorders of the retinal pigment epithelial cell layer and thus an early indicator of diseases like age-related macular degeneration (Section 3.4) [26].

Advanced glycation end products AGE is an end product of the glucose metabolism which accumulates in proteins of the eye lens. Since these proteins are long-living, AGE is stored in these molecules for a long period of time. Studies have shown that AGE concentration in diabetic patients is considerably higher than in people of same age without diabetes. As a consequence, AGE may be useful as a biomarker for diabetes screening [27, 28].

Selective detection of fluorophores A selective detection of fluorophores on the fundus by means of multispectral techniques is generally very difficult. As mentioned in Section 6.7.7, this is due to the following reasons [29]:

- The fluorescence intensity of the endogenous fluorophores is very low. The signal cannot be increased by raising the excitation intensity, as we have to regard the limits given by the maximum permissible exposure (MPE) values of the eye (Section 4.4). The signal-to-noise ratio can be improved through longer measuring times, but requires additional tools to compensate for eye motion.
- As a result of the transmittance properties of the eye (Figure 2.12 in Section 2.1.10), the fluorescence examination can only be performed in a spectral range between 400 and 900 nm. As the excitation maxima of the relevant endogenous fluorophores often lie below 400 nm, the fluorescence signal cannot be maximized.

- Fluorescence excitation and emission spectra are very broad and do partly overlap so that multiple excitations are possible. This makes it more difficult to distinguish between different fluorophores.
- Autofluorescence from other eye structures (e.g., the eye lens) impairs the measurement, as it overlays the desired fluorescence signal of the fundus. With confocal scanning methods (Section 6.8), however, it is possible to at least partly solve this issue.

All these limitations mean that retinal metabolism-relevant fluorophores, with the exception of lipofuscin, currently cannot be detected at all or not selectively enough with multispectral techniques. However, fluorescence lifetime measurement (FLIM) methods can be used for selective detection of retinal fluorophores. By excitation with picosecond laser pulses (Section B.4.4), the characteristic autofluorescence lifetime t_f is measured for every point of the fundus. For this purpose, a confocal scanning laser ophthalmoscope (Section 6.8.1) with a picosecond laser source is used in combination with time-correlated single photon counting (TCSPC) [17, 26, 30].

Some systems based on this concept are presently in the research phase. They could potentially have the following advantages over spectral methods:

- Fluorescence lifetime measurements only require minimal excitation intensities and allow two-dimensional analysis with sufficiently high lateral resolution.
- FLIM is not affected by the fluorescence intensity. Hence, a fluorophore with a weak signal can also be detected when a strongly light-emitting fluorophore is excited simultaneously as long as the characteristic fluorescence lifetimes are different.

8.2.3 Prospects

In the course of the development of new pharmaceutical drugs for the treatment of eye diseases and the optimization of drug treatment strategies, new applications are emerging for the described methods which are used to characterize the metabolism at the cellular level [16, 31]. The key focus in this research application is the development of *reporting biomarkers*. With such markers, the physiological effects of drugs on the body might be visualized *in vivo* which, in turn, may simplify the development of new treatment options.

8.3 Recommended Reading

For further information about functional diagnostics, please refer to the following references:

- Visual Field Examination (Section 8.1): [1, 2, 5, 9, 12, 13, 32–34]
- Metabolic Mapping (Section 8.2): [16–19, 21, 29, 35, 36]

8.4 Problems

P8.1. Differential light sensitivity In the ZEISS Humphrey Field Analyzer HFA II-i, the maximum luminance of the light stimulus is $\mathcal{L}_{\max} = 10\,000\,\text{asb}$. For the Haag–Streit Octopus, we have $\mathcal{L}_{\max} = 1000\,\text{asb}$. In both devices a luminance $\mathcal{L}_{\text{b}} = 10\,\text{cd/m}^2$ is used for background illumination. With the HFA II-i, the examiner measures a DLS value of 30 dB. What would be the corresponding DLS value of the Octopus perimeter?

P8.2. Luminance Consider a hot tungsten platelet with a luminating area of $1\,\text{mm}^2$. The platelet has a temperature of 3500 K. Under the assumption that the platelet can be considered a black body, what is its luminance? Compare this result with the maximum luminance of $\mathcal{L}_{\max} = 10\,000\,\text{asb}$ for the ZEISS Humphrey Field Analyzer HFA II-i.

References

1 Schiefer, U., Pätzold, J., and Dannheim, F. (2005) Konventionelle Perimetrie – Teil I: Einführung – Grundbegriffe. *Der Ophthalmologe*, **102**, 627–646.

2 Heijl, A., Patella, V.M., and Bengtsson, B. (2012) *The Field Analyzer Primer: Effective Perimetry*, Carl Zeiss Meditec, Inc., 4. edn.

3 Johnson, C.A., Wall, M., and Thompson, H.S. (2011) A history of perimetry and visual field testing. *Optom. Vis. Sci.*, **88**, 8–15.

4 Gloor, B.P. (2009) Franz Fankhauser: the father of the automated perimeter. *Surv. Ophthalmol.*, **54**, 417–425.

5 Weijland, A., Fankhauser, F., Bebie, H., and Flammer, J. (2004) *Automated Perimetry – Visual Field Digest*, Haag–Streit Group, 5. edn.

6 Bizios, D., Heijl, A., and Bengtsson, B. (2011) Integration and fusion of standard automated perimetry and optical coherence tomography data for improved automated glaucoma diagnostics. *BMC Ophthalmol.*, **11**, 20.

7 Medeiros, F.A., Lisboa, R., Weinreb, R.N., Girkin, C.A., Liebmann, J.M., and Zangwill, L.M. (2012) A combined index of structure and function for staging glaucomatous damage. *Arch. Ophthalmol.*, **130**, 1107–1116.

8 van der Schoot, J., Reus, N.J., Colen, T.P., and Lemij, H.G. (2010) The ability of short-wavelength automated perimetry to predict conversion to glaucoma. *Ophthalmology*, **117**, 30–34.

9 Johnson, C.A. (2011), Short wavelength automated perimetry (swap). http://www.perimetry.org/PerimetryHistory/Short-wavelength-automated-perimetry.htm (accessed 26 August 2013).

10 Kelly, D.H. (1966) Frequency doubling in visual responses. *J. Opt. Soc. Am. A*, **56**, 1628–1633.

11 Johnson, C.A. and Samuels, S.J. (1997) Screening for glaucomatous visual field loss with frequency-doubling perimetry. *Invest. Ophthalmol. Vis. Sci.*, **38**, 413–425.

12 Spry, P.G. and Johnson, C.A. (2008) *Primer for FDT perimetry using Humphrey Matrix*, Welch Allyn.

13 Zeppieri, M. and Johnson, C.A. (2011), Frequency doubling technology (FDT) perimeter. http://webeye.ophth.uiowa.edu/ips/perimetryhistory/FDP/index.htm (accessed 26 August 2013).

14 Ramachandran, V.S. and Rogers-Ramachandran, D. (1991) Phantom contours: a new class of visual patterns that selectively activates the magnocellular pathway in man. *Bull. Psychon. Soc.*, **29**, 391.

15 Quaid, P.T. and Flanagan, J.G. (2005) Defining the limits of flicker defined form: effect of stimulus size, eccentricity and number of random dots. *Vis. Res.*, **45**, 1075–1084.

16 Schmitz-Valckenberg, S., Cordeiro, M.F., Fitzke, F.W., and Holz, F.G. (2010) Toward molecular imaging, in *Medical Retina*, Springer, pp. 214–220.

17 Schweitzer, D. (2010) Metabolic mapping, in *Medical Retina*, Springer, pp. 107–124.

18 Riva, C.E., Geiser, M., and Petrig, B.L. (2010) Ocular blood flow assessment using continuous laser Doppler flowmetry. *Acta Ophthalmol.*, **88**, 622–629.

19 Harris, A., Kagemann, L., Ehrlich, R., Rospigliosi, C., Moore, D., and Siesky, B. (2008) Measuring and interpreting ocular blood flow and metabolism in glaucoma. *Can. J. Ophthalmol.*, **43**, 328–336.

20 Wang, Y., Fawzi, A.A., Varma, R., Sadun, A.A., Zhang, X., Tan, O., JA, J.A.I., and Huang, D. (2011) Pilot study of optical coherence tomography measurement of retinal blood flow in retinal and optic nerve diseases. *Invest. Ophthalmol. Vis. Sci.*, **52**, 840–845.

21 Harris, A., Dinn, L., Kagemann, R.B., and Rechtman, E. (2003) A review of methods for human retinal oximetry. *Ophthalmic Surg. Laser Imaging*, **34**, 152–164.

22 van Assendelft, O.W. (1970) *Spectrophotometry of Haemoglobin Derivatives*, Royal Vangorcum Ltd.

23 Prahl, S. (1999) Optical absorption of hemoglobin. http://omlc.ogi.edu/spectra/hemoglobin (accessed 26 August 2013).

24 Nelson, D.A., Krupsky, S., Pollack, A., Aloni, E., Belkin, M., Vanzetta, I., Rosner, M., and Grinvald, A. (2005) Special report: Noninvasive multi-parameter functional optical imaging of the eye. *Ophthalmic Surg. Lasers Imaging*, **36**, 57–86.

25 Chance, B. (1967) Pyridine nucleotide as an indicator of the oxygen requirements for energy-linked functions of mitochondria. *Circ. Res.*, **38**, 131–138.

26 Schmitz-Valckenberg, S., Fleckenstein, M., Scholl, H.P., and Holz, F.G. (2009) Fundus autofluorescence and progression of age-related macular degeneration. *Surv. Ophthalmol.*, **54**, 96–117.

27 Yu, N.T., Krantz, B.S., Eppstein, J.A., Ignotz, K.D., Samuels, M.A., Long, J.R., and Price, J. (1996) Development of a non-invasive diabetes screening device using the ratio of fluorescence to Rayleigh scattered light. *J. Biomed. Opt.*, **1**, 280–288.

28 FreedomMeditech website (2012) http://www.freedom-meditech.com (accessed 26 August 2013).

29 Schweitzer, D., Schenke, S., Hammer, M., Schweitzer, F., Jentsch, S., Birckner, E., Becker, W., and Bergmann, A. (2007) Towards metabolic mapping of the human retina. *Microsc. Res. Tech.*, **70**, 410–419.

30 Becker, W. (2005) *Advanced Time-Correlated Single Photon Counting Technique*, Springer.

31 Eter, N. (2010) Molecular imaging in the eye. *Br. J. Ophthalmol.*, **94**, 1420–1426.

32 Comer, G.W. (2006) Visual-field screening and analysis, in *Borish's Clinical Refraction*, 2nd edn, (ed. W.J. Benjamin), Butterworth-Heinemann/Elsevier, pp. 544–618.

33 Midena, E. (2006) *Perimetry and the Fundus: An Introduction to Microperimetry*, Slack Inc.

34 International Organization for Standardization (1999/2008) Ophthalmic instruments – Perimeters, ISO 12866:1999 + Amd 1:2008.

35 Riva, Ch.E. and Petrig, B.L. (2003) Laser Doppler techniques in ophthalmology – principles and applications, in *Ophthalmology – Basics, Diagnostic and Surgical Aspects*, (eds F. Fankhauser and S. Kwasniewska), Kugler Publications, The Hague, pp. 1–59.

36 Bahram Khoobehi, B. and Beach, J.M. (2011) Hyperspectral image analysis for oxygen saturation automated localization of the eye, in *Computational Analysis of the Human Eye with Applications*, World Scientific Publishing Co. Pte. Ltd., Singapore, pp. 123–185.

Part Three

9
Laser–Tissue Interaction

In this section, we will consider the interaction of light with biological tissue with a focus on ophthalmic applications. More specifically, we want to explain how the energy of incident laser light (Appendix B) can be used to modify the shape, structure, and function of eye tissue in a controlled manner. Generally, when light is incident on a medium, it is either reflected from the surface, absorbed, scattered inside the medium, or transmitted without interaction. For homogeneous media, the energy transfer mostly happens via absorption. But biological tissues usually consist of many small, randomly distributed components so that the incident photons are deflected from their straight path of travel. This effect is referred to as *scattering*. In case of *inelastic scattering*, the medium generally gains energy in the form of heat, whereas the scattered photons change their wavelength. In the case of *elastic scattering*, photons are only deflected and do *not* transfer any amount of energy to tissue.

Although absorption and scattering are fundamentally different kinds of light–matter interaction, it is hard to distinguish between them in experiments with tissue. To detect the amount of light energy deposited in biological matter, we typically use an arrangement like that shown in Figure 9.1. We can use this to determine the intensity I_0 emitted by the light source and the intensity I arriving at the detector after passage through the medium. Since the detector has a limited field of view, we cannot retrace whether scattering or absorption has led to the measured loss of photons. A new quantity, the *absorbance*[1] (also known as *extinction*)

$$A = \ln\left(\frac{I_0}{I}\right) , \tag{9.1}$$

is thus introduced which takes all causes for photon loss into account (for a certain "collecting" solid angle). Then, we use scattering models to calculate back which portion of loss is attributed to which type of interaction.

1) It is also common to define the absorbance via the common logarithm.

Optical Devices in Ophthalmology and Optometry, First Edition. Michael Kaschke, Karl-Heinz Donnerhacke, and Michael Stefan Rill.

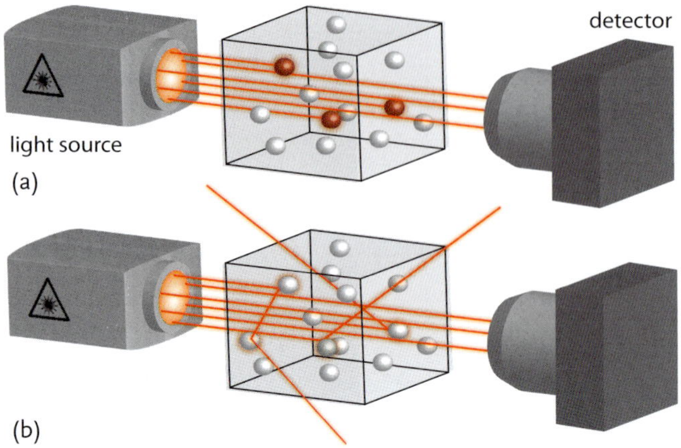

Figure 9.1 Light attenuation due to (a) absorption and (b) scattering. Although the interactions are fundamentally different, the intensity signal measured by the detector is equal, as it has only a limited field of view.

9.1 Absorption

The absorption of a medium is characterized by the absorption coefficient $\mu_a(\lambda)$ ($[\mu_a] = 1/\text{cm}$). The dependence of the absorption coefficient on the wavelength of light is described by the absorption spectrum (see e.g., Figure 9.3 in Section 9.3). When light passes through a purely absorbing homogeneous medium, we may calculate the intensity distribution along the light's traveling path according to Lambert–Beer's law

$$I = I_0 e^{-\mu_a(\lambda) L} . \tag{9.2}$$

I_0 denotes the intensity of the incident light and L the passed distance. So, Eq. (9.2) accounts for the local distribution of absorbed energy. We see that most of the light energy is deposited directly at the surface and rapidly decreases inside the medium. Alternatively to the absorption coefficient, we may also consider the *absorption length* or *depth of penetration*

$$\delta_a = \frac{1}{\mu_a(\lambda)} \tag{9.3}$$

at which the intensity of the incident light beam has been reduced to $1/e = 37\%$.

Absorption means that incident photons are annihilated and transfer all of their energy to atoms/molecules. For classic light sources or low laser light intensities, we usually observe (linear) *one-photon absorption* for which the energy of one incident photon has to match the energy difference between two atomic/molecular levels (Figure 9.2a).

If the intensity of the incident light beam is high enough, it is possible that two or more photons will be annihilated simultaneously when interacting with the same

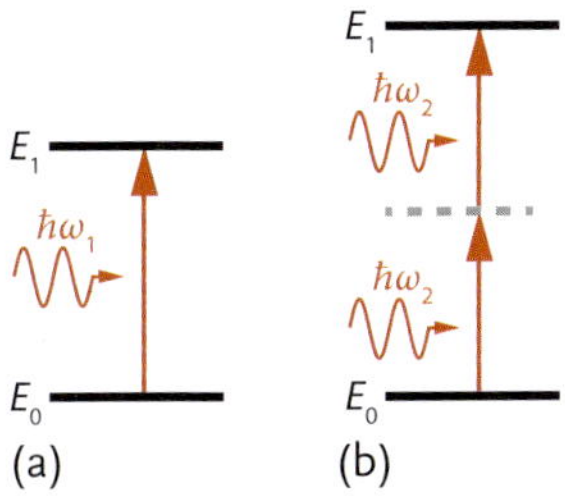

Figure 9.2 Energy scheme of one- and two-photon absorption. (a) In the case of one-photon absorption, one incident photon with an energy of $E = E_1 - E_0 = \hbar\omega_1$ excites an atom from the ground state E_0 to a higher energy level E_1. (b) If the laser intensity is high enough, two photons may simultaneously transfer their energy to an atom. For this so-called two-photon process, an energy of $2\hbar\omega_2 = E_1 - E_0$ is required.

atom/molecule. Since more than one photon is involved, such a process is referred to as *nonlinear multiphoton absorption*. For classic light sources, the attainable intensity is normally too low to induce multiphoton absorption. The required light intensity for multiphoton absorption can only be achieved in the focal volume of pulsed laser sources (Section B.4.4)[2]. In this case, the sum of all photon energies is available for the atom to be excited to a much higher energy state. In Figure 9.2b, this is shown for the example of two-photon absorption.

Multiphoton absorption processes can be used to cut transparent tissue (e.g., cornea or eye lens) in refractive surgery (see also Section 9.4.4). If the considered tissue is transparent for a certain wavelength range this means that the energy of each incident photon is too low to excite the atoms/molecules by one-photon absorption. A laser beam with low intensity thus simply passes through the tissue without interaction. If we now increase the laser intensity considerably and focus the beam into the transparent tissue, the intensity in the focal volume becomes high enough to induce multiphoton absorption in a spatially confined region. In contrast to one-photon absorption, most of the light energy is now deposited in a specific volume and not along the entire traveling path of the laser beam.

9.2 Elastic Scattering

The loss in intensity due to scattering is described by a similar relation as absorption in Eq. (9.2), that is,

$$I = I_0 e^{-\mu_s(\lambda) L} , \tag{9.4}$$

where $\mu_s(\lambda)$ is the scattering coefficient of the medium. The formalism to quantify the scattering of light at particles is much more complicated than for absorption. Indeed, different models are used depending on the relative size of particles and the incident wavelength.

2) Other nonlinear optical effects observed for intense laser light are discussed in Section 9.5.

9.2.1
Rayleigh Scattering

Rayleigh scattering describes the interaction of a small spherical particle embedded in a homogeneous medium. In order for Rayleigh's model to apply, the particle diameter d must be at least $10\times$ smaller than the wavelength of the scattered wave ($d \leq 10\lambda$). Thus, when illuminated with visible light, Rayleigh scattering occurs in biological tissue at cell membranes, macromolecular proteins, and collagen fibers. For Rayleigh scattering, the exact shape of the scattering center is irrelevant and can be treated as a sphere of equivalent volume in any case. For unpolarized light (Section A.2.1.4), the angle-dependent light intensity distribution reads

$$I(\phi) = I_0 \left(\frac{1+\cos^2\phi}{2L_{\text{scd}}^2}\right)\left(\frac{n^2-1}{n^2+2}\right)^2\left(\frac{2\pi}{\lambda}\right)^4\left(\frac{d}{2}\right)^6 , \qquad (9.5)$$

where d denotes the diameter of the spherical particle with refractive index n. L_{scd} is the distance from the scattering center to the detector and ϕ the scattering angle. From Eq. (9.5) it follows that Rayleigh scattering is inversely proportional to the fourth power of wavelength. Blue light of 400 nm wavelength is scattered $9.4\times$ more strongly than red light at 700 nm.[3)]

9.2.2
Mie Scattering

For biological samples, we have the situation that the wavelength of the incident light is comparable to the particle size ($d \approx \lambda$). In this case, the exact shape of the scattering center becomes crucial. A theory for this type of scattering was developed by Gustav Mie (1868–1957) [1]. In general, the intensity of Mie-scattered light shows a weaker dependence on the wavelength than Rayleigh scattering according to

$$I(\lambda) \propto \lambda^{-x} \quad \text{with} \quad 0.4 \leq x \leq 0.5 . \qquad (9.6)$$

Mie scattering preferably takes place in the forward direction, whereas forward and backward scattered intensities are equal in the Rayleigh regime. When illuminated with visible light, Mie scattering occurs in biological tissue for the cell core, lysosomes, vesicles, and mitochondria.

9.3
Optical Properties of Biological Tissue

For turbid media like tissue, all types of light–matter interaction are concurrently present. However, for certain spectral ranges, one of them might be more relevant

3) Rayleigh scattering is the primary cause of the blue color of the Earth's sky on a sunny day. Here, the shorter wavelengths of sunlight are more strongly scattered by the atmosphere than the longer wavelengths.

for absorbance than the others. Thus, it is common to define a ratio of scattering and absorption coefficient

$$\mathcal{A} = \frac{\mu_s}{\mu_a + \mu_s} \tag{9.7}$$

known as the *albedo*. For $\mathcal{A} = 0$ only absorption is present, while for $\mathcal{A} = 1$ only scattering processes are present.

Figure 9.3 shows the wavelength-dependent absorption coefficient μ_a of the main components and chromophores of eye tissue (i.e., water (black), protein (orange), melanin (gray), and hemoglobin (red and violet)). The diagram reveals that absorption dominates in the ultraviolet (UV) spectral range, mainly due to the contribution of proteins, and in the infrared (IR) spectral range, mainly due to water. In the visible (VIS) range, the spectral properties of biological tissue are mainly influenced by chromophores like melanin, xanthophyll, and the blood pigments oxy- and deoxyhemoglobin. In the human eye, melanin can be found in the iris, the trabecular meshwork, and the retinal pigment epithelium. Xanthophyll is located in the macula, while oxyhemoglobin (arterial blood) and deoxyhemoglobin (venous blood) are mainly present in choroidal and retinal tissue. Figure 9.3 also shows the wavelength of lasers which are typically used for treatment of diseases or disorders in ophthalmology. We see that lasers which emit UV and IR radiation are best suited for treatments at the tissue surface. Interventions in deeper tissue layers are better performed with near-infrared (NIR) laser light in the wavelength

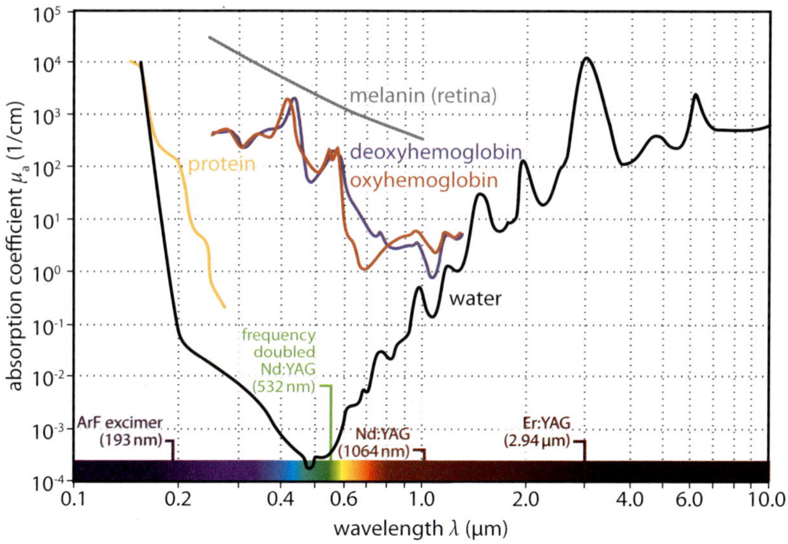

Figure 9.3 Absorption coefficients of the main components and chromophores of eye tissue (i.e., water, protein, melanin, and hemoglobin) as a function of wavelength (see also Problem P9.1). The wavelengths of some therapy laser sources are indicated as well (see also Section B.5). Adapted from [2].

region about 1 μm. For selective treatment of blood-rich tissue, laser wavelengths in the green-yellow range can be used (Problem P9.1).

9.4
Interaction of Irradiated Biological Tissue

When laser radiation interacts with biological tissue, we can observe different interaction effects which can be used for therapeutic purposes. The laser-induced changes in tissue depend on the chosen radiation parameters and certain optical and thermal tissue properties. Typical radiation parameters are wavelength λ, energy E, intensity I, and exposure time t_{exp} (duration of light exposure). Relevant tissue properties are described by the reflectance R (Section A.1.1), the absorption μ_a, and scattering coefficients μ_s (optical parameters), as well as the heat capacity and heat conductivity (thermal parameters).

The possible interaction effects can be classified into photochemical interactions, photothermal interactions, photoablation, plasma-induced ablation, and photodisruption. In Figure 9.4, the different types of interaction are depicted in a double-logarithmic diagram. The horizontal axis is assigned to the exposure time, and the vertical axis is assigned to the applied light intensity. The gray lines in the diagram

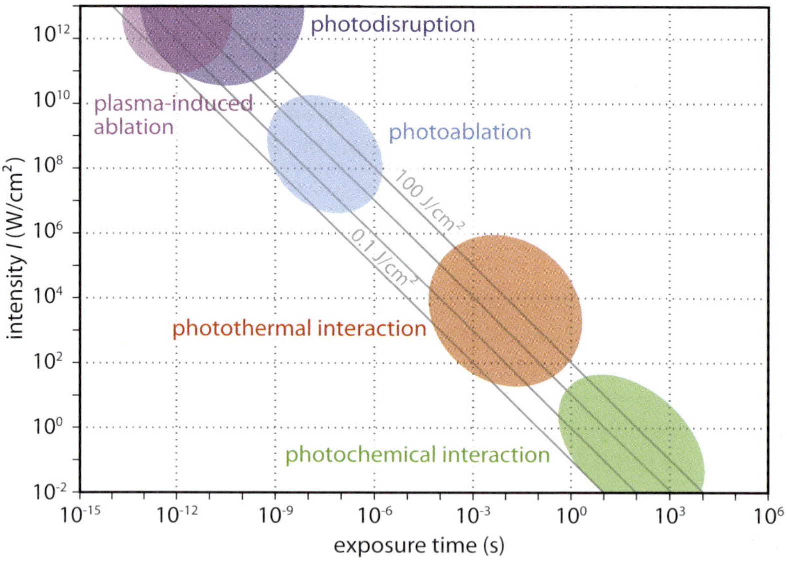

Figure 9.4 Double-logarithmic map of the basic interaction types grouped by exposure. We distinguish between photochemical interaction (green), photothermal interaction (red), photoablation (blue), photodisruption (violet), and plasma-induced ablation (pink). Adapted from [3].

represent constant values of *exposure*[4] Φ ($[\Phi]$ = Ws/cm^2 = J/cm^2). Let us now discuss the laser–tissue interactions starting from long exposure times of several seconds (lower right corner of Figure 9.4) to short exposure times (upper left corner) in the femtosecond (fs) range (Problems P9.2 and P9.3).

9.4.1 Photochemical Response

Photochemical effects mainly occur at intensities between 10^{-2} and 50 W/cm^2 by using low-power lasers (mW range) at long exposure times (some seconds to hours). During exposure, molecules are optically excited so that chemical reactions can be triggered. Well-known examples of photoinduced chemical reactions are photosynthesis in plants, the bleaching of dyes, and the tanning effect of sunlight on human skin. During the visual process in the human eye, the so-called *photoisomerization*[5] of the visual pigment rhodopsin in rods and cones is one of the first steps of light perception, as it leads to the activation of photoreceptors as a whole.

In ophthalmology, photochemical effects are used in *photodynamic therapy* (PDT; Section 10.1). In PDT, special chromophores called *photosensitizers* are used which can be excited by photon absorption to a long-lived (triplet) state. The absorbed energy stored in the triplet state of the photosensitizer is then transferred by interaction with adjacent molecules (e.g., oxygen molecules) in which a chemical reaction is induced. After the energy transfer, the photosensitizer falls back to the ground state. The toxic products generated by the chemical reaction react with adjacent tissue and thus lead to the desired change of the tissue structure. The corresponding chemical reaction chain of the so-called *photosensitized oxidation* is illustrated in Figure 9.5. After the absorption of laser photons, photosensitizer S is transferred

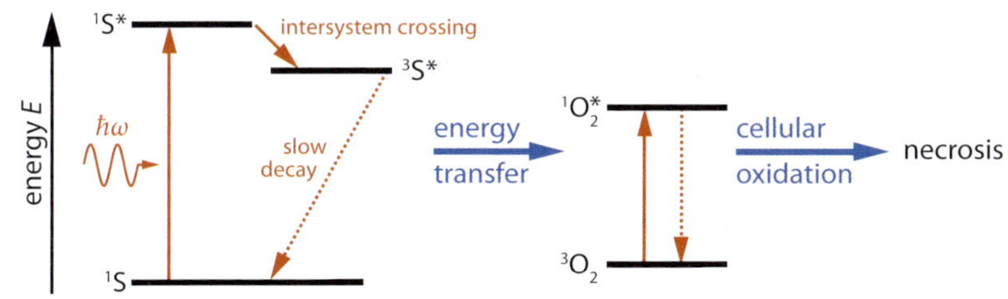

Figure 9.5 Sequence of photosensitized oxidation in photodynamic therapy. After absorption of laser photons, the photosensitizer S is first excited from the ground state ^1S to $^1S^*$. Intersystem crossing then transits the molecule to an excited triplet state $^3S^*$. The state $^3S^*$ directly interacts with an adjacent triplet oxygen molecule $^3O_2$, which is then transferred to an excited singlet state $^1O_2{}^*$. The excited singlet oxygen is highly reactive so that tissue cells can be oxidized. This finally leads to the desired tissue necrosis.

4) In literature, the exposure is sometimes also referred to as the *fluence* or *energy density* (see also Section A.2.1.5).
5) Two molecules that differ in shape but consist of the same atoms are termed *isomers*. A change of the shape of a portion of the molecules is called *isomerization*.

from the ground state ^1S to an excited singlet state $^1S^*$. Besides fluorescent decay (Section B.1.2) to the ground state, intersystem crossing to an excited triplet state $^3S^*$ occurs. In so-called *type II* reactions, the triplet state $^3S^*$ of the photosensitizer directly interacts with an adjacent oxygen molecule which is in the triplet state $^3O_2$ which is then transferred to an excited singlet state $^1O_2^*$ according to

$$^3S^* + {}^3O_2 \rightarrow {}^1S + {}^1O_2^* \,. \tag{9.8}$$

Excited singlet oxygen $^1O_2^*$ is a very reactive form of oxygen which leads to cellular oxidation and a type of irreversible tissue damage called *necrosis.*

In ophthalmology, PDT (Section 10.1) is used to selectively damage sub-foveal abnormal blood vessels in retinal diseases like wet age-related macular degeneration (wet AMD; Section 3.4). For this purpose, the photosensitizer verteporfin[6] is injected into the patient's vein, which then accumulates primarily inside the walls of newly grown vessels. It is activated by irradiation with red laser light of 689 nm wavelength. Since the introduction of a very efficient anti-vascular endothelial growth factor (anti-VEGF) drug therapy in 2006, however, the importance of PDT for treatment of wet AMD has strongly decreased.

9.4.2
Photothermal Response

Tissue shows a photothermal response when exposed to light intensities in the 10–10^6 W/cm^2 range for some microseconds to seconds. In this regime, the chromophores, proteins, and water of the irradiated cells absorb the incident light energy and convert it to heat. Clearly, if the cell temperature rises just slightly above body temperature for a short time, we do not expect any notable reaction. But if the temperature rises above a critical value, the tissue cells will be irreversibly damaged so that their repair mechanism does not work anymore, that is, necrosis. The threshold behavior is illustrated in the diagram of Figure 9.6, which has been derived from several empirical observations. As expected, no measurable response can be found when tissue cells are slightly heated to $T \leq 42\,°C$, that is, less than 5 K above body temperature. First conformational changes of cell molecules by bond destruction and membrane alterations appear when tissue is kept at $T > 42.5\,°C$ for some minutes. These effects are referred to as *hyperthermia.* In cells kept at $T > 50\,°C$ for several seconds, the enzyme activity is reduced and certain repair mechanisms are disabled.

Tissue coagulates at $T > 60\,°C$. This means that proteins and collagen inside the cells denaturate. However, the desired therapeutic effect is only generated by the biological reaction (healing response) to the thermal damage. The biological processes that take place and eventually lead to the therapeutic effect are not yet completely understood. It is generally assumed that photocoagulation either starts or prevents the secretion of growth factors. For example, in proliferative di-

6) Verteporfin is a photosensitive dye which is commercially available under the trade name Visudyne™.

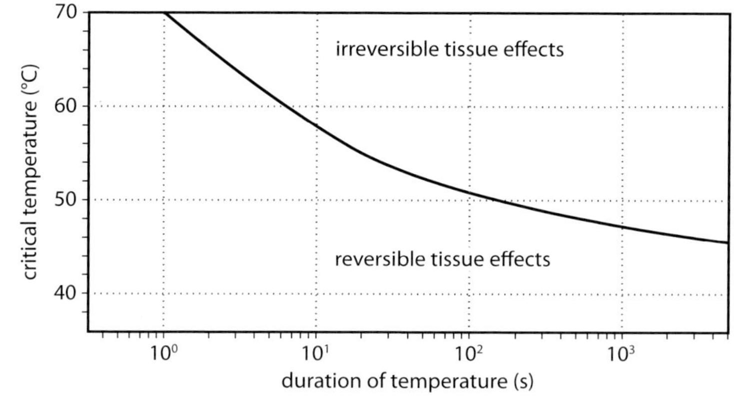

Figure 9.6 Critical temperature versus the duration of temperature at which the tissue is kept. "Irreversible tissue effect" means that the cells are considerably damaged by the radiation so that the repair mechanism is disabled (necrosis). In contrast, the cells may regenerate from "reversible tissue effects". Adapted from [3].

abetic retinopathy (Section 3.5), the destruction of oxygen consuming photoreceptors by pan-retinal photocoagulation (Section 10.2.3.1) downregulates the secretion of growth factors and thus stops the undesirable formation of new abnormal blood vessels (neovascularization) in the macular region. In other eye diseases like diabetic macular edema, it rather stimulates the regeneration of retinal pigment epithelium cells, which in turn leads to the desired therapeutic effect, for example, disappearance of macular edema.

If the treatment of eye diseases shall be planned based on photocoagulation with lasers, we have to set a few parameters which influence the therapeutic outcome. The most important process parameters are:

1. Irradiation parameters:
   - laser wavelength λ,
   - exposure time t_{exp},
   - light intensity I, and
   - spot size of the laser beam $2w_0$.
2. Tissue parameters:
   - absorption coefficient μ_a,
   - scattering coefficient μ_s,
   - thermal penetration depth L_{tpd}, and
   - thermal relaxation time t_r.

In the following, we will consider some of these parameters in detail.

Laser wavelength A very important process parameter in photothermal interactions is the wavelength of the used laser light. In Sections 9.1 and 9.2, we noticed that the absorption and scattering coefficients of ocular tissue depend on the wavelength. The choice of the laser wavelength therefore determines in which part of the eye the interaction takes place.

Exposure time The heat conduction, and thus the volume of the coagulated area, can be set by the exposure time t_{exp}. For pulsed laser sources, t_{exp} is given by the pulse duration (Section B.4.4). The extent of thermal damage is estimated by means of the thermal penetration depth L_{tpd}.

Thermal penetration depth Due to thermal conduction, the deposited energy spreads out and heats up adjacent, unexposed tissue cells as well. The resulting temperature distribution during the exposure time is characterized by the time-dependent thermal penetration depth

$$L_{tpd} = \sqrt{4\kappa t_{exp}}\,, \tag{9.9}$$

where κ represents the thermal diffusion constant[7] ($[\kappa] = m^2/s$) and t_{exp} the exposure time. L_{tpd} is the distance at which the temperature has decreased to $1/e$ of its maximum value T_{max} and is a measure for the speed at which heat spreads out inside the tissue.

Eye tissue consists mainly of water. With regard to the thermal properties, we may thus approximate κ_{tissue} by $\kappa_{water} = 1.4 \times 10^{-7}\,m^2/s$. If eye tissue is irradiated for $t_{exp} = 100$ ms, adjacent tissue at a radial distance of $L_{tpd} = 240\,\mu m$ can also be thermally damaged.

Thermal relaxation time Directly related to the thermal penetration depth is the thermal relaxation time t_r. This parameter represents the time required for the peak temperature rise ΔT_{max} in a heated tissue region to decrease to $1/e$ of the total rise. The thermal relaxation time is determined by [4]

$$t_r = \frac{\delta_a^2}{4\kappa} = \frac{1}{4\mu_a^2\kappa}\,. \tag{9.10}$$

With $t_{exp} = t_r$ and Eq. (9.9), the thermal penetration depth can be written as $L_{tpd} = 1/\mu_a = \delta_a$. Hence, the thermal relaxation time can also be considered as the exposure time at which L_{tpd} equals the optical depth of penetration δ_a of the laser radiation (Section 9.1). For exposure times $t_{exp} < t_r$ and single-pulse irradiation, the size of the damaged tissue volume only depends on the optical penetration depth. The thermal damage of unexposed tissue due to heat diffusion is negligible.[8]

When photothermal effects are used to treat tissue (photocoagulation, photoablation), the exposure time must be chosen adequately to minimize undesired thermal

7) In literature, the thermal diffusion constant is also called *thermal diffusivity* or *temperature conductivity*.

8) If laser pulses with high repetitions rates are applied to the same treatment area, the thermal damage volume will be enlarged even at exposure times $t_{exp} < t_r$ due to energy accumulation effects.

damage of adjacent tissue structures. A correct consideration of t_r is thus crucial for the therapy. As a reference, the shortest thermal relaxation time of approximately 1 μs occurs for the absorption peak of water at wavelengths around 3 μm. Thermal damage of adjacent structures around the exposed region is thus irrelevant for pulse durations $\tau < 1$ μs and low repetition rates in water-containing tissues.

Intensity With the laser beam intensity[9], we are able to control the peak temperature rise ΔT_{max}. This parameter must be adapted to the absorption coefficient $\mu_a(\lambda)$.

Spot size On the one hand, the spot size of the laser beam determines the area of directly irradiated tissue. Thus, for a given laser power, it determines the incident intensity of the laser beam. On the other hand, the spot size determines the temperature distribution in the irradiated tissue.

Since all process parameters are interrelated, the radiation parameters must be carefully chosen depending on the goal of treatment, the treatment location, and the corresponding tissue properties so that an optimal treatment effect is obtained with minimum damage of adjacent tissue (Problem P9.4).

9.4.3 Photoablation

If laser light with a pulse length < 1 μs and an intensity between 10^7 and 10^{10} W/cm^2 (Figure 9.4) interacts with strongly absorbing tissue above a certain threshold exposure, near-surface tissue layers are removed in an explosive manner. This effect is called *photoablation*. Since the absorption of eye tissue is particularly high in the UV and IR spectral range (Figure 9.3), short-pulse lasers with a corresponding wavelength can be used for this purpose. Depending on the wavelength of the laser light, we may distinguish between two different interaction processes [5]:

1. *Photodecomposition ablation:* With interactions of short-pulsed UV-lasers, organic molecular bonds in proteins and collagen are broken up by means of the absorption of high energy UV photons (dissociation). The photon energy (see Info Box B.1 in Appendix B) of UV laser light must be higher than the dissociation energy E_{diss} of the molecular bonds, that is,

$$\frac{h}{2\pi}\omega = \frac{hc}{\lambda} > E_{diss} . \tag{9.11}$$

   The dissociation energies E_{diss} of most organic molecular bonds are in the range between 3 and 7 eV[10]. For example, to break up the bonds of a carbon–carbon (C–C) molecule ($E_{diss} = 3.6$ eV) in proteins and collagen, UV light with

9) In literature, the intensity is also called irradiance or power density.
10) 1 eV (electron volt) corresponds to about 1.602×10^{-19} J.

a wavelength below 350 nm is required. The energy exceeding the dissociation energy contributes to the internal or kinetic energy of the dissociation products. This creates an expanding plume consisting of material removed from the ablation site. Photodecomposition ablation with UV lasers is never a purely photochemical process, as thermal processes take place as well. The contribution of thermal processes to photoablation is significantly smaller for shorter wavelengths. At the wavelength of the ArF excimer laser ($\lambda = 193$ nm), photoablation is dominated by photochemical decomposition.

2. *Photothermal ablation:* When short-pulsed laser light in the mid- and far-infrared range is used, water contained in tissue evaporates instantly due to its high absorption coefficient and rips the tissue open. The tissue fragments are then ejected like during an explosion.

Compared to photothermal ablation, it is possible to generate significantly smoother ablation profiles with primarily photochemically induced ablation using UV lasers in the wavelength range of ≈ 200 nm, as this process is less disruptive. In ophthalmology, photoablation is used to correct refractive errors by reshaping the corneal surface (Section 10.3.3.1).

For surface reshaping in the sub-µm range, a high precision is required which can only be achieved with a small ablation depth per pulse as well as minimal thermal and mechanical side effects. The tissue layer thickness ablated with each laser pulse, that is, *ablation depth* L_{abl}, is related to the exposure Φ via the *blow-off model*. For pulse durations $\tau < 100$ ns, the ablation depth is given by [5]

$$L_{\text{abl}} = \frac{1}{\mu_{\text{a}}(\lambda)} \ln\left(\frac{\Phi}{\Phi_{\text{th}}}\right) . \tag{9.12}$$

Φ_{th} is the threshold exposure required to achieve ablative material removal, and $\mu_{\text{a}}(\lambda)$ denotes the absorption coefficient of the tissue to be treated. In Figure 9.7, L_{abl} is plotted versus Φ for different absorption coefficients μ_{a}. The higher μ_{a}, the lower is Φ_{th} and the thinner are the tissue layers which can be ablated per pulse. Above the threshold exposure Φ_{th}, the ablated depth can be adjusted precisely by changing Φ of the laser pulse. In the case of low absorption (low values of μ_{a}), Φ_{th} increases, and L_{abl} rises much faster due to a steeper slope.

The ArF excimer laser (Section B.5.1.2) is preferably used for photoablative corneal procedures in refractive surgery, as the absorption coefficient of the corneal stroma is maximal ($\mu_{\text{a}} = 29\,000$/cm) for the emitted wavelength of $\lambda = 193$ nm [5]. Due to the short pulse duration of the ArF excimer laser ($\tau = 20\text{ ns} < t_{\text{r}}$), thermal diffusion, and thus the thermal damage to adjacent tissue structures, is negligible during exposure.

9.4.4
Plasma-Induced Ablation and Photodisruption

In the focus of a laser beam with a short-pulse duration and intensities above 10^{11} W/cm^2, a phenomenon called *optical breakdown* occurs. This phenomenon is often accompanied by a bright, visible plasma and a sparking noise. It occurs in

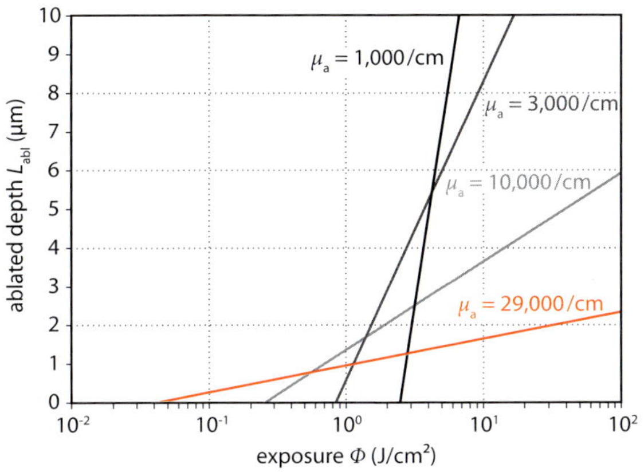

Figure 9.7 Ablated (etch) depth L_{abl} versus exposure Φ. The gray tone curves are calculated with the blow-off model for three different absorption coefficients $\mu_a(\lambda)$. The red line is a best-fit curve of measured data for corneal stroma at the emission wavelength of an ArF excimer laser ($\lambda = 193$ nm). Adapted from [5].

tissues with strong optical absorption as well as in materials which are transparent for the used laser wavelength at low intensities. Due to the high absorption coefficient of the laser-induced plasma, light energy can also be deposited in biological tissue which is transparent for the laser wavelength, such as cornea and eye lens.

9.4.4.1 Laser-Induced Optical Breakdown in Biological Tissue

In biological tissue which is transparent at low light intensities, the formation of a laser-induced optical breakdown happens in two steps:

1. *Multiphoton ionization:* At first, (quasi-)free "seed" electrons are generated by nonlinear absorption (Section 9.1) in the focus of the laser beam. The rate at which the multiphoton ionization occurs is proportional to I^n, where I is the laser intensity in the focus and n the number of photons required to ionize the tissue molecules [3].
2. *Avalanche ionization:* The seed electrons absorb any amount of photon energy in a nonresonant process and are thus accelerated. This process takes place in the presence of a third particle (e.g., an ion or molecule) and is called *inverse bremsstrahlung*[11)]. After repeated absorption events, the accelerated electrons gain more and more kinetic energy so that they are eventually able to ionize other molecules by impact ionization. In turn, this creates other free electrons (Figure 9.8). If the incident laser light is intense enough to overcome the losses of the free electrons, for example, through diffusion or recombination, an *avalanche* effect occurs. This leads to a very short rise time of the free electron density N_{el} in the order of picoseconds (ps).

11) A third particle (ion/atom) is needed to fulfill energy and momentum conservation.

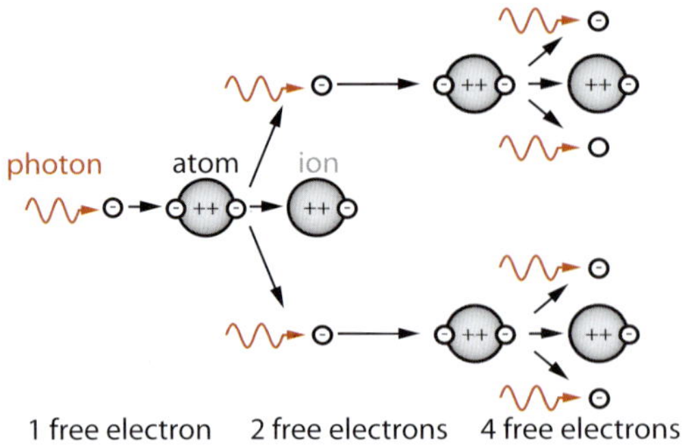

Figure 9.8 Scheme of an electron avalanche ionization. The electrons are represented by particles with a minus sign, and ions are indicated by "++". Adapted from [3].

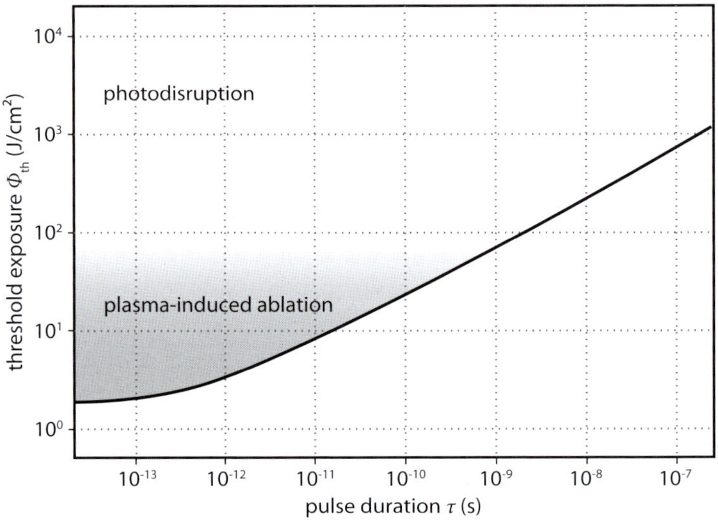

Figure 9.9 Threshold parameters for the onset of plasma-induced ablation (gray area) and photodisruption in corneal tissue. Adapted from [3].

If the free electron density exceeds a critical value of approximately 10^{18}–$10^{20}/\text{cm}^3$, a "cloud" of ions and free electrons is formed in the focus of the laser beam which is referred to as a *plasma*. In this case, an optical breakdown has been generated.

As shown in Figure 9.9, the threshold exposure Φ_{th} needed to create an optical breakdown decreases when we decrease the pulse duration τ of laser light. In the ps to µs range, this behavior can be approximated by [3]

$$\Phi_{th} \propto \sqrt{\tau} \, . \tag{9.13}$$

For pulse durations shorter than 200 fs, the threshold exposure Φ_{th} is nearly constant.

If the treated tissue has a high absorption coefficient μ_a for the wavelength of the used laser light, the threshold for plasma formation with ns to ps pulses is considerably reduced, as the seed electrons are provided by *thermionic emission*[12]. For fs pulses, however, μ_a of the exposed target is irrelevant, since multiphoton ionization dominates because of the short interaction time.

9.4.4.2 **Effect of Moving Optical Breakdown**

For light intensities right above the threshold value I_{th} at which an optical breakdown occurs, the formed plasma is located within the focus of the laser beam (beam waist). But if the intensity in the focal region considerably exceeds this threshold ($I \gg I_{th}$), the plasma expands from the beam waist towards the direction of light incidence. This effect can be described by the *moving breakdown model* (Figure 9.10) [6] which assumes that the optical breakdown occurs at all locations with $I > I_{th}$, independently of any preceding plasma formation. This assumption is fulfilled for pulse durations in the fs to ps range. In the nanosecond (ns) range, however, the plasma expands further than expected from the model, as the threshold of the optical breakdown is now also influenced by the formed plasma.

The maximum plasma expansion Δz_{max} from the focus towards the incoming laser beam is reached for the peak intensity of the laser pulse and is given by [3]

$$\Delta z_{max} = z_R \sqrt{\beta - 1}\,. \tag{9.14}$$

z_R denotes the Rayleigh length (Section A.2.2.1) and $\beta = I_{peak}/I_{th}$ the ratio between the peak pulse intensity I_{peak} and threshold intensity I_{th} for laser-induced breakdown. According to (A83) in Section A.2.2.1, the Rayleigh length is related to the radius of the beam waist w_0 via

$$z_R = \frac{\pi w_0^2}{\lambda} \tag{9.15}$$

so that we obtain

$$\Delta z_{max} = \frac{\pi w_0^2}{\lambda} \sqrt{\beta - 1}\,. \tag{9.16}$$

In the region behind the beam waist, no plasma is formed at all, since this region is effectively "shielded" by light absorption in the plasma (Figure 9.10b–d). The light intensity after the peak does *not* lead to any further expansion of the plasma cloud, but only heats it up. According to (9.16), the plasma expansion can be minimized with a tightly focused laser beam and when we set the laser intensity slightly above I_{th}.

9.4.4.3 **Plasma Absorption and Shielding**

The plasma absorption coefficient $\mu_{a,plasma}$ is related to the electron density N_{el} via

$$\mu_{a,plasma} \propto N_{el}^2\,. \tag{9.17}$$

12) Thermionic emission is the flow of electrons (or ions) from a material surface caused by excessive heating. In light sources such as fluorescent tubes, electrons are created by a thermionic cathode. If the thermal energy is high enough, it overcomes the binding energy of the atom so that electrons (or ions) are "ripped off" the material. In older literature, electrons and ions were referred to as *thermions*, which lead to the name of the effect.

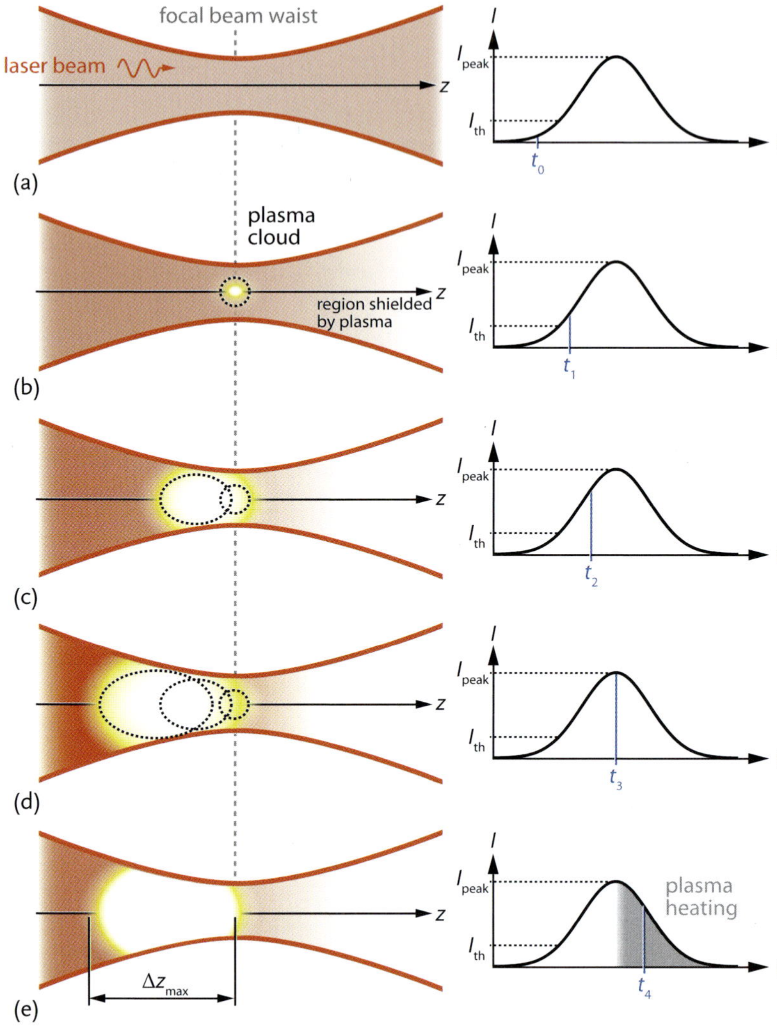

Figure 9.10 Formation of a cascade of laser-induced plasma clouds by focusing a laser beam into a transparent medium. The moving breakdown model describes how the plasma gradually expands after an optical breakdown. Left column: Scheme of Gaussian laser focus (red) and formed plasma cloud (yellow). Right column: Diagram which illustrates the light intensity I at certain time steps t_i while the laser pulse passes by. (a) Below the threshold intensity I_{th}, no plasma is formed. (b) The threshold intensity I_{th}, at which an optical breakdown occurs, is first exceeded in the beam waist so that a plasma cloud is formed. Due to plasma absorption, the region behind the plasma cloud is shielded from laser radiation. (c) At time step t_2, I_{th} has already been exceeded at planes which lie in front of the beam waist. Thus, the plasma cloud moves towards the incident direction of the laser beam. (d) When the maximum intensity I_{peak} is reached in the focus (time step t_3), the plasma cloud has maximally expanded to Δz_{max}. Its maximum size depends on the ratio $\beta = I_{peak}/I_{th}$ and the focusing conditions determined by the Rayleigh length z_R. (e) The portion of the light pulse after the peak ($t > t_3$) just heats up the plasma, while the size of the plasma cloud remains unchanged until the whole pulse has passed by. Adapted from [6].

Table 9.1 Typical parameters in plasma-induced ablation and photodisruption for nanosecond (ns), picosecond (ps), and femtosecond (fs) laser pulses. Data taken from [6, 7].

Parameter	ns range	ps range	fs range
Threshold exposure Φ_{th} (J/cm^2)	10–100	2–10	1–3
Maximum pressure at a distance of 1 mm from the laser focus (bar)	100–500	10–100	1–5
Diameter of cavitation bubble (μm)	100–2000	200–500	< 80
Conversion of absorbed light energy to cavitation bubble energy (%)	22	11	3

$\mu_{a,plasma}$ is thus a nonlinear function of the absorbed energy [3]. Due to the high plasma absorption, eye tissue (e.g., retinal tissue) behind the plasma cloud is protected. The so-called *plasma shielding* is an important safety issue during laser surgery. Depending on the exposure and pulse duration, typical values for $\mu_{a,plasma}$ range between 100 and 400/cm [6].

9.4.4.4 **Plasma Energy Density**

Plasma formation, vaporization, chemical dissociation of tissue within the plasma, shock wave formation, and cavitation bubbles are typical phenomena associated with an optical breakdown. With increasing energy density (energy per volume) of the plasma, the mechanically disruptive effects (e.g., shock waves and cavitation bubbles) become more and more significant.

The laser energy transferred to the plasma increases with increasing pulse duration, which can be deduced from Eq. (9.13) and Figure 9.9. Hence, depending on the pulse duration and the deployed laser energy/fluence, the laser-induced plasma may have two different effects on the tissue in the focal region, that is,

- plasma-induced ablation in the fs range and
- photodisruption in the ps to ns range.

In Table 9.1, typical parameters of plasma-induced interaction processes are listed with reference to the laser pulse duration.

Plasma-Induced Ablation For fs pulses, Φ_{th} above which an optical breakdown occurs is low. Since the energy density of the plasma is low as well, mechanical disruptive effects during the laser–tissue interaction can be ignored. The duration of the interaction is extremely short so that almost the entire deposited laser energy is kept inside the plasma cloud. Within the plasma volume, the tissue is destroyed by thermal vaporization due to the high plasma temperature and chemical dissociation by electrons.[13] The reaction products resulting from this process (i.e., hydrogen, oxygen, hydrocarbons, and carbon dioxide) form a gas bubble which gradually

13) The electrons have sufficient energy to break up the bonds of organic molecules.

disappears through tissue resorption. Analogous to photoablation with ultraviolet laser light (Section 9.4.3), this process is referred to as *plasma-induced ablation*. In photoablation (Section 9.4.3) and plasma-induced ablation, the tissue is split into volatile components. However, we should keep in mind that the underlying laser–tissue interactions are different (i.e., one-photon versus multiphoton absorption and optical breakdown).

By means of plasma-induced ablation, we can realize very clean and well-defined cuts in transparent tissue. The ablated region is spatially confined to the optical breakdown region and shows almost no thermal or mechanical side effects. This technique is thus well-suited for cornea-based refractive surgery (e.g., LASIK flap creation or fs lenticule extraction; Section 10.5.4.1) and cataract surgery (e.g., laser-assisted capsulorhexis; Section 10.5.4.2).

Photodisruption For pulse durations in the ps to ns range, Φ_{th} above which an optical breakdown occurs is relatively high (Figure 9.9). Hence, the light energy transferred to the plasma cloud is high as well. In this regime, the optical breakdown is generally accompanied by a popping sound which is an indicator of additional mechanical reactions. When these mechanical secondary effects dominate, photodisruption (from Latin: *ruptus* = ruptured) is induced, which is characterized by the following side effects:

- *Shock wave:* During the laser-induced plasma formation, the free electrons gain a lot of kinetic energy and diffuse into the surrounding tissue. As the ions immediately follow, the plasma cloud expands radially at supersonic speed. This creates a destructive shock wave which pushes the surrounding tissue away from its center (Figure 9.11). During this, the tissue structure is ruptured by emerging shear forces. After about 50 ns, the expansion slows down to ordinary acoustic speed. The maximum diameter of a spherical shock wave lies in the order of a few micrometers, whereas it is smaller for shorter pulse durations.
- *Cavitation:* Because of the high plasma temperature, tissue within the focal volume is instantaneously evaporated. The pressure of the formed gas of water and carbon dioxide is much higher than atmospheric pressure. Thus, the gas volume increases and pushes the surrounding tissue away (Figure 9.11). After some microseconds, the pressure inside the resulting cavitation bubble drops below atmospheric pressure. The bubble collapses and creates another shock wave. This happens periodically within some milliseconds. The cavitation volume, which corresponds to the damaged tissue volume, is proportional to the laser pulse energy, and the diameter can be as large as a few millimeters.

The characteristic time scale at which all these phenomena happen is plotted for a 30 ps laser pulse in Figure 9.12. The most relevant applications of ns photodisruption in ophthalmology are laser posterior capsulotomy (Section 10.4.4.1) and laser peripheral iridotomy (Section 10.4.4.2; see also Problem P9.5).

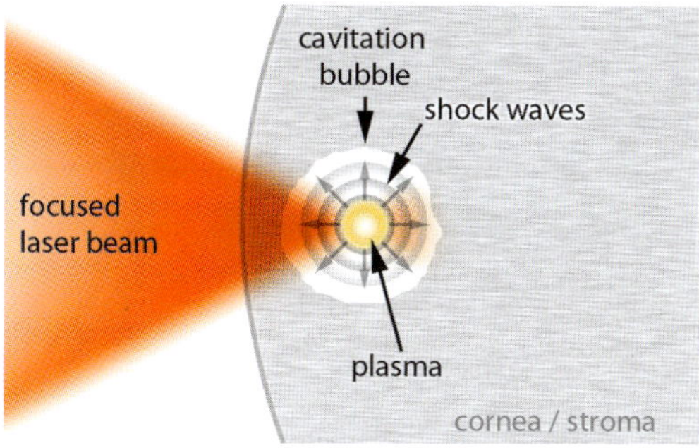

Figure 9.11 Photodisruption in corneal tissue. The high exposure of a tightly focused laser beam creates a plasma, acoustic shock waves, and cavitation bubbles.

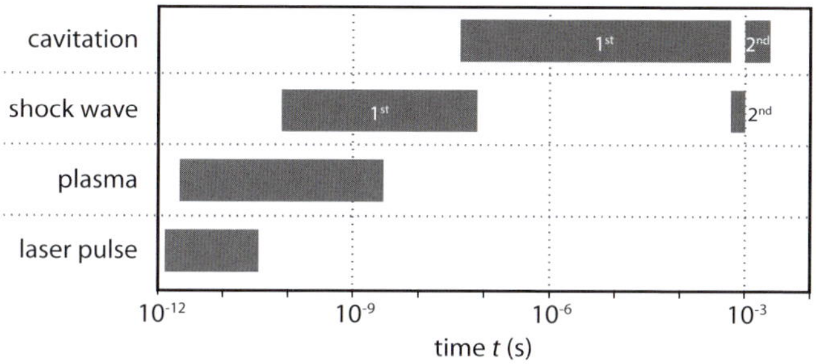

Figure 9.12 Time scale for processes which contribute to photodisruption. For this diagram, a laser pulse length of 30 ps is assumed. Adapted from [3].

9.5
Propagation of Femtosecond Pulses in Transparent Media

To obtain an optical breakdown with pulse durations in the range of 100 fs, intensities between 10^{12} and 10^{13} W/cm^2 are required [3]. When fs pulses with such high intensities pass through a transparent medium, we do not only observe multiphoton absorption (Section 9.1), but also other nonlinear effects. For the generation and application of fs laser pulses, *self-focusing* (Section 9.5.1) and *self-phase modulation* (Section 9.5.2) are of particular interest. As fs laser pulses are polychromatic (Section A.2.4), the *group velocity dispersion* (Section 9.5.3) must be considered as well.

Self-focusing and self-phase modulation are based on the fact that, for very intense laser pulses, the refractive index n of a medium depends on the light intensity I. This optical response of a medium is known as the *optical Kerr effect* for which we have

$$n(I) = n_0 + n_2 I \,, \tag{9.18}$$

where n_0 is the linear and n_2 the second-order nonlinear component of the refractive index. For pulse durations in the range of 100 fs, the nonlinear component of the refractive index n_2 of optical media takes values with a magnitude of $10^{-16}\,\mathrm{cm^2/W}$.

In the following, we will describe these nonlinear effects in brief so that the principles and applications in Section 10.5 can be understood. More details about nonlinear optics are presented in [8, 9].

9.5.1 Self-Focusing

For sufficiently high laser intensities, the intensity distribution along the beam waist leads to a local variation of the refractive index $n(I)$. If we assume a Gaussian beam profile (Section A.2.2.1), the refractive index is effectively larger in the center than in the periphery of the beam. As a consequence, the wavefront is deformed in a similar way as for a positive lens which, in turn, leads to a focusing effect.

Self-focusing is necessary for the generation and amplification of fs laser pulses. For example, it is used to form a so-called *Kerr lens*, which is one of the most common passive mode locking techniques (Section B.4.4.2) to generate very short laser pulses (Section 10.5.3). However, self-focusing is undesirable for laser beam amplification, since the high intensities may destroy the gain medium. To avoid this, the laser pulse must be broadened before it passes through the gain medium. The corresponding method, referred to as *chirped pulse amplification*, is discussed in Section 9.5.3.

9.5.2 Self-Phase Modulation

Self-phase modulation is a nonlinear optical effect which leads to a spectral broadening of an intense laser pulse (Problem PB.8). The effect is again based on the intensity dependence of the refractive index (optical Kerr effect). When a laser pulse travels through a medium, the light intensity rises at a certain location in the medium and then falls when the pulse passes by. At this location, the refractive index thus changes in time which causes a phase and frequency shift of the pulse. Let us consider the time-dependent phase change $\alpha(t)$ of the laser beam after it has traveled a distance L in the medium. In this case, we have

$$\alpha(t) = \omega_0 t - \boldsymbol{k}\boldsymbol{x} = \omega_0 t - \frac{2\pi}{\lambda_0} n(I) L \,, \tag{9.19}$$

where $\boldsymbol{k}$ denotes the wave vector, ω_0 the initial angular frequency, and λ_0 the light wavelength in vacuum. With $n(I) = n_0 + n_2 I(t)$, we obtain for the angular frequency

$$\omega(t) = \frac{\mathrm{d}\alpha}{\mathrm{d}t} = \omega_0 - \frac{2\pi}{\lambda_0} n_2 L \frac{\mathrm{d}I}{\mathrm{d}t} \,. \tag{9.20}$$

It is evident from Eq. (9.20) that $\omega(t)$ decreases when the laser intensity increases ($\mathrm{d}I/\mathrm{d}t > 0$). On the other hand, the frequency $\omega(t)$ increases at the end of the pulse

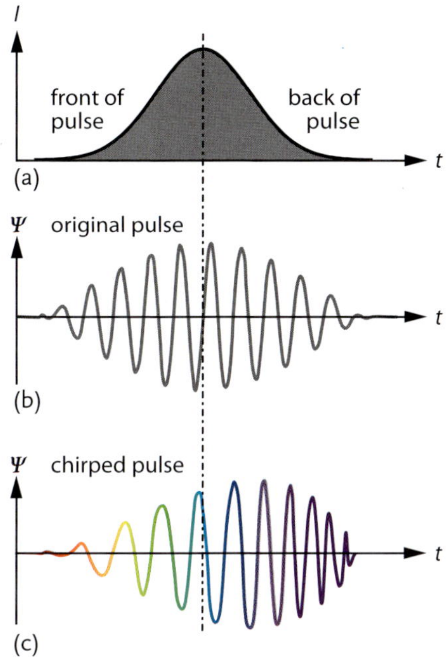

Figure 9.13 (a) An intense laser pulse transits through a transparent medium. (b) Original pulse shape. (c) Due to self-phase modulation, the pulse undergoes a frequency shift. The front of the pulse is shifted to lower frequencies and the back to higher frequencies. As a consequence, the pulse is spectrally broadened.

when the intensity decreases ($\mathrm{d}I/\mathrm{d}t < 0$). As shown in Figure 9.13c, the laser pulse front is thus shifted to lower frequencies (longer wavelengths) and the pulse back to higher frequencies (shorter wavelengths). As a result, the spectral bandwidth of the laser pulse is broadened, while the pulse duration is maintained. However, the effects of group velocity dispersion in a medium will act on the pulse shape.

9.5.3
Group Velocity Dispersion

Group velocity dispersion (GVD) is a phenomenon which is caused by the fact that the group velocity c_g (Section A.2.4) of light in a transparent medium depends on the wavelength. Dispersion (Section A.1.1) has an important impact on the propagation of fs pulses, because according to (B38) and (A114), a laser pulse always has a finite spectral bandwidth. For Gaussian beams, the spectral bandwidth $\Delta\lambda$ is related to the pulse duration τ via

$$\Delta\lambda\tau = 0.44\frac{\lambda^2}{c}\,, \tag{9.21}$$

where λ is the center wavelength of the laser light and c the speed of light (see also Section B.4.4.2).[14] The GVD is given by the second derivative of the refractive index with respect to the wavelength, that is,

$$\mathrm{GVD} = \frac{\lambda^3}{2\pi c^2} \frac{\mathrm{d}^2 n}{\mathrm{d}\lambda^2} \, . \tag{9.22}$$

Positive (negative) values of GVD ([GVD] = $\mathrm{fs}^2/\mathrm{mm}$) correspond to normal (anomalous) chromatic dispersion (Section A.1.1). In the case of positive GVD, the long-wavelength portions of the pulse spectrum are advanced with regard to the short-wavelength portions. This effect is referred to as *positive chirp*. For GVD < 0, the short-wavelength portions of the pulse spectrum are advanced with regard to the long-wavelength portions. This effect is then referred to as *negative chirp*. In any case, a nonzero GVD ultimately leads to a temporal spreading of the initial pulse. For example, at a wavelength of $\lambda = 800\,\mathrm{nm}$, the group velocity dispersion of fused silica is $+35\,\mathrm{fs}^2/\mathrm{mm}$.

In applications, we usually try to minimize the path lengths of pulses in optical media (e.g., glass) in order to keep the temporal spreading as low as possible. If necessary, optical arrangements with negative GVD (e.g., prism or grating pairs) must be added to the beam path to compensate for undesired effects. For example, if we use complex focusing optics with long path lengths in glass, a negative *pre-chirp* is required such that the focusing optics "compresses" the time-spread pulse and thus recovers the original pulse duration. It is also important to keep in mind that the path lengths of glass are different along the beam waist. This may also cause a spatial deformation of the pulse.

The pulse spreading due to GVD is intentionally used for *chirped pulse amplification* (CPA) of fs laser pulses (Section 10.5.3; Problem PB.9). The pulse to be amplified is temporally spread by a medium with positive GVD (e.g., optical fiber stretcher) so that its intensity is effectively reduced. With this method, self-focusing is reduced which might destroy the gain medium. After the laser pulse has passed through the gain medium, the pulse duration is compressed again by optical media or systems with negative GVD.

9.6 Ophthalmic Laser Safety

A variety of laser systems is used in ophthalmic diagnosis and therapy. In each application, potential laser hazards to the patient, physician, as well as the attending and service personnel exist. In diagnostic applications, the potential hazards are minimal, since eye-safe laser systems must be used.[15] But during ophthalmic therapies, patients are intentionally exposed to intense laser light to modify shape,

14) For example, a Gaussian laser pulse with a duration of $\tau = 100\,\mathrm{fs}$ and a center wavelength of 800 nm has a spectral bandwidth of $\Delta\lambda = 9\,\mathrm{nm}$.

15) General remarks regarding optical hazards that apply to all optical diagnostic and imaging methods are outlined in Section 4.4.

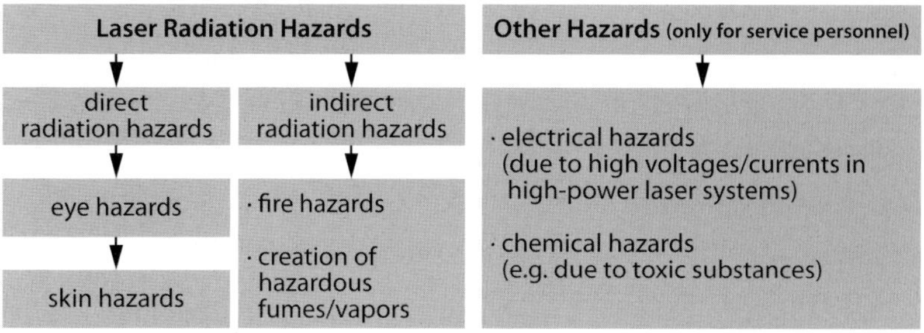

Figure 9.14 Potential hazards from laser equipment used in ophthalmic therapy.

structure, and function of eye tissue. Hence, an accidental exposure or misuse of laser light may lead to severe injuries.

In the following, we will restrict our discussion to laser safety during ophthalmic therapies. The basic potential hazards (Figure 9.14) from laser therapy equipments can be categorized into direct or indirect laser radiation hazards and other hazards.

Direct laser radiation hazards In practice, direct laser radiation hazards are most important. Both eye and skin can be easily injured by lasers and other intense optical light sources. The hazard potential depends on the wavelength, power, and duration of radiation exposure (Table 9.2). The normal aversion response to heat sensation of exposed skin will normally limit potentially hazardous thermal exposure of the skin. Eye tissue is, however, considerably more at risk than skin tissue due to the potential loss of vision. Even moderately intense laser beams may lead to serious irreversible injuries. In particular, if Q-switched (Section B.4.4.1) or mode-locked lasers (Section B.4.4.2) are used, operators and assistants must be very careful, as these lasers usually provide extremely high light intensities, and diffuse reflections from a surface may also cause severe eye damage.

The spectral range of the laser determines which part of the eye is damaged in the case of exposure (Figure 9.3). In particular, laser light with wavelengths in the (retinal hazard) range between 400 and 1400 nm is extremely dangerous. When this light is directly focused onto the retina, it may cause thermal and photochemical damage.

Indirect laser radiation hazards During laser treatment, hazardous fumes or vapors may be generated which must be evacuated. In addition, there is always a risk of fire when continuous wave (cw) lasers (Section B.4.4) with an output power above 0.5 W are used in the presence of highly flammable substances.

Other hazards Many laser systems have built-in high voltage components. Pulsed lasers are particularly dangerous because of the stored energy in the capacitor banks. Hence, most lethal occurrences are caused by electric shocks during servicing. In addition, some gain materials used in ophthalmic lasers are toxic (e.g., fluorine gases in ArF excimer lasers; Section B.5.1.2). However, these kinds of haz-

Table 9.2 Potential optical radiation hazards to eye and skin from lasers and other intense radiation sources. The hazard potential depends on the wavelength, power, and duration of radiation exposure (* This type of injury is limited to long-term exposure.).

Wavelength (nm)	Type of hazard	Cause of hazard
180–315	Photokeratitis of cornea and conjunctiva	Light absorption by actin in cornea*
280–315	Erythema (sunburn), accelerated skin aging	Light absorption by skin tissue*
280–480	Increased pigmentation, long-wave erythema	Light absorption by skin tissue*
315–400	Photochemical cataract	Absorption of UV radiation by lens*
400–550	Photochemical damage of retina (blue-light retinal hazard)	Light absorption by retinal tissue*
400–1400	Thermal damage of retina	Light absorption by retinal tissue
400–10^6	Skin burns	Light absorption by skin tissue
1400–3000	Photothermal cataract, corneal burn	Absorption of IR radiation by lens or cornea*
3000–10^6	Corneal burn	Light absorption by cornea

ards are only relevant for service personnel if, for example, protective enclosures are removed or gas bottles are exchanged.[16)]

Due to the potential risks for eye and skin during laser operation, the safe use of lasers has been regulated by official safety standards and guidance. The basic approach of nearly all laser safety standards is the classification of lasers (Section 9.6.1) by their hazard potential. Then, appropriate control measures are deduced. The standards define requirements for an adequate system design, work environment (safety systems), and usage. Table 9.3 lists some international standards which are relevant for general laser operation and the use of lasers for medical applications. Tables 9.4 and 9.5 specify corresponding standards for European countries and the United States, respectively. These standards include engineering, procedural, and administrative controls necessary for the safety of patients and health care professionals.

9.6.1 Laser Classes

The standards classify lasers according to the amount of radiation accessible during "normal" usage, that is, the so-called *accessible emission limits* (AEL). They do not refer to the emitted radiation during periods of service or maintenance! The maximum output energy or power determines the laser classification, even though

16) Today, ArF excimer lasers are well designed, and only premixed gases are used.

Table 9.3 International safety standards relevant for general and medical usage of laser systems.

Standard	Content and description
IEC 60825-1 ed2.0 (2007)	Requirements for the safety of laser products. Part 1: Equipment classification and requirements.
ISO 6161:1981 (2008)	Requirements regarding filters and personal eye-protectors.
IEC 60601-2-22 ed3.0 (2007-05)	Requirements for medical electric equipment. Part 2-22: Particular requirements for basic safety and essential performance of surgical, cosmetic, therapeutic and diagnostic laser equipment.

Table 9.4 Safety standards relevant for general and medical usage of laser systems in European countries.

Standard	Content and description
EN 60825-1 (equiv. to IEC 825-1)	Classification of laser (diode) systems emitting at wavelengths from 180 nm to 1 mm regarding potential risks for skin and eye. The standard provides only limited information about hazards arising from electricity and emission of toxic substances. Included topics: • Requirements for manufacturers regarding laser classification, indication, and technical properties. • Regulations regarding the operation of laser systems.
EN 60825-4	Requirements regarding safeguards and protective devices.
EN 207/EN 208	Technical specification for testing of filters and laser protective goggles.
EN 165–EN 171	General requirements regarding eye protection.
EN 60601-2-22 (equiv. to IEC 601-2-22)	Particular requirements for the safety of diagnostic and therapeutic laser equipment.
EN ISO 15004-1	Fundamental requirements and test methods applicable to all ophthalmic instruments.
EN ISO 15004-2	Requirements for optical radiation safety for ophthalmic instruments that direct radiation into or at the eye for diagnostic, cosmetic, preventive, and therapeutic purposes.

normal operation may require less than the maximum output. In addition, the classification relates to the potential for injury from the beam itself and not from related hazards like the high-voltage power supply. The lasers are classified according to their ability to produce damage in exposed people. For each laser class, the

Table 9.5 Safety standards relevant for general and medical usage of laser systems in the United States of America.

Standard	Content and description
ANSI Z136.1	Classification of laser (diode) systems emitting at wavelengths between 180 nm and 1 mm regarding potential risks for skin and eye. Included topics: • Requirements for manufacturers regarding laser classification, indication and technical properties. • Regulations and recommendations regarding the operation of laser systems.
ANSI Z136.3	Guidance for safe use of lasers for diagnostic, cosmetic, preventive, and therapeutic applications in health care facilities. The latter are defined as any location where a laser is applied to humans to alter bodily structure or function or relieve symptoms.
Code of Federal Regulations "21", Parts 1040.10 and 1040.11, Medical Devices Act	Details about the enforcement and implementation of laws and regulations applying to radiation-producing medical devices and other electronic products by the Federal Food and Drug Administration (FDA) Center for Devices and Radiological Health (CDRH).

standards instruct specific manufacture requirements and user precautions. In the following, we provide an overview of existing laser classes:[17)]

Class 1: Lasers of this class are eye-safe under all foreseeable operating conditions, including the use of optical instruments for intrabeam viewing.

Class 1M: Lasers of this class are safe for viewing directly with the naked eye, but can be hazardous if the user employs optical instruments for intrabeam viewing. In general, the use of magnifying glasses increases the hazard from a widely-diverging beam (e.g., LEDs), and binoculars or telescopes increase the hazard from a wide, collimated beam. Radiation in Classes 1 and 1M can be visible or invisible. Class 1M lasers usually emit beams which are divergent or have a large diameter.

Class 2: Lasers of this class emit visible radiation in the wavelength range from 400–700 nm for which eye protection is usually afforded by natural aversion responses to very bright light (e.g., eye lid closure reflex). Laser radiation of this class is safe for accidental viewing under all operating conditions. However, it may not be safe for a person who deliberately stares into the laser beam for longer than 0.25 s. Class 2 lasers are limited to an output power of 1 mW for cw operation (pulse length $\tau > 0.25$ s).

17) A detailed description of laser classes can be found in the IEC 60825-1 (Table 9.3) or ANSI Z136.1 (Table 9.5) standards.

Class 2M: Lasers of this class emit visible radiation in the wavelength range from 400–700 nm for which eye protection is normally afforded by natural aversion responses to very bright light (e.g., eye lid closure reflex). Laser radiation of this class is safe for accidental intrabeam viewing, as with Class 2, but may be hazardous (even for accidental viewing) when viewed with a visual aid or optical instruments (similar to Class 1M). Class 2M lasers usually emit beams which are divergent or have a large diameter.

Class 3R: Lasers of this class emit light in the wavelength range between 302.5 and 10^6 nm. Radiation in this spectral range is considered low risk, but potentially hazardous. The class limit for 3R is $5 \times$ the applicable limit for Class 1 (invisible radiation) or Class 2 (visible radiation). Hence, cw visible lasers emitting between 1 and 5 mW are normally assigned to Class 3R. Important user precautions: Only trained personnel are allowed to operate lasers of Class 3R. Appointment of a laser safety officer (LSO) is required for Class 3R lasers that emit invisible radiation.

Class 3B: Lasers of this class are medium-powered light sources which emit either in the visible or invisible spectral range. For a continuous wave laser, the maximum output power is limited to 500 mW. Class 3B lasers are potentially hazardous if a direct beam or specular reflection is viewed by the unprotected eye. However, viewing diffuse reflections is normally safe. Laser radiation in this class may cause skin injuries. Important manufacture requirements and user precautions: Lasers of Class 3B must be equipped with a key switch and a safety interlock. The appointment of an LSO is required, and only trained personnel are allowed to operate lasers of Class 3B.

Class 4: This is the highest and most dangerous laser class. Lasers of this class can cause severe eye injury from the direct beam, its specular reflections, and/or from diffuse reflections. Class 4 laser beams may cause skin injuries and could also create a burning. Their use thus requires extreme caution. Important manufacture requirements and user precautions are the same as for Class 3B.

Any laser product of a given class may contain "embedded" laser systems which are greater than the class assigned to the product. In these cases, engineering controls (protective housings and interlocks) ensure that during normal operation, the user cannot access radiation in excess of the assigned product class.

9.6.2
Safe Use of Ophthalmic Laser Systems

Laser systems used in ophthalmic therapy virtually all belong to Classes 3B and 4. Thus, according to the laser standards, only trained personnel is allowed to operate these systems. In addition, the appointment of an LSO is required. The LSO should be a trained person, who is familiar with laser safety guidance and terminology. He or she should have a balanced view of what might be perceived as a very hazardous

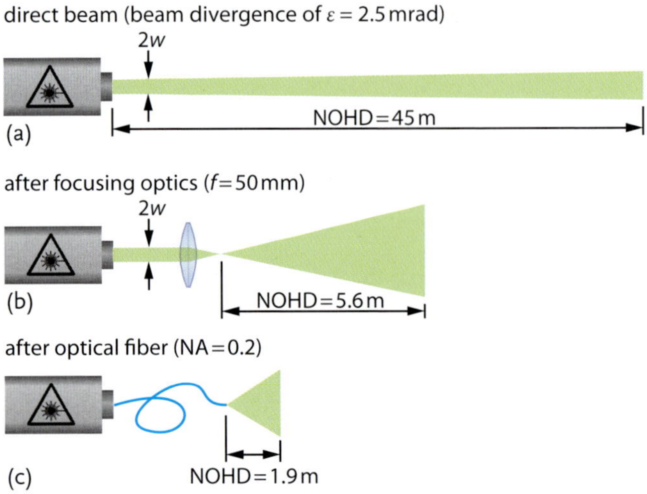

Figure 9.15 Nominal ocular hazard distance analysis for the unintended look into a laser beam with a power of 1 W, a wavelength of 532 nm, and a beam diameter of $2w = 2$ mm. Outside the NOHA, an unintended look into the laser beam is not dangerous. Please note that the schematics are not to scale. (a) Direct laser beam with a beam divergence of $\varepsilon = 2.5$ mrad. (b) Laser beam focused by a lens with a focal length of $f = 50$ mm. (c) Laser beam emitted from an optical fiber with a numerical aperture of NA = 0.2.

laser-related situation and what procedures are necessary to take appropriate precautions. For example, the LSO should be able to estimate which area around the laser source is potentially hazardous. Such a qualified safety analysis is based on the maximum permissible exposure (MPE) values in W/cm^2 or J/cm^2 which are set below known hazard levels. In the IEC 60825-1 or ANSI Z136.1, the MPE values are given separately for eye and skin tissue. They depend on the light wavelength and exposure time (or pulse length).

The area in which the MPE values for eyes are exceeded is referred to as the *nominal ocular hazard area* (NOHA) or *nominal ocular hazard distance* (NOHD). Only in this area, suitable protective measures (e.g., protective goggles) are necessarily required. Figure 9.15 shows a safety analysis (Problem P9.6) for an unintended look into green laser beams with different divergence (Section B.4.2). As the eyelid closure reflex limits the exposure time to 0.25 s, we obtain an MPE value of 25.5 W/m^2. This analysis will also reveal that for lasers with small beam divergence (e.g., laser pointer), eye damage may occur even at large distances.

The potential hazards are quite different for patients, physicians, attending people, and service staff. In the following, they are presented with corresponding safety instructions:

Patient Most laser safety regulations do not directly apply to the exposure of patients at the target site. However, accidental exposure of patients due to misdirection of the beam or due to unintentional launching of the system may lead to severe injury of the skin and eye. Thus, physicians should always be on the alert when the laser system is turned on.

Service personnel For maintenance and regular servicing work, the cover of the laser system has to be removed to have free access to the resonator. If laser light unintentionally leaves the system (e.g., due to reflections at tuning mirrors or prisms), service personnel may be exposed to collimated beams with small divergence. This means that the laser intensity decreases very slowly with distance. Even after the beam has traveled more than several tens of meters, it is still extremely hazardous for the human eye! In contrast, focused laser light or light emitted from a fiber tip rapidly diverges and is thus only dangerous within a few meters around the focal point (Figure 9.15). Service personnel must therefore *always* wear adequate protective eyeglasses during laser operations which filter most of the emitted light intensity. Special care should be taken with invisible IR and UV laser radiation.

Physicians In contrast to service personnel, physicians "merely" have to handle focused laser beams and/or beams emitted from fiber applicators. In both cases, the beam divergence is so high that the hazard area is limited to a few meters. Moreover, physicians view the target through the optics of a slit lamp, microscope or endoscope. Due mandatory safety requirements, reflections are safely attenuated by internal filters. Nevertheless, if the laser is accidentally turned on when the physician is not looking through the designated port, he or she will be as much at risk as any other unprotected person in the room.

Special care should be taken with hand-held laser delivery systems like fiber optical waveguides and headworn indirect ophthalmoscopes. Since the physician's hand is closest to the beam, no reflective items (watch, ring, etc.) should be worn! Reflections at these items may harm users, attending personnel, and patients.

Attending personnel Nurses, surgical assistants, and other assisting staff are potentially exposed to misdirected laser light. This can happen, for example, when laser light is reflected from contact lenses and surgical instruments in the beam path. As, unlike the physician, the assisting staff is not protected by the device's built-in filters, the emitted laser light is potentially hazardous within a few meters around the laser focus. Hence, attending staff must wear protective eyeglasses if they are located close to the laser source. To minimize the unintended exposure of bystanders, the laser should always be directed at a wall to terminate the beam in case the patient is not in position.

9.7 Recommended Reading

More detailed information about laser–tissue interactions can be found in [3, 10, 11]. Further details about ophthalmic laser safety are presented in [12, 13].

9.8 Problems

P9.1. Penetration depth of lasers Calculate the penetration depths for the following lasers in water and in blood-rich tissue: Nd:YAG laser, Nd:YAG (frequency doubled) laser, and ArF excimer laser.

P9.2. Light–tissue interaction

1. Which light–tissue interaction can only be realized by the use of lasers? For which interaction mechanisms can conventional (thermal) light sources be used alternatively?
2. What are the primary reasons for laser usage in medical therapy?

P9.3. Intensity

1. Calculate the time average of the product of two trigonometric functions with equal frequencies. The time average is defined by

$$\langle a(t)b(t)\rangle = \frac{1}{T}\int_0^T a(t)b(t)\mathrm{d}t \tag{9.23}$$

in which

$$a(t) = \mathrm{Re}\left(A\mathrm{e}^{\mathrm{i}\omega t}\right)\ ; \quad b(t) = \mathrm{Re}\left(B\mathrm{e}^{\mathrm{i}\omega t}\right) \tag{9.24}$$

and $T = 2\pi/\omega$.
2. Show that the time average of the square of a function

$$a(t) = \mathrm{Re}(V(t)) = \mathrm{Re}\left(A_1\mathrm{e}^{\mathrm{i}\omega t} + A_2\mathrm{e}^{\mathrm{i}\omega t}\right) \tag{9.25}$$

is given by

$$\langle a^2(t)\rangle = \frac{1}{2}\{V(t)V^*(t)\}\,. \tag{9.26}$$

3. A red diode laser ($\lambda = 700$ nm) with 10 mW electrical power and 10% electrical efficiency is focused on a 100 μm² spot. Which photon fluence (number of photons per time and area) does this correspond to?
4. Compare the result of 3. to the achievable intensity with a 20 W light bulb (assumed to be a Planck black body with 3000 K temperature), optical efficiency of 2% for the same spot size and a bandpass filter at 700 nm with bandwidth of $\Delta\lambda = 20$ nm. How would you design the collection optics? What electrical power would be needed to achieve the same spot intensity as in 3.?

P9.4. Photocoagulation Suppose you cook an egg in the pan. Why does the egg white turn from transparent to white?

P9.5. Plasma-induced ablation and photodisruption Plasma-induced ablation and photodisruption are two kind of a laser–tissue interaction occurring at short irradiation times and high intensities. What do they have in common and what is different in the two processes?

P9.6. Laser safety Verify the nominal ocular hazard distance safety analysis depicted in Figure 9.15 for an unintended look into a laser beam with a power of 1 W, a wavelength of 532 nm, and a beam diameter of $2w = 2$ mm for

1. a direct laser beam with a beam divergence of 2.5 mrad and
2. a laser beam focused by a lens with a focal length of 50 mm.

References

1 Bohren, C.F. and Huffmann, D.R. (2010) *Absorption and Scattering of Light by Small Particles*, John Wiley & Sons, Inc.

2 Mainster, M.A. (1986) Wavelength selection in macular photocoagulation. tissue optics, thermal effects, and laser systems. *Ophthalmology*, **93**, 952–958.

3 Niemz, M.H. (2007) *Laser-Tissue Interactions: Fundamentals and Applications*, Springer, 3rd edn.

4 Choi, B. and Welch, A.J. (2001) Analysis of thermal relaxation during laser irradiation of tissue. *Lasers Surg. Med.*, **29**, 351–359.

5 Vogel, A. and Venugopalan, V. (2011) Pulsed laser ablation of soft biological tissue, in *Optical-Thermal Response of Laser-Irradiated Tissue*, Springer, pp. 551–615, 2nd edn.

6 Vogel, A. (2003) Mechanism of short-pulsed plasma-mediated laser ablation and disruption, in *Lasers in Ophthalmology – Basics, Diagnostic and Surgical Aspects*, Kugler Publications, pp. 99–113.

7 Lubatschowski, H., Maatz, G., Heisterkamp, A., Hetzel, U., Drommer, W., Welling, H., and Ertmer, W. (2000) Application of ultrashort laser pulses for intrastromal refractive surgery. *Graefes Arch. Clin. Exp. Ophthalmol.*, **238**, 33–39.

8 Boyd, R.W. (2008) *Nonlinear Optics*, Academic Press, 3rd edn.

9 Diels, J.C. and Rudolph, W. (2006) *Ultrashort Laser Pulse Phenomena: Fundamentals, Techniques, and Applications on a Femtosecond Time Scale*, Academic Press, 2nd edn.

10 Welch, A.J. and van Gemert, M.J.C. (eds) (2011) *Optical-Thermal Response of Laser-Irradiated Tissue*, Springer, 2nd edn.

11 Fankhauser, F. and Kwasniewska, S. (eds) (2003) *Lasers in Ophthalmology – Basics, Diagnostic and Surgical Aspects*, Kugler Publications.

12 Sliney, D.H. (2003) Ophthalmic Laser Safety, in *Lasers in Ophthalmology – Basics, Diagnostic and Surgical Aspects*, Kugler Publications, The Hague, pp. 1–10.

13 Sliney, D.H. and Trokel, S.L. (1993) *Medical Lasers and Their Safe Use*, Springer.

10
Laser Systems for Treatment of Eye Diseases and Refractive Errors

The use of lasers in ophthalmic therapy was one of the very first applications of lasers at all and is up to now the most significant medical laser application. The basic concept of laser therapy methods in ophthalmology is the selective change of tissue properties by means of different types of laser–tissue interaction (Section 9). As laser light can be used to treat all optically accessible areas in the eye, it forms the basis for many noninvasive[1)] or at least minimally invasive noncontact techniques. Today, noncontact laser methods are used with great success for the treatment of many different eye diseases (Table 10.1) and to correct refractive errors of the eye (Table 10.2).

In the following sections, we will group the laser systems not by application, but by the employed laser–tissue interaction, that is,

- photochemical interactions (photoactivation lasers or lasers for photodynamic therapy),
- photothermal interactions (photocoagulation lasers),
- photoablation (photoablation lasers), and
- photodisruption (nanosecond lasers).
- plasma-induced ablation (femtosecond lasers).

Tables 10.1 and 10.2 provide also a link to the application-oriented categorization.

History The pioneering work of Gerhard Meyer-Schwickerath (1920–1992) in the area of photocoagulation [1] laid the foundation for the successful implementation of lasers in eye surgery. As a light source for photocoagulation, he first used the sun (in 1948) and later (in the early 1950s) carbon arc lamps. Finally, he developed together with ZEISS scientist Hans Littmann the xenon lamp photocoagulator LIKO. The ZEISS LIKO (first manufactured in 1957) was the first commercial light-based surgical instrument which allowed noncontact and noninvasive interventions inside the eye. With the rapid development of lasers from 1960 onwards (Appendix B), new radiation sources became available which soon replaced the "classic" light sources and offered additional treatment options. The most important milestones for laser applications in ophthalmology are listed in Table 10.3.

1) Here, "noninvasive" refers to an intraocular surgery without incisions.

Optical Devices in Ophthalmology and Optometry, First Edition. Michael Kaschke, Karl-Heinz Donnerhacke, and Michael Stefan Rill.

Table 10.1 Laser-based procedures for the treatment of certain eye diseases. The sections in which the procedures are discussed are indicated in brackets.

Disease	Laser procedure (section)	Laser–tissue interaction (section)
Retinal diseases	• Standard photocoagulation (pan-retinal, grid, focal, etc.) (10.2.3.1) • Short-pulse treatment (10.2.3.2) • Transpupillar thermotherapy (TTT) (10.2.3.3)	Photothermal (10.2)
	• Photodynamic therapy (PDT) (10.1.1)	Photochemical (10.1)
Glaucoma	• Argon laser trabeculoplasty (ALT) (10.2.5) • Selective laser trabeculoplasty (SLT) (10.2.5) • Micropulse laser trabeculoplasty (MLT) (10.2.5) • Cyclophotocoagulation (CPC) (10.2.5)	Photothermal (10.2)
	• Laser peripheral iridotomy (LPI) (10.4.4.2)	Photodisruption (ns laser pulses) (10.4)
Secondary cataract (posterior capsular opacification)	• Laser posterior capsulotomy (LPC) (10.4.4.1)	Photodisruption (ns laser pulses) (10.4)
Cataract (laser-assisted cataract surgery) (10.5)	• Laser anterior capsulotomy (10.5.4.2) • Clear corneal incisions (CCI) (10.5.4.2) • Lens fragmentation (10.5.4.2) • Arcuate keratectomy (10.5.4.2)	Photodisruption (ps laser pulses) (10.4)/plasma-induced ablation (fs laser pulses) (10.5)
Corneal diseases		
Keratoconus	• fs laser-assisted keratoplasty (FLAK) (10.5.4.1) • Tunnel resections for intrastromal corneal ring segments (ICRS) (10.5.4.1)	Plasma-induced ablation (fs laser pulses) (10.5)
Erosions	• Surface ablation (10.3.3.1)	Photoablation (10.3)

10.1
Laser Systems Based on Photochemical Interactions

Photodynamic therapy (PDT) is based on a photochemical response of tissue due to the joint action of photosensitizer and laser radiation (Section 9.4.1). In ophthalmology, this method is used for selective treatments of sub-foveal choroidal neovascularization (CNV) which affects patients with age-related macular degeneration (AMD) (Section 3.4) or high myopia (Section 3.1). In industrialized countries, neovascular (wet) AMD is the main cause of irreversible blindness for persons older than 50 years, and its incidence increases with age ([4, 5]; see also Section 3.9). Un-

Table 10.2 Overview of laser-assisted procedures in refractive surgery. The sections in which the procedures are discussed are indicated in brackets.

Refractive surgery technique	Laser procedure (section)	Laser–tissue interaction (section)
Cornea-based refractive surgery (ablation)	Surface ablation (10.3.3.1): • Photorefractive keratectomy (PRK) • Laser epithelial keratomileusis (LASEK) • Epithelial LASIK (epiLASIK) Intrastromal ablation (10.3.3.2): • Laser *in situ* keratomileusis (LASIK)	Photoablation (10.3)
Cornea-based refractive surgery (incision)	• Flap creation in laser *in situ* keratomileusis (fs LASIK) (10.5.4.1) • Refractive lenticule extraction (ReLEx®) (10.5.4.1) • Arcuate keratectomy in astigmatism (AK) (10.5.4.1) • Presbyopia intrastromal treatment (INTRACOR®) (10.5.4.1) • Tunnel resections for intrastromal corneal ring segments (ICRS) (10.5.4.1) • Creation of intracorneal pockets for corneal inlays (10.5.4.1)	Plasma-induced ablation (fs laser pulses) (10.5)
Lens-based refractive surgery Refractive lens exchange Implantation of premium IOLs	• Laser anterior capsulotomy (10.5.4.2) • Clear corneal incisions (CCI) (10.5.4.2) • Lens fragmentation (fs phaco) (10.5.4.2)	Plasma-induced ablation (fs laser pulses) (10.5)/ photodisruption (ps laser pulses) (10.4)

til the mid-1990s, focal photocoagulation (Section 10.2) of newly formed abnormal blood vessels was the only available therapy to avoid imminent blindness caused by wet AMD. But since photoreceptors (Section 1.2) are also destroyed by conventional photocoagulation (thus generating absolute scotoma), the possibilities to treat the macular region were very limited. These difficulties were overcome by PDT, which was introduced by Ursula Schmidt-Erfurth *et al.* and Gilbert Kliman *et al.* in 1994 (Table 10.3). PDT allowed, for the first time, destruction of abnormal blood vessels in a selective manner without affecting the photoreceptor layer. Treatments in the foveal region thus became possible as well.

PDT does not cure wet AMD, but just slows down the progressive impairment of central vision. Nevertheless, it became established as the method of choice to treat sub-foveal CNV within few years after introduction. PDT with the photosensitizer

Table 10.3 Important milestones in the development of therapy laser procedures for ophthalmology. Data taken from [2, 3].

Year	Milestone	Researchers
1962/1963	Ruby laser photocoagulation	Koester, Campbell *et al.*; Zweng *et al.*
1968	Argon laser photocoagulation	L'Esperance
1979	Argon laser trabeculoplasty for glaucoma	Wise, Witter
1979	Laser photodisruption	Aron-Rosa *et al.*; Fankhauser *et al.*
1983/1986	Excimer laser photoablation of corneal surface (photorefractive keratectomy)	Trokel, Srinivasan *et al.*; Seiler, Wollensack *et al.*
1986	Photorefractive keratotomy	Marshall, Trokel *et al.*
1991	Laser *in situ* keratomileusis (LASIK)	Pallikaris *et al.*
1992/2000	Selective retina therapy (SRT) or short-pulse treatment	Roider, Birngruber *et al.*
1994	Photodynamic therapy	Kliman *et al.*; Schmidt-Erfurth, Birngruber *et al.*
1995	Selective laser trabeculoplasty for glaucoma	Latina, Park
1998/2001	fs laser-assisted corneal flap cutting for LASIK (fs LASIK)	Ratkay-Traub, Juhasz, Kurtz *et al.*
2003	fs laser intrastromal cutting and ablation for myopia and hyperopia	Ratkay-Traub, Juhasz, Kurtz *et al.*
2009	fs laser-assisted cataract surgery	Nagy, Takacs *et al.*

(photochemical substance) verteporfin[2)] is less stressful for patients, relatively cost-effective, and can be repeated at regular intervals. In 2006, a very efficient anti-VEGF drug therapy was introduced which strongly diminished the importance of PDT (Section 10.1.4).

10.1.1
Basics of Photodynamic Therapy

The fundamental concepts of photochemical laser–tissue interaction have already been described in Section 9.4.1. Let us now discuss the required process steps for PDT treatments with verteporfin:

1. *Application of photosensitizer:* The photosensitizer verteporfin is injected to the vein.
2. *Selective accumulation of the photosensitizer in the target region:* After injection, verteporfin preferentially attaches to the walls of newly formed abnormal vessels. Hence, its concentration in the chorio-retinal tissue changes with time.

2) Verteporfin is a photosensitive dye which is commercially available under trade name Visudyne.

About 5 min after the injection, the concentration difference of verteporfin between abnormal vessels and adjacent tissue has reached the maximum. This is the right time to start the local laser treatment.

3. *Activation of verteporfin by means of laser light:* For the treatment, laser light with a wavelength of $\lambda = 689$ nm is used for which verteporfin has a relative absorption maximum. The absorption of the most relevant endogenous chromophores (Section 6.7.7), such as melanin and xanthophyll, is relatively low in this spectral range. As a consequence, the laser light is able to penetrate deeply into the choroid, where the CNV develops. In addition, the scattering losses in the anterior eye segment are also low for the used laser wavelength.
4. *Photoinduced dissociation and CNV tissue necrosis:* As mentioned in Section 9.4.1, the excited photosensitizer molecule transits to a long-lived energy state. In type II reactions, the triplet state ${}^3\text{S}^*$ of the photosensitizer directly interacts with molecular triplet oxygen ${}^3\text{O}_2$ which is already present in the tissue. The oxygen molecule is then excited to the singlet state O_2^* according to

$$ {}^3\text{S}^* + {}^3\text{O}_2 \rightarrow {}^1\text{S} + {}^1\text{O}_2^* \,. \tag{10.1}$$

The excited singlet oxygen ${}^1\text{O}_2{}^*$ is highly reactive thus leading to cellular oxidation and necrosis of the vessel wall of CNV.

PDT is based on the joint action of photosensitizer and laser radiation. The chemical substance or the laser radiation alone have actually no effect. For this reason, the photosensitizer verteporfin must be used in combination with specially approved PDT laser instruments. Out of the three instruments worldwide that have met these requirements, only the ZEISS VISULAS 690^{plus} laser system (Figure 10.1) is still on the market today (in 2013).

10.1.2 Technical Equipment Concepts

The setup of a PDT laser system (Figure 10.1) essentially consists of

- a laser source with control unit (Section 10.1.2.1),
- a beam transmission system from the laser source to the beam delivery device, and
- a beam delivery device consisting of a laser link adapted on a diagnostic slit lamp (Section 6.4).

In PDT, solutions for beam delivery and beam application have been adopted which proved themselves in photocoagulation. The corresponding techniques are discussed in detail in Sections 10.2.4.2 and 10.2.4.3. In the following, we thus restrict ourselves to the discussion of the laser source and some characteristics of the beam delivery system.

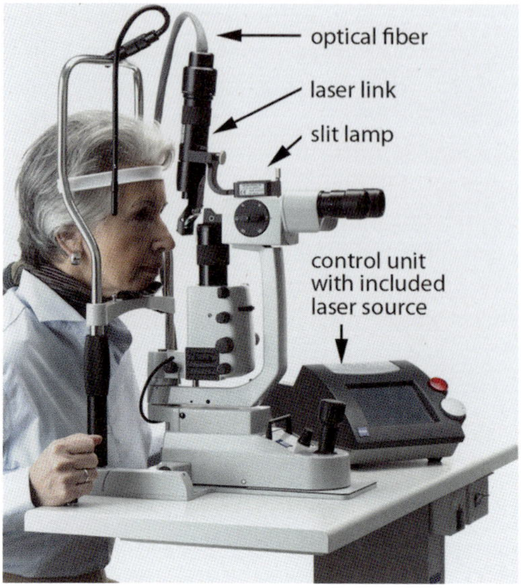

Figure 10.1 Photograph of the ZEISS VISULAS 690^{plus} PDT laser system with patient. Light emitted by the laser source is guided by an optical fiber to a laser link which is mounted onto a standard diagnostic slit lamp. The end surface of the optical fiber is imaged by the optical system of the laser link via a link mirror and a contact lens (not shown) onto the patient's retina to form a laser spot of desired size. The slit lamp is used for the visual treatment control only. Courtesy of Carl Zeiss.

10.1.2.1 **Laser Source for Photodynamic Therapy**

The following procedure is prescribed for PDT with Visudyne:

- Duration of irradiation: 83 s
- Exposure: 50 J/cm^2
- Wavelength: 689 nm $\pm$ 3 nm

If the laser spot has a diameter of 5 mm on the fundus, a constant output power of approximately 120 mW must be applied. To meet this requirement, a high-performance laser diode with a maximum output power of 400 mW is typically used. The wavelength of the diode laser is stabilized by a closed-loop control which keeps the device constant in temperature. The laser's output power is monitored by means of three independent photodiodes, whereas one of them is located directly at the beam exit of the laser link. The entire laser source is usually integrated into the control unit (Problems P10.1 and P10.2).

10.1.2.2 **Beam Delivery Device**

A laser link mounted onto a standard diagnostic slit lamp is used as the beam delivery device (Figure 10.1). The end surface of the fiber transmission system is imaged by the optical system of the laser link via a link mirror and a contact lens (not shown) onto the retina to obtain a laser spot of desired size. The spot size can be set by the zoom system of the laser link in the range from 800–5000 μm. The

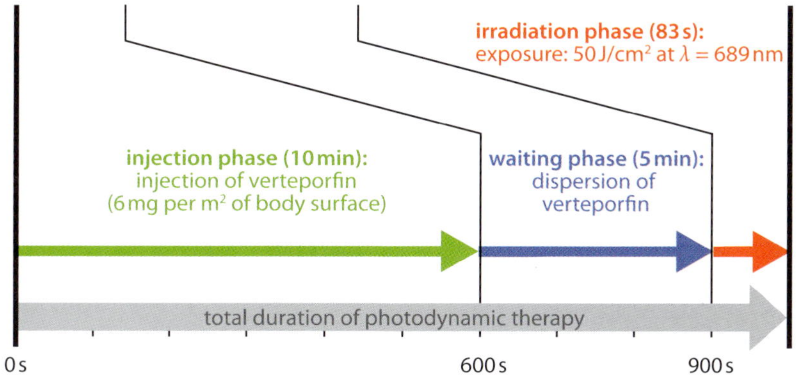

Figure 10.2 Typical treatment procedure in photodynamic therapy with verteporfin including the time scale.

slit lamp is used for visual treatment control only. Further details are presented in Section 10.2.4.3.

10.1.3 Treatment Procedure

The steps of the PDT procedure are illustrated in Figure 10.2. The required laser spot size on the fundus results from the greatest linear diameter of the lesion to be treated plus an additional safety margin of 0.5 mm. The lesion diameter can be determined from fluorescence angiography images of the eye fundus (Section 6.7.6). The laser spot size is then automatically calculated from this value with a special software tool. Eventually, the laser output power is automatically adjusted to apply the standard dose, while taking the magnification factor of the used laser contact lens (Section 6.4.4.1) into account. As the photosensitizer remains active over a period of 24–48 hours, the treated eyes should not be exposed to bright light, and the patient should wear sunglasses for eye protection.

10.1.4 Prospects

Since the introduction of drug-based therapy for wet AMD in 2006 which uses anti-vascular endothelial growth factor (anti-VEGF) substances[3] (e.g., *pegaptanib* available as Macugen® and *ranibizumab* available as Lucentis®), the significance of PDT has strongly decreased. The drug-based therapy revolutionized treatment of wet AMD, as it stops the loss of vision and, above all, even improves the visual acuity. However, the anti-VEGF substances must be injected several times a year to ensure a therapeutic success. Because of their high price and since multiple injections are quite stressful for the patient, researchers try to find ways to increase

3) These substances are directly injected into the vitreous.

the required treatment cycle, for example, by combining the drug therapy with conventional PDT. However, such attempts have not been successful yet [6]. Thus, the application options for PDT in ophthalmology are currently limited [7].

10.2
Laser Systems Based on Photothermal Interactions

Photocoagulation lasers are primarily used to treat chorio-retinal diseases (Sections 3.4–3.6). In serious conditions such as proliferative diabetic retinopathy, diabetic macular edema, and retinal vein occlusion, photocoagulation is the method of choice to prevent loss of vision. In the anterior eye segment, photothermal interactions are used to decrease the intraocular pressure (IOP) in different types of glaucoma (Section 3.3).

History Photocoagulation has been used very successfully for more than 50 years for the noninvasive treatment of eye diseases. Since the first experiments in the late 1940s [1], this method has been steadily improved (particularly by the use of laser sources), and the fields of application have considerably increased. Today, judging by the global instrument base ($\approx$ 50 000 units in 2012) and the volume of performed procedures, photocoagulation has become the most important laser application in ophthalmology.

10.2.1
Functional Principle

In photothermal interactions, laser light is absorbed by biological tissue and converted into heat. The deposited thermal energy increases the tissue temperature in the absorption region and, because of thermal conduction, in the surrounding volume. Depending on the degree and duration of the temperature increase, the tissue is more or less thermally damaged. However, the desired therapeutic effect is only produced by the biological reaction (healing response) to the damage (Section 9.4.2). The biological processes that take place and eventually lead to the therapeutic effect are not yet completely understood. It is generally assumed that photocoagulation either starts or prevents the secretion of growth factors. For example, in proliferative diabetic retinopathy (Section 3.5), the destruction of oxygen consuming photoreceptors by pan-retinal photocoagulation stops the undesirable formation of new abnormal blood vessels (CNV). In other eye diseases like diabetic macular edema, it rather stimulates the regeneration of retinal pigment epithelial (RPE) cells [8, 9].

10.2.2
Process Parameters

Some important process parameters to obtain the desired photothermal effect include:

1. Irradiation parameters:

   - Laser treatment wavelength (λ).
   - Exposure time t_{exp} in continuous wave (cw) lasers (Section B.4.4).
   - Pulse duration τ and/or repetition rate $1/t_{\mathrm{per}}$ in pulsed lasers (Section B.4.4).
   - Intensity (Section A.2.1.5).

2. Tissue parameters:

   - Absorption coefficient $\mu_{\mathrm{a}}(\lambda)$ (Section 9.1).
   - Scattering coefficient $\mu_{\mathrm{s}}(\lambda)$ (Section 9.2).
   - Thermal penetration depth L_{tpd} (Section 9.4.2).

Let us now consider some process parameters which are particularly important for the photocoagulation of the eye fundus. The specific process parameters for the different treatment modes are then discussed in Section 10.2.3.

An important process parameter in photothermal interactions is the wavelength of the used laser light. In the ideal case, the radiation penetrates the anterior eye segment and is only absorbed by the tissue to be treated. Referring to the transmittance properties of the anterior ocular media (Figure 2.12 in Section 2.1.10), it is in principle possible to use laser wavelengths in both the visible (VIS) and the near-infrared (NIR) spectral ranges for photocoagulation on the fundus. In Figure 10.3, the absorption coefficients $\mu_{\mathrm{a}}(\lambda)$ of relevant chromophores are plotted for this spectral range as a function of the wavelength. In addition, the wavelengths used in commercial coagulation lasers are shown. As we can see in Figure 10.3, melanin has the highest absorption coefficient in the relevant spectral range. In the posterior segment, this substance is primarily embedded in the RPE and, to a lesser extent, in the choroid (*choroidal melanocytes*). When the retinal tissue is

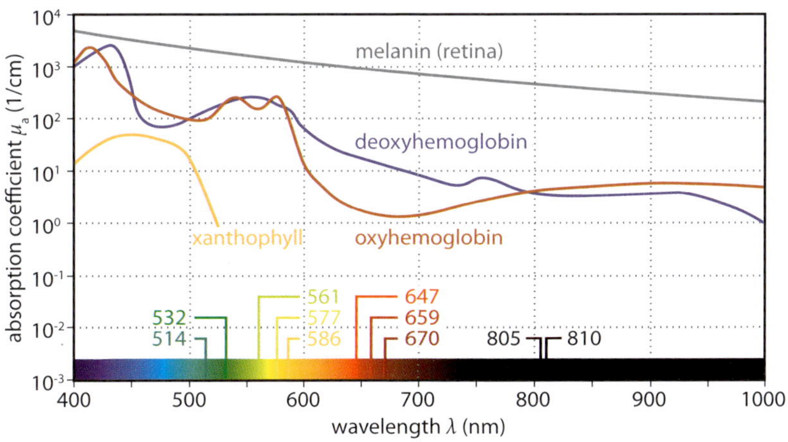

Figure 10.3 Absorption coefficients of retinal melanin (gray), macular xanthophyll (orange), oxyhemoglobin (red), and deoxyhemoglobin (violet) for wavelengths between 400 and 1000 nm. Adapted from [10].

coagulated with laser light which is efficiently absorbed by melanin, the energy of the laser beam is thus mainly deposited in the RPE. In the anterior segment, only the iris pigment epithelium and the trabecular meshwork contain a considerable amount of melanin.

If tissue shall be coagulated with VIS and ultraviolet (UV) laser light, we have to make sure that it does not lie behind a layer of blood. Oxyhemoglobin and deoxyhemoglobin are chromophores contained in blood which strongly absorb light at these wavelengths and may thus lead to undesirable effects or loss of performance in the desired treatment effect. For this reason, the high absorption coefficient of hemoglobin in the VIS spectral range is only used for some special applications in photocoagulation.

The pigment xanthophyll is embedded in the inner layers of the retina in an area of approximately 800 μm around the macula. Since the unintended coagulation of structures in the macular region (e.g., of the nerve fiber layer) may lead to severe visual impairment or visual loss, we usually avoid wavelengths for which the xanthophyll absorption is high, especially when macular diseases shall be treated.

Depending on the application and the desired photothermal tissue effect, we have to select a suitable laser wavelength from the four different spectral ranges indicated in Figure 10.3, that is, green, yellow, red, and NIR.

Green wavelength range (514 nm, 532 nm) The general purpose of retinal tissue coagulation is to cause a selective thermal damage (denaturation). In the green spectral range, the absorption of melanin and hemoglobin is very high (Figure 10.3). Hence, green laser light can be used for different applications and is the proven standard method for the treatment of retinal diseases. It is, however, beneficial to work with other wavelengths in the case of

- opacities in the anterior eye segment (e.g., cataract), as scattering losses increase with decreasing wavelengths (Section 9.2),
- treatments in the macular area (undesirable absorption by xanthophyll), and
- treatments of tissue structures behind a blood or fluid layer (undesirable absorption by hemoglobin and/or performance loss due to light scattering in the fluid film).

Yellow wavelength range (561 nm, 577 nm, 586 nm) As compared to the green wavelengths, yellow laser light shows the following properties (compare with Figure 10.3):

- absorption maximum for oxyhemoglobin,
- lower melanin absorption,
- lower absorption of xanthophyll, and
- lower scattering losses when light passes through turbid optical media (e.g., in the case of cataract) or areas filled with a fluid (e.g., in the case of edema).

Consequently, yellow laser light has certain advantages for the treatment

- in the macular area,
- of local and diffuse vessel leakages (focal or grid laser coagulation),
- of patients with a weakly or irregularly pigmented retina[4], and
- of patients with corneal or lens opacities (turbid cornea or eye lens).

Red wavelength range (647 nm, 659 nm, 670 nm) As compared to the green and yellow wavelengths, red laser light shows the following properties (compare with Figure 10.3):

- Minimal absorption of oxyhemoglobin, which means a maximum light transmittance through blood.
- Lower absorption by melanin, which results in a higher depth of penetration into the choroid.
- Maximum difference between melanin and oxyhemoglobin absorption.
- Negligible absorption by xanthophyll.
- Lower losses due to scattering.

Red laser light thus has certain advantages for the treatment

- of tissue structures behind a blood or fluid layer and
- of pigmented lesions of the choroid (e.g., choroidal tumors).

Near-infrared wavelength range (805 nm, 810 nm) For near-infrared (NIR) laser wavelengths, there are basically no differences with regard to applications as compared to red wavelengths (Figure 10.3). However, since the absorption of melanin becomes lower for long wavelengths, NIR laser light is able to penetrate even deeper into the choroid. This can be advantageous for the treatment of distinct choroidal diseases. But patient discomfort (pain) seems to be greater for NIR diode laser photocoagulation than for VIS light. From a technical point of view, NIR laser sources are substantially smaller and less expensive than laser sources which emit red light. In the anterior eye segment, NIR laser coagulators are primarily used for treatment of glaucoma (Section 10.2.5.2).

10.2.3 Treatment Modes

We may distinguish between three treatment modes based on photothermal laser–tissue interaction which are relevant in ophthalmology, that is,

- *standard (threshold) photocoagulation* (Section 10.2.3.1),
- *short-pulse treatment* (Section 10.2.3.2), and
- *transpupillary thermotherapy* (long-pulse sub-threshold photocoagulation; Section 10.2.3.3).

4) High hemoglobin absorption in the choriocapillaris (Figure 1.8 in Section 1.2), which is rich in blood vessels, ensures more regular coagulation effects.

10.2.3.1 Standard Photocoagulation

Most ophthalmic laser procedures are performed with standard photocoagulation. In this method, continuous wave (cw) lasers are used to apply single laser spots to separate locations of the retina by means of a single- or multipulse method. In the *single-pulse method*, discrete laser spots are applied in a sequential manner under visual control. In the *multipulse method* or *multipulse pattern scan treatment*, a set of discrete pulses is automatically applied in quick succession with a predefined point pattern (Figure 10.8). This method was introduced in 2006, and its technical implementation is described in Section 10.2.4.2.

In standard photocoagulation, the following exposure parameters are typically used (Problem P10.3):

- exposure time in single-pulse method: 50–200 ms,
- exposure time in multipulse method: 10–50 ms,
- laser power: 50–300 mW, and
- spot diameter: 50–500 μm.

The laser light is primarily absorbed in the RPE and leads to a locally confined coagulation. However, with the typical exposure times, neighboring tissue areas will also be thermally damaged because of heat conduction. In particular, adjacent structures of the retina and choroid are affected.

Protein denaturation caused by photocoagulation leads to an increased scattering of the tissue, and the usually reddish retinal surface turns grayish-white. In practice, this discoloration is used as a dosimetry control for successful photocoagulation. During a treatment, the preselected exposure parameters (i.e., laser wavelength, spot diameter, and exposure time) are more or less kept fixed. The *threshold power* determined by the discoloration is then controlled by adjusting the laser power. This approach is referred to as the *threshold method*, and standard photocoagulation is thus often called *threshold photocoagulation*.

The heat conduction and thus the spatial expansion of the coagulated area can be set by the exposure time. The extent of thermal damage is estimated by means of the thermal penetration depth L_{tpd} according to Eq. (9.9) in Section 9.4.2. For example, an exposure time of 100 ms leads to $L_{tpd} \approx 240\ \mu m$. This means that for a standard exposure time, inner layers of the retina and portions of the choroid are thermally damaged in addition to the RPE. In particular, the strongly oxygen-consuming photoreceptors are also destroyed so that the desired therapeutic result (e.g., stopping the undesired growth of new abnormal retinal vessels (Section 3.4)) is achieved [9].

Standard photocoagulation is an established method that has been used with great success for the treatment of different retinal pathologies for many years. However, retinal tissue is destroyed to achieve the therapeutic effect, and anatomical changes and functional losses thus cannot be avoided. For example, in panretinal photocoagulation, [5] about 15–30% of the peripheral retinal regions outside of the macula are coagulated with several thousand laser spots. Accordingly, func-

5) This method is used to treat diabetic retinopathy in advanced stages (proliferative diabetic retinopathy; Section 3.5).

tional losses such as reductions in visual field, color vision, night vision, and contrast sensitivity must carefully be taken into account in the treatment planning. For this reason, standard photocoagulation is only applied in the case of advanced disease stages, in which an otherwise imminent loss of vision justifies the significant side effects.

10.2.3.2 Short-Pulse Treatment

Standard photocoagulation cannot be used for treatments in the macular region (or only under certain conditions), since the large coagulation zone may damage the adjacent photoreceptor layer which, in turn, may lead to an impairment or loss of vision (central scotoma). To decrease the volume of the damaged zone, we could, in principle, reduce the pulse duration (exposure time). If, however, the pulse duration is reduced, the probability increases that an energy overdose causes an undesirable photomechanical tissue destruction (e.g., ruptures, hemorrhages). As a consequence, the suitable parameter range for exposure (the so-called *therapeutic window*) becomes smaller [11]. This issue can be solved if the laser energy is delivered in a burst of low-energy micropulses, rather than a single pulse. To prevent further undesirable damage to adjacent retinal structures, the time intervals between the individual pulses t_{per} of a sequence must be greater than the thermal relaxation time t_r (Section 9.4.2). In this way, tissue is able to cool down before the next pulse arrives.

In short-pulse treatment, the laser energy is delivered with a train of repetitive short pulses within an "envelope" whose width is typically in the range of 100–500 ms. Depending on the duration of the individual pulses in the train, we distinguish between three modes of short-pulse treatment, that is, micropulse treatment, selective retina therapy, and retinal rejuvenation therapy.

Micropulse treatment In micropulse treatment mode, the duration of the pulses is in the range of typically 100–300 μs, which results in a *thermal* damage of the RPE. Due to the short pulse duration, the thermal damage of adjacent tissue structures is reduced [12]. For example, according to Eq. (9.9), L_{tpd} is approximately 10 μm for a pulse duration of $\tau = 200$ μs.

In micropulse treatment mode, a laser wavelength of 810 nm was initially used. More recently, laser wavelengths of 532 and 577 nm are used as well.

Selective retina therapy In selective retina therapy (SRT), laser pulses with durations in the range of several microseconds (1–2 μs) are used to cause *nonthermal* damage of the RPE. If the RPE is exposed to laser pulses with a duration < 50 μs, the high melanin absorption leads to an explosive intracellular vaporization [13]. As a consequence, small steam bubbles are created which destroy the RPE cells.

If tissue is exposed to single microsecond laser pulses, heat conduction can be ignored. For example, according to Eq. (9.9), L_{tpd} is approximately 1 μm for a pulse duration of 1 μs. As the strongly light absorbing RPE layer has a thickness of 4–5 μm, the damage is locally confined, and the adjacent photoreceptor layer remains intact. Again, the time intervals between the individual pulses of a sequence must be greater than the thermal relaxation time ($t_{per} > t_r$).

Retinal rejuvenation therapy In retinal rejuvenation therapy (Ellex 2RT™), 3 ns-laser pulses are used to cause *nonthermal* damage of the RPE. Compared to SRT, the formed microbubbles affect only internal areas of the RPE cells (i.e., the cytoskeleton) and preserve the cell walls [14, 15].

The short-pulse treatment modes allow selective damage of the RPE and thus stimulate regeneration of the RPE cell layer. Hence, these methods can be used for RPE-related disorders such as macular edema. They have the same therapeutic effect as conventional photocoagulation, but the thermal damage to adjacent tissue structures can be reduced or eliminated. In this way, the disease can be treated at an earlier stage and/or with less pain and discomfort for the patient [8, 12, 13].

In contrast to standard photocoagulation, no visible tissue changes are generated during the treatment. Thus, no threshold for online dosimetry control (i.e., *during* the therapy) is ophthalmoscopically visible during the treatment,[6] Since online dosimetry was not available for a long time, the introduction of short-pulse techniques slowly progressed into clinical practice, although the techniques had been known for many years. Currently (in 2013), the short-pulse methods SRT and 2RT are still in the clinical introduction phase. Recently, however, an objective online dosimetry technique has made it to the trial phase [16]. It is based on the detection of sound waves generated by explosive heating or bubble formation (Section 10.2.6; see also Problem P10.4).

10.2.3.3 **Transpupillary Thermotherapy**

In transpupillary thermotherapy (TTT), the fundus is exposed to low-intensity light of a NIR diode laser ($\lambda = 810\,\text{nm}$) for a relatively long time (e.g., 60 s). This generates local hyperthermia ($\Delta T \leq 10\,\text{K}$) which damages cells without visible structural changes and eventually leads to the desired therapeutic effect. The underlying mechanism of this method is actually unclear. Hence, TTT is rarely used and basically applied in the treatment of small pigmented choroid melanomas. In this case, the NIR laser light is primarily absorbed in the choroid because of its significant penetration depth [17].

10.2.4
Technical Equipment Concepts

The setup of a photocoagulation laser (Figure 10.4) essentially consists of

- a laser source (Section 10.2.4.1) with power supply and control unit,
- a beam transmission system from the laser source to the beam delivery device (Section 10.2.4.2), and
- a beam delivery device.

6) The lesions caused by the short laser pulses can be detected after the treatment (*offline*) via fluorescein angiography (strong fluorescence, that is, hyperfluorescence) or autofluorescence imaging (weak or no fluorescence, that is, hypoautofluorescence).

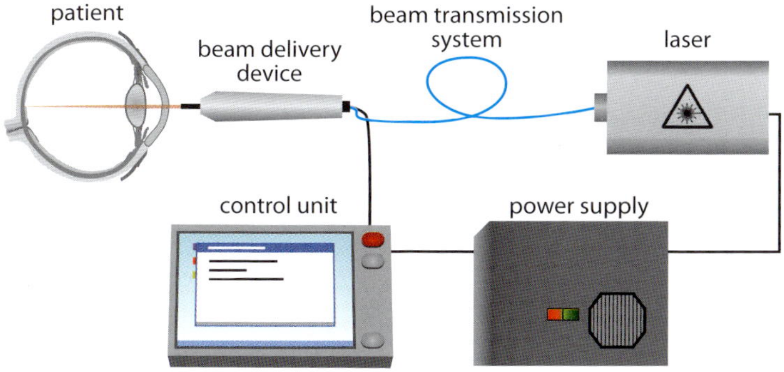

Figure 10.4 Principle setup of a photocoagulator. Light emitted by the laser source is guided by a beam transmission system to a beam delivery device. Different delivery devices are used for the targeted application of laser radiation. The treatment parameters (e.g., intensity, treatment time) are set by a control unit.

Different delivery devices are used for the targeted application of laser radiation, that is,

- slit lamps (Section 10.2.4.3),
- laser indirect ophthalmoscopes (Section 10.2.4.4 and Problem P10.6), and
- endophotocoagulation probes (Section 10.2.4.5).

10.2.4.1 **Laser Sources for Photothermal Treatment**

The most important laser source for photocoagulation (Table 10.4) is the frequency-doubled, optically pumped Nd:YAG laser with a wavelength of $\lambda = 532\,\text{nm}$ (Section B.5.3.2). This laser source has replaced the argon ion laser (Section B.5.1.1) which dominated in the early years of photothermal therapy. During the last few years, additional laser sources have been introduced to the market with wavelengths in the yellow, red, and NIR spectral ranges. As discussed in Section 10.2.2, the different wavelengths offer certain advantages for specific applications. Thus,

Table 10.4 List of laser sources with typical wavelengths used for photocoagulation of eye tissue.

Laser type	Wavelengths (nm)
Argon ion gas laser	488, 514
Optically pumped solid-state laser (Nd:YAG (Section B.5.3.2), Nd:YVO_4 (Section B.5.3.3)) with resonator internal frequency doubling	532, 560, 660, 670
Optically pumped semiconductor diode laser (Section B.5.2) with resonator internal frequency doubling	514, 532, 577
Diode-pumped solid-state laser	586
Semiconductor diode laser (Section B.5.2)	647, 805, 810

many modern laser photocoagulation systems allow selection between different laser wavelengths.

In addition to the actual photocoagulation laser, a visible low-power laser source is required to aim at the location to be treated. Mostly, laser diodes are used for this purpose whose emitted light is coupled into the beam path of the coagulation laser. In the course of this discussion, by "laser beam" or "laser light" we will always mean the combination of therapy and aiming laser beams.

10.2.4.2 **Beam Transmission Systems**

To transmit laser light from the laser source to the beam delivery device, a flexible optical multimode fiber (Figure 10.5) is used. The optical fiber consists of a fused quartz core with a refractive index n_c and a surrounding cladding made of doped fused quartz with a lower refractive index $n_{cl} < n_c$. Due to different refractive indices, total reflection (Section A.1.1) occurs at the interface between fiber core and cladding for light rays incident at an angle $> \gamma_{crit}$ (Figure 10.5a). In this way, the light beam can be guided inside the cable with low losses. Coatings on the fiber cladding and the outer jacket protect the optical fiber from mechanical stresses.

The critical angle γ_{crit}, above which total reflection occurs, is given by

$$\gamma_{crit} = \arcsin\left(\frac{n_{cl}}{n_c}\right) . \tag{10.2}$$

γ_{crit} also determines the maximum acceptance angle α_{max} under which light can enter the optical fiber at the entrance surface. The maximum acceptance angle is

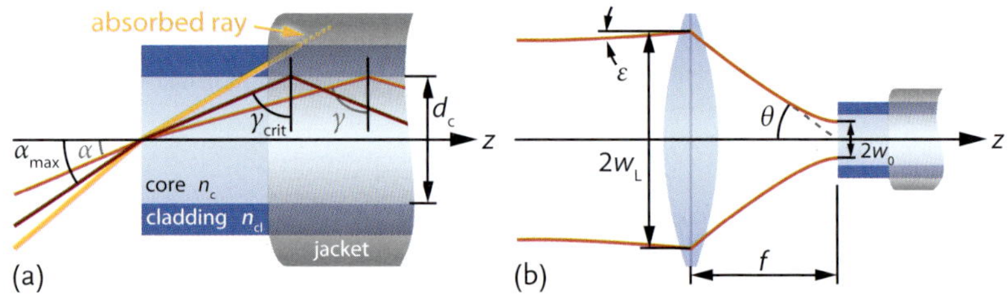

Figure 10.5 Requirements to couple a laser beam into an optical multimode fiber. (a) The principle of beam transmission in a step-index optical fiber is based on total reflection (Section A.1.1) which occurs at the core–cladding interface if the refractive index of the fiber cladding n_{cl} is smaller than the refractive index of the fiber core n_c. For low-loss transmission, the convergence angle θ (see also Eq. (10.5)) of the focused beam bundle must be smaller than the maximum acceptance angle of the fiber α_{max}. γ denotes the reflection angle and γ_{crit} the critical angle below which no total reflection occurs. (b) The diameter of laser focus $2w_0$ must be smaller than the core diameter of the fiber d_c. f denotes the focal length of the lens, w_L the laser beam radius at the focusing lens, and ε the beam divergence of a (nearly collimated) laser beam. If one of the coupling conditions in (a,b) is not fulfilled, laser light may enter the cladding of the optical fiber and is absorbed in the jacket (dotted orange line). This leads to transmission losses and, with standard laser powers in the range of 1 W, to an immediate thermal destruction of the fiber.

determined by

$$\alpha_{\max} = \frac{1}{n_0} \arcsin \sqrt{n_c^2 - n_{cl}^2} \,, \tag{10.3}$$

where n_0 is the refractive index of the surrounding medium, that is, in most cases air ($n_0 = 1$). Instead of $\alpha_{\max}$, the numerical aperture NA (Section A.1.4) is often used as a measure for the range of ray angles which can enter the optical fiber. It is given by

$$\mathrm{NA} = n_0 \sin \alpha_{\max} = \sqrt{n_c^2 - n_{cl}^2} \,. \tag{10.4}$$

Typical characteristics for a multimode, step-index optical fiber used in commercial laser photocoagulators are NA ≤ 0.2–0.3 and a core diameter of $d_c = 50$–160 μm. To couple a laser beam into the optical fiber, the laser beam must be focused onto the fiber core with a lens L (Figure 10.5b). Here, we have to take the following conditions into account:

1. The convergence angle θ of a focused Gaussian beam (Section A.2.2.1) must be smaller than $\alpha_{\max}$. We have

$$\theta = \arctan\left(\frac{w_L}{f}\right) < \alpha_{\max} \,, \tag{10.5}$$

   where f is the focal length of the focusing lens and w_L the beam radius on the lens.
2. The diameter of the laser focus $2w_0$ must be smaller than the core diameter of the fiber d_c (Figure 10.5b) so that

$$2w_0 \approx 2f\varepsilon \leq 0.7 d_c \,. \tag{10.6}$$

If one of the coupling conditions (1) or (2) is not fulfilled, laser light may enter the jacket of the optical fiber, where it is absorbed. With standard laser powers in the range of 1 W, this leads to immediate thermal destruction of the fiber. Furthermore, we have to make sure that the laser intensity on the entrance surfaces of the fiber is smaller than the damage threshold of the core material. For this reason, optical fibers cannot be used for transmission of high power laser pulses. With an accurate adjustment, losses in transmittance (typically about $2 \times 4\%$ of the total transmittance) only happen due to surface reflections at both end surfaces of the optical fiber (Problem P10.6). Losses inside the fiber can usually be neglected for the standard fiber lengths of a few meters.

10.2.4.3 **Laser Beam Delivery with a Slit Lamp**

Photocoagulation with lasers is most often performed under visual control with a slit lamp (Section 6.4). As discussed in Section 6.4.4.1, the posterior segment of the human eye can only be visualized with a slit lamp by means of auxiliary lenses. Hence, for retinal laser treatments, suitable laser contact lenses are additionally required. Auxiliary (mirror contact) lenses are also needed to treat eye structures of the anterior segment which are otherwise not directly accessible (e.g., the iridocorneal angle).

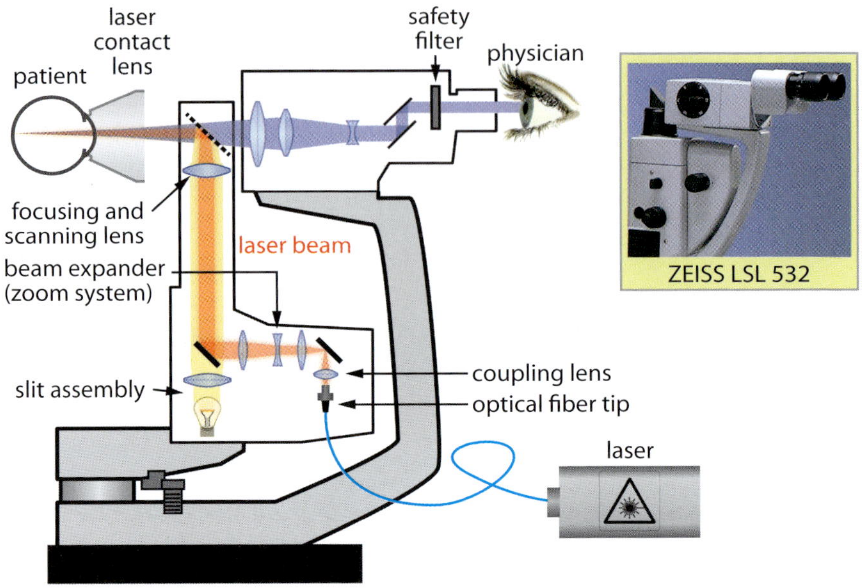

Figure 10.6 Schematic setup of a laser slit lamp with a laser beam transmission and imaging system integrated to the illumination beam path. The illumination beam of the slit lamp is shown in yellow, the laser beam in red, and the observation path in blue. To visualize and treat the patient's eye fundus, a laser contact lens is placed on the cornea (Section 6.4.4.1). An additional safety filter is added to the observation path which protects the physician's eye from backscattered and reflected laser light. Inset: Photograph of the ZEISS LSL 532 laser slit lamp. Courtesy of Carl Zeiss.

Implementation The laser beam is coupled into the illumination or observation beam path of the slit lamp by means of a laser beam transmission and imaging system. The latter can be connected to the slit lamp either by direct integration (so-called *laser slit lamp*) (Figure 10.6) or in the form of an *external slit lamp adapter* or *laser link* which is mounted onto a standard diagnostic slit lamp as an auxiliary unit (Figure 10.7).

In photocoagulators, special laser slit lamps with an integrated laser beam imaging system are usually employed. External solutions (for already installed slit lamps) are less expensive than laser slit lamps. However, as an additional link mirror is required, the working distance between eye and instrument is smaller.

Laser beam imaging The imaging system for the laser beam images the end surface (output) of the optical fiber with variable magnification into the focal plane of the slit lamp microscope and additionally allows precise lateral motion of the laser spot within the field of observation. The laser spot diameter should be adjustable within a range of 50 μm to approximately 1000 μm. If the laser beam is used to coagulate parts of the fundus, the beam diameter in the anterior segment should be as large as possible in order to prevent unintended thermal damage of cornea and

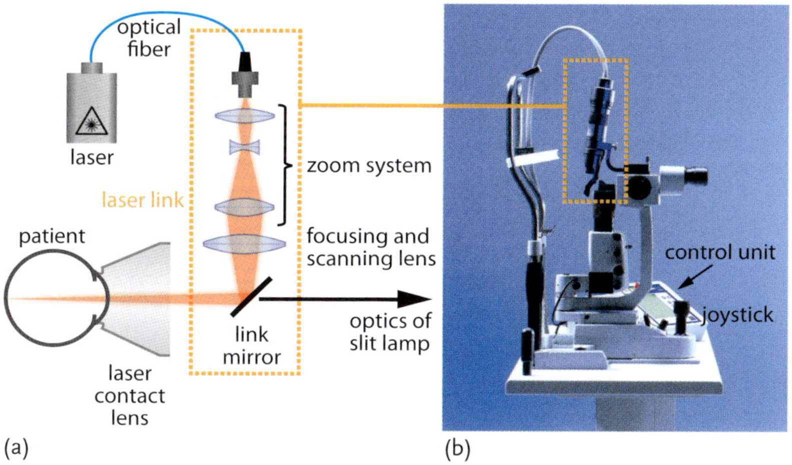

Figure 10.7 External laser link for diagnostic slit lamps (see also Problem P10.7). (a) Schematic setup of a laser link. (b) Photograph of the ZEISS VISULINK® (inside orange frame) mounted onto a ZEISS SL 130 slit lamp. Courtesy of Carl Zeiss.

eye lens. These requirements are best fulfilled by specially designed zoom systems (Section 6.2.3.4).

The effective spot size of a laser beam on the retina is determined by the product of the laser spot size in the intermediate image plane of the slit lamp microscope and the magnification of the laser contact lens.

Laser spot motion/deflection In *single-pulse treatments,* individual laser spots are sequentially aimed at desired locations on the fundus. In the simplest case (diagnostic slit lamp with laser link), the laser spots are manually moved with a joystick on the mechanical base. In laser slit lamps, the laser spot position can be set with a more precise micromanipulator. Here, the laser beam transmission and imaging system contains a beam deflection system which allows a very exact change of the lateral xy position of the laser spot in the targeted eye structure. The beam deflection is realized either by two galvanometric scanners or by an electromechanical xy shift of the focusing lens (Figure 10.6).

The beam deflection system may also be used for automatic generation of an arbitrary sequence of spot positions (scan pattern). In the *multipulse pattern scan treatment,* a sequence of laser pulses is applied in a predefined pattern at the push of the release button (Figure 10.8a–e). By lateral displacement of these predefined patterns, arbitrary two-dimensional coagulation patterns can be realized on the fundus (Figure 10.8f). The multipulse pattern scan treatment was first introduced in 2006 by the company OptiMedica (Pattern Scan Laser PASCAL®). Compared to the traditional single-pulse treatment, the multipulse method has the following advantages:

- Time savings (approximately by a factor of 5 for larger coagulation areas).
- Regularly spaced spots, that is, no overlaps with risk of overdosed coagulation.
- Shorter exposure per spot (20–50 ms as compared to approximately 100–250 ms). As a consequence, the thermally damaged zone becomes smaller.
- Less treatment pain.

Because of these advantages, more advanced commercial laser photocoagulators offer multipulse pattern treatment.

Laser safety To ensure that the physician's eye is not exposed to reflected or scattered laser light and to prevent eye damage due to high laser light intensities, the observation beam paths of the slit lamp microscope must be equipped with laser safety filters. *Active safety filters* automatically swing into the observation beam path when the therapy laser beam is activated. The examined/treated area is thus unrestrictedly visible as long as the therapy laser beam is blocked. *Passive safety filters* are color filters which always remain in the beam path. They have special coatings which block the wavelength of the therapy laser, while color vision is only slightly impaired.

In laser slit lamps, the safety filters are permanently installed. However, all bystanders within the laser hazard area should wear laser protective goggles and must be aware of common safety rules (Section 9.6).

In the case of external laser link systems, a special coating on the link mirror (Figure 10.7a) prevents reflected or scattered laser light from entering the observation beam path of the microscope.

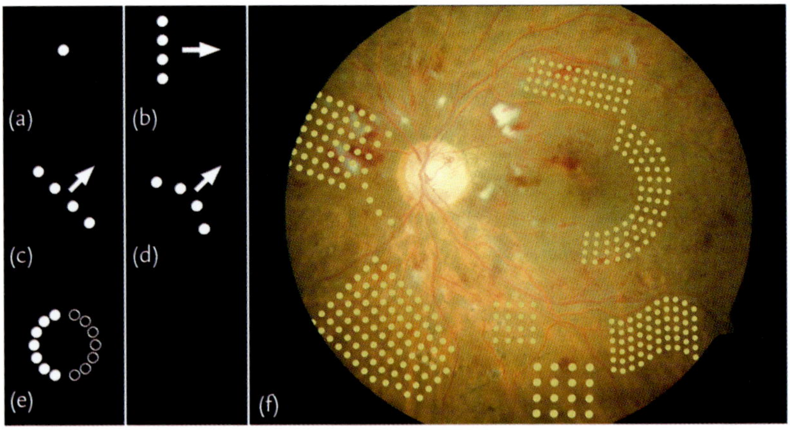

Figure 10.8 Multipulse method. A selection of predefined scan sequences is shown in (a–e). (a) Single spot. (b) Horizontal or vertical shifting of a linear spot sequence generates rectangular, square, or arbitrarily shaped scan patterns. (c) Shifting of a linear spot sequence with variable starting orientation generates rectangular, square, or arbitrarily shaped scan patterns. (d) Shifting a curved spot sequence with variable starting orientation generates quadrants and circular segments as scan patterns. (e) Mirroring of a curved spot sequence of variable starting orientation for open circular arc scan patterns. (f) Simulation of multipulse scan patterns applied to a patient's fundus. Courtesy of Carl Zeiss.

10.2.4.4 **Laser Beam Delivery with a Laser Indirect Ophthalmoscope**

Laser treatments of the fundus can also be carried out with a laser indirect ophthalmoscope (LIO) (Section 6.6.3). In this case, the laser beam is coupled into the beam path of a head-mounted indirect ophthalmoscope via one or more link mirrors, similar to a slit lamp (Figure 10.9). Analogous to the diagnostic indirect ophthalmoscope, an ophthalmoscopy lens is used to form an intermediate aerial image of the fundus (blue plane in the inset of Figure 10.9). The end surface of the optical fiber (output) is imaged with adjustable magnification into the laser spot image plane (red plane in the inset of Figure 10.9) which lies approximately 30–40 cm in front of the head-mounted indirect ophthalmoscope. The distance of the laser spot image plane therefore corresponds to a comfortable viewing distance. In the standard setup, a laser spot with a diameter of approximately 1 mm is formed at a laser spot image plane distance of 35 cm. The diameter of the laser spot on the fundus $2w_f$ is determined by the spot diameter $2w_i$ in the intermediate aerial image plane for the fundus (Problem P10.5). By changing the distance between physician (head-mounted ophthalmoscope) and ophthalmoscopy lens, w_i can be varied. w_f is minimal if the laser spot image plane coincides with the intermediate aerial fundus image.

The radius of the laser spot on the fundus w_f can be determined from the magnification β of an indirect ophthalmoscope (given by Eq. (6.73)) via

$$w_f = \frac{w_i}{\beta} = -\frac{w_i \mathcal{D}_{oph}}{\mathcal{D}_{eye}} . \tag{10.7}$$

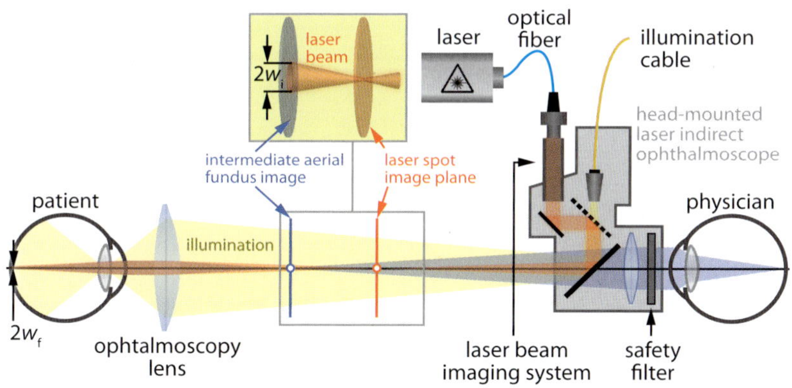

Figure 10.9 Schematic setup of a head-mounted laser indirect ophthalmoscope (LIO). The laser beam (red) of the beam delivery system is coupled via a semi-transparent mirror into the optical path of the indirect ophthalmoscope. The end surface of the optical fiber (output) is imaged with adjustable magnification into a plane that lies approximately 30–40 cm in front of the physician's eye (comfortable viewing distance) and finally by the ophthalmoscopy lens onto the fundus to obtain a laser spot of desired size w_f. The ophthalmoscopy lens forms an intermediate image of the retina which is used for visual control of the laser spot application. The illumination (observation) beam path of the indirect ophthalmoscope is shown in yellow (blue).

$\mathcal{D}_{eye}$ denotes the refractive power of the patient's eye and $\mathcal{D}_{oph}$ the refractive power of the ophthalmoscopy lens. In practice, w_f can be set either in discrete steps by changing the refractive power of the ophthalmoscopy lens or continuously by changing the distance between the physician's eye and the ophthalmoscopy lens (Problem P10.8).

Application notes The LIO is an ideal beam delivery system for patients who cannot be treated with slit lamp photocoagulators (bedridden patients, children, etc.). As it is more difficult to exactly position the laser spot with a head-mounted LIO than with a mechanically stable slit lamp, the LIO is mostly used to treat peripheral areas of the retina. In this way, the risk of unintended coagulation in the macular region (leading to a loss of central vision) is minimized.

Laser safety In the LIO, safety filters for the physician are permanently installed to the observation beam path. As no beam control is available due to the head-mounted design, special care must be taken to prevent injury to attending persons. For this reason, all bystanding people within the laser hazard area (Section 9.6.2) should wear laser protective goggles.

10.2.4.5 Laser Beam Delivery with Endophotocoagulation Probes and Contact Tips

Endophotocoagulation probes are specially designed optical fibers with a handpiece and an injection cannula which contains the end surface of the beam delivering fiber. These devices are used intraoperatively after surgical removal of the vitreous (*vitrectomy*). The divergent laser beam emitted by the fiber is guided to the retinal area to be treated (Figure 10.10). Straight, angled, aspirating, and illuminating endophotocoagulation probes with different diameters (so-called *gauge sizes*) are available for distinct applications. Special contact tips are used for transscleral[7] photocoagulation in glaucoma therapy (Figure 10.12b).

10.2.5 Clinical Applications

10.2.5.1 Posterior Segment

Table 10.5 lists relevant retinal diseases which can be treated with photocoagulation lasers and the corresponding therapeutic effects. The biophysical principles for the individual treatment modes have been discussed in Section 10.2.3. As a supplement, Figure 10.11 shows the typical laser photocoagulation patterns for selected treatment procedures.

7) "Transscleral" means that the laser light passes through the sclera and is then absorbed by the tissue structure to be treated.

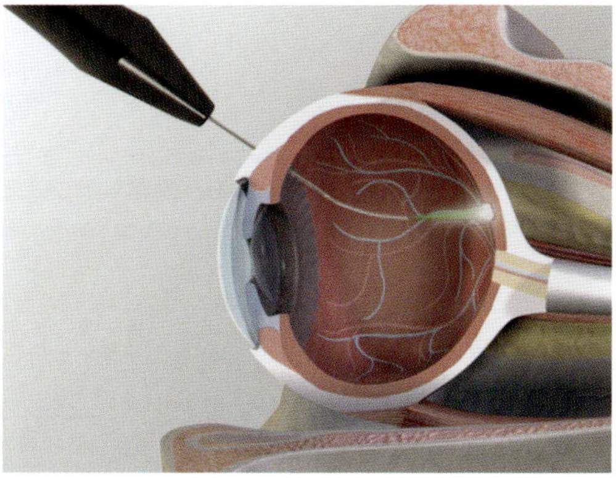

Figure 10.10 Illustration of an angled endophotocoagulation probe whose injection cannula is inserted into the patient's eye. At the probe tip, laser light is emitted which can be locally applied to the retinal tissue (white glowing spot).

Table 10.5 List of common retinal conditions which can be treated with photocoagulation.

Disease	Laser procedure	Rationale
Proliferative diabetic retinopathy	Standard pan-retinal photocoagulation (PRP)	Destruction of peripheral retinal tissue (Figure 10.11a) decreases stimulus for neovascularization
Diabetic macular edema due to leaking microaneurysms	Standard focal laser treatment (FLT)	Treatment of leaking microaneurysms and stopping of leakage by vascular thrombosis
Diabetic macular edema due to diffuse leakage	Standard macular grid treatment (MGT)	Treatment of diffuse capillary leakage except for the foveal region (Figure 10.11b)
Macular edema in branch retinal vein occlusion	Standard macular grid treatment (MGT)	Treatment of edematous areas except for the foveal region (Figure 10.11b)
Macular edema	Short-pulse treatment	Treatment of edematous areas except for the foveal region. Selective thermal damage of the RPE triggers regeneration processes. Eventually, the impaired "pump" function of the RPE is restored.
Retinal tears (detachment)	Laser retinopexy	Retinal tissue around the detached area is coagulated (Figure 10.11c). In this way, a formation of scar tissue is stimulated which eventually adheres the detached retina to the underlying tissue.

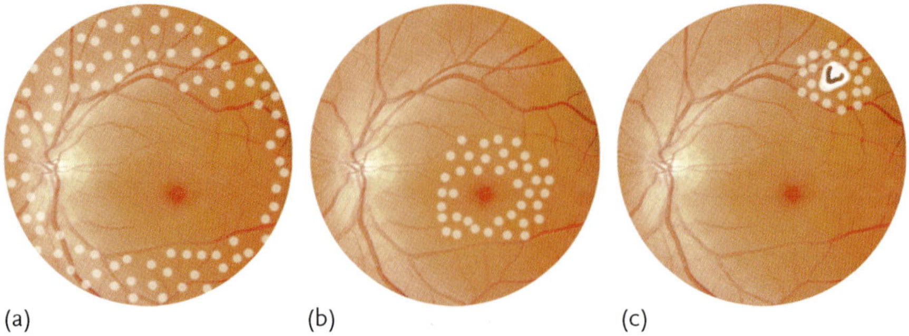

Figure 10.11 Typical photocoagulation patterns. (a) Pan-retinal or scattering laser photocoagulation for proliferative diabetic retinopathy. "Pan-retinal" means that only areas outside the internal, large retinal vessel arcs are treated. (b) Standard macular grid treatment for macular edema due to diffuse leakage. Only the retinal area close to the macula is treated. (c) Conventional retinopexy laser treatment for retinal tears (retinal detachment). Coagulation around the detached area stimulates the formation of scar tissue which eventually adheres the detached retina to the underlying tissue.

10.2.5.2 Anterior Segment

Besides the mentioned applications in the posterior eye segment, photocoagulation can also be applied in the anterior segment. For example, it can be used to reduce the IOP in different types of glaucoma (Section 3.3). The IOP is determined by the pressure of the aqueous humor which is produced by the epithelium of the ciliary body and enters the anterior chamber of the eye through the opening between iris and eye lens. The liquid is then drained via the trabecular meshwork located at the iridocorneal angle through Schlemm's canal into the (epi)scleral venous system (Figure 3.5a in Section 3.3). The IOP rises if more aqueous humor is produced than can be drained. Except for few cases, an elevated IOP is related to an increased outflow resistance. Although an increased IOP is *not* the only risk factor for the development of glaucoma, it is indeed the only one that can be influenced by therapeutic measures (e.g., medication, surgery, or laser therapy).

A number of different photothermal laser procedures are used to lower the IOP in glaucoma [18, 19], that is,

- argon laser trabeculoplasty (ALT),
- selective laser trabeculoplasty (SLT),
- micropulse laser trabeculoplasty (MLT), and
- cyclophotocoagulation (CPC).

The standard exposure parameters used for the individual procedures are listed in Table 10.6. Glaucoma conditions which can be treated with these methods are shown in Table 10.7, together with the corresponding photothermal response of tissue, and the resulting therapeutic effects.

Laser trabeculoplasty In all trabeculoplasty methods, the trabecular meshwork is photothermally damaged on a spot-by-spot basis to improve the drainage. Approxi-

Table 10.6 Typical laser parameters for standard photothermal procedures used in glaucoma therapy. The mentioned treatment modes are discussed in Section 10.2.4.3.

Procedure	Wavelength	Spot size	Duration	Treatment mode
Argon laser trabeculoplasty (ALT)	514 nm or 532 nm	50 μm	100 ms	Standard photocoagulation
Selective laser trabeculoplasty (SLT)	532 nm	400 μm	3 ns (burst mode)	Short-pulse treatment
Micropulse laser trabeculoplasty (MLT)	810 nm	200 μm	100–300 μs (burst mode)	Short-pulse treatment
Cyclophotocoagulation (CPC)	810 nm	–	1–2 s	Standard photocoagulation

Table 10.7 Glaucoma conditions which can be treated with photocoagulation.

Disease	Laser procedure	Rationale
Open-angle glaucoma	Argon laser trabeculoplasty (ALT)	Focal photocoagulation of trabecular meshwork improves aqueous outflow which reduces the intraocular pressure.
Open-angle glaucoma	Selective laser trabeculoplasty (SLT)	Due to short pulse durations, SLT interacts only with pigmented cells while trabecular structures are not affected. SLT lowers the outflow resistance of the aqueous humor which reduces the intraocular pressure.
Open-angle glaucoma	Micropulse laser trabeculoplasty (MLT)	Similar to SLT.
Glaucoma which is difficult to treat	Cyclophotocoagulation (CPC)	Tissue of the ciliary body is destroyed which, in turn, decreases the "production" of aqueous humor.

mately 50–100 coagulation spots are applied within an angular range of 180°–360°. The pigment melanin is the main absorber in the trabecular meshwork. Similar to standard laser coagulation, ALT generates grayish-white tissue denaturation. In short-pulse methods (SLT, MLT), the trabecular meshwork is damaged in a selective manner by a sequence of individual pulses. The development of small steam bubbles is often used as the dosimetry criterion to set the pulse energy. The therapeutic effect, that is, the reduction of the IOP, is quite similar for all procedures. However, the underlying functional principle is not fully understood.

In trabeculoplasty, the laser beam is applied by means of a slit lamp (Section 10.2.4.3) and a gonioscopic lens (Figure 10.12a; see also Section 6.4.4.1). Formerly, argon ion lasers were used in ALT. But today, the photocoagulation is usually

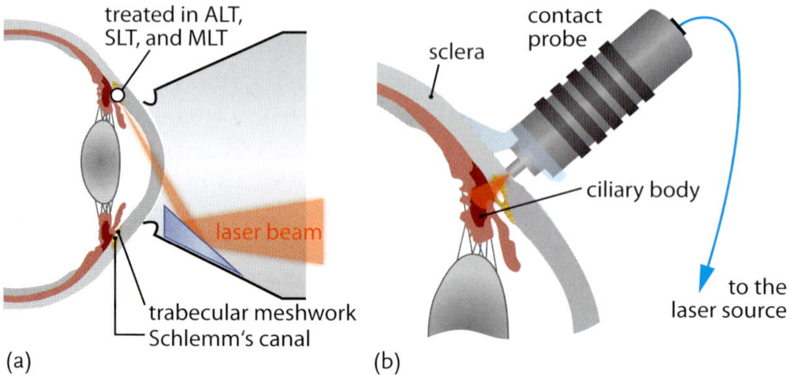

Figure 10.12 Photothermal laser procedures used in glaucoma therapy. (a) Laser trabeculoplasty with gonioscopic lens. The incident laser beam (red) is deflected to the trabecular meshwork by the mirror of a gonioscopic lens. The treated tissue zones for argon laser trabeculoplasty (ALT), selective laser trabeculoplasty (SLT), and micropulse laser trabeculoplasty (MLT) are indicated as well. (b) Transscleral cyclophotocoagulator with contact probe. Laser light (red) passes through the sclera and then interacts with the tissue of the ciliary body.

performed with frequency-doubled Nd:YAG lasers ($\lambda = 532$ nm) (Section B.5.3.2). SLT requires special Q-switched (Section B.4.4.1) and frequency-doubled Nd:YAG lasers which generate single pulses with a duration of 3 ns and low duty cycle. With such laser sources, the overall exposure time is in the range of 100 ms. To cause a selective damage with MLT, single laser pulses with a duration of 100–300 μs, a spot diameter of typically 200 μm, and a wavelength of 810 nm are used. These laser pulse parameters are provided by laser diodes so that, in contrast to SLT, no special laser equipment is required for MLT.

Cyclophotocoagulation In *cyclophotocoagulation* (CPC), the epithelium of the ciliary body is coagulated on a spot-by-spot basis to reduce the production of aqueous humor and thus to lower the IOP. For *transscleral photocoagulation* (TPC), a special contact probe is placed on the sclera, as shown in Figure 10.12b. Because of its significant penetration depth and the relatively long exposure time, the emitted NIR light of a diode laser ($\lambda = 810$ nm) generates extensive tissue necrosis from the sclera all the way to the highly pigmented ciliary body. About 17–20 coagulation spots are typically applied in an angular range of 270°. Alternatively, CPC can also be performed with special endoscopes (*endoscopic cyclophotocoagulation* (ECP)).

10.2.6
Prospects

Laser systems based on photothermal laser–tissue interaction are successfully used for the treatment of a number of different eye diseases. In the last few years, the introduction of new treatment strategies (such as pattern scanning) and the short-pulse (sub-threshold) treatment allowed major progress towards faster and safer treatments with fewer side effects. In addition, a trend towards image-supported, computer-assisted, and semi- or fully-automated treatment methods exists. By integrating a fundus imaging system into laser photocoagulators, the advantages of digital fundus imaging can be combined with a target-locked laser treatment. In this case, the coagulation procedure is performed automatically by means of a preselected program which has to be compiled before the treatment based on the previously acquired fundus images. The first example of this solution is the Navigated Laser Photocoagulator (Navilas®) by the company OD-OS. This system promises image-guided laser spot placement with high precision [20].

Associated with these trends, there is a fundamental demand for (online) dosimetry methods which can be used during a short-pulse treatment. With such dosimetry methods, the applied laser power could be automatically adjusted in accordance with a predefined threshold criterion. In turn, this would allow a reliable control of sub-threshold treatments. Online dosimetry methods could also be used for semi- or fully- automated pattern scan treatments, as they would ensure that the minimum exposure required to achieve a desired therapeutic effect is actually applied, regardless of tissue pigmentation and transparency or scattering properties of ocular media.

A promising approach towards objective online dosimetry methods is the evaluation of optoacoustic effects which are generated during photothermal interactions of pulsed laser light with biological tissue. When laser pulses are applied to the fundus, thermoelastic pressure waves are emitted due to a thermal expansion of the heated tissue. If a constant pulse energy is used, the amplitude of the pressure wave is proportional to the tissue temperature [16]. In this way, it is possible to perform real-time temperature measurements in standard (threshold) and short-pulse (sub-threshold) treatments. The intensity of the ultrasound threshold can be measured with pressure sensors in the laser contact lens, and the treatment laser is automatically switched off as soon as the temperature of the tissue has reached a preselected value. Alternatively, one could also think of other optoacoustic online dosimetry methods based on the formation of microbubbles in the nonthermal short-pulse treatment methods (SRT, 2RT).

10.3 Laser Systems Based on Photoablation

Laser-induced photoablation (Section 9.4.3) is used to correct refractive errors of the eye by reshaping the front surface of the cornea. At present, refractive corneal surgery with excimer lasers is the most often performed elective surgical procedure.

History José Barraquer can be considered as one of the most important pioneers in corneal refractive surgery. As early as 1964, he demonstrated that (contrary to the doctrines held at the time) a mechanically reshaped cornea will maintain its new form. Barraquer's insight paved the way for correction of refractive errors by corneal reshaping. He called his method *keratomileusis*[8] (keras, kerat- = cornea; mileusis = carving). But only the employment of laser-induced photoablation rendered his envisioned refractive surgery useful in clinical practice. In 1983, the non-contact laser-based photoablation of corneal tissue was demonstrated by Stephen Trokel (born 1934) and Rangaswamy Srinivasan (born 1929). They found out that light of an ArF excimer laser (Section B.5.1.2) can be used to remove corneal tissue very precisely and without any thermal side effects. In the following years, laser-supported refractive correction steadily continued to improve with the introduction of new refractive surgery techniques such as PRK and LASIK (Table 10.3). The precision of these methods has increased continuously so that photoablation with excimer lasers became widely accepted in clinical practice.

10.3.1 Basics of Photoablation Treatments

The precision of surface reshaping depends on the transversal and axial treatment accuracy, whereas the latter is determined by the ablation depth per pulse. In addition, the thermal and mechanical side effects must be minimized. As discussed in Section 9.4.3, short-pulsed UV lasers with a wavelength of approximately 200 nm must be used to achieve the required performance. In this wavelength range, the ablation process is primarily photochemically induced and thus less disruptive. The higher the absorption coefficient at a certain laser wavelength, the lower is the ablation threshold, and the thinner are the tissue layers which can be removed (Figure 9.7 in Section 9.4.3). When we compare all available UV laser sources, the absorption coefficient of the corneal stroma is maximal for the wavelength of the ArF excimer laser which emits light with a wavelength of $\lambda = 193$ nm.

The tissue ablation per laser pulse L_{abl} can be estimated from the blow-off model via Eq. (9.12). For $\mu_a^{-1}(193\,\mathrm{nm}) = 0.35\,\mu\mathrm{m}$ and $\Phi_{\mathrm{th}} = 50\,\mathrm{mJ/cm^2}$, an ablation rate between 0.3 and 0.5 μm per pulse is obtained if exposure values between 120 and $250\,\mathrm{mJ/cm^2}$ are used. No other surgical method allows such a thin layer removal.

To avoid thermal damage to adjacent tissue structures, the pulse duration must be shorter than the thermal relaxation time t_r. The shortest thermal relaxation time of approximately 1 μs is found at the absorption peak of water ($\lambda \approx 3\,\mu\mathrm{m}$). As the pulse duration of the ArF excimer laser lies in the range of $\tau = 20\mathrm{ns}$, thermal damage to adjacent structures can be ignored for low repetition rates.[9] Moreover, mutagenic cell damage can be largely ruled out for a wavelength of 193 nm, because proteins in the cell matrix strongly absorb radiation at 193 nm before it reaches the cell nucleus containing the DNA [21]. Summing up, the ArF excimer laser is the preferred laser source for refractive corneal surgery (Problem P10.9).

8) Barraquer cut a central front part of the cornea with a specially developed mechanical microkeratome. The frozen corneal part was then reshaped with a turning lathe and sewed on after defrosting.

9) For high repetition rates, a secondary thermal damage of corneal tissue is still an issue, which may lead to haze formation.

10.3.1.1 Ablation Profiles for Correction of Refractive Errors

The refractive correction is determined by the ablation profile used, as it determines the exact shape and size of tissue to be removed. In the following, we will briefly discuss the most important patient-specific ablation strategies (Problems P10.10–P10.12). In addition to the presented profiles, some other special ablation profiles exist (e.g., PresbyLASIK, Q-factor customized ablation, etc.) which are described in the specialized literature, for example, [22–24].

Munnerlyn profile In 1988, Charles Munnerlyn (born 1940) proposed the first ablation profile for refractive corneal surgery [25]. When myopia shall be corrected, a convex-concave corneal tissue volume with a central thickness a_0 is ablated from the stroma, as shown in Figure 10.13a. Removal of this corneal material is equivalent to insertion of a negative lens which corrects myopia. In this way, the radius of curvature of the cornea before the treatment r_{pre} is increased to r_{post}. The desired refractive correction $\Delta\mathcal{D}$ (degree of myopia in diopters[10)]) is related to the radii via

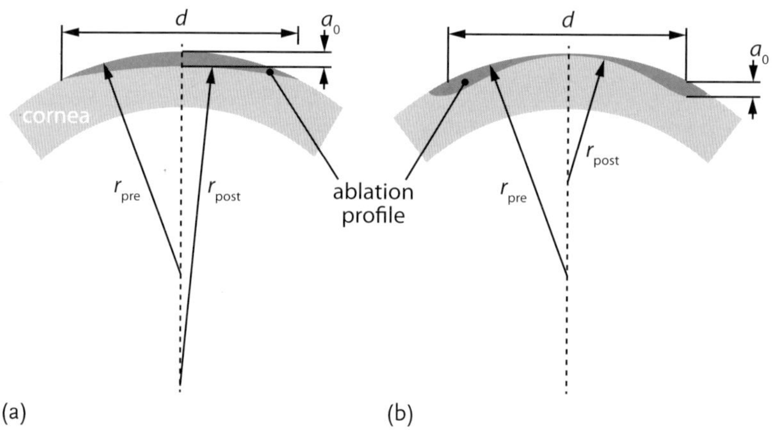

Figure 10.13 Munnerlyn profile used for refractive corneal surgery. (a) Ablation profile to correct myopia (dark gray). In myopia, the refractive power of the cornea is too high. For correction, the central part of the cornea is ablated such that the radius of curvature increases. We thus have $r_{pre} < r_{post}$, where r_{pre} is the radius of curvature before the treatment (preoperative) and r_{post} the radius of curvature after the treatment (postoperative). a_0 is the maximum ablation depth at the corneal vertex, and d is the diameter of the optical zone. (b) Ablation profile to correct hyperopia (dark gray). In the case of hyperopia, the corneal radius of curvature must be steepened so that $r_{post} < r_{pre}$. a_0 now refers to the maximum optical depth in the periphery, whereas the ablated depth at the corneal vertex is nearly zero. Existing astigmatism can be treated by angular-dependent correction with toric ablation profiles.

10) In subjective refraction methods based on phoropters and trail lens frames (Chapter 5), the refractive status of the eye is determined via trial lenses which are placed at a defined distance (usually 12 mm) in front of the corneal vertex. However, since in corneal surgery the refractive correction happens directly in the plane of the corneal vertex, we have to convert the measured refractive values according to the new (zero) distance (see also Section 5.5).

$$\Delta\mathcal{D} = (n_c - 1)\left(\frac{1}{r_{\text{post}}} - \frac{1}{r_{\text{pre}}}\right) \tag{10.8}$$

with the corneal refractive index $n_c = 1.377$. The required central ablation depth a_0 in micrometers can be calculated via (derivation presented in Section 10.3.1.4)

$$a_0 = \frac{\Delta\mathcal{D}d^2}{8(n_c - 1)} \approx \frac{1}{3}\Delta\mathcal{D}d^2 , \tag{10.9}$$

where d is the diameter of the optical zone in millimeters[11]. According to Eq. (10.9), approximately 12 μm of tissue must be removed from the central part of the cornea for each diopter of refractive correction if $d = 6$ mm.

The *Munnerlyn profile* can also be used to correct hyperopia (Section 3.1). In this case, the ablated profile aims at steepening the radius of curvature of the corneal front surface (Figure 10.13b). Existing astigmatism can be treated by angular-dependent correction with toric ablation profiles.

When the cornea is reshaped according to the Munnerlyn profile, the natural asphericity of the cornea is modified. In particular, in the case of myopia correction, spherical aberration is created which increases with $\Delta\mathcal{D}$ and d (Problems P10.14 and P10.15).

Wavefront-optimized ablation *Wavefront-optimized aspherical ablation profiles* have been introduced to solve the mentioned drawbacks of the Munnerlyn profile. They differ from the conventional Munnerlyn profile mainly in the periphery of the optical zone and are designed to avoid an increase of spherical aberrations [26].

As for the Munnerlyn profile, the refractive status must be determined by subjective refraction methods (Chapter 5) before a treatment. To calculate the diameter of the optical zone, the mesopic pupil diameter is measured as well. For most refractive surgery treatments, the wavefront-optimized ablation is the preferred method to correct spherocylindric refractive errors (Section 3.1).

Wavefront-guided ablation In some patients' eyes, significant higher-order aberrations (Section A.1.8.2) impair vision additionally to spherocylindric refractive errors. In these cases, the determination and correction of the spherocylindric errors alone may not be sufficient, and *wavefront-guided ablation* is required. To obtain a high quality of vision, the measured wavefront aberrations (Sections 5.3.1.2 and A.1.7) must be corrected to the greatest possible extent. The targeted ablation profile $a(x, y)$ results directly from the determined wavefront aberration function $\mathcal{W}(x_p, y_p)$ and/or the optical path difference $\text{OPD}(x_p, y_p)$ (both measured in mi-

11) In general, the diameter of the optical zone is smaller than the treated ablation zone. In this way, we obtain a smoother transition to the zone which is not ablated.

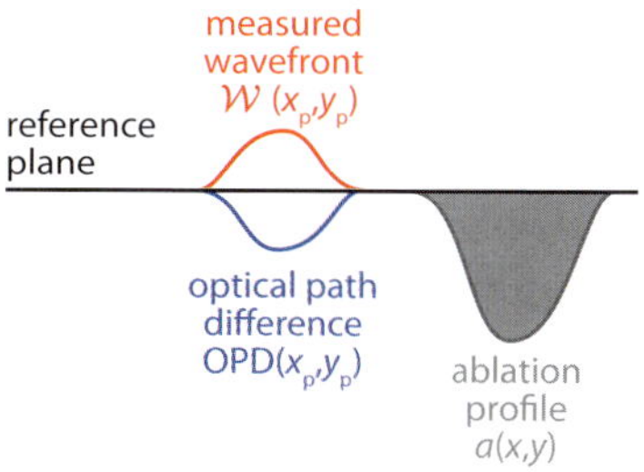

Figure 10.14 Scheme to illustrate the calculation of the ablation profile $a(x, y)$ (gray) from the measured wavefront aberration $\mathcal{W}(x_\text{p}, y_\text{p})$ (red; Section 5.3.1.2) and/or the optical path difference $\text{OPD}(x_\text{p}, y_\text{p})$ (blue; Section 5.3.1.3) of a myopic eye (Problem P10.16).

crometers)[12] as

$$a(x, y) \approx -\frac{\mathcal{W}(x_\text{p}, y_\text{p})}{n_\text{c} - 1} = \frac{\text{OPD}(x_\text{p}, y_\text{p})}{n_\text{c} - 1} . \tag{10.10}$$

x_p and y_p are the horizontal and vertical coordinates in the pupil plane, x and y are the horizontal and vertical coordinates on the corneal surface, and n_c is the refractive index of the cornea. Figure 10.14 shows how the ablation profile is in general derived from the measured wavefront aberration.

In wavefront-guided ablation, the cornea must be reshaped with a high degree of precision, which can only be achieved with the *flying-spot* ablation technique (Section 10.3.1.2). For this purpose, a thin, computer-controlled laser beam scans the cornea while the desired surface profile is formed by a sequence of laser ablations (Problem P10.16).

Topography-guided ablation *Topography-guided ablation* profiles are primarily used to correct pronounced irregularities of the corneal front surface, for example, due to irregular astigmatism after a corneal or cataract surgery. Here, the ablation profile $a(x, y)$ is determined by (Figure 10.15)

$$a(x, y) = e(x, y) - e_\text{target}(x, y) , \tag{10.11}$$

where $e(x, y)$ denotes the actual elevation profile of the cornea (see also Figure 6.32) which is measured before the treatment procedure, for example, with a Placido disk-based video keratoscope (Section 6.3.2) or a Scheimpflug system (Section 6.5.2). $e_\text{target}(x, y)$ is the desired target profile.

A critical point in topography-guided ablation is the definition of the target profile $e_\text{target}(x, y)$. For severely irregular corneas, a full refractive correction is sometimes not possible, since the corneal tissue is not thick enough and/or since reliable preoperative refractive data is not available. Stepwise approaches are typically used in these cases.

12) As discussed in Section A.1.7, the coordinates of the wave aberration function are usually defined in the $x_\text{p} y_\text{p}$ pupil plane. However, the ablation profile must be referred to the corneal surface plane. As the eye's pupil plane and the corneal surface plane are located relatively close to each other, the approximation $a(x, y) \approx a(x_\text{p}, y_\text{p})$ is justified.

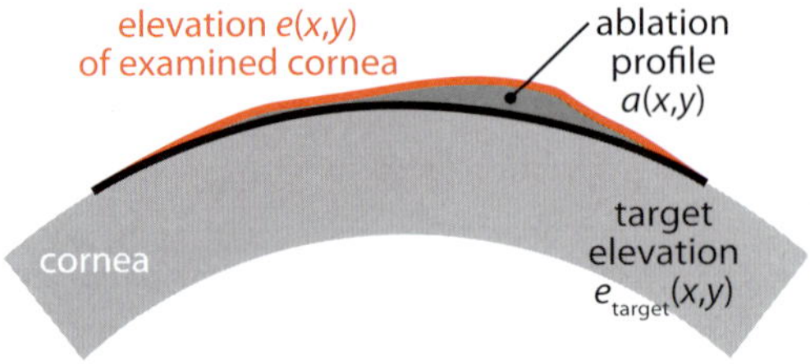

Figure 10.15 Principle of topography-guided ablation. The corneal topography $e(x, y)$ before the treatment (red line) shall be reshaped to obtain the targeted elevation profile $e_{\text{target}}(x, y)$ (black line). For this purpose, the calculated profile $a(x, y)$ (dark gray) has to be removed with photoablation.

10.3.1.2 Ablation Techniques

To generate the desired ablation profile, three different techniques can be used, that is, the *broad-beam, scanning-slit*, and *flying-spot* (or *scanning-spot*) ablation techniques (Figure 10.16).

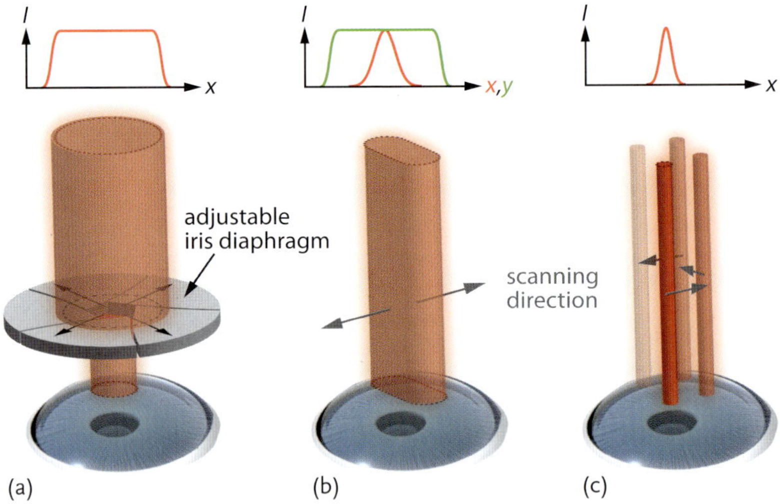

Figure 10.16 Different techniques used to reshape the corneal surface by photoablation. (a) Broad-beam ablation technique. The patient's cornea is exposed to a locally fixed laser beam with a *top hat* intensity profile (see $I(x)$ diagram). To attain the desired ablation profile, the diameter of the treatment beam is gradually increased by widening the iris diaphragm. (b) Scanning-slit ablation technique. The light beam has a Gaussian beam profile along one direction (here x direction; highlighted in red) and a top hat profile along the perpendicular direction (here y direction; highlighted in green). To obtain the desired ablation profile, the laser beam is scanned along the x direction. (c) Flying-spot ablation technique. The corneal surface is scanned with a small circular laser beam with a Gaussian beam profile. The desired ablation profile is achieved by applying the laser spots locally with an appropriate overlap.

Broad-beam ablation technique In the broad-beam ablation technique, the cornea is exposed to a locally fixed laser beam with a *top hat* intensity profile. To obtain the desired ablation profile, the beam is masked by a circular aperture stop which is able to vary the spot diameter for each pulse (Figure 10.16a). By gradually changing the spot diameter, the center of the cornea is exposed to a higher laser dose than the periphery which, in turn, leads to a greater ablated depth.

This simple ablation technique was used by the first laser systems for refractive corneal surgery. To date, it has been entirely replaced by other advanced techniques, since the broad-beam ablation technique does not allow the realization of more complex profiles needed for wavefront or topography-guided ablation. In addition, this technique requires a higher laser pulse energy than the flying-spot technique, since (for constant exposure) the laser energy is proportional to the square of the beam diameter.

Scanning-slit ablation technique In the scanning-slit ablation technique, the cornea is exposed to a slit-shaped beam bundle with a Gaussian intensity distribution in one direction of the beam spot and a top hat intensity distribution in the perpendicular direction (Figure 10.16b). To obtain a smooth ablation profile, the slit-shaped beam is scanned across the cornea in the direction of the Gaussian intensity distribution. The slit orientation is then changed after each line scan so that potential inhomogeneities in the intensity distribution of the top hat profile are compensated. Size, length, and position of the rotating slit-shaped beam are changed stepwise by means of a computer-controlled aperture stop to generate the desired ablation profile.

Flying-spot ablation technique Today, the flying-spot technique is the one most commonly used for corneal photoablation. Here, the corneal surface is scanned with a small circular laser beam which has a fixed diameter.[13] The desired ablation profile is then realized by a defined distribution of locally applied laser spots with an appropriate overlap (Figure 10.16c). From Figure 10.17b, it follows that the laser beam must have a Gaussian profile to obtain a smooth, homogeneous ablation profile. A top hat beam profile (Figure 10.17a) is rather unsuitable for this purpose.

As the flying-spot technique allows ablation of arbitrary, high-precision profiles, it is well-suited for customized treatments such as wavefront-guided, wavefront-optimized, or topography-guided photoablation. The precision of the surface ablation profiles depends on the transversal and axial treatment accuracy, whereas the latter is determined by the ablation depth per pulse L_{abl}. According to Eq. (9.12), we have $L_{\mathrm{abl}} \approx 0.3\,\mu\mathrm{m}$ for a typically used exposure of $150\,\mathrm{mJ/cm^2}$. The corresponding wavefront retardation inside corneal tissue results as $L_{\mathrm{abl}}(n_{\mathrm{c}} - 1) \approx 0.11\,\mu\mathrm{m}$. Hence, the L_{abl} value achieved for an ArF laser meets the require-

13) A modification of the flying-spot ablation technique is the *variable spot scanning method* (VSS) at which both the diameter and the position of the beam can be changed.

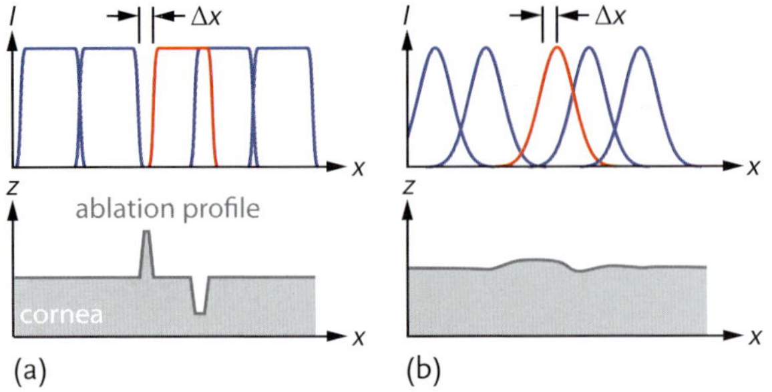

Figure 10.17 Influence of small laser spot displacements Δx on the smoothness of the ablated corneal surface generated by the flying-spot ablation technique. (a) Laser spots with a top hat intensity profile produce more surface irregularities than (b) spots with a Gaussian profile at equal spot displacements from the ideal spot location (red).

ments of Rayleigh's quarter wavelength rule.[14] The transversal treatment accuracy depends on the laser spot diameter and the positioning accuracy of the laser spots. The demands made on the precision of the ablation profiles increase with the order of the aberrations to be corrected. For example, computer simulations have shown that a Gaussian beam diameter of approximately 2 mm is sufficient to correct spherocylindric refractive errors (i.e., Zernike polynomials with radial order $n = 2$). But to correct higher-order aberrations (i.e., Zernike polynomials with radial order $2 < n \leq 6$), laser beams with a diameter ≤ 0.6 mm are required [22, 28]. Since we can usually ignore aberrations related to Zernike polynomials with $n > 5$, a Gaussian laser beam with a diameter of ≤ 1 mm is suitable for most wavefront-guided ablation processes. In any case, we have to make sure that the transition at the edges of the optical zone and the untreated cornea are as smooth as possible.

The system requirements with regard to speed and positioning accuracy of the laser spots increase for smaller beam diameters. On the one hand, a fast scanning system needs a more accurate positioning system for the laser spots including fast eye tracking, which makes the technical realization more complex. On the other hand, for small spot treatments, we need high repetition rates between 200 and 750 Hz to achieve acceptable treatment times (e.g., in the range of 3–5 s per diopter for correction of myopia). In fact, the treatment duration is an important issue, as dehydration of the stroma influences the effective ablation depth per pulse.

10.3.1.3 Physical and Biological Side Effects

For a good treatment result, two aspects must be considered:

1. An appropriate ablation method (e.g., wavefront-optimized ablation) must be chosen which fits the actual situation. Then, a suitable ablation profile must be calculated based on patient-specific data.

14) An optical system can be considered as "nearly diffraction-limited", if the amplitude of maximum wavefront aberration in the exit pupil deviates from the ideal reference surface by less than $\lambda/4$. This is the so-called *Rayleigh quarter wavelength rule* [27] according to which the wavefront deviation must be ≤ 0.14 µm for a wavelength of 550 nm (at which the sensitivity of the eye is maximal).

2. It must be ensured that the laser system transfers the calculated ablation profile to the cornea with the required degree of precision. Indeed, a few physical factors (e.g., changes of the ablation rate) have to be taken into account, which may lead to deviations between the desired and the obtained ablation profile. In addition, biomechanics and healing effects may influence the refractive correction after a surgery.

All of these factors have to be identified and quantified. The corresponding effects must then be compensated by adjusting the treatment parameters so that the desired ablation profile is finally realized.

Changes of the ablation rate According to the blow-off model Eq. (9.12), the ablation rate per laser pulse depends on the applied exposure. As the surface of the cornea is curved, both the laser spot size and the reflectance change as a function of the location on the corneal surface. The corresponding variation of exposure is dominated by the change of the spot size [23]. Without correction, the ablation rate would gradually decrease from the corneal center to the periphery, as the effective exposure drops (Problem P10.15).

The ablation rate may also change during the treatment due to the time-dependent dehydration of the stroma. In this case, the eye is "overcorrected", as the ablation rate increases with the degree of dehydration. We have to compensate this effect by means of empirical data, since the dehydration rate increases (nearly) proportional to the total treatment time (for flap creation and ablation). Laser systems with short treatment times look particularly advantageous. However, temperature increase of the cornea limits the speed, as it may cause thermal haze.

The ablation rate is also influenced by laser light absorption of the ablated stroma (called a *laser plume*) in front of the treatment area. This effect can be eliminated by removing the plumes by active air flow management.

Tissue response and biomechanical changes After a refractive surgery, wound healing processes of the corneal epithelium and biomechanical changes in the stroma can lead to deviations of the intended correction. This must also be taken into account by empirical data.

10.3.1.4 Excursus: Derivation of the Munnerlyn Formula

In Section 10.3.1.1, we introduced the Munnerlyn ablation profile whose central ablation depth is given by Eq. (10.9). In the following, the corresponding Munnerlyn formula shall be derived from geometric considerations. For this purpose, we need to know the *sagitta* of a spherical surface (Figure 10.18) given by

$$s = r - \sqrt{r^2 - \left(\frac{d}{2}\right)^2} \qquad (10.12)$$

with the radius r of the spherical surface at a distance $d/2$ from the sphere vertex.

In the case of myopia, the radius of curvature must be increased by photoablation so that the corneal surface profile becomes flatter. This situation is depicted in

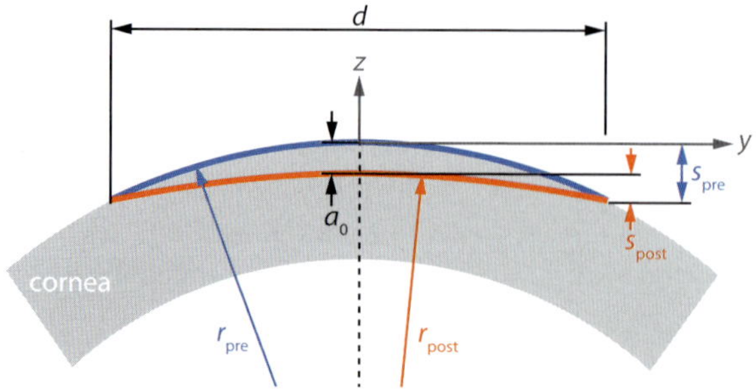

Figure 10.18 Geometry used for the derivation of the Munnerlyn formula (10.9). a_0 is the central ablation depth, d the diameter of the optical zone, r_{pre} (blue) the corneal radius of curvature before photoablation, and r_{post} (red) the corneal radius of curvature after the treatment. s_{pre} and s_{post} denote the sagittae before and after the treatment, respectively.

Figure 10.18, where the cornea before and after the treatment is represented by the blue and red curves, respectively. According to Eq. (A11) in Section A.1.2, a spherical surface which separates two volumes with refractive indices n and n' has a refractive power of

$$\mathcal{D} = \frac{n' - n}{r} . \tag{10.13}$$

The required ablation depth can now be calculated as the height difference of the corneal vertex before and after ablation. But note that the height difference disappears at the outer rim of the ablation zone, since a smooth transition to the residual cornea is required (Section 10.3.1.1). If we assume that the cornea has a spherical surface before the treatment and that the medium outside the cornea is air ($n \approx 1$), we can use Eqs. (10.12) and (10.13).

For the parameters before the treatment, (indicated by "pre"), we have

$$\mathcal{D}_{\text{pre}} = \frac{n_{\text{c}} - 1}{r_{\text{pre}}} , \tag{10.14}$$

$$s_{\text{pre}} = r_{\text{pre}} \left(1 - \sqrt{1 - \left(\frac{d}{2 r_{\text{pre}}} \right)^2} \right) , \tag{10.15}$$

with the refractive index of the corneal tissue $n_{\text{c}} = 1.377$. The parameters after the treatment (indicated by "post") read

$$\mathcal{D}_{\text{post}} = \mathcal{D}_{\text{pre}} + \Delta\mathcal{D} , \tag{10.16}$$

$$r_{\text{post}} = \frac{n_{\text{c}} - 1}{\mathcal{D}_{\text{pre}} + \Delta\mathcal{D}} , \tag{10.17}$$

$$s_{\text{post}} = r_{\text{post}} \left(1 - \sqrt{1 - \left(\frac{d}{2 r_{\text{post}}} \right)^2} \right) . \tag{10.18}$$

As the central ablation depth is equal to the difference of the sagittae ($a_0 = s_{pre} - s_{post}$), we can use Eqs. (10.14)–(10.18), which yields

$$a_0 = \frac{n_c - 1}{\mathcal{D}_{pre}}\left(1 - \sqrt{1 - \left(\frac{d\mathcal{D}_{pre}}{2(n_c - 1)}\right)^2}\right) - \frac{n_c - 1}{\mathcal{D}_{pre} + \Delta\mathcal{D}}\left(1 - \sqrt{1 - \left(\frac{d(\mathcal{D}_{pre} + \Delta\mathcal{D})}{2(n_c - 1)}\right)^2}\right). \tag{10.19}$$

This is the exact conditional equation for the minimum central ablation depth in the case of myopia. For Eq. (10.19), we may use the approximation that the quadratic terms of the roots are $\ll 1$. If the diameter of the ablation zone d and $\Delta\mathcal{D}$ are small and if

$$1 - \sqrt{1 - x} \approx \frac{x}{2} + \frac{x^2}{8}, \tag{10.20}$$

we finally obtain

$$a_0 \approx \frac{d^2\Delta\mathcal{D}}{8(n_c - 1)} + \frac{d^4(3\mathcal{D}_{pre}^2\Delta\mathcal{D} + 3\mathcal{D}_{pre}\Delta\mathcal{D}^2 + \Delta\mathcal{D}^3)}{128(n_c - 1)^3} \approx \frac{1}{3}\Delta\mathcal{D}d^2 \tag{10.21}$$

which is equal to Eq. (10.9). This formula also holds for hyperopia corrections. However, in this case, a_0 represents the ablation depth at the outer rim of the treated zone, whereas the corneal center remains unchanged (Figure 10.13b).

10.3.2 Technical Equipment Concepts

A laser platform for refractive corneal surgery essentially consists of

- a laser source (Section 10.3.2.1),
- a beam delivery system for beam shaping, focusing, and scanning (Section 10.3.2.2),
- an observation system for direct treatment control (Section 10.3.2.3),
- an alignment tool for the therapy laser (Section 10.3.2.4),
- an eye tracking system (Section 10.3.2.5),
- devices for energy monitoring and control (Section 10.3.2.6), and optionally
- a diagnostic unit (wavefront analyzer or corneal topography system) used to gather relevant data for therapy planning (Section 10.3.2.7).

The general setup of a photoablation laser system is shown in Figure 10.19a. For reference, Figure 10.19b shows a photograph of the therapy laser system ZEISS MEL 90™.

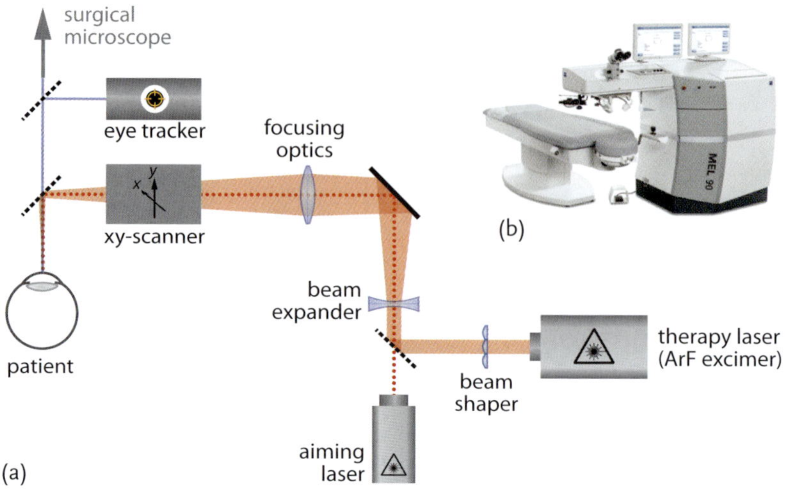

Figure 10.19 (a) Principle of a laser system used for refractive corneal surgery. At first, the intensity profile of the therapy laser is modified by a beam shaper. The shaped laser beam is then expanded, focused, and finally deflected by an *xy* scanner such that it is incident on the desired position on the corneal surface. As the therapy laser is invisible, an aligned aiming laser beam (red dotted line) is used to set the focus position. The *xy* scanner is automatically readjusted by an eye tracker which rapidly identifies the lateral position of the pupil center of the patient's eye. The physician is able to adjust and follow the treatment procedure with a surgical microscope. (b) Photograph of the therapy laser system ZEISS MEL 90 which is designed according to the presented scheme. Courtesy of Carl Zeiss.

10.3.2.1 **Laser Sources for Photoablation**

For photoablation, excimer lasers are mostly used (Table 10.8) whose general setup, functional principle, and properties are described in Section B.5.1.2. From the biophysical point of view, the ArF excimer laser is best suited for refractive corneal surgery (Section 10.3). Although excimer lasers are relatively large and maintenance-intensive compared to solid-state lasers (e.g., changing gas bottles, replacing laser optics), they are safe and reliable light sources without real alternatives.[15)]

10.3.2.2 **Beam Shaping, Focusing, and Scanning**

After the laser beam is emitted, it passes through a delivery system in which it is homogenized, focused, and guided to the targeted tissue. If the flying-spot technique (Section 10.3.1.2) is used, a circular laser beam with a Gaussian intensity is required to ensure a smooth ablation profile. However, beams from excimer lasers usually have a squared cross-section due to the transversally excited gain medium (Figure B.17 in Section B.5.1.2). In addition, a uniform intensity profile cannot be obtained due to the short excitation times in the ns range. Thus, the laser beam

15) At present (in 2013), only the companies Katana Technologies and CV Laser Pty Ltd. use a Nd:YAG laser with a wavelength of $\lambda = 210\,\text{nm}$ instead of the common ArF excimer laser.

Table 10.8 Specifications of currently available excimer laser systems. d_{beam} is the beam diameter and $1/t_{per}$ the repetition rate of the laser pulses. $1/t_s$ and t_l are the sampling rate and the latency time of the image-based eye tracking system (video camera), respectively. The laser systems above the separation line are based on an ArF excimer laser. The systems below the line are based on a Nd:YAG solid-state laser.

Company	Name	Working principle	d_{beam} (mm)	Beam profile	$1/t_{per}$ (Hz)	Eye tracker $1/t_s$ (Hz)	Eye tracker t_l (ms)
Alcon	WaveLight® EX500	Flying spot	0.68	Gaussian	500	1050	2
Abbott Medical Optics (AMO)	STAR S4 IR™	Variable-spot scanning (VSS)	0.65 to 6.5 (variable)	Top hat	6–20	60	20
Carl Zeiss	MEL 90	Flying spot	0.7	Gaussian	250/500	1050	2
Bausch+Lomb	Technolas® 217P	Flying spot	2 or 1	Quasi-Gaussian	100	240	7
Nidek	EC-5000 CX III	Scanning slit or flying spot	2 × 9 or 1.0 (beam array)	Quasi-Gaussian	5–50	200 (1000)	6 (3)
Schwind	Amaris® 1050RS	Flying spot	0.54	Super-Gaussian	1050	1050	0
CV Laser Pty Ltd.	Pulzar™ Z1	Flying spot	0.6	Quasi-Gaussian	300–400	2500	1
Katana Technologies	LaserSoft	Flying spot	0.2	Quasi-Gaussian	2000–4000	1050	1

must be homogenized and converted to a Gaussian beam before it can be guided to the interaction volume. For this purpose, special microlens arrays are used.

The reshaped beam is then expanded and focused onto the cornea. The desired focal position is set by an xy scanner placed after the focusing optics. The scanner consists of two computer-controlled galvanometric or piezoelectric mirrors which allow scanning of the laser beam across the corneal surface in a precise manner. In particular, for systems based on the flying-spot technique, the requirements concerning speed and positioning accuracy of the beam are very high. To obtain a sufficiently smooth ablation profile with a Gaussian beam, the actual beam position should not deviate more than 1/100 of the spot diameter from the ideal position. For a spot diameter of 1 mm and a distance of 20 cm between corneal surface and scanning mirror, the beam position must therefore be controlled with an accuracy in the range of 50 μrad or about 10″.

10.3.2.3 Observation System

The beam of the therapy laser formed in the delivery system is coupled into the observation path of a surgical microscope (Section 6.2) with a beam splitter (similar to the light fiber in Figure 6.20 in Section 6.2.3.5). The surgical microscope is used for optimal alignment of the focused laser beam relative to the patient's eye and to monitor the entire treatment process.

10.3.2.4 Alignment System

For a precise focal positioning of the invisible therapy laser beam, different types of alignment systems can be used, such as crossed aiming laser beams or projected slit images. In crossed beam systems, the therapy beam is enclosed by at least two visible aiming beams (see also Section 10.4.3) which cross in one point when the target is exactly located in the focus of the therapy laser beam. Before treatment, the axial position of the focus is aligned to the corneal surface by means of a surgical microscope. After the presetting, an eye tracking system takes over the automatic lateral readjustments.

10.3.2.5 Eye Tracking System

Ablation profiles are usually generated by a series of overlapping laser pulses which must be guided very precisely to a target position on the corneal surface. Small displacements of successive laser pulses may lead to an irregular surface shape, and a decentered ablation profile may cause a considerable impairment of the patient's vision.[16] As the patient's eye is not mechanically fixated during the treatment, potential head and/or eye movements shift the treatment area relative to the laser beam. Head movements can be avoided by fixating the patient's head in a headrest on the examination table. The headrest can be precisely adjusted in all directions by means of a joystick. Coarse eye movements can be prevented by means of a fixation target which is coaxially coupled into the beam path of the excimer laser. However, involuntary eye movements, so-called *saccades*, still occur during fixation. Moreover, it is often difficult for the patient to look at the fixation target during the treatment, as vision becomes blurred when the corneal tissue is removed.

Fixations-related saccades take place at high angular velocities of maximal $\approx 100°/s$ but with relatively small amplitudes ($< 1°$) [29]. During the laser treatment, such small but fast eye movements can be compensated with an eye tracking system. Eye tracking systems determine the eye position at the beginning and then recognize relative shifts during the treatment. The offset data is then sent to the controller of the xy scanner which adjusts the beam position before the laser pulse is applied. The time required for the whole process is called *response* or *latency* time and depends on the maximum speed of eye movements, the repetition rate of laser pulses, and the laser spot diameter on the cornea. In turn, the spot diameter on the cornea defines the maximum permissible spot shift. The *sampling rate* indicates how often the tracking system measures the eye position. Latency times

16) In fact, this is one of the most serious complications in refractive corneal surgery!

and sampling rates of eye trackers used in commercial laser systems are listed in Table 10.8.

Image-based tracking systems The majority of eye tracking systems determine eye movements from acquired images of the pupil and the iris of the patient's eye. For this purpose, NIR video cameras with a high sampling rate are used. If the patient's eye is illuminated by an array of displaced IR light spots, the pupil area appears to be black and thus can be detected with high precision. Fast image processing algorithms then retrieve the position of the pupil center (or more precisely, the center of the entrance pupil).

Since the pupil center may shift as the pupil diameter changes, the border line between the dark colored iris and the white sclera should be also detected and used for eye tracking to avoid potential errors. Some tracking systems only rely on this method. Other systems additionally evaluate and monitor the iris structure (*iris registration*) so that possible rotations of the eye around its visual axis (*cyclotorsion*) are determined. Clearly, cyclotorsion is only relevant for the treatment with ablation profiles *without* rotational symmetry.

Video-based eye trackers used for flying-spot laser systems usually have a latency time between 1 and 10 ms. Eye movements can thus be detected very fast, and the precision is sufficient to avoid negative influence of eye movements on treatment results.

Analog tracking systems Besides image-based eye tracking systems, analog *photoelectric-based* tracking systems are also used in some clinical instruments. For example, four light slits are used which are rotated against each other by an angle of 90° and are projected onto the border line between the colored iris and the white sclera. The intensity of the reflected light is then measured by photodetectors. If the patient's eye moves, the recorded detector signals change depending on the moving direction of the transition line between the dark iris and the light surface of the sclera. The differential signals from opposite sites are then used to calculate the eye movement by comparing the previous signals with the current ones. Since analog signal processing is used, sampling rates in the kHz range can be achieved a lot easier. However, analog tracking systems are merely able to identify relative pupil shifts. So, analog tracking is less reliable than image-based tracking with video systems. For this reason, analog trackers are used as supporting or backup systems for video-based eye trackers.

10.3.2.6 Energy Monitoring and Control

The treatment programs resulting from the calculated ablation profiles assume a constant light exposure. To correct potential fluctuations during the treatment or to completely switch off the entire system if the laser exposure exceeds or falls below permissible limit values, multiple detectors are used. They monitor the laser intensity at the input and output ends of the optical delivery system. In addition, the entire system should be calibrated with appropriate exposure tests on a regular basis. For this purpose, a test target (plastic-coated metal foil or PMMA disk) is placed in the processing plane. A test routine then checks whether a specified ablation rate is attained for a given exposure value.

A possible source of error which influences constant light exposure is ozone formed from oxygen in the surrounding air by intense UV laser light. Ozone absorbs the laser light and erodes optical surfaces in the beam path. To prevent this, the beam path of the optical delivery system is usually flushed with a protective gas (e.g., nitrogen).

10.3.2.7 Diagnostic Unit for Therapy Planning Data Capture

For therapy planning, commercial photoablation lasers are often connected to diagnostic instruments such as wavefront analyzers (Section 5.3) and/or corneal topography systems (Sections 6.3 and 6.5). Many manufacturers offer such joint refractive surgery platforms. These solutions simplify the clinical workflow and allow the transfer of relevant data from the diagnostic unit to the laser system in a reliable way.

10.3.3 Surgical Ablation Techniques

To attain a desired refractive correction, one of two surgical approaches can be used, that is, either surface ablation (Section 10.3.3.1) [30] or intrastromal ablation (Section 10.3.3.2).

10.3.3.1 Surface Ablation

Photorefractive keratectomy In *photorefractive keratectomy* (PRK), the corneal epithelium (Figure 1.7 in Section 1.1) is removed either mechanically with a blunt instrument, such as a spatula or rotating brush (Figure 10.20a), or with an excimer laser. After the stromal tissue has been removed via photoablation (Figure 10.20b), the eye is covered with a "bandage" soft contact lens. Within a few days, the epithelium regenerates and closes the wound on the surface (Figure 10.20c). Since in PRK only the epithelium is removed before the treatment, more stromal tissue is available for refractive correction than in the LASIK procedure (Section 10.3.3.2).

The primary side effects of PRK during the healing phase are pain and restricted vision. In addition, the risk of scar formation on the stromal surface increases for an increasing degree of refractive correction, and the success of the intervention cannot be conclusively evaluated until approximately one month after the surgery.

PRK was introduced by Stephen Trokel and John Marshall in 1986 (Table 10.3) and is nowadays the most often performed *surface* ablation method in corneal refractive surgery.

Laser sub-epithelial keratomileusis and epithelial laser *in situ* keratomileusis *Laser sub-epithelial keratomileusis* (LASEK) and *epithelial laser in situ keratomileusis* (epi-LASIK) are refractive correction methods similar to PRK with the objective of short-

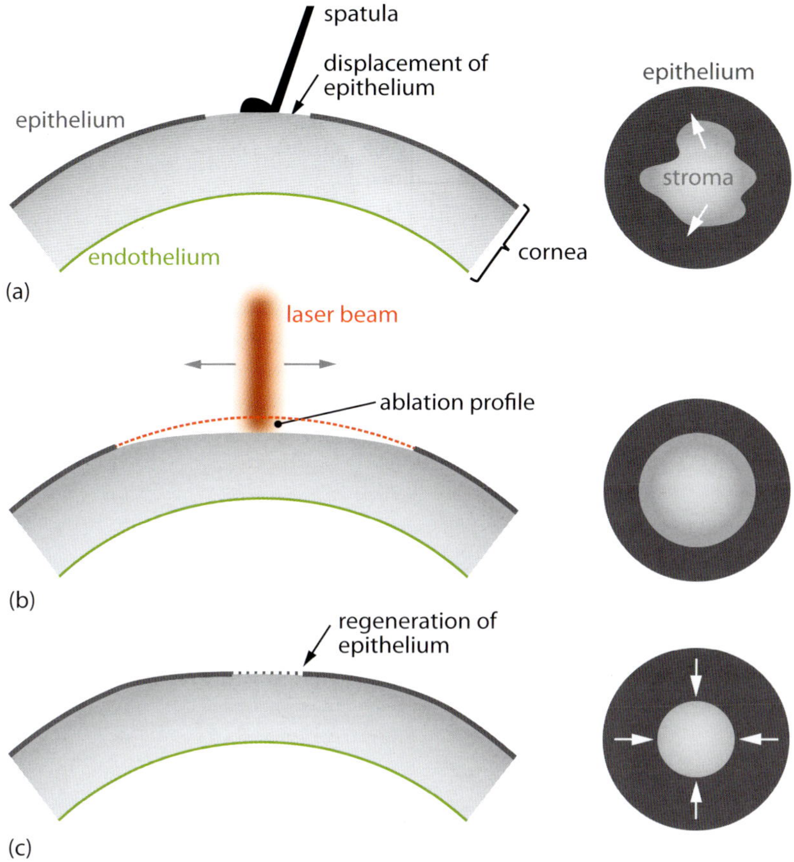

Figure 10.20 Procedure of photorefractive keratectomy (PRK). In the left column, the cross-section of the cornea is presented. The right column shows the corresponding top-view scheme. (a) The corneal epithelium (dark gray line) is mechanically removed with a spatula or rotating brush. (b) The stromal tissue is then exposed to a laser beam (red). Within the optical zone, tissue is removed via photoablation. (c) After a few days, the corneal epithelium regenerates on the reshaped surface (healing process).

ening the healing process and reducing potential side effects. In contrast to PRK, the epithelial layer is *not* destroyed in the treatment area, but detached as a whole sheet and put back in place once the photoablation has been completed. In epi-LASIK, a special microkeratome (epi-keratome) is used to separate the thin epithelial sheet from the remaining cornea. In LASEK, a microsurgical instrument called a *trephine* is used to create a circular cut in the epithelial tissue. The circular epithelial flap is then separated from the underlying tissue by shortly applying a diluted alcohol solution. Despite of potential advantages both methods are rarely used in clinical practice.

10.3.3.2 Intrastromal Ablation (LASIK)

In *laser* in situ *keratomileusis* (LASIK), a corneal flap (slice) with a thickness of approximately 80–180 μm is partly cut by means of a microkeratome and then flipped over like a cover (Figure 10.21b). The cut can also be performed with a focused femtosecond (fs) laser, as described in Section 10.5.4.1. In the stroma bed, the tissue is then removed by photoablation with a laser beam. Afterwards, the corneal flap is repositioned. As it attaches again to the stroma by suction due to cohesion forces, the corneal flap does not need to be sutured. The epithelial layer is only cut in the periphery (circular cut) so that the healing process is substantially faster than in PRK. Vision is thus only impaired within the first few hours after the surgery, and the treatment is less painful than PRK.

LASIK was developed by Ioannis Pallikaris (born 1947) in 1991 and is nowadays the most often applied ablation technique in corneal refractive surgery.

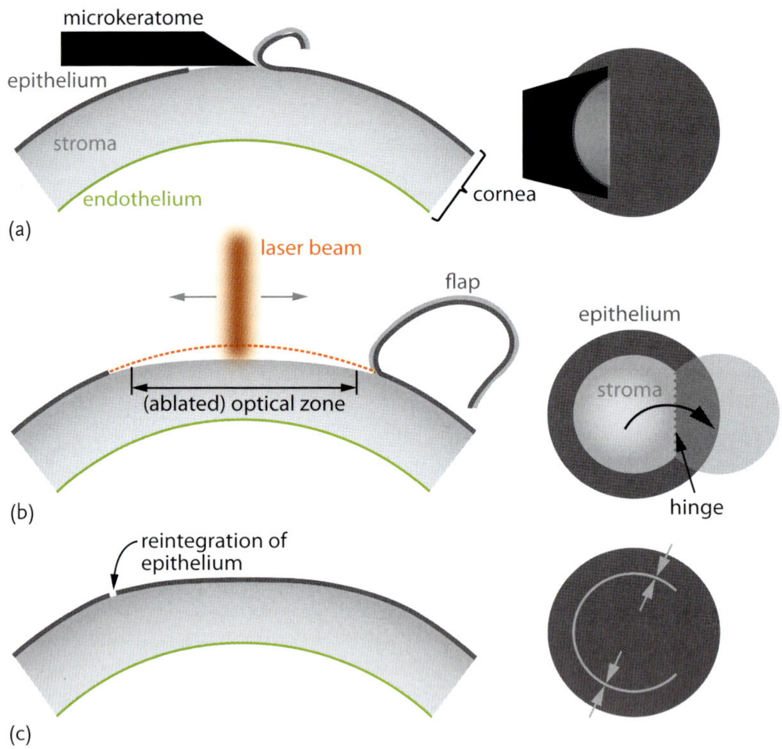

Figure 10.21 Procedure of laser *in situ* keratomileusis (LASIK). In the left column, the cross-section of the cornea is presented. The right column shows the corresponding top-view scheme. (a) The top of the cornea is cut with a microkeratome or by a femtosecond laser (not shown). (b) The formed LASIK flap is flipped over like a cover, and the exposed stromal tissue is ablated. (c) After the laser treatment, the flap is folded back onto the ablated surface. As the epithelium has not been removed, the healing process is considerably faster than for photorefractive keratectomy.

10.3.3.3 Limitations of Application and Refractive Result

We can estimate the maximum achievable correction of refractive power $\Delta\mathcal{D}_{\max}$ by means of the Munnerlyn formula (10.9). With

$$a_0 \approx \frac{1}{3}\Delta\mathcal{D}d^2 , \tag{10.22}$$

we obtain the conditional equation

$$\Delta\mathcal{D}_{\max} = \frac{3a_{0,\max}}{d^2} , \tag{10.23}$$

where $a_{0,\max}$ is the maximum possible central ablation depth in micrometers and d the diameter of the optical zone (diameter of lenticule) in millimeters. For given d, $\mathcal{D}_{\max}$ is only determined by $a_{0,\max}$. In the case of myopia correction with LASIK, we thus have

$$a_{0,\max} = L_{\mathrm{cc}} - L_{\mathrm{flap}} - L_{\mathrm{rs}} . \tag{10.24}$$

L_{cc} denotes the central corneal thickness, L_{flap} the flap thickness, and L_{rs} the minimum residual stromal thickness. To keep the cornea stable after surgery, a minimum residual stromal thickness of $L_{\mathrm{rs}} \approx 250\ \mu\mathrm{m}$ must be preserved. The central corneal thickness L_{cc} can be determined before the intervention by means of different measurement methods (Table 4.2 in Section 4.2).

Example 10.1

Maximum Possible Correction with LASIK For a central corneal thickness of 530 μm and a flap thickness of 140 μm, we can calculate from Eq. (10.24) a maximum central ablation depth of $a_{0,\max} = 140\ \mu\mathrm{m}$. The maximum correction of refractive power in the case of myopia is determined by Eq. (10.23). In the case of an optical zone diameter of $d = 6\ \mathrm{mm}$, we have $\Delta\mathcal{D}_{\max} \approx 12\ \mathrm{D}$.

From these considerations, it follows that a thinner flap thickness allows a greater refractive correction. As surface ablation methods such as PRK do not require a flap at all, they are advantageous for patients with a small corneal thickness. However, the theoretically maximum possible refractive correction is usually not exhausted in clinical practice, since with rising refractive correction values $\Delta\mathcal{D}$, the obtained refraction increasingly deviates from the target value, and the risk of side effects rises. In particular, for surface ablation techniques, the formation of scars is a critical issue so that the central ablation depth should not exceed a value of approximately 100 μm. The recommended correction limits are [31]:

- For surface ablation (PRK, LASEK, and epiLASIK):
  - Myopia correction: up to −8 D
  - Correction of astigmatism: up to 6 D
  - Hyperopia correction: up to +3 D

- For LASIK:
  - Myopia correction: up to −10 D
  - Correction of astigmatism: up to 6 D
  - Hyperopia correction: up to +4 D

For patients with a preoperative corneal thicknesses of less than 480 μm, corneal refractive surgery is not recommended at all (regardless of the desired refractive correction). Within the recommended range of application, the outcomes are virtually identical for intrastromal and surface ablation surgery.

10.3.4 Prospects

In clinical practice, refractive *corneal* surgery is the predominant refractive surgical procedure. However, it represents only one possibility to correct refractive errors (Section 3.1) and treat presbyopia (Section 2.1.4). At present, the conventional noninvasive correction methods like eyeglasses and contact lenses are much more prominent. Since a major part of the population has spherocylindrical refractive errors and people are affected by presbyopia when they become older, there is a significant market potential for refractive corneal surgery. This is the obvious motivation to develop advanced laser treatment systems (Section 10.5) which have the potential to increase public acceptance.

10.4 Laser Systems Based on Photodisruption with Nanosecond Pulses

Photodisruption with ns laser pulses is most often used for *laser posterior capsulotomy* (Section 10.4.4.1). Laser posterior capsulotomy is a procedure that is frequently required some months or years after cataract surgery (Section 3.2) to create a hole in the turbid back surface of the lens capsule (called *posterior capsular opacification* (PCO) or *secondary cataract*). In fact, the probability that such a treatment becomes necessary within five years approaches 10–30% [32]. As approximately 20 million cataract surgeries are performed worldwide each year, the number of required laser posterior capsulotomy treatments is quite significant.

Another application of laser-induced photodisruption is *laser peripheral iridotomy* (Section 10.4.4.2), that is, a procedure in which the iris is perforated for treatment of angle-closure glaucoma (Section 3.3). Laser-induced photodisruption also allows removal of disturbing tissue membranes inside the eye in a noninvasive manner (*laser membranectomy*).

Since the first successful experiments by Danièle Aron-Rosa (born 1934) [33] and Franz Fankhauser (born 1924) in the late 1970s (Table 10.3), laser-induced photodisruption has emerged as the method of choice for precise and noninvasive microsurgery inside the living eye. Today (in 2013), judging by the globally installed instrument base (approximately 25 000 units) and the volume of interventions performed with them, photodisruption with ns laser pulses is the second most important laser application in ophthalmology after photocoagulation.

10.4.1 Functional Principle

In ophthalmology, photodisruption with ns laser pulses is used for noninvasive cutting or perforation of transparent or nontransparent tissue inside the eye. The functional principle is based on the fragmentation of tissue in the laser focus as a result of a laser-induced optical breakdown (Section 9.4.4). Tissue structures behind the focal volume are effectively shielded from laser damage because of plasma shielding (Section 9.4.4.3). If tissue membranes shall be cut, rupture lines are created by means of laterally shifted perforation points along which the tissue eventually breaks up because of internal tension.

10.4.2 Process Parameters

In photodisruption with ns pulses, the threshold exposure Φ_{th} (in J/cm^2) needed to achieve an optical breakdown is in the order of 10–100 J/cm^2. Due to the high energy input, primarily tissue-destroying mechanical effects occur in the form of shock waves and cavitation (Table 9.1; Section 9.4.4.4). The processes leading to the formation of a laser-induced plasma are subject to statistical fluctuations. The threshold exposure Φ_{th} is usually defined as the exposure at which the probability for an optical breakdown is 50%. Thus, an exposure well above Φ_{th} is required to ensure a reliable process in clinical practice. However, for exposure values $\Phi > \Phi_{\mathrm{th}}$, an expansion of the laser-induced plasma from the focal region towards the incoming laser beam is observed (Section 9.4.4.2). In addition, the size of the cavitation bubble and the amplitude of the shockwaves increase with the laser energy, altogether enlarging the damage zone. For this reason, the task is to find suitable process parameters to achieve a reliable optical breakdown with the lowest level of energy possible. In the ideal case, the desired therapeutic effect (tissue fragmentation) is confined to a small area around the focal point (high processing accuracy) without damaging surrounding tissue. The tighter the beam can be focused, the less laser energy is required. With a tightly focused laser beam and a laser intensity slightly above I_{th}, the undesired plasma expansion can be minimized according to Eq. (9.16).

To obtain a small focus diameter, ns lasers used in photodisruptors are usually operated in the fundamental Gaussian mode (TEM_{00}; Section B.4.1). In some photodisruptors (e.g., ZEISS VISULAS YAG III), ns pulses with a *super-Gaussian* intensity profile are generated by means of a special laser resonator design [34]. For equal pulse energies, super-Gaussian beams achieve a higher peak intensity in the focal plane than Gaussian laser beams (see Info Box 10.1). In this way, the threshold energy can be further reduced as compared to a standard Gaussian profile.

Info Box 10.1: Super-Gaussian Profile

Super-Gaussian profiles are intensity profiles between standard Gaussian and top hat profiles (Figure 10.22). The super-Gaussian intensity profile is determined by

$$I(r) = I_0 \exp\left(-2\left|\frac{r}{w_0}\right|^m\right) \tag{10.25}$$

with the radial coordinate r, the beam waist radius w_0[17], and the super-Gaussian order $m > 2$. For $m = 2$, we obtain the classic Gaussian profile and for $m \gg 2$ a quasi-top hat profile. When focused, super-Gaussian beams achieve a higher peak intensity in the focal plane than Gaussian laser beams (Section B.4). In Figure 10.23, the maximum intensity in the focal plane is shown in arbitrary units for super-Gaussian beam profiles with varying exponent m and beam waist radius w_0 relative to the circular, limiting aperture radius a. The peak intensity I_{peak} is normalized with reference to the special case of a Gaussian profile (red curve in Figure 10.22) and a beam waist radius of $w_0 = a$.

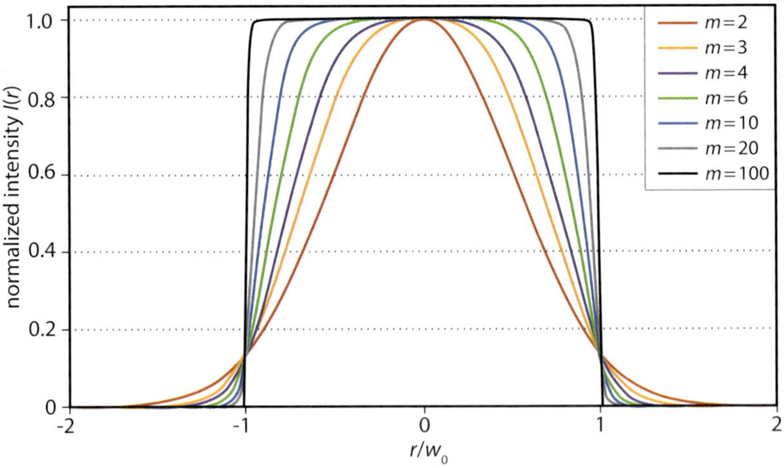

Figure 10.22 Normalized intensity profiles of Gaussian ($m = 2$) and super-Gaussian ($m > 2$) beams. Courtesy of Herbert Gross.

For beam radii smaller than the limiting aperture radius ($w_0 < a$), the effective numerical aperture NA (Section A.1.4) decreases. So, the focus is broadened which, in turn, leads to a reduced I_{peak}. If w_0 is too large ($w_0 > a$), the truncation loss at the circular aperture results in a reduction of I_{peak} as well. If m is small, the outer pupil zone is "underfilled", and the peak intensity is again reduced. The most suitable setup is a beam with aperture radius ratio of $w_0/a \approx 1$ and a uniformly filled pupil according to the top hat profile (Problem P10.18).

The generation of a coherent top hat beam profile with a laser resonator is not easy. According to Figure 10.23, a super-Gaussian profile with a high value of m is used to approach the theoretical best case of a top hat profile. For laser beams with such a profile, it is possible to obtain a peak intensity which is 20% higher than for Gaussian profiles ($m = 2$). Hence, an optical breakdown in the focal plane can already be achieved with a lower laser energy.

In Figure 10.24, the corresponding intensity profiles are shown in the plane of the focusing optics, the focal plane, and on the retina for a Gaussian and a super-Gaussian beam.

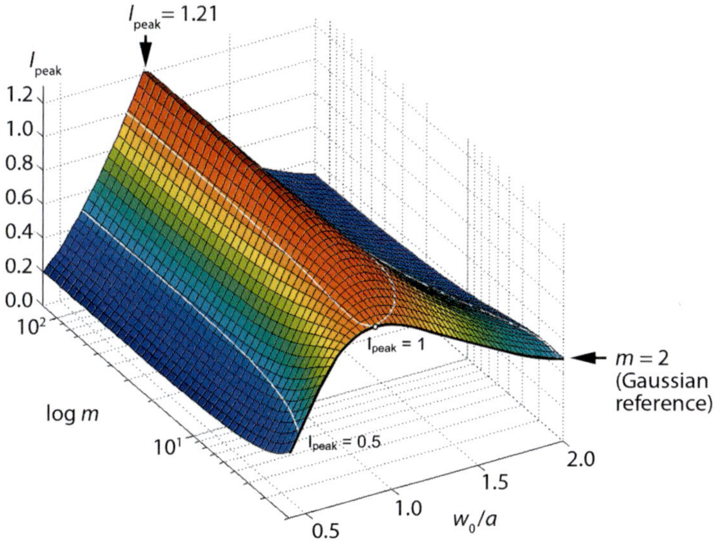

Figure 10.23 Maximum intensity I_{peak} in the beam waist in arbitrary units for super-Gaussian beam profiles with varying exponent m and beam waist radius w_0 relative to the limiting aperture a of the optical system. Courtesy of Herbert Gross.

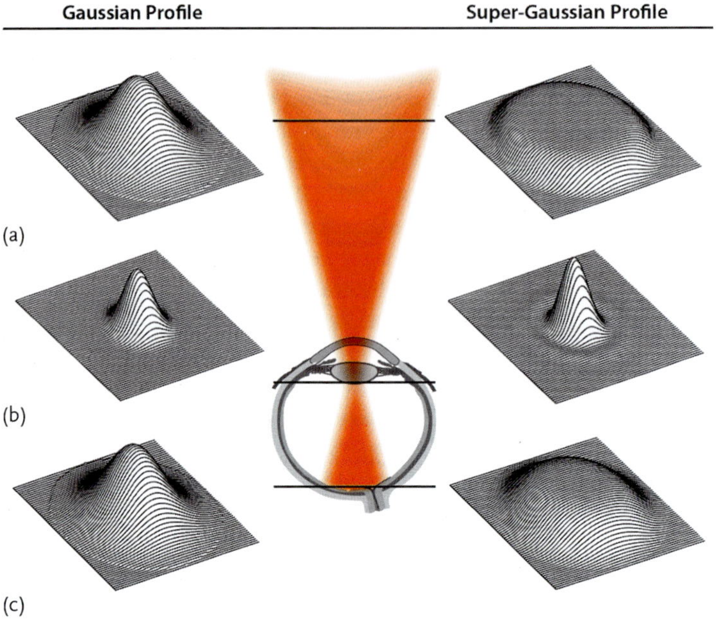

Figure 10.24 Comparison of Gaussian and super-Gaussian intensity profiles in (a) the plane of the focusing optics, (b) the focal plane, and (c) on the retina. When lasers with a super-Gaussian beam profile are used for laser posterior capsulotomy, not only is the peak intensity in the focal plane larger, but also the radiation exposure of the retina lower. Adapted from Carl Zeiss.

10.4.3
Technical Equipment Concepts

In commercial ns photodisruptor systems (Figure 10.27), Q-switched (Section B.4.4.1) Nd:YAG lasers ($\lambda = 1064$ nm) are commonly used as a light source. The laser beam is usually guided to the tissue with the optics of a slit lamp (Section 6.4). For this purpose, the pulsed laser beam is projected into the observation beam of a slit lamp and focused onto the treated tissue by means of the slit lamp objective (Figure 10.25a). To minimize the focus diameter, the beam diameter is increased by means of a beam expander so that the full numerical aperture of the objective lens can be used (Section A.2.2.3). In addition, a specific laser contact lens (capsulotomy or iridectomy lens) with integrated focusing optics is employed to further reduce the diameter of the laser focus in the eye by a factor of approximately 1.5 (Figure 10.29b).

17) Here, we define the beam radius analogous to Section A.2.2.1, that is, by the radial distance at which the maximum beam intensity decreases by a factor $1/e^2 = 0.135$.

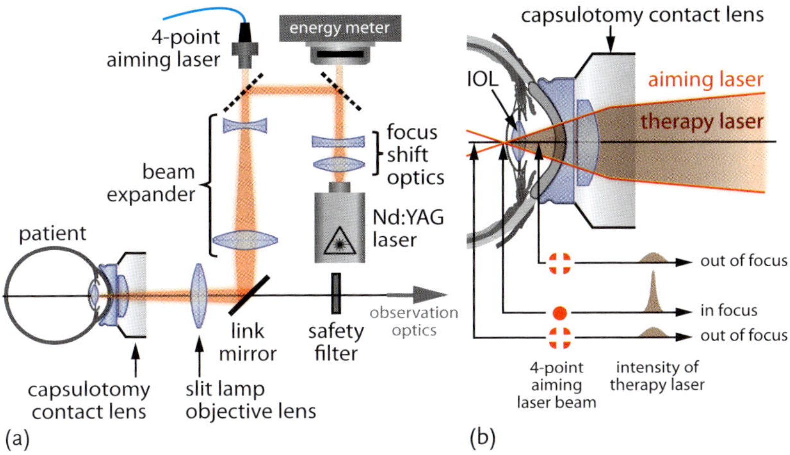

Figure 10.25 (a) Setup of a photodisruption laser. With a link mirror, both the four-point aiming laser and the therapy laser are coupled into the beam path of a slit lamp. (b) Functional principle of the four-point aiming laser. In the optical system, the aiming laser (red line) is arranged such that the four laser spots are separated behind and in front of the therapy laser focus (dark red area). If the four individual aiming laser spots merge to one point, the therapy laser focus setting is at the desired position. The pattern of the aiming laser and the corresponding intensity of the therapy laser are shown below the scheme.

Aiming lasers As the therapy laser beam emits invisible NIR light, it is surrounded by four peripheral aiming laser beams emitting visible red light. The aiming beams are used to verify that the therapy laser is accurately focused (Figure 10.25b) and that it is directed towards the tissue to be treated. All aiming beams merge in one point when the target lies exactly in the focal plane. With this arrangement, we can optimally set the axial adjustment and correct astigmatic refractive errors in the case of transverse alignment errors (e.g., laser beam is off-center). Besides this "four-dot arrangement", other arrangements of aiming lasers are available, for example, with two rotating partial beams facing each other and enveloping the therapy laser beam.

Variable focus shift To shift the axial position of the therapy laser focus in a defined manner, a focus shift unit is implemented in photodisruptor systems. The optics of the focus shift unit changes the divergence/convergence of the therapy laser beam so that its focal position is modified relative to the fixed focal position of the aiming lasers.

A variable focus shift is required for many applications. Due to the relatively large size of cavitation bubbles, eye structures outside the actual focal point can also be damaged. As the interaction range around the focal point is quite large in ns photodisruption (Table 9.1), it is necessary to have a controlling unit for adaption of the focal position to the actual application. For example, in posterior capsulotomy, the focal plane of the therapy laser lies behind that of the aiming lasers (posterior focus

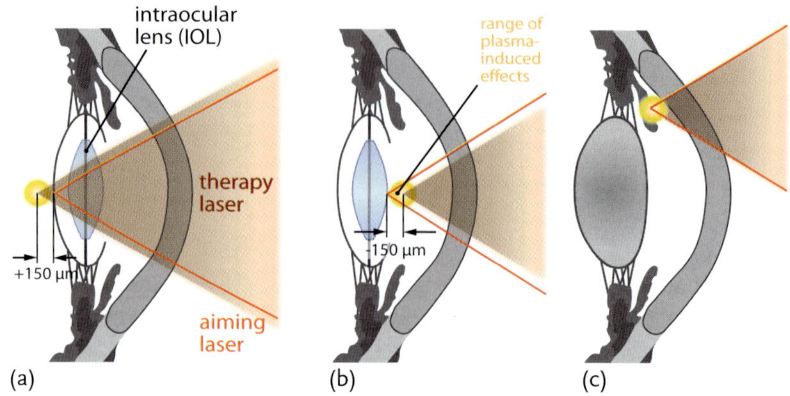

Figure 10.26 Principle of variable focus shift for different applications. In general, the focus shift is set according to the applied laser pulse energy. The shown value of 150 µm only represents the typical value range. (a) Posterior focus shift used in the case of laser posterior capsulotomy. The aiming laser beams (red line) are focused onto the surface of the posterior lens capsule. However, due to the focus shift, the focus of the therapy laser (dark red area) is located in the vitreous. The range of the plasma-induced mechanical effects (shockwaves, cavitation – orange) is now limited to the posterior lens capsule. The posterior lens capsule can thus be perforated without pitting the surface of the intraocular lens (IOL). (b) Anterior focus shift sometimes used to clean the surface of the implanted IOL. The aiming lasers beams are focused onto the surface of the IOL. The therapy laser is effectively focused into the anterior chamber so that only the deposits on the surface of the IOL are removed while avoiding lens pitting. (c) For laser peripheral iridotomy, a focus shift is not necessary, as no sensitive objects or tissue structures are located next to the targeted location.

shift; Figure 10.26a). The focus shift is set such that the posterior lens capsule is perforated as desired, but the back surface of the implanted intraocular lens (IOL) cannot be damaged. If the focus shift is too small, this would lead to a pitted lens surface. Another application is the front surface cleaning of an implanted IOL, for example, by removing cell layers which have grown over it. In this case, an anterior focus shift is used (Figure 10.26b). The amount of focus shift is chosen such that the deposits are removed by photodisruption while avoiding lens pitting. In iridotomy, however, a focus shift is not necessary, since no sensitive eye structures are located next to the targeted location (Figure 10.26c; Problem P10.19).

At present, a number of manufacturers offer laser photodisruptors. In Figure 10.27, commercial ns photodisruptor systems by Ellex and Lumenis are shown. Table 10.9 provides an overview of typical device parameters.

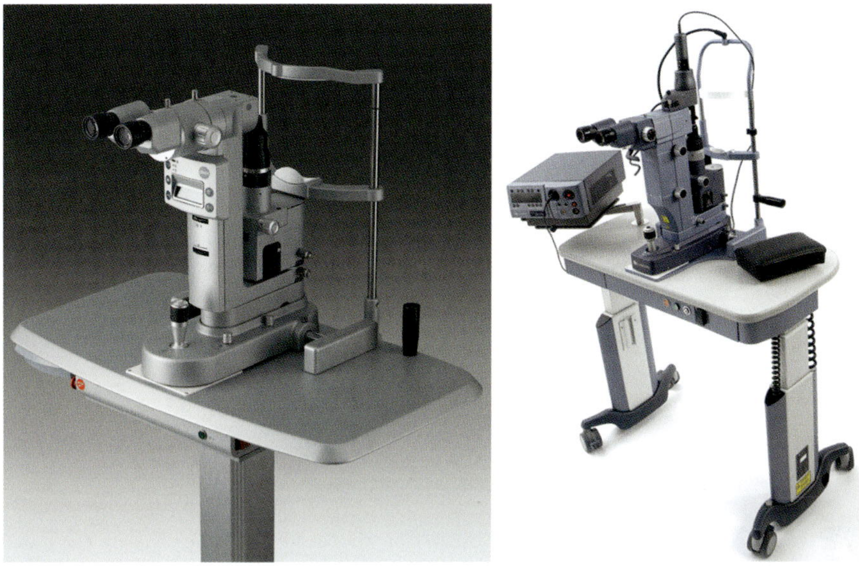

(a) (b)

Figure 10.27 Commerical photodisruptors based on ns laser pulses. (a) Photograph of Ellex Super Q®. Courtesy of Ellex Deutschland GmbH. (b) Photograph of Lumenis Selecta® Trio™ (YAG photodisruptor, SLT, and photocoagulator in one platform). Courtesy of LUMENIS Inc.

10.4.4 Clinical Applications

The most important clinical applications of ns photodisruptors are

- laser posterior capsulotomy (Section 10.4.4.1) and
- laser peripheral iridotomy (Section 10.4.4.2).

10.4.4.1 Laser Posterior Capsulotomy

One of the first steps of cataract surgery is the creation of a central round hole in the anterior eye lens capsule (capsulorhexis). Through this hole, the surgeon gains access to the lens core. In the next step, the turbid nucleus of the eye lens is liquefied by means of ultrasound (a process referred to as *phacoemulsification*) and then withdrawn by suction. This procedure basically preserves the capsule of the eye lens. Eventually, an IOL is inserted through the anterior central hole into the empty lens capsule bag. However, some months or years after the surgery, the rear side of the lens capsule often becomes turbid so that this posterior capsular opacification (PCO; or secondary cataract) must be removed. PCO is caused by the proliferation of epithelial cells of the eye lens which remain in the equator of the lens capsule after phacoemulsification. These epithelial cells migrate to the posterior part of the capsule behind the IOL and cause a second opacification. As a consequence, the patient suffers again from a progressive decrease of the visual acuity and a loss of

Table 10.9 Typical device parameters of laser photodisruptors based on ns laser pulses.

Treatment laser	
Type/wavelength	Passive Q-switched Nd:YAG with $\lambda = 1064\,\text{nm}$
Intensity profile	Gaussian or super-Gaussian
Pulse duration	4 ns
Pulse energy (single pulse)	Maximal 10 mJ (continuously or variable in steps)
Pulse energy (burst mode)	Double pulse: maximal 20 mJ. Triple pulse: maximal 30 mJ (continuously or variable in steps)
Pulse repetition rate	Single pulse: 3 Hz Burst mode: 1 Hz
Spot size	8 μm (in air)
Cone angle	16°
Optical breakdown threshold	2–3 mJ (in air)
Focal shift (posterior or anterior)	0–500 μm (continuously or variable in steps)
Aiming laser	
Type/wavelength	Diode laser with $\lambda = 635\,\text{nm}$ or $670\,\text{nm}$
Output power	0–200 μW (continuously or variable in steps)
Aiming method	Rotating partial beams or four-dot aiming system

contrast sensitivity. To treat this visual impairment, the central part of the turbid posterior lens capsule membrane is perforated via photodisruption (Figure 10.28). This procedure is called *laser posterior capsulotomy*.

10.4.4.2 **Laser Peripheral Iridotomy**

When the flow of the aqueous humor from the posterior to the anterior chamber of the eye is blocked, photodisruption is used to perforate the iris root. In this way, the risk of primary (acute) angle-closure glaucoma (Section 3.3) due to a pupillary block is reduced. The corresponding procedure is called *laser peripheral iridotomy* (LPI).

With a pupillary block, the aqueous humor can no longer drain into the anterior chamber. The increased pressure in the posterior chamber bends the iris towards the cornea (upper half of Figure 10.29a). If the iridocorneal angle is closed, the aqueous humor can also not drain through the trabecular meshwork, which results in a considerable increase of the IOP. After a successful LPI treatment, the hole created in the iris bypasses the pupillary block, and the IOP again decreases. As a consequence, the iris resumes its normal position, and the iridocorneal angle increases (lower half of Figure 10.29a).

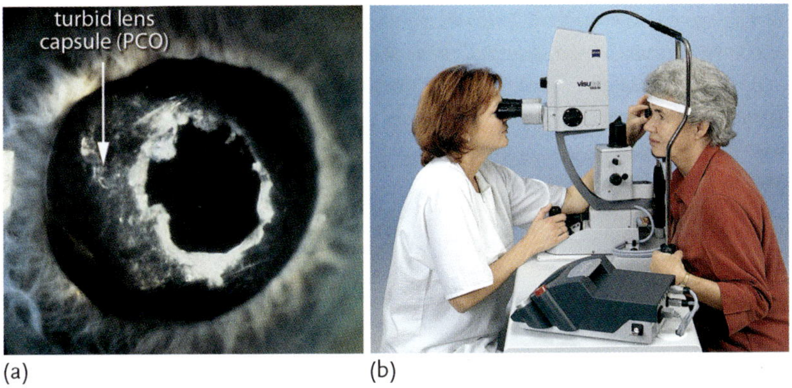

Figure 10.28 Laser posterior capsulotomy. (a) A sequence of therapy laser pulses perforates and opens up the turbid posterior lens capsule (PCO) by photodisruption. (b) ZEISS VISULAS YAG III in practical application. The laser capsulotomy contact lens held by the physician has two purposes. First, it additionally focuses the laser beam. Second, it stabilizes the patient's eye and keeps the eye lid out of the beam path. Courtesy of Carl Zeiss.

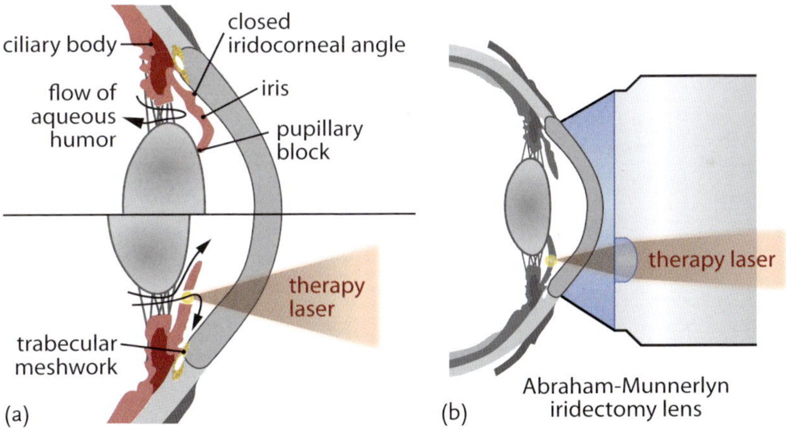

Figure 10.29 Laser peripheral iridotomy (LPI). (a) LPI is used to treat pupillary block glaucoma (shown in upper half of scheme). For the sake of simplicity, the iridectomy contact lens is not shown in this scheme. With a therapy laser beam (dark red area), a small hole is created in the iris tissue such that the aqueous humor can flow into the anterior chamber and the IOP decreases. The iris resumes its normal position, and the iridocorneal angle increases. (b) For LPI treatment, special contact lenses, called *iridectomy lenses*, are used. For example, the Abraham-Munnerlyn iridectomy lens increases the convergence angle of the incident laser beam by a factor of 1.5.

10.4.5 Prospects

New designs and materials of IOLs as well as improved implantation techniques have reduced the incidence of PCO. However, as the number of cataract surgeries performed worldwide steadily increases, the amount of interventions seems to compensate the reduced incidence of PCO. Nanosecond photodisruptors will thus remain an essential tool for cataract surgeons in the next years.

10.5 Laser Systems Based on Plasma-Induced Ablation with Femtosecond Pulses

Nanosecond laser pulses are not well suited to create precise cuts in tissue. Because of the relatively high laser energy needed to induce an optical breakdown, a relatively large volume is damaged due to dominant plasma-induced mechanical effects. If, however, the pulse duration is reduced down to the fs range, the optical breakdown takes place at substantially lower pulse energies (Figure 9.9), and the plasma-induced mechanical effects (e.g., cavitation) become very small (Table 9.1). Compared to ns pulses, the interaction volume is thus basically confined to the focus of the laser pulse.

History In 2001, the first commercial fs laser system IntraLase™ FS was presented ([35]; Table 10.3). The main application of this instrument was the cutting of corneal tissue to create LASIK flaps in refractive surgery (Section 10.5.4.1). As it allowed three-dimensional cuts with very high precision, the fs laser-based method (fs laser keratom) largely replaced the manual, mechanical cutting tools called *microkeratomes*. Following the acceptance of fs lasers for flap cutting, numerous new applications have been introduced to clinical practice. Today, fs laser-supported methods are used in

- refractive corneal surgery (Section 10.5.4.1),
- treatment of corneal diseases (Section 10.5.4.1), and
- laser-assisted cataract surgery (Section 10.5.4.2).

10.5.1 Functional Principle

The functional principle of ophthalmic fs lasers is based on plasma-induced ablation [21, 36] of tissue in the laser focus as a result of a laser-induced optical breakdown (Section 9.4.4.1).[18] As the laser–tissue interaction is confined to the laser

18) The term "plasma-induced ablation" describes the interaction processes of fs laser pulses correctly. However, the term "photodisruption", typically used for ns pulses, is also commonly used for fs lasers (e.g., fs photodisruption laser).

focus, ophthalmic fs lasers allows noncontact tissue fragmentation with high precision inside the eye. By scanning a preset pattern with a rapid sequence of laser pulses, precise tissue cuts in the μm range can be realized without thermal or mechanical side effects.

10.5.2
Process Parameters

The adequate parameter range for plasma-induced ablation is shown as a gray area in Figure 9.9. The lower limit of possible exposures is determined by the exposure threshold Φ_{th} above which an optical breakdown occurs. Φ_{th}, in turn, depends on the pulse duration τ. Above an exposure of about 50 J/cm^2 (regardless of τ), the energy input is so high that mechanical effects due to shock waves and cavitation bubbles dominate, and cuts on the μm scale are no longer possible.[19)]

Let us consider the size of the interaction volume (and the related cutting precision) for plasma-induced ablation in detail. The axial extension of the interaction volume from the focus towards the incoming laser beam is determined by the maximum expansion of the laser-induced plasma Δz_{max}. According to Eq. (9.14), we have

$$\Delta z_{max} = z_R \sqrt{\frac{I_{peak}}{I_{th}} - 1} \,, \tag{10.26}$$

with the Rayleigh length z_R (Section A.2.2.1; Problem P10.20), the peak pulse intensity I_{peak}, and the threshold intensity I_{th} for laser-induced optical breakdown. To achieve a high axial cutting precision (i.e., small values of Δz_{max}), we thus have to use a light intensity right above the threshold intensity I_{th}, and the laser beam must be focused tightly. On the one hand, a small beam waist diameter $2w_0$ is useful, since (according to Eq. (A83)) the Rayleigh length z_R is proportional to w_0^2, and with a small Rayleigh length, we increase the axial cutting precision. On the other hand, a small laser focus requires less energy to induce an optical breakdown which, in turn, leads to a smaller plasma volume. Although a small laser focus also means a slower cutting speed, this drawback can be largely compensated by a higher repetition rate.

Commercially available instruments for ophthalmic applications generate laser pulses with a duration of 200–800 fs (Table 10.10), as shorter pulses are not useful due to the following reasons:

1. According to Eqs. (B38) and (9.21), shorter pulses also mean a broader spectral bandwidth $\Delta\lambda$ of the emitted laser light. This fact must be taken into account when designing the optical imaging system of the laser. It becomes much more

19) However, recent research by Vogel *et al.* has revealed that plasma-induced ablation is still possible with pulse durations in the ns range, if laser sources with temporally smooth laser pulses without spikes and wavelengths below 1000 nm are used [37]. This technique is implemented in the SCHWIND SmartTech laser.

complicated to minimize chromatic aberrations (Section A.1.9) and group velocity dispersion (Section 9.5.3).

2. As shown in Figure 9.9, the exposure threshold required to obtain an optical breakdown approaches a nearly constant value for pulse durations < 200 fs. Hence, for a given focus diameter, the energy input into the tissue, and thus the volume of the plasma cloud, remains nearly constant. A reduction of the pulse duration beyond this point does not improve the precision or speed of the process.
3. The shorter the pulse duration, the higher the peak intensity for a constant pulse energy. Hence, the probability increases that the laser–tissue interaction is accompanied by undesirable nonlinear optical effects, which may change the propagation and focusing conditions (Section 9.5). Such undesirable nonlinear effects already appear for standard laser parameters used in commercial sys-

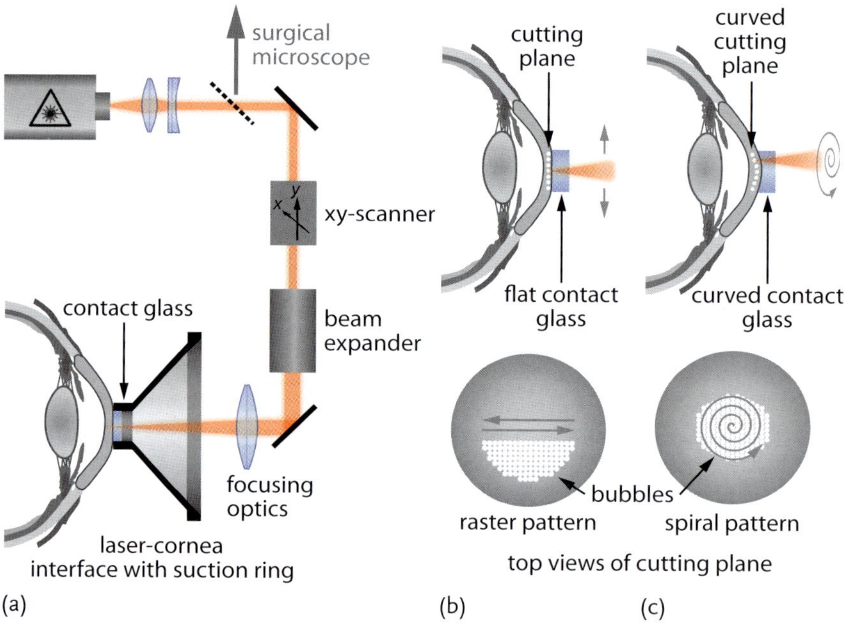

Figure 10.30 (a) Setup of a fs laser system used for corneal surgery. For visualization, the laser system is usually coupled to a surgical microscope. In the systems used for laser-assisted cataract surgery, a 3D imaging device is employed instead of (or in addition to) the surgical microscope. To fixate the position of the eye relative to the treatment beam, a laser–cornea interface is used which is connected to the cornea or the limbus by a peripheral suction ring. Before treatment, the patient-side part of the laser system with a contact glass on its front-side is attached to the cone-shaped opening of the laser–cornea interface. (b) Arrangement in the case of a flat contact glass. (c) Arrangement in the case of a curved contact glass. To cut the tissue, the laser spots are applied in a plane parallel to the contact glass surface in a zigzag pattern or along a spiral. Each "shot" generates an opaque gas bubble in the laser focus that is gradually resorbed by the surrounding tissue.

tems and must be compensated in the optical design (e.g., the focusing optics) by appropriate measures.

10.5.3
Technical Equipment Concepts

An ophthalmic fs laser system (Figure 10.30a) essentially consists of

- a laser source (Section 10.5.3.1),
- an xyz scanner with focusing optics (Section 10.5.3.2),
- a patient interface (Section 10.5.3.2),
- an imaging system for treatment planning and real-time control, and
- a control unit.

Table 10.10 summarizes the most relevant technical data of fs laser systems used for corneal or cataract surgery. Currently (in 2013), a number of manufacturers offer fs laser systems for ophthalmic surgery some of which are shown in Figure 10.31.

10.5.3.1 Laser Sources for Ophthalmic Femtosecond Laser Systems

To generate sufficiently intense laser pulses, most ophthalmic fs laser systems use diode-pumped amplifiers which are based on *chirped pulse amplification* (CPA) (Figure 10.32) [38]. In a small laser resonator cavity (Section B.3), a sequence of fs pulses is generated via (passive) Kerr lens mode locking (Section B.4.4.2). This method is based on self-focusing (Section 9.5.1) of short pulses with a high in-

Table 10.10 Technical features of commercially available fs laser systems for corneal treatment and laser-assisted cataract surgery.

Parameter	fs corneal surgery	fs cataract surgery
Laser concept	fs amplifier	fs amplifier
Wavelength	$\approx$ 1050 nm	$\approx$ 1050 nm
Pulse duration	200–800 fs	400 fs
Pulse energy	$< 1\,\mu$J	$\approx 10\,\mu$J
Repetition rate	80 kHz to > 5 MHz	12–160 kHz
Spot size	$\approx 1\,\mu$m	$\approx 10\,\mu$m
Treatment patterns	Zigzag, spirals	Crosses, pie cut, circles
Laser–cornea coupling	Flat or curved laser–cornea interface	Fluid-filled or curved laser–cornea interface
Imaging system	Surgical microscope and/or video microscope	Surgical microscope or video microscope and confocal structured illumination (Scheimpflug principle) or optical coherence tomography

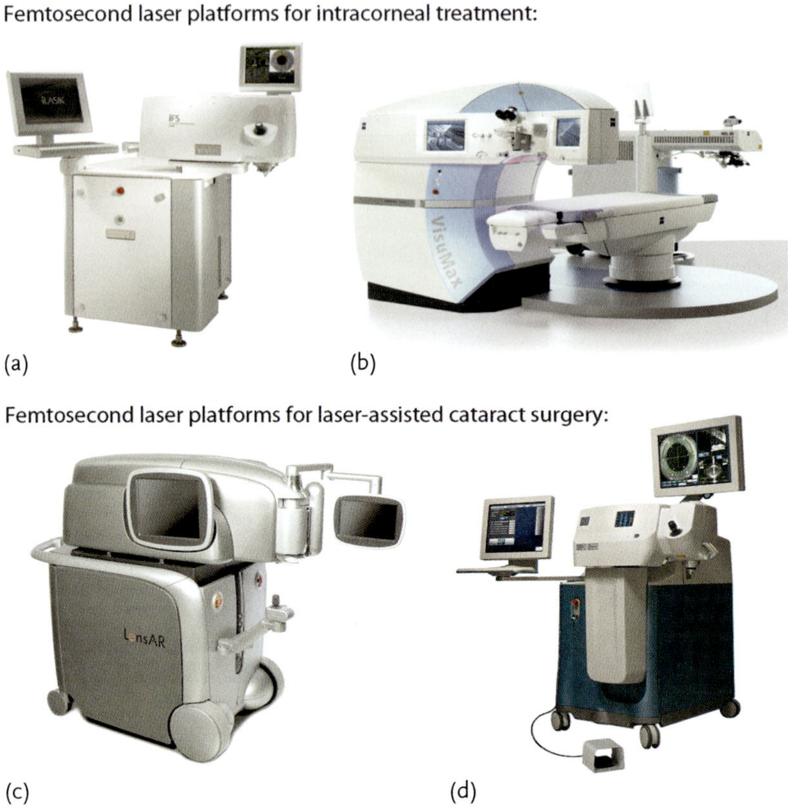

Figure 10.31 (a) Femtosecond laser platform IntraLase iFS™ by Abbott Medical Optics (AMO) for intracorneal treatments. Courtesy of AMO Germany GmbH. (b) Femtosecond laser platform ZEISS VisuMax™ for intracorneal treatment. Courtesy of Carl Zeiss. (c) Femtosecond laser platform LensAR™ for laser-assisted cataract surgery. Courtesy of Topcon Deutschland GmbH. (d) Femtosecond laser platform Alcon LenSx 550™ for laser-assisted cataract surgery. Courtesy of ALCON PHARMA GmbH.

tensity in the gain medium. Only the tightly focused laser modes are able to pass through a small aperture stop in the resonator. Since the focusing effect of the Kerr lens depends on the intensity, very short laser pulses are generated by mode locking if both the light amplification and bandwidth are high enough. The pulses to be amplified are temporally expanded by a medium with positive group velocity dispersion (optical fiber stretcher) so that its intensity is effectively reduced. In this way, self-focusing is reduced, which might destroy the gain medium. After the laser pulse has passed through the gain medium (fiber amplifier), the pulse duration is compressed again by optical media or systems (pulse compressor) with negative group velocity dispersion.

Ziemer FEMTO LDV systems use only a diode-pumped laser oscillator without subsequent amplification. As their corresponding pulse energy is much lower (in the nJ range), focusing optics with a large numerical aperture must be used to

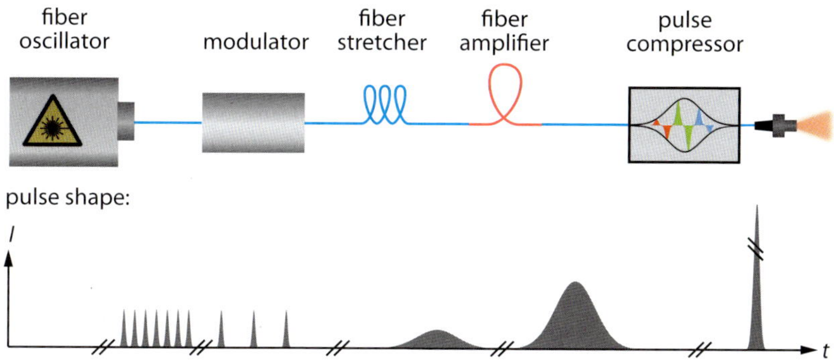

Figure 10.32 Principle of chirped pulse amplification (CPA), which is used to generate intense fs laser pulses. The shape of the laser pulses after passing the individual components is shown in the diagram below the schematic setup. Pulsed laser light of a mode-locked laser passes through a modulator by which the repetition rate is reduced. The pulses are then temporally expanded by a fiber stretcher (i.e., a long optical fiber) to prevent damage of the gain medium in the following fiber amplifier. After amplification, the pulses are eventually recompressed to durations of a few hundred fs (pulse compressor). As the energy of the laser pulse in the compressor is conserved, its peak intensity considerably rises after compression.

achieve radiant exposures which are high enough to induce an optical breakdown. In addition, because of the extremely small interaction volume, repetition rates in the MHz range are required to obtain an adequate cutting speed.

10.5.3.2 **Patient Interface**

The laser beam emitted by the laser source is expanded, focused, and guided to the treatment area via an xyz scanner. To follow a preset pattern, the beam deflection performed by the xyz scanner is controlled by a computer system.

To obtain the required treatment precision, a patient interface (PI) is used to mechanically couple the laser system to the eye. It is fixed to the cornea or the limbus with a peripheral suction skirt. The required underpressure is set manually or via the computer system. The cone-shaped opening of the PI is used to attach the patient-side part of the laser system (Figure 10.30a). For docking, either the patient-side part of the laser system or the patient bed can be precisely moved in all directions with a motor-controlled system. At the patient-side front end, we have a contact glass with a flat or spherical contact surface (Figure 10.30b or c) which serves as a reference plane for the calculation of the incision geometry. In the case of a spherical contact surface, the cornea is minimally deformed. Logically, the undesirable increase of the IOP during the treatment is considerably lower than for a flat contact surface. However, the requirements regarding the scanning geometry are higher (3D versus 2D scanning patterns).

Figures 10.30b,c show how the cornea is exposed to a sequence of laser spots. For tissue cutting, the laser spots are applied in a plane parallel to the glass surface in a zigzag pattern or along a spiral. Each "shot" generates an opaque gas bubble in the laser focus that cleaves the surrounding tissue before it is gradually resorbed. Small closely spaced spots allow easy separation of remaining tissue bridges and provide a smooth surface. The treatment is performed under visual control by means of an integrated surgical microscope (Section 6.2) or a video microscope.

10.5.3.3 Solutions for Laser-Assisted Cataract Surgery

Plasma-induced ablation as well as photodisruption with ps pulses can be used for laser-assisted cataract surgery. In this case, a few optical adaptations are required, as the object to be treated lies in the anterior chamber behind the eye pupil. This means that, with a given depth position of the object, the maximum size of the laser system's exit pupil is limited by the size of the entrance pupil of the patient's eye. Although the eye pupil is always widened with medication, the focus diameter of the laser beam is still increased approximately by a factor of 10 compared to corneal surgeries due to the limited aperture. The larger focus diameter[20] has to be compensated by increasing the pulse energy (Table 10.10). Because of the higher energy input and the larger focus diameter, the cutting precision of fs lasers in cataract surgery is lower compared to corneal surgery.

As the position and thickness of the eye lens is different for every patient, the corneal surface or a PI surface alone cannot be used as a reference for treatment planning. Before the treatment, the three-dimensional geometry of the anterior segment of the patient's eye must therefore be captured with a suitable imaging method. In this way, the necessary data about the relative positions (with regard to the contact glass surface) of the corneal back surface, the iris, as well as the front and back surfaces of the eye lens are determined. The used imaging methods are optical coherence tomography (OCT; Section 7.4) or confocal structured illumination methods based on the Scheimpflug principle (Section 6.5.2). With the acquired image data, a three-dimensional map of the anterior segment can be created which then serves as a basis for the treatment planning. During cataract surgery, the anterior segment is visualized by means of an integrated video microscope. This approach allows intuitive treatment planning and visual control of all process steps.

10.5.4 Clinical Applications

In the following, we will briefly discuss the most important clinical applications of fs laser photoablation in intracorneal and cataract surgery. Further details about the corresponding topics can be found in the referenced literature.

20) According to Eq. (A87), the focus diameter w_0 is inversely proportional to the divergence angle ε of the focused laser beam, and, according to Eq. (A96), it is directly proportional to the quotient of focal length f' and beam radius on the focusing lens w_L. If the laser beam is, however, focused on the back surface of the eye lens, the maximum divergence angle is limited by the ratio between the diameter of the entrance pupil and the distance between focal point and entrance pupil center ($\overline{F'E}$).

10.5.4.1 Intracorneal Surgery

LASIK flap creation The creation of corneal flaps for laser *in situ* keratomileusis (LASIK; Section 10.3.3.2) was the first and still is the most important clinical application of fs lasers in ophthalmology [39–41]. Although microkeratomes (mechanical cutting devices) are small, flexible, and low-cost tools, at present a significant portion of flaps in LASIK worldwide are created with fs lasers. The advantages of the fs laser-based method are as follows:

- Corneal flaps can be prepared with high precision and reproducibility in any desired geometry (diameter, thickness, side-cut angle, hinge position, and length).
- Considerably reduced complications due to irregular flap cuts (e.g., buttonholes and free caps).
- Patient acceptance is higher for bladeless flap creation.

In flap creation with fs lasers, the deeper part of the stroma is cut at first, as the generated gas bubbles may impair the subsequent process steps (Figure 10.33a). The lamellar cut is performed at a preset depth by applying laser spots in zigzag or spiral patterns. Then, the laser focus is vertically displaced and further laser spots

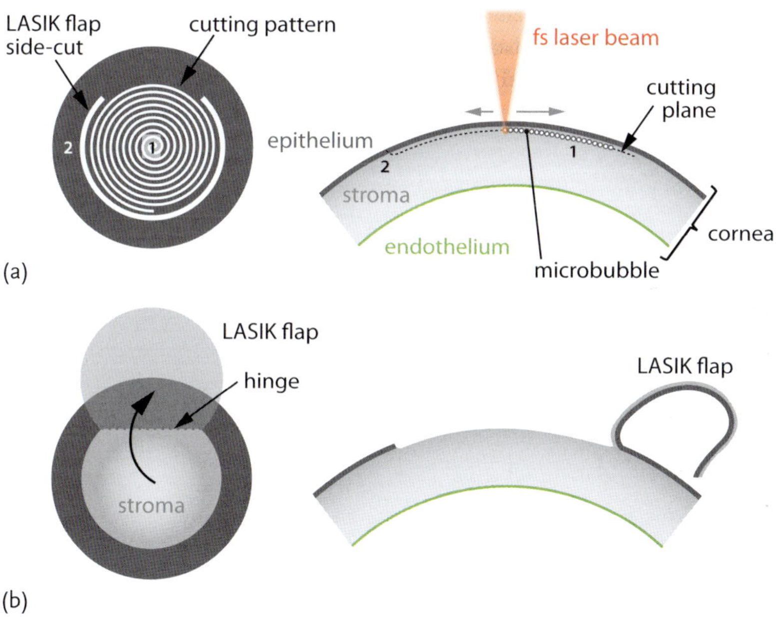

Figure 10.33 Formation of a LASIK flap with a fs laser. The left column shows a top view onto the cornea. In the right column, we can see the corresponding cross-section of the cornea (side view). (a) (1) At first, a lamellar cut is performed at a preset depth in the stromal tissue by applying fs laser spots in a spiral pattern. For this purpose, a series of small gas bubbles are lined up. (2) Next, a circular LASIK flap is cut that is connected to the corneal front surface. Importantly, the side-cut must be carried out such that a hinge is formed. (b) After the incisions have been created, the surgeon can flip over the flap, and the stromal tissue is exposed.

are applied along a circular spiral path. In each cutting cycle, a certain section is left "untouched" which serves later as a hinge for the flap. With a spatula, the flap is released from the stroma bed and folded backwards at the hinge (Figure 10.33b). The vertical side cut allows repositioning of the flap after the refractive surgery and reduces the risk of a lateral shift of the flap.

Refractive lenticule extraction Refractive lenticule extraction (ReLEx) is a refractive surgery technique which allows correction of myopia and myopic astigmatism (Section 3.1) in a single step without excimer lasers [42–44]. For this purpose, a lamellar cut inside the stroma is performed by scanning the fs laser beam in a spiral pattern (Figure 10.34a). A second cut prepares a thin disk-shaped piece of corneal tissue (*lenticule*) with the desired shape, depending on the refractive error of the patient's eye. The lenticule is then removed via standard LASIK flap cutting (ReLEx flex) or through a small lateral cut (blue line in Figure 10.34). The latter procedure is referred to as *small incision lenticule extraction* (ReLEx smile). As the surface of the cornea changes according to the removed volume of the lenticule, the desired refractive error correction is achieved.[21)]

Currently, the ReLEx technique is only implemented in the ZEISS VisuMax® fs laser system. The obvious advantage is that all process steps in refractive surgery can be performed with a single laser. The patient does not need to be moved from one laser platform to the other so that time is saved in the overall procedure. Since the traditional shot-by-shot tissue removal with excimer laser photoablation (Section 10.3) is no longer required, the treatment time is considerably reduced, especially for high refractive corrections. In addition, the surgical results are independent of corneal hydration and environmental conditions (Section 10.3.1.3), because in contrast to excimer laser surface ablation the cuts are performed inside the cornea.

Femtosecond laser-assisted keratoplasty When diseases affect the central part of the cornea such that vision is considerably impaired, corneal transplants may be required. In *penetrating keratoplasty* (PKP), the central portion of the cornea is completely exchanged (full-thickness corneal transplant). For this purpose, a mechanical circular blade cutting tool called *trephin* is used.

The manual cutting procedure can be improved by means of fs lasers, as the recipient and the donor cornea can be prepared with high precision by a computer-controlled process. When cuts are performed with *femtosecond laser-assisted keratoplasty* (FLAK) [40, 45], arbitrary cutting geometries can be realized (examples shown in Figure 10.35). In FLAK, a fs laser first creates a penetrating 360° cut in the patient's cornea. Then, the cut volume of the central cornea is removed and the donor corneal graft, created with the identical cutting pattern, is placed in the patient's cornea with or without sutures. As arbitrary cutting geometries can be realized, the fixation of the graft into the patient's eye is better than the smooth

21) The lenticule shape corresponds to the ablation profile generated by excimer lasers (Section 10.3.1.1).

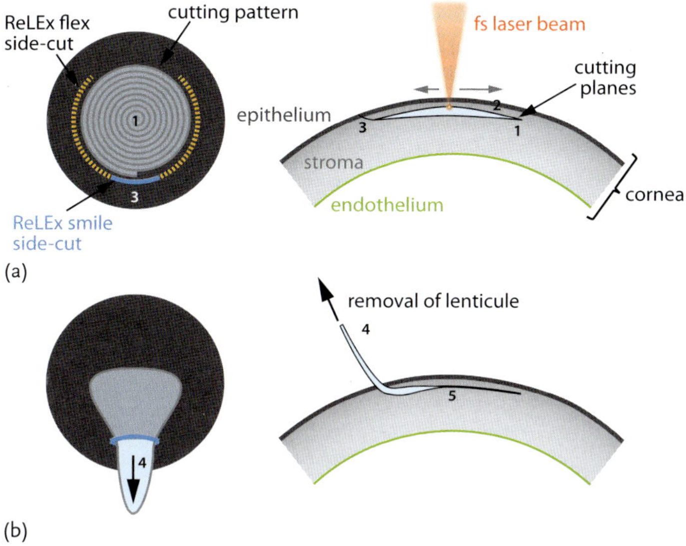

Figure 10.34 Principle of refractive lenticule extraction (ZEISS ReLEx). The left column shows a top view onto the cornea. In the right column, we can see the corresponding corneal cross-section (side view). (a) (1) A lamellar intrastromal cut is performed by scanning the laser beam in a spiral pattern. (2) A second cut prepares a refractive lenticule (disk-shaped piece of stromal tissue) with the desired shape whose actual form depends on the eye's refractive error. (3) A small side-cut (blue) of less than 4 mm is created (ReLEx smile). (b) (4) The prepared lenticule is removed manually through the small incision. (5) The surface of the cornea changes according to the removed volume of the lenticule which, in turn, leads to the desired change in refractive power. In the alternative ReLEx flex procedure, a LASIK flap-like side-cut (orange) is created. After the flap has been opened, the lenticule can be removed.

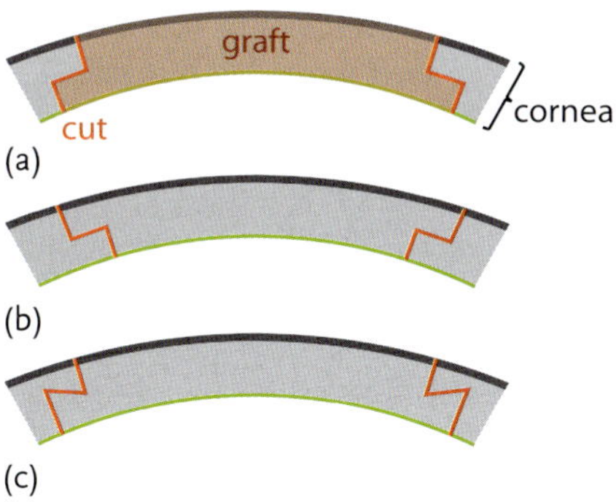

Figure 10.35 Examples of cutting patterns which can be realized with fs laser-assisted keratoplasty (FLAK). (a) Top hat pattern. (b) Mushroom pattern. (c) Zig-zag pattern.

cylindrical cuts performed with a mechanical trephin. As a consequence, fewer sutures are required after fs laser cuts, and the risk of suture-induced astigmatism is reduced.

Arcuate keratectomy in astigmatism Natural astigmatism (Section 3.1.2) or surgically induced astigmatism, for example, following a corneal transplant or cataract surgery, can be corrected by means of circular cuts into the peripheral parts of the cornea (*limbal relaxing incisions*; Figure 10.36). Cuts of a specific depth and length running perpendicular to the principal meridian (Section 3.1.2) with stronger curvature lead to a flattening along this meridian. Correspondingly, we have a steepening effect along the other (perpendicular) principal meridian. Compared to the traditional manual procedure, this intervention can be applied more accurately and at lower risk with a fs laser [41].

Other intracorneal surgical applications Other fields of intracorneal application for fs lasers in ophthalmology are [41]:

- *Tunnel resections for intrastromal corneal ring segments* (ICRSs): ICRSs are sickle-shaped PMMA implants which have defined diameters and arc lengths according to the intended degree of refractive correction. They are placed in the peripheral part of the corneal stroma. The achieved change of the corneal shape is used for the correction of mild to moderate myopia and for the treatment of keratoconus (Section 3.1.6). The fs laser is required to rapidly create a corneal channel necessary for implantation with a well-defined geometry. Compared with manual techniques, the tunnel dimensions (width, diameter, and depth) can be formed with much higher accuracy.
- *Intracorneal pockets for corneal inlays*: Corneal inlays are used to treat presbyopia (Section 2.1.4) in that they are inserted centrally into the stroma of the nondominant eye to improve near vision. The inlays consist of a biocompatible material and are either realized as pinhole diaphragms (Kamra inlay) or small lenses (PresbyLens). Femtosecond lasers can be used to create pockets which exactly match the inlay shape and size to be implanted.
- *Presbyopia intrastromal treatment*: Presbyopia intrastromal treatment is a minimally invasive method for the treatment of presbyopia which was developed by Luis Ruiz [46]. A corresponding procedure called INTRACOR® is offered by Technolas Perfect Vision. With a fs laser, five concentric rings are cut within the stroma around the optical axis. All cuts end shortly below the corneal front and back surfaces and lead to a reorganization of the biomechanical forces inside the cornea. The underlying functional principle is, however, not fully understood. After this procedure, the cornea has a multifocal hyperprolate topography which eventually causes an increased depth of field with improved near vision. This method can only be used for emmetropic or slightly hyperopic eyes (Section 3.1.1). To date, long-term safety and stability results are not yet available.

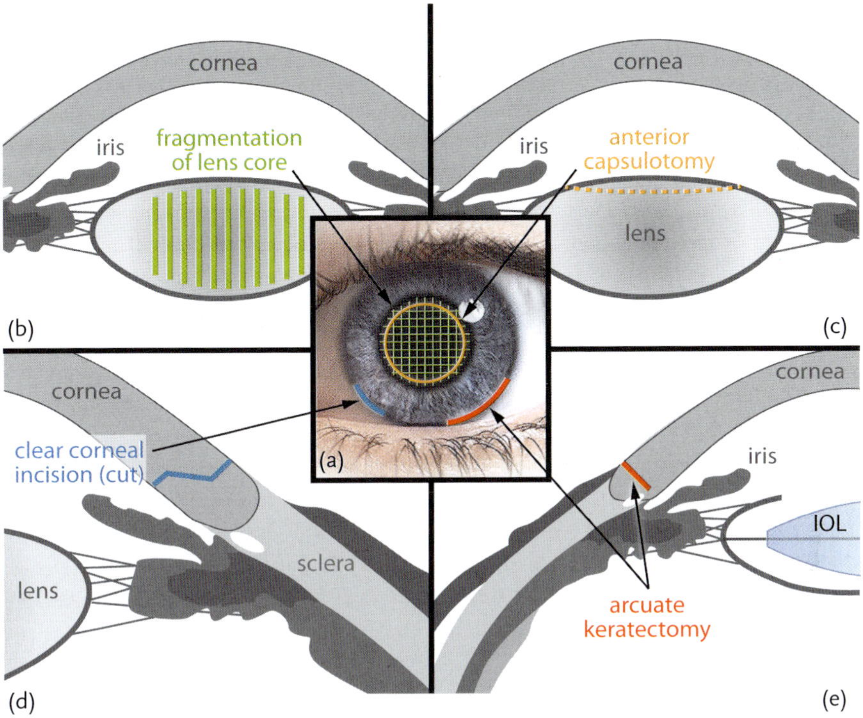

Figure 10.36 Treatment procedures in fs laser-assisted cataract surgery. (a) Front view of the patient's eye which shows the different cut positions and patterns. (b) The lens core is fragmented with special treatment patterns (green) to allow an easy removal of soft cataracts or to soften harder cataracts. To prevent capsule rupture due to laser damage, a safety distance to the lens capsule must be adhered. (c) Laser anterior capsultotomy (LAC) is used to create an exactly central and round hole in the anterior eye lens capsule. Through this hole, the physician gains access to the lens core for subsequent phacoemulsification and IOL implantation. (d) Clear corneal incision is a common method to access the anterior chamber in cataract surgery. With fs lasers, special cut designs can be realized to obtain a perfect self-sealing incision. (e) If necessary, arcuate keratectomy is performed by a limbal relaxation incision to correct natural or surgery-induced astigmatism if necessary.

10.5.4.2 Femtosecond Laser-Assisted Cataract Surgery

In cataract surgery, fs lasers can be used to simplify critical process steps during the intervention [47–49]. The actual application options are illustrated in Figure 10.36.

Laser anterior capsulotomy One of the most critical steps of phacoemulsification is the creation of a central round hole in the anterior eye lens capsule, that is, a *continuous circular capsulorhexis*. Through this hole, the surgeon gains access to the lens core. Conventional capsulorhexis is indeed a demanding manual procedure, since a correct hole diameter with reference to the IOL diameter is important. In laser anterior capsulotomy, the central part of the anterior eye lens capsule is circularly cut and then removed with a suitable surgical tool. With fs lasers, it is possible to create a precisely sized, shaped, and centered hole in the anterior eye lens capsule which is more resistant to tearing while manipulating the IOL during surgery.

Lens fragmentation The fragmentation of the lens nucleus is a very demanding process step of phacoemulsification, as it requires many intraocular surgical maneuvers. The interaction of fs lasers with lens tissue can be used to cut patterns into the lens nucleus in order to allow an easy removal of soft cataracts or to soften harder cataracts. In the latter case, fs laser pretreatments of the nucleus reduce the required amount of ultrasound energy to be applied by the phacoemulsification probe. In this way, surgical side effects such as endothelial injury of the cornea and complications concerning the lens capsule are reduced. In addition, the process time for manual phacoemulsification decreases when using this method.

Clear corneal incisions (CCI) *Self-sealing clear corneal incision* is a prominent method to access the anterior chamber during cataract surgery. With a fs laser, defined cuts with required dimensions are performed. This process is independent from placement or patient factors such as deep-set eyes and tissue thickness. The characteristics of the cut (e.g., size and length) can be easily adapted to the present situation. With special cut designs, one can also ensure a complete seal at the end of the surgery.

Arcuate keratectomy (AK) The method and advantages of fs laser-supported arcuate keratectomy for astigmatism correction have already been described in Section 10.5.4.1. The requirements for the correction of remaining astigmatism after a cataract surgery are particularly high for premium IOLs. For example, in the case of multifocal IOLs, good results can only be obtained if the postoperative astigmatism is less than 0.5 D.

Typical workflow Generally, in all fs laser procedures, deeper structures in the treated tissue are cut at first, since the gas bubbles may impair the subsequent process steps. Hence, compared to conventional cataract surgery, the treatment procedure in fs laser-assisted cataract surgery is actually reversed. At first, the capsulotomy is performed to gain access to the lens core. This is followed by the lens fragmentation. Then, corneal incisions are created to gain access to the anterior chamber. Once all incisions are completed, the surgeon uses the suction handpiece of the *phaco unit* to remove the pretreated lens core and to clean the inner surface of lens capsule. Eventually, the IOL is implanted into the empty lens capsular bag. If necessary, an arcuate keratectomy is performed to correct present and/or surgery-induced astigmatism.

10.5.5
Prospects

Femtosecond lasers have considerably enhanced many corneal and cataract surgery methods. Over the last few years, their range of application has constantly expanded. At present, however, the main hurdles to overcome (in particular with regard to cataract surgery) are the relatively high costs and the large footprint of the systems. In the forthcoming years, we expect to see a cost-reduction of the required components, the development of more modular instrument designs with standard platforms for different applications in corneal and cataract surgery, and a further expansion of the field of applications.

Currently, new laser-supported methods for the correction of presbyopia and/or for the restoration of the accommodative power of the eye lens are being tested [50]. If these clinical trials are successful, a new field of application will arise for fs lasers with significant market potential. In [51], it is also discussed that fs lasers could be used to customize the refractive power of an implanted IOL in a noninvasive manner.

10.6 Recommended Reading

For further information about laser systems for the treatment of eye diseases and refractive errors, please refer to the following references:

- Laser Systems Based on Photochemical Interactions (Section 10.1): [7, 52].
- Laser Systems Based on Photothermal Interactions (Section 10.2): [8, 11–13, 15, 18].
- Laser Systems Based on Photoablation (Section 10.3): [23, 53–56].
- Laser Systems Based on Photodisruption with Nanosecond Pulses (Section 10.4): [57].
- Laser Systems Based on Plasma-Induced Ablation with Femtosecond Pulses (Section 10.5): [36, 40, 41, 44, 45, 48, 49].

10.7 Problems

P10.1. Photodynamic therapy I Calculate the number of photons deposited into tissue using the parameters in Section 10.1.2.1.

P10.2. Photodynamic therapy II The absolute spectral absorption maximum of the photosensitizer verteporfin is at a wavelength of 400 nm. Hence, an excitation at this wavelength would be very efficient. Why is a wavelength of $\lambda = 689$ nm used instead to destroy choroidal neovascularizations during a photodynamic therapy (PDT)? After the photosensitizer has been injected, we have to wait a while until the laser therapy can be started. Why so?

P10.3. Photothermal effects Estimate the temperature increase in tissue using the parameters in Section 10.2.3.1.

P10.4. Retinal photocoagulation I Compare for typical parameters of standard and short-pulse photocoagulation the temperature increase, the diameter of coagulated tissue, and the deployed energy.

P10.5. Retinal photocoagulation II For photocoagulation on the retina, an indirect ophthalmoscope can be used. With a 20 D ophthalmoscopy lens, a real (aerial) image of the retina is obtained. What is the size of the laser spot on the retina for an emmetropic eye if the laser beam diameter in the plane of the aerial image is 1 mm?

P10.6. Step-index fiber

1. Consider a step-index fiber with a numerical aperture of NA = 0.2 and a refractive index of the fiber core of $n_c = 1.5$. What is the refractive index of the cladding?
2. An expanded and collimated laser beam with a waist radius of $w_0 = 0.5$ cm and a beam divergence of $\varepsilon = 0.2$mrad shall be coupled into an optical fiber with NA = 0.2 and a core diameter of $d_c = 50$ μm. What are the minimum and maximum focal lengths of the fiber coupling lens to achieve a loss-free coupling-in? Why is it more advantageous for the instrument design to use lenses with a shorter focal length? Coupling losses due to Fresnel reflection at the surface shall be considered here as second-order. Is that justified?

P10.7. Laser link Calculate a suitable layout for the laser link shown in Figure 10.7 using a diode laser and as parameters the data given in Section 10.2.3.1. Assume an optical fiber diameter of 100 μm.

P10.8. Head-mounted laser indirect ophthalmoscope (LIO)

1. Calculate the laser focus diameter on the retina of the head-mounted laser indirect ophthalmoscope shown in Figure 10.9. Assume a standard ophthalmoscopy lens (20 D) is to be used and an optical fiber diameter of 100 μm. Design the laser beam imaging system so that a laser spot image of 1 mm is obtained in the (intermediate) aerial image plane of the fundus of an emmetropic eye. Consider typical distances (arm lengths etc.).
2. What is the pointing stability of the laser spot on the retina if the head of the ophthalmologist moves by 1°. What happens if the ophthalmoscopy lens is tilted by 5° during treatment?

P10.9. Photoablation (blow-off model) Calculate the ablation rate per laser pulse L_{abl} in corneal stroma according to the blow-off model for an ArF excimer laser ($\lambda = 193$ nm) and a KrF excimer laser ($\lambda = 248$ nm). The absorption coefficients of stroma are $\mu_a(193\,\text{nm}) = 29\,000/\text{cm}$ and $\mu_a(248\,\text{nm}) = 290/\text{cm}$. The corresponding exposure thresholds are $\Phi_{th}(193\,\text{nm}) = 50\,\text{mJ/cm}^2$ and $\Phi_{th}(248\,\text{nm}) = 500\,\text{mJ/cm}^2$. In both cases, the laser exposure is assumed to be $\Phi = 600\,\text{mJ/cm}^2$.

P10.10. Photoablation (thermal effects) An ArF excimer laser has a typical pulse length of 20 ns. Why are thermal effects of this laser negligible when it is used for photoablation of corneal stroma (thermal diffusion constant of stroma $\kappa = 1.5 \times 10^{-7}\,\text{m}^2/\text{s}$; absorption coefficient $\mu_a(193\,\text{nm}) = 29\,000/\text{cm}$)?

P10.11. Photoablation (refractive change) In refractive corneal surgery, an excimer laser is used to increase the central radius of curvature of the corneal front surface by +3%.

1. Let us assume that the treated eye can be described by the Exact Gullstrand Eye model. Calculate the change of the refractive power of the corneal front surface.
2. For what kind of refractive error is such a treatment useful?

P10.12. Photoablation (required precision) Let us consider a laser system which allows photoablation with a precision of $\pm 5\,\mu\text{m}$.

1. What is the maximum precision during refractive surgery which can be achieved with this instrument for a typical ablation zone diameter of 6 mm?
2. Does it make sense to have an ablation precision of $\pm 1\,\mu\text{m}$?

P10.13. Photoablation (LASIK I) A myopic eye has a central corneal thickness of 530 μm and shall be treated with LASIK. The flap thickness is 160 μm and the diameter of the optical zone 6 mm. Calculate the maximum possible correction of the refractive power under consideration of the stability limit of the corneal thickness.

P10.14. Photoablation (LASIK II) A myopic patient desires a LASIK refraction correction. The required refractive power of the corrective lens is determined preoperatively by subjective refraction using a phoropter. The trial lenses of the phoropter are placed in front of the eye at a distance of $L = 12\,\text{mm}$ from the corneal vertex. The patient achieves the best far distance visual acuity with a trial lens of back vertex power $\mathcal{D}'_v = -6\,\text{D}$. What is the necessary ablation depth in the center if the physician chooses the Mullerlyn-ablation profile with an optical zone diameter of 6 mm for myopia correction?

P10.15. Photoablation (exposure correction) The photoablation rate per laser pulse depends on the applied exposure (see Eq. (9.12)). As the surface of the cornea is curved, the beam waist diameter changes as a function of the location on the corneal surface. Without correction, the ablation rate would gradually decrease from the corneal center to the periphery, as the effective exposure drops. Calculate the necessary correction factor for the laser energy as a function of the laser spot decentration from the corneal vertex.

P10.16. Photoablation (OPD) and wavefront aberrations The aberrations of the eye can be described, for example, by means of the optical path difference OPD (Section 5.3.1.3). From the OPD, we immediately obtain the data for the necessary ablation profile in a wavefront-guided refractive corneal surgery. The OPD map shows directly at which locations of the cornea how much stromal material needs to be removed in order to achieve a reduction of the optical path length and thus for the correction of the aberrations. Verify the "rule of three" which states that for the correction of +1 μm OPD or (−1 μm wavefront aberration) approximately 3 μm corneal stroma (refractive index $n_c = 1.376$) must be removed.

P10.17. Aberration with Gaussian beams Simulate with a mathematical software tool (e.g., Matlab) the ablation of a corneal tissue slab (thickness of 200 μm) with a Gaussian beam (beam diameter of 0.7 mm and parameters of Section 10.3.1 and Table 10.8). Optimize the step width and the total procedure time. Assume a fixed repetition rate of $1/t_{rep} = 250\,\text{MHz}$.

P10.18. Super-Gaussian beams Assume a Gaussian and a super-Gaussian beam ($m = 20$, $w_0 = 1\,\text{mm}$) with equal energy, that is,

$$\int_{-\infty}^{\infty} I_{m=2}(r)\mathrm{d}r = \int_{-\infty}^{\infty} I_{m=20}(r)\mathrm{d}r\,. \tag{10.27}$$

Calculate the peak intensity for both beam profiles in the focus of a lens ($f = 10\,\text{cm}$) with a pupil radius of 1 mm.

P10.19. YAG Laser Capsulotomy Calculate for typical values of f_{obj} and f_{cl} ("cl" means capsulotomy lens) the spot size for a YAG laser beam assuming that the aperture of the objective lens is fully used. How long must the focus shift be to cover two thirds of the diameter of the backside capsule (same for front side capsule)? Use the parameters from Table 10.9.

P10.20. Laser-induced plasma Calculate the maximum expansion Δz_{max} of the laser-induced plasma for a laser pulse with constant energy with the following parameter set:

- Pulse length $\tau = 200$, 500, and 1000 fs
- Wavelength $\lambda = 1$ μm
- Beam waist $w_0 = 1$ μm

Assume that for $\tau = 1000$ fs we have $I_{peak} = I_{th}$. Also calculate the total energy deposition.

References

1 Meyer-Schwickerath, G. (1949) Koagulation der Netzhaut mit Sonnenlicht. *Ber. Dtsch. Ophthalmol. Ges.*, **55**, 256.

2 Palanker, D.V., Blumenkranz, M.S., and Marmor, M.F. (2011) Fifty years of ophthalmic laser therapy. *Arch. Ophthalmol.*, **129**, 1613–1619.

3 Krueger, R.R., Rabinowitz, Y.S., and Binder, P.S. (2010) The 25th anniversary of excimer lasers in refractive surgery: Historical review. *J. Refract. Surg.*, **26**, 749–760.

4 Bressler, N.M. (2004) Age-related macular degeneration is the leading cause of blindness. *J. Am. Med. Assoc.*, **291**, 1900–1901.

5 Pascolini, D. and Mariotti, S.P. (2012) Global estimates of visual impairment – 2010. *Br. J. Ophthalmol.*, **96**, 614–618.

6 Larsen, M., Schmidt-Erfurth, U., Lanzetta, P., Wolf, S., Simader, C.,

Tokaji, E., Pilz, S., and Weisberger, A. (2012) Verteporfin plus ranibizumab for choroidal neovascularization in age-related macular degeneration: Twelve-month MONT BLANC study results. *Ophthalmology*, **119**, 992–1000.
7 Berger, L., Wolf-Schnurrbusch, U., Brinkmann, C., and Wolf, S. (2010) Current indications for ocular photodynamic therapy – A review of the literature and two case reports. *Med. Laser Appl.*, **25**, 235–241.
8 Roider, J., Brinkmann, R., and Birngruber, R. (2003) Selective retinal pigment epithelium laser treatment – Theoretical and clinical aspects, in *Lasers in Ophthalmology – Basics, Diagnostic and Surgical Aspects*, Kugler Publications, pp. 119–129.
9 Stefànsson, E. (2003) From physical energy to biological effect: how retinal laser treatment affects diabetic retinopathy, in *Lasers in Ophthalmology – Basics, Diagnostic and Surgical Aspects*, Kugler Publications, pp. 73–78.
10 Mainster, M.A. (1986) Wavelength selection in macular photocoagulation. tissue optics, thermal effects, and laser systems. *Ophthalmology*, **93**, 952–958.
11 Framme, C., Roider, J., Brinkmann, R., Birngruber, R., and Gabel, V.P. (2008) Basic principles and clinical application of retinal laser therapy. *Klin. Monatsbl. Augenheilkd.*, **225**, 259–268.
12 Sivaprasad, S., Elagouz, M., McHugh, D., Shona, O., and Dorin, G. (2010) Micropulsed diode laser therapy: evolution and clinical applications. *Surv. Ophthalmol.*, **55**, 516–530.
13 Brinkmann, R., Roider, J., and Birngruber, R. (2006) Selective retina therapy (SRT): A review on methods, techniques, preclinical and first clinical results. *Bull. Soc. Belge Ophthalmol.*, **302**, 51–69.
14 Ellex (2012), Retinal rejuvenation therapy. http://www.ellex.com/assets/files/2rt/Ellex2RT_mediakit_iss2.pdf (accessed 8 August 2013).
15 Wood, J.P., Plunkett, M., Previn, V., Chidlow, G., and Casson, R.J. (2011) Nanosecond pulse lasers for retinal applications. *Lasers Surg. Med.*, **43**, 499–510.
16 Brinkmann, R., Koinzer, S., Schlott, K., Ptaszynski, L., Bever, M., Baade, A., Luft, S., Miura, Y., Roider, J., and Birngruber, R. (2012) Real-time temperature determination during retinal photocoagulation on patients. *J. Biomed. Opt.*, **17**, 061219.
17 Fuisting, B. and Richard, G. (2010) Transpupillary thermotherapy (TTT) – review of the clinical indication spectrum. *Med. Laser Appl.*, **25**, 214–222.
18 Meyer, J.J. and Lawrence, S.D. (2012) What's new in laser treatment for glaucoma? *Curr. Opin. Ophthalmol.*, **23**, 111–117.
19 Lin, S.C. (2008) Endoscopic and transscleral cyclophotocoagulation for the treatment of refractory glaucoma. *J. Glaucoma*, **17**, 238–247.
20 OD-OS (2012) Navilas® – A new era in retinal disease management. http://www.od-os.com (accessed 28 August 2013).
21 Niemz, M.H. (2007) *Laser-Tissue Interactions: Fundamentals and Applications*, Springer, 3rd edn.
22 Huang, D. (2004) Physics of customized corneal ablation, in *Wavefront Customized Correction: The Quest for Supervision II*, Slack Inc., Thorofare, pp. 171–180.
23 Dai, G. (2008) *Wavefront Optics for Vision Correction*, SPIE Bellingham.
24 Mrochen, M., Koller, T., and Seiler, T. (2011) Ablationsprofile, in *Refraktive Chirurgie*, Springer, pp. 98–113, 1st edn.
25 Munnerlyn, C.R., Koons, S.J., and Marshall, J. (1988) Photorefractive keratectomy: A technique for laser refractive surgery. *J. Cataract Refract. Surg.*, **14**, 46–52.
26 Mrochen, M., Donitzky, C., Wüllner, C., and Löffler, J. (2004) Wavefront-optimized ablation profiles: theoretical background. *J. Cataract Refract. Surg.*, **30**, 775–785.
27 Born, M. and Wolf, E. (1997) *Principles of Optics: Electromagnetic Theory of Propagation, Interference and Diffraction of Light*, Cambridge University Press.
28 Huang, D. and Arif, M. (2002) Spot size and quality of scanning laser correction

of higher-order wavefront aberrations. *J. Cataract Refract. Surg.*, **28**, 407–416.

29 Yee, K. (2004) Active eye tracking for excimer laser refractive surgery, in *Aberration-Free Refractive Surgery*, Springer, pp. 125–140.

30 Taneri, S., Weisberg, M., and Azar, D.T. (2011) Surface ablation techniques. *J. Cataract Refract. Surg.*, **37**, 392–408.

31 Kohnen, T., Knorz, M., and Neuhann, T. (2007) Bewertung und Qualitätssicherung refraktiv-chirurgischer Eingriffe durch die DOG und den BVA. *Der Ophthalmologe*, **104**, 719–726.

32 Schaumberg, D.A., Dana, M.R., Christen, W.G., and Glynn, R.J. (1998) A systematic overview of the incidence of posterior capsule opacification. *Ophthalmology*, **105**, 1213–1221.

33 Aron-Rosa, D., Aron, J.J., Griesemann, M., and Thyzel, R. (1980) Use of the neodymium-YAG laser to open the posterior capsule after lens implant surgery: A preliminary report. *J. Am. Intraocul. Implant Soc.*, **6**, 352–354.

34 Gross, H., Bauer, H., Lasser, T., and Wiechmann, M. (1995) Unstable laser resonator for generating a stable fundamental mode beam profile, Patent US 5406578 A.

35 Ratkay-Traub, I., Juhasz, T., Horvath, C., Suarez, C., Kiss, K., Ferincz, I., and Kurtz, R. (2001) Ultra-short pulse (femtosecond) laser surgery: Initial use in LASIK flap creation. *Ophthalmol. Clin. North Am.*, **14**, 347–355.

36 Vogel, A. (2003) Mechanism of short-pulsed plasma-mediated laser ablation and disruption, in *Lasers in Ophthalmology – Basics, Diagnostic and Surgical Aspects*, Kugler Publications, pp. 99–113.

37 Vogel, A., Linz, N., and Freidank, S. (2008) Method for laser machining transparent materials, Patent US 2010/0163540 A1.

38 Perry, M.D. and Mourou, G. (1994) Terawatt to petawatt subpicosecond lasers. *Science*, **264**, 917–924.

39 Vaddavalli, P.K. and Yoo, S.H. (2011) Femtosecond laser in-situ keratomileusis flap configurations. *Curr. Opin. Ophthalmol.*, **22**, 245–250.

40 Soong, H.K. and Malta, J.B. (2009) Femtosecond lasers in ophthalmology. *Am. J. Ophthalmol.*, **147**, 189–197.

41 Kim, P., Sutton, G.L., and Rootman, D.S. (2011) Applications of the femtosecond laser in corneal refractive surgery. *Curr. Opin. Ophthalmol.*, **22**, 238–244.

42 Sekundo, W., Kunert, K., Russmann, C., Gille, A., Bissmann, W., Stobrawa, G., Sticker, M., Bischoff, M., and Blum, M. (2008) First efficacy and safety study of femtosecond lenticule extraction for the correction of myopia: Six-month results. *J. Cataract Refract. Surg.*, **34**, 1513–1520.

43 Shah, R., Shah, S., and Sengupta, S. (2011) Results of small incision lenticule extraction: All-in-one femtosecond laser refractive surgery. *J. Cataract Refract. Surg.*, **37**, 127–137.

44 Shah, S. and Shah, M. (2013) ReLEX – A journey towards no blade no excimer no flap femtosecond laser vision correction, in *Ready Reckoner of Modern Refractive Surgery*, AJR Medi Solutions, New Delhi, pp. 201–209.

45 Farid, M. and Steinert, R.F. (2010) Femtosecond laser-assisted corneal surgery. *Curr. Opin. Ophthalmol.*, **21**, 288–292.

46 Ruiz, L.A., Cepeda, L.M., and Fuentes, V.C. (2009) Intrastromal correction of presbyopia using a femtosecond laser system. *J. Refract. Surg.*, **25**, 847–854.

47 Uy, H.S., Edwards, K., and Curtis, N. (2012) Femtosecond phacoemulsification: the business and the medicine. *Curr. Opin. Ophthalmol.*, **23**, 33–39.

48 He, L., Sheehy, K., and Culbertson, W. (2011) Femtosecond laser-assisted cataract surgery. *Curr. Opin. Ophthalmol.*, **22**, 43–52.

49 Krueger, R.R, Talamo, J.H., Lindstrom, R.L. (eds) (2013) *Textbook of Refractive Laser Assisted Cataract Surgery (ReLACS)*, Springer.

50 Lubatschowski, H., Schumacher, S., Fromm, M., Wegener, A., Hoffmann, H., Oberheide, U., and Gerten, G. (2010) Femtosecond lentotomy: Generating gliding planes inside the crystalline lens to regain accommodation ability. *J. Biophotonics*, **3**, 265–268.

51 Bille, J.F., Sahler, R., Aguilera, R., Zhou, S., and Schanzlin, D.J. (2011) Generation and in situ modification of customized IOLs, in *ASCRS Symposium on Cataract, IOL and Refractive Surgery*, www.aarenic.com/pdfs/bille_ris.pdf (accessed 28 August 2013).

52 Chan, W.M., Lim, T.H., Pece, A., Silva, R., and Yoshimura, N. (2010) Verteporfin PDT for non-standard indications – a review of current literature. *Graefes Arch. Clin. Exp. Ophthalmol.*, **248**, 613–626.

53 Kohnen, T. (ed.) (2011) *Refraktive Chirurgie*, Springer.

54 Krueger, R.R., Applegate, R.A., and MacRae, S.M. (eds) (2004) *Wavefront Customized Visual Correction: The Quest for Super Vision II*, Slack Inc., Thorofare.

55 Kugler, L.J. and Wang, M.X. (2010) Lasers in refractive surgery: History, present, and future. *Appl. Opt.*, **49**, F1–F9.

56 American National Standards Institute (2007) Laser systems for corneal reshaping (ANSI Z80.11-2007).

57 Krauss, M. (2003) Principles of photodisruption, in *Lasers in Ophthalmology – Basics, Diagnostic and Surgical Aspects*, Kugler Publications, pp. 303–313.

A
Basics of Optics

The concepts of the ophthalmic and optometric devices presented in this book can only be understood with a well-founded knowledge in geometric and wave optics. Hence, we append this chapter to ensure a common knowledge base for all readers. The intention was not to present the content of an entire textbook of optics. However, the following sections cover all topics that are relevant for the discussions in the ophthalmic system parts of this book.

Although "light" is something we are familiar with in our daily lives, its nature is hard for us to imagine. Light is nothing we can directly touch or manipulate. We can only learn about its properties via its interactions with matter.

A first elaborate description of light was developed in the seventeenth century. Classical geometric optics (ray optics) describes the propagation of light through and around objects by rays. In this model, light behaves like a projectile (or particle) whose flight path is influenced by interactions with media. So, this model focuses on the location and direction of light rays, while the intrinsic properties of light are ignored.

But already in the seventeenth century interference and diffraction effects had been observed in experiments and could not be explained in terms of geometric optics. Scientists like Christiaan Huygens (1629–1659) proposed that light has a wave-like character. With this approach, the laws of reflection and refraction could be explained. Wave optics was further improved by Thomas Young (1773–1829) and Augustin Fresnel (1788–1827) who introduced the interference principle for superposition of waves.

In the nineteenth century, Michael Faraday (1791–1867) and James Maxwell (1831–1879) extended the wave model. They discovered that light is an electromagnetic wave. This had far-reaching consequences particularly for the explanation of light–matter interaction. In this way, for instance, it was possible to explain the wavelength dependence of refraction. By this theory, visible (VIS) light became essentially the same physical phenomenon as microwaves, infrared (IR) radiation, ultraviolet (UV) radiation, and X-rays. The only difference is the corresponding wavelength. In 1888, the electromagnetic wave theory was experimentally verified by Heinrich Hertz (1857–1894).

Optical Devices in Ophthalmology and Optometry, First Edition. Michael Kaschke, Karl-Heinz Donnerhacke, and Michael Stefan Rill.

It is quite interesting that the question of whether light is a particle or a wave is a recurring topic in history. Within the framework of quantum theory, Albert Einstein (1879–1955) proposed that light must have a momentum like small "solid" particles (photons). This assumption was based on the photoelectric effect which was a puzzling experiment that seemed to contradict the classic wave concept. In the early twentieth century, it became evident that the apparent dichotomy of the particle and wave descriptions in the macroscopic world is removed on the atomic scale. This is also known as the "wave-particle duality". Quantum optics reveals that an intuitive understanding of light's nature is beyond our imagination.

Four different approaches therefore exist to describe the behavior of light in optical systems, that is, geometric, wave, electromagnetic, and quantum optics. Depending on the given problem or application, we have to choose the adequate formalism which is most intuitive and feasible to describe the considered problem or application. For almost all purposes, geometric optics (Section A.1) and wave optics (Section A.2) are sufficient to understand most topics in ophthalmology and optometry. However, photons play a major role in the context of lasers (Chapters 9, 10, and B).

A.1 Geometric Optics and Optical Imaging

In geometric optics, light is treated as a bundle of rays which travels in optical media and obeys a set of geometric laws. As an example, let us consider a transparent, flat glass slide which is embedded in air as shown in Figure A.1a. When a light

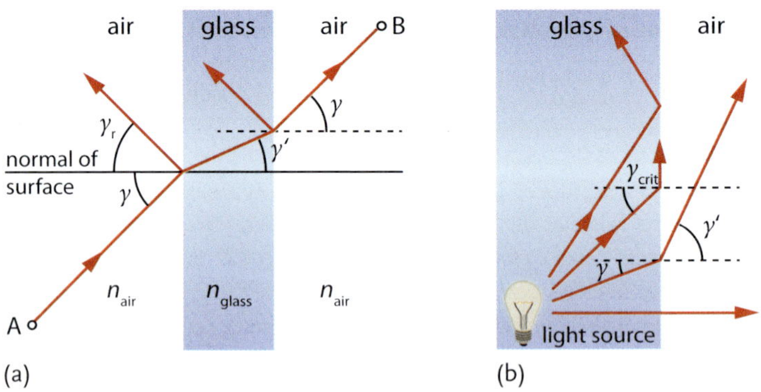

Figure A.1 (a) Reflection, propagation, and transmission of a light ray which is incident from air on a flat glass slide. The ray is deflected at the air–glass interface towards the normal of the glass surface due to refraction. When the ray exits the medium, it is again deflected, but this time away from the normal of the surface. (b) Let us consider a light source inside glass. For $\gamma < \gamma_{crit}$, the ray is again deflected as in (a). For $\gamma = \gamma_{crit}$, the ray is deflected such that it travels along the glass surface. If $\gamma > \gamma_{crit}$, we have total reflection at the air–glass interface, that is, the ray does not leave the glass volume anymore. Instead, it bounces back and forth between both interfaces.

beam is incident on the left air–glass interface at an angle γ, some of it is reflected at an angle of

$$\gamma_r = \gamma \, , \tag{A1}$$

while the rest enters the glass body. The incoming, transmitted, and reflected rays all lie in the plane of incidence. The ray which enters the body now also changes its direction ($\gamma' \neq \gamma$). This effect is referred to as *refraction*. At the other interface, the ray is again partly reflected and partly transmitted. The transmitted portion is deflected to the other side so that it travels again at the angle γ.

A.1.1
Refraction and Dispersion

Why do incident rays change their direction at both air–glass interfaces? Obviously, it has something to do with the inherent properties of the medium. Here, Fermat's[1] principle gives a possible clue: light rays always choose the path which can be traversed in the least time. Keeping this in mind and looking again at Figure A.1a, we see that the light ray tries to minimize the traveling time in glass. Hence, glass seems to slow down the light rays.[2] As a consequence, the fastest route from point A to point B requires the angle of travel to be changed (in our example: $\gamma' < \gamma$).

The highest speed of light is found in vacuum, where light is as fast as $c_0 = 299\,792\,458$ m/s. For all other media, no matter if they are gaseous, liquid, or solid, the speed decreases corresponding to the related refractive index n, that is, a dimensionless number greater than one. In an arbitrary medium with refractive index n_m, the speed of light is $c_m = c_0/n_m$. For example, we have $c_{glass} = c_0/1.52 \approx 2 \times 10^8$ m/s in glass.

The angles of travel outside and inside a medium (γ and γ', respectively) are directly related to the refractive index via Snell's[3] law. For the general case of an arbitrary medium (primed parameters) embedded in a host medium, Snell's law reads

$$n \sin\gamma = n' \sin\gamma' \, . \tag{A2}$$

As long as γ is smaller than a critical angle

$$\gamma_{crit} = \arcsin\left(\frac{n}{n'}\right) \, , \tag{A3}$$

the situation is similar to the example given in Figure A.1a. For $\gamma = \gamma_{crit}$, the refracted ray is exactly tangent to the interface. And if the angle of incidence exceeds the critical angle, the ray can no longer enter the adjacent medium. In the example of Figure A.1b, the ray is kept inside the glass slide in that it is reflected back at

1) Pierre de Fermat (1607–1665).
2) The reason why the speed of light c is altered in a medium can only be explained by atomic physics.
3) Willebrord van Roijen Snell (1580–1626).

both interfaces without "losing" intensity. This so-called *total (internal) reflection* is actually a lossless effect which is used for optical fibers to transfer light over long distances (Section 10.2.4.2).

At the interface of two different media, the refractive index determines not only the ray deflection, but also the fraction of light that is transmitted and reflected. For a normal incident light ray ($\gamma = 0°$) and neglected absorption (Section 9.1), the reflected fraction (reflectance) is given by

$$R = \left(\frac{n - n'}{n + n'}\right)^2 \tag{A4}$$

and the transmitted fraction (transmittance) by

$$T = \frac{4nn'}{(n + n')^2} \,. \tag{A5}$$

The relations (A4) and (A5) are referred to as the *Fresnel equations.*[4] A more general form of the Fresnel equations for oblique incidence can be found in Section A.2.1.4.

Summing up, refraction occurs at every interface between two media with different refractive indices. The refractive index is in turn a function of the wavelength (color) of the incident light, which is referred to as *dispersion*. This phenomenon is caused by interactions (scattering and absorption) of light with atoms or molecules and is thus specific for each material.

A well-known experiment to show the effect of dispersion is illustrated in Figure A.2, where a beam of white light shines onto a glass prism. Besides the overall refraction, the incident beam is split into the colors of the rainbow. Since the refractive index depends on the wavelength λ (Figure A.3), we have different refraction angles for each wavelength. As white light is composed of all visible colors, rays with different wavelengths are fanned out by the prism and eventually hit the wall at different positions. A medium whose refractive index decreases with increasing wavelengths ($\mathrm{d}n/\mathrm{d}\lambda < 0$) is said to have *normal* dispersion (as shown in Figure A.3), whereas the contrary case ($\mathrm{d}n/\mathrm{d}\lambda > 0$) is said to have *anomalous* dispersion.

Shining the light of an arc lamp onto a prism does not lead to a continuous color fan. In this case, we rather see discrete spectral lines which are separated by totally dark regions. Some of these material-specific emission lines are used to define a measure for dispersion in the visible spectral range, that is, the so-called *Abbe number*[5]

$$\nu = \frac{n_\mathrm{e} - 1}{n_{\mathrm{F}'} - n_{\mathrm{C}'}} \,. \tag{A6}$$

4) Both the reflectance R and transmittance T are numbers between 0 and 1. In case of $R = 0$ ($T = 0$), absolutely no light is reflected (transmitted) at the interface of the considered media. In contrast, the incident ray is completely reflected (transmitted) if $R = 1$ ($T = 1$). In the absence of absorption, we have $R + T = 1$.

5) Ernst Abbe (1840–1905).

Figure A.2 A typical experiment to demonstrate dispersion: white light which contains all color components of the visible spectrum is split into the colors of the rainbow by a glass prism.

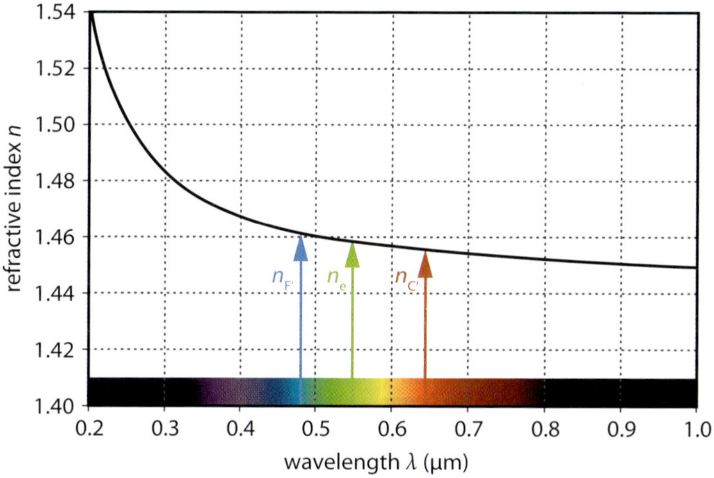

Figure A.3 Refractive index of fused silica (SiO_2) in the near-infrared, visible, and ultraviolet spectral range (color spectrum is shown for reference). The refractive index $n(\lambda)$ changes considerably from violet (e.g., at 0.4 μm = 400 nm) to near-infrared wavelengths (e.g., at 1.0 μm = 1000 nm). To determine the Abbe number of fused silica, n is measured for certain wavelengths: $n(\lambda_{F'}) = 1.4638$, $n(\lambda_e) = 1.4601$, $n(\lambda_{C'}) = 1.4567$. From Eq. (A6), we obtain the Abbe number $\nu_{silica} = 64.8$.

$n_{F'}$, n_e, and $n_{C'}$ correspond to the refractive indices of a considered medium at wavelengths $\lambda_{F'} = 480\,\text{nm}$, $\lambda_e = 546\,\text{nm}$, and $\lambda_{C'} = 644\,\text{nm}$, respectively.[6] Following Eq. (A6), we see that media with low dispersion have high values of ν. Abbe numbers are particularly used to classify different sorts of glass and other optically transparent materials ([1]; Problem PA.1).

6) These wavelengths correspond to spectral emission lines of mercury (λ_e) and cadmium ($\lambda_{F'}$, $\lambda_{C'}$) vapor lamps.

A.1.2
Imaging by Spherical Surfaces

So far, we have only considered the case of refraction at a plane surface. The situation becomes a little more complicated if the surface is curved. This is very relevant in practice as most standard optical systems can be regarded as a sequence of curved surfaces. For the sake of simplicity, we will initially limit ourselves to monochromatic light, that is, light with one wavelength, so that dispersion can be ignored. In more concrete terms, we will trace light rays which are incident on an embedded spherical medium with a refractive index $n' > n$, where n is the refractive index of the surrounding medium. A cross-section of the arrangement is shown in Figure A.4[7]. The rays are emanated in all directions from object point Q on the left. Here, we will concentrate on just one of these rays which forms an angle γ with the optical axis (horizontal symmetry axis). When the ray hits the spherical surface at point A, it is refracted by obeying Snell's law (A2) and crosses the optical axis again at point Q′. The spherical interface thus creates a point-to-point projection from Q to Q′ at a distance $|s| + |s'|$.

In the following, we will derive the basic equation which links the distances s and s' of the point-to-point imaging to the refractive indices n, n', and the radius of curvature r' of the spherical surface. From the triangle $\overline{Q'AC'}$ in Figure A.4, we

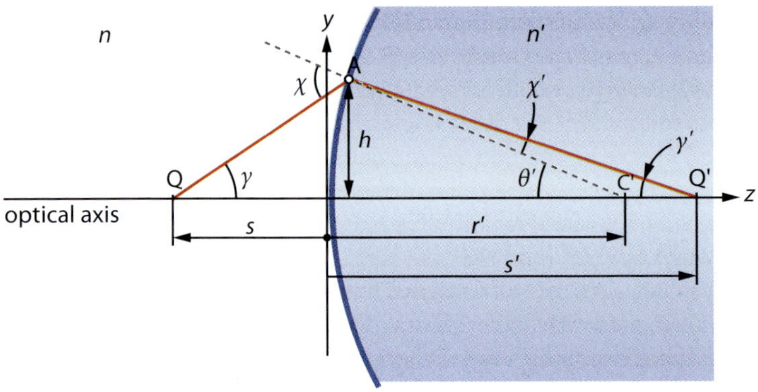

Figure A.4 Point-to-point projection by a spherical surface. A light ray is emanated from the on-axis object point Q at an angle γ. The considered space is divided by a spherical surface. On the left, the refractive index is n, while we have a different refractive index $n' > n$ on the right. An incident ray $\overline{QA}$ is thus refracted by obeying Snell's law. At point Q′, the ray $\overline{AQ'}$ crosses the optical axis again at an angle γ'. The center of curvature of the spherical surface is denoted as C′, and r' is the radius. s is the horizontal distance from Q to the surface boundary (object distance), and s' the distance from the boundary to Q′ (image distance). h is the vertical distance of the ray from the optical axis to the point of intersection A (ray height).

7) Primed variables describe optical design parameters and points within the image space, that is, between the first refracting surface and the image. Unprimed variables are used for the corresponding object space.

see that

$$\theta' = \gamma' + \chi' . \tag{A7}$$

It is useful to express the angle χ' in terms of χ. For this purpose, we will apply Snell's law to point A. Although Snell's law appears to be simple at first sight, the sine terms would make the following calculations quite complicated. For this reason, we will select a ray which travels close to the optical axis. This is the so-called *paraxial approximation* of small angles which allows substitution of $\sin\gamma$ and $\tan\gamma$ by γ. Analogous relations also hold for χ, χ', γ', and θ'. For these paraxial rays, we may use the simplified Snell's law of refraction given by $n\chi \approx n'\chi'$. The relation (A7) now becomes

$$\theta' \approx \gamma' + \frac{n}{n'}\chi . \tag{A8}$$

In addition, from the triangle $\overline{\mathrm{QAC'}}$ we deduce

$$\chi = \gamma + \theta' , \tag{A9}$$

which yields

$$\theta'(n' - n) = n\gamma + n'\gamma' . \tag{A10}$$

Next, we retrieve image distance s', object distance s, and height h from the angles. For this purpose, we use the paraxial approximations $\theta' \approx h/r'$, $\gamma \approx -h/s$, and $\gamma' \approx h/s'$ so that we obtain[8)]

$$\frac{n'}{s'} - \frac{n}{s} = \frac{n' - n}{r'} . \tag{A11}$$

For a given object distance s, the image distance s' thus no longer depends on angle γ' in the case of paraxial approximation. The important consequence of this result is that *all* paraxial rays emanated from Q intersect at the same point Q′. Point Q′ is referred to as the *image point*[9)] (Problem PA.2).

A.1.2.1 **Thin Lenses**

In the next step, we add a second spherical surface to the system, as shown in Figure A.5. We assume that the centers of both spheres lie on the optical axis so that the whole system is rotationally symmetric. If the sum of radii $|r_1| + |r_2|$ is greater than the center-to-center distance, the intersecting volume forms a positive,

8) Distances from the spherical surface to the left (i.e., towards point Q) are set to negative values. Distances from the spherical surface to the right (i.e., towards point Q′) are set to positive values. This sign definition is represented by the pointing direction of the arrows.

9) Please keep in mind that such a definite point-to-point projection does *not* work for rays which form large angles with the optical axis. Since the paraxial approximation cannot be applied in such a case, s' and the location of Q′ vary for every ray. As a result, imaging errors (aberrations) appear which we will examine later in Section A.1.6.

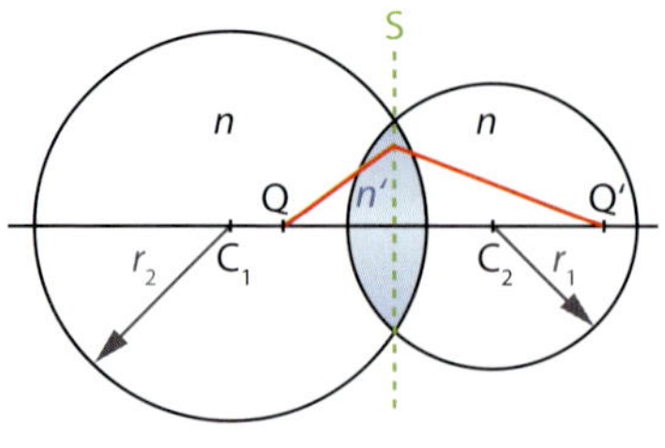

Figure A.5 Geometric formation of a lens by two spherical volumes. For $|r_1| + |r_2| > \overline{C_1C_2}$, a positive (biconvex) lens is formed by the intersecting volume of both spheres (blue area). To observe refraction at the interface, the refractive index of the intersecting volume n' must not be equal to the refractive index of the exterior volume n.

biconvex lens. We further assume that the lens is "thin" so that an incident ray emerges at about the same height at which it enters the lens. In terms of geometric design, we consider the thin lens as a flat refracting sheet S (green dashed line in Figure A.5). The imaging equation of a thin lens, also known as the *lens maker's equation* (Problem PA.3), is given by

$$\frac{1}{s'} - \frac{1}{s} = \frac{n' - n}{n} \left(\frac{1}{r_1} - \frac{1}{r_2} \right) . \tag{A12}$$

The derivation of Eq. (A12) can be performed either by using the matrix approach to be discussed in Section A.1.3 or by applying Eq. (A11) to both spherical surfaces (Problem PA.4).

The rules of imaging are not only valid for an on-axis object point. We can also apply them to extended objects as long as the paraxial approximation is valid. In this context, it is helpful to regard an extended object as a collection of object points which are distributed around the optical axis (off-axis) and all lie in one common object plane O. We define h_O as the object height. For the optical imaging, we select a single point from the object plane and trace all emanated paraxial rays. As before, the rays converge at a respective image point after passage through the lens. If we do this for all the other object points, we will recognize that the whole collection of image points lies in a common image plane I.[10] The largest distance of the image points from the optical axis is defined as the image size h'_I. On the whole, we replaced the former point-to-point by a plane-to-plane imaging. This means that we may see a sharp image of an extended object by placing a flat screen in the image plane (Figure A.6).

For the optical calculation of extended objects, we select the outermost object point O_1 in Figure A.6 and trace the first ray which is parallel to the optical axis (i.e., ray 1). After refraction by the lens at A_1, the ray crosses the optical axis at focal point F′. The second ray starts from O_1 as well, but passes through the lens center C under oblique incidence without any deflection from the straight traveling path. The position of image plane I is then determined by the intersection of both rays. The geometric design allows the roles of object plane and image plane to be

10) This is again only true for paraxial approximation.

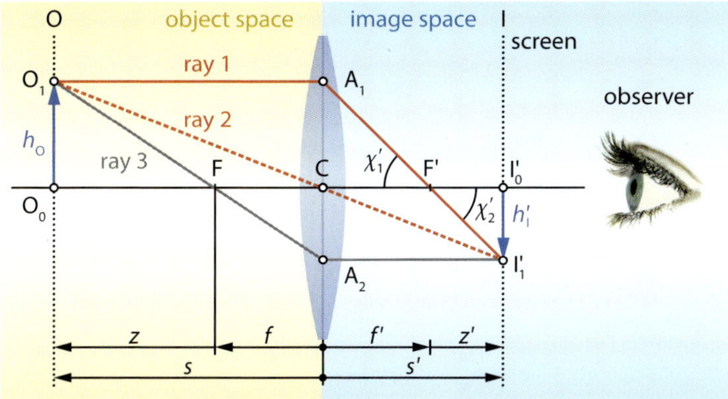

Figure A.6 Optical imaging of an extended object using a thin lens. For the geometric design, we consider two specific rays. Ray 1 travels parallel to the optical axis at a distance h_O (which is effectively the height of the considered object). After refraction by the lens, ray 1 crosses the optical axis on the image side at focal point F′. Ray 2 has the same origin as ray 1, but is obliquely incident on the lens surface. As a consequence, ray 2 passes through the center of the lens C and, simultaneously, crosses the optical axis without being deflected. Image plane I is determined by the geometric intersection of both rays.

exchanged. This brings us to ray 3 (gray line in Figure A.6). The corresponding object-side focal point is represented by F.

As paraxial image formation by lenses is based on elementary geometric rules, one can easily translate the geometric design to corresponding imaging equations. A closer look at Figure A.6 reveals that the triangles $\overline{\mathrm{FI}_1'\mathrm{I}_0'}$ and $\overline{\mathrm{F}'\mathrm{A}_1\mathrm{C}}$ are similar. This is because χ_1' and its opposite angle χ_2' are equal. In addition, the triangles $\overline{\mathrm{O}_0\mathrm{CO}_1}$ and $\overline{\mathrm{I}_0'\mathrm{CI}_1'}$ are similar[11)] so that

$$\frac{h_I'}{h_O} = \frac{s' - f'}{f'} = \frac{s'}{s} \tag{A13}$$

from which we derive the *lens equation*

$$\frac{1}{s'} - \frac{1}{s} = \frac{1}{f'} \,. \tag{A14}$$

f' is focal length between lens center C and focal point F′. In ophthalmology and optometry, f' is often replaced by the reciprocal refractive power $\mathcal{D}' = n'/f'$, where n' is the refractive index of the medium on the image side. Note that the refractive power has the dimension of inverse meters (diopters) $[\mathcal{D}] = \mathrm{D} = 1/\mathrm{m}$.

We can also calculate the ratio between object and image size, which is referred to as the *magnification*

$$\beta = \frac{s'}{s} = \frac{h_I'}{h_O} \tag{A15}$$

11) Distances from the symmetry plane of the lens to the left (i.e., towards object plane O) are set to negative values. Distances from the symmetry plane of the lens to the right (i.e., towards image plane I) are set to positive values. This is also illustrated by the pointing direction of the arrows.

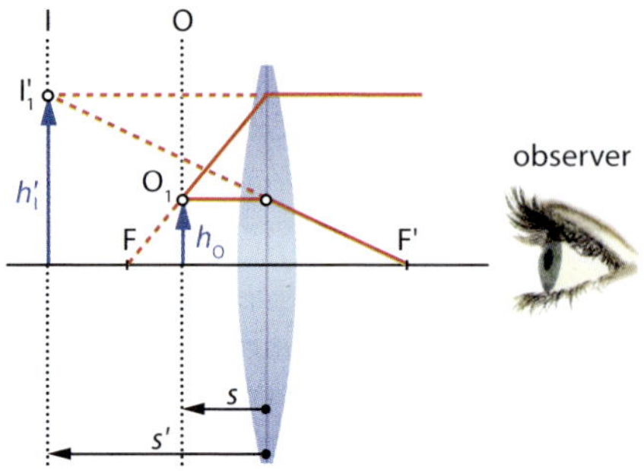

Figure A.7 If the object is placed between object-side focal point F and lens, an upright virtual image is formed which is larger than the object ($h_I' > h_O$).

of the optical system. β is positive if both object and image are upright, and negative if the image is flipped upside down. For example, we see a flipped image in Figure A.6 which is smaller than the original object. In this case, the magnification β is a number between -1 and 0. The image is said to be *real*, since we could place a screen in image plane I to observe a sharp image of the object.

For some magnifying visual devices (magnifying loupes, telescopes, and so on), the object is placed between object-side focal point F and lens (Figure A.7). In contrast to Figure A.6, the parallel and focal rays diverge on the image side. Thus, no sharp image is formed there. But when the ray paths are extended in the reverse direction (dashed lines), an upright *virtual* image is obtained that appears to be located on the object side. In contrast to real images, virtual images can only be detected by using an optical imaging system (e.g., eye or camera).

A.1.2.2 **Thick Lenses**

At the beginning of our discussion about thin lenses (Section A.1.2.1), we assumed that all incident rays emerge from the lens at about the same height at which they entered it. But if the lens thickness L is comparable to its radii of curvature, that is, a "thick" lens, we have to take a closer look at the ray deflection at each interface. In Figure A.8a, we trace an incident ray which is parallel to the optical axis. This ray is refracted at point A by the left lens surface such that it emerges at point B. Refraction at the right lens surface changes the traveling path of the ray again so that it eventually crosses image-side focal point F′. In practice, we usually do not care about the exact ray path inside the lens. Instead, we are interested in the effective change in direction. Therefore, it is sufficient to use a simplified geometric design which gives the same results. For this purpose, the incoming ray path is extended in a forward direction (orange line in Figure A.8a) and the outgoing ray path in a backward direction (green line). The point at which both extended ray paths intersect lies on the image-side principal plane K′. At this plane, the incident parallel rays are "effectively" refracted to the focal point. Accordingly, if an incident light ray crosses the object-side focal point F at an oblique angle, it is "effectively" refracted at object-side principal plane K. Thus, it emerges parallel to the optical axis on the image side. For positive lenses, the positions of the corresponding prin-

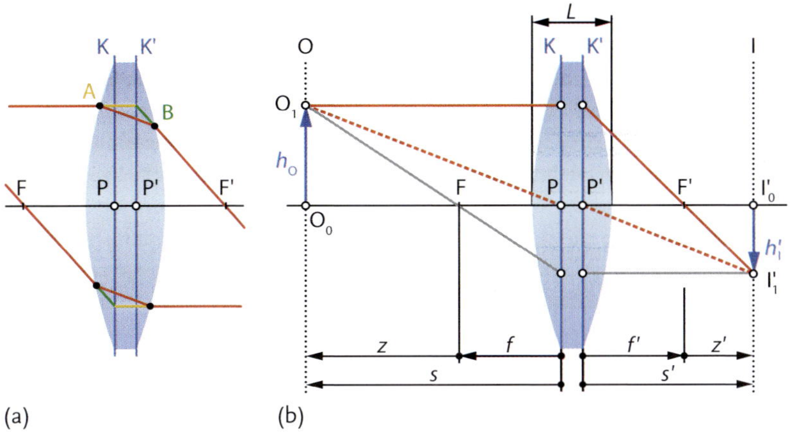

Figure A.8 Imaging with a thick spherical lens. (a) Usually, refraction would lead to deflection of rays as shown for the red path. However, extending the ray path outside the lens results in the orange and green lines. Their points of intersection define the principal planes K and K′. By design, we may thus consider refraction as a ray deflection which occurs at principal planes. (b) Imaging diagram of an extended object by a thick lens.

cipal points P and P′ are determined by the object- and image-side focal lengths $f = \overline{\mathrm{O_0P}} - \overline{\mathrm{O_0F}} < 0$ and $f' = \overline{\mathrm{O_0F'}} - \overline{\mathrm{O_0P'}}$, respectively. The imaging equation of a thick lens is given by

$$\frac{1}{s'} - \frac{1}{s} = \frac{n' - n}{n}\left(\frac{1}{r_1} - \frac{1}{r_2} + \frac{(n' - n)L}{nn'r_1r_2}\right), \qquad \text{(A16)}$$

where L is the lens thickness, n' the refractive index of the lens medium (usually glass), and n the refractive index of the exterior medium. r_1 and r_2 denote the object-side and image-side lens radii, respectively. For $L = 0$, Eq. (A16) passes into the lens maker's equation (A12) (Problem PA.3).

A.1.2.3 **Types of Lenses**

So far, we have explored positive, biconvex lenses formed by the *intersecting* volume of two spheres. The added radii of both spheres are greater than the distance between sphere centers (Figure A.9a). For positive lenses, parallel incident light rays converge at focal point F′ (or F in the reverse direction). A similar behavior is also found for plano-convex and meniscus-shaped lenses. Due to their differing geometry and refractive power, the object-side and image-side principal planes K and K′ have to be shifted as shown in Figure A.9b,c. All of these types of positive lenses bend light rays towards the optical axis, and they are said to have a *positive* refractive power.

In addition to positive lenses, concave or *negative* lenses (Figure A.9d,e) also exist. They can be designed by virtual connection of two *spatially separated* spheres (Figure A.10). Negative lenses cause an incident bundle of parallel rays to diverge after passage, and converging rays are made less convergent. Converging rays become parallel rays if their point of convergence coincides with the focal point of the

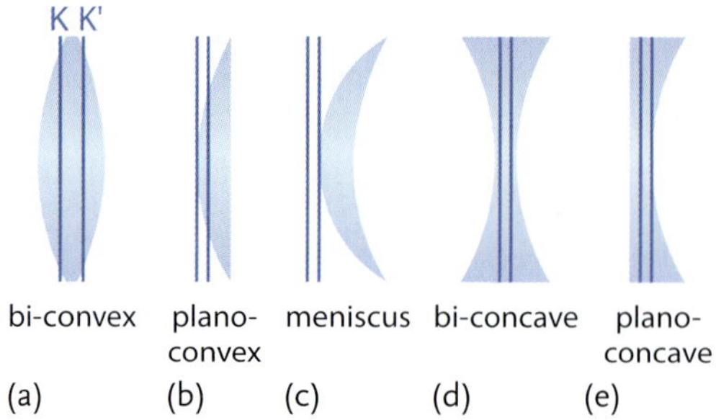

Figure A.9 Various types of positive and negative lenses. (a) Biconvex, (b) plano-convex, and (c) meniscus lenses refract incident light in basically the same manner. But due to their differing geometry, the position of the principal planes K and K′ must be adapted. (d) Biconcave and (e) plano-concave lenses.

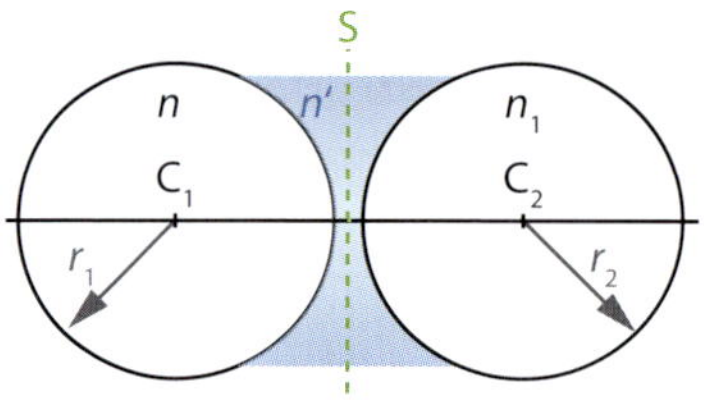

Figure A.10 Geometric formation of a negative, biconcave lens by "virtual" connection of two spatially separated spherical volumes ($|r_1| + |r_2| < \overline{C_1 C_2}$). For thin negative lenses, refraction effectively occurs at the flat refracting sheet S (see also Figure A.5).

lens. The lens equation (A14) can also be used for negative lenses when we choose a negative focal length $f' < 0$. With a single negative lens, only virtual images can be formed.

A.1.3
The Ray Tracing Approach to Paraxial Optical Systems

We will now continue with somewhat more complex optical systems like a set of thin lenses. For this purpose, it is sufficient to restrict ourselves to systems for which all optical components are centered at the optical axis. The imaging rules that we derived for a single lens can thus also be applied to more complex optical systems. For example, let us consider a simple microscope which consists of two positive lenses, as shown in Figure A.11. The lens next to the object forms an image that is either real or virtual. This image now serves as an object for the next lens, which forms another image. In general, we can thus simply repeat the imaging procedure for an arbitrary number of lenses.

Optical devices in ophthalmology normally require more than just one or two lenses. They may also contain curved mirrors or combinations of negative and positive lenses. A "brute-force" calculation of such optical systems may become a difficult and time-consuming undertaking. For this reason, more sophisticated methods are required which can also be used for computer-aided simulations. We already realized that every ray emanated by an extended object is characterized by two values: ray height h and angle γ. Every component of an optical system modifies these quantities. Thus, we will trace a ray which travels along the z axis and is incident on an arbitrary optical component (Figure A.12). If it is a paraxial

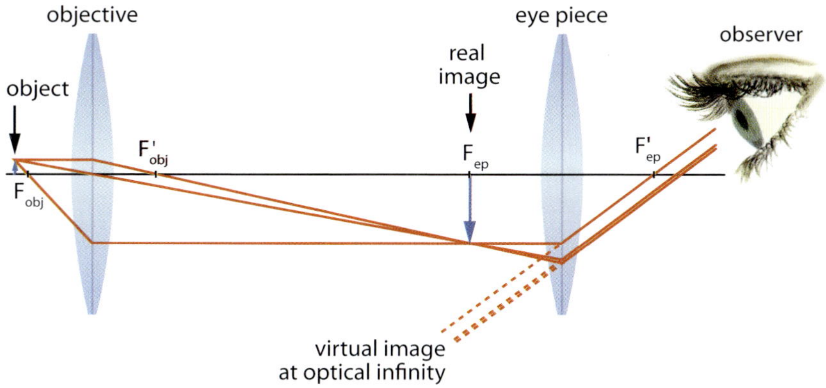

Figure A.11 Magnified imaging by two lenses. Since the image plane of the right lens (eye piece) is on the object side, the resulting virtual image (here situated at optical infinity) is magnified. Continuous lines represent the traveling path of real light rays. Dashed lines represent "virtual" rays which are emanated from the virtual image focused by the observer.

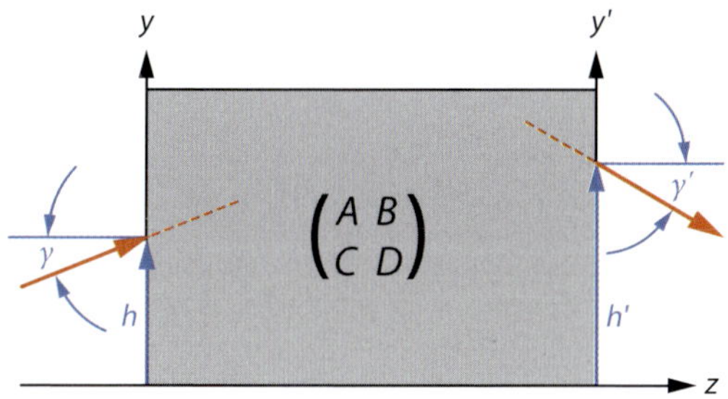

Figure A.12 Underlying concept of the ray tracing approach. The optical path of a ray is parameterized by ray height h and the angle relative to the optical axis γ. A, B, C, and D are functions uniquely defined for optical components (Table A.1) which determine the deviations from a straight path. Adapted from [2].

ray, the sine terms in Snell's law can be replaced by the angle itself ($\sin\gamma \approx \gamma$). In this case, h, h', γ, and γ' are directly related via

$$h' = Ah + B\gamma\,, \tag{A17}$$

$$\gamma' = Ch + D\gamma\,. \tag{A18}$$

Here, we introduced the parameters A, B, C, and D which describe how the optical component acts on an incident ray.[12] In the next step, we combine Eqs. (A17) and (A18) in a matrix operator equation such that

$$\begin{pmatrix} h' \\ \gamma' \end{pmatrix} = \begin{pmatrix} A & B \\ C & D \end{pmatrix} \begin{pmatrix} h \\ \gamma \end{pmatrix} \equiv \underline{\mathbf{M}} \begin{pmatrix} h \\ \gamma \end{pmatrix}\,. \tag{A19}$$

12) For matrix calculations, we use a different sign convention than used in Section A.1.2. Rays pointing downwards are described by an angle with negative value.

Table A.1 ABCD matrices of several optical components. n' denotes the refractive index of the refracting medium (e.g., lens material). n is the refractive index of the exterior medium through which the *incident* light rays travel, and n'' is the refractive index of the exterior medium behind the optical system through which the outgoing rays travel. L denotes the slab or lens thickness, f' the focal length, and r_i the radii of curvature ($r > 0$ for convex surfaces; $r < 0$ for concave surfaces). For media with quadratic radial index profiles: $n(r) = n_0(1 - \varepsilon^2 r^2)$, $\varepsilon =$ const. Regarding the ABCD matrix of an afocal telescope, please refer to Problem PA.4.

Optical component	ABCD matrix
Free space (propagation)	$\underline{\mathbf{M}} = \begin{pmatrix} 1 & L \\ 0 & 1 \end{pmatrix}$
Slab (propagation)	$\underline{\mathbf{M}} = \begin{pmatrix} 1 & L/n' \\ 0 & 1 \end{pmatrix}$
Flat surface (refraction)	$\underline{\mathbf{M}} = \begin{pmatrix} 1 & 0 \\ 0 & n/n' \end{pmatrix}$
Spherical surface (refraction)	$\underline{\mathbf{M}} = \begin{pmatrix} 1 & 0 \\ \frac{n-n'}{n' r} & \frac{n}{n'} \end{pmatrix}$
Thin lens (refraction)	$\underline{\mathbf{M}} = \begin{pmatrix} 1 & 0 \\ \frac{n'-n}{n}\left(\frac{1}{r_2} - \frac{1}{r_1}\right) & 1 \end{pmatrix} = \begin{pmatrix} 1 & 0 \\ -1/f' & 1 \end{pmatrix}$
Thick lens (refraction)	$\underline{\mathbf{M}} = \begin{pmatrix} 1 - \frac{L(n'-n)}{n' r_1} & \frac{Ln}{n'} \\ \frac{n'-n}{n r_2}\left(1 - \frac{L(n'-n)}{n' r_1}\right) - \frac{n'-n}{n r_1} & 1 + L\left(\frac{n'-n}{n' r_2}\right) \end{pmatrix}$
Thick lens (embedded between two different media)	$\underline{\mathbf{M}} = \begin{pmatrix} 1 + \frac{L(n-n')}{n' r_1} & \frac{Ln}{n'} \\ \frac{n'-n''}{n'' r_2} + \frac{L(n'-n'')(n-n')}{n'' n' r_1 r_2} - \frac{n'^2-n' n}{n'' n' r_1} & \frac{n}{n''} + Ln\left(\frac{n'-n''}{n'' n' r_2}\right) \end{pmatrix}$
Quadratic radial index profile	$\underline{\mathbf{M}} = \begin{pmatrix} \cos(\varepsilon L) & \sin(\varepsilon L)/\varepsilon \\ -\varepsilon \sin(\varepsilon L) & \cos(\varepsilon L) \end{pmatrix}$
Planar mirror (reflection)	$\underline{\mathbf{M}} = \begin{pmatrix} -1 & 0 \\ 0 & -1 \end{pmatrix}$
Spherical mirror (reflection)	$\underline{\mathbf{M}} = \begin{pmatrix} 1 & 0 \\ -2/r & 1 \end{pmatrix}$

We join all parameters to one *ABCD matrix* which contains all relevant information for the imaging process. ABCD matrices of various typical components are listed in Table A.1. If light rays travel backwards through an optical system (inversion of the direction), we simply have to invert the matrix. In such a case, the imaging relation is given by

$$\begin{pmatrix} h \\ \gamma \end{pmatrix} = \begin{pmatrix} D & -B \\ -C & A \end{pmatrix} \begin{pmatrix} h' \\ \gamma' \end{pmatrix} \equiv \underline{\mathbf{M}}^{-1} \begin{pmatrix} h' \\ \gamma' \end{pmatrix} . \tag{A20}$$

For instructive purposes, let us derive the matrix expressions for a spherical surface and a thin lens just to understand the concept of matrix optics.

Example A.1

Spherical Surface When a ray hits a spherical surface, it does not change its height at the interface ($h = \text{const}$). The first conditional equation thus simply reads

$$h' = 1h + 0\gamma \quad \Rightarrow A = 1, B = 0\,. \tag{A21}$$

However, the ray is deflected due to refraction. Referring to Figure A.4, we apply Snell's law in paraxial approximation. With

$$\chi' = \gamma' + \theta' \tag{A22}$$

(χ' is pointing downwards) and

$$\chi = \gamma + \theta'\,, \tag{A23}$$

we obtain

$$\gamma' = \left(\frac{n-n'}{n'r'}\right)h + \left(\frac{n}{n'}\right)\gamma \quad \Rightarrow C = \frac{n-n'}{n'r'}, D = \frac{n}{n'}\,. \tag{A24}$$

Finally, we merge the results of Eqs. (A21) and (A24) to the matrix equation

$$\begin{pmatrix} h' \\ \gamma' \end{pmatrix} = \begin{pmatrix} 1 & 0 \\ \frac{n-n'}{n'r'} & \frac{n}{n'} \end{pmatrix} \begin{pmatrix} h \\ \gamma \end{pmatrix} \equiv \underline{\mathbf{M}} \begin{pmatrix} h \\ \gamma \end{pmatrix}\,. \tag{A25}$$

For $r' = r$, $\underline{\mathbf{M}}$ corresponds to the matrix of the forth row in Table A.1.

Example A.2

Thin Lens It was already mentioned that a thin lens is nothing more than a combination of two spherical surfaces. As a consequence, we have to apply Eq. (A25) twice while taking the different curvatures into account. So, we have

$$\begin{aligned} \begin{pmatrix} h' \\ \gamma' \end{pmatrix} &= \underbrace{\begin{pmatrix} 1 & 0 \\ \frac{n'-n}{n r_2} & \frac{n'}{n} \end{pmatrix}}_{\text{right surface}} \underbrace{\begin{pmatrix} 1 & 0 \\ \frac{n-n'}{n' r_1} & \frac{n}{n'} \end{pmatrix}}_{\text{left surface}} \begin{pmatrix} h \\ \gamma \end{pmatrix} \\ &= \begin{pmatrix} 1 & 0 \\ \frac{n'-n}{n}\left(\frac{1}{r_2} - \frac{1}{r_1}\right) & 1 \end{pmatrix} \begin{pmatrix} h \\ \gamma \end{pmatrix}\,. \end{aligned} \tag{A26}$$

With reference to Eqs. (A12) and (A14), we can express Eq. (A26) in a simpler way as

$$\begin{pmatrix} h' \\ \gamma' \end{pmatrix} = \begin{pmatrix} 1 & 0 \\ -1/f' & 1 \end{pmatrix} \begin{pmatrix} h \\ \gamma \end{pmatrix} \equiv \underline{\mathbf{M}} \begin{pmatrix} h \\ \gamma \end{pmatrix}\,. \tag{A27}$$

This expression corresponds to the fifth row in Table A.1.

Arbitrary optical systems To calculate the imaging of an arbitrary optical system, we proceed as for the thin lens example. This means that we decompose the whole system into individual optical elements for which we have a simple matrix expression (some of them given in Table A.1). The resulting arrangement of k elements can be regarded as a chain of matrix operations that must be multiplied like (Problem PA.4)

$$\underline{\underline{\mathbf{M}}}_{\text{system}} = \underline{\underline{\mathbf{M}}}_k \underline{\underline{\mathbf{M}}}_{k-1} \cdots \underline{\underline{\mathbf{M}}}_2 \underline{\underline{\mathbf{M}}}_1 \,. \tag{A28}$$

A.1.4
Aperture Stops, Field Stops, and Pupils

Optical systems always have at least one component that limits the solid angle at which rays may pass through. Up to now, we have assumed (without explicitly mentioning it) that the only limiting factor is the diameter of the used lenses or the inner diameter of their mounts. Rays which pass by a lens do not contribute to the imaging process and are thus ignored. Additionally, we can introduce other limiting elements like diaphragms which do not change the path of rays. We mainly distinguish between two types of diaphragm:

1. *Aperture stop*: Aperture stops are diaphragms introduced to an imaging system which narrow the solid angle of ray bundles emanated by the object. For this reason, they reduce the light intensity that arrives at the image plane.
2. *Field stop*: Field stops reduce the size of an image or object. They are used to define the boundary and field size (i.e., the visible image size).

The position in the optical system determines whether a diaphragm acts like an aperture or a field stop.[13] To find out which "role" a diaphragm takes up, it is useful to consider rays which are able to pass through the whole optical system. For an on-axis object point (O_0), the outermost limiting rays are referred to as *marginal rays* (violet lines in Figure A.13). Wherever the marginal rays cross the optical axis ($y_{mr} = 0$), an image is formed. When we place the diaphragm at or near this position, it "cuts off" the outer parts of the corresponding image, and thus acts as a field stop.[14] We may also regard the field stop as an "additional overlaid image" which effectively limits the projected field size of the original object.

The diaphragm acts as an aperture stop if placed at or near a pupil plane ($y_{mr} \approx$ const), as it blocks some incident rays which are emanated by the object. Although the aperture stop regulates the amount of rays that can arrive at the image plane, the object is completely projected to the image side. Hence, we have *no* cut-off effect, but the brightness and resolution (as we will see in Section A.2.1.6) of the whole image is reduced instead.

13) In fact, the definitions of aperture and field stops describe only two special cases. In real optical systems with vignetting (Section A.1.4.1), we often cannot assign a defined role to each diaphragm.
14) A typical example for a field stop is the finite size of a CCD chip in a camera system.

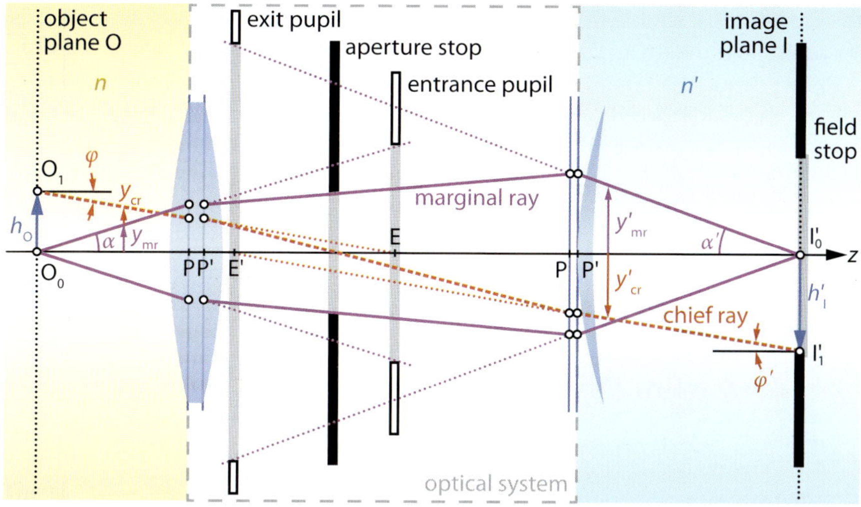

Figure A.13 Geometry of optical imaging. The aperture stop blocks rays emanated from the object so that the resulting image brightness is reduced. The field stop limits the field size of the image. Entrance and exit pupils are the images of the aperture stop in object and image space, respectively. The marginal rays (violet) are characterized by the maximum emission angle of rays α and the maximum convergence angle α'. The marginal ray height at each position along the z axis is represented by y_{mr}. The chief ray (dashed red line) is characterized by the angle φ and its height y_{cr} at a certain z position. n and n' are the refractive indices of the exterior media in object space (orange subspace) and image space (blue subspace), respectively. The primed parameters refer to the image space.

The image of the aperture stop formed by the optical components between object plane and aperture stop is referred to as the *entrance pupil*. Its position and diameter determines the maximum emission angle α of rays from the on-axis object point O_0 which contributes to the optical imaging. The extensions of the object-side part of the marginal rays graze the edges of the entrance pupil (Figure A.13).[15] A measure for the range of ray angles, which can enter the optical system, is the numerical aperture

$$\mathrm{NA} = n \sin \alpha \,, \tag{A29}$$

where n is the refractive index of the medium in the object space (yellow subspace in Figure A.13) into which the system is embedded.

The image of the aperture stop formed by the optical components between aperture stop and image plane is referred to as the *exit pupil*. It determines the maximum convergence angle α' of rays in the image space (blue subspace in Figure A.13). The extension of the image-side part of the marginal ray, which converges at the on-axis image point I'_0, grazes the edges of the exit pupil. Summing

15) If no lens is placed between object and aperture stop, the aperture stop itself may serve as the entrance pupil. However, this only holds as long as no other optical component exists which limits α even more!

up, the marginal rays define the diameters of entrance and exit pupils as well as the position of the images.

To determine the image size and the positions of entrance and exit pupils, we use the so-called *chief ray* (dashed red line in Figure A.13). The chief ray emerges from the off-axis object point O_1 and passes through the center of the aperture stop. The pupil centers are located at the intersections of the object- and image-side extensions of the chief ray with the optical axis, that is, points E and E′. The image size h'_1 of an object is given by the chief ray height y_{cr} in the image plane.

If the optical system only consists of a thin lens (e.g., as shown in Figure A.6), the lens forms the aperture stop, and both entrance and exit pupils coincide with this aperture stop. In more complex optical systems (as shown in Figure A.13), however, the pupils are often different from the aperture stop and must not necessarily be physical diaphragms.

A.1.4.1 **Vignetting**

The aperture stop determines shape and size of the ray bundle emanated from on-axis object point O_0. However, when the bundle of rays is emanated from an off-axis object point, a part of it may be blocked by the lens mount or simply passes by the lenses and thus does not contribute to the imaging (Figure A.14). This "shading effect" is called *vignetting* and becomes more severe the larger the distance between object point and optical axis. Vignetting thus decreases the brightness toward the outer zone of an image. To reduce this effect in cascaded optical systems, the exit pupil of the first sub-system E'_1 has to match the entrance pupil of the second (downstream) sub-system E_2 as exactly as possible. If E'_1 and E_2 are not located at the same position on the optical axis and do not have the same diameter, some incident light rays cannot enter the second sub-system and the resulting image has a nonuniform brightness.

Vignetting can be removed when no lens rims or mounts act as the limiting elements for any object point. Alternatively, we can also cut off the shadowed parts of the image with a field aperture. In the latter approach, we obviously crop the visible image, but the remaining part has a homogeneous brightness.

A.1.4.2 **Helmholtz–Lagrange Invariant**

For any optical system, invariants exist which are constant throughout the entire optical path. The *Helmholtz–Lagrange invariant* [2] is given by

$$\mathcal{H}^2 = (n\varphi y_{mr} - n\alpha y_{cr})^2 \,. \tag{A30}$$

n denotes the refractive index, α the marginal ray angle, and y_{mr} the marginal ray height. φ is the chief ray angle and y_{cr} the chief ray height at a selected position z along the optical axis (Figure A.13). To calculate $\mathcal{H}^2$, we can select any arbitrary plane along the optical axis which is *not* located inside the optical system (subspace in Figure A.13 framed by the gray dashed lines). In particular, we may select as one plane the object plane O and as another one the image plane I (dotted black lines in Figure A.13). When we take the possibly different refractive indices of the object- and image-side media into account ($n \neq n'$), note that the height of the marginal

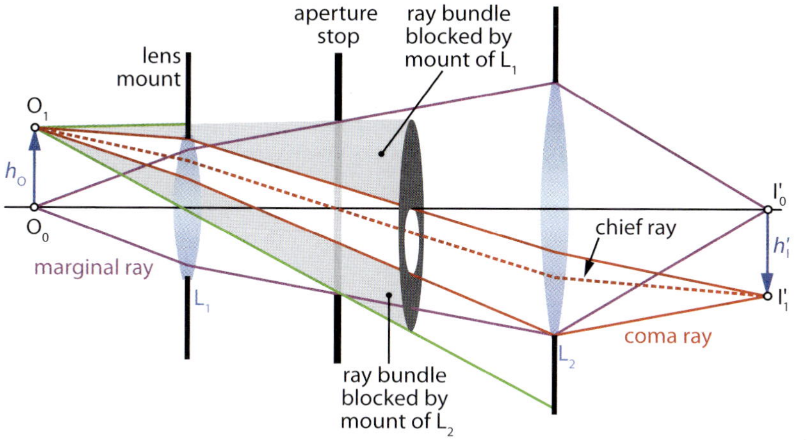

Figure A.14 The origin of vignetting in an optical system which consists of two lenses (L_1, L_2) and an aperture stop. The green lines represent the outermost rays which graze the edge of the aperture stop. A portion of the ray bundle emanated from off-axis object point O_1 is blocked. As a consequence, the outer zone of the image becomes darker.

ray y_{mr} is zero in both the object and image plane, and use $y_{cr} = h_O$, it follows from Eq. (A30) that

$$\mathcal{H}_O^2 = (n\alpha h_O)^2 = (n'\alpha' h_I')^2 = \mathcal{H}_I^2 \,. \tag{A31}$$

The primed and unprimed parameters refer to the image and object space (Figure A.13), respectively.

From Eq. (A31), we arrive at another invariant which is referred to as the *throughput* or *étendue* G. The étendue characterizes the amount of light passing through an optical system. As above, G is determined by the area of the entrance pupil times the solid angle subtended by the light source as seen from the pupil (see also [2, 3]). For any arbitrary optical system which fulfills Fermat's principle (Section A.1.1), the étendue is constant in every pupil plane of the considered optical system[16)] and proportional to the square of the Helmholtz–Lagrange invariant (Problem PA.5). So, we have

$$G \propto \mathcal{H}^2|_{\text{pupil}} = (n\varphi y_{mr})^2 \,. \tag{A32}$$

A.1.5
Limitations of the Paraxial Beam Approximation

One of the first assumptions in this chapter was the paraxial approximation for small refraction angles: $\sin\gamma \approx \tan\gamma \approx \gamma$. Here, we assumed that rays emanating from an object point travel close to the optical axis. When considering the

16) When two optical systems are combined, the étendues of all sub-systems must be equal to maximize the amount of light passing through the whole system.

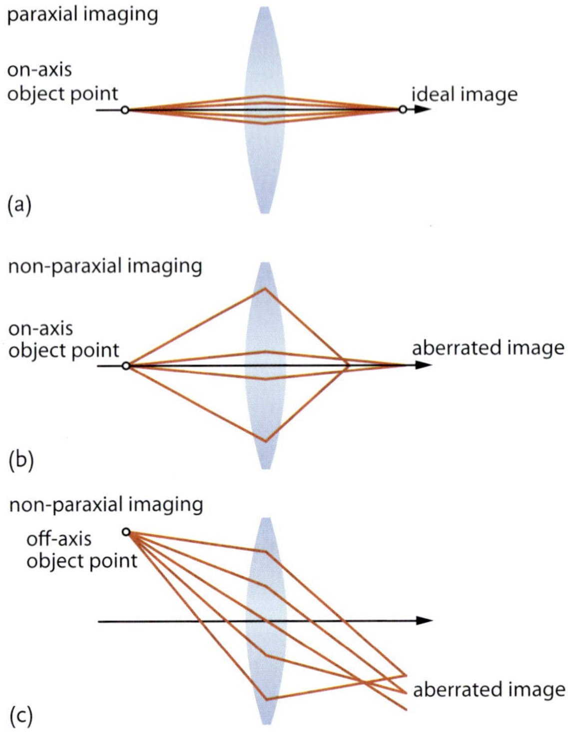

Figure A.15 Schematics of (a) ideal point-to-point imaging, (b) aberrated imaging due to a wide opening angle of incident monochromatic rays, and (c) aberrated imaging due to oblique incidence of monochromatic rays from an off-axis object point. The graphics show the limitations of the paraxial approximation which are usually assumed in geometric optics. Deviations from the ideal case must always been considered for real optical systems.

imaging behavior of a positive lens, we realized that all paraxial, monochromatic rays converge at one single image point (Figure A.15a). In this case, a sharp image of the object point was obtained. But what happens with rays for which the paraxial approximation is *not* valid anymore? So far, we have simply ignored the nonparaxial rays. But in reality, they are usually present as well. For example, Figure A.15b shows how nonparaxial rays from an on-axis object point converge at different points after passage through the lens. Rays at different angles thus form different image points along the optical axis. Figure A.15c shows that rays from an off-axis object point form different image points in the image plane. As we do not have a defined point-to-point projection in both cases, the image becomes blurred. Any deviation from the ideal paraxial case leads to so-called *aberrations*. All these facts also hold for every point of an extended object.

If aberrations are present in an optical system, the image of an object point looks "spread out". To quantify this behavior, we may use the *point-spread function* PSF, which is the image of an object point formed by a given optical system. A full understanding of the PSF is, however, only possible in the context of wave optics, as diffraction effects have to be considered for image formation (Section A.2.1.6). For now, we may think of this in a simplified manner in that we regard the intensity distribution of a point image as the density of rays homogeneously emanated from the object point into any solid angle passing through the image plane. Each ray corresponds to the same amount of energy. This concept is often called *geometric spot*. When diffraction effects can be neglected, the geometric spot is similar to the PSF. We can use it at this stage to get an understanding of the concept of aberrations.[17]

A.1.6 Aberrations

In practice, we often want to image a given object as accurately as possible. Correction of ray aberrations is thus an extremely important topic for the development of optical systems and devices. Since each type of aberration can have a positive or negative sign, it is in principle possible to cancel out all kinds of deviation from the ideal image with a suitable arrangement of optical components. Nevertheless, a lot of effort and experience is needed to optimize the image quality of optical systems to a sufficient degree.

We distinguish between longitudinal and transverse ray aberrations:

- *Longitudinal aberrations* refer to different intersection points along the chief ray for paraxial and marginal rays. For example, if we use a positive lens, the image-side position at which rays from an on-axis object point cross the optical axis depends on the ray angle γ (Figure A.15b).
- *Transverse aberrations* refer to the shift of ray intercepts in the transverse direction in a fixed reference plane (Figure A.15c). This description is particularly important for considerations of lateral blur and resolution. In the special case of afocal systems, that is, the image is infinitely far away, transverse aberrations must be described as angular deviations.

Before we go through the maths of aberration theory, it is instructive to get a quick overview of the imaging consequences of the most relevant types of aberration.

A.1.6.1 Spherical Aberration

We consider a simple optical system which consists of an on-axis object point and a positive (biconvex) lens, as depicted in Figure A.16a. For now, we do not restrict ourselves to paraxial rays. We also trace rays which pass through the outer zone of the lens. When Snell's law is applied to all rays without any approximation, rays passing through the outer zone of the lens intersect with the optical axis

17) The point images plotted in Figures A.16 and A.17 are the point-spread functions and *not* the geometric spots.

further to the left than the paraxial rays. As refraction depends on the angle of incidence γ, no unique image point exists. The concentric blurring is called *spherical aberration.*

In Figure A.16b, corresponding images of the point object (i.e., the point-spread functions) are depicted for the object plane and the plane of axis intersection of paraxial rays. Since all rays are centered at the optical axis, the PSF is rotationally symmetric.

From this, we conclude that spherical aberration is particularly strongly pronounced for outer ring illumination of lenses with short focal lengths, such as microscope lenses and focusing systems. For correction of spherical aberrations, among other options, several positive and negative lenses are combined in a suitable manner, or aspheric lenses are used.

A.1.6.2 **Coma**

Let us look at another imaging aberration called *coma*, which appears only for off-axis object points or if components of an optical system are misaligned. The ray diagram in Figure A.17a shows how obliquely incident rays are refracted by a positive lens. Rays from an off-axis point passing through a particular zone of the lens intersect the image plane in a circle. The size and lateral shift of these circles increases with the diameter of the zone. Thus, the rings produce the characteristic coma tail (see picture of "image plane" in Figure A.17b).[18]

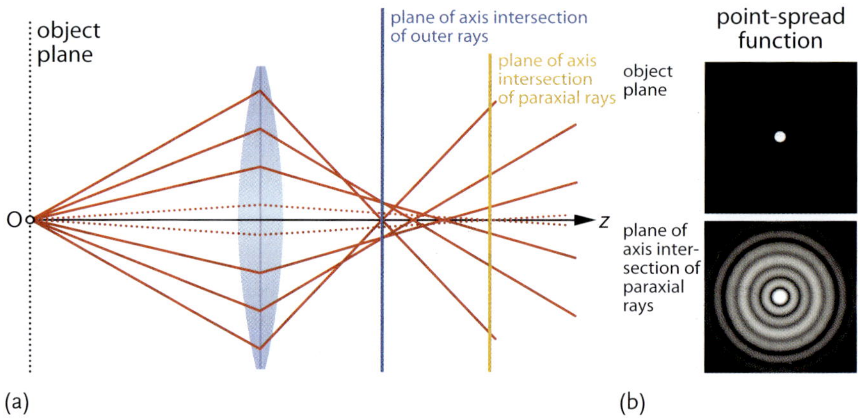

Figure A.16 The origin of spherical aberration for a single lens. (a) Light rays refracted in the outer zone of the lens have the shortest focal length, and rays refracted close to the optical axis (dotted line) are focused farther away from the lens. (b) Point-spread function of an optical system with spherical aberration in the plane of axis intersection of paraxial rays. Note that diffraction effects are taken into account in this image. So, interference patterns are also included which would not appear in the geometric spot image.

18) Coma is very easy to demonstrate experimentally. Simply tilt a magnifying glass while bright light, for example, sun light, is shining through it. The image projected onto the floor will be elongated and broadened at one end.

In complex optical systems, off-axis coma can be compensated by a suitable optical design. Optical systems are called *aplanatic* if they are free of spherical aberration and corrected for coma (at least for near-axis points).

A.1.6.3 **Astigmatism**

If two rays emanating from an off-axis object point pass through a lens at different angles, they will "see" different effective radii of the surface curvature. This already occurs in thin ray bundles. As a consequence, we find two different focal positions for the sagittal (xz) planes and meridional/tangential (yz) planes (Figure A.18) which, in turn, leads to a blurred image. Between the sagittal and meridional image planes, we have a *disk of least confusion* at which the deviations from the ideal image in vertical and horizontal directions are best compensated for. However, the resolution is worse than in the case of ideal imaging, since all fine object details are softened.

In rotationally symmetric systems, astigmatism can only occur for off-axis object points, as it is an effect of broken symmetry. Systems with asymmetric or off-axis optical components can also have astigmatic errors for on-axis object points. Logically, aspheric, toric, or any other free-form surfaces exhibit an even more pronounced astigmatism.

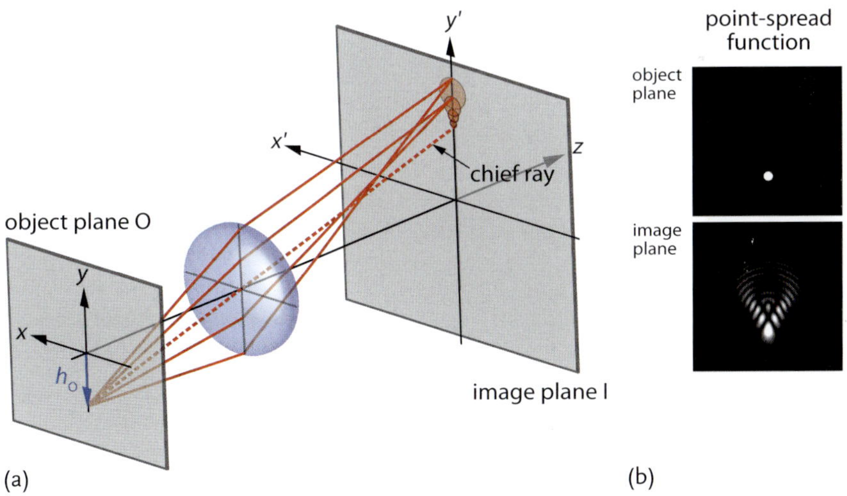

Figure A.17 The origin of coma aberration. (a) Ray diagram of an off-axis object point. Due to the oblique angle of incidence, the image of an object point is deformed to a comet tail as is shown for the point-spread function in (b). (b) Point-spread function of an optical system with coma aberration in the image plane. Diffraction effects are also included which would not appear in the geometric spot image.

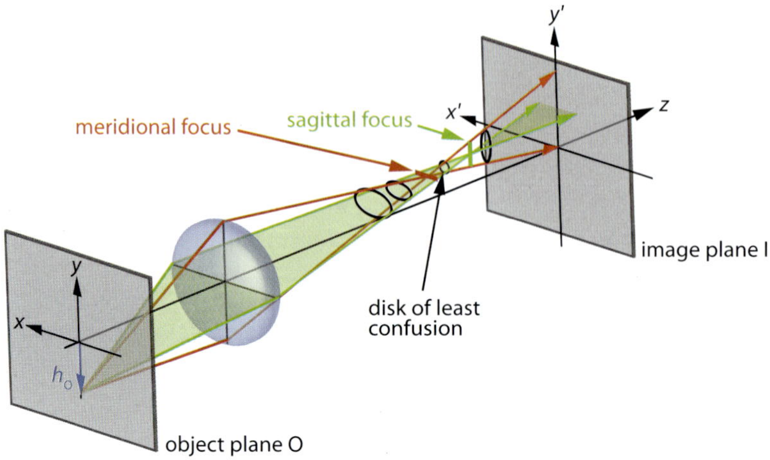

Figure A.18 The origin of astigmatism. Light rays which travel in the meridional (red) and sagittal (green) planes are refracted differently. Both sets of rays thus intersect at different positions (i.e., the meridional and sagittal foci) and produce elongated spot shapes ranging from linear to elliptical. In between, the geometric spot recovers a circular geometry. This is why it is called the "disk of least confusion".

A.1.6.4 **Field Curvature**

As we can see in Figure A.19, a positive lens projects extended objects onto a curved surface. With a flat screen, we could therefore scan between the image planes formed by paraxial and coma rays to see either sharp outer zones or a sharp center. An example is shown in the insets on the right.

Departures of an ideal image due to field curvature are quite difficult to eliminate and even more severe at low magnifications. To correct this aberration, we need combinations of lenses which add positive and negative refractive powers to flatten the image field. As a consequence of the design principles, in photographic systems, positive lenses with a larger diameter and high refractive power but low refractive index, as well as negative lenses with a smaller diameter are used. For stereo microscopes, field curvature is also an important issue (Section 6.2). The simplest solution would be to capture the image on a curved screen. Unfortunately, this is often not feasible for technical systems.

A.1.6.5 **Distortion**

Distortion is the only type of aberration which is not related to image blurring. It rather leads to a change of magnification for extended images (of extended objects) with increasing distance of the image point from the optical axis. To understand the origin of distortion, we consider an extended object which is imaged by a thin lens (Figure A.20). In addition, we place a circular aperture stop in front of the lens. The position of each image point is determined by the chief ray that passes through the center of the aperture stop. When the aperture stop touches the lens, the chief ray passes through the lens without any deflection, and the image shows

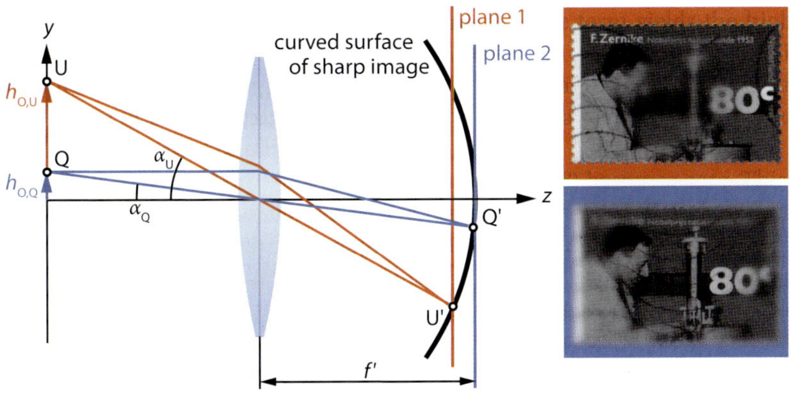

Figure A.19 The origin of field curvature: We place an ideal "point-imaging" lens at a distance f' from a screen, where f' is the image-side focal length of the lens. In this case, image points near the optical axis will be located in plane 2. However, image points from rays further away from the optical axis will be formed in front of the focal plane. The z position of the image point drops off by the cosine of angle α. A sharp image can thus only be obtained on curved surfaces, which is often technically demanding. α_U and α_Q are the field angles related to the object points U and Q, respectively. If a planar photodetector is placed at plane 1 (red), we will observe an image that is sharp near the edges (see image with red frame), since off-axis rays are focused. If the photodetector is placed at plane 2 (blue), the central part of the image will become sharp (see image with blue frame).

no distortion at all (orthoscopic system). We now place the aperture stop between lens and object plane, as depicted in Figure A.20a. The chief ray again crosses the center of the aperture stop. But this time, it passes through the outer zone of the lens. The image size (i.e., the magnification) thus decreases in a "nonlinear" manner with distance from the optical axis in that the image corners decrease more than the central part. In this effect called *barrel distortion*, straight lines of an object appear bent in the image. When the aperture stop is placed behind the lens on the image side, we see the contrary effect (Figure A.20b). The image size increases with distance from the optical axis, that is, the magnification is larger toward the image corners. This is called *pincushion distortion*. In contrast to position, the diameter of the aperture stop does not influence distortion, since the chief ray does not change its path when we make the aperture smaller or larger.

Pronounced distortion can be found for thick "fish-eye" camera lenses which allow viewing of a wide field angle. In this extreme configuration, distortion cannot be optically corrected anymore. In general, optical systems can be corrected for distortion by a symmetric arrangement of the optical components. For example, in simple optical systems, we achieve this by centering the aperture stop between two equal lenses or by putting the aperture stop directly onto a single lens.

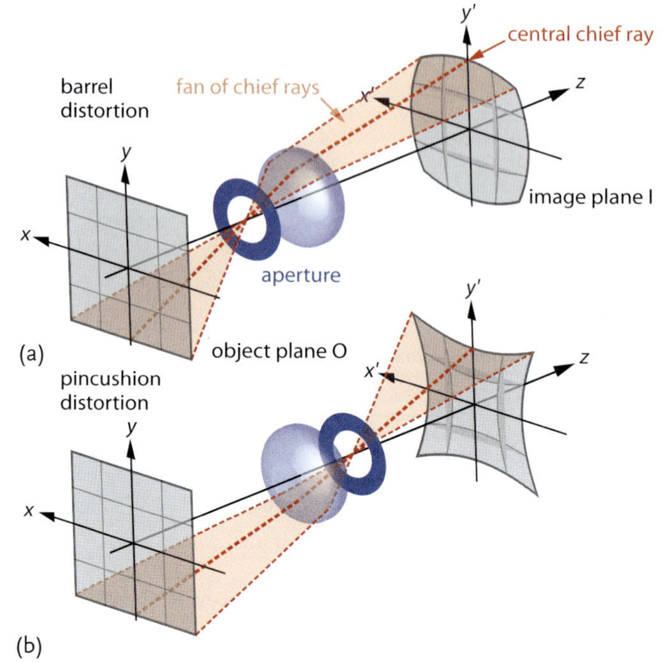

Figure A.20 The origin of image distortion. (a) Barrel distortion which leads to the well-known fish-eye effect: a magnification of the image center. (b) Pincushion distortion which leads to a relative magnification at the edges of the image.

A.1.7
Wavefront Aberration and Image Quality

Longitudinal and transverse ray aberrations are a very useful concept to design and characterize optical systems. We will now look at a related formalism which provides an alternative description of aberrations. Wavefront aberrations also provide a more convenient way to quantify aberrations and also take the scalability with wavelength into account.[19)]

How are wavefront aberrations linked to longitudinal and transverse aberrations (Sections A.1.6.1–A.1.6.5)? At first sight, it may seem to be confusing to talk about "wavefronts" in geometric optics. In fact, we will deal with light as a wave phenomenon only later in Section A.2. For now, it is sufficient to consider that when light propagates in three dimensions, it can be regarded as a wave with wave crests and troughs forming wavefronts. Rays that we have considered so far are nothing more than normals of these wavefronts (Section A.2.1.3). As shown in Figure A.21a, they point along the traveling direction of the light wave.

19) In particular, this theory is used to quantify refractive errors in human eyes (Section 3.1) and to optimize the image quality of diagnostic devices (e.g., Sections 5.3 and 6.3.2) and therapeutic systems (Chapter 10).

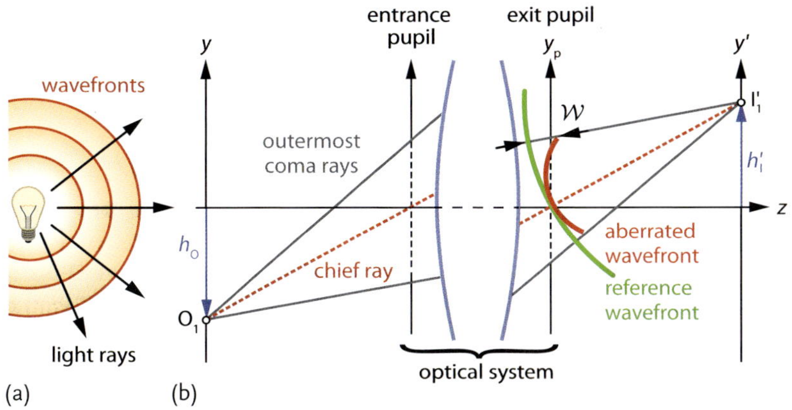

Figure A.21 (a) Rays can be interpreted as vectors which point along the local traveling direction of light waves. They are thus perpendicular to the wavefront. (b) In the nonparaxial case, aberrations can either be described in terms of a "ray model" (geometric optics) or a "wave model" (wave optics) of light. These are indeed two different approaches to the same optical effects. Adapted from [2].

Let us have a look at the arrangement of Figure A.21b, where rays are emanated from an off-axis point O_1 of an extended object. The emission angles of these *coma rays* are restricted by an exit pupil[20]. As rays and wavefronts are directly related to each other, a reference wavefront exists (bold green line in Figure A.21b) whose normal is the chief ray and crosses the optical axis in the center of the exit pupil. If the diameter of the exit pupil is small, the generated wavefront (bold red line in Figure A.21b) on the image side will hardly deviate from the reference wavefront. If, however, the pupil diameter is increased, the generated wavefront will generally have a complex surface shape.

Wave aberration function All deviations of the generated wavefront from the reference wavefront are referred to as *wavefront aberrations* and can be described by means of the wave aberration function $\mathcal{W}(x_p, y_p)$. $\mathcal{W}$ is defined in the exit pupil plane ($x_p y_p$ plane in Figure A.22) such that its mean value is zero for the reference wavefront. We have

$$\langle \mathcal{W}(x_p, y_p) \rangle = \frac{1}{S_{ep}} \iint \mathcal{W}(x_p, y_p) \mathrm{d}x_p \mathrm{d}y_p = 0 \,, \tag{A33}$$

where S_{ep} denotes the area of the exit pupil. By convention, $\mathcal{W} > 0$ if the wavefront locally travels ahead of the reference wavefront, and $\mathcal{W} < 0$ if it locally lays behind. The wavefront offset is chosen such that

$$\mathcal{W}(x_p = 0, y_p = 0) = 0 \,. \tag{A34}$$

20) Since the exit pupil is the image of the entrance pupil, we do not have to take the entrance pupil into account.

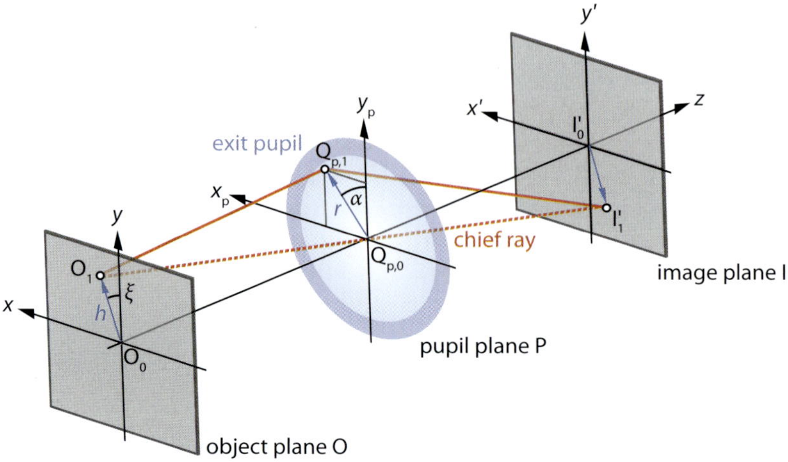

Figure A.22 Coordinate system used for the definition and decomposition of the wave aberration function $\mathcal{W}$. O_0 is the origin of the $(x\,y)$ object plane at which the optical axis passes through. $O_1(x, y)$ is an arbitrary object point in the object plane. $Q_{p,1}(x_p, y_p)$ represents an arbitrary point in the $(x_p\,y_p)$ pupil plane through which the ray passes. This point can also be characterized by (r, α) polar coordinates. Primed coordinates (x', y') refer to the image plane.

It must be pointed out that $\mathcal{W}(x_p, y_p)$ only describes aberrations for a fixed object point. To obtain the complete information about present aberrations of an optical system, we thus have to consider the whole "object field" instead of a distinct object point. If an arbitrary object point is characterized by the object plane coordinates (x, y), the wave aberration function will depend on four variables $\mathcal{W}(x, y, x_p, y_p)$. This general form of the wave aberration function holds for arbitrary optical systems. However, in practice, we usually consider rotationally symmetric optical systems for which adapted variables can be found that are invariant with respect to rotation about the optical z axis. For rotationally symmetric systems, $\mathcal{W}$ is equal for all object points with the same radial distance $h = \sqrt{x^2 + y^2}$ (Figure A.22). Thus, it becomes a function of

1. the square of the object height $h_O^2 = h^2$,
2. the square of the ray's radial distance from the optical axis in the pupil plane $r^2 = x_p^2 + y_p^2$, and
3. the azimuthal angle difference in the field and pupil plane $hr\cos(\alpha - \xi)$.

In the following, we only consider object points which lie on the y axis ($x = 0$, $\xi = 0$) such that the number of free variables can be reduced by one and the wave aberration function in rotationally symmetric optical systems can be expressed by

$\mathcal{W}(y, x_p, y_p)$ (Cartesian coordinate system) or by $\mathcal{W}(h, r, \alpha)$ (polar coordinate system[21]).

Root mean square wavefront error The wave aberration function $\mathcal{W}$ contains all information about the monochromatic aberrations of an optical system. However, in practice, suitable metrics (Section 5.4.1) are used for wavefront analysis. The most frequently used metric in the pupil plane is the *root mean square wavefront error* which is defined by

$$\text{RMS}_{\text{wfe}} = \sqrt{\frac{1}{N}\sum_i \left(\mathcal{W}(h, r_i, \alpha_i) - \langle\mathcal{W}\rangle\right)^2} \,. \tag{A35}$$

RMS_{wfe} is the *standard deviation* of N measured values of $\mathcal{W}$ at varying positions (r_i, α_i) within the pupil plane. It indicates how strongly the measured wavefront deviates from the reference wavefront. $\langle\mathcal{W}\rangle = 1/N \sum_i \mathcal{W}(h, r_i, \alpha_i)$ is the (arithmetic) mean value of the wave aberration function which equals zero for the reference wavefront (see also Eq. (A33)). Equation (A35) thus simplifies to

$$\text{RMS}_{\text{wfe}} = \sqrt{\frac{1}{N}\sum_i \left(\mathcal{W}(h, r_i, \alpha_i)\right)^2} \,. \tag{A36}$$

Point-spread function The wave aberration function can be determined either directly by measurements (Section 5.3) or indirectly from the image of an object point, that is, the point-spread function PSF. In analogy to the geometric spot definition (Section A.1.5), the PSF characterizes noticeable aberrations in the $(x'y')$ image plane and is linked to $\mathcal{W}$ via

$$\text{PSF}(x', y') \propto \left| \mathcal{F}_{x',y'} \left\{ p(x_p, y_p) e^{2\pi i/(\lambda)\mathcal{W}(x_p, y_p)} \right\} \right|^2 \,. \tag{A37}$$

$\mathcal{F}$ is the Fourier[22] transform (see Info Box A.2 in Section A.2.4) and λ the wavelength of the incident light. p represents the apodization function which describes size, shape, and amplitude transmission of the exit pupil.[23]

Strehl ratio We can also introduce at this point the *Strehl ratio* $\mathcal{S}$ which is another measure of the effect of aberrations on the image quality. It is expressed by the reduction of the maximum value of the PSF. The Strehl ratio is defined as (Prob-

21) We can easily transform between polar and Cartesian coordinates by means of the relations $y_p = r \sin\alpha$, $x_p = r\cos\alpha$, and $r = \sqrt{x_p^2 + y_p^2}$.

22) Joseph Fourier (1768–1830).

23) A simple example for the apodization function: Consider a circular diaphragm centered at $r = 0$ with a hole diameter of $2r_{\text{ph}}$. Inside the hole, we have $p = 1$ for $r < r_{\text{ph}}$. Outside the hole, we have $p = 0$ for $r > r_{\text{ph}}$.

lem PA.6)

$$\mathcal{S} = \frac{I_{\text{max,aberr}}}{I_{\text{max,noaberr}}} , \tag{A38}$$

where $I_{\text{max,aberr}}$ and $I_{\text{max,noaberr}}$ denote the maximum light intensity values of the aberrated and nonaberrated point-spread functions, respectively. As aberrations spread out the PSF, $\mathcal{S}$ is always a number equal or less than one.[24] The greater the effect of aberration, the lower the value of $\mathcal{S}$ and the poorer the image quality.

A.1.8
Classification and Expansion of the Wave Aberration Function

Once the wave aberration function $\mathcal{W}$ is determined for each object point, we have in principle all information about the aberrations of an optical system at hand. However, we face the problem that uncorrected optical systems usually suffer from a number of different aberrations to different degrees. As the correction can only be understood individual aberration by individual aberration (e.g., spherical aberration, coma, distortion, and so on), it is necessary to decompose $\mathcal{W}$ into the respective contributions. In this way, we may link each type of aberration treated in Sections A.1.6.1–A.1.6.5 to the relevant component of the expansion of the wavefront aberration. Mathematically speaking, $\mathcal{W}$ is expressed as a weighted sum, where each term describes solely *one* type of aberration.

A.1.8.1 Taylor Expansion of the Wave Aberration Function

The most straightforward way to decompose $\mathcal{W}$ into the individual types of aberration is the Taylor expansion.[25] When we apply the Taylor expansion to $\mathcal{W}$ in a strictly mathematical manner, we will also obtain terms which are optically irrelevant. According to Eq. (A34), $\mathcal{W}$ is defined in such a way that it vanishes in the center of the exit pupil ($x_\text{p} = y_\text{p} = 0$). As a consequence, all coefficients in the expansion which do *not* depend on the pupil coordinates (x_p, y_p) are zero.

As we only consider optical systems with rotational symmetry and a circular exit pupil, we can expand $\mathcal{W}$ in terms of h^2, r^2, $hr\cos\alpha$, and powers thereof.[26] The Taylor expansion of the wave aberration function in polar coordinates then reads (Problem PA.7)

$$\begin{aligned}\mathcal{W}(h, r, \alpha) = &\sum_{l,n,m} \mathcal{W}_{lnm} h^l r^n \cos^m \alpha \\ = &\mathcal{W}_{200} h^2 + \mathcal{W}_{020} r^2 + \mathcal{W}_{111} h r \cos\alpha \\ &+ \mathcal{W}_{400} h^4 + \mathcal{W}_{040} r^4 + \mathcal{W}_{131} h r^3 \cos\alpha + \mathcal{W}_{220} h^2 r^2 \\ &+ \mathcal{W}_{222} h^2 r^2 \cos^2\alpha + \mathcal{W}_{311} h^3 r \cos\alpha + \ldots \, .\end{aligned} \tag{A39}$$

24) A value of 1 is obtained for the ideal image without aberrations.

25) Brooke Taylor (1685–1731) stated that any arbitrary function can be expressed as a sum of its derivatives.

26) Note that we restrict our discussion to object points located on the y axis for which $\xi = 0$ and $x = 0$.

Table A.2 Schematic representation of the Taylor expansion of optical aberrations. The terms in the cells which are highlighted in red describe the image position. The terms of the cells which are highlighted in blue and green describe primary and secondary aberrations, respectively.

radius on exit pupil r^n / **object height h^l**	h^0	h^1	h^2	h^3	h^4	h^5	
r^1		$hr\cos\alpha$ tilt		$h^3r\cos\alpha$ primary distortion		$h^5r\cos\alpha$ secondary distortion	distortion
r^2	r^2 defocus		$h^2r^2\cos^2\alpha$ h^2r^2 astigm./field c.		$h^4r^2\cos^2\alpha$ h^4r^2		
r^3		$hr^3\cos\alpha$ primary coma		$h^3r^3\cos^3\alpha$ $h^3r^3\cos\alpha$			
r^4	r^4 primary spherical		$h^2r^4\cos^2\alpha$ h^2r^4 sec. astigm.				
r^5		$hr^5\cos\alpha$ secondary coma					
r^6	r^6 secondary spherical						
	spherical aberration	coma	astigmatism				

$\mathcal{W}_{lnm}$ are weighting factors which tell us how great the influence of each type of aberration actually is. l, n, and m are the powers of the expansion variables used.

Let us now correlate each term of the Taylor expansion in Eq. (A39) with an individual type of aberration. In the horizontal direction of Table A.2, the resulting terms are categorized by the power l of the object height h. In the vertical direction, the terms are categorized by the power n of the radius r. The resulting diagonals (with different color codes) represent the aberration "classes". The terms related to the red tiles are not aberrations in the usual sense. They describe a lateral (tilt) or axial (defocus) shift of an image point. The terms related to the blue tiles are called *primary aberrations* and correspond to the aberrations which we already discussed in a qualitative manner in Sections A.1.6.1–A.1.6.5. The *secondary aberrations* in the diagonal with green tiles are used to describe more complex imaging errors. Usually, their contribution to the total wavefront aberration is much smaller than for primary aberrations. Logically, even higher orders of aberration also exist which are not considered here.

In principle, the described Taylor expansion can be used to decompose any complicated wavefront aberration function into the individual types of aberration. But, disadvantageously, the coefficients $\mathcal{W}_{lnm}$ depend on each other and the number of terms used in the expansion. Mathematically speaking, the terms of the Taylor expansion are *not* orthonormal (see [4]).

A.1.8.2 Zernike Expansion of the Wave Aberration Function

Zernike[27] polynomials $\mathcal{Z}_n^m(r, \alpha)$ are much better suited for the decomposition of the wavefront aberration function, because they are orthonormal on a circular area. This means that each term has its own best-fit coefficient which does not influence the others. The Zernike expansion of the wave aberration function, which corresponds to Eq. (A39), reads

$$\begin{aligned} \mathcal{W}(h, r, \alpha) &= \sum_n \sum_{m=-n}^{n} c_n^m(h) \mathcal{Z}_n^m(r, \alpha) \\ &= \sum_n \sum_{m=-n}^{-1} c_n^m(h) \left(-\mathcal{N}_n^m \mathcal{R}_n^{|m|}(r) \sin(m\alpha)\right) \\ &\quad + \sum_n \sum_{m=0}^{n} c_n^m(h) \left(\mathcal{N}_n^m \mathcal{R}_n^{|m|}(r) \cos(m\alpha)\right) . \end{aligned} \tag{A40}$$

n and m are the integer radial and azimuthal indices, respectively. They are either both even or both odd and always fulfill $m \leq n$. Thus, for a given n, the azimuthal index m is constrained to $-n, -n+2, -n+4, \ldots, n$. The order of a Zernike coefficient is determined by the value of its radial index n. The Zernike coefficients $c_n^m(h)$ in Eq. (A40) represent weighting factors to which a dependence on a fixed object height h is assigned. In ophthalmic optics, this is a common representation, since the Zernike polynomials have a simple form in this case (Table A.3). However, in practice, we must determine $c_n^m(h)$ for every considered object height h. In [5], an alternative approach is presented for which the object height is directly included in the Zernike polynomials.

In addition to the standard double indexing used in Eq. (A40), Zernike coefficients can also be identified with just one index j. According to [6], we may "translate" the indices n and m with the relation

$$j = \frac{n^2 + 2n + m}{2} . \tag{A41}$$

In this way, a simpler notation for the Zernike expansion in Eq. (A40) is obtained, that is,

$$\mathcal{W}(h, r, \alpha) = \sum_j c_j(h) \mathcal{Z}_j(r, \alpha) . \tag{A42}$$

27) Frits Zernike (1888–1966).

Info Box A.1: Explicit Form of Zernike Polynomials

Zernike polynomials can be separated into a radial function $\mathcal{R}_n^m(r)$, an angular function $\exp(i m\alpha)$, and a normalization factor $\mathcal{N}_n^m$ such that

$$\mathcal{Z}_n^m(r,\alpha) = \begin{cases} \mathcal{N}_n^m \mathcal{R}_n^{|m|}(r)\cos(m\alpha) & \text{for } m \geq 0 \\ -\mathcal{N}_n^m \mathcal{R}_n^{|m|}(r)\sin(m\alpha) & \text{for } m < 0 \end{cases} . \quad \text{(A43)}$$

r is the radial distance from the pupil center normalized to the pupil radius (range of values: $0 \leq r \leq 1$). The azimuthal angle (meridional parameter) α is defined for an angular range of 0–2π. The normalization factor (keeping $\mathcal{Z}_n^m$ inside a unit circle) is given by

$$\mathcal{N}_n^m = \sqrt{(2-\delta_{0,m})(n+1)}\,, \quad \text{(A44)}$$

where $\delta_{0,m} = 1$ for $m = 0$ and $\delta_{0,m} = 0$ for $m \neq 0$. The explicit form of the radial polynomials reads

$$\mathcal{R}_n^{|m|}(r) = \sum_{l=0}^{\frac{n-|m|}{2}} \left(\frac{(-1)^l (n-l)!}{l!\left(\frac{n+|m|}{2}-l\right)!\left(\frac{n-|m|}{2}-l\right)!} \right) r^{n-2l} . \quad \text{(A45)}$$

Both the radial function (A45) and the normalization factor (A44) are somewhat complex. For this reason, the explicit expressions of relevant Zernike terms are shown in the outer right column of Table A.3. In the fourth column, the Zernike terms are related to the corresponding types of aberration. In Table A.3, we have also classified the types of aberration by *lower-order* (or *spherocylindric*) *aberrations* with $n = 2$ ($j = 3, 4, 5$) and *higher-order aberrations* with $n > 2$ ($j \geq 5$), which is very common in ophthalmology [6, 7].[28)]

Figure A.23 shows graphical plots of the formulas in the outer right column of Table A.3. This is an illustrative way to visualize the deformations of the reference wavefront caused by each type of aberration. The reference wavefront of an ideal spherical wave is represented as a flat, green surface in the $x_p y_p$ plane[29)]. The deviations caused by the respective type of aberration are represented by crests and

28) For a long time, defocus and astigmatism were the only known types of aberration in ophthalmology. All other refractive imperfections of eyes were grouped together as irregular astigmatism and generally set aside as being too difficult to measure and correct with eye glasses and contact lenses. However, laser-supported refractive corneal surgery (Section 10.3) has gained acceptance during recent years so that there was a need for precise aberration analysis techniques beyond defocus and astigmatism. The formerly known aberrations were termed "lower-order" and the newly discovered aberrations were termed "higher-order". Tilt ($n = 1$) describes a mere shift of an image point in the image plane. Hence, in technical and ophthalmic optics, it is not considered to be an optical aberration in the usual sense.

29) Think of it as the plane of the exit pupil.

Table A.3 Different types of aberration (nomenclature according to ISO 24157 standard) and their corresponding representation by Zernike polynomials according to Eq. (A43). The aberrations are sorted by the single index *j* which can be used instead of *n* and *m*. The outer right column shows the explicit mathematical expressions. The graphical plots of $\mathcal{Z}_n^m(r, \alpha)$ are depicted in Figure A.23. Please note that "quatrefoil" is sometimes also called "quadrofoil".

j	*n*	*m*	Type of aberration	Imaging consequence	$\mathcal{Z}_n^m(r, \alpha)$
0	0	0	Piston	None	1
1	1	−1	Vertical tilt	Shift of image point (distortion)	$2r \sin \alpha$
2	1	1	Horizontal tilt	Shift of image point (distortion)	$2r \cos \alpha$
			Lower-order aberrations/Spherocylindric aberrations		
3	2	−2	Primary (oblique) astigmatism	Orientation-dependent shift of focus	$\sqrt{6}r^2 \sin(2\alpha)$
4	2	0	Defocus	Axial focus shift	$\sqrt{3}(2r^2 - 1)$
5	2	2	(With/against the rule) astigmatism	Orientation-dependent shift of focus	$\sqrt{6}r^2 \cos(2\alpha)$
			Higher-order aberrations		
6	3	−3	(Oblique) trefoil	Imaging anomalies with threefold symmetry	$\sqrt{8}r^3 \sin(3\alpha)$
7	3	−1	Primary (vertical) coma	Image asymmetry and vertical shift of image	$\sqrt{8}(3r^3 - 2r) \sin \alpha$
8	3	1	Primary (horizontal) coma	Image asymmetry and horizontal shift of image	$\sqrt{8}(3r^3 - 2r) \cos \alpha$
9	3	3	(Horizontal) trefoil	Imaging anomalies with threefold symmetry	$\sqrt{8}r^3 \cos(3\alpha)$
10	4	−4	(Oblique) quatrefoil	Imaging anomalies with fourfold symmetry	$\sqrt{10}r^4 \sin(4\alpha)$
11	4	−2	Secondary (oblique) astigmatism	Orientation-dependent shift and degradation of focus	$\sqrt{10}(4r^4 - 3r^2) \sin(2\alpha)$
12	4	0	Primary spherical aberration	Blurring depending on pupil zone	$\sqrt{5}(6r^4 - 6r^2 + 1)$
13	4	2	Secondary (with/against the rule) astigmatism	Orientation-dependent shift and degradation of focus	$\sqrt{10}(4r^4 - 3r^2) \cos(2\alpha)$
14	4	4	(Horizontal) quatrefoil	Imaging anomalies with fourfold symmetry	$\sqrt{10}r^4 \cos(4\alpha)$

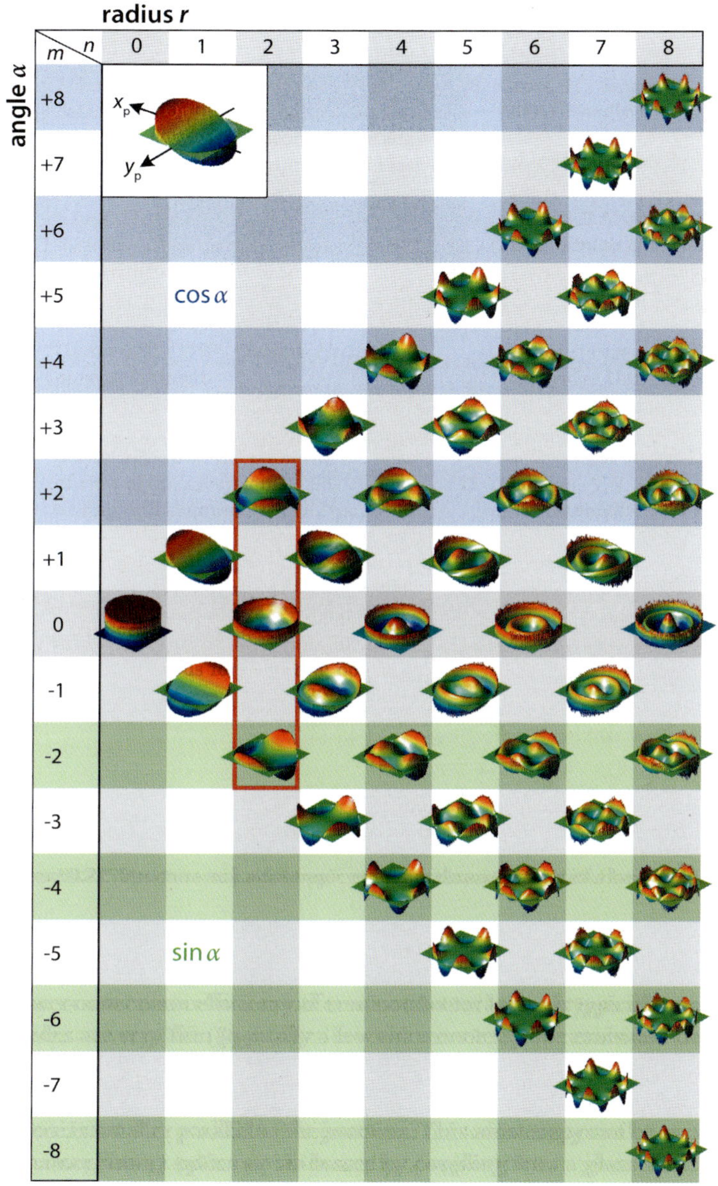

Figure A.23 Three-dimensional representation of Zernike polynomials. Deviations from the reference wavefront (green plane) are shown in red elevations or blue sinks. In other words, red, orange, and yellow regions represent advancing parts of the wavefront with respect to the reference wavefront. Retarded parts are shown in blue. Lower-order aberrations (see also Table A.3) are highlighted by a red frame. All types of aberration which are depicted to the right of the red frame are classified as higher-order aberrations. Adapted from [8].

troughs in the graphs. With some practice and creativity, one may also imagine the shape of the geometric spot from these graphs.

For illustrative purposes, let us now discuss the wavefront aberrations for spherical aberration and coma in more detail.

Example A.3

Spherical Aberration ($n = 4, m = 0, j = 12$) The graph in Figure A.23 which represents spherical aberration shows that the aberrated wavefront is considerably advanced at the outer zone of the pupil compared to the paraxial wavefront.[30] We now translate this to the schematic in Figure A.24. Wavefronts emanating from the object point Q are perfectly spherical (black). For clarification, we also included the corresponding rays (gray) which are normal to the wavefront. The paraxial part of the wavefronts converges close to the paraxial focal point Q'_{pa}, whereas the nonparaxial parts converge closer to the lens. The aberrated wavefront can be approximated by the spherical wavefront with minimum RMS_{wfe} (blue; Section A.1.7) which converges at image point Q'_{RMS}.

Example A.4

Coma ($n = 3, m = 1, j = 8$) For off-axis object points or decentered lenses, the radial symmetry of the optical system is broken. As a consequence, a nonsymmetric wavefront is formed which consistently leads to a nonsymmetric image. We can observe this general behavior in the graphical representation of the Zernike term for primary coma. As depicted in Figure A.25, the cross-section of the aberrated wavefront (red) crosses the paraxial wavefront (green) on the optical axis in the center of the exit pupil. Hence, we obtain by the retarded and advanced parts of the aberrated wavefront the typical tail-shaped image shown in Figure A.17b.

The Zernike polynomial expression of the wave aberration function is a valuable tool to describe and quantify image errors. In ophthalmology, the wavefront aberration is directly measured with suitable devices (Section 5.3). An alternative approach in technical optics is to detect and then analyze the point-spread function PSF of a real optical system (see Eq. (A37)). Subsequently, the obtained wavefront can be decomposed by a suitable Zernike polynomial fitting algorithm (see Eq. (A40)).[31] As a result, the Zernike coefficients c_n^m are determined which, in turn, provide useful information on how to optimize an optical system's image quality and performance (Problem PA.8).

30) In this example, we use the basic waveform of the corresponding Zernike term. In fact, the Zernike coefficient may also be negative so that the waveform appears inverted.

31) Because of the square in Eq. (A37), this is a nontrivial task requiring some form of retrieval algorithm.

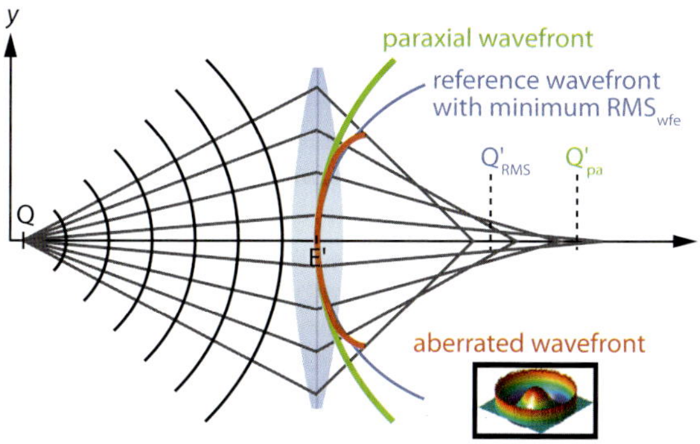

Figure A.24 The influence of spherical aberration ($\mathcal{W}_4^0 > 0$) on the shape of the wavefront. In paraxial approximation, the positive lens would refract an incident plane wavefront such that it forms a sphere around the focal point Q'_{pa} (green). In contrast, the wavefront influenced by spherical aberration (red) is advanced at the edges of the exit pupil so that no definite image point can be found. The spherical wavefront with the lowest RMS$_{wfe}$ value (blue; Section A.1.7) approximates the aberrated wavefront (red) best and converges at point Q'_{RMS}. E′ is the center of the exit pupil. Inset: Graphical representation of the corresponding Zernike polynomial $\mathcal{Z}_4^0$.

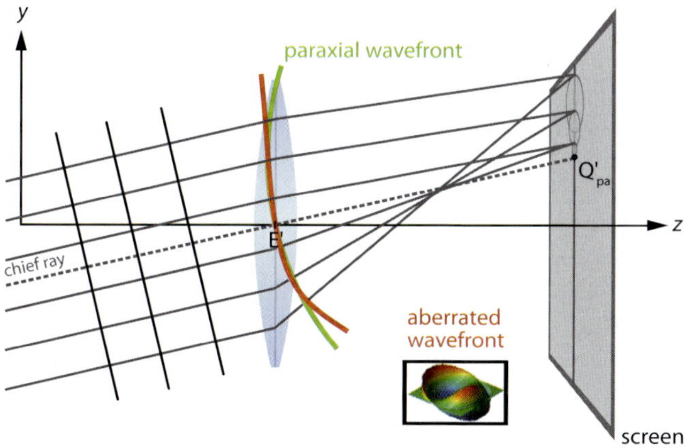

Figure A.25 The influence of coma ($\mathcal{W}_3^1 > 0$) on the shape of a wavefront. An aberration-free positive lens would refract an obliquely incident plane wave such that it forms spherical wavefronts (green) which converge at the paraxial image point Q'_{pa}. In contrast, the wavefront generated by a lens with coma aberration (red) is deformed such that it generates a typical comet-like image. E′ is the center of the exit pupil. Inset: Graphical representation of the respective Zernike polynomial $\mathcal{Z}_3^1$.

A.1.9
Chromatic Aberration

Until now, we have discussed imaging with monochromatic light. In real life, such conditions can be provided by a laser beam. There are also situations, however, where the light beam consists of rays with multiple colors (polychromatic light). Even if we restrict ourselves to paraxial rays, things become more complex in this case, since dispersion must be taken into account (Section A.1.1).

The refraction by a lens differs for each color and thus leads to different image points. If we place a screen behind the lens, as shown in Figure A.26, only one color will deliver a sharp image, whereas the others generate a blurred image. For an on-axis point, the axial dependence of the focus location is called *axial chromatic aberration*. The image location of off-axis points moves laterally as well as axially with the wavelength. The lateral shift in image location for off-axis object points is called *lateral chromatic aberration* and leads to different magnifications for each color.

A.2
Wave Optics

The concept of geometric optics only applies when looking at effects on length scales much larger than the considered wavelength. Since the wavelength of visible light is shorter than one thousandth of a millimeter, this condition is fulfilled in many cases. However, if we use special optical components or try to understand how light is bent around very small opaque objects, we have to use the wave picture of light. In fact, wave optics provides a more general description of light from which we may deduce geometric optics as a special case.

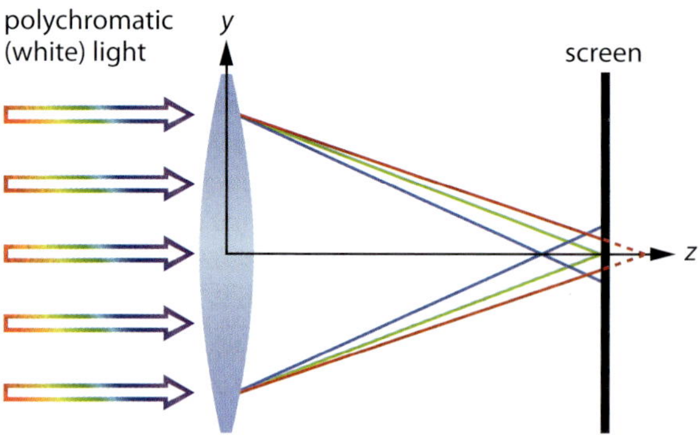

Figure A.26 The origin of axial chromatic aberration caused by the dispersive character of lenses. Here, only the green light component is sharply imaged. The other colors form a blurred image in the screen plane, as their focal planes do not coincide.

In wave optics, light is described as a transversal wave[32] which behaves quite similar to (mechanical) water waves. But, in contrast, light is absolutely massless and able to travel in a vacuum. Nevertheless, we may use the same wave equation of motion

$$\nabla^2 \psi - \frac{1}{c^2}\frac{\partial^2 \psi}{\partial t^2} = 0 \tag{A46}$$

for both types of waves. In the case of optical light waves, c denotes the speed of light, t the time, $\psi = \psi(\mathbf{x}, t)$ the wave function, and $\mathbf{x} = (x, y, z)$ the spatial coordinate. The wave function describes the local and temporal behavior of the light wave amplitude. $\nabla^2 \psi$ is the second spatial derivative of the wave function. Its explicit expression in Cartesian coordinates is

$$\nabla^2 \psi = \frac{\partial^2}{\partial x^2}\psi + \frac{\partial^2}{\partial y^2}\psi + \frac{\partial^2}{\partial z^2}\psi \,. \tag{A47}$$

A.2.1
Monochromatic Harmonic Waves

In principle, any wave function $\psi(\mathbf{x}, t)$ that satisfies Eq. (A46) represents a possible wave form for light. Let us start, as it is also common for mechanical waves, with so-called *harmonic* waves for which a maximum amplitude ψ_0 and a sinusoidal phase factor $\varphi(\mathbf{x})$ are defined. The solution of Eq. (A46) then reads

$$\psi(\mathbf{x}, t) = \psi_0(\mathbf{x}) \sin(\varphi(\mathbf{x}) - \omega t) \,, \tag{A48}$$

where ω denotes the angular frequency (oscillations per second, $[\omega] = 1/\mathrm{s}$). It should be mentioned that any arbitrary wave form can be generated by a sum (superposition) of harmonic waves (Sections A.2.3 and A.2.4).

For the sake of simplicity, we replace the phase factor $\varphi(\mathbf{x})$ in Eq. (A48) using Euler's[33] formula

$$\mathrm{e}^{\mathrm{i}\alpha} = \cos\alpha + \mathrm{i}\sin\alpha \,, \tag{A49}$$

with the imaginary unit $\mathrm{i} = \sqrt{-1}$ and obtain

$$\psi(\mathbf{x}, t) = \psi_0(\mathbf{x})\mathrm{e}^{\mathrm{i}(\varphi(\mathbf{x}) - \omega t)} \,. \tag{A50}$$

The standard rules of exponential functions allow us to split Eq. (A50) into a spatial part

$$\mathcal{X}(\mathbf{x}) = \psi_0(\mathbf{x})\mathrm{e}^{\mathrm{i}\varphi(\mathbf{x})} \tag{A51}$$

and a temporal part

$$\mathcal{T}(t) = \mathrm{e}^{-\mathrm{i}\omega t} \tag{A52}$$

32) Specifically, light is a transversal wave with oscillating electromagnetic fields [9, 10].

33) Leonhard Euler (1707–1783).

so that $\psi(\mathbf{x}, t) = \mathcal{X}(\mathbf{x})\mathcal{T}(t)$. If we insert this into Eq. (A46), the Helmholtz equation for the spatial dependence of the wave function results as

$$\left(\nabla^2 + k^2\right) \mathcal{X}(\mathbf{x}) = \left(\nabla^2 + k^2\right) \psi_0(\mathbf{x}) e^{i\varphi(\mathbf{x})} = 0 . \tag{A53}$$

Here, we used the wave number $k = \omega/c = 2\pi/\lambda$. We now consider two periodic wave forms which are of particular interest for our discussions.

A.2.1.1 Spherical Waves

Let us first consider a point source of light which emits spherical harmonic waves uniformly in all directions. Due to the spherical symmetry, we can replace the phase factor $\varphi(\mathbf{x})$ by the product of wave number k and radius r. r measures the distance from the light source. A solution of the wave equation (A46) which fulfills these characteristics is given by

$$\psi_{\text{sph}}(r, t) = \frac{\psi_0}{r} e^{i(kr - \omega t)} . \tag{A54}$$

The wavefronts of the spherical wave are separated by multiples of the wavelength (Figure A.27a) with equal amplitudes so that

$$r\psi(r, t_0) = (r + m\lambda)\psi(r \pm m\lambda, t_0) , \tag{A55}$$

where m is an integer. Similarly, we have

$$\psi(r_0, t) = \psi(r_0, t \pm t_{\text{per}}) , \tag{A56}$$

where $t_{\text{per}} = 2\pi/\omega$ represents the temporal periodicity.

The wavefronts travel with a phase velocity of

$$c_p = \frac{c_0}{n} = \frac{\omega}{k} = \frac{\lambda}{t_{\text{per}}} \tag{A57}$$

and, at the same time, increase their total surface area, like for the inflation of a balloon.

A.2.1.2 Plane Waves

Another solution of the wave equation (A46) is the so-called *plane wave*. In general, a plane wave travels along a specific direction. Hence, the wave number k, that we introduced earlier, should be replaced by the wave vector $\mathbf{k}$ (with the absolute value $|\mathbf{k}| = k$). The wave function of a plane wave is given by

$$\psi_{\text{pw}}(\mathbf{x}, t) = \psi_0 e^{i(\mathbf{kx} - \omega t)} . \tag{A58}$$

Wavefronts of plane waves with equal amplitude and phase are separated by multiples of the wavelength λ so that we have $\psi(r, t_0) = \psi(r \pm m\lambda, t_0)$.

Plane waves can also be deduced from spherical waves for the case that the point source is placed "far away" from the optical system.[34] If the size of the involved

34) For example, a star is a point source which is very far away. Its light thus arrives on the Earth as a plane wave.

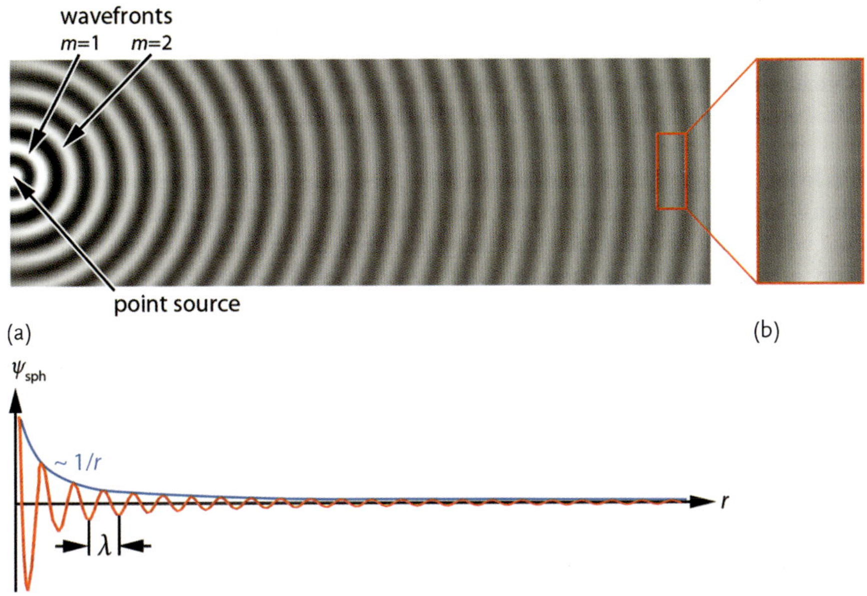

Figure A.27 (a) Spherical wave emitted from a point source. Bright regions represent high positive amplitudes (wave crests) and dark regions negative amplitudes (wave troughs). The wave function ψ is plotted against the radius r from the light source. (b) When we cut out a small part of the wave (red box) "far" away from the light source, it becomes clear that a spherical wavefront can be approximated by a plane wavefront.

optical components (e.g., diameter of lenses and apertures) is small compared to the distance between the point source and the optical system, we may simply ignore the curvature of the spherical wavefronts (Figure A.27b).

A.2.1.3 Eikonal Equation: The Link Between Geometric and Wave Optics

A comparison of the wave functions (A54) and (A58) reveals that for arbitrary wavefronts the spatial part can be generalized to

$$\mathcal{X} = \psi_0(\mathbf{x})\mathrm{e}^{\mathrm{i}k_0 S(\mathbf{x})} . \tag{A59}$$

Here, we assume that the vector property of $\mathbf{k}$ and the refractive index n of the medium are included in the three-dimensional surface function $S(\mathbf{x})$, the so-called *eikonal*. k_0 is the wave number in vacuum related to the vacuum wavelength via $\lambda_0 = 2\pi/k_0$. The wavefronts are determined by $S(\mathbf{x}) = \text{const}$, and the wavefront normals point along the direction of the gradient ∇S.

We will now assume that the wave travels in a homogeneous medium for which the refractive index $n(\mathbf{x})$ varies very slowly with position. Similarly, the amplitude $\psi_0(\mathbf{x})$ changes sufficiently slowly with respect to the location so that it can be regarded as constant within a distance of λ_0. In mathematical terms this means that $\lambda_0^2 \nabla^2 \psi_0/\psi_0 \ll 1$. From this, we can derive the eikonal equation

$$|\nabla S(\mathbf{x})|^2 = n(\mathbf{x})^2 . \tag{A60}$$

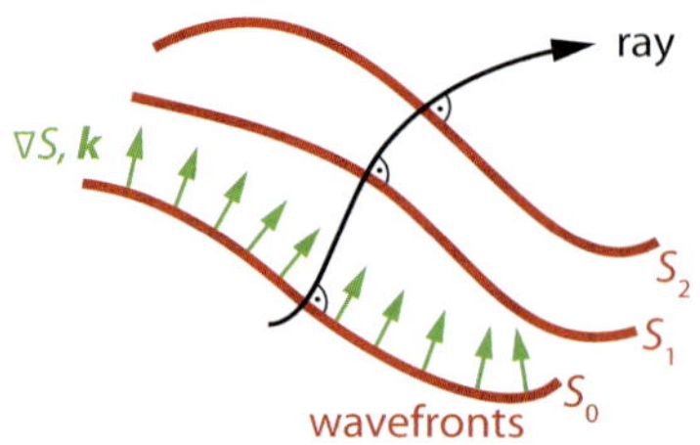

Figure A.28 The link between wave and geometric optics. Rays (green) are normal to the surfaces of equal phase (wavefronts; red) and thus point along the propagation direction of light waves (black arrow).

In principle, this equation can also be derived from geometric optics (from Fermat's principle [10], to be exact). With Eq. (A60) we can thus identify light rays as wave vectors which are perpendicular to the wavefronts and point in the traveling direction of the wave (Figure A.28).

A.2.1.4 **Polarization**

Since light is a transversal wave, the amplitude vector cannot point in any arbitrary direction, but is kept in the plane of oscillation which is normal to the wave vector. In this plane, the direction of the amplitude vector can be freely chosen. This direction is called the *polarization*[35]. For a plane wave, we must therefore write the wave function more precisely as

$$\psi(\mathbf{x}, t) = \hat{\mathbf{x}}\psi_{0x}\mathrm{e}^{\mathrm{i}(k_z z-\omega t)} + \hat{\mathbf{y}}\psi_{0y}\mathrm{e}^{\mathrm{i}(k_z z-\omega t+\xi)}\,, \qquad \text{(A61)}$$

where $\hat{\mathbf{x}}$ and $\hat{\mathbf{y}}$ are orthogonal unit vectors pointing in the x and y directions, respectively. z denotes the propagation direction of the wave. As both vectors are orthogonal, the relation $\hat{\mathbf{x}} \cdot \hat{\mathbf{y}} = 0$ is fulfilled. The wave amplitude obviously has two amplitude components ψ_{0x} and ψ_{0y} which are retarded by a phase factor $\exp(i\xi)$.

The emitted light of many "classical" sources (sun, light bulbs, gas-filled tubes, and so on) is unpolarized. This means that light is composed of different waves with arbitrary and fluctuating values for ψ_{0x}, ψ_{0y}, and ξ. But for some applications (e.g., filtering and light manipulation) it is sometimes required to single out a certain polarization.

The polarization must also be considered when light passes through a *birefringent* (anisotropic) medium. In such a medium, the refractive index depends on the polarization and traveling direction of light. Let us have a look at the nerve fiber layer of the retina as an example (Section 6.8.3). On the micrometer scale, the nerve fibers can be regarded as a layer of long parallel cylinders which are perpendicular to the retinal surface. This means that the fibers are oriented along a

35) The human eye is just weakly sensitive to the polarization of light. Interestingly, some humans are able to perceive polarized light in the center of the visual field as a yellowish horizontal bar with fuzzy ends, for example, when looking through polarized sunglasses onto a bright background. This phenomenon was first described by Wilhelm von Haidinger (1795–1871) and is thus called *Haidinger's brush*. Other animals like bees, octopuses, squids, cuttlefishes, and mantis shrimps are able to detect light polarization quite precisely for navigation and communication purposes.

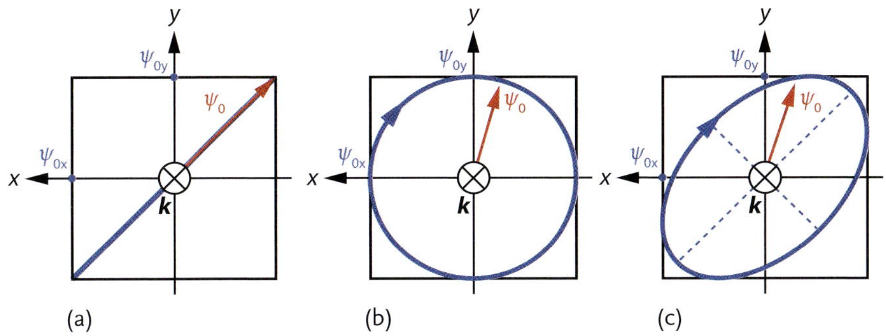

Figure A.29 Different states of polarization. (a) Linear, (b) right circular, and (c) right elliptical polarization of a light wave. In this scheme, we look at the amplitude vector ψ_0 from the light source.

preferred direction. Let us assume that the fibers are parallel to the x direction. When unpolarized light passes through these fibers, the incident wave (described by Eq. (A61)) alters its state of polarization and breaks up into two components with $\boldsymbol{\psi}_1(\mathbf{x}, t) = \hat{\mathbf{x}}\psi_{0x} \exp(\mathrm{i}k_z z - \mathrm{i}\omega t)$ and $\boldsymbol{\psi}_2(\mathbf{x}, t) = \hat{\mathbf{y}}\psi_{0y} \exp(\mathrm{i}k_z z - \mathrm{i}\omega t + \mathrm{i}\xi)$. The change of polarization happens because of the maximum refractive index for the polarization component parallel to the fibers and the minimum refractive index for the polarization component perpendicular to the fibers.

In the course of this book, the vector property of the amplitude is only considered in very special cases. We rather ignore the polarization elsewhere to keep the discussions simple and clear. In the latter cases, we substitute the vector $\boldsymbol{\psi}$ by its scalar ψ.

Linear polarization Linear polarization is obtained if the x and y components of the wave function show no relative retardation ($\xi = 0$). Here, the amplitude oscillates along a straight line in the xy plane (Figure A.29a). The wave function for linear polarization thus reads

$$\boldsymbol{\psi}_{\mathrm{lin}}(\mathbf{x}, t) = \left(\hat{\mathbf{x}}\psi_{0x} + \hat{\mathbf{y}}\psi_{0y}\right) \mathrm{e}^{\mathrm{i}(k_z z - \omega t)} . \tag{A62}$$

Due to the properties of the complex exponential function, linear polarization also results for $\xi = \pm m\pi$, where m is an integer.

In the context of linear polarization, we should also mention the general Fresnel equations which describe the transmittance T and reflectance R at the interface of two media with different refractive indices n and n' (see also Section A.1.1 for the special case of normal incidence). For a light wave polarized perpendicular to the plane of incidence (s-polarization), we have

$$R_{\mathrm{s}} = \frac{I_{\mathrm{r,s}}}{I_{\mathrm{i}}} = \left(\frac{n\cos\gamma - n'\cos\gamma'}{n\cos\gamma + n'\cos\gamma'}\right)^2 = \frac{\sin^2(\gamma - \gamma')}{\sin^2(\gamma + \gamma')} , \tag{A63}$$

$$T_{\mathrm{s}} = \frac{I_{\mathrm{t,s}}}{I_{\mathrm{i}}} = \left(\frac{n'\cos\gamma'}{n\cos\gamma}\right)\left(\frac{2n\cos\gamma}{n\cos\gamma + n'\cos\gamma'}\right)^2 . \tag{A64}$$

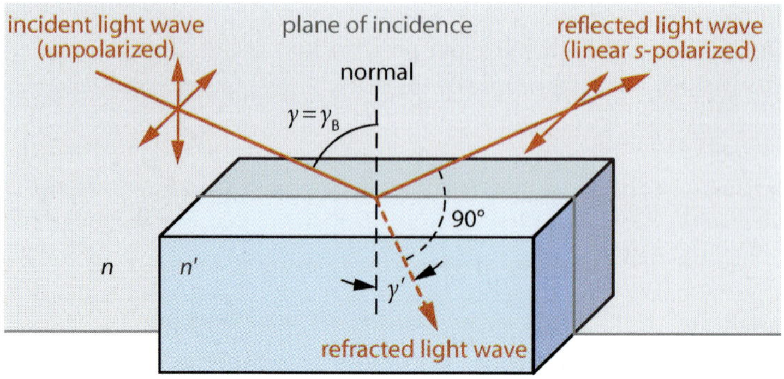

Figure A.30 Light incident on a glass slide at the Brewster angle γ_B is reflected with s-polarization. This corresponds to a linear polarization orthogonal (normal) to the plane of incidence. The refracted portion of light is a composite of both linear polarization components (*s* and *p*).

For light which is polarized parallel to the plane of incidence (*p*-polarization), the Fresnel equations are given by

$$R_p = \frac{I_{r,p}}{I_i} = \left(\frac{n' \cos\gamma - n \cos\gamma'}{n' \cos\gamma + n \cos\gamma'}\right)^2 = \frac{\tan^2(\gamma - \gamma')}{\tan^2(\gamma + \gamma')} , \quad \text{(A65)}$$

$$T_p = \frac{I_{t,p}}{I_i} = \left(\frac{n' \cos\gamma'}{n \cos\gamma}\right)\left(\frac{2n \cos\gamma}{n' \cos\gamma + n \cos\gamma'}\right)^2 . \quad \text{(A66)}$$

$I_{r,s}$ and $I_{r,p}$ are the intensities of the reflected *s*- and *p*-polarized portion of the light wave, respectively. $I_{t,s}$ and $I_{t,p}$ are the intensities of the transmitted *s*- and *p*-polarized portion of the light wave, respectively. I_i denotes the intensity of the incident wave.

If light is incident on the interface at the Brewster angle (Figure A.30 and Problem PA.9)

$$\gamma_B = \arctan\left(\frac{n'}{n}\right) , \quad \text{(A67)}$$

only the *s*-polarized portion is reflected. The remaining *p*-polarized portion is refracted and travels through the medium. As a consequence, the reflected beam becomes linearly polarized.

Circular polarization If the amplitudes in the x and y direction are equal ($\psi_{0x} = \psi_{0y}$) and the phase retardation is set to $\xi = \pm\pi/2$, we have the special case of circular polarization for which the wave function reads

$$\boldsymbol{\psi}_{cp}(\mathbf{x}, t) = \hat{\mathbf{x}}\psi_{0x} e^{i(\mathbf{kx}-\omega t)} \pm \hat{\mathbf{y}} i \psi_{0y} e^{i(\mathbf{kx}-\omega t)} . \quad \text{(A68)}$$

In the case of $\xi = -\pi/2$, the amplitude vector rotates clockwise when we look from the light source. This is referred to as "right circular polarization" (RCP; Figure A.29b).[36] Correspondingly, the amplitude vector rotates counterclockwise for $\xi = +\pi/2$ which is then called "left circular polarization" (LCP). Circularly polarized light can be generated from linear polarized light with a quarter-wave plate (see [9, 10] and Info Box 6.4).

Elliptical polarization In the most general case of fully polarized light, the tip of the amplitude vector $\boldsymbol{\psi}_0$ lies on an ellipse (Figure A.29c). Like for circular polarized light, ψ_0 either rotates clockwise or counterclockwise during propagation.

A.2.1.5 Radiometric and Photometric Quantities

In geometric optics, we identified the density of rays as the intensity. Indeed, this was just an auxiliary interpretation. Now, we will link the amplitude ψ_0 of the wave function to the optical intensity I which is defined as the power per unit area ($[I] = \mathrm{W/m^2}$, "watt per square meter") and related to the wave function via

$$I(\mathbf{x}) = \frac{1}{2t_{\mathrm{per}}} \int_{-t_{\mathrm{per}}}^{t_{\mathrm{per}}} |\psi(\mathbf{x}, t)|^2 \mathrm{d}t \,. \tag{A69}$$

Here, we have averaged over the period t_{per} of the oscillation. In contrast to the wave function, the intensity is a directly measurable quantity.

With given intensity, the power $P(t)$ ($[P] = \mathrm{W}$, "watt") flowing into an area $\mathbf{S}$ that is perpendicular to the traveling direction can be calculated by

$$P = \int_{\mathrm{S}} I(\mathbf{x}) \mathrm{d}\mathbf{S} \,. \tag{A70}$$

The energy E ($[E] = \mathrm{J}$, "joule") accumulated during the time interval $\mathrm{d}t$ is determined by

$$E = \int P \mathrm{d}t \,. \tag{A71}$$

I, P, and E are so-called *radiometric* quantities, which are measured in terms of absolute power. At this point, we also introduce *photometric* quantities which describe the radiometric quantities weighted by the wavelength-dependent sensitivity function $\mathcal{V}(\lambda)$ of the human eye (Figure 2.17 in Section 2.3). Typically, one uses $\mathcal{V}(\lambda)$ of the retinal cones, which is the averaged sensitivity function of S, M, and L cones. Table A.4 gives an overview of the relations and definitions of radiometric and photometric quantities.

The *luminous power* ϕ_{v} ($[\phi_{\mathrm{v}}] = \mathrm{lm}$, "lumen") is the photometric measure for the light power perceived by the eye. From this, we can derive the photometric "brightness" of a light source, that is, the so-called *luminous intensity* $\mathcal{I}_{\mathrm{v}}$, which is defined as follows: A monochromatic light source which emits an optical power of

36) In contrast to the convention used in this book, circular polarization may also be defined the other way round, that is, we look from the detector at the tip of the wave vector $\mathbf{k}$.

Table A.4 Comparison of radiometric and photometric quantities. ε is the angle between the normal of the considered surface S and the specified direction. Ω denotes the solid angle and $\mathcal{V}(\lambda)$ the wavelength-dependent sensitivity function of the human eye. $K_m = 1/683$ W is an optical conversion power which, in principle, defines the unit "candela" (cd). In the SI system, the "candela" is one of the basic physical units. Unit abbreviations: lumen (lm), candela (cd), lux (lx), watt (W), joule (J), steradiant (sr), meter (m), second (s).

Radiometric quantity	Unit	Photometric quantity	Unit	Conversion/Definition
Power P	W	Luminous power ϕ_v	lm	$\phi_v = K_m \int_{380\text{nm}}^{780\text{nm}} \mathcal{V}(\lambda) \frac{dP}{d\lambda} d\lambda$
Radiant intensity	W/sr	Luminous intensity $\mathcal{I}_v$	cd = lm/sr	$\mathcal{I}_v = \frac{d\phi_v}{d\Omega}$
Radiance	W/(sr m^2)	Luminance $\mathcal{L}_v$	cd/m^2	$\mathcal{L}_v = \frac{d^2\phi_v}{d\Omega\, dS \cos\varepsilon}$
Intensity I (irradiance)	W/m^2	Illuminance $\mathcal{E}_v$	lx = lm/m^2	$\mathcal{E}_v = \int_{2\pi\text{sr}} \mathcal{L}_v \cos\varepsilon\, d\Omega$
Energy E	J = Ws	Luminous energy Q_v	lm s	$Q_v = \int \phi_v dt$
Exposure Φ (fluence, energy density)	J/m^2	Luminous exposure H_v	lx s	$H_v = \frac{dQ_v}{dS}$

$K_m = 1/683$ W (watts) at a wavelength of 555 nm into a solid angle of 1sr (steradian) has a luminous intensity of 1 cd (candela).

For light sources with a certain surface area S, we may also use the *luminance* $\mathcal{L}_v$. It is the ratio of the luminous intensity emitted in a distinct direction divided by the surface area projected into that direction.[37] Thus, the unit of the luminance is "candela per square meter" ($[\mathcal{L}_v] = \text{cd/m}^2$).

When the apparent "brightness" shall be characterized from a receiver's point of view, we use the *illuminance* $\mathcal{E}_v$ ($[\mathcal{E}_v] = \text{lx}$, "lux"). This quantity tells us how strongly the incident light illuminates a certain surface. In practice, one often has to determine the effectiveness of a uniform, extended light source which overfills the eye's field of view. In principle, we could use the standard definition of the illuminance, but the distance between the pupil and retina varies for every person and is usually not available. For practical reasons, the *retinal illuminance* $\mathcal{L}_{ret} = \mathcal{L}_v S_{pupil}$ ($[\mathcal{L}_{ret}] = \text{td}$, "troland"[38]) is introduced for which we only need to know the source luminance and the area of the entrance pupil. Note that the retinal illuminance does not accurately describe the subjective sensation of brightness, as the imaging properties of ocular media (Chapter 2) are not considered.

37) The projected surface area is given by $S_{proj} = S_{surface} \cos\varepsilon$, where ε denotes the angle between the considered direction and the surface normal.

38) The "troland" is defined as the retinal illumination for a pupil area of $S_{pupil} = 1\,\text{mm}^2$ produced by a radiating surface with a luminance of $\mathcal{L} = 1\,\text{cd/m}^2$.

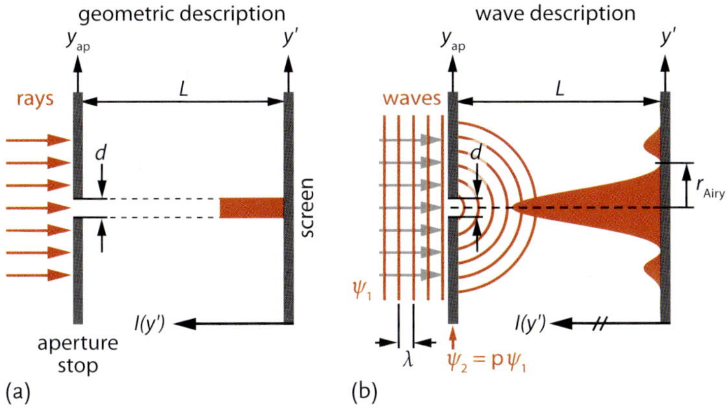

Figure A.31 Light incident on an aperture stop. (a) Geometric description: In geometric optics, only those rays which are directly in sight arrive at the screen. As a consequence, a sharp intensity distribution $I(x', y')$ is obtained. This is the image that we observe if the size of the aperture stop is much larger than the wavelength of light ($a \gg \lambda$). If $a \approx \lambda$, we must switch to the wave description. (b) Wave description: A plane wave ψ_1 passes a circular aperture stop. After the wave has traveled a distance L in free space, it forms an Airy intensity pattern on the screen. The distance between the central maximum and the first minimum of the Airy pattern is defined as the radius r_{Airy} of the so-called Airy disk.

A.2.1.6 Diffraction and Angular Resolution

When light passes through an aperture stop, geometric optics would lead us to expect a bright spot on the screen with a sharp shadow cast as shown in Figure A.31a. But in reality, light is "bent around" the edge of the aperture stop. This is a purely wave optical effect called *diffraction*. Diffraction patterns can be more or less pronounced, depending on the size of the aperture stop relative to the wavelength of light λ and its distance to the screen L.

In the simplest model of diffraction, the so-called *Fraunhofer diffraction*[39], we assume that the screen is far away from the aperture ($L \gg \lambda$). As depicted in Figure A.31b, an incident plane wave ψ_1 passes through an aperture stop ($x_{ap} y_{ap}$ plane) which modifies the wave function to $\psi_2 = p\,\psi_1$ with the apodization function

$$p(x_{ap}, y_{ap}) = \begin{cases} 1 & \text{inside the aperture} \\ 0 & \text{outside the aperture} \end{cases} . \tag{A72}$$

On the screen, the intensity $I(x', y')$ forms a so-called *Airy pattern*[40]. The pattern consists of a central peak (Airy disk) surrounded by concentric rings. The radius of the Airy disk is given by $r_{Airy} = 1.22\lambda/d$ (Figure A.32b).

Let us consider a light wave which is diffracted by an aperture stop or by any obstacle of arbitrary shape. We assume that the incident light is monochromatic

39) Joseph von Fraunhofer (1787–1826).
40) George Airy (1801–1892).

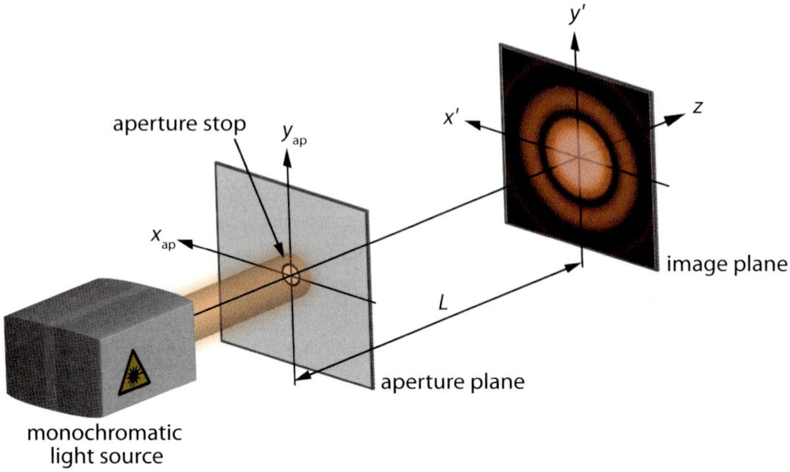

Figure A.32 Geometry used to describe Fraunhofer diffraction at a circular aperture stop.

(with wavelength λ) and not polarized. In addition, the Fresnel number

$$F = \frac{a^2}{L\lambda} \tag{A73}$$

shall be relatively large, where a represents the characteristic size of the aperture stop (e.g., the diameter in the case of a circular aperture stop). Under these conditions, the intensity distribution is given by

$$I(x', y') \approx \frac{1}{(\lambda L)^2} \times \left| \int_{-\infty}^{\infty} \int_{-\infty}^{\infty} p(x_{ap}, y_{ap}) \psi_1 \exp\left(\frac{-2\pi i (x' x_{ap} + y' y_{ap})}{\lambda L} \right) \mathrm{d}x_{ap} \mathrm{d}y_{ap} \right|^2 . \tag{A74}$$

The integration from minus to plus infinity leads to a finite result, as the wave is "confined" by the aperture stop. For the special case of a circular aperture stop (Figure A.32), the intensity distribution on the screen reads

$$I_{circ}(x', y') = I_1 \left(\frac{\pi d^2}{2\lambda L} \right)^2 \left(\frac{\mathcal{J}_1(\kappa)}{\kappa} \right)^2 , \quad \kappa = \frac{\pi d \sqrt{x'^2 + y'^2}}{\lambda L} . \tag{A75}$$

d is the aperture diameter and $\mathcal{J}_1$ the Bessel[41] function of first order [4]. Diffraction appears not only for "empty" aperture stops but also for lenses and mirrors since every optical component has a finite diameter.

As a consequence of diffraction, the maximum angular resolution of an optical system is limited. To understand this behavior, let us imagine two point sources

41) Friedrich Bessel (1784–1846).

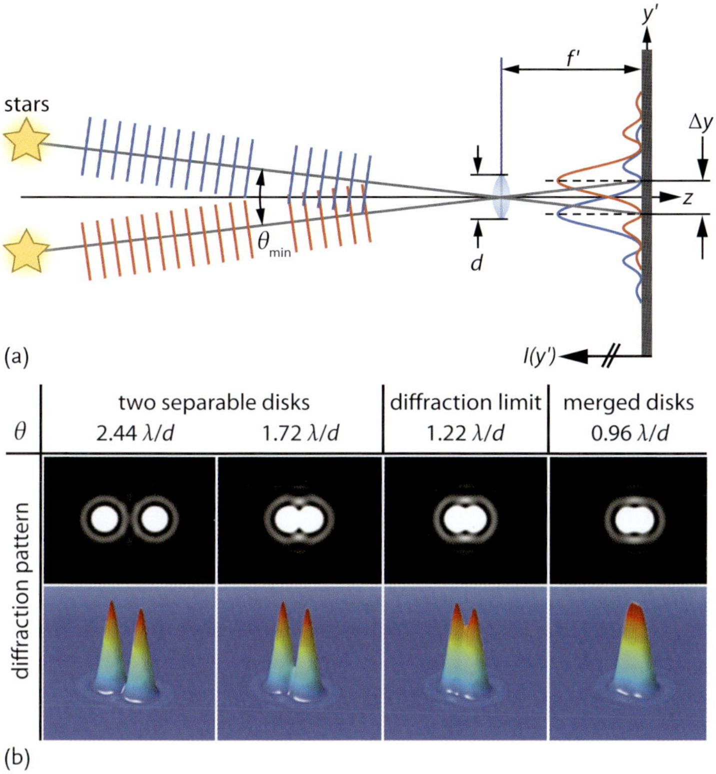

Figure A.33 Resolution limit of an optical system. (a) Two distant point sources (e.g., stars) emit plane waves at an angle $\theta_{\text{min}}/2$ which are incident to the same lens forming an aperture with a diameter d. After passage through the lens, the Airy intensity patterns of both light sources overlap. (b) Two-dimensional (top) and three-dimensional (bottom) plots of the intensity pattern $I(x', y')$ (overlapping Airy diffraction patterns) of two adjacent point sources. Down to $\theta_{\text{min}} = 1.22\lambda/d$ the patterns are well-separated, whereas they merge together for $\theta_{\text{min}} < 1.22\lambda/d$, that is, below the diffraction limit.

which emit plane waves like shown in Figure A.33a. The waves are incident on a lens and include an angle θ. After passage through the lens, they create two overlapping Airy patterns. For small values of θ, the distance between the central crests of both intensity patterns can be approximated by

$$\Delta y \approx f'\theta \,. \tag{A76}$$

In the x direction, the corresponding condition $\Delta x \approx f'\theta$ holds. For symmetry reasons, the lateral optical resolution will be represented by $\Delta(x, y)$ in the following. If we reduce the angle θ, the images are gradually approaching each other. Some example images are depicted in Figure A.33b as two- and three-dimensional intensity plots. Above a threshold distance (diffraction limit), we can clearly distinguish the images of both point sources.

The Rayleigh[42] criterion specifies that the two images can be regarded as just resolved when the central peak (maximum) of one image coincides with the first trough (minimum) of the other, as shown for the blue and red curves in Figure A.33a. If the distance between both central crests is greater, the two point sources are well-separated. But if the distance is smaller, the points merge together. Based on this condition, the maximum angular resolution is defined by the inverse of the minimum resolvable angle θ_{min} as

$$\theta_{\text{min}} = \arcsin\left(1.22\frac{\lambda}{d}\right) \approx 1.22\frac{\lambda}{d}\,, \tag{A77}$$

where d is the aperture diameter.

For a microscope, the minimum lateral distance between two resolvable objects in both the x and y direction can be derived directly from Eq. (A77) and is often specified by the Abbe diffraction limit

$$\Delta(x, y) = 1.22\left(\frac{\lambda}{2\text{NA}}\right)\,, \tag{A78}$$

in which NA is the numerical aperture $\text{NA} = n \sin\alpha$ defined by the medium's refractive index in front of the lens aperture and the maximum acceptance angle α of the microscope objective (Section A.1.4). The relations (A77) and (A78) pose fundamental limitations to the resolution of small objects, which are also relevant for imaging in ophthalmology.

A.2.2
Paraxial Solutions of the Wave Equation

We will now reconsider the Helmholtz equation (A53) under paraxial approximation that also plays a very important role in geometric optics (Section A.1.2). The simple laws which can be derived for the paraxial case help us to understand the general behavior of imaging.

Let us first establish a link between paraxial rays and paraxial waves.[43] For this purpose, we consider the spatial part of a plane wave which travels in the z direction. The wave function is given by $\mathcal{X}(z) = \psi_0(\mathbf{x})\exp(ik_z z)$, where we assumed a modulated amplitude $\psi_0(\mathbf{x})$ (envelope) which changes slowly within the distance of a wavelength (Figure A.34). As rays are normal to the wavefront, we will restrict ourselves to those parts of the wavefront which form small angles to the optical axis. The wavefront thus bends just slightly and forms a paraxial beam. Since $\mathcal{X}(z)$ must satisfy Eq. (A53), the modulated amplitude $\psi_0(\mathbf{x})$ has to fulfill

$$\left(\frac{\partial^2\psi_0}{\partial x^2} + \frac{\partial^2\psi_0}{\partial y^2} - 2ik_z\frac{\partial\psi_0}{\partial z}\right) \approx 0 \tag{A79}$$

42) John Rayleigh (1842–1919).

43) We will see later that both approaches can be treated with the same elegant matrix formalism introduced in Section A.1.3.

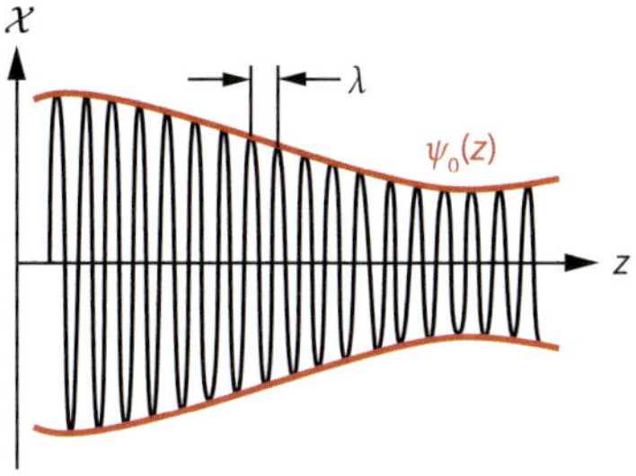

Figure A.34 Paraxial wave function $\mathcal{X}(z)$ with a modulated amplitude $\psi_0(z)$.

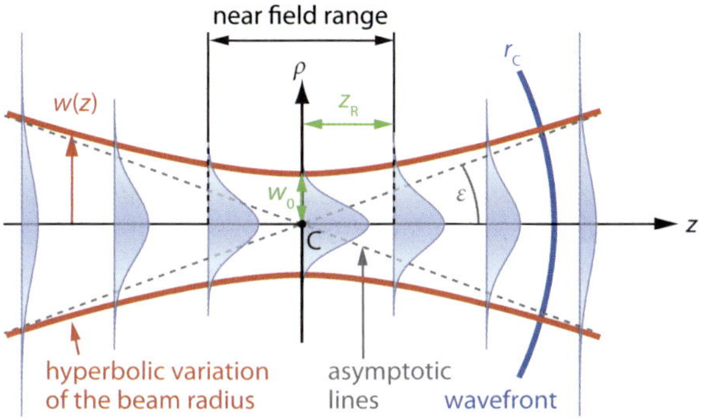

Figure A.35 Geometric parameters of the Gaussian beam. $w(z)$ denotes the beam radius, w_0 the waist radius (focus), and ε the divergence angle. At a large distance from the focus, the spherical wavefront of the Gaussian beam with radius of curvature $r_C(z)$ is centered around the waist center C. Gaussian beams thus approximate spherical waves in the far field range.

which is often called the *slowly varying envelope approximation* of the Helmholtz equation.

A.2.2.1 **Gaussian Beams**

The question now is, what do the solutions of Eq. (A79) look like? One of the practically most relevant solutions are Gaussian[44] beams which are particularly used to describe laser beams. The modulated solution of Eq. (A79) for the case of a Gaussian beam reads (see also Problems PA.10 and PB.3)

$$\psi_0(z) = \frac{\mathcal{A}}{q(z)} \exp\left(-ik_z \frac{\rho^2}{2q(z)}\right) . \tag{A80}$$

$\mathcal{A}$ is a constant and $\rho = \sqrt{x^2 + y^2}$ denotes the radial distance from the beam axis (z axis in Figure A.35). $q(z)$ is the complex beam parameter which we will discuss next.

44) Carl Friedrich Gauss (1777–1855).

Complex beam parameter and wavefront curvature The parameters of a Gaussian beam are described by means of the complex beam parameter $q(z)$ via

$$\frac{1}{q(z)} = \frac{1}{r_C(z)} - \frac{i\lambda}{\pi w^2(z)} \,. \tag{A81}$$

Here, we used the beam radius $w(z)$, the wavelength λ, and the radius of curvature

$$r_C(z) = z\left(1 + \frac{z_R^2}{z^2}\right) \,. \tag{A82}$$

The *Rayleigh length*

$$z_R = \frac{\pi w_0^2}{\lambda} \tag{A83}$$

is a measure for the depth of focus and defines the near field range $-z_R \le z \le +z_R$. In the near field range, the beam radius is $w_0 < w(z) < \sqrt{2}w_0$ with the beam radius at the beam waist $w_0 = w(z = 0)$[45]. From Eq. (A81) we conclude that the radius of curvature r_C and the beam radius $w(z)$ are associated with the real and imaginary parts of $q(z)$, respectively. At the waist position $z = 0$, the beam has a minimum radius of w_0 and a planar wavefront. This means that $r_C(z = 0) \to \infty$ and $q(z)$ is purely imaginary. As a consequence, $\psi_0(z = 0)$ becomes real (see Eq. (A80)). "Far away" from the waist (*far field* range), the curvature of the wavefront may be approximated by a sphere around the waist center point C, since $q(z) \approx r_C(z)$. Point C lies on the beam axis at the waist position $z = 0$.

The spatial evolution of the complex beam parameter is given by

$$q(z) = z + i z_R \,. \tag{A84}$$

Intensity and beam profile The intensity of the Gaussian beam is given by[46]

$$\begin{aligned} I(\rho, z) &\propto \mathcal{X}\mathcal{X}^* \\ &\propto I_0 \left(\frac{w_0}{w(z)}\right)^2 \exp\left(-\frac{2\rho^2}{w^2(z)}\right) \,, \end{aligned} \tag{A85}$$

where $\mathcal{X}(\rho, z) = \psi_0(\rho, z) \exp(i k_z z)$ is the term describing the spatial evolution of a plane wave. $I(\rho, z)$ is maximal on the beam axis ($\rho = 0$) and decreases successively when moving away from the axis. In this context, the beam radius $w(z)$ is defined by the radial distance at which the maximum beam intensity decreases by a factor 0.135 (i.e., $I(w, z) = I_{peak}(z)/e^2$). If we move along the z direction, the intensity profile maintains the Gaussian shape, but the beam radius changes by

$$w(z) = w_0 \sqrt{1 + \left(\frac{z}{z_R}\right)^2} \,. \tag{A86}$$

45) $2w_0$ is also called the *spot size* of a beam.

46) $\mathcal{X}^*$ is the complex conjugate of $\mathcal{X}$.

Beam divergence From Eq. (A86) it follows that the hyperbolic variation of $w(z)$ has an asymptotic behavior in the far field (dashed gray lines in Figure A.35). This property can be used to define the divergence[47] of a Gaussian beam. The divergence angle ε is determined by the slope of the asymptotes via

$$\tan\varepsilon = \frac{w_0}{z_R} = \frac{\lambda}{\pi w_0}\,. \tag{A87}$$

Equation (A87) tells us that a smaller radius at the beam waist means a larger beam divergence and vice versa. In many practical situations, the beam divergence is small so that we may use the (paraxial) approximation $\tan\varepsilon \approx \varepsilon$.

A.2.2.2 Gaussian Beams in Paraxial Optical Systems

In the context of paraxial approximation in geometric optics (Section A.1.3), ABCD matrices have been used to calculate how light is imaged by an arbitrary optical system. In fact, we can use the same approach to describe the optical imaging of focused Gaussian beams.

We already mentioned that spherical waves are a good approximation for Gaussian beams far away from the beam waist. We will thus initially consider a spherical wave emanated from an on-axis object point O (Figure A.36). In paraxial approximation, the wave's object-side radius of curvature is $r_{C1} = h/\gamma$. After passage through the lens, the outgoing wave's radius of curvature is $r_{C2} = h'/\gamma'$. Next, we insert both radii into Eqs. (A17) and (A18) and derive the relation

$$r_{C2} = \frac{Ah + B\gamma}{Ch + D\gamma} = \frac{Ar_{C1} + B}{Cr_{C1} + D}\,. \tag{A88}$$

The radius of curvature of a Gaussian beam is related to the complex beam parameter $q(z)$ via (A81). One can show that an analogous Gaussian ABCD law

$$q_2 = \frac{Aq_1 + B}{Cq_1 + D} \tag{A89}$$

results by replacing r_C with q. That Eq. (A89) holds for large z can be easily seen from Eq. (A81). It can be shown that Eq. (A89) also holds for all values of z.

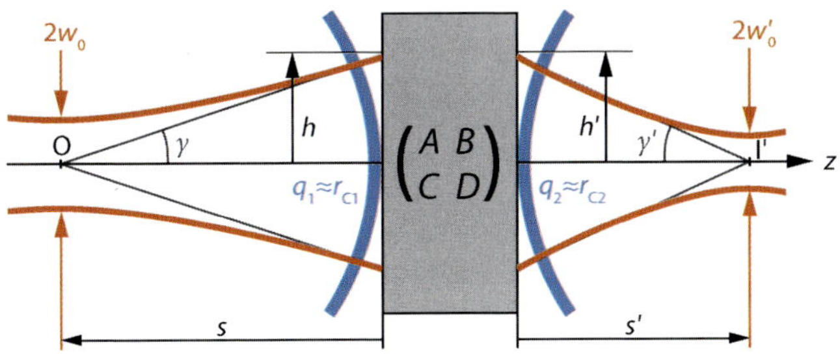

Figure A.36 Paraxial imaging of a Gaussian beam by an arbitrary optical system.

47) The beam divergence is a result of the wave nature of light. It can be considered as the diffraction of a beam at its own aperture, that is, the beam waist.

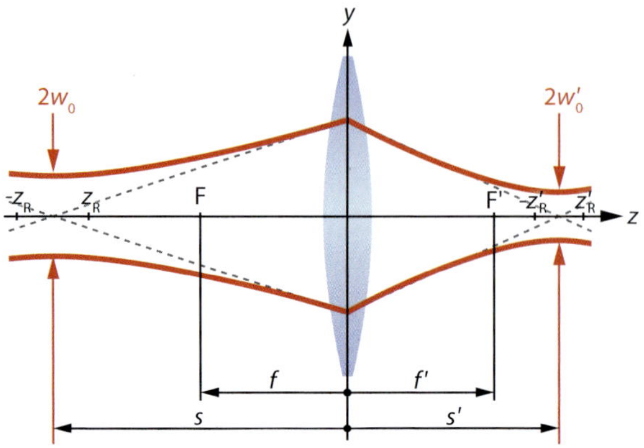

Figure A.37 Imaging of a Gaussian beam by a thin lens. Unprimed (primed) quantities refer to geometrical parameters on the object (image) side.

A.2.2.3 **Gaussian Beam Focusing**

We will now apply Eq. (A89) to a Gaussian beam which passes through a thin lens (Figure A.37). Specifically, we consider the imaging of a Gaussian beam waist. The corresponding ABCD matrix is given by

$$\underline{\mathbf{M}} = \begin{pmatrix} A & B \\ C & D \end{pmatrix} = \begin{pmatrix} 1 - \frac{s'}{f'} & s' - s\left(1 - \frac{s'}{f'}\right) \\ -\frac{1}{f'} & 1 + \frac{s}{f'} \end{pmatrix}, \tag{A90}$$

where s and s' denote the distances between the beam waists and the lens on the object and image side, respectively. f' denotes the distance between the lens and its image-side focal point F′. When we insert Eq. (A90) into Eq. (A89) and take into account that both q_1 and q_2 are imaginary at the beam waist, we obtain

$$s' = \frac{f'(s^2 + sf' + z_R^2)}{(s + f')^2 + z_R^2} \tag{A91}$$

and for the image-side waist radius

$$w'_0 = \frac{w_0 f'}{\sqrt{(s + f')^2 + z_R^2}} . \tag{A92}$$

Equation (A91) can also be written as

$$\frac{1}{s'} = \frac{1}{f'} + \frac{1}{s}\left(1 - \frac{z_R^2}{z_R^2 + sf' + s^2}\right) \tag{A93}$$

and thus passes into the lens equation (A14) for $z_R = 0$. Starting from Eqs. (A91) and (A92), let us consider two special cases:

1. If the beam waist on the object side is placed exactly on the focal point of the lens ($s = -f'$), the projected waist lies in the image-side focal plane[48] ($s' = f'$). The image-side waist radius is given by

$$w_0' = \frac{\lambda f'}{\pi w_0} . \tag{A94}$$

A large object-side waist radius is thus related to a small waist radius on the image side. This means that the beam is effectively focused.

2. If the beam waist on the object side is placed very far away from the lens ($s \gg -f', z_R$), the projected waist once again lies close to F′ ($s' \approx f'$). The image-side waist radius is now approximated by

$$w_0' \approx \frac{w_0 f'}{|s|} . \tag{A95}$$

In practice (e.g., for laser resonators), s is usually unknown. But we can calculate this value indirectly from the beam radius at the lens surface w_L with $s \approx w_L w_0 \pi/\lambda$. Accordingly, we obtain

$$w_0' \approx \frac{\lambda f'}{\pi w_L} . \tag{A96}$$

Please note the similarity between Eqs. (A94) and (A96) (Problem PA.10).

A.2.3 Monochromatic Superposition of Harmonic Waves

A major difference between geometric and wave optics is the treatment of the interaction and superposition of light beams. For this purpose, let us consider two bundles of light with the same wavelength which are generated by two spatially separated, coherent (Section A.2.3.2) point sources (Figure A.38). In geometric optics, we correlate the ray density with the light intensity and assume that the individual rays do not interact with each other. In this simple picture, all emanated rays would be homogeneously distributed all over the screen. However, experiments show a periodic, complex intensity pattern which is related to the wave nature of light and can only be explained by wave optics.

A.2.3.1 Interference

If two or more waves, called *partial waves*, "collide" and penetrate each other, their amplitudes superpose (overlap) linearly at every point where they are simultaneously present. This phenomenon is called *interference*. Depending on the relative phase difference[49], the resulting amplitude $\psi_{0,\mathrm{tot}}$ may be either smaller or larger

48) In this case, Gaussian beams behave differently from point sources. The image of a point source would be infinitely far away due to plane wavefronts (parallel rays) formed on the image side.

49) If two identical waves do not exactly line up, there is a phase difference between them. The phase difference can be quantified in radians, degrees, or fractions of a wavelength. A phase difference

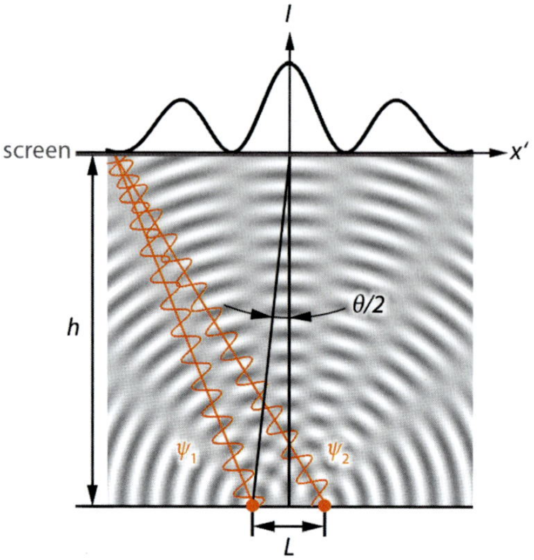

Figure A.38 Interference of two partial waves with equal wavelength. Two spatially separated, coherent (Section A.2.3.2) point sources (e.g., created by a double slit) emit spherical waves which overlap and form an interference pattern. Light regions (dark regions) correspond to positive (negative) values of the amplitude. On a screen, we may observe the intensity pattern $I(x')$ along the x axis.

than the amplitudes of the "original" partial waves $\psi_{0,m}$. For an arrangement like that shown in Figure A.38, the wave function of two superposed harmonic waves is given by[50)]

$$\begin{aligned}\psi(\mathbf{x}, t) &= \psi_1 + \psi_2 \\ &= \psi_{0,1}\mathrm{e}^{\mathrm{i}(\mathbf{kx}-\omega t+\xi_1)} + \psi_{0,2}\mathrm{e}^{\mathrm{i}(\mathbf{kx}-\omega t+\xi_2)} \\ &= \underbrace{\left(\psi_{0,1}\mathrm{e}^{\mathrm{i}\xi_1} + \psi_{0,2}\mathrm{e}^{\mathrm{i}\xi_2}\right)}_{=\psi_{0,\mathrm{tot}}} \mathrm{e}^{\mathrm{i}(\mathbf{kx}-\omega t)} ,\end{aligned} \tag{A97}$$

where $\xi_1(\mathbf{x}, t)$ and $\xi_2(\mathbf{x}, t)$ are relative phases.

The intensity of the interference pattern

$$\begin{aligned}I &= \frac{1}{2t_{\mathrm{per}}} \int_{-t_{\mathrm{per}}}^{t_{\mathrm{per}}} |\psi(\mathbf{x}, t)|^2 \mathrm{d}t \\ &= I_1 + I_2 + 2\psi_{0,1}\psi_{0,2}\cos(\Delta\xi)\end{aligned} \tag{A98}$$

is derived from Eq. (A97) with the relative phase difference $\Delta\xi = \xi_2 - \xi_1$. The third term in Eq. (A98) is the interference term which is a periodic function of $\Delta\xi$.

50) For the sake of simplicity, we assume that the wave function $\psi(\mathbf{x}, t)$ is a scalar quantity. The following discussions and results would be similar for a vectorial wave function $\boldsymbol{\psi}(\mathbf{x}, t)$.

For $\Delta\xi = 2\pi m$ (m is an integer), the interference term yields the highest absolute values because of the cosine. In this case, the partial waves are shifted by multiples of the wavelength. Hence, the crests (troughs) overlap in-phase and thus further increase (decrease). Consequently, we obtain an enhanced amplitude contrast which is called *constructive interference*.

For $\Delta\xi = (2m + 1)\pi/2$, the cosine factor in Eq. (A98) becomes zero. This means that the partial waves are shifted by *half* the wavelength such that the crests of one partial wave overlap with the troughs of the other. Hence, linear superposition of both waves cancels the amplitudes. This is referred to as *destructive interference*.

Interference does not only appear when multiple light sources are present. Due to the Huygens–Fresnel[51)] principle, each point in space which is penetrated by light may act as a new point source for spherical waves. Diffraction can thus be regarded as a kind of "self-interference". Considering Figure A.31b once again, this new approach states that the incident wave ψ_1 can be virtually split into numerous partial waves at every point within the aperture. The diffraction pattern given by Eq. (A74) is then generated by interference of the newly emanated spherical waves.

A.2.3.2 **Coherence**

Now why is interference only observable in some experiments and not in all? Interference patterns appear only for a fixed, defined phase difference. If, however, $\Delta\xi$ changes in time, the interference term in Eq. (A98) will vanish due to temporal averaging. For this reason, we have to introduce a measure for the "possibility to show interference" that is called *coherence*. The delay over which the phase difference changes by a significant amount ($\Delta\xi > 2\pi$) is defined as the coherence time t_c. An associated quantity is the coherence length L_c which is defined as the distance the wave travels during t_c. L_c can thus be understood as the maximum path length difference of two superposed partial waves which allows observation of an interference pattern.

For monochromatic light, the coherence time is infinite. However, in reality, the phase difference is never perfectly fixed in time. In particular, for light sources with a broad spectrum (Section A.2.4.1), the coherence time is considerably reduced down to some femtoseconds. When the considered time interval (distance) exceeds t_c (L_c), the fixed phase correlation disappears. Regarding the former example of two superposed sources, the intensity distribution of the incoherent wave changes to

$$I_{inc} = I_1 + I_2 \,. \qquad \text{(A99)}$$

In this case, the interference term disappears and the individual intensities of the partial waves are simply added.

A.2.4
Polychromatic Superposition of Waves

The perfectly monochromatic light source is a theoretical concept. In reality, we always have to deal with interactions of waves which have different wave-

51) Christiaan Huygens (1629–1695).

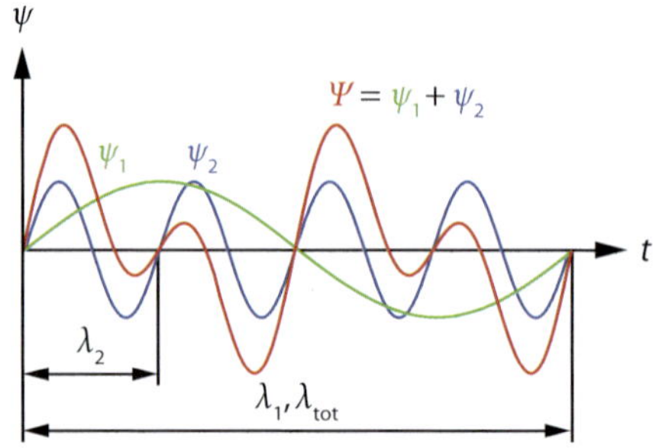

Figure A.39 Superposition of two harmonic wave functions ψ_1 and ψ_2 with different wavelengths (colors). Analogous to the superposition of monochromatic harmonic waves, the resulting function is given by $\Psi = \psi_1 + \psi_2$.

lengths/frequencies[52]. The resulting polychromatic wave function $\Psi(\mathbf{x}, t)$ can either be described as a wave function with *non*harmonic time dependence or as a superposition of harmonic monochromatic waves. In the latter case, we simply add the partial wave functions like for a monochromatic superposition:

$$\Psi(\mathbf{x}, t) = \psi_1(\mathbf{x}, t) + \psi_2(\mathbf{x}, t) + \psi_3(\mathbf{x}, t) + \ldots \tag{A100}$$

This approach is illustrated for two partial harmonic waves in Figure A.39.

In contrast to monochromatic superposition, adding more and more partial waves with different frequencies continuously lowers coherence. As long as the added frequencies are distributed closely around a central frequency ω_0, wave packets are created. Each wave packet forms a train of amplitudes and can be regarded as a separate pulse (Figure A.41) which travels with the group velocity

$$c_g = \frac{d\omega}{dk} \,. \tag{A101}$$

Light which is emitted as pulse trains is referred to as *low-coherence light* and is used, for example, in optical coherence tomography (Chapter 7).

In the most general case, an infinite number of harmonic waves with different wavelengths generates a random, incoherent wave function (i.e., white light). Here, we may replace the discrete sum in Eq. (A100) by an integral across all appearing frequencies ranging from zero to infinity, that is,

$$\Psi(\mathbf{x}, t) = \int_0^\infty \psi_0(\omega) e^{i(\mathbf{kx} - \omega t)} d\omega \,. \tag{A102}$$

Since $\Psi(\mathbf{x}, t)$ is a random function of time and position, the intensity as defined by Eq. (A69) is also random. If we want to measure a constant intensity signal, the integration time of the photo detector must be considerably increased. For this

52) Wavelength λ and frequency ω are related via Eq. (A57): $\lambda = 2\pi c/\omega$.

purpose, we introduce the total average intensity

$$\hat{I} = \lim_{\tau \to \infty} \frac{1}{2\tau} \int_{-\tau}^{\tau} |\Psi(\mathbf{x}, t)|^2 dt = \langle |\Psi(\mathbf{x}, t)|^2 \rangle_\tau \,, \tag{A103}$$

where we have chosen an integration time τ much longer than the periodicity t_{per} of any partial wave.

Info Box A.2: Fourier Transformation

The Fourier transform is defined as

$$U(\omega) = \int_{-\infty}^{\infty} u(t) e^{i\omega t} dt \equiv \mathcal{F}\{u(t)\} \,, \tag{A104}$$

where the coefficient $u(t)$ is, in turn, given by the inverse Fourier transform

$$u(t) = \frac{1}{2\pi} \int_{-\infty}^{\infty} U(\omega) e^{-i\omega t} d\omega \equiv \mathcal{F}^{-1}\{U(\omega)\} \,. \tag{A105}$$

The basic idea of Fourier transformation is to express an arbitrary function $u(t)$ by a series of sinusoidal (harmonic) functions. Each term of the series represents a single-frequency component weighted by a constant amplitude coefficient $U_{0,m}(m\omega_0)$, where m is an integer. To improve the approximation by each additional term, the frequency ω_0 is successively increased like

$$u(t) = U_{0,0} + U_{0,1} e^{-i\omega_0 t} + U_{0,2} e^{-2i\omega_0 t} + \ldots = \sum_{m=0}^{\infty} U_{0,m} e^{-im\omega_0 t} \,. \tag{A106}$$

To reconstruct an arbitrary *periodic* function, a (possibly infinite) sum is needed[53]. However, to reconstruct an arbitrary, *nonperiodic* function, a transition from the discrete form in Eq. (A106) to the continuous integral form of Fourier transformation in Eq. (A105) is required.

Summing up, the Fourier transformation allows analysis of a given function $u(t)$ in terms of its frequency components $U(\omega)$ and vice versa. From a time-dependent signal, we may thus derive the frequency-domain spectrum.

The same formalism can also be used for functions which depend on space coordinates to determine their spatial frequencies (and vice versa). For example, the optical resolution can be defined by a spatial frequency. By applying the Fourier transform, we can calculate back how this quantity transfers to a minimum local distance (Section A.2.1.6). Another example is the retrieval of a point object from its point-spread function. Some important functions and their respective frequency spectra are listed in Table A.5.

Table A.5 List of certain functions $u(t)$ and their Fourier transforms $U(\omega)$. $\delta(t)$ is the Dirac delta function, which has nonzero values only for $t = 0$. $\prod(t/2t_0)$ is the Heaviside function which is constant within the time interval $[-t_0, t_0]$ and zero outside.

Initial Function	**Fourier Transform**
$u(t) = \cos(\omega_0 t)$ 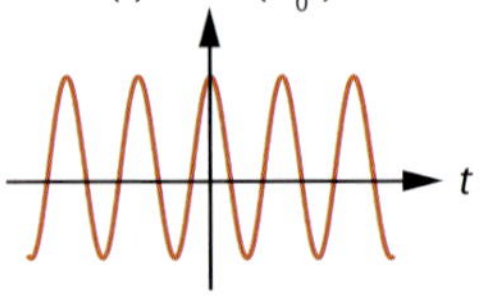	$U(\omega) = \pi\,[\delta(\omega+\omega_0)-\delta(\omega-\omega_0)]$ 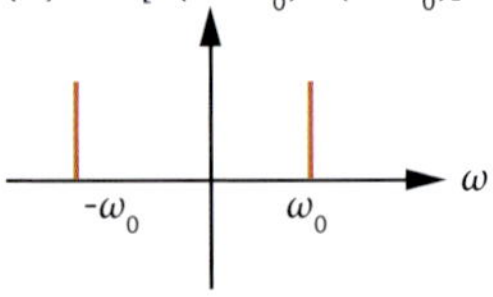
$u(t) = \Pi(t/2t_0)$ 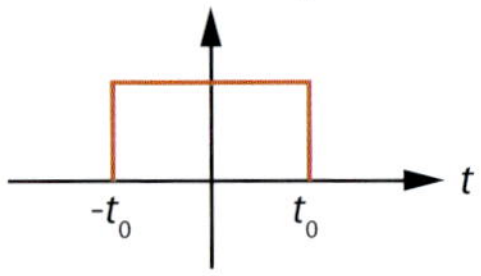	$U(\omega) = 2t_0\,\mathrm{sinc}(\omega t_0) = 2t_0 \sin(\omega t_0)/\omega t_0$ 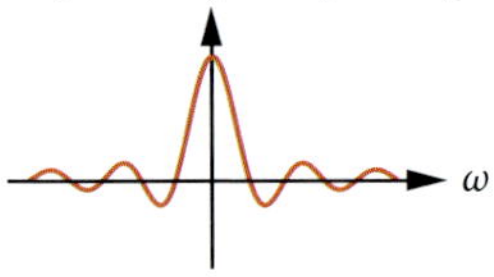
$u(t) = \delta(t)$ 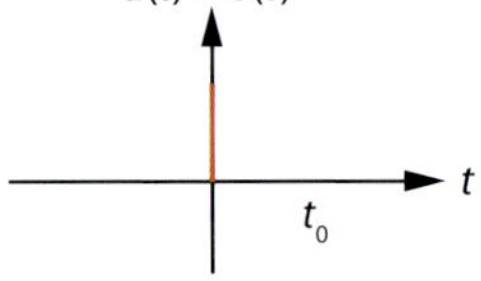	$U(\omega) = 1$ 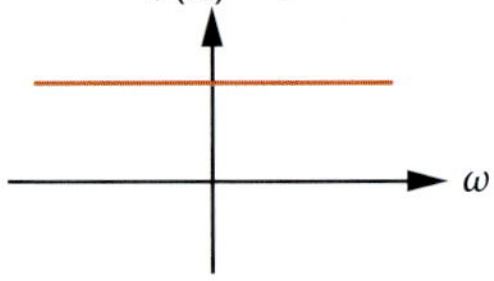
$u(t) = \exp(-t^2/2\sigma^2)$ 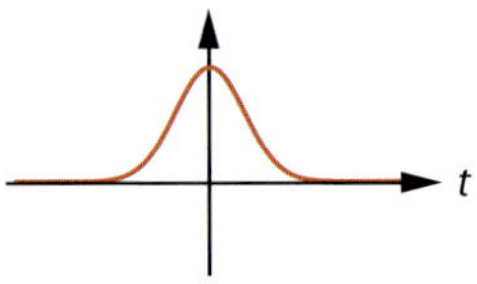	$U(\omega) = \sqrt{2\pi}\,\sigma \exp(-\omega^2\sigma^2/2)$ 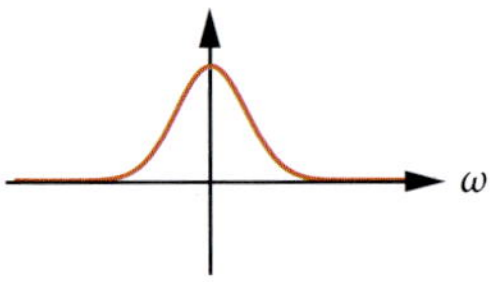

A.2.4.1 Spectral Distribution of Light Sources

Let us consider the average spectrum of a polychromatic light source. At first, we determine the amplitude of one partial wave with frequency ω by Fourier transformation[54] of Eq. (A102). With $\Psi(\mathbf{x}, \omega) = \psi_0(\omega)\exp(\mathrm{i}\mathbf{kx})$, we obtain

53) For example, the wave function Ψ in Figure A.39 is fully recovered for $m = 4$. By following the approach in Eq. (A106), it can be expressed via

$$\Psi(t) = \underbrace{\psi_{0,0}}_{=0} + \psi_{0,1}\,\mathrm{e}^{-\mathrm{i}\omega t} + \underbrace{\psi_{0,2}\,\mathrm{e}^{-2\mathrm{i}\omega t}}_{=0} + \underbrace{\psi_{0,3}\,\mathrm{e}^{-3\mathrm{i}\omega t}}_{=0} + \psi_{0,4}\,\mathrm{e}^{-4\mathrm{i}\omega t},$$

with $\lambda_{tot} = \lambda_1 = \frac{2\pi c}{\omega}$ and $\lambda_2 = \frac{2\pi c}{(4\omega)}$

54) For the Fourier transformation, we have to extend the integration range to $-\infty$. This is possible since negative values of time relate to the past and negative values of frequency do not contribute to the integral: $\int_{-\infty}^{0} \psi_0(\omega)\exp(\mathrm{i}\mathbf{kx} - \mathrm{i}\omega t)\mathrm{d}\omega = 0$.

$$\Psi(\mathbf{x}, \omega) = \int_{-\infty}^{\infty} \Psi(\mathbf{x}, t) e^{i\omega t} dt \,, \tag{A107}$$

where $\Psi(\mathbf{x}, t)$ is the polychromatic wave function (see Eq. (A100)). The average energy per area can now be calculated by averaging the partial waves (frequency components) over the interval between ω and $\omega + d\omega$, that is, $\langle |\Psi(\mathbf{x}, \omega)|^2 \rangle d\omega$. Here, $\langle |\Psi(\mathbf{x}, \omega)|^2 \rangle$ represents the energy spectral density of light.

In Eq. (A107), we integrated across a random function $\Psi(\mathbf{x}, t)$ extending to infinity ($t \to \infty$). As a consequence, the energy carried by the wave becomes infinite! To solve this issue, we "cut out" a limited time frame τ (Figure A.40) and obtain the truncated Fourier transform

$$\Psi_\tau(\mathbf{x}, \omega) = \int_{-\tau/2}^{\tau/2} \Psi(\mathbf{x}, t) e^{i\omega t} dt \,. \tag{A108}$$

The average energy per area for the time frame τ is given by $\langle |\Psi_\tau(\mathbf{x}, \omega)|^2 \rangle d\omega$. Extending the duration of the time frame to infinity ($\tau \to \infty$) yields the power spectral density

$$\sigma(\omega) = \lim_{\tau \to \infty} \frac{1}{\tau} \langle |\Psi_\tau(\mathbf{x}, \omega)|^2 \rangle \,. \tag{A109}$$

$\sigma(\omega)$ is nonzero only for positive frequencies. By integration over all frequencies, we find the total average intensity

$$\hat{I} = \int_0^{\infty} \sigma(\omega) d\omega \,. \tag{A110}$$

For the sake of clarity, let us compare intensity I in Eq. (A69) of a monochromatic wave with the total average intensity $\hat{I}$ in Eq. (A110) of a polychromatic wave. Since a monochromatic wave has a well-defined period, the light intensity I can be determined by acquiring the signal for a duration of at least one period τ. We may thus consider the definition (A110) as a kind of "measuring instruction". For polychromatic light, the contribution of all partial waves must be taken into account. At first, the intensity of each partial wave has to be considered, which is the power spectral density $\sigma(\omega)$ given by Eq. (A109). $\sigma(\omega)$ is comparable to the Fourier transform of the monochromatic light intensity I. Then, we sum up the contributions of all partial waves in order to obtain the total average intensity $\hat{I}$.

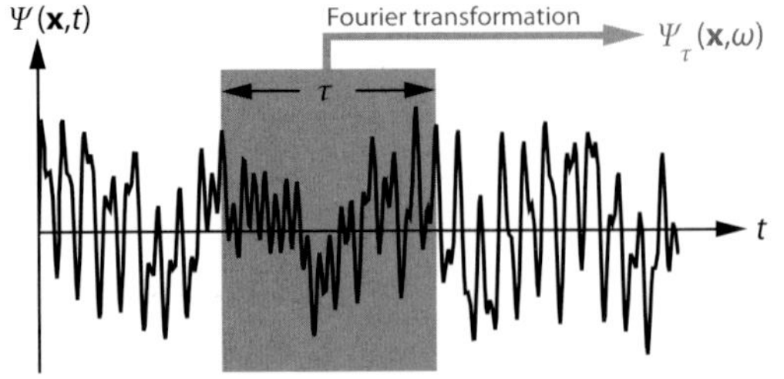

Figure A.40 Random wave function $\Psi(\mathbf{x}, t)$. To calculate the power spectral density, we consider a limited time frame τ before Fourier transformation.

A.2.4.2 Wiener–Khinchin Theorem

With the autocorrelation function defined by

$$\mathcal{G}(\Delta t) = \lim_{\tau\to\infty} \frac{1}{2\tau} \int_{-\tau}^{\tau} \Psi^*(t)\Psi(t+\Delta t)\mathrm{d}t$$
$$= \langle \Psi^*(t)\Psi(t+\Delta t)\rangle_\tau \quad \text{(A111)}$$

we may express the power spectral density $\sigma(\omega)$ of Eq. (A109) in an alternative way by

$$\sigma(\omega) = \int_{-\infty}^{\infty} \mathcal{G}(\Delta t)\mathrm{e}^{\mathrm{i}\omega\Delta t}\mathrm{d}(\Delta t)\,. \quad \text{(A112)}$$

In other words, the power spectral density is the Fourier transform of the autocorrelation function. This relation is also known as the *Wiener–Khinchin theorem.*[55]

The autocorrelation function $\mathcal{G}$ is a mathematical tool used to find either repeating patterns or nonrandomness of an arbitrary wave function. For example, it allows us to find a periodic signal which is supposedly hidden by noise. This is possible because the original wave function is correlated to the time-shifted version $(t+\Delta t)$ of itself. The autocorrelation can also be used to express the coherence time as

$$t_\mathrm{c} = \int_{-\infty}^{\infty} \left|\frac{\mathcal{G}(\Delta t)}{\mathcal{G}(0)}\right|^2 \mathrm{d}(\Delta t)\,, \quad \text{(A113)}$$

where $\mathcal{G}(0) = \hat{I}$. With Eq. (A112) and the most common definition of the spectral width $\Delta\omega$ (see [10]), we may derive the bandwidth theorem

$$t_\mathrm{c} = \frac{\text{const}}{\Delta\omega}\,. \quad \text{(A114)}$$

55) The definition in Eq. (A112) of the power spectral density is normally more relevant in practice than Eq. (A109), since the Fourier transform may not exist for random wave functions. In this case, $\sigma(\omega)$ can still be derived from the autocorrelation function.

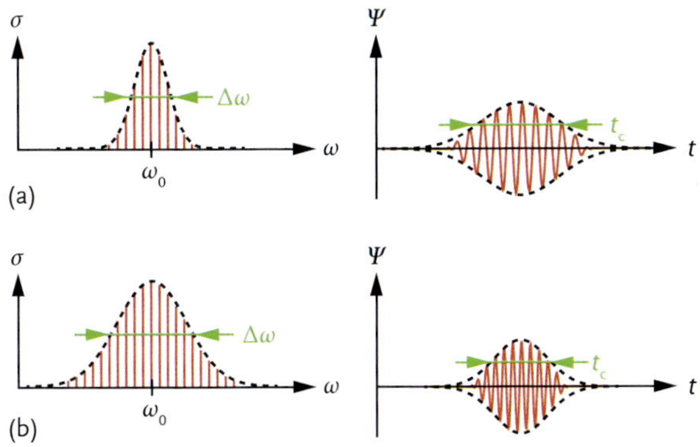

Figure A.41 Dependence of spectral width $\Delta\omega$ (here defined by the full width at half maximum) and coherence time t_c. ω_0 denotes the central frequency and the dashed curve illustrates the envelope of amplitudes. The coherence time is longer in (a) than in (b), since the spectral width is smaller in (a).

It states that a large spectral width of a light source means a short coherence time (Figure A.41) and vice versa. This fact is relevant for the specification of optical devices which use a low-coherence light source (see, e.g., Section 7.3; Problem PA.11).

A.3 Recommended Reading

In the course of this chapter, a brief overview of geometric and wave optics is given which is most relevant for the understanding of basic ophthalmic concepts. Recommendations for advanced literature and references to original publications are provided at the respective passage in the text. To gain a deeper insight into general topics in optics, please refer to standard optical textbooks [9–12]. A profound discussion of optical imaging and aberration theory can be found in [8, 13, 14].

A.4 Problems

PA.1. Optics on a summer day It is a warm day and you are standing on the edge of a swimming pool. You are wearing Polaroid sunglasses which obviously reduce the glare from sunlight reflected from the water surface ($n_{\text{water}} = 1.33$).

1. Which polarization direction is blocked by your sunglasses?
2. If the water surface is observed at a certain angle, the glare of the sunlight is perfectly filtered out by the sunglasses. Derive the conditional equation for this angle from the Fresnel equations (A63)–(A66).

3. When looking at the opposite side of the swimming pool, you realize that the bottom edge appears to be at an angle of 30° below the horizontal. But when you sit on the pool edge, the bottom edge of the opposite side of the pool appears to be at an angle of 14° below the horizontal. Determine length and depth of the pool. *Hint:* Estimate the height of your eyes above the surface of the water when standing and sitting.
4. You now gaze directly into the sun. Estimate the time it takes for a photon to travel from the surface of the sun to your retina. What is the additional time delay if you are wearing sunglasses?
5. You replace your Polaroid sunglasses with a special type of filter glasses. Only light with a wavelength of 550 nm can pass through these glasses. Calculate the number of photons that enter your eye if you look for 0.1 s at the sun. What energy is absorbed by your eye during that time (all photons shall be absorbed). *Hints:* You need to use some results of Appendix B; power output of the sun: $P_{\mathrm{sun}} = 4.2 \times 10^{26}$ W; distance between Earth and sun: $d_{\mathrm{ES}} = 1.5 \times 10^{11}$ m; pupil diameter: $d_{\mathrm{p}} = 1.5$ mm.
6. In order to cool down, you are jumping into the pool. As discussed in Chapter 1, human eyes perceive color via three types of cones. Explain why the color of an object that appears blue in air also appears blue underwater even though the speed of light (and hence its wavelength) is shortened.

PA.2. Atmospheric refraction Imagine you stand at the ocean shore to watch the setting sun. Interestingly, you can already observe the first rays, although the sun is actually at an angle α below the horizon.

1. Calculate the angle $\alpha(r, n, h)$. Assume that the Earth's atmosphere has a uniform refractive index ($n = 1.0003$) and extends to a height of $h = 20$ km. Beyond the atmosphere, there is vacuum. The Earth's radius is $r = 6378$ km.
2. Why does the Sun appear to be flattened during the setting?

PA.3. Lens maker's equation Derive the lens maker's equation for a thin and a thick lens.

PA.4. Galilei's telescope In 1610, Galileo Galilei discovered the four moons of Jupiter using an afocal telescope. Although such an optical system does not alter the divergence of an incident bundle of rays, it *does* alter the width of the beam and thus increases the magnification. Galilei's device consisted of a thin, negative lens (ocular lens with focal length: $f_1 = -5$ cm) and a thin, positive lens (objective lens with focal length: $f_2 = 80$ cm).

1. Derive the imaging equation of a thin lens (lens maker's equation) by applying Eq. (A11) twice and by using the ABCD matrix of a spherical surface in Table A.1.
2. Use the ABCD matrix approach to derive the imaging equation for a thick lens (A16).

3. Derive the ABCD matrix for the Galilei telescope. What is the condition for an afocal optical system? Was the observed image of Jupiter's moons upright or inverted?
4. Just one year later, Johannes Kepler showed that telescopes can also be made of two positive lenses. The lenses were separated by the sum of their focal lengths. Check, using the ABCD matrix approach, whether the observed image was upright or inverted.
5. We now use the Galilei's telescope as a laser beam expander. For this purpose, we consider a Nd:YAG laser ($\lambda = 1064$ nm) which emits a Gaussian beam with a waist radius of $w_0 = 1.3$ mm. Calculate the resulting beam diameter after passage through the afocal Galilei telescope.
6. The expanded laser beam shall be focused with another lens so that the peak intensity does not fall below 80% within a distance of $\Delta z = 1$ mm. What is the minimum focal length of this lens? How large is the beam diameter at the focal point?

PA.5. LED–Fiber Coupling Consider a GaAs LED ($n_{\text{GaAs}} = 3.4$) with a flat surface. The setup can be considered as a point source which is located close to the GaAs–air surface. At a distance of 2 mm, we place a silica fiber ($n_{\text{silica}} = 1.46$; core diameter $d_c = 1$ mm) which has a maximum acceptance angle of 14° in air.

1. What fraction of light (percentage) emitted by the active region of the LED can be coupled into the fiber? How does this value change if we fill the volume between fiber and GaAs LED with water? Neglect the reflection losses at both media interfaces.
2. By looking at the fiber, you realize that it has a cladding. For your setup, you need to know what kind of material has been used for the fiber. In a specification list, you found three possible options: $n_{c1} = 1.493$, $n_{c2} = 1.440$, or $n_{c3} = 1.430$. Which cladding material has been used in this case?
3. How can the LED fiber coupling be understood in the concept of the Helmholtz–Lagrange invariant?

PA.6. Point-spread function and Strehl ratio In practice, the Strehl ratio (or definition brightness) is defined as the normalized ratio of the point-spread intensity on the axis for the real system to the ideal system, that is,

$$\mathcal{S} = \frac{I_{\text{PSF}}^{\text{real}}(0,0)}{I_{\text{PSF}}^{\text{ideal}}(0,0)} \,. \tag{A115}$$

This relation is used to characterize diffraction-limited optical systems. Write the Strehl ratio for (Fraunhofer) far field conditions by assuming a circular, evenly illuminated pupil as a function of the wave aberration $\mathcal{W}(x_p, y_p)$.

PA.7. Taylor expansion of wavefront aberrations Show that only terms with the products h^2, r^2, and $hr\cos\alpha$ are physically relevant for the Taylor expansion of the wavefront aberration function $\mathcal{W}(h, r, \alpha)$.

PA.8. Zernike expansion of wavefront aberrations According to Zernike, the wave aberrations for defocus and spherical aberrations are defined in normalized coordinates as

$$\mathcal{W}_{\text{def}}(r) = c_2^0(2r^2 - 1) , \tag{A116}$$

$$\mathcal{W}_{\text{sph}}(r) = c_4^0(6r^4 - 6r^2 + 1) . \tag{A117}$$

1. For the spherical aberration across a circular pupil, calculate the peak-valley value, the mean value, and the RMS_{wfe} value for $c_4^0 = 1$.
2. When determining the Zernike coefficient, one always assumes a normalized pupil radius. This reference plays a key role in the result. In reality, the size of the pupil often cannot be determined very accurately. How does the determination of a spherical Zernike coefficient change if the assumed pupil size deviates by 5%? What is the associated error in the specification of the defocus?
3. Factorize an aspherical cylinder surface into Zernike polynomials using the equation $F(x, y) = y^4$.

PA.9. Brewster angle Derive the equation for the Brewster angle γ_B from Eqs. (A63)– (A66).

PA.10. Gaussian Beams ASTRA is a television satellite which travels in a geostationary orbit (distance to sea level $h = 35.8$ km). The signal is transmitted via radiation with a wavelength of $\lambda = 2.7$ cm and a transmission power of $P = 100$ W. The radiation can be considered as a Gaussian beam with the spatial envelope function (A80).

1. Show that the emitted beam fulfills the paraxial approximation.
2. Calculate power and intensity received by a parabolic antenna (diameter $d =$ 1 m) located at sea level. *Hint:* Use the approximation $1 - e^{-x} \approx x$.
3. What would happen if the satellite could transmit the digital television signal with visible light (e.g., by using an argon ion laser)?

PA.11. Autocorrelation function, spectral density, and coherence length Calculate the autocorrelation function $\mathcal{G}$, the spectral density $\sigma(\omega)$, and coherence length L_c for various pulse forms and spectral distributions:

1. Gaussian pulse
2. Rectangle pulse
3. Lorentz spectrum
4. sech^2 pulse

Use the definitions for $\mathcal{G}(\Delta t)$ and $\sigma(\omega)$ as given in Eqs. (A111) and (A109), respectively. For the coherence length as well as the spectral density use appropriate definitions, such as the full width at half maximum (FWHM), second momentum of a normalized function, or second momentum of a squared normalized function.

References

1 Schott AG (2010) Optical glass – all datasheets, Online Product Specification. http://www.schott.com/advanced_optics/english/download/index.html (accessed 28 August 2013).

2 Gross, H. (2005) *Handbook of Optical Systems – Fundamentals of Technical Optics*, vol. 1, John Wiley & Sons.

3 Liang, R. (2010) *Optical Design for Biomedical Imaging*, SPIE Press.

4 Greiner, W. (1998) *Classical Electrodynamics*, Springer.

5 Agurok, I. (1998) Double expansion of wavefront deformation in Zernike polynomials over the pupil and the field-of-view of optical systems: Lens design, testing, and alignment. *SPIE Proc.*, **3430**, 80.

6 International Organization for Standardization (2008) Reporting aberrations of the human eye (ISO 24157:2008(e)).

7 American National Standards Institute (2004), Methods for reporting optical aberrations of eyes (ANSI Z80.28-2004).

8 Gross, H., Singer, W., and Totzeck, M. (2005) *Handbook of Optical Systems – Physical Image Formation*, vol. 2, John Wiley & Sons, Inc.

9 Hecht, E. (1999) *Optics*, Oldenbourg.

10 Saleh, B.E.A. and Teich, M.C. (2007) *Fundamentals of Photonics*, John Wiley & Sons, Inc.

11 Born, M. and Wolf, E. (1997) *Principles of Optics: Electromagnetic Theory of Propagation, Interference and Diffraction of Light*, Cambridge University Press.

12 Smith, F.G., King, T.A., and Wilkins, D. (2007) *Optics and Photonics – An Introduction*, John Wiley & Sons, Inc.

13 Mahajan, V.N. (2001) *Optical Imaging and Aberrations, Part II – Wave Diffraction Optics*, SPIE Publications.

14 Gross, H., Zügge, H., Peschka, M., and Blechinger, F. (2007) *Handbook of Optical Systems – Correction of Optical Systems*, vol. 3, Wiley-VCH GmbH & Co. KG.

B
Basics of Laser Systems

In ophthalmology, the unique optical properties of lasers are used in many diagnostic (Chapters 5–7) and therapeutic (Chapter 10) methods. Because of this special role, we added this separate chapter about the theoretical and technical basics of laser systems to ensure a common knowledge base for all readers. This chapter serves as a quick introduction and reference and does not intend to cover all details about laser physics.

In 1960, the first laser[1)] was built by Theodore Maiman (1927–2007) based on pioneering work by Charles Townes (born 1915) and Arthur Schawlow (1921–1999). In those days, nobody was really aware of the extraordinary importance of this special kind of light source. Today, lasers are considered to be one of the paramount achievements of the twentieth century. Their applications range from information and medical technology to mechanical processing, metrology, and chemical sensing.

The success story of lasers actually dates back to the development of quantum mechanics in the early twentieth century. In particular, Albert Einstein's (1879–1955) publications on light–matter interaction [1, 2] and Nils Bohr's (1885–1962) atom model paved the way for this unique device. These models take into account the observation that the energy of an atom (or a molecule) cannot be continuously altered. Only certain discrete energy states are "allowed". For example, an atom may transit from a discrete lower state $|1\rangle$ with energy E_1 to an excited state $|2\rangle$ with energy E_2 by absorbing light which carries exactly the difference energy $E = E_2 - E_1$ (Figure B.1a). The transition also works the other way round in that the excited atom "falls back" (is *depleted/de-excited*) to a lower state, whereupon the energy difference is released by emission of light (Figure B.1b,c). To describe this behavior, Einstein proposed a simple formalism which we will use in the following discussions.

1) The term "laser" was coined by Gordon Gould (1920–2005). It is an acronym for "*Light Amplification by Stimulated Emission of Radiation*".

Optical Devices in Ophthalmology and Optometry, First Edition. Michael Kaschke, Karl-Heinz Donnerhacke, and Michael Stefan Rill.

Info Box B.1: Photons: The Elementary Particles of Light

In the context of light–atom interaction, it is common to treat light as a massless particle called a *photon*. Similar to the wave concept of light (Section A.2), the photon is characterized by the frequency $\omega = 2\pi c/\lambda$, where λ is the wavelength and c is the speed of light. The photon energy is directly related to ω via

$$E = \frac{h}{2\pi}\omega = \hbar\omega \,, \tag{B1}$$

where $h \approx 6.6261 \times 10^{-34}$ Js denotes Planck's constant.

B.1
Einstein's Two-Level Model of Light–Atom Interaction

Einstein considered two energy states of an atom[2] and postulated three types of atomic transitions which are illustrated in Figure B.1, that is, absorption, spontaneous emission, and stimulated emission. For each process, he introduced an individual *Einstein coefficient* which characterizes the respective transition rate (number of transition events per unit time).

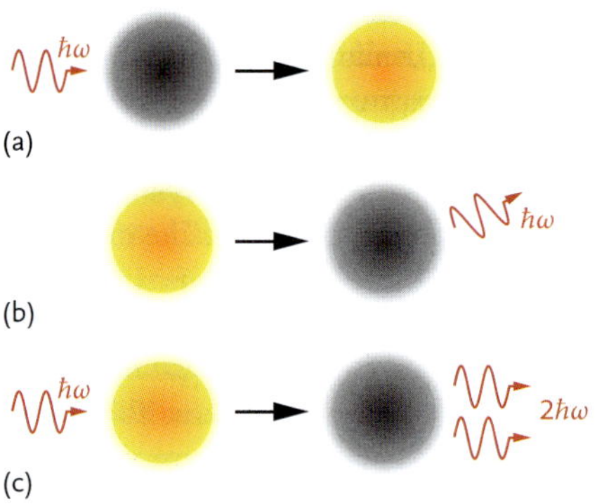

Figure B.1 Energy transitions of an atom. (a) Absorption: By absorption of a photon (wiggly line), the atom in the lower energy state $|1\rangle$ (darker) is excited to a higher state $|2\rangle$ (lighter). (b) Spontaneous emission: An excited atom transits to $|1\rangle$ by releasing a photon. (c) Stimulated emission: An incident photon stimulates the excited atom to transit to $|1\rangle$ by emitting a second photon which has the same frequency, polarization, and direction as the incident one.

2) In principle, these considerations also apply to ions and molecules. However, in molecules, as is the case for atoms in a less complex way too, other radiative and nonradiative transition processes must be considered as well.

B.1.1
Absorption

An atom in the lower energy state $|1\rangle$ is excited to a higher state $|2\rangle$ by absorption of a photon with the energy $E_2 - E_1 = \hbar\omega_{21}$ (Figure B.1a). As a consequence, the photon is annihilated as its complete energy is transferred to the atom. The rate at which the number of atoms in the lower state N_1 changes in time is given by

$$\frac{\mathrm{d}N_1}{\mathrm{d}t} = -B_{12} N_1 \rho(\omega) \,, \tag{B2}$$

where B_{12} is the Einstein coefficient for absorption and $\rho(\omega)$ the spectral energy density of light (energy per volume per frequency interval) with $[\rho] = \mathrm{Js/m^3}$. The subscript "12" of the Einstein coefficient indicates that the transition starts with the energy level $|1\rangle$ and ends up in $|2\rangle$.

B.1.2
Spontaneous emission

An excited atom spontaneously transits to the lower state after an average lifetime τ_{21}. At this point, a photon of energy $\hbar\omega_{21}$ is released and travels along a random direction (Figure B.1b). Spontaneous emission is a statistical process and arises from the atom's natural tendency to lose its excess energy. The rate at which the population of excited atoms N_2 changes in time is given by

$$\frac{\mathrm{d}N_2}{\mathrm{d}t} = -\frac{N_2}{\tau_{21}} = -A_{21} N_2 \,. \tag{B3}$$

$A_{21} = 1/\tau_{21}$ is the Einstein coefficient for spontaneous emission.

If the lifetime τ_{21} of an atom in the excited state is in the range of some nanoseconds (ns, 10^{-9}s), the emission of photons is referred to as *fluorescence*. Atoms remain virtually forever in the ground state, since this is the only stable energy state.

B.1.3
Stimulated emission

In case of stimulated emission, an incident photon of energy $\hbar\omega_{21}$ triggers an excited atom to transit to a lower state by generating a second photon (Figure B.1c). Interestingly, the second photon is a "clone" of the incident one in that it has the same frequency, polarization, and travels along the same direction. Both emitted photons are thus said to be *coherent* (Section A.2.3.2). Analogous to Eqs. (B2) and (B3), the transition rate of stimulated emission is given by

$$\frac{\mathrm{d}N_2}{\mathrm{d}t} = -B_{21} N_2 \rho(\omega) \,. \tag{B4}$$

B.1.4
Relation of Einstein Coefficients

Although the Einstein coefficients A_{21}, B_{12}, and B_{21} are associated to different transition processes, they are all directly related to each other. If we know one of them, we can work out the rest. Let us consider mutually noninteracting atoms inside an isolated box. Under steady-state conditions with light present, the rates of atomic excitations and depletions are exactly balanced, that is,

$$\underbrace{B_{12} N_1 \rho(\omega)}_{\text{absorption}} = \underbrace{A_{21} N_2 + B_{21} N_2 \rho(\omega)}_{\text{emission}} \; . \tag{B5}$$

Starting from Eq. (B5), Einstein derived ([2], Problem PB.1) that

$$B_{12} = B_{21} \tag{B6}$$

indicating that the probabilities for absorption and stimulated emission are equal. In addition, he found that

$$A_{21} \propto \omega^3 B_{21} \; . \tag{B7}$$

Compared to stimulated emission and absorption, the contribution of spontaneous emission becomes more significant the higher the light frequency. This is why it is generally more difficult to build a laser at high frequencies (or short wavelengths) of light. Furthermore, transitions with high absorption probability also tend to have a high emission rate, for both spontaneous and stimulated processes.[3)]

B.2
Light Amplification by Stimulated Emission

Lasers have some outstanding features compared to "classic" light sources like the sun, light bulbs, and gas-discharge lamps which make them useful for many applications in science, technology, and medicine:

- The brightness of laser light (emitted light intensity per solid angle) is extraordinarily high. Even small laser pointers provide a brightness which is about 50 × higher than that of the sun.
- Light emitted by laser sources usually has a very narrow spectral bandwidth (Section A.2.4.2). For some types of lasers, the emitted light is nearly monochromatic. Classic light sources attain similar monochromatism only through extreme filtering which, in turn, leads to drastic losses in light intensity. Laser sources also exist which emit in a broad spectral range (supercontinuum lasers) and thus have a low temporal coherence (Section A.2.3.2).

3) An exact quantum mechanical treatment of atomic transitions (see, for example, [3]) reveals that Eqs. (B6) and (B7) also hold true if the system is *not* in a steady state. This makes Einstein's approach also applicable to the description of dynamic systems like the laser.

- A laser beam can be made very directive (low divergence) even for long traveling distances. It can also be focused to a very small spot, both being features of high spatial coherence.
- Laser light with a narrow spectral bandwidth has a long coherence length and time (Section A.2.3.2).

After reviewing the previously mentioned types of atomic transitions, it comes as no surprise that stimulated emission is the cause of these laser properties. All of the very special laser properties rely on *coherent amplification* of light. Loosely speaking, this means that photons of an incident light beam are "cloned" many times. For stimulated emission to happen, the incident light beam has to pass through an adequate *gain medium* where the cloning or amplification happens. The gain medium consists of atoms/molecules for which the energy difference of allowed atomic states exactly matches the energy carried by the incident photons.

Under normal environmental conditions, most atoms occupy a lower or even the lowest (ground) state so that photons are mostly absorbed when traveling through the gain medium. Hence, if more photons are to be generated by stimulated emission than annihilated by absorption, we have to "force" the majority of atoms to occupy an excited state. More specifically, we have to find a method to invert the population of atoms in the lower and the excited state so that the steady-state condition (B5) can be replaced by the optical gain condition

$$B_{12} N_1 \rho(\omega) < A_{21} N_2 + B_{21} N_2 \rho(\omega) \,. \tag{B8}$$

With Eq. (B6) and by neglecting spontaneous emission in Eq. (B8), we obtain

$$N_1 < N_2 \,. \tag{B9}$$

To fulfill population inversion, atoms of the gain medium must be actively transferred to the excited state, which is called *pumping*. For this purpose, energy (e.g., in the form of photons or electrons) is added to the system via an external energy source, the so-called *pump source*[4].

B.2.1
Conditions for Population Inversion

Let us consider atoms exposed to polychromatic light given by the situation shown in Figure B.2 as an energy level scheme. The energy states of the atomic system are represented by horizontal bars. The states $|1\rangle$ and $|2\rangle$ are called the *lower* and *upper laser levels*, respectively. The vertical arrows correspond to possible transition processes. By photon absorption, atoms are excited to the upper laser level $|2\rangle$ at a rate $\mathbb{P}_2$ per second per volume. After a lifetime τ_{21}, they transit back to $|1\rangle$ because of spontaneous emission. State $|1\rangle$ is not necessarily the ground state of the atoms. Another level $|0\rangle$ might exist which is beneath the lower laser level $|1\rangle$. For this

4) As soon as the pump source is switched off, the atoms immediately fall back to a lower (or the lowest) energy level and, eventually, no stimulated emission appears anymore.

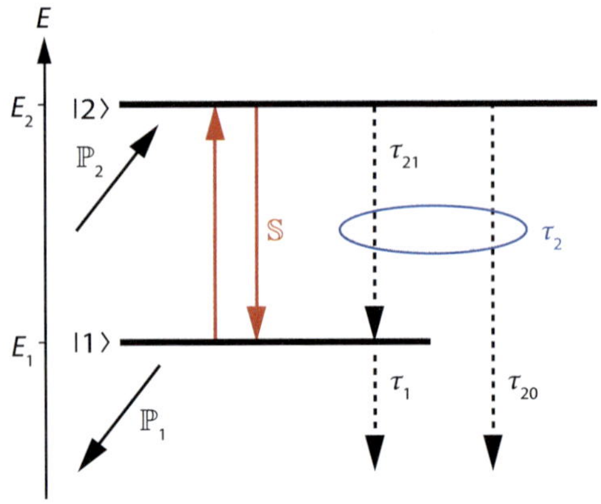

Figure B.2 Illustration and overview of parameters which are introduced to describe multilevel atomic transitions. The lifetimes are denoted by τ, the pump rates by $\mathbb{P}$, and stimulated emission by $\mathbb{S}$. Horizontal bars correspond to energy levels, and arrows represent the atomic transitions. Red arrows relate to stimulated transitions.

reason, an excited atom may also transit spontaneously to the ground state $|0\rangle$ after a lifetime τ_{20}. The total depletion rate by spontaneous emission from the excited state $|2\rangle$ is then given by $1/\tau_2 = 1/\tau_{21} + 1/\tau_{20}$. Pump rate $\mathbb{P}_1$ and lifetime τ_1 of the lower laser level are defined in an analogous way, whereas $\mathbb{P}_1$ does *not* include spontaneous emission from $|2\rangle$ to $|1\rangle$.

To understand the dynamics of a laser system, let us assume a low pump intensity at first. In this case, only the contributions of the pump source and spontaneous emission are taken into account so that the population of atoms in state $|1\rangle$ and $|2\rangle$ changes in time like

$$\frac{\mathrm{d}N_1}{\mathrm{d}t} = -\mathbb{P}_1 + \frac{N_2}{\tau_{21}} - \frac{N_1}{\tau_1}\,, \tag{B10}$$

$$\frac{\mathrm{d}N_2}{\mathrm{d}t} = +\mathbb{P}_2 - \frac{N_2}{\tau_2}\,, \tag{B11}$$

respectively. The relations (B10) and (B11) are the rate equations of a nonamplifying system.

If pumping and de-excitation are exactly balanced, the population of states remains constant despite dynamic processes. For this stationary state ($\mathrm{d}N_1/\mathrm{d}t = \mathrm{d}N_2/\mathrm{d}t = 0$), the population difference[5] $N_2 - N_1$ between the upper and lower laser levels can be calculated from Eqs. (B10) and (B11) as

$$\Delta N_0 = N_2 - N_1 = \mathbb{P}_2\tau_2\left(1 - \frac{\tau_1}{\tau_{21}}\right) + \mathbb{P}_1\tau_1\,. \tag{B12}$$

If the factor $(1-\tau_1/\tau_{21})$ is positive, we may deduce the following necessary requirements for population inversion:

5) The population difference $N_2 - N_1$ is a direct measure for the degree of inversion.

1. The pump rates $\mathbb{P}_1$ and $\mathbb{P}_2$ are large.
2. The depletion rate $1/\tau_2$ from the excited state is low. This means, on average, that the atoms stay for a long time in state $|2\rangle$.
3. The lifetime in the lower laser level is short. This means that the atoms remain only a short time in state $|1\rangle$ so that this state becomes quickly depleted.

In the next step, we increase the pump beam's intensity. As a consequence, intensity amplification by stimulated transitions (red arrows in Figure B.2) must be included. For this purpose, we will introduce an additional transition rate $\mathbb{S}$ (with $\mathbb{S} \propto B_{21}\rho(\omega)$; Section B.1.4) and expand the rate equations (B10) and (B11) to

$$\frac{dN_1}{dt} = -\mathbb{P}_1 + \frac{N_2}{\tau_{21}} - \frac{N_1}{\tau_1} \underbrace{+N_2\mathbb{S} - N_1\mathbb{S}}_{\text{stimulated transitions}} , \tag{B13}$$

$$\frac{dN_2}{dt} = +\mathbb{P}_2 - \frac{N_2}{\tau_2} \underbrace{-N_2\mathbb{S} + N_1\mathbb{S}}_{\text{stimulated transitions}} . \tag{B14}$$

Equation (B14) reveals that the population of the upper laser level is reduced by stimulated emission. However, since stimulated emission increases the population of $|1\rangle$, the amount of absorption events also rises (see fourth term of Eq. (B14)). In the steady state, the population difference in the presence of stimulated emission is

$$\Delta N_S = N_2 - N_1 = \frac{\Delta N_0}{1 + \mathbb{S}\,(\tau_2 + \tau_1 - \tau_1\tau_2/\tau_{21})} , \tag{B15}$$

where we used Eq. (B12). As expected for a laser, the number of atoms in the excited state and, consequently, the degree of inversion are reduced by the stimulated transition rate $\mathbb{S}$ and increased by the pump rate $\mathbb{P}_2$.

B.2.2 Multilevel Optical Pumping

Let us now discuss some properties of optical pumping in detail. For this purpose, we consider atomic systems with different numbers of available laser levels and their respective conditions for population inversion.

Two-level system Up to now, we have restricted ourselves to two-level atomic systems as described by Einstein's model (Figure B.3a). The limitations of this model become clear when we regard high pump intensities. Under these conditions, the contribution of spontaneous emission can be ignored. Due to the Einstein relation (B6), each incident photon has the same probability of being absorbed or becoming a trigger for stimulated emission. Thus, it is *not* possible to achieve population inversion $\Delta N > 0$ by optical pumping of a two-level system. However, population inversion can be achieved by other methods, for example, by electrical pumping (Section B.5.2) and by using free electron beams [4].

Three-level system For light amplification by optical pumping, an atomic system with at least three energy levels is required, as is schematically shown Figure B.3b. The lifetime in the auxiliary pumping level $|3\rangle$ should be much shorter than in the upper laser level $|2\rangle$, that is, $\tau_{21} \gg \tau_{32}$. In this case, atoms are pumped from $|1\rangle$ to $|3\rangle$ at a rate $\mathbb{P}$. The rapid transition from $|3\rangle$ to $|2\rangle$ effectively pumps state $|2\rangle$ so that $\mathbb{P} \approx \mathbb{P}_2$. State $|3\rangle$ is thus basically not populated ($N_3 \approx 0$), while the population of atoms grows in the (metastable) laser level $|2\rangle$ with increased pumping. The three-level system shown in Figure B.3b can be considered as a special case of Figure B.2 with $\mathbb{P}_1 = \mathbb{P}_2 = \mathbb{P}$, $\tau_1 = \infty$ (i.e., lower laser level corresponds to the ground state), and $\tau_2 = \tau_{21}$. With Eqs. (B13) and (B14), we derive the corresponding rate equation

$$\mathbb{P} - \frac{N_2}{\tau_{21}} - N_2\mathbb{S} + N_1\mathbb{S} = 0 \,. \tag{B16}$$

By introducing the total number of atomic states $N_{\text{tot}} \approx N_1 + N_2$ and the pumping transition probability $\mathbb{W} \approx \mathbb{P}/(N_{\text{tot}} - N_2 + N_1)$, we obtain the population difference of a three-level laser system

$$\begin{aligned} \Delta N_{3\text{L}} &\approx N_2 - N_1 \\ &= N_{\text{tot}} \left(\frac{2\mathbb{W}\tau_{21} - 1}{(1 + 2\mathbb{W}\tau_{21})(1 + \tau_{\text{s}}\mathbb{S})} \right) \end{aligned} \tag{B17}$$

with

$$\tau_{\text{s}} = \frac{2\tau_{21}}{1 + 2\tau_{21}\mathbb{W}} \,. \tag{B18}$$

Four-level system For a four-level system (Figure B.3c), the lower laser level $|1\rangle$ has a slightly higher energy than the absolute ground state $|0\rangle$ and is rapidly emptied after a very short time τ_{10} via nonradiative transitions. The auxiliary level $|3\rangle$ is again used for optical pumping. It is also hardly populated due to fast transitions from $|3\rangle$ to $|2\rangle$. Consequently, we may set τ_{32} and τ_{10} to zero and $N_{\text{tot}} \approx N_0 + N_2$. With $\tau_{20} \gg \tau_{21} \gg \tau_{10}$, the population difference of a four-level system is then given by

$$\begin{aligned} \Delta N_{4\text{L}} &\approx N_2 - N_0 \\ &= N_{\text{tot}} \left(\frac{\mathbb{W}\tau_{21}}{1 + \mathbb{W}\tau_{21} + \mathbb{S}\tau_{21}} \right) \,. \end{aligned} \tag{B19}$$

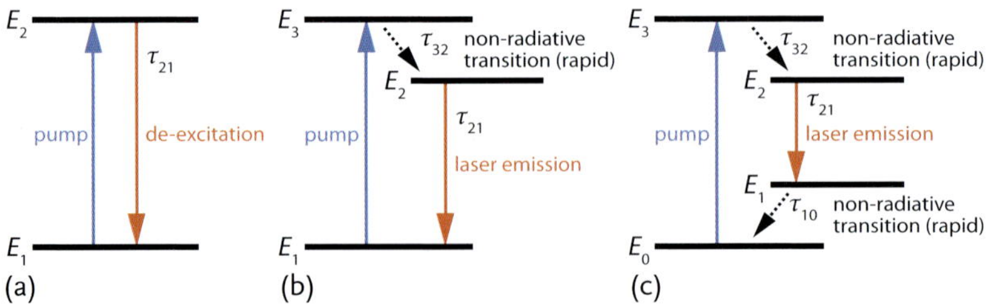

Figure B.3 Transitions in (a) two-level, (b), three-level, and (c) four-level atomic systems.

Comparison of three- and four-level system Let us now compare the conditions for inversion of three-level and four-level systems. For this purpose, we consider the relative inversion $\Delta N_{j\mathrm{L}}/N_{\mathrm{tot}}$ of a *j*-level system versus the product of pumping transition probability $\mathbb{W}$ and lifetime τ_{21} in accordance with Eqs. (B17) and (B19). $\Delta N_{j\mathrm{L}}/N_{\mathrm{tot}} = 0$ defines the threshold above which population inversion is obtained. If $\Delta N_{j\mathrm{L}}/N_{\mathrm{tot}} = -1$, all atoms are in the lower laser level. Accordingly, $\Delta N_{j\mathrm{L}}/N_{\mathrm{tot}} = +1$ represents the case at which all atoms have been excited to the upper laser level. Since $\mathbb{W} \propto \mathbb{P}$, $\mathbb{W}\tau_{21}$ is a measure for the pumping power of the laser system. Note that we set $\mathbb{S}\tau_{21} = \mathrm{const}$, as this factor does not depend on the pump rate.

For the transition $\mathbb{W}\tau_{21} \rightarrow 0$, the relative inversion of a three-level system approaches

$$\lim_{\mathbb{W}\tau_{21}\rightarrow 0} \frac{\Delta N_{3\mathrm{L}}}{N_{\mathrm{tot}}} = \frac{-1}{1+\mathrm{const}} . \tag{B20}$$

Without pumping, more atoms are present in the lower laser level than in the upper level. Only for $\mathbb{W}\tau_{21} > 0.5$ does the relative inversion of a three-level system become positive. In the case of a four-level laser system, however, the relative inversion equals zero for the transition $\mathbb{W}\tau_{21} \rightarrow 0$. Thus, already for low pumping powers we find a population inversion ($\Delta N_{4\mathrm{L}}/N_{\mathrm{tot}} > 0$).

In Figure B.4, the functions of Eqs. (B17) and (B19) are plotted for comparison. As expected, the relative inversion $\Delta N_{j\mathrm{L}}/N_{\mathrm{tot}}$ of both systems rises with increasing pumping power, and we can see that the relative inversion of a four-level system (blue curve) has always higher values than that of a three-level system (red curve).

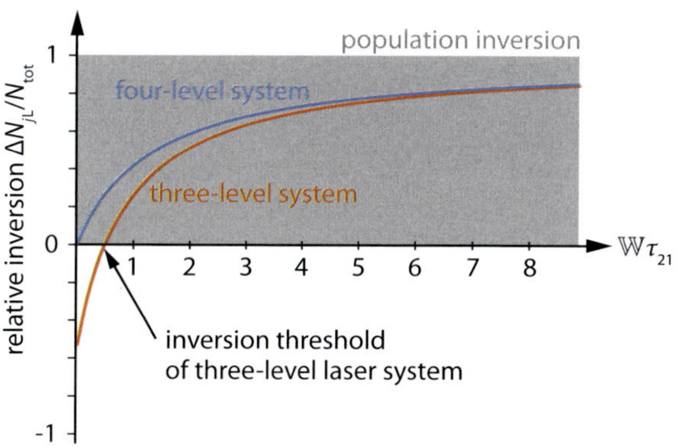

Figure B.4 Relative inversion versus the product of pumping transition probability $\mathbb{W}$ and lifetime τ_{21} for optically pumped three- (red line) and four-level (blue line) laser systems according to Eqs. (B17) and (B19). $\mathbb{W}\tau_{21}$ is a measure for the pumping power of the laser system. The higher degree of population inversion of the four-level system is clearly visible.

B.3
Laser Oscillator

So far, we have only looked at the atomic states to support coherent light amplification. However, a feasible gain medium is only half of the story. For example, consider a rod-shaped gain medium and remember that the traveling directions of spontaneously emitted photons are distributed in a statistical manner. All photons exiting the medium sideways do not effectively contribute to light amplification. By chance, those photons which travel parallel to the rod axis induce more stimulated emission events than all the others so that the resulting radiation is preferably enhanced along the rod axis. If the efficiency of stimulated emission is high enough, an intense light beam is formed. If the rod length is extended, the output power will increase even more and the beam divergence will be reduced. Clearly, such an arrangement already exhibits all typical characteristics of a laser. But losses inside the gain medium lower the total light output. This can only be compensated for by even longer traveling distances of photons inside the gain medium. However, long gain media are not really a practicable solution, because of mechanical difficulties and cost considerations of the very pure materials. To optimize light amplification, we thus let photons pass multiple times through a shorter gain medium by adding a reflecting unit, the so-called *resonator*. As we will see, this setup leads to a superposition of light waves which again results in a constructive interference effect (Section A.2.3).

The combination of gain medium and resonator is called a *laser oscillator*. Based on this concept, various designs have been proposed during the past few decades (Figure B.5). Depending on the oscillator geometry, the beam shape differs in diameter, divergence, and composition.

B.3.1
Inversion Threshold

Let us first take a look at the consequences of such an oscillator setup on the laser light intensity. Let us consider a gain medium of length L_g and volume V_g and enclose it with two facing mirrors as depicted in Figure B.6. This arrangement allows the light beam to bounce back and forth. The resulting intensity inside the resonator is then determined by

$$I = I_+ + I_- \, . \tag{B21}$$

I_+ and I_- denote the intensities of wavefronts inside the resonator which travel to the right and to the left, respectively. For light amplification, the intensity I must be at least maintained after each round-trip. This means that the intensity enhancement by the gain medium must outweigh the round-trip losses due to absorption, scattering, and so on. Following the beam along the resonator brings us to

$$T_i G R_1 T_i G R_2 I_+ \geq I_+ \, , \tag{B22}$$

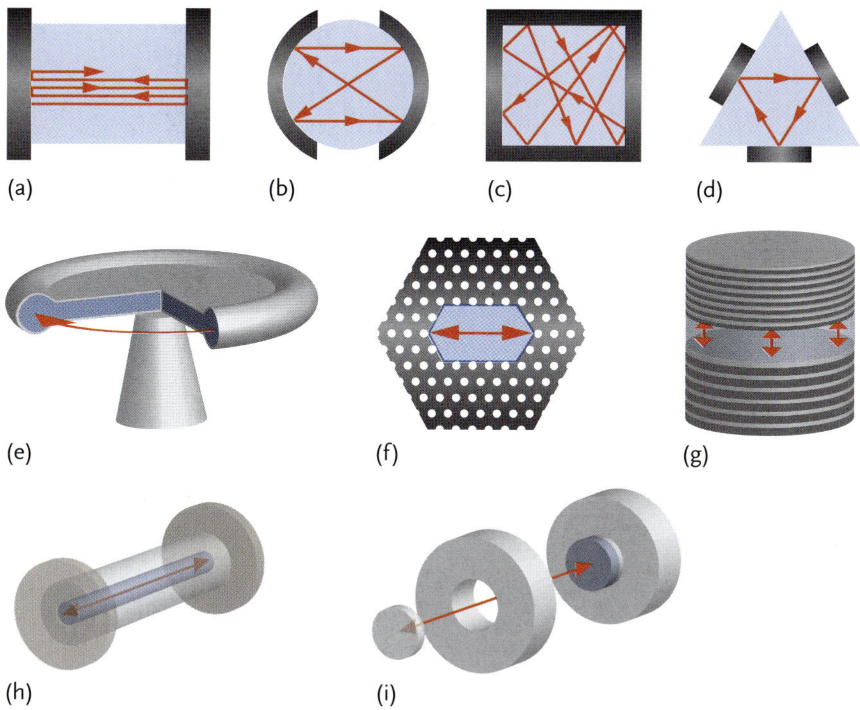

Figure B.5 Different designs of laser oscillators. The gain medium is blue, the reflector unit of the resonator is shown in gray, and the path traveled by the photons is represented by red arrows. (a) Two plane mirrors facing each other. (b) Concentric resonator design consisting of two opposite spherical mirrors. (a) and (b) are special cases of the Gaussian oscillator design (Section B.4). (c) Rectangular resonator cavity. (d) Triangular ring resonator. (e) Microdisk resonator. (f) Photonic crystal oscillator for which the resonator itself is made of the gain material. Reflections around the central "defect" region arise, since photons cannot penetrate the periodic hole array. (g) Micropillar oscillator consisting of a disk-shaped gain medium embedded between two Bragg gratings ("DBRs") that act as mirrors. (h) Fiber laser resonator. The gain medium is enclosed by a cladding to keep the beam inside the resonator (total internal reflection at cladding, Section A.1.1). At both medium–air interfaces the beam is partially reflected. (i) Disk laser resonator (see also Section B.5.3.1).

where T_i is the internal transmittance, R_1 and R_2 the reflectances of the resonator mirrors (Figure B.6), and G the gain medium amplification factor (gain factor). The latter is determined by

$$G = \mathrm{e}^{\sigma(N_2 - N_1)L_\mathrm{g}} \tag{B23}$$

and thus grows exponentially with the population inversion $\Delta N = N_2 - N_1$. In Eq. (B23), σ denotes the wavelength-dependent optical gain cross-section that describes the strength with which light interacts with atoms via the transition. A large value of σ corresponds to a strong interaction and, hence, to a rapid growth of the light intensity.

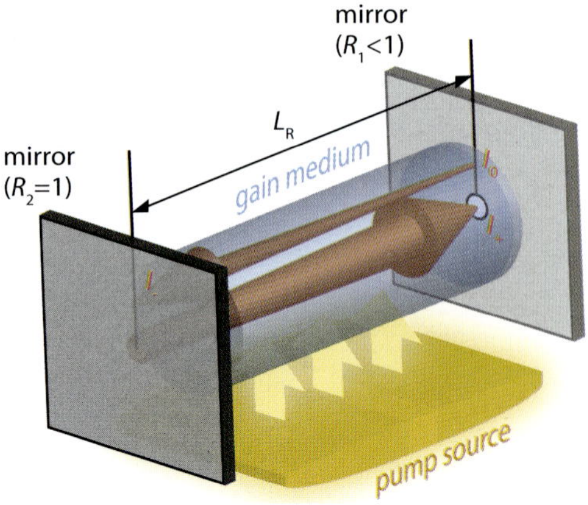

Figure B.6 Scheme of a typical laser oscillator. The gain medium (blue) is encapsulated by two mirrors (gray) which are placed at a distance L_R. The left mirror is highly reflective ($R_2 = 1$), whereas the right mirror has a lower reflectance ($R_1 < 1$) and thus acts as an output coupler. Energy is introduced to the oscillator by a pump source (orange). The photons inside the resonator pass the gain medium multiple times (shown as red arrows) while the beam intensity (represented by arrow thickness) is increased.

With Eq. (B22) and with the mean value of the mirror reflectances $R = \sqrt{R_1 R_2}$, the threshold condition for an oscillator follows as

$$R T_i G \geq 1 \,. \tag{B24}$$

If $R T_i G < 1$, the number of photons added by pumping is too small for laser operation. This means that the intrinsic losses of the gain medium "suppress" all photons generated by stimulated emission. Accordingly, the population inversion of real laser systems must be higher than previously discussed in Section B.2.2, where losses have been neglected. From Eqs. (B23) and (B24), the threshold inversion for laser operation results as

$$\Delta N_{th} = -\frac{\ln(T_i R)}{\sigma L_g} \,. \tag{B25}$$

Related to the threshold inversion, we can define a threshold pumping power P_{th} which depends on atomic parameters like optical gain cross-section σ and lifetime of the laser level. If the pumping power exceeds the threshold ($P_{pump} > P_{th}$), a net amplification of the light intensity is obtained. The output power of a laser then reads

$$P_{out} = \eta_{sl}(P_{pump} - P_{th}) \tag{B26}$$

with the slope efficiency

$$\begin{aligned} \eta_{sl} &= \frac{\Delta P_{out}}{\Delta P_{pump}} \\ &= \left(\frac{\eta_{pump}}{\ln(T_i \sqrt{R})}\right)\left(\frac{R-1}{R+1}\right) \,. \end{aligned} \tag{B27}$$

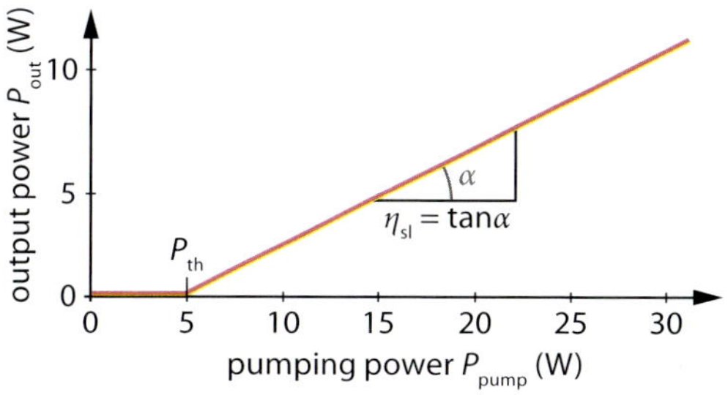

Figure B.7 The gain medium amplification factor and absorption as well as the reflectance of the resonator mirrors determine the minimum threshold pumping power P_{th}. Below P_{th} we have no laser operation. Above P_{th}, the net output power $P_{out} = \int I\,d\mathbf{S}$ (Eq. (A70)) increases linearly. The efficiency η of the system is given by the slope of the curve. Hence, for the example given in this diagram, the threshold pumping power is $P_{th} = 5$W, and the slope efficiency is $\eta_{sl} = 50\%$.

The slope efficiency depends on the internal transmittance T_i, the reflectance of the resonator mirrors, and the pump efficiency[6] η_{pump}. Alternatively, the output power can be written as

$$\begin{aligned} P_{out} &= I_s \frac{V_g}{L_g}\left(\frac{1-R}{1+R}\right)\left(\frac{P_{pump}}{P_{th}} - 1\right) \\ &= \left(\frac{R-1}{R+1}\right)\left(\frac{\eta_{pump}}{\ln(T_i\sqrt{R})} P_{pump} + \frac{V_g}{L_g} I_s\right) , \end{aligned} \tag{B28}$$

where I_s is the saturation intensity. It is apparent from Eq. (B26) that the laser output power rises linearly with P_{pump} starting from P_{th} (please refer to the example shown in Figure B.7). The optimization of the output power for a given pump power usually involves a trade-off between high slope efficiency and low threshold pumping power.

Besides the differential slope efficiency, frequent use is made of the power-conversion efficiency

$$\eta_{pc} = \frac{P_{out}}{P_{pump}} , \tag{B29}$$

which is the electrical-to-optical power efficiency. It is defined as the ratio of output power to supplied pumping power. η_{pc} does *not* include the additional power required for cooling and controlling the laser system (see also Problem PB.4).

B.3.2
Standing Wave Condition

In the previous section, we looked at the light intensity inside a laser resonator and derived a condition for a self-sustained oscillation. This can also be understood as

6) The pump efficiency is given by the ratio of power transferred to the gain medium by pumping and power required by the pump source.

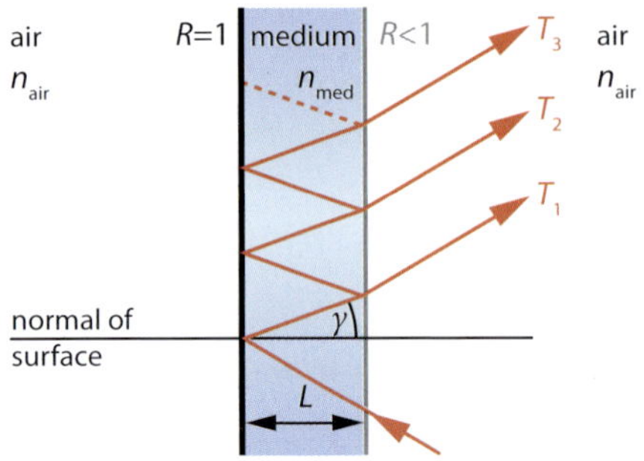

Figure B.8 The Fabry–Pérot interferometer is the simplest optical system to obtain constructive self-interference of a light wave. It consists of one fully reflecting mirror (bold black line) and one partially reflecting mirror (gray) facing each other at a distance L.

the "amplitude amplification" of constructively interfering (Section A.2.3.1) standing waves inside the resonator. Now, we shall also consider the phase relation of the bouncing waves. Placing two mirrors around the gain medium reminds us of a Fabry–Pérot interferometer (Figure B.8) which is the simplest arrangement to induce constructive interference. In analogy to the resonator in Figure B.6, a Fabry–Pérot interferometer involves multiple reflections between two mirrors. One mirror is fully reflecting, whereas the other is partially transparent. So, a portion of the photons is transmitted each time they reach the medium–air surface of the transparent mirror. The emerging beams interfere constructively if the phase difference is zero or a multiple of 2π for adjacent beams. This condition is fulfilled for

$$2Ln_{\text{med}} \cos\gamma = m\lambda \,, \tag{B30}$$

where L is the mirror distance, γ the beam's angle of incidence, n_{med} the refractive index of the medium, and m a positive integer. Since the laser resonator can be seen as a Fabry–Pérot interferometer with zero angle of incidence ($\gamma = 0$), the corresponding standing wave condition is given by

$$\frac{m\lambda}{2n_{\text{med}}} = L_{\text{R}} \,, \tag{B31}$$

where L_{R} is the resonator length. Here, we assumed that the medium fills the entire resonator cavity.

The standing light waves which build up inside the resonator are referred to as *modes*. Here, we distinguish between longitudinal modes which, according to Eq. (B31), only differ in the wavelength; and transverse modes which differ in both the wavelength and the intensity pattern along the beam cross-section. Depending on the type of operation, lasers can be operated in *multimode* or *single-mode*. In multimode operation, laser light consists of multiple modes with equal difference in wavelength

$$\delta\lambda = 2n_{\text{med}} L \tag{B32}$$

between adjacent modes. In single-mode operation, only one longitudinal mode is present in the resonator so that the spectral bandwidth of the laser is extremely narrow and the coherence length is long (Section A.2.4.2).

Info Box B.2: Modes

Modes are standing light waves in an optical resonator. Longitudinal modes differ only in the wavelength. Transverse modes also show different intensity patterns along the beam cross-section (see, for example, Figure B.10).

B.4 The Gaussian Oscillator

Let us continue with a more detailed examination of resonator designs and the resulting laser modes. Although a huge variety of resonator designs exists, commercial laser systems very often use a so-called *Gaussian oscillator*. As shown in the examples of Figure B.5a,b, it consists of a gain medium enclosed by two reflective mirrors. Inside the resonator, photons bounce back and forth between the mirrors by forming Gaussian beams (Section A.2.2.1). At least one of the resonator mirrors has a reflectance $R < 1$ to couple the beam out.

B.4.1 Resonator Stability Condition

For efficient amplification, the geometric design of the resonator must be optimized to reduce losses. The aim is to "keep" the photons inside the resonator. For this purpose, we need a "design recipe" which tells us how to construct such a "stable" laser oscillator.

Curved mirrors can focus light beams like positive lenses, and the beam shape between two identical optical components is approximately the same. From these facts we conclude that a resonator which consists of two facing mirrors is comparable to a periodic sequence of thin lenses separated by a distance L_R (Figure B.9). We can now apply the ABCD matrix approach for paraxial rays (Section A.1.3) in which

$$\begin{pmatrix} A & B \\ C & D \end{pmatrix}$$

describes a full round-trip in the cavity or, equivalently, a periodic sequence of m lenses (with $m \to \infty$). With this, we can derive a geometric stability condition by demanding that the rays stay within the resonator (or the lens sequence), which requires (Problem PB.2)

$$\frac{|A + D|}{2} \leq 1 \,. \tag{B33}$$

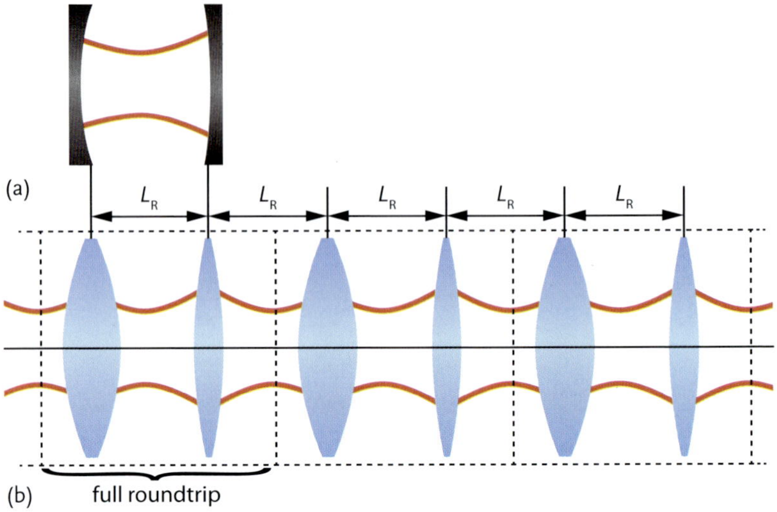

Figure B.9 In the ABCD matrix formalism, the resonator with two facing mirrors (a) is comparable to an infinite periodic sequence of thin lenses (b) which are separated by a distance L_R.

This condition implies that the paraxial beam travels along the optical system without being "deflected" out of the optical system.

Gaussian beams are special solutions of the paraxial wave equation (A79). Thus, a condition which is equivalent to Eq. (B33) can be derived for the Gaussian resonator with the Gaussian beam formalism (Section A.2.2.1). In this approach, the wavefronts of the Gaussian beam resonator have to coincide with the mirror radii r_1 and r_2, as such waves have minimal diffraction losses at the mirror surface. Laser beams which fulfill this condition support Gaussian modes (also called *transversal electromagnetic modes* or TEM modes; Figure B.10) as described in Section A.2.2.1.

To evaluate Eq. (B33) further, we will define the *g parameters*

$$g_1 = 1 + \frac{L_R}{r_1}\,, \tag{B34}$$

$$g_2 = 1 + \frac{L_R}{r_2}\,, \tag{B35}$$

which allow us to rewrite the stability condition (Problem PB.2) by

$$0 \leq g_1 g_2 \leq 1\,. \tag{B36}$$

The g parameters include the geometry of the laser resonator via the resonator length L_R and the mirror radii r_i ($r_i < 0$ for a concave mirror surface; $r_i > 0$ for a convex mirror surface). In Figure B.11, the stable parameter range for Gaussian resonators is highlighted in blue. It is evident that typical resonator designs (planar, confocal, concentric) are special cases of the stability regime.

If the product $g_1 g_2$ is outside the stable regime, modes with random beam profiles appear. Compared to the stable regime, the beam diameter is then larger and

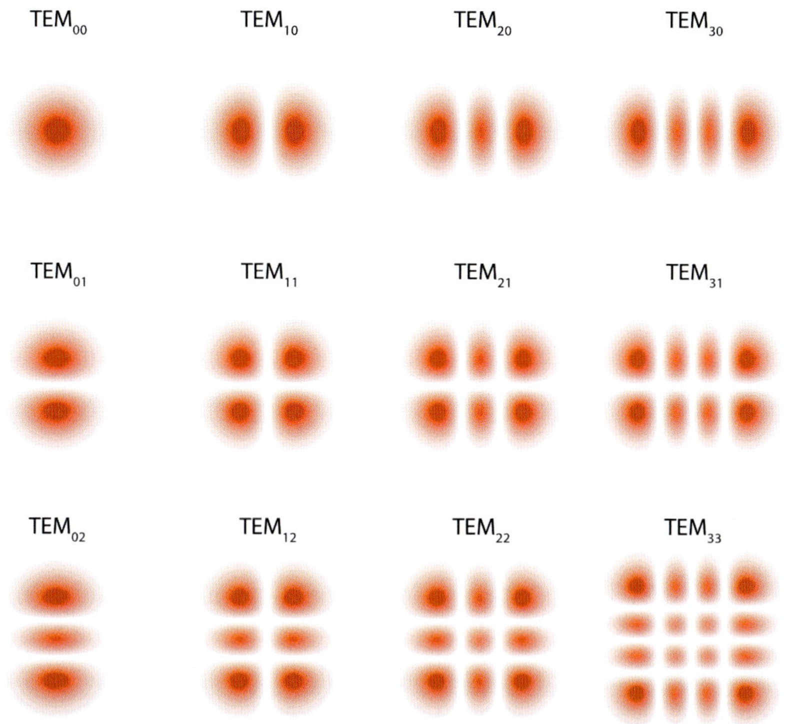

Figure B.10 Intensity profiles of the cross-sections of the Hermite–Gaussian resonator modes. The modes are denominated by TEM_{nm} which refers to "*t*ransversal *e*lectromagnetic *m*ode". The integers m and n represent the number of nodes or zeros of intensity in the vertical and horizontal directions. TEM_{00} is thus the fundamental mode with a Gaussian intensity profile and maximum intensity on the beam axis.

the oscillator becomes lossy. However, it is sometimes meaningful to use an unstable resonator design ([5], for example, to generate super-Gaussian beam profiles (see Info Box 10.1 in Section 10.4.2) or for high-power applications (e.g., pumping of lasers and industrial laser welding). The increased beam diameter reduces the thermal load of the gain medium which, in turn, could lead to material degeneration or even destruction (Problems PB.5 and PB.6).

B.4.2
Divergence

For the fundamental Gaussian mode (TEM_{00} mode; see also Figure B.10), the beam divergence (Section A.2.2.1) is minimal which leads to maximum brightness and minimum waist radius (Section A.2.2.3). If higher Gaussian modes are also present in the resonator, the divergence angle ε and the waist radius w_0 increase. We may

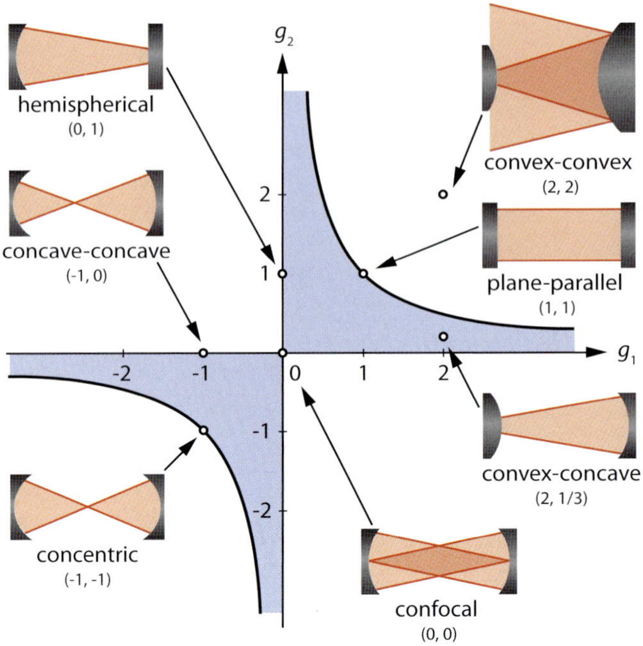

Figure B.11 Stability diagram of Gaussian laser resonators. Various design types are denoted by (g_1, g_2). For example, $(g_1, g_2) = (0, 0)$ stands for a confocal arrangement. The blue area indicates the nominal stable arrangements.

quantify this behavior with the M factor which is determined by

$$\varepsilon w_0 = M^2 \frac{\lambda}{\pi} \; . \tag{B37}$$

In the case of $M^2 = 1$, Eq. (B37) passes into the approximation of Eq. (A87) for small ε ($\tan \varepsilon \approx \varepsilon$). We thus have a situation where only the fundamental Gaussian mode is present. When higher modes are also present, both divergence angle and waist radius are rising, which means that $M^2 > 1$.

B.4.3
Polarization

Each mode generated inside the resonator can, in principle, have two degrees of freedom that correspond to two independent orthogonal polarizations (Section A.2.1.4). Since losses in the resonator usually vary for each polarization direction, the respective amplification is also different. As a consequence, one polarization direction is amplified and the other suppressed. Light emitted by a laser source is thus very often linearly polarized, which is a desired feature for many applications.

B.4.4
Pulsed Laser Operation

Throughout the former sections, we looked at lasers with continuous operation above threshold and constant output power, so-called *continuous wave* (cw) lasers. A laser can also be operated in a pulsed mode, where the optical power may, for example, appear in periodic packets of some duration τ (Figure B.12a). The advantage compared to cw operation is the possibility to provide very high output intensities on a short time scale. By adjusting the repetition rate $1/t_{\mathrm{per}}$ (i.e., the number of pulses per second), the off-time between two consecutive pulse trains is controlled during which light emission is stopped. In this way, we can reduce the thermal load of irradiated objects. The total thermal energy transferred to the object depends on the output power divided by the repetition rate. Spectral width $\Delta\omega$ and pulse duration τ of a pulse are related via the time-bandwidth product (see also Section A.2.4.2)

$$\tau\Delta\omega > 2\pi \ . \tag{B38}$$

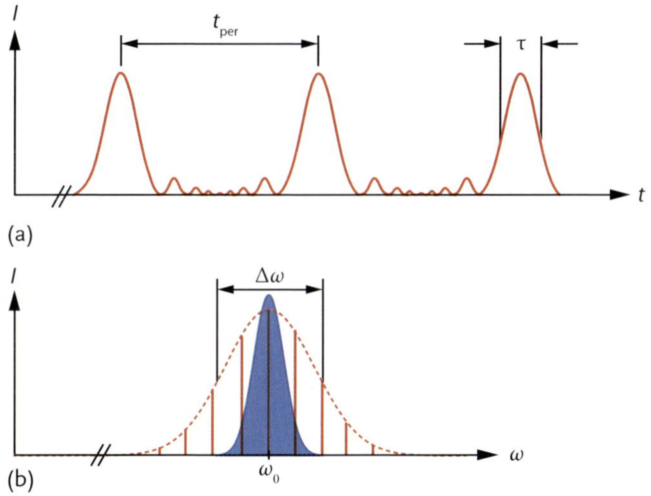

Figure B.12 Output intensity of pulsed laser light. (a) Intensity I versus time t of a pulsed laser beam (mode-locked laser). t_{per} denotes the periodicity of pulses and $1/t_{\mathrm{per}}$ the repetition rate. Note that different definitions of the pulse duration exist. Here, the pulse duration τ is defined as the period at which the intensity $I = \psi\psi^*$ is higher than half the maximum peak value (also known as the full-width half-maximum condition). (b) Emission of laser pulses is always related to a broadened spectral distribution. The diagram illustrates the intensity I versus frequency ω. In the case of a mode-locked laser (Section B.4.4.2), the emitted laser light consists of many modes with an equidistant frequency distribution (red peaks). The corresponding frequency range supported by the gain medium is represented by the dashed curve. The corresponding spectral bandwidth $\Delta\omega$ can be defined by the full width at half maximum (FWHM). In the case of a Q-switched laser (Section B.4.4.1), a continuous light spectrum around a central frequency ω_0 is emitted (blue curve).

In the case of extremely short pulses, the spectrum of laser light is thus considerably broadened (Figure B.12b), whereas cw lasers are in most cases virtually monochromatic.

The simplest way to generate laser pulses is by switching the pump source on and off (gain switching). However, this technique is quite inefficient, as energy is wasted during the off-times, and the pulses' peak power cannot exceed the steady power in cw mode. Q-switches or mode-locking techniques, which we will discuss in the next sections, are more versatile approaches to generate short pulses with a high peak intensity.

B.4.4.1 **Q-switches**

The basic idea of Q-switching[7] [4] is to suppress light oscillations inside the resonator until population inversion is maximized. For this purpose, a controllable switch[8] is added which initially causes high losses inside the resonator. Amplification by stimulated emission is thus suppressed while the number of excited atoms gradually increases during pumping. When maximum population inversion ΔN_{max} is attained, the resonator is "switched on". As a consequence, the losses are reduced and amplification by stimulated emission can occur so that all excited atoms transit to the lower laser level within a very short time frame t_{de} (Figure B.13). Since the energy stored in the gain medium is extremely high in this case, a short and very intense laser pulse is emitted with a maximum output power (Problem PB.7) of

$$P_{\text{max}} \approx \frac{\hbar\omega T V \Delta N_{\text{max}}}{2 t_{\text{rt}}} . \tag{B39}$$

T is the transmittance of the decoupling mirror, V the volume of the resonator, and $t_{\text{rt}} = 2L_{\text{R}}/c$ the round-trip time. P_{max} thus directly depends on the "excess" population inversion $\Delta N_{\text{max}} > \Delta N_{\text{th}}$ when the resonator is switched on and is inversely proportional to resonator length L_{R}. Shorter resonators are thus preferable for high-power systems.

If the system is very strongly pumped, the pulse duration equals the round-trip time, as all excited atoms are de-excited by stimulated emission in one round-trip. Typically, pulse durations in the order of ns (10^{-9} s) are obtained with a maximum output power in the GW (10^9 W) range. The applications of Q-switched laser in ophthalmology are discussed in Sections 10.2.5 and 10.4.3 (Problem PB.7).

B.4.4.2 **Mode Locking**

As we have seen in Section B.4.1, amplification by resonant feedback works only for light waves which fulfill the standing wave condition (B31). In reality, a laser

7) "Q" stands for "quality".

8) Many different approaches exist to increase and decrease the resonator performance in a fast manner. Some optical switches are based on electro-optical effects, where the gain medium's refractive index is controlled by an external electric circuit. Other switches use acousto-optical effects to control diffraction via ultrasound. In addition, passive absorbers exhibiting intensity-dependent filtering are used. A detailed discussion of possible techniques can be found in [4, 6].

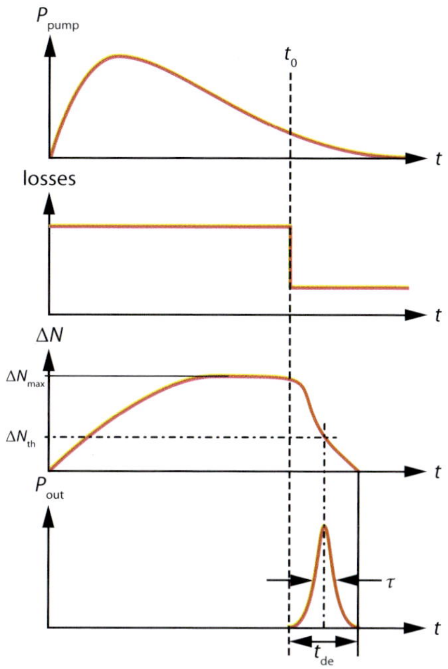

Figure B.13 Temporal behavior of a Q-switched laser oscillator. Pumping power P_{pump}, losses, population inversion ΔN, and output power P_{out} are plotted versus time t. t_0 indicates the point in time when the resonatorquality (internal losses, transmittance of mirrors etc.) is switched on and t_{de} is the time it takes to deplete the upper laser level. ΔN_{th} is the threshold population value of the population inversion. The Q-switch process can be repeated many times.

oscillator supports a number of modes (all having slightly different wavelengths; see Eq. (B32)) which all meet the standing wave condition, as exemplary shown in Figure B.14. The number of modes is not only determined by the resonator geometry, but also by the gain bandwidth $\Delta\lambda$. The latter describes the wavelength range supported by the gain medium (dashed curve in Figure B.12b). Usually, the modes are independent of each other, in terms of each having a different phase in time. Since constructive and destructive interferences (Section A.2.4) occur at the same time, we obtain strong variations in the intensity (mode beating). In the case of a broad gain bandwidth with a large number of modes, the interference effects average to a near-constant output intensity. If, however, all modes are forced by some method to oscillate with a fixed phase relation, constructive interference occurs periodically. This generates very intense light pulses with a rate of $c/(2L_R)$ after each round-trip of photons inside the resonator. In this case, the modes are said to be *locked* or *coupled* [6]. The maximum output power of a mode-locked laser is given by

$$P_{max} \propto (2m + 1)^2 . \tag{B40}$$

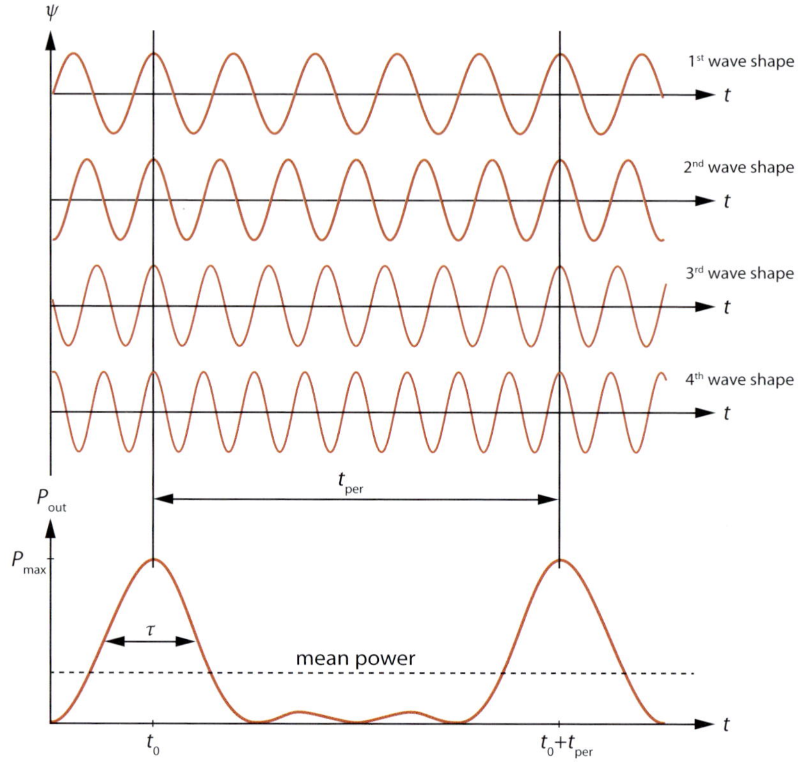

Figure B.14 Principle of mode locking for the example of four modes with different frequencies.

It is thus proportional to the square of the number of modes $m = 2\Delta\lambda L_R/c^2$. By comparison, P_{max} is directly proportional to $2m + 1$ in the case of nonlocked modes. Hence, the more modes that are locked, the higher the achievable output peak power.

The duration of each pulse is determined by the number of coupled modes, the gain bandwidth $\Delta\lambda$, and the pulse shape. For Gaussian beams, the minimum pulse duration is given by

$$\tau_{min} = \frac{0.44 \cdot 2L_R}{mc} = \frac{0.44\Delta\lambda}{c} \; . \tag{B41}$$

Typical values for τ_{min} of mode-locked lasers are in the order of picoseconds (ps) to femtoseconds (fs) (10^{-15}–10^{-12} s). These very short pulses are commonly used in ophthalmology for corneal incisions, cutting flaps, and, more recently, for incisions in the eye lens (Section 10.5).

The methods for mode locking can be classified as *active* or *passive*. Active methods typically involve external signals to trigger the process. For example, an acousto-optic modulator acts as a fast shutter which opens and closes with a rate synchronized to the round-trip time of photons inside the resonator. Modes with a certain phase relation are continuously amplified, whereas the others are attenuated. Passive mode-locking techniques, on the other hand, rely on placing special optical elements with nonlinear characteristics of either refraction or absorption inside the resonator (Problems PB.8 and PB.9).

B.5
Technical Realization of Laser Sources

Laser light has many attractive features which can be advantageous for particular applications. The laser properties are mainly determined by the employed gain medium, the resonator design, and the mode of operation (cw or pulsed). The gain medium determines the generated wavelengths and determines whether the laser can be operated continuously or in pulses. The resonator design primarily defines the mode structure, while the choice of the mode of operation determines the spectral range and the pulse duration. Due to pioneering developments in laser technology, nearly all relevant parameters are tunable to the range of interest. Table B.1 provides an overview of most relevant laser parameters and classifications.

Table B.1 Overview of the most relevant laser parameters and classifications.

Parameter	Types	Remarks
Operation mode	Continuous	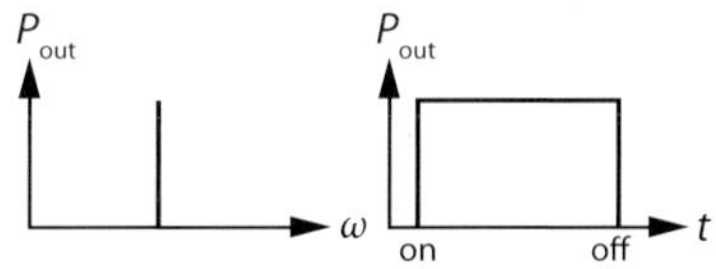
	Q-switched (pulsed)	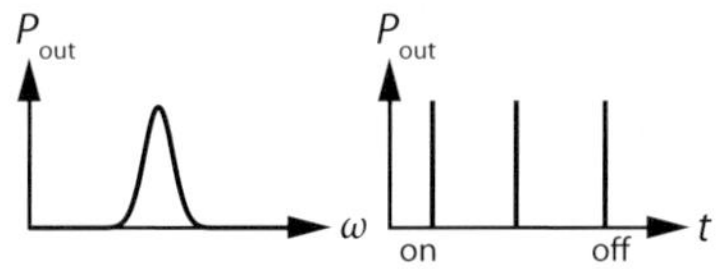
	Mode-locked (pulsed)	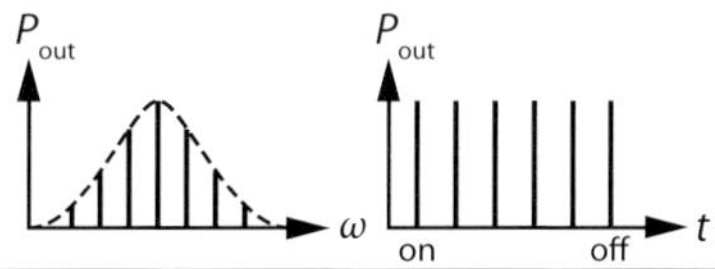
Spectral width	Single-mode Multimode	
Emitted wavelength(s)	Ultraviolet (UV) Visible (VIS) Infrared (IR)	
Beam shape	Divergence Gaussian/Super-Gaussian/Top-hat	

Lasers are classified according to their gain medium into solid-state, dye, semiconductor, or gaseous. The available wavelengths are, on the one hand, set by the intrinsic laser levels of the gain medium and, on the other hand, expandable by means of nonlinear effects like frequency-doubling [7]. Thus, laser radiation is available throughout the whole spectrum from the far-infrared (FIR) up to the ultraviolet (UV) regime.

In the following section, we will discuss the technical aspects of gas, semiconductor, and solid-state lasers. For each case, some specific systems will be highlighted which are most relevant for ophthalmic applications.

B.5.1 Gas Lasers

Gaseous gain media are enclosed by a cylindric tube and pumped by UV light, electron beams, electric current, or chemical reactions. In many cases, lasing is induced by a gas discharge. For this purpose, a high voltage is applied inside the tube by a pair of electrodes which leads to an electric current flowing through the gas. During this, charge carriers collide with each other and with atoms/molecules. In the latter case, these atoms/molecules are either excited or ionized. However, only a small part of the energy provided by the gas discharge is converted to laser light, since it is mainly dissipated as heat. Hence, high-power gas lasers have to be cooled by water, whereas air cooling is mostly sufficient for low-power systems. A list of commonly used gas lasers can be found in Table B.2.

B.5.1.1 Gas-Ion Lasers

Typical gain media of gas-ion lasers are noble gases like argon, neon, or krypton. To make noble gas a suitable gain medium, two electron–atom collisions are needed.

Table B.2 Selection of gas lasers and their properties. For ophthalmic applications, the ArF excimer and the Ar^+ ion laser are often used. In the past, the helium–neon laser served as an optical alignment tool. Abbreviations for the mode(s) of operation: continuous-wave (cw), pulsed (p). The last column provides the magnitude of typical cw output power in watts. For pulsed lasers, the typical power per pulse is given in joules.

Category	Chemical Formula	Typical Operation Mode	Typical Pulse Duration	Common Wavelength	Output Power or Pulse Energy
Excimer/ exciplex laser	ArF	p	10–20 ns	193 nm	10^{-2} J
	KrF	p	10–20 ns	248 nm	10^{-2} J
	XeCl	p	20–300 ns	308 nm	10^{-1} J
Helium–neon laser	Ne	cw	–	633 nm	10^{-3} W
Gas-ion laser	Ar^+	cw	–	488 nm	10^{0} W
	Ar^+	cw	–	514 nm	10^{0} W
	Kr^+	cw	–	647 nm	10^{-1} W

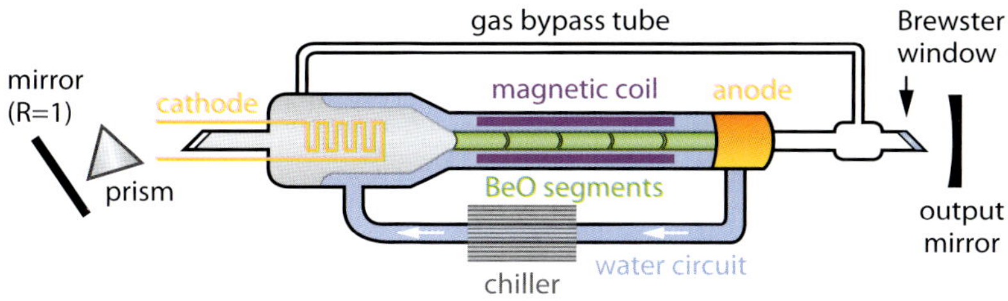

Figure B.15 Schematic setup of an argon ion-laser with beryllium oxide (BeO) bore. The argon gas is ionized between cathode and anode. The resulting ion plasma is then kept in the center of the gas tube by a magnetic coil. The prism on the left-hand side is used to select one of the available laser wavelengths. The Brewster window on the right is used to polarize the laser light parallel to the plane of incidence (Section A.2.1.4). In contrast, the portion of light inside the resonator which is polarized perpendicular to the plane of incidence is suppressed.

The first electron–atom collision ionizes the gas atom, whereas the second collision excites this ion to the upper laser level. For this purpose, gas-ion lasers use high-current gas discharge tubes which consume a lot of electric power (several tens of kilowatts) and produce substantial heat. To handle the high temperatures, the gain medium is filled into an beryllium oxide (BeO) ceramics. This material has a good thermal conductivity and is sufficiently robust against the impact of reactive ions. A magnetic coil keeps the ion plasma in the center of the gas tube to reduce both thermal and electric load. The gas tube is cooled with flowing water. If the system has a closed-loop cooling system, an efficient chiller must be integrated. The total power-conversion efficiency of gas-ion lasers is usually below 0.1%.

Argon ion lasers were the first ophthalmic laser sources used for coagulation of retinal tissue (Figure 10.11 in Section 10.2.5). Beyond that, they can be used to coagulate the trabecular meshwork (trabeculoplasty; Section 10.2.5.2) or for iridotomy (Section 10.4.4.2). Argon lasers can be selected to operate at different wavelengths such as 514 nm (green), 488 nm (blue-green), and 351 nm (UV) by using a prism inside the resonator (Figure B.15) or filters at the output.

B.5.1.2 Excimer Lasers

The term *excimer* is an abbreviation for "excited dimer" and refers to molecules which exist merely in the excited state for a short time. If an excimer transits back to the ground state, the molecule decomposes to atoms by emission of a photon (Figure B.16). Since the unstable ground state is always "empty", excimers are gain media which allow output powers of as high as 200 W.

The gain medium of an excimer laser consists of a noble gas (e.g., xenon and krypton) which does not usually form any chemical compound. However, when

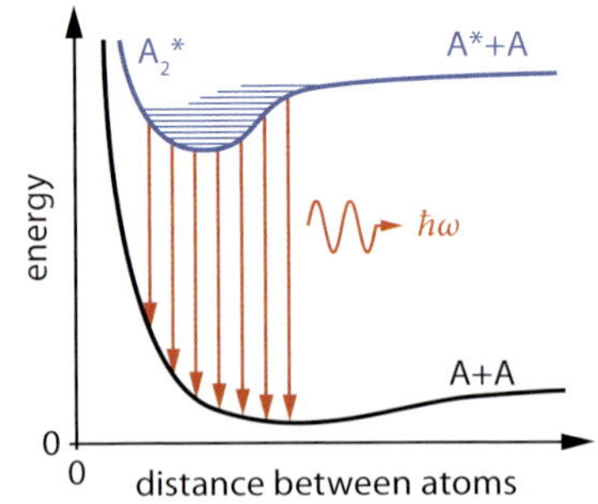

Figure B.16 Energy diagram of excited dimers (excimers). Two atoms A + A are excited and form a molecule A_2^* whose energy depends on the distance between both atoms (see progression of blue curve). However, this molecule is unstable and decomposes again: $A_2^* \rightarrow A + A$. Since the energy of both individual atoms (black curve) is lower than the energy of the molecule (blue), the difference energy is emitted as a light wave. In other words, the excimer is de-excited to the ground state.

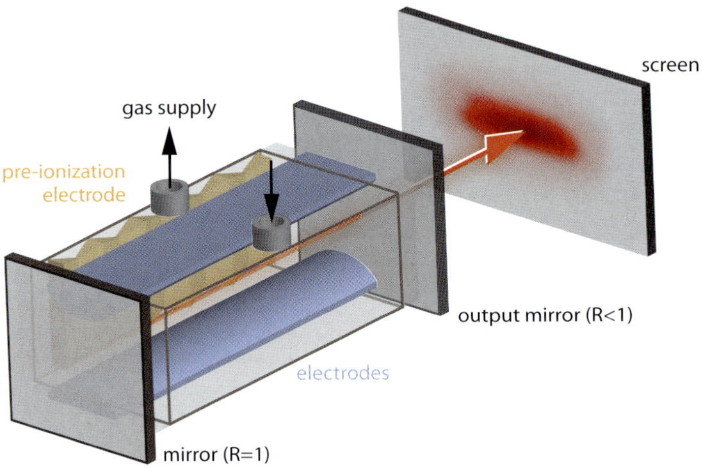

Figure B.17 Schematic setup of an excimer laser. Due to the typical electrode geometry the emitted beam profile is nearly rectangular.

excited by an electrical discharge or high-energy electron beams, noble gas atoms form temporarily-bound molecules with each other (dimers) or with added halogens[9] (complexes) such as fluorine and chlorine.

The resonator of an excimer laser usually consists of a plane dielectric mirror with high reflectance and a plane calcium fluoride (CaF_2) or magnesium fluoride (MgF_2) front mirror with a reflectance of 0.05 to 0.30.

9) If complexes of noble gases and halogens are used (e.g., ArF, KrF, and so on), the system shall rather be referred to as an *exciplex* laser which is the abbreviation for "excited complex", but the term "excimer" is normally used for halogen-containing molecules such as XeCl as well..

The technical realization of excimer lasers is quite demanding. Gas in the resonator is nearly at atmospheric pressure, which makes it hard to obtain a homogeneous gas discharge. For this purpose, a preionization electrode is implemented (Figure B.17). In addition, the gas inside the resonator must be exchanged between consecutive laser pulses to avoid discharge instabilities. Excimer lasers can thus only be operated in pulsed mode. They deliver pulses in the UV spectral range with pulse durations from 10–100 ns at repetition rates of up to 2 kHz. After around $10^{(}$ pulses, a complete gas exchange becomes necessary, since the highly reactive excimer forms compounds with the material of the gas tube. The abrasion products from the electrode material are deposited on the resonator mirrors so that they have to be cleaned regularly.

The power-conversion efficiency ranges from 0.2–4%. Compared to solid-state and semiconductor laser sources (Sections B.5.2 and B.5.3, respectively), this is quite low. It should be noted, however, that spontaneous emission considerably increases for short wavelengths (Eq. (B7)) so that not many alternatives exist.

The short wavelengths emitted by excimer lasers are strongly absorbed by water and proteins so that light cannot penetrate deep into the eye tissue. For this reason, excimer lasers are primarily used in ophthalmology for tissue ablation in refractive surgery (Section 10.3).

Info Box B.3: The Band Model of Solids

Solids consist of a huge amount of atoms which are periodically arranged and densely packed in a crystalline structure [7]. Since the energy levels of the individual atoms overlap, they strongly interact so that the discrete levels transform into continuous energy bands (Figure B.18) which are separated by the energy gap E_{gap}.

Solids are classified depending on the electron population of their energy bands. The conduction band of metals is only partially populated which explains their high electrical conductivity. In contrast, the conduction band of insulators is free of electrons and energetically separated from the completely filled valence band. Semiconductors have a relatively small energy gap between valence and conduction bands. Thus, at low temperatures, they behave like insulators, whereas the conduction band is populated at room temperature. The conduction by thermally excited electrons and holes is called *intrinsic* (Figure B.19a). By adding impurities to the semiconductor crystal, new energy levels are generated. In the case of *n-type* doping, impurities with extra electrons are introduced to the crystal. In terms of energy, these *donor* levels lie next to the conduction band (Figure B.19b) so that the energy required to excite electrons to the conduction band is significantly decreased. Similarly, we can add impurities which give rise to a deficit of electrons. This so-called *p-type* doping (Figure B.19c) creates *acceptor* levels close to the valence band.

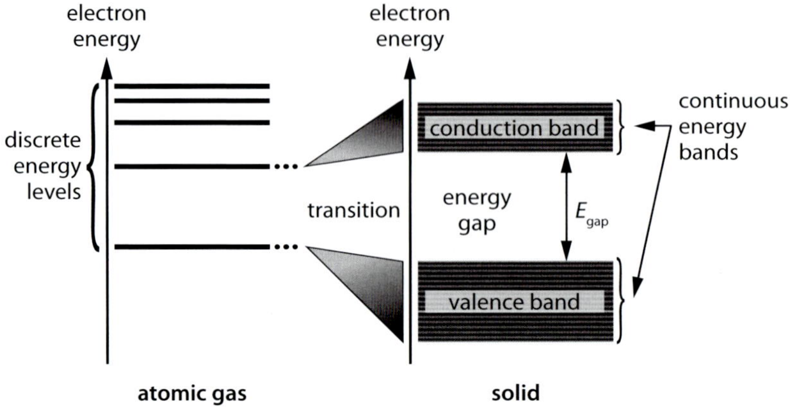

Figure B.18 Formation of electronic energy bands in solids. As the separated atoms (left) are arranged periodically to form a solid (right), the energy levels start to overlap with each other. This causes strong interactions and, thus, the formation of broad energy bands. Optical transitions between them appear over a continuous spectral range rather than for discrete wavelengths. For the description of optical properties of solids, the highest energy bands (valence and conduction bands) are of particular interest.

Much like semiconductors, we can also introduce impurities to insulators to create new energy states which are comparable to atomic states. Hence, we can once again find upper and lower laser levels. Doped insulators are the gain media of solid-state lasers.

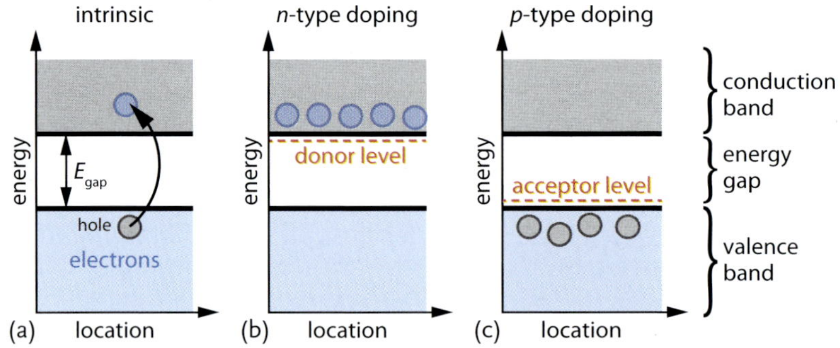

Figure B.19 Energy band structure of (a) a pure (intrinsic), (b) *n*-doped, and (c) *p*-doped semiconductor. The blue circles represent electrons, whereas the gray circles refer to holes. Additionally, the donor and acceptor levels are schematically shown (red).

B.5.2
Semiconductor Lasers

Let us consider the energy levels of doped or undoped semiconductors in Figure B.19. It is obvious that conduction and valence bands serve as upper and lower laser levels, respectively. As already mentioned, semiconductors are electrically conductive under standard conditions. When we integrate the crystal into an electric circuit, the flowing current (of injected electrons) pumps the atoms directly. This process replaces the optical excitation.

A pure semiconductor is not useful as a gain medium, since no vacancies are present in the valence band. Hence, no stimulated emission can happen. We actually need a junction between *n*- and *p*-type doped semiconductors. Due to electron diffusion, an intrinsic ("i") region builds up between both doped regions, where excess electrons and holes compensate for each other. All in all, we thus obtain a "p-i-n" structure which is also known as a *diode*.

In Figure B.20, the corresponding energy levels are plotted versus the location. If no bias voltage is applied, the holes of the *p*-type semiconductor are at the same energy level as the excess electrons of the *n*-type semiconductor (Figure B.20a). As a consequence, it is impossible to identify an upper or lower laser level. But if a forward bias voltage $V_0 > 0$ is applied across the junction, the voltage "bends" the energy levels as shown in Figure B.20b and the intrinsic (depletion) region shrinks. Hence, the energy levels of the doped regions show a spatial overlap. In this case, the donor levels of the *n*-type semiconductor may act as an upper laser level, whereas the acceptor levels of the *p*-type semiconductor serve as a lower laser level.

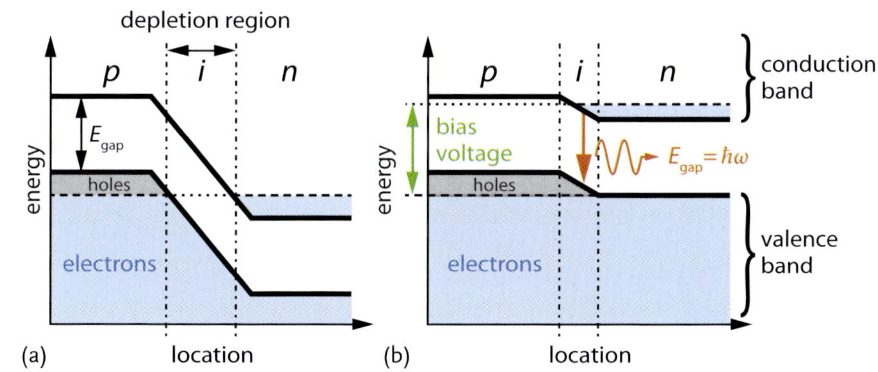

Figure B.20 Energy scheme of a diode ("p-i-n structure") versus location. (a) Band structure if no bias voltage is applied. (b) With a positive bias voltage, the band structures of *n*- and *p*-type semiconductors are bent and brought into line. In the active region between them, electrons in the conduction band coexist with holes in the valence band. Hence, electrons can transit from the conduction band to the valence band (red arrow) by emitting photons of energy $E_{gap} = \hbar\omega$. This is actually a stimulated emission process required for laser operation.

Table B.3 Semiconductor lasers which are commonly used in ophthalmic systems. Abbreviations for the mode(s) of operation: continuous-wave (cw), pulsed by gain switching (p). The last column lists the magnitude of typical cw output power in watts. Lasers with composite gain media (stacks of semiconductors) may deliver a much higher output power in the range of $> 10^2$ W.

Category	Gain medium	Operation mode	Wavelength	cw Output power (W)
Gallium–arsenide laser	GaAlAs	cw/p	650–880 nm	10^1
	InGaAsP	cw/p	630–2000 nm	10^1
	InGaAs	cw/p	904–1065 nm	10^1
	InGaAs	cw/p	1.27–1.33 μm	10^1
	InGaAs	cw/p	1.43–1.57 μm	10^1
Gallium–nitride laser	InGaN	cw/p	370–493 nm	10^{-1}
Gallium–phosphite laser	InGaAlP	cw/p	630–685 nm	10^{-1}

A p-i-n structure can be either used as a laser (Table B.3 for typical representatives) or a light-emitting diode (LED). In contrast to an LED, a laser diode is additionally embedded in a resonator.[10)]

Info Box B.4: Light-Emitting Diode (LED)

The light-emitting diode (LED) is a forward-biased p-i-n diode whose light emission is based on spontaneous processes. The wavelength (color) of the emitted light is determined by the energy gap of the used semiconductor. Typically, the spectral width is broader than that of a corresponding laser source. Compared to "classic" light sources (e.g., light bulbs), LEDs have a lower energy consumption, longer lifetime, improved physical robustness, and smaller size.

The output power of a laser diode depends on the injection current flowing through the device. If the amperage is below a certain threshold I_{th}, the emitted light is mainly characterized by spontaneous emission and behaves like an LED. Above I_{th}, the losses of the diode are compensated and stimulated emission predominates (Figure B.21). The output efficiency then increases considerably, and the diode operates in laser mode. Since the emitted light stems from spontaneous and stimulated processes, it can have a relatively broad spectral width of around 25 nm.

The center of the spectrum shifts to longer wavelengths with increasing temperature. In the case of a GaAlAs diode, the range of change is about 0.25–0.30 nm/K.

10) There exist several feasible designs for semiconductor lasers, for example homo- and hetero-structures, quantum-wells, distributed-feedback designs, VCSELs, and so on. Further information about laser diode designs can be found in standard textbooks on laser technology, for example, [4, 6, 7].

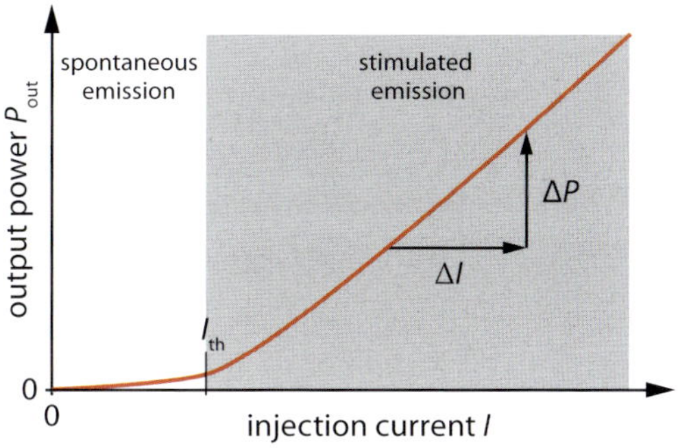

Figure B.21 Output power of light emitted by a laser diode versus the injected current. Below the threshold current I_{th}, spontaneous emission is the most relevant process. Above I_{th}, photons are predominantly generated by stimulated emission.

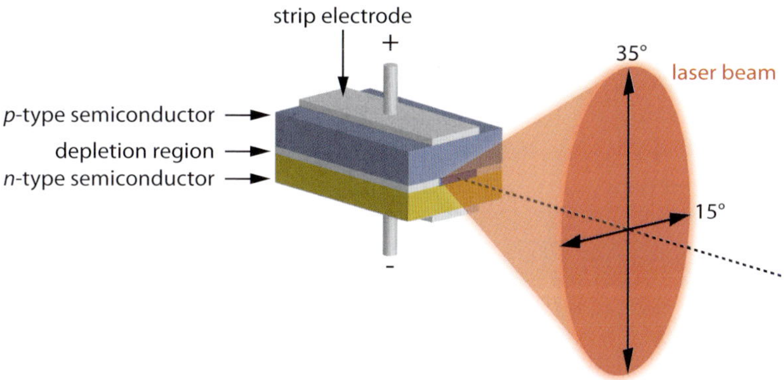

Figure B.22 Structure and anisotropic emission characteristics of a laser diode.

Active temperature stabilization with Peltier elements is thus required for laser applications, where the emitted wavelength must be as stable as possible. The power-conversion efficiency of semiconductor lasers is typically around 25%. Since diodes are very thin (typically a few micrometers), the emission characteristics are anisotropic like shown in Figure B.22. Perpendicular to the p-i-n structure, the emitted radiation has a large angle of divergence ($\approx 20°-40°$), whereas the divergence is smaller parallel to the junction. This anisotropy can be corrected by special (anamorphotic) optics or via losses by coupling into a glass fiber. Coupled into a fiber, laser light can be easily guided to the treatment region (Section 10.2.4.2).

Semiconductor lasers have many advantages compared to gas and solid-state lasers (Section B.5.3). They are usually much cheaper, more compact, show high beam stability due to their rigidity, and are very easy to handle (e.g., no toxic gases or hygroscopic materials). Moreover, a huge variety of gain media exists so that almost any infrared to visible output wavelength is available (Table B.3). This is the reason why semiconductor lasers have replaced gas and solid-state systems for many applications during the recent decades.

Table B.4 Solid-state lasers which are relevant for ophthalmic applications. Abbreviations for the mode(s) of operation: continuous-wave (cw), Q-switched (qs), mode-locked (ml). The last column lists the typical cw output power in watts.

Category	Gain medium	Operation mode	Typical pulse duration	Wavelength	Typical cw output power (W)
Glass laser	$Nd:SiO_2$	ml	0.1–100 ns	1.06 μm	–
Ruby laser	$Cr:Al_2O_3$	cw/qs	1–250 μs	694 nm	10^0
Titanium–sapphire laser	$Ti:Al_2O_3$	cw/qs/ml	0.05–100 ps	670–1130 nm	10^0
Vanadate laser	$Nd:YVO_4$	cw/qs/ml	0.1–10 ns	1.06 μm	10^{+1}
YAG laser	$Yb:Y_3Al_5O_{12}$	cw/qs/ml	0.5–1000 ns	1.05 μm	10^{+1}
	$Nd:Y_3Al_5O_{12}$	cw/qs/ml	0.1–250 μs	1.06 μm	10^{+2}
	$Nd:Y_3Al_5O_{12}$	cw/qs/ml	0.1–250 μs	1.12 μm	10^{+0}
	$Nd:Y_3Al_5O_{12}$	cw/qs/ml	0.1–250 μs	1.32 μm	10^{+1}
	$Ho:Y_3Al_5O_{12}$	cw/ml	0.1–250 μs	2.08 μm	10^{+1}
	$Er:Y_3Al_5O_{12}$	cw/ml	0.1–250 μs	2.94 μm	10^{+1}

B.5.3 Solid-State Lasers

Solid-state lasers are based on active media which consist of ions doped into an insulating host material. The excitation of ions is usually carried out by optical pumping via classic light sources (flash lamps), semiconductor lasers, or other laser sources. The power-conversion efficiency of solid-state lasers depends significantly on the choice of the pumping system. It varies from $< 1\%$ for flashlamps to $> 25\%$ for systems pumped with laser diodes. Since the lifetime of ions in the upper laser level is relatively long and the bandwidth of the gain medium often relatively broad, a large amount of energy can be stored in solid gain media. For this reason, high output powers and very short pulse durations down to some fs (10^{-15} s) can be achieved (Table B.4).

B.5.3.1 Optimized Shapes of Solid Gain Media

As solid-state lasers have high output intensities, a considerable portion of light is absorbed by the gain medium as well. If the gain medium is a massive slab- or rod-shaped crystal, it rapidly heats up and cooling becomes a serious problem (Figure B.23). The mere use of a closed-loop water cooling system is often not sufficient. Therefore alternative geometric shapes have been proposed for which excess heat is easily released. The advantages and disadvantages of both fiber and disk lasers are summarized in Table B.5.

Table B.5 Advantages and challenges of fiber and disk lasers with reference to a standard rod-shaped geometry of solid gain media.

Laser Type	Advantages	Challenges
Fiber laser	• Efficient cooling • Compact design • Few mechanical components • Large bandwidth • High pump efficiency • High beam quality	• Output power limited by non-linear effects • Fiber might be destroyed at high power • Chromatic dispersion
Disk Laser	• Efficient cooling • Compact design • High pump efficiency • Beam quality and pump efficiency independent from output power	• Substantial losses due to spontaneous emission in transverse direction

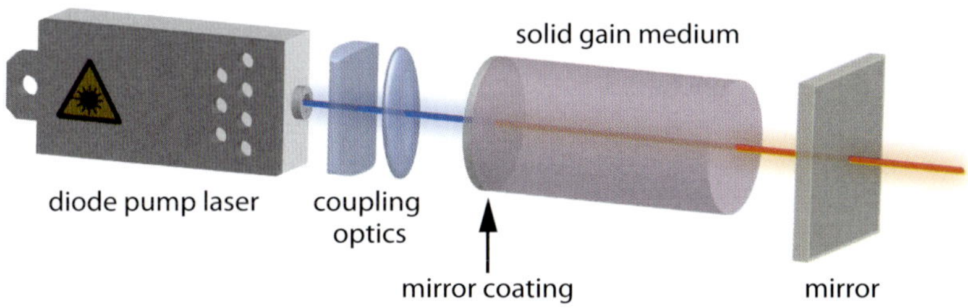

Figure B.23 Schematic setup of a diode-pumped solid-state laser. The mirror coated onto the terminal face of the solid gain medium (violet) is transparent for light emitted by the diode pump laser (blue line) and highly reflective for light generated by stimulated emission (red line).

Fiber laser Some solid gain media can be drawn to fibers (Figure B.5h), particularly those with silica glass as a host material. The fibers are then pumped by diodes at the end surfaces. The generated laser modes are guided inside the fiber because of total internal reflection. Since the surface area is very large compared to the volume of the gain medium, excessive heat can be easily released. Fiber lasers are able to emit light of the fundamental Gaussian mode with $M^2 \approx 1$, which means low divergence and a small spot size (Section B.4.2).

Disk laser Another possibility to build efficient solid-state lasers is a thin (about 100 μm) disk-shaped gain medium (Figures B.5i and B.24). In the case of an optimized oscillator geometry, the laser beam may cross the gain medium multiple times so that a high amplification and efficiency is attained. The reflecting surface of the crystal is glued to a heat sink and, therefore, experiences only little thermally-induced mechanical tension. Excessive heat is dissipated by the entire basal plane of the disk.

B.5.3.2 Neodymium–YAG Laser

A neodymium-doped yttrium aluminum garnet (Nd:YAG) crystal suitable for laser operation is obtained by replacing 1% of yttrium (Y^{3+}) ions of the colorless host crystal $Y_3Al_5O_{12}$ with neodymium ions (Nd^{3+}). The gain medium is optically pumped at a wavelength of 808 nm either by flashtubes or laser diodes. Depending on the pumping technique used, the power-conversion efficiency ranges from 3–5% for flashlamps and 25–50% for laser diodes. The wavelength of emitted photons is usually around 1064 nm, but laser transitions also exist at 940, 1123, 1319, and 1440 nm.

The oscillator may be constructed in different ways. In most cases, the gain medium itself serves as the optical resonator in that the mirrors are coated directly onto the terminal faces of the crystal (Figure B.23). However, open systems are also available which allow the inclusion of switches and modulators to modify the beam geometry. Nd:YAG lasers operate in both pulsed and continuous mode. In a mode-locked system, the pulse durations may go down to the ps range, whereas the output power can be some kilowatts.

Due to their high flexibility, Nd:YAG lasers have a variety of applications in ophthalmology (Problem PB.5). For example, Q-switched and mode-locked systems are used for posterior capsulotomy (Section 10.5.4.2) and iridotomy (Section 10.4.4.2). The near-infrared laser light can also be frequency-doubled by a nonlinear optical crystal. The resulting green light (532 nm wavelength) is barely absorbed by water, whereas the absorption in hemoglobin and melanin is high (Figure 9.3). Coagulation of retinal tissue (Section 10.2.5) and glaucoma therapy are thus also common fields of application. Indeed, frequency-doubled cw systems have replaced argon ion lasers due to their compact design and relatively low power consumption.

B.5.3.3 Neodymium–Vanadate Laser

The emitted laser wavelength of neodymium-doped yttrium vanadate (Nd:YVO_4) crystals is similar to Nd:YAG, that is, 1064 nm. So, the typical applications in ophthalmology are quite similar. Compared to Nd:YAG, Nd:YVO_4 shows a much higher pump absorption and slope efficiency. Moreover, both the emitted gain bandwidth and the wavelength range for pumping are broader. Nd:YVO_4 is well suited for mode-locked lasers with very high pulse repetition rates up to 160 GHz. However, the crystal has a lower thermal conductivity so that it is often designed as a thin disk.

A simplified arrangement of a vanadate disk laser with folded geometry is shown in Figure B.24. Real-world setups are actually a bit more complicated, since the pump beam has to pass multiple times through the thin gain medium to increase efficiency.

B.5.3.4 Ytterbium–YAG Laser

Like neodymium, ytterbium (Yb) is a chemical element which belongs to the group of rare earth metals. But in contrast to neodymium, the gain bandwidth of laser transitions is typically quite large. Ytterbium-doped (Yb^{3+}) yttrium aluminum garnet crystals (Yb:YAG) act as a three-level laser system (Section B.2.2) for a wave-

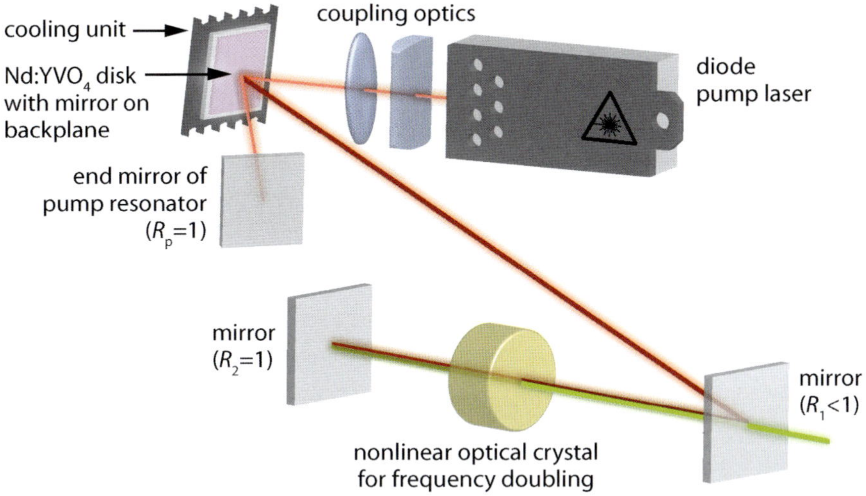

Figure B.24 Simplified schematic setup of a frequency-doubled Nd:YVO_4 disk laser (*folded* geometry). The gain medium (violet) is pumped by a semiconductor laser with a wavelength of about 810nm. Due to anisotropic emission characteristics, the pump beam is focused by a cylindrical and a spherical lens. The laser light emitted by the Nd:YVO_4 crystal (1064 nm wavelength) is guided to the "output" resonator formed by the lower pair of mirrors. An integrated non-linear optical crystal (yellow) doubles the frequency of the incident near-infrared light so that green light (532 nm wavelength) is eventually emitted at the output.

length of 1030 nm. Hence, some emitted photons are reabsorbed by the atoms and heat up the whole crystal. Like Nd:YVO_4, the gain medium has to be cooled efficiently and is thus often designed as a disk.

B.6 Recommended Reading

Since lasers have become one of the most important tools in science and industry during recent decades, a vast amount of books, papers, and reviews has been published in this field. Some of them are outstanding regarding their didactic content. For a detailed discussion about the general working principles of lasers, we recommend [4, 7–13]. For beginners, [5, 14] provide an easy introduction to the fundamental concepts.

B.7 Problems

PB.1. Einstein relations Starting from Eq. (B5), Einstein found that the probabilities for absorption and stimulated emission are equal for a two-level system. So, he supposed that the number of atoms in the ground state and in the excited state (N_1 and N_2, respectively) are given by the Boltzmann distribution

$$\frac{N_2}{N_1} = \exp\left(-\frac{\hbar\omega_{21}}{k_B T}\right),$$

where k_B is the Boltzmann constant. In addition, he used Planck's formula for the energy density

$$\rho(\omega) = \frac{\hbar\omega^3}{\pi^2 c^3}\left(\exp\left(\frac{\hbar\omega}{k_B T}\right) - 1\right)^{-1}.$$

Follow Einstein's approach and show that Eq. (B6) can be derived from Eq. (B5).

PB.2. Stability condition A paraxial light beam bouncing forth and back in a resonator can be considered as if the beam would pass through a periodic sequence of lenses. Each component is described by an ABCD matrix (Section A.1.3).

1. To derive the general stability condition (B33) for resonators, we calculate

   $$\begin{pmatrix} h_2 \\ \gamma_2 \end{pmatrix} = \begin{pmatrix} A & B \\ C & D \end{pmatrix}\begin{pmatrix} A & B \\ C & D \end{pmatrix}\begin{pmatrix} h_0 \\ \gamma_0 \end{pmatrix}. \tag{B42}$$

   Solve the resulting quadratic equation by using the ansatz

   $$h_m = h_0 K^m \tag{B43}$$

   with $K = \text{const}$ and $m = 0, 1, 2, \ldots$. In the case of lens systems, $\det \underline{\mathbf{M}} = AD - BC = 1$. A periodical and stable solution for Eq. (B43) is obtained if the linear combination $K = K_+ - K_-$ is a real number.
2. Let us now consider the special case of a Gaussian resonator. The corresponding ABCD matrix can be calculated from the matrices of a thin lens and free space according to

   $$\begin{pmatrix} A & B \\ C & D \end{pmatrix} = \begin{pmatrix} 1 & 0 \\ 2/r_1 & 1 \end{pmatrix}\begin{pmatrix} 1 & L_R \\ 0 & 1 \end{pmatrix}\begin{pmatrix} 1 & 0 \\ 2/r_2 & 1 \end{pmatrix}\begin{pmatrix} 1 & L_R \\ 0 & 1 \end{pmatrix}. \tag{B44}$$

   Derive the condition (B36) from Eq. (B33) with the dimension parameters given in Eqs. (B34) and (B35).

PB.3. Gaussian laser beams We consider a helium–neon laser beam in a TEM_{00} mode which has a waist radius of $w_0 = 1.3$ mm. The beam shall be expanded and subsequently focused by an optical system.

1. Calculate the ABCD matrix of an optical system which consists of a negative and a positive lens which have a distance of L. What is the condition for an afocoal Gailiei telescope? How does the matrix change in this case?
2. The laser beam shall be expanded to a diameter of $2w_0 = 8$ mm by using a Gailiei telescope which consists of thin lenses with a face-to-face length of $L = 50$ mm. Calculate the focal lengths of the lenses in the Galilei system.

3. Next, the collimated expanded beam shall be focused so that we obtain a depth of field of $\Delta z = 1$ mm. Here, the depth of field is defined as the range at which the beam intensity does not fall below 80% of the maximum intensity at the waist. What is the minimum focal length to achieve this? How large is the diameter of the focus?
4. We assume that the focusing lens, with a minimum focal length as calculated in 3., is placed along the laser path such that the waist (diameter of 8 mm) lies 300 mm in front of the lens. Calculate the waist position behind the lens relative to the image-side focal point F′ of the lens. Does the waist lie in front of or behind the focal point F′?
5. Consider the change of the Gaussian beam parameters when the beam passes through an afocal Kepler- and Galilei-type telescope system. Such an optical system can be used to expand or compress the beam diameter. Calculate the minimum beam diameter and the divergence angle. How are these parameters related to each other? Consider also the product of minimum beam diameter and divergence angle.

PB.4. Laser power Calculate the cw power of a Nd:YAG laser as a function of the degree of reflectance of the output mirror for various pumping power values of 1 kW, 2 kW, and 4 kW. Determine the optimal decoupling degree $T = 1 - R$. Use MathCAD or a similar program to calculate the profile of the output power for various output mirrors first and then attempt to find an analytical solution for the optimal degree of out-coupling (i.e., maximum output power). Use the following values:

- Saturation intensity: $I_s = 2.2\,\text{kW/cm}^2$.
- Laser rod dimensions: 0.5 cm (diameter), 10 cm (length).
- Transmittance in resonator: $T_i = 0.95$.
- Pump efficiency $\eta_{pump} = 5.5\%$.

PB.5. Nd:YAG laser For photocoagulation, a frequency-doubled cw Nd:YAG laser is used. The Nd:YAG resonator consists of a concave mirror with a radius of curvature of 250 mm and a flat decoupling mirror. What is the maximum distance L_{max} between the mirrors to obtain a stable configuration? Is it possible to design a stable resonator made of two convex mirrors with the same radius of curvature and same distance L_{max}?

PB.6. Thermal lens An active laser medium within a laser resonator usually generates a so-called *thermal lens* due to thermal effects. Consider whether this will be a convergent or a divergent lens. For this purpose, look up the temperature dependence of the refractive index of laser materials on the internet! For building a model, let us consider a confocal-planar resonator of length L_R. For simplification, let the thermal lens with focal length f (sign!) be positioned exactly in the middle. The planar mirror is the output mirror. Calculate the stability condition as a function of the focal length of the thermal lens. How is the decoupled Gauss bundle changed by the lens (waist, divergence)?

PB.7. Q-switch Consider a Q-switched Nd:YAG laser with a resonator/gain medium length of $L_g = 30\,\text{cm}$. The rod-shaped gain medium has a diameter of $d = 1\,\text{cm}$ and a gain cross-section of $\sigma = 3 \times 10^{-19}\,\text{cm}^{-3}$. It is actively used at $\approx 50\%$. Estimate the peak power, pulse duration, and pulse energy. We start with the following assumptions:

1. The number of excited atoms n_i is $4\times$ the threshold inversion multiplied by the active volume.
2. The resonator quality is switched by changing the transmission from 0–100% and by using mirrors with a reflectance of $R = \sqrt{R_1 R_2} = 0.5$. The peak power is approximately given by

$$P_{\text{peak}} \approx \frac{\Delta N_i \hbar\omega}{2\tau_{\text{res}}}\,, \tag{B45}$$

where the photons' lifetime in the resonator is $\tau_{\text{res}} = L/(c - cR)$ and c is the speed of light. ΔN_i is the initial inversion before switching and the switching time is taken to be very short. The equation can be understood as a process at which the laser level is completely emptied within the resonator attenuation time. Note that losses due to decoupling are predominant. The pulse duration is approximately $3\tau_{\text{res}}$.

PB.8. Ultra-short light pulses and self-phase modulation The optical Kerr effect (Section 9.5), that is, the dependence of the refractive index on the intensity, is a third-order nonlinear effect. It is described by $n = n_0 + n_2 I$, where the nonlinear refractive index n_2 has the following values:

- Glass (BK7): $n_2 = 5 \times 10^{-15}\,\text{cm}^2/\text{W}$
- Water: $n_2 = 10^{-16}\,\text{cm}^2/\text{W}$
- Doped fiber: $n_2 = 10^{-10}\,\text{cm}^2/\text{W}$.

1. Show that a thin plate of BK7 with a thickness of 5 mm has the effect of a lens with a refractive index of n_l in the case of a 100 fs pulse at a wavelength of approximately 550 nm and an approximately plane wave front (Gaussian mode, waist radius 0.5 mm). Calculate the focal length as a function of realistic pulse energies (in the range from 1 nJ–10 μJ). Make use of the fact that the effect of a lens on a plane wave can be described by the phase term

$$\exp\left(\frac{ik(x^2 + y^2)}{2n_l}\right)\,. \tag{B46}$$

2. The phase of the pulse also changes during passage through the medium according to

$$\varphi(t) = -2\pi\frac{L}{\lambda}n_2 I(t)\,. \tag{B47}$$

Compare a Gaussian pulse and a sech² pulse with a pulse duration of τ_0 for which the time dependence of the field strength amplitudes shall be given by $\exp(-t^2/\tau_0^2)$ and $\mathrm{sech}(t/\tau_0)$, respectively. Calculate the frequency response $\omega(t) = \mathrm{d}\varphi(t)/\mathrm{d}t$ for both pulse forms after passage through 10 mm of glass. Neglect self-focusing and group velocity dispersion. Assume the bundle to be approximately collimated when it travels through the material. Do we have an up- or a down-chirp?

3. Calculate the spectrum of the Gaussian pulse upstream and downstream of the material including the full widths at half maximum of the spectra. Regarding the spectra after passage through the material, let us consider the chirp effect to be a small correction that can be approximated to simplify the calculation. Does the product of pulse duration and spectral width give you any hints?

PB.9. Ultra-short light pulses in the presence of dispersion Considering the propagation of ultra-short light pulses, it is necessary to take nonlinear effects and the effect of group velocity dispersion into account (Section 9.5). An ultra-short light pulse is given by

$$\psi(z,t) = \psi_0(z,t)\mathrm{e}^{2\pi i \nu_0 t - ikz} \, . \tag{B48}$$

In the space-time domain, ν_0 is the center frequency of the pulse and k the propagation constant determined by

$$k(\nu) = \frac{2\pi\nu n(\nu)}{c_0} \tag{B49}$$

with the speed of light in vacuum c_0 and the refractive index $n(\nu)$. In the so-called *SVE approximation* (slowly varying envelope), the propagation equation can be written as

$$\frac{\partial\psi_0}{\partial t^2} + \frac{4\pi i}{D_\nu}\frac{\partial\psi_0}{\partial z} = 0 \, , \tag{B50}$$

where the dispersion coefficient is given by

$$D_\nu = \frac{1}{2\pi}\frac{\mathrm{d}^2 k}{\mathrm{d}\nu^2} = \frac{\mathrm{d}}{\mathrm{d}\nu}\left(\frac{1}{c_\mathrm{g}}\right) \tag{B51}$$

with the group velocity

$$\frac{1}{c_\mathrm{g}} = \frac{1}{2\pi}\frac{\mathrm{d}k}{\mathrm{d}\nu} \, . \tag{B52}$$

The SVE approximation shows formal similarity to the paraxial wave equation (A79), which was used to derive the Gaussian modes. Accordingly, the solution for $\psi_0(z,t)$ will formally correspond to the solution for Eq. (A80) after appropriate substitution. We have

$$\psi_0(z,t) = A_0\sqrt{\frac{-iz_0}{z - iz_0}}\exp\left(\frac{i\pi\left(t - \frac{z}{c_\mathrm{g}}\right)^2}{D_\nu(z - iz_0)}\right) \, . \tag{B53}$$

Table B.6 Dispersion parameters of different sorts of glass at a wavelength of 546 nm.

Material	Refractive Index	$dn/d\lambda$ (mm^{-1})	$d^2n/d\lambda^2$ (mm^{-2})
BK7	1.51872	−51.3514	269 835
SF6	1.81265	−204.024	1 305 289
FK54	1.43815	−30.748	161 347

From this can be derived the pulse duration $\tau(z)$ and the so-called *chirp parameter* $a(z)$ which are given by

$$\tau(z) = \tau_0\sqrt{1 + \left(\frac{z}{z_0}\right)^2} \tag{B54}$$

and $a(z) = z/z_0$ with the dispersion length $z_0 = \pi\tau_0^2/D_\nu$.

1. Describe the analogy between Gaussian pulses and Gaussian bundles. Which variables are equivalent? Which variables correspond to the bundle width, wavefront curvature, angle of divergence, and Rayleigh length?
2. Calculate the intensity of the pulse as a function of z.
3. Calculate the pulse width and chirp of an originally bandwidth-limited pulse of 100 fs at a wavelength of approximately 546 nm after passage through different BK7 glass rods with lengths of 5, 10, and 50 mm.
4. What happens to a pulse that propagates with a down-chirp into a medium with positive group velocity? Determine the optimal length of the glass block made of BK7 after which a pulse with a pulse duration of τ_1 and a negative chirp of a_1 has become a bandwidth-limited pulse. How short will the pulse have become by then?

Use the material parameters for a wavelength of 546 nm from Table B.6.

References

1 Einstein, A. (1905) Über einen die Erzeugung und Verwandlung des Lichtes betreffenden heuristischen Gesichtspunkt. *Ann. Phys., Lpz.*, **17**, 132.

2 Einstein, A. (1917) On the quantum theory of radiation. *Phys. Z.*, **18**, 121.

3 Fox, M. (2007) *Quantum Optics: An Introduction*, Oxford University Press.

4 Hooker, S. and Webb, C. (2010) *Laser Physics*, Oxford University Press.

5 Koechner, W. and Bass, M. (2003) *Solid-State Lasers: A Graduate Text*, Springer.

6 Saleh, B.E.A. and Teich, M.C. (2007) *Fundamentals of Photonics*, John Wiley & Sons, Inc.

7 Fox, M. (2001) *Optical Properties of Solids*, Oxford University Press.

8 Diels, J.C. and Rudolph, W. (2006) *Ultrashort Laser Pulse Phenomena: Fundamentals, Techniques, and Applications on a Femtosecond Time Scale*, Academic Press, 2nd edn.

9 Eichler, J. and Eichler, H.J. (2010) *Laser Bauformen, Strahlführung, Anwendungen*, Springer, 7th edn.
10 Dausinger, F., Lichtner, F., and Lubatschowski, H. (eds) (2004) *Femtosecond Technology for Technical and Medical Applications*, Springer.
11 Ebersbach, D. and Gottschalk, F. (eds) (2013) *Laser Technology, Applications and Future Prospects*, Nova Science Publishers, Inc.
12 Siegman, A.E. (1986) *Lasers*, University Science Books.
13 Svelto, O. (2009) *Principles of Lasers*, Springer, 4th edn.
14 Hecht, J. (2008) *Understanding Lasers: An Entry-Level Guide*, John Wiley & Sons, Inc.

C
Summary of Used Variables and Abbreviations

This chapter provides an overview list of symbols and variables introduced throughout this book. Since we adapted the nomenclature to standard literature, some variables are multiply defined. Hence, the following tables can be used as a backup to avoid misunderstandings.

C.1
Chapters 1–3

AMD	Age-related macular degeneration
ARM	Age-related maculopathy
BRVO	Branch retinal vein occlusion
CNV	Choroidal neovascularization
CRVO	Central retinal vein occlusion
GDP	Gross domestic product
IOP	Intraocular pressure
MAR	Minimum angle of resolution
NA	Numerical aperture
OCT	Optical coherence tomography
ONH	Optic nerve head
PD	Interpupillary distance
RNFL	Retinal nerve fiber layer
RPE	Retinal pigment epithelium
RVO	Retinal vein occlusion
WHO	World Health Organization

$\overline{XY}$	distance between the two points X and Y
C	Rotation center of eye
E	Center of entrance pupil
E′	Center of exit pupil
F	Object-side focal point
F′	Image-side focal point
I′	Image point

Optical Devices in Ophthalmology and Optometry, First Edition. Michael Kaschke, Karl-Heinz Donnerhacke, and Michael Stefan Rill.

N	Object-side nodal point
N′	Image-side nodal point
O	Object point
P	Object-side principal point
P′	Image-side principal point
$\mathrm{Q}_{\mathrm{far}}$	Far point
$\mathrm{Q}_{\mathrm{near}}$	Near point
V	Vertex of cornea
A_{far}	Far point refraction
A_{near}	Near point refraction
ΔA_{max}	Amplitude of accommodation
L	Distance
ΔL	Stereoscopic depth perception
Q	Asphericity parameter
V	Visual acuity
V_{S}	Snellen visual acuity
d_{pupil}	Pupil diameter
h_{O}	Object height
h'_{I}	Image height
n	Refractive index of object-side medium
n'	Refractive index of image-side medium
r	Radius (in Section 2.1.7: distance of light ray from pupil center)
r_{C}	Radius of curvature
s_{far}	Far point distance
s_{near}	Near point distance
s_{nv}	Typical near viewing distance
s_{P}	Parallax distance
$\Delta(x, y)$	Lateral resolution
Δz_{dof}	Depth of field
$\mathcal{D}$	Refractive power
$\mathcal{D}'_{\mathrm{c}}$	Refractive power of cornea
$\mathcal{D}'_{\mathrm{cv}}$	Corneal back vertex power
$\mathcal{D}'_{\mathrm{eye}}$	Total refractive power of eye
$\mathcal{D}'_{\mathrm{l}}$	Refractive power of eye lens
α	Object-side field angle of marginal ray
α'	Image-side field angle of marginal ray
β	Magnification
ε	Stereo angle
η	efficiency
θ	Angle of resolution
κ	Angle between visual axis and optical axis
λ	Wavelength of light
ν	Abbe number
φ	Object-side field angle of chief ray
φ'	Image-side field angle of chief ray

C.2
Chapters 4 and 5

A	Aperture stop
BCVA	Best corrected visual acuity
CCD	Charge-coupled device
cSLO	Confocal scanning laser ophthalmoscope
HOA	Higher-order aberration
L	Lens
LED	Light-emitting diode
LOA	Lower-order aberration
M	Test mire/aperture
MRI	Magnetic resonance imaging
NIR	Near-infrared
MPE	Maximum permissible exposure
OPD	Optical path difference
OPL	Optical path length
PSF	Point-spread function
RD	Ring diaphragm
RPE	Retinal pigment epithelium
HS-WFS	Hartmann–Shack wavefront sensor
VIS	Visible

D'	Image of detector element
E	Center of entrance pupil
E'	Center of exit pupil
F	Object-side focal point
F'	Image-side focal point
I	Image plane
I'	Image point
N	Object-side nodal point
N'	Image-side nodal point
O	Object point
P	Object-side principal point
P'	Image-side principal point
$\mathrm{Q}_{\mathrm{far}}$	Far point
A_{far}	Far point refraction
I	Intensity (distribution)
L	Distance
L_{wd}	Working distance
ΔL	Spot shift
N	Amount
V_j	Base function
a	Distance from detector elements to optical axis
c_n^m, c_j	Zernike coefficients

d	Distance/diameter
d_0	Reference distance
d_{pupil}	Pupil diameter
f	Object-side focal length
f, g	Arbitrary functions (used in Sections 5.3 and 5.4)
f'	Image-side focal length
h_{O}, h	Object height
h'_{I}	Image height
i, j	Integers/indices
n	Refractive index
n, m	Indices (used in Sections 5.3 and 5.4)
r	Radius/radial coordinate
s'	Image distance (distance between principal plane of lens and image plane)
s_{far}	Far point distance
Δt	Time period
v_{lb}	Speed of eye movement
v_{fr}	Speed of fundus reflexes
z'	Axial shift
$\mathcal{D}$	Refractive power
$\mathcal{D}'_{\text{eye}}$	Total refractive power of eye
$\mathcal{D}_{\text{rl}}$	Refractive power of retinoscopy lens
$\mathcal{D}'_{\text{corr}}$	Refractive power of corrective lens
$\mathcal{D}'_{\text{v}}$	Back vertex power
$\Delta\mathcal{D}$	Theoretical measuring accuracy
$\mathcal{W}$	Wave aberration function
$\mathcal{Z}_n^m, \mathcal{Z}_j$	Zernike polynomial
α	Angle/angular coordinate
λ	Wavelength of light
φ	Angular ray deviations

C.3 Chapter 6

2D	Two-dimensional
3D	Three-dimensional
APD	Avalanche photo diode
BIO	Binocular indirect ophthalmoscopy
BS	Beam splitter
CCD	Charge-coupled device
cSLO	Confocal scanning laser ophthalmoscope
DP	Detector plane
DS	Double slit diaphragm

ECC	Enhanced corneal compensation
EP	Eye piece
F	Focus plane
FAF	Fundus autofluorescence
FAG	Fluorescence angiography
FWHM	Full width at half maximum
GAT	Goldmann applanation tonometer
HFL	Henle fiber layer
I	Image plane
ICG	Indocyanine green
ICGA	Indocyanine green angiography
IOL	Intraocular lens
K	Collimator lens
L	Loupe/lens (plane)
LASIK	Laser *in situ* keratomileusis
LED	Light-emitting diode
LSM	Laser scanning microscope
M	Test mire/aperture

NA	Numerical aperture
O	Objective lens/object plane
OD	Right eye (*oculus dexter*)
OS	Left eye (*oculus sinister*)
P	Prism
PD	Interpupillary distance
PMT	Photo multiplier tube
PSF	Point-spread function
RTA	Retinal Thickness Analyzer
SL	Scheimpflug line
SLP	Scanning laser polarimeter
T	Tube lenses
TBUT	Tear breakup time
VCC	Variable corneal compensator

C	Center point
E	Center of entrance pupil
E′	Center of exit pupil
F	Object-side focal point
F′	Image-side focal point
I	Image plane
I′	Image point
K	Principal plane
L	Lens plane
N	Object-side nodal point
N′	Image-side nodal point

O	Object point/object plane
P	Object-side principal point
P′	Image-side principal point
Q_{far}	Far point
R	Reflection point
V	Vertex of cornea
Z_0	Reference plane
A, B, C, D	Coefficients of ABCD matrix
I	Light intensity
K_a	Axial curvature ($K_a = 1/r_a$)
K_m	Meridional curvature ($K_m = 1/r_m$)
L, y	Distance
L_{pp}	Distance between patient's and physician's pupil
L_{wd}	Working distance
ΔL	Stereoscopic depth perception (in Section 6.5: Image shift)
M	Zoom factor
S_{ep}	Area of exit pupil
T	Transmittance
a	Slope of function
b	Stereo base (distance between observation channels)
c	Speed of light
d	Diameter
d_{fov}	Field of view (diameter)
d_{phys}	Diameter of physician's pupil
d_{pupil}	Pupil diameter
f	Focal length
h_O, h	Object height
h'_I	(Retinal) image height
$\Delta h(x, y)$	Distance from reference plane
n^*	Keratometric index
r	Radius
r_a	Axial radius of curvature
r_C	Radius of curvature
r_m	Meridional radius of curvature
s	Object distance
s'	Image distance
s_{far}	Far point distance
s_{near}	Near point distance
s_{nv}	Typical near viewing distance
s_{ref}	Reference viewing distance
$\Delta s(x, y)$	Lateral shift
$\Delta s'$	Longitudinal chromatic aberration (wavelength-dependent change of image distance)
$\Delta(x, y)_{cSLO}$	Lateral resolution of confocal scanning laser ophthalmoscope
Δz_{cSLO}	Axial resolution of confocal scanning laser ophthalmoscope

Δz_{dof}	Depth of field
$\mathcal{C}$	Instrument constant
$\mathcal{E}_{\text{v}}$	Illuminance
$\mathcal{D}$	Refractive power
$\mathcal{D}'_{\text{c}}$	Refractive power of cornea
$\mathcal{D}'_{\text{cv}}$	Corneal back vertex power
$\mathcal{D}_{\text{L}}$	Refractive power of loupe
$\mathcal{D}_{\text{oph}}$	Refractive power of ophthalmoscopy lens
$\mathcal{L}_{\text{v}}$	Illuminance
Γ	Magnification of zoom system
α_{fov}	Angular field of view
β	Magnification
β_{um}	Usable magnification
γ	(Visual) field angle
ε	Stereo angle (in Section 6.2.3.4: scaling factor)
θ	Angle of resolution
λ	Wavelength of light
ν	Abbe number
$\Delta\xi$	Phase difference
ψ	Wave function

C.4 Chapter 7

1D	One-dimensional
2D	Two-dimensional
3D	Three-dimensional
A/D	Analog/digital
a.u.	Arbitrary unit
BS	Beam splitter
CCD	Charge-coupled device
CMOS	Complementary metal-oxide semiconductor
const	Constant number
DOCT	Doppler OCT
FD-OCT	Frequency-domain optical coherence tomography
FFT	Fast-Fourier transformation
FWHM	Full width at half maximum
GVD	Group velocity dispersion
LCI	Low-coherence interferometry
MEMS	Microelectromechanical systems
NA	Numerical aperture
OCDR	Optical coherence domain reflectometry
OCT	Optical coherence tomograph/tomography

PSF	Point-spread function
PS-OCT	Polarization-sensitive OCT
R	Reference arm
Re()	Real part of function
RSOD	Rapid scanning optical delay
S	Signal arm
seFD	Spatially encoded frequency-domain
seTD	Spatially-encoded time-domain
SS-OCT	Swept-source optical coherence tomography
SNR	Signal-to-noise ratio
teFD	Time-encoded frequency-domain
teTD	Time-encoded time-domain
TD-OCT	Time-domain optical coherence tomography
VCSEL	Vertical cavity surface emitting lasers

B	Electronic bandwidth of detector
I	Intensity
L	Length/distance
L_c	Coherence length
ΔL	Length difference
N	Amount/number of channels
P	Power
R	Reflectance
R_S	Average reflection/backscatter coefficient in sample arm
R_R	Average reflection/backscatter coefficient in reference arm
S	(Gaussian) spectral distribution
$U(t)$	Detector signal
c	Speed of light
c_g	Group velocity
c_p	Phase velocity
f	Frequency of photodiode signal
i	Imaginary unit
k	Wave number
n	Refractive index
t	Time
t_c	Coherence time
t_g	Round-trip group delay
t_p	Pulse parameter/round-trip phase delay
Δt_R	Temporal range
$\Delta(x, y)_{cSLO}$	Lateral resolution of confocal scanning laser ophthalmoscope
$\Delta(x, y)_{OCT}$	Lateral resolution of optical coherence tomograph
z_{max}	Maximum scan depth
δz	Axial broadening of point-spread function
Δz	Axial displacement
Δz_{cSLO}	Axial resolution of confocal scanning laser ophthalmoscope

Δz_{OCT}	Axial resolution of optical coherence tomograph
$\mathcal{F}$	Fourier transform
$\mathcal{G}$	Autocorrelation function
δ	Delta function
η	Quantum efficiency of detector
λ	Wavelength of light
λ_0	Center wavelength of spectrum
$\Delta\lambda$	Spectral bandwidth
ν	Frequency of light ($\nu = \omega/2\pi$)
ν_0	Center frequency of spectrum
$\Delta\nu$	Full spectral width
$\Delta\nu_{Doppler}$	Doppler frequency shift
$\Delta\xi$	Phase difference
σ	Power/energy spectral density
τ	Pulse duration
τ_{int}	Integration time in measurement
ψ	Wave function
ω	(Angular) frequency of light
ω_0	Center frequency of spectrum
$\delta\omega$	Spectral width (Set of discrete values (channels))

C.5 Chapter 8

AGE	Advanced glycation end products
CT	Computer tomography
CRT	Cathode ray tube
DLS	Differential light sensitivity
FDF	Flicker-defined form (edge) perimetry
FDT	Frequency-doubling technology
FLIM	Autofluorescence lifetime measurement
MD	Mean deviation
MPE	Maximum permissible exposure
MRI	Magnetic resonance imaging
OS	Oxygen saturation
SAP	Standard automated (white-on-white) perimetry
SWAP	Short-wavelength automated perimetry
TCSPC	Time-correlated single photon counting
VFI	Visual field indices

A	Absorbance
C	Substance concentration
I	Light intensity

$\mathcal{L}$	Luminance
$\mathcal{L}_b$	Background luminance
$\Delta\mathcal{L}_{th}$	Threshold differential luminance
λ	Wavelength of light
μ_a	Absorption coefficient
$\mu_{a,mol}$	Molar absorption coefficient
ω	(Angular) frequency of light

C.6
Chapters 9 and 10

AEL	Accessible emission limits
AK	Arcuate keratectomy
ALT	Argon laser trabeculoplasty
AMD	Age-related macular degeneration
ANSI	American National Standards Institute
CCI	Clear corneal incisions
CNV	Choroidal neovascularization
CPA	Chirped pulse amplification
CPC	Cyclo-photocoagulation
ECP	Endoscopic cyclo-photocoagulation
EN	European standard
epiLASIK	Epithelial laser *in situ* keratomileusis
FDA	Food and Drug Administration
FLAK	Femtosecond laser-assisted keratoplasty
flex	Femtosecond lenticule extraction
FLT	Focal laser treatment
fs	Femtosecond
GVD	Group velocity dispersion
ICRS	Intrastromal corneal ring segments
IEC	International Electrotechnical Commission
IOP	Intraocular pressure
IR	Infrared
ISO	International Organization for Standardization
L	Lens
LASEK	Laser epithelial keratomileusis
LASIK	Laser *in situ* keratomileusis
LED	Light-emitting diode
LIO	Laser indirect ophthalmoscope
LPC	Laser posterior capsulotomy
LPI	Laser peripheral iridotomy
LSO	Laser safety officer
MGT	Macular grid treatment

MPE	Maximum permissible exposure
MLT	Micropulse laser trabeculoplasty
NIR	Near-infrared
NOHA	Nominal ocular hazard area
NOHD	Nominal ocular hazard distance
ns	Nanosecond
OPD	Optical path difference
PCO	Posterior capsular opacification
PDT	Photodynamic therapy
PKP	Penetrating keratoplasty
PMMA	Polymethyl methacrylate (acrylic glass)
PRK	Photorefractive keratectomy
PRP	Pan-retinal photocoagulation
ps	Picosecond
ReLEx	Refractive lenticule extraction
RPE	Retinal pigment epithelium
SBK	Sub-Bowman keratomileusis
SLT	Selective laser trabeculoplasty
smile	Small incision lenticule extraction
SRT	Selective retina therapy
TEM	Transversal electromagnetic mode
TPC	Transscleral photocoagulation
TTT	Transpupillar thermotherapy
UV	Ultraviolet
VEGF	Vascular endothelial growth factor
VIS	Visible

A	Absorbance
E	Energy (level)
E_{diss}	Dissociation energy
I	Light intensity
I_0	Light intensity emitted by light source
I_{peak}	Peak intensity
L	Length/distance
L_{abl}	Ablation depth per pulse
L_{cc}	Central corneal thickness
L_{flap}	Flap thickness
L_{rs}	Residual stromal thickness
L_{scd}	Distance from scattering center to detector
L_{tpd}	Thermal penetration depth
N_{el}	Free electron density
T	Temperature
ΔT_{max}	Peak temperature rise
a	Ablation profile (in Section 10.4.2: clear aperture radius)
a_0	Maximum ablation/optical depth

c	Speed of light
d	Diameter
f	Focal length
$\hbar$	Planck's constant ($\hbar = h/(2\pi)$)
k	Wave number
$\mathbf{k}$	Wave vector
n	Refractive index
r	Radius (of curvature)
s	Sagitta of spherical surface
t	Time
t_{exp}	Exposure time
t_{per}	Pulse period (Repetition rate: $1/t_{\mathrm{per}}$)
t_{r}	Thermal relaxation time
w	Beam radius
w_0	Waist radius of Gaussian beam
w_{f}	Beam radius on eye fundus
w_{L}	Beam radius on lens
z_{max}	Maximum expansion
z_{R}	Rayleigh length
$\mathcal{A}$	Albedo
$\mathcal{D}$	Refractive power
$\mathcal{W}$	Wave aberration function
$\Delta\mathcal{D}$	Refractive correction
Φ	Exposure
Φ_{th}	Threshold exposure
Ψ	Polychromatic wave function
α_{max}	Maximum acceptance angle
β	Magnification (in Section 9.4.4.2: intensity ratio)
γ	Field angle
γ_{crit}	Critical angle (critical angle of total internal reflection)
δ_{a}	Depth of penetration
ε	Beam divergence angle
θ	Convergence angle
κ	Thermal diffusion constant
λ	Wavelength of light
$\Delta\lambda$	Spectral bandwidth
μ_{a}	Absorption coefficient
$\mu_{\mathrm{a,plasma}}$	Plasma absorption coefficient
μ_{s}	Scattering coefficient
τ	Pulse duration
ϕ	Scattering angle
ω	Frequency of light
ω_0	Center frequency of spectrum

C.7 Appendix A

const	Constant number
NA	Numerical aperture
PSF	Point-spread function
RMS	Root mean square
C	Center point of lens
E	Center of entrance pupil
E′	Center of exit pupil
F	Object-side focal point
F′	Image-side focal point
I	Image plane
I_0'	On-axis image point
I_1'	Off-axis image point
K	Principal plane
N	Object-side nodal point
N′	Image-side nodal point
O	Object plane
O_0	On-axis object point
O_1	Off-axis object point
P	Object-side principal point
P′	Image-side principal point
Q_p	Point in pupil plane
A, B, C, D	Coefficients of ABCD matrix
E	Energy
F	Fresnel number
G	Étendue (throughput)
H_v	Luminous exposure
I	Light intensity
$\hat{I}$	Total average intensity
K_m	Optical conversion power
L	Lens/slab thickness, distance
L_c	Coherence length
$\underline{\mathbf{M}}$	ABCD matrix
P	Power
Q_v	Luminous energy
R	Reflectance
$S(\mathbf{x})$	Eikonal
S_ep	Area of exit pupil
S_pupil	Pupil area
T	Transmittance
a	Characteristic size of aperture stop
c	Speed of light

c_0	Speed of light in air/vacuum
c_g	Group velocity
c_n^m	Zernike coefficients
d	Diameter
h	Ray height
h_O	Object height
h'_I	Image height
f	Object-side focal length
f'	Image-side focal length
i	Imaginary unit
k	Wave number
$\mathbf{k}$	Wave vector
n	Refractive index
p	Apodization function
q	Complex beam parameter
r	(Lens surface) radius
r_C	Radius of curvature
s	Object distance
s'	Image distance
t	Time
t_c	Coherence time
t_{per}	Temporal periodicity
w	Beam radius
w_0	(Gaussian) waist radius
$\mathbf{x}$	Space vector
$\hat{\mathbf{x}}$	Unit vector
$\Delta(x, y)$	Lateral resolution
y_{cr}	Chief ray height
y_{mr}	Marginal ray height
z_R	Rayleight length
$\mathcal{D}$	Refractive power
$\mathcal{E}_v$	Illuminance
$\mathcal{F}$	Fourier transform
$\mathcal{G}$	Autocorrelation function
$\mathcal{H}$	Helmholtz-Lagrange invariant
$\mathcal{I}_v$	Luminous intensity
$\mathcal{J}$	Bessel function
$\mathcal{L}_v$	Luminance
$\mathcal{N}$	Normalization factor of Zernike polynomial
$\mathcal{R}$	Radial polynomial of Zernike function
$\mathcal{S}$	Strehl ratio
$\mathcal{T}$	Temporal part of wave function
$\mathcal{V}$	Sensitivity function
$\mathcal{W}$	Wave aberration function
$\mathcal{X}$	Spatial part of wave function

$\mathcal{Z}_n^m$, $\mathcal{Z}_j$	Zernike polynomial
Φ	Exposure
Ψ	Polychromatic wave function
Ψ_τ	Truncated Fourier transform of polychromatic wave function
α	Object-side field angle of (marginal) ray
α'	Image-side field angle of (marginal) ray
β	Magnification
γ	Field angle
γ_{B}	Brewster angle
γ_{crit}	Critical angle (critical angle of total internal reflection)
γ_{r}	Reflection angle
δ	Delta function
ε	Beam divergence angle
θ'	Convergence angle
θ_{res}	Angle of resolution
λ	Wavelength of light
ν	Abbe number
ξ	Phase retardation
ρ	Radial coordinate (used in Section A.2.2.1)
σ	Power spectral density
τ	Integration time
ϕ_{v}	Luminous power
φ	Object-side field angle of chief ray (in Section A.2: phase factor)
φ'	Image-side field angle of chief ray
ψ	Wave function
$\nabla^2\psi$	Second spatial derivative of wave function
ω	Frequency of light
$\Delta\omega$	Bandwidth

C.8
Appendix B

const	Constant number
cw	Continuous wave
fs	Femtosecond
IR	Infrared
FIR	Far-infrared
LASER	Light amplification by stimulated emission of radiation
LED	Light-emitting diode
NA	Numerical aperture
ns	Nanosecond
ps	Picosecond
TEM	Transversal electromagnetic mode

UV	Ultraviolet
VIS	Visible

$\|X\rangle$	(Energy) state X of atom
A_{21}	Einstein coefficients for spontaneous emission
B_{12}	Einstein coefficients for absorption
B_{21}	Einstein coefficients for stimulated emission
E	Energy
E_{gap}	Energy gap
G	Gain (medium amplification) factor
I	Light intensity (in Section B.5.2: Amperage)
L	Length/distance
L_{R}	Resonator length
M	M factor (related to beam divergence)
N	Population
ΔN	Population difference
P	Power
P_{out}	Output power
P_{pump}	Pumping power
P_{th}	Threshold pumping power
R	Reflectance
T	Transmittance
V	Volume (in Section B.5.2: Voltage)
c	Speed of light
g	g parameter (related to laser stability condition)
h	Planck's constant
n	Refractive index
r	(Mirror) radius
t	Time
t_{per}	Pulse period (repetition rate: $1/t_{\text{per}}$)
w_0	Radius of (Gaussian) beam waist
$\mathbb{P}$	Pump rate
$\mathbb{S}$	Stimulated transition rate
$\mathbb{W}$	Pumping transition probability
ε	Beam divergence angle
η	Efficiency
η_{pc}	Power-conversion efficiency
η_{pump}	Pump efficiency
η_{sl}	Slope efficiency
λ	Wavelength of light
ρ	Spectral energy density
σ	Optical gain cross-section
τ	Pulse duration (in Sections B.1–B.2.2: lifetime of atomic state)
ω	Frequency of light
$\Delta\omega$	Bandwidth

Index

Optical Devices in Ophthalmology and Optometry, First Edition. Michael Kaschke, Karl-Heinz Donnerhacke, and Michael Stefan Rill.

Q

R